T0299888

Bioprocessing and Biotreatment of Coal

edited by

Donald L. Wise
Northeastern University
Boston, Massachusetts

CRC Press
Taylor & Francis Group
Boca Raton London New York

CRC Press is an imprint of the
Taylor & Francis Group, an **informa** business

Library of Congress Cataloging--in--Publication Data

Wise, Donald L. (Donald Lee)
  Bioprocessing and biotreatment of coal / Donald L. Wise.
  p. cm.
  Includes bibliographical references and index.
  ISBN 0-8247-8305-0 (alk. paper)
  1. Coal--Cleaning. 2. Coal--Biotechnology. I. Title.
TP325.W76   1990
662.6'23-- --dc20   90-3916
                                                        CIP

This book is printed on acid-free paper.

MARCEL DEKKER, INC.
270 Madison Avenue, New York, New York 10016

Current printing (last digit):
10 9 8 7 6 5 4 3 2 1

PRINTED IN THE UNITED STATES OF AMERICA

Preface

This book includes the most comprehensive set of chapters possible for the newly emerging area of biotechnology applied to fossil fuels. Although many people believe the energy crisis is over, there is still a strong worldwide interest in coal research. A new feature of this research on coal is the rapidly emerging interest in applications of modern biotechnology. For example, this book includes a chapter on pioneering work in genetically producing sulfatase enzymes for removal of organic sulfur from coal (to prevent "acid rain"). Other research work is included which builds on recent discoveries that certain "lignases" (actually, peroxidase enzymes) cleave the lignin-type lattices in coal producing a type of liquid fuel. Only a few years ago, the biological (i.e., enzymatic) breakdown of coal was simply not considered. While there is also some immediate commercial interest in using these lignase enzymes for "biopulping" and "biobleaching" of paper, as well as selected pollution control situations ("bioremediation"), the major funding support appears to be for bioliquefaction of coal. Note that much of the research interest in coal pertains to lower-rank coals, including "lignite" in the United States and "brown coal" in other parts of the world. In summary, this book focuses on current research work in biotechnology applied to fossil fuels. The promise for commercialization appears to be quite strong.

Readers may appreciate learning of background and technical status of the work leading up to publication. Many of the contributors were involved several years ago in a pioneering research effort by a major American electrical utility firm, Houston Lighting and Power Company, to investigate applications of biotechnology to fossil fuels. Based on a number of related publications, I was invited to explore the technical feasibility of and economical potential for what was essentially a "coal biorefinery." The major focus of the "coal biorefinery" was to explore the potential for producing methane from lignite. After conducting a preliminary process engineering economic evaluation, an experimental feasibility program was launched. It became quite clear that a number of key technical questions needed to be addressed, in addition to continuing with the work then underway.

It was recommended that we embark on a novel University Grant Program, which I administered, in which experts in selected areas of modern biotechnology would be invited to participate. It was understood in initiating this University Grants Program that few, if any, of these experts in biotechnology had previously worked with coal—the objective of each research topic was to experimentally explore selected applications of modern biotechnology applied to fossil fuels. For example, one research worker, who was investigating the genetic engineering of sulfatase enzymes as applied to certain polysaccharides found in seaweeds, was invited to extend his research work and investigate the enzymatic removal of organic sulfur from coal. We believe this work presented the first definitive concept for using enzymes (not microorganisms) to remove organic sulfur from coal. In a similar manner, chapters from other research projects selected and conducted under this University Grants Program are included.

As I continued to coordinate the University Grants Program, as well as carry out my own experimental investigations and economic evaluations on producing methane and BTX-type liquid fuels from lignite and lower ranked coals, other research workers were invited to join the University Grants Program. At one time ten different research groups were carrying out investigations on the bioprocessing and biotreatment of coals. Further, in the course of attending technical meetings and presenting papers on the overall "coal biorefinery" work, I met a number of other research workers who were also in the forefront of applying modern biotechnological concepts to the bioprocessing and biotreatment of coal. The need for a focused reference text in this newly emerging field was clear. Thus, based on the work of those colleagues who participated in the University Grants Program, and the work of those colleagues who were independently pursuing research in this field, our present book evolved. We therefore present a very deliberately organized set of chapters describing efforts that have principally matured together. That is, this text is definitely not a collection of chapters based on an invited symposium, although such special symposia texts are useful and highly regarded; this text simply evolved from a focused effort specifically directed to the bioprocessing and biotreatment of coal.

In this volume we discuss bioprocessing, bioconversion, and biosolubilization as the several themes of research, and later chapters deal with biosystems, biocleaning, and sulfur removal. While much of the work presented does pertain to research being carried out in the United States, we recognize the major contribution—and keen interest—in coal bioprocessing and biotreatment on an international basis. For example, Dr. R. M. Fakoussa of West Germany is now generally recognized as the first research worker to consider the biological "breakdown" of coal. (A number of earlier workers recognized the potential for microbiological removal of pyritic sulfur.) On the other hand, Dr. Fakoussa's Ph.D. thesis was in German and unfortunately because of this did not get the attention it deserved. Stating this impressive contribution from Dr. Fakoussa does not in any way detract from the truly astounding impact in the United States by the work of Dr. Martin S. Cohen on the enzymatic solubilization of coal.

Just as the work on coal biosolubilization has had recent impact on the thinking of those involved in coal utilization, so has the question of using modern biotechnology to remove sulfur from coal. A number of workers have for many years diligently pursued research on microbial methods for (principally) pyritic sulfur removal from coal and their most recent work is included. On the other hand, a more basic approach has recently been pursued by, for example, Professor Steven Krawiec at Lehigh University (initiated by Professor Fikret Kargi, now at Washington University), as well as others.

In addition to chapters describing experimentally based work, a number of people interested in the bioprocessing and biotreatment of coal have contributed significantly to this text by providing an in-depth insight into research opportunities, economic factors, and regional perspectives. Because I value these nonexperimental inputs so highly, several chapters are included on these key topics.

Without providing an exhaustive overview of each individual chapter, which might spoil the reader's enjoyment of searching on his own, I close this introduction by wishing the reader to join me, along with all the chapter authors, in this exciting new field.

Donald L. Wise

In addition to chapters describing experimentally based work, a number of people interested in the bioprocessing and biotreatment of coal have contributed significantly to this text by providing an in-depth insight into research opportunities, economic factors, and regional perspectives. Because I value these nonexperimental inputs so highly, several chapters are included on these key topics.

Without providing an exhaustive overview of each individual chapter, which might spoil the reader's enjoyment of sampling on his own, I close this introduction by wishing the reader to join me, along with all the chapter authors, in this exciting new field.

Donald L. Wise

Contents

Contributors

Anthony S. Atkins   Staffordshire Polytechnic, Staffordshire, England

S. Barik   University of Arkansas, Fayetteville, Arkansas

Roger M. Bean   Battelle, Pacific Northwest Laboratories, Richland, Washington

Ernst Beier   DMT Fachhochschule Bergbau, Bochum, West Germany

M. Luisa Blazquez   Complutense University of Madrid, Madrid, Spain

D. Chandra   Indian School of Mines, Dhanbad, India

Charles C. Y. Chen   Ohio State University, Columbus, Ohio

E. C. Clausen   University of Arkansas, Fayetteville, Arkansas

Martin S. Cohen   University of Hartford, West Hartford, Connecticut

Gordon R. Couch   IEA Coal Research, London, England

Michael D. Dahlberg   Pittsburgh Energy Technology Center, U. S. Department of Energy, Pittsburgh, Pennsylvania

Henrique C. G. Do Nascimento[*]   University of Southern California, Los Angeles, California

Carlos Dosoretz   MIGAL-Galilee Technological Center, Kiryat-Shmona, Israel

René M. Fakoussa   University of Bonn, Bonn, West Germany

James G. Ferry   Virginia Polytechnic Institute and State University, Blacksburg, Virginia

G. Fuchs   University of Ulm, Ulm, West Germany

J. L. Gaddy   University of Arkansas, Fayetteville, Arkansas

Benedict J. Gallo   Boston University School of Medicine, Boston, Massachusetts

---

[*]*Current affiliation:* Engineering Division, South Coast Air Quality Management District, El Monte, California

Celal F. Gokcay   Middle East Technical University, Ankara, Turkey

Hans E. Grethlein   Michigan Biotechnology Institute, Lansing, Michigan

Kathleen Gribschaw   Houston Lighting and Power Company, Houston, Texas

John D. Haddock   Virginia Polytechnic Institute and State University, Blacksburg, Virginia

S. B. Harrison   University of Arkansas, Fayetteville, Arkansas

Herbert J. Hatcher   Idaho National Engineering Laboratory, Idaho Falls, Idaho

Rebecca S. Hsu-Chou*   University of Southern California, Los Angeles, California

Holger W. Jannasch   Woods Hole Oceanographic Institution, Woods Hole, Massachusetts

Vijay T. John   Tulane University, New Orleans, Louisiana

E. R. Johnson   University of Arkansas, Fayetteville, Arkansas

La Mar J. Johnson   Idaho National Engineering Laboratory, Idaho Falls, Idaho

Fikret Kargi   Washington University, St. Louis, Missouri

Robert M. Kelly   The Johns Hopkins University, Baltimore, Maryland

Ernest E. Kern   Houston Lighting and Power Company, Houston, Texas

John J. Kilbane II   Institute of Gas Technology, Chicago, Illinois

Steven Krawiec   Lehigh University, Bethlehem, Pennsylvania

A. Kröger   J. W. Goethe-University, Frankfurt, West Germany

Don Lapin   University of Houston, Houston, Texas

Mark J. Laquidara   Dynatech Scientific, Inc., Cambridge, Massachusetts

Kwangil Lee†   University of Southern California, Los Angeles, California

Alfred P. Leuschner‡   Dynatech Scientific, Inc., Cambridge, Massachusetts

Susan N. Lewis   Oak Ridge National Laboratory, Oak Ridge, Tennessee

Uri Marchaim   MIGAL–Galilee Technological Center, Kiryat-Shmona, Israel

Current affiliations:
   *Gtel Environmental Labs, Inc., Southwest Region, Torrence, California
   †Cal Science Engineering, Cypress, California
   ‡Remediation Technologies, Inc., Concord, Massachusetts

Judith K. Marquis Arthur D. Little, Inc., Cambridge, Massachusetts

Annette S. Martel * Dynatech Scientific, Inc., Cambridge, Massachusetts

Jack V. Matson University of Houston, Houston, Texas

Ronald G. L. McCready Canadian Centre for Mineral and Energy Technology (CANMET), Ottawa, Ontario, Canada

A. K. Mishra Bose Institute, Calcutta, India

Gregory J. Olson[†] National Institute of Standards and Technology, Gaithersburg, Maryland

Aharon Oren The Institute of Life Sciences, The Hebrew University of Jerusalem, Jerusalem, Israel

S. Prieto University of Arkansas, Fayetteville, Arkansas

Leah Reed Sheladia Associates, Inc., Rockville, Maryland

Sirirat Rengpipat Michigan State University, East Lansing, Michigan

Charles D. Scott Oak Ridge National Laboratory, Oak Ridge, Tennessee

Duane R. Skidmore Ohio State University, Columbus, Ohio

Gerald W. Strandberg Oak Ridge National Laboratory, Oak Ridge, Tennessee

Rudolf K. Thauer University of Marburg, Marburg, West Germany

J. L. Vega University of Arkansas, Fayetteville, Arkansas

John W. Wang[‡] University of Southern California, Los Angeles, California

Bailey Ward The University of Mississippi, University, Mississippi

Bary W. Wilson Battelle, Pacific Northwest Laboratories, Richland, Washington

Teh Fu Yen University of Southern California, Los Angeles, California

Reyhan Yurteri Middle East Technical University, Ankara, Turkey

J. G. Zeikus Michigan Biotechnology Institute, Lansing, Michigan

Current affiliations:
   *Remediation Technologies, Inc., Concord, Massachusetts
   †Pittsburgh Energy Technology Center, Pittsburgh, Pennsylvania
   ‡Management Division, Department of Public Works, Los Angeles, California

1

Research Opportunities in Coal Bioprocessing

LEAH REED   *Sheladia Associates, Inc., Rockville, Maryland*

1.1 INTRODUCTION

Within the past year, Sheladia Associates, Inc. has investigated the types of research necessary for the development of processes to gasify coal by microbiological means. The ultimate aim of this work was to propose a detailed research plan for use by the Department of Energy in program management. The application of new developments in biotechnology provides an innovative approach to coal gasification which can make a significant contribution to the product of high-value products from coal. This contribution may take the form of lower processing costs, less rigorous processing conditions, regenerable catalysts, better control of toxic constituents, or a combination of these and other factors. However, before any of these novel processes can be demonstrated at the commercial level, much research is necessary into basic mechanisms, metabolic pathways, reactor design, and other facets of the end process. In this chapter the research path developed by Sheladia is presented together with specific research initiatives that may contribute to the development of such processes. The initiatives were developed during the course of the study by scientists and engineers working within the pertinent technical areas. We have provided objectives as well as a short discussion of the work entailed for each initiative as a way of demonstrating possible research ideas within the proposed research path.

1.2 APPROACH

An approach was developed that took into consideration the multidisciplinary nature of biogasification processes and the fact that current research in this area is at a very preliminary stage. We then defined the research areas that were most likely to affect progress in coal biogasification. Each of these areas was investigated in detail by technical staff members. Research reports were compiled in each of the six areas as appendices to the final report. Based on these investigations, research recommendations were made. The approach we used to develop this work included the following steps:

1. Investigate the current level of development in pertinent research areas.
2. Define topics where research is needed.
3. Construct an independent panel representative of necessary research disciplines.
4. Present the results of the investigation to the panel.
5. Develop further research ideas with panel members.
6. Organize and enlarge on all pertinent ideas.
7. Design a research path to include all initiatives.

The research initiatives were organized into nine major categories based on subject and level of development toward the ultimate process. These nine categories were then arranged in a set of five sequential stages from basic research to commercialization.

1.3 RESEARCH AREAS

The six major research areas investigated in this work were coal structure and reactivity, gasification technology (nonbiological), genetic engineering, coal bioprocessing, bioprocess technology, and regulatory implications of biotechnology processes. New developments in any of these areas can potentially affect the development of coal biogasification processes. The last topic, although not technical in nature, was added because any change in regulatory requirements or restrictions may cause corresponding changes in process requirements or may even restrict certain avenues of research.

Coal structure and reactivity was an obvious choice since coal is the substrate for the majority of the processes considered. (Upgrading of synthesis gas from conventional coal gasifiers was also considered.) Coal is a very complex material composed primarily of carbon, hydrogen, oxygen, and nitrogen in various functional groups, along with lesser amounts of other organic and inorganic constituents. The ways in which these elements are bonded affects the stability and reactivity of the material as well as the ability of microorganisms to attach to and degrade the particles. Toxic constituents in the coal also affect biological processes.

Nonbiological gasification technology was investigated as a benchmark for comparison of biological processes in terms of cost, mechanism, conditions, and other selected parameters. Retrofit technology, where part of an existing process could be replaced by a biological reactor, was also examined as a way of introducing microbial processes to industry without requiring the construction of an entire facility.

One of the most exciting areas we investigated was the field of genetic engineering. Through manipulation of the inheritable material within a cell it will be possible to produce pronounced effects on the reaction of microorganisms. These effects may cause faster processing times, alteration of specificities, combining of several traits within one organism, as well as other benefits. Approaches and techniques developed in other industries can be applied to modifying the microorganisms active in coal conversion processes. These techniques can be used in ways that do not raise issues related to human health or the environment.

Current work in coal bioprocessing is, of course, the starting point for speculation about future research in the bioconversion of coal. Bacterial production of sulfuric acid from coal (now called bacterial desulfurization) has been examined since the late 1940s. Production of methane by bacterial has been a commercial process for much longer;

sewage treatment and landfill gas production are two such processes. The newest area in coal bioprocessing is the production of water-soluble polymers from insoluble coal particles. Expansions and combinations of these three sets of biological processes can provide new processes for future production of clean, upgraded fuel from coal.

The development of bioprocess technology is the key to efficient, cost-effective commercial-scale microbial processes. Bioprocesses have been used since the first bread was baked and the first wine fermented. Large-scale production of foodstuffs, pharmaceuticals, solvents, and so on, has occurred for more than a century. In the last few decades, however, advances have been made not only in the alteration of pertinent organisms, but in the design of reactors. Continuous-flow bioreactors will replace batch processes, and immobilized cells and enzymes will supplant ordinary microorganisms. All of these changes will provide better control of reactions through improved mass transfer, simplified separations, control of toxicity, and other factors related to commercial production of microbial products. Again, advances produced in other biological industries can be applied to the problems of coal conversion and utilization.

Along with the foregoing technical areas are two other factors that should be considered when designing a research plan. Research is always performed in some context, with goals or objectives defined at the outset. In the case of coal conversion and utilization, that context is set within the objectives of the Department of Energy. These objectives include:

1. Increasing the contribution of coal by:
   a. Improving the environmental, technical, and economic performance of coal-based systems
   b. Increasing the numbers of areas of application and flexibility of use of coal-based systems
2. Increasing the effective resource base for premium gas and liquid fuels through:
   a. Enhanced resource recovery
   b. Production of liquid and gaseous fuel analogs from coal, shale, and tar sands
3. Coupling process technology research with appropriate environmental research for the economic management of environmental impacts related to processing and using fossil fuels
4. Coupling process technology research to advanced research activities needed to understand fundamental scientific and engineering mechanisms

An additional facor of concern is the potential regulation of the biotechnology industry, especially as concerns the use of genetically modified organisms. At the present time, any such work is monitored primarily by the Environmental Protection Agency through the Toxic Substances Control Act (TSCA). Several other agencies are reviewing their regulations and a White House Task Force has been formed to investigate this issue. By the time that most of the coal bioprocessing work has reached the commercialization stage, we anticipate that precedents on containment procedures and other pertinent issues will have been set by researchers and regulatory agencies in areas such as agriculture.

1.4 RESEARCH STAGES

The format used for the proposed research plan is based on a typical flow diagram that provides a logical progression of events for meeting program objectives. A graphic presentation of the proposed plan, showing research stages as well as individual research

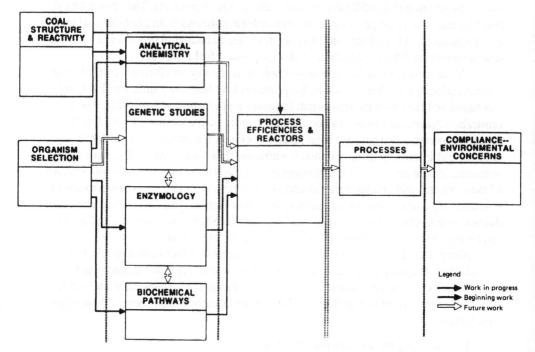

Figure 1.1 Research path for the development of coal biogasification processes. General research categories are presented with the research stage noted at the bottom of the figure.

categories, is shown in Fig. 1.1. The plan is divided into research stages rather than years because the research in coal bioprocessing has essentially just begun. At such an early stage of development, predictions of time are arbitrary and misleading. The research plan as proposed contains five major research stages: foundation, biochemical research, process research, process development, and environmental impact. The first three stages consist of basic and applied research, while the fourth stage contains demonstration projects. The final stage addresses research required to answer regulatory questions about the processes developed.

The *foundation* stage and its two categories address the data acquisition and basic research necessary to support the work to be carried out at later stages. The two categories, *coal structure and reactivity* and *organism selection*, are designed to identify and characterize the substrate and catalyst to be used in the bioprocesses. "Catalyst," of course, refers to the microorganisms.

The *biochemical research* stage provides for the development of the catalyst microorganisms through three separate paths. Research on *biochemical pathways* will define the responses of the organisms to their environment, including both nutrient and toxic effects, and to delineate the sequence of enzymatic reactions used by selected strains to perform coal conversion or beneficiation reactions. Through research in *enzymology*, individual enzymes from these pathways will be isolated and characterized. Attempts can be made at this point to develop purely enzymatic reactors for coal processing. To further develop selected cultures, *genetic studies* will be carried out to discover the inheritance of requisite traits and methods for genetic improvement of strains. A fourth category at this stage consists of research in *analytical chemistry* to investigate the interaction of the organisms with the coal itself. Both effects of the organism on the substrate (e.g., bond breaking) and coal structure as it affects microbial reactivity are of concern.

At the stage of *process research*, new concepts in bioreactor design and control will be adapted or developed. All the research at this stage has been assigned to the category, *process efficiencies and reactors*. Applicable research is concerned with macroscopic interaction of the substrate, coal or its intermediate conversion products, with the catalyst microorganisms or enzymes. Such parameters include mass transfer, particle size, agitation rates, and stability of catalyst, together with on-line monitoring and control of reactions. Optimization of laboratory- and pilot-scale reactors is also consistent with research at this stage.

Initiatives listed under the *process development* stage are expected to take the processes designed and optimized at the last research stage and develop them into commercializable demonstration-scale activities. The specific activities under development comprise the category *processes*. Each has the potential to be a feasible technology, although this list of initiatives does not contain all possible technologies. For example, microbial liquefaction of coal is another activity that could be considered in this list of technologies. Liquefaction may be a necessary intermediate step in the biogasification of coal, but in this context it is not an end process.

The last research stage, *environmental impact*, deals with the concept of compliance. No commercial process can be developed without a concurrent investigation of adverse effects on the environment. In the case of microbial processes where the techniques of genetic engineering may be employed, compliance requirements are extremely stringent at present and may be more severe in the future. The list of initiatives in the category *compliance–environmental concerns* represents current concerns about the effects of engineered organisms and their potential release into the environment.

Some of the work described above is already in progress; other work is just beginning. Much of the work, however, has yet to be implemented. Work in progress as well as work soon to be undertaken is designated in Fig. 1.1. The following sections of the document explain each research stage in greater detail and discuss relevant research at each state.

1.5 FOUNDATION

The two categories comprising the first stage of the research plan, "coal structure and reactivity" and "organism selection," contain key research elements that must be addressed and resolved at the outset of bioprocessing work. These categories deal with the identification and characterization of the starting materials, coal and microorganisms, through which all other work will be developed. It is of critical importance that certain of these issues be investigated in advance to facilitate later work. For example, it is essential to develop analytical methods (standard laboratory procedures for analyzing substrates and products) and reference materials by which to compare experimental results. In many cases, appropriate methods of analysis do not exist and time must be spent developing assays before conversion or beneficiation work can begin. In like manner, different coals must be characterized in greater depth than the accepted proximate and elemental analyses in order to determine the effect of sulfur, nitrogen, oxygen, or metal functionalities on the microorganisms or on microbial processing. This work would also contribute to research on the effects of microbial processing on the coal particles. The relative reactivities of the organic functional groups contained in the coal are again of importance in investigating coal/microbial reactivity as well as in providing means for comparing data generated from reactions with different coals.

1.5.1 Coal Structure and Reactivity

Figure 1.2 presents the specific research initiatives proposed under "coal structure and reactivity" as well as the relationship of these initiatives to later categories. Much of the work described in this first category has no biological component whatsoever. The emphasis is placed on the physical and chemical properties of coal itself, not on the effects these properties may have on microorganisms or microbial components. This research has traditionally been carried out in many nonbiological coal programs. Some data are available, although not necessarily complete or in the desired format. These initiatives are included here to demonstrate the areas in which coal structure and/or reactivity may affect microbial processing and the kinds of data that are specific to this biological research.

Initiative: Development of Analytical Methodologies, Audits, and Reference Materials
Objective: To standardize analytical testing methods and reference materials, and to research methods for reducing analytical errors
Discussion: Several ongoing research programs are involved in the development of methods for coal analysis. However, virtually no work has been undertaken on audits and reference materials. With regard to the individual coals, coal banks have been situated at the Argonne National Laboratory and the Pennsylvania State University to provide several grades and volumes of samples for coal researchers. There is no agreement, how-

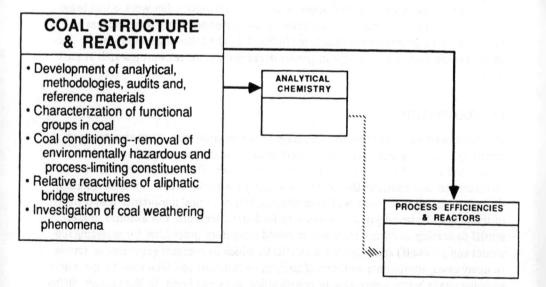

Figure 1.2 Research initiatives in the category of *coal structure* and *reactivity* that relate to biological processes. These studies are considered fundamental and are already under way.

ever, about standard coals or analytical methods. Specific analytical methods that are lacking include those for precise sulfur analysis, especially a determination of the organic sulfur content in coal.

Initiative: Characterization of Functional Groups in Coal
Objective: To develop direct methods to quantify the presence of sulfur, nitrogen, and oxygen functionalities and metal constituents in coal
Discussion: Some research has been carried out to determine functional groups pertinent to coal reactivity and to quantify these constituents by techniques such as molecular spectroscopy. The data generated for coal liquid mixtures, cleaned coal, or for powdered coals may or may not be applicable to biologically treated coals. Of primary interest to bioprocessing is an understanding of oxygen functionalities in raw and biologically treated coals.

Initiative: Coal Conditioning: Removal of Environmentally Hazardous and Process-Limiting Constituents
Objective: To develop clean coal technologies for the cost-effective removal of sulfur, nitrogen, oxygen, metals, and ash-forming minerals from coal, so as to improve chemical reactivity and biodegradability
Discussion: The removal of these constituents from coal is environmentally desirable and advantageous in terms of processing efficiency and operating costs. Some processes have been examined for physical or chemical means of removal, such as pulverization, solvent extraction, froth flotation, or treatment with alkali at high temperatures, most of which are energy intensive. Future work calls for new approaches to the removal of these materials, including the use of catalysts to break down oxygen- and nitrogen-containing functional groups.

Initiative: Relative Reactivities of Aliphatic Bridge Structures
Objective: To compare the relative bond strengths between key pairs of nuclei, particularly in aliphatic bridge structures, with a view to understanding and predicting coal reactivity
Discussion: The utilization of coal is constrained by a lack of understanding of structure and functional characteristics at the molecular level. This is especially true for biological processes, in which specific sites are targeted for attack by microorganisms or by their enzyme extracts. This initiative would be a complementary study to already existing coal structure investigations.

Initiative: Investigation of Coal Weathering Phenomena
Objective: To assess the impact of exposure parameters such as moisture content and temperature on the rate and degree of coal oxidation, and to develop laboratory methods to estimate the extent of weathering
Discussion: The oxidative state of coal was shown to be a critical parameter in microbial coal desulfurization as well as in microbial liquefaction. Some research into this phenomenon has been carried out with respect to coal stockpiles, but the level and direction of research does not meet the requirements of a bioprocessing program. Research could be aimed at developing a chemical classification system for coal that can be correlated with the mineralogical history of the deposit.

1.5.2 Organism Selection

In the area of organism selection it is essential that suitable organisms with an affinity for coal be identified and isolated, as these strains form the basis for further process development. Initiatives comprising this category are shown in Fig. 1.3. A collection of organisms should be developed for each facet of coal conversion or beneficiation. As a part of this process, individual strains must be compared and ranked in terms of their capabilities to perform specific transformations and also in terms of the speed and specificity of transformation. These steps will provide a spectrum of organisms of varying capabilities from which to choose in the later development of bioprocesses.

Initiative: Isolation of Organisms That Grow at the Expense of Coal
Objective: To identify microorganisms capable of producing liquid or gaseous coal
             products that can be used as fuels, chemical feedstocks, or process additives
Discussion: This initiative can draw on research conducted in related areas for the past
30 to 40 years. Organisms have been isolated from mines, mine wastes, and coal piles, as well as from noncoal areas such as oil-contaminated soil, deep-sea thermal vents, and hot springs. Similar work has been conducted in the fermentation of biomass where polymer degrading and methane-producing organisms have been isolated. Microorganisms that habituate weathered refuse and overburden coal piles and low-rank coals are of particular interest because they enhance the prospect of minehead disposal by conversion to useful products. Although a large number of cultures have been isolated, it is still necessary to maintain a certain level of investigation into the existence of unknown useful organisms.

Initiative: Selection of Cultures for Unit Conversions
Objective: To identify and characterize those bacterial and fungal species that efficient-
             ly degrade coal and that exhibit maximum tolerance over a broad range of
             physical/chemical conditions

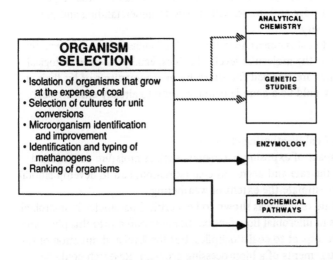

Figure 1.3 Research initiatives in the category of *organism selection* for coal processing. Collection of active cultures is basic to the development of coal bioprocesses. Each facet of organism reaction with coal is represented by a different array of cultures.

Discussion: A unit conversion is a single operation along the route of the total transformation desired in a process. The purpose of selecting unit conversions is to produce a well-elucidated and well-controlled series of operations that can be combined into an efficient process. Such selection has been undertaken in the past using an empirical approach. At present there is no work at the level of detail or organization contemplated in this research plan.

Initiative: Microorganism Identification and Improvement
Objective: To identify candidate microorganisms for the cost-efficient conversion of coal to more desirable products
Discussion: A compendium of defined microorganisms is available from the American Type Culture Collection as well as other microbial collections. Relating newly isolated organisms to members of these collections decreases the amount of research necessary to define metabolic requirements. Typing and species identification are under way for recent isolates active in coal conversion or beneficiation processes, as is improvement of cultures by mutagenesis and by selective culturing.

Initiative: Identification and Typing of Methanogens
Objective: To collect and isolate organisms from methane-producing cultures and to determine the physical and chemical requirements for maximum productivity
Discussion: Methanogens provide the final conversion step in the progression from coal to fuel gas. As a class, these organisms are well used but not well known in terms of genus, species, and so on. A larger variety of methane-producing cultures would provide greater selection in terms of nutritional requirements and tolerances, conversion rates, and other process parameters.

Initiative: Ranking of Organisms
Objective: To determine the relative abilities of various microorganisms to alter raw and conditioned coals
Discussion: Organisms are ultimately ranked in terms of conversion rate, efficiency, or specificity. Alternative parameters include tolerance to toxic materials, utilization of inexpensive substrates, growth temperature or pH, and so on. The ranking of suitable microorganisms should encompass testing on raw coals as well as treated coals and may involve the use of model compounds.

1.6 BIOCHEMICAL RESEARCH

This stage of the research plan contains four groups of research elements—*analytical chemistry, biochemical studies, enzymology*, and *genetic studies*—which comprise the key biological research areas. Few of these initiatives have been investigated at present; definitive information is available on even fewer. A much greater emphasis should be placed on initiatives at this research stage because these results will define the parameters and constraints for future bioprocesses. Through these investigations, means can be developed for creating the most efficient catalyst—enzyme or organism—for interaction with coal or its conversion products. It is vital that the small amount of work in progress on enzymology, in particular, be expanded and that further enzymes be identified and isolated from active strains.

1.6.1 Analytical Chemistry

The category "analytical chemistry" consists of investigations of coal/microorganism interactions. There are aspects of the coal "molecule" that inhibit as well as contribute to microbial reactions. Some of these inhibitory factors are contaminants, such as heavy metals; others are toxic organic compounds bound into the organic structure. Contributory factors might resemble compounds necessary for microbial growth and proliferation. Another emphasis of this category includes the ways in which natural biochemical pathways are related to reactions that might improve coal. A determination of the chemical changes to the coal matrix after microbial processing is also necessary in support of this work. Figure 1.4 presents research initiatives pertinent to these investigations.

Initiative: Coal Structure and Microbial Reactivity

Objective: To determine the critical linkages in coal that pertain to its degradation by microbes, and to identify structural and chemical features that limit or enhance microbial reactivity

Discussion: The work entailed in this research initiative follows closely upon the results of initiatives in coal structure and reactivity. However, the functional groups of interest are only those that promote the mechanistic understanding of microorganism interaction with coal. A small amount of work has been performed on microbial reactivity in relation to coal rank as well as in relation to coal weathering. More work is needed on microbial access to and reactions with specific functionalities identified in the coal matrix.

Initiative: Coal as a Substrate: Structure Versus Bioactivity

Objective: To determine the specificity of various microorganisms in attacking sites in coal macerals, and the extent to which sulfur and aromatic functionalities may inhibit biological reactivity.

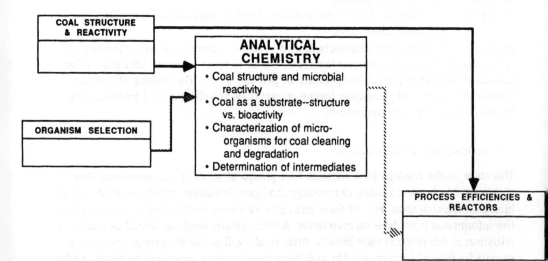

Figure 1.4 Research initiatives related to coal/microorganism reactions. Research in this category draws heavily on advances in coal structure research. Work must be repeated for new strains of microorganisms.

Discussion: This work will expand on activities described in the previous initiative by examining microbial pathways that transform molecules related to organic functionalities present in coal. Because of the importance of phenols and other aromatics as inhibitors of biological reactivity, biological pathways for conversion of these materials may be significant. This and the previous initiative should be given more prominence because knowledge of substrate/catalyst interactions is fundamental to efficient process design.

Initiative: Characterization of Microorganisms for Coal Cleaning and Degradation
Objective: To determine the mechanisms of microbial action on coal, and to optimize the physical and chemical conditions for rapid microbial attack at specific coal structure sites
Discussion: The characterization of microorganisms for coal beneficiation or coal degradation will be based on three broad mechanisms by which bacteria and/or fungi alter coal: altering the wettability preference of ash and pyrite particles, direct chemical attack at molecular sites, and secretion of enzymes onto the coal substrate. A study has been initiated to elucidate the mechanisms of bond breaking by microorganisms, using both model compounds and coal. Additional work is needed in all areas (desulfurization, liquefaction, and gasification) to determine any alterations in the physical and chemical structures of coals of various ranks after bioprocessing.

Initiative: Determination of Intermediates
Objective: To identify possible mechanisms in the microbial degradation of coal by the separation and analysis of intermediates
Discussion: Microorganisms that liquefy coal can be classed in terms of the structure and functionality of the organic intermediates produced. To assist in this characterization, liquid samples should be fractionated and analyzed by methods such as electrophoresis, mass spectrometry, nuclear magnetic resonance, and infrared spectroscopy. These intermediates can be classified according to their acidic/basic character, polarity, as well as the effect of pretreatment of the coal on conversion yield. Similar studies can be performed for the intermediates in desulfurization and gasification processes. The liquefaction work is under way, although no specific molecular structures have been identified. Functional groups have been characterized and quantified by several methods, although the results from different methods are not always comparable. For gasification research, degradation studies of model compounds related to lignin and peat were undertaken, but not coal directly. Identification of the intermediates in coal bioprocesses will allow for better coupling of subsequent microbial processes.

1.6.2 Biochemical Pathways

The actual mechanisms by which the organisms use coal or adapt to a coal-containing environment are investigated in the category "biochemical pathways." Figure 1.5 presents initiatives leading to a greater understanding of these pathways. Organisms are capable of adapting to a wide variety of environments, whether they contain toxic materials, new forms of nutrients, or physical factors such as pH or pressure alterations. The means of adaptation and the new pathways developed or enhanced for adaptation are appropriate topics of research for microbial applications to coal processing. In addition, investigation and identification of biochemical pathways will enhance and accelerate research in enzymology.

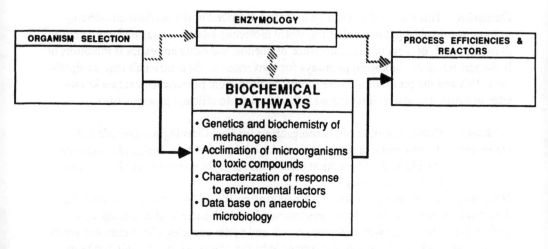

Figure 1.5 Research initiatives to investigate the series of biological mechanisms by which organisms use coal or adapt to a coal-containing environment. Pathways are specific to classes of microorganism and research is supported by advances in enzymology.

Initiative: Genetics and Biochemistry of Methanogens

Objective: To elucidate the biochemical pathways of methane production from coal, determine the enzymes that catalyze the reactions, and identify the genes that stimulate the production of these enzymes

Discussion: Methanogens are the most important class of organisms from the standpoint of biogasification because of their ability to convert simple substrates to methane. Unfortunately, the mechanisms of this conversion have not been determined, nor have other biosynthetic pathways used by these organisms. This additional information will support genetic engineering of these organisms to maximize the production and potential use of the pertinent enzymes rather than whole cells. Use of enzymes may be an important area of research because of the difficulties inherent in culturing methanogens. Biochemical data related to methanogens is also necessary to the optimization of reactor conditions. Some data are available from sewage and biomass fermentation research. However, in comparison to most industrial microorganisms, the methanogens are not well characterized.

Initiative: Acclimation of Microorganisms to Toxic Compounds

Objective: To characterize the response of coal-degrading microorganisms to toxic coal constituents, and to determine tolerance limits and the extent to which toxicity may be mitigated by acclimation

Discussion: Coal is an extremely complex material containing many toxic organic compounds as well as heavy metals and other inorganic materials that can poison the environment for microorganisms. Toxic materials must be identified in relation to each set of microorganisms proposed for coal processing because a large variation in response is possible. Cultures can be selected for resistance to some of these compounds or tolerance can be developed through acclimation to increasingly greater concentrations of the element or compound in question. Organisms will acclimate to the presence of noxious compounds by several biological mechanisms, including degradation of the compound to remove it from the environment. The acclimation response has been investigated in the area of coal gas upgrading for the compounds carbon monoxide, hydrogen sulfide, and carbonyl sulfide. Further investigations should be undertaken using coal extracts and coal.

Initiative: Characterization of Response to Environmental Factors
Objective: To study the effects of physical and chemical variables on microbial growth and to interpret the responses to different stimuli in terms of reaction path mechanisms
Discussion: Microbial growth rates and the production of inducible enzymes can vary widely depending on chemical and physical factors in the growth medium. At a certain pH, temperature, or in the presence of certain substrates, an entire enzyme system may be absent. Alternatively, with promoters or inducers added, activity may exceed expectations. Other environmental parameters may include aeration or susceptibility to damage from factors such as high shear rates. These factors need to be investigated for potential cultures while research is at the bench or small pilot scale as a precursor to the optimization of reactor conditions.

Initiative: Data Base on Anaerobic Microbiology
Objective: To develop a data base to catalog information pertaining to the various reaction mechanisms, intermediates, end products, and enzymes associated with the anaerobic conversion of coal to methane
Discussion: The result of this initiative would be a compilation of existing information on anaerobes as well as new data generated through this and other initiatives. Anaerobic organisms have been studied much less extensively than aerobic organisms, due to greater difficulty in culturing. As a result, much less is known about their growth requirements, enzyme systems, metabolic pathways, and so on. These organisms have great potential for energy conversion because of their inherent efficiency. This translates into a 95% conversion into product for anaerobes versus a 50% conversion for aerobic cultures. The balance is converted to cell mass, so the disposal volumes (and therefore cost) for spent biomass (sludge) are comparably less for anaerobes as well.

1.6.3 Enzymology

As demonstrated in Fig. 1.6, the enzymology category promotes research into the basic biochemical mechanisms by which microorganisms transform substrates into products. Necessary activities along with the isolation and characterization of enzymes capable of desulfurization, liquefaction, or gasification reactions are the development of more reactive enzymes and more efficient means for using enzymes. This type of work includes research on immobilized enzymes and carriers as well as the synthesis of catalysts that mimic enzymatic reactions.

Initiative: Isolation and Characterization of Enzymes for Coal Desulfurization, Liquefaction, and Methanation
Objective: To extract enzymes that improve or degrade coal and define their physicochemical parameters
Discussion: Enzymes are the central factor in all biological processes. They determine the speed and specificity of biological reactions as well as whether certain conversions will occur at all. Efficient application of enzymes as catalysts requires knowledge of their

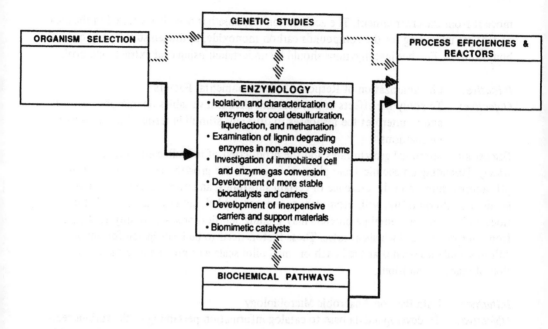

Figure 1.6 Research initiatives related to alterations of coal by specific microbial enzymes. Although this work is related to the categories of *biochemical pathways* and *genetic studies*, the research is not developed to the point of smooth information flow.

physicochemical parameters, such as binding constants, reaction rates, effects of immobilization, and so on. Prior to these characterization studies, isolation and purification methods must be devised, including assays using coal substrates or model compounds. Data generated in this effort can then be used to support genetic studies, the investigation of metabolic pathways, and reactor design. As such, this is one of the most important initiatives within the proposed research plan. Research has begun in the characterization of enzymes responsible for coal liquefaction, but no other enzyme work has been initiated.

Initiative: Examination of Lignin-Degrading Enzymes in Nonaqueous Systems
Objective: To determine whether or not oxidative enzymes capable of degrading lignin can degrade coal in nonaqueous media
Discussion: Nonaqueous environments are currently under investigation for microbial systems because it is possible to attain greater stability and versatility with enzyme-catalyzed reactions when they occur in organic solvents. Nonpolar products, which would have a higher fuel value, might also result from reactions carried out under conditions where the water concentration is limited. The extracellular enzymes produced by lignin-degrading fungi can liquefy coal and can operate in nonaqueous environments. It is not yet known whether coal liquefaction could be achieved by these enzymes in nonaqueous environments, but the potential is high.

Initiative: Investigation of Immobilized Cell and Enzyme Gas Conversion
Objective: To biologically produce and convert gases such as methane and other hydrocarbon gases to products of greater value

Discussion: Immobilized cell and enzyme technology has received attention lately because of additions to operational stability, continuous operation capability, high cell/ enzyme concentrations, and decreased aeration requirements. A good foundation of work has been performed on the production of methane from synthetic coal gas (primarily carbon monoxide), including feasibility studies using immobilized cells. Work has just begun on the conversion of coal gas to liquid fuels using microorganisms. The conversion of methane to products such as methanol can be performed catalytically; equivalent bioprocesses have yet to be developed.

Initiative: Development of More Stable Biocatalysts and Carriers
Objective: To examine the structure and function of enzymes and cells to identify molecular features that can be correlated with catalyst reactivity and stability
Discussion: Immobilized enzymes and immobilized cell processes are key developments in modern microbial process designs. The use of these techniques provides a higher density of reactive sites and higher throughput rates along with decreased costs in areas such as agitation and aeration. Advances are needed in terms of enzyme/cell stability after immobilization and in designing carriers that are stable to temperature, pressure, medium components, and so on. Some information in this area can be drawn from research in the food and the pharmaceutical industries. No work related to coal processes is currently under way.

Initiative: Development of Inexpensive Carriers and Support Materials
Objective: To examine the mechanisms of bacterial attachment to solid surfaces to determine what qualities are essential to good support materials
Discussion: Large-scale immobilized cell reactors will require high-surface-area materials in large quantities for cell and enzyme attachment. To reduce the process costs, these materials should be inexpensive and easy to manufacture. In the past, ordinary materials such as rubber rings, glass beads, and gravel have been used as adsorptive supports, as well as polystyrene or agarose beads for chemical binding of cells or enzymes. It is necessary to investigate the mechanisms of bacterial attachment so that more efficient support materials can be chosen or so that new materials can be designed. As in the previous initiative, some information can be found in food and pharmaceutical research.

Initiative: Biomimetic Catalysts
Objective: To study the chemistry of enzymatic coal conversion to methane in order to develop biomimetic catalyst systems
Discussion: The identification of coal reactive enzymes and the examination of their reaction mechanisms is necessary to permit the design of larger-scale biological processes. The development of catalysts with better properties than enzymes may also be part of the process design. Biomimetic catalysts need not be enzymes themselves, but need only possess the functional characteristics that promote specific reactions. Quantum mechanical and semiempirical molecular modeling tools can be used to predict the functional character and reactivity of coal and to examine the reactive sites of pertinent enzymes. Such analysis can be used to design nonprotein catalysts for coal conversion processes, which might have a smaller size and greater specificity than enzymes. Because the enzymes useful to coal conversion have not been well purified or characterized, no work is yet possible in this area.

1.6.4 Genetic Studies

Research constituting the genetic studies category is concerned with elucidating the chromosome organization and control mechanisms of bacteria and fungi that are active in coal conversion processes. Once these mechanisms are known, cloning and transfer of genes becomes more efficient. In addition, alteration of control regions on native DNA can be used to increase the production of pertinent enzymes. Suggested work in these areas is presented in Fig. 1.7.

Initiative: Genetic Analysis and Study of Chromosome Organization
Objective: To identify and map genes responsible for coal-degrading enzymes and to estimate the size and organization of chromosomes
Discussion: Lignin-degrading enzymes such as laccase have tentatively been identified in coal solubilization experiments. Enzymes for other coal alteration processes, such as desulfurization, have yet to be identified. Once these enzymes are isolated, the corresponding genes can be found. At that point in time, techniques of recombinant DNA technology can be applied in conjunction with classical genetic techniques to map the pertinent genes and determine the interaction between structural genes and control regions on the chromosome. Knowledge of this interaction is necessary to the construction of organisms for large-scale processing. No work has been initiated in this area, with the exception of identifying plasmids in organisms pertinent to coal research.

Initiative: Cloning and Characterization of Genes for Coal Desulfurization, Liquefaction, and Gasification
Objective: To isolate pertinent genes for transfer to cloning organisms for *in vivo* or *in vitro* production of enzymes

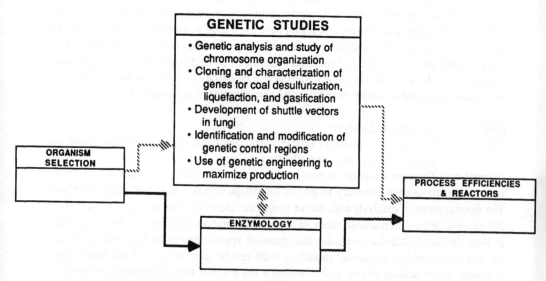

Figure 1.7   Research initiatives concerned with chromosome organization and microbial control mechanisms. Advances in this area may speed related research but are dependent on basic research in microbial physiology.

Discussion: As the enzymes and corresponding genes for microbial desulfurization, liquefaction, and gasification processes become known, it will be necessary to locate these genes on the pertinent plasmids or chromosomes and to identify control regions so that one can amplify the production of genetic material for further characterization. Once the gene has been characterized, further modifications can be made to enhance the activity or production of the protein product. This method of genetic alteration provides more control and specificity with a greater assurance of producing desired traits than do classical genetic techniques. One or two reports have appeared documenting the discovery of plasmids in organisms pertinent to coal processing. However, information related to coal conversion functions may not be coded on these plasmids.

Initiative: Development of Shuttle Vectors in Fungi
Objective: To isolate episomes from filamentous fungi for synthesis with bacterio-phages, thereby creating elements capable of mobilizing and transferring genetic material from a given source, bacterial or fungal, to coal-gasifying microorganisms

Discussion: Shuttle vectors are the prime tool of the genetic engineer in mobilizing genes from one organism to another, possible unrelated organism. Shuttle vectors have the capacity to replicate in two or more hosts and are especially useful for transfers from fungi to bacteria and back. Because fungi are critical to the solubilization of coal, yet difficult to handle in submerged culture such as in a large bioreactor, shuttle vectors may be used to transfer pertinent genetic material to a more amenable host. No shuttle vectors have been developed for the organisms pertinent to coal bioprocessing.

Initiative: Identification and Modification of Genetic Control Regions
Objective: To identify and modify the promoter and operator regions of genes to stimulate greater production of enzymes in organisms that produce methane from coal

Discussion: The control regions on a plasmid or chromosome determine how much of an enzyme is produced and under what conditions. In other commercial areas, such as the food industry, microorganisms have been forced to overproduce a desired product through manipulation of these genetic control regions. The modern techniques of recombinant DNA technology serve as enhanced tools for these manipulations. Enzyme activities must be identified and mapped onto specific plasmid or chromosome fragments as inputs to this initiative.

Initiative: Use of Genetic Engineering to Maximize Production
Objective: To genetically engineer microorganisms to produce cultures capable of improved conditioning and gasification of coal

Discussion: Because energy production is a high-volume, low-product-value process, efficient organisms and processes will be necessary for economic competitiveness. Recombinant DNA technology is at a stage of development where it can be applied to soil bacteria and other microorganisms of interest to energy production. Plasmid DNA, which is necessary for the transfer of genetic material, has been discovered in *Thiobacillus* and *Pseudomonas* species (desulfurization), certain filamentous fungi (liquefaction), as well as other organisms. DNA from the methane producer *Methanococcus vanielli* (gasification) has been cloned in yeast. As yet, little work has been initiated on identifying the genetic elements that code for and control the production of desired energy products.

1.7 PROCESS RESEARCH

This stage of the proposed research plan is responsible for core research in novel reactor concepts and supporting reactor characterization studies. The group of research initiatives is contained within the single category "process efficiencies and reactors," shown in Figure 1.8. No specific research projects related to these initiatives have begun, although some work in this area is under way as part of other projects. On an overall basis, this stage can be considered as being at a detailed planning stage, awaiting the identification of specific applications.

Two separate facets of bioreactor research are considered in this category. On the one hand, it is necessary to design and develop new concepts in reactors and to investigate the controlling parameters of these new systems. On the other hand, characterization and optimization of any bioprocessing system should be examined long before it has progressed to the demonstration stage. Accordingly, one section of these initiatives deals with the subject of reactor design, while the next considers reactor characterization and optimization.

Initiative: Development of Low-Temperature Nonbiological Gasification Processes
Objective: To develop novel catalysts for existing gasification processes that would permit lower operating temperatures
Discussion: Biological reactivity is limited to temperatures below approximately 150°C, although some rare microorganisms are known to exist at even higher temperatures. Therefore, the application of microbial systems to the treatment of preheated feed streams or hot effluent streams from conventional (nonbiological) gasifiers may not be possible until catalytic gasification is achieveable at 200°C or lower. Progress has already

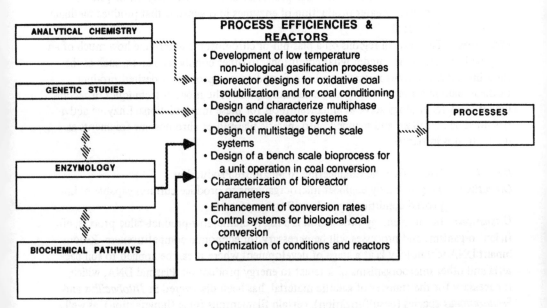

Figure 1.8 Research initiatives related to reactor design and characterization. Research in this category must integrate coal research with advances in biochemical engineering. Work is dependent on the development of hardy, efficient strains of microorganisms.

been made in synthesizing novel catalysts for low-temperature liquefaction of coal, and this is encouraging in regard to direct gasification research. Development of this technology would facilitate the addition of biological gas cleaning and upgrading steps without excessive loss of operating efficiency resulting from the need to cool the process gases to suitable organism exposure temperatures.

Initiative: Bioreactor Designs for Oxidative Coal Solubilization and for Coal Conditioning

Objective: To design biological systems capable of catalyzing oxidative coal solubilization, sulfur and metals removal, and methanation, possibly involving symbiotic microorganisms

Discussion: The effectiveness of established aerobic cultures in coal desulfurization, demineralization, and liquefaction should be compared in a number of pilot-scale bioreactors to identify the optimal design for the process. Pretreatment of coals may also show some significant effects in these environments. Although not defined as a specific project, there is ongoing work in this area. Several research groups have proposed reactor designs for coal desulfurization (inorganic and organic), coal liquefaction, and upgrading of coal gas. These reactors are primarily bench scale, although designs for pilot-scale bioreactors have been devised.

Initiative: Design and Characterization of Multiphase Bench-Scale Reactor Systems

Objective: To examine the use of various solvent/microbe compositions in facilitating the mass transfer of biocatalysts to molecular sites on coal substrates

Discussion: Microbial reactions—especially those with an insoluble substrate such as coal—require the contact of reactants present in the gas, liquid, and solid phases. The effectiveness of microbes in degrading coal can be severely limited by the lack of connectivity in the pore structure of the coal substrate, which results in lack of mass transfer within the coal particle. Furthermore, the cross-sectional area of the pore throat space available to flow is a function of the wettability preference of the surface as well as the viscosity of the permeating liquid. The passage of microorganisms into these pores might therefore be restricted in certain solvent systems. The path of biocatalysts to their target sites can also be restrained by molecular geometry. Some work on the use of organic solvents with enzymes for coal liquefaction has recently begun, and several new approaches are expected soon.

Initiative: Design of Multistage Bench-Scale Systems

Objective: To conduct bench-scale tests to determine the effects of staged processing on biological coal gasification kinetics and product quality

Discussion: Microbial coal cleaning and processing is highly desirable because external energy requirements for processing are low. Microorganisms may metabolize inorganic sulfur for energy or may oxidize organic sulfur, organic functional groups, or metals in response to their toxicity. In so doing, some direct their acquired energy to the production of useful gaseous products, fuel, or fuel intermediates. Because of the multiplicity of plausible reaction pathways, each of which is associated with a specific thermodynamic potential, in theory a number of feasible approaches to biological gasification are possible. There is, however, a need to identify the most efficient routes. Again, some of this work has been integrated into the experimentation of other projects. There is no focused approach to research in this area.

Initiative: Design of a Bench-Scale Bioprocess for a Unit Operation in Coal Conversion
Objective: To immobilize and test several enzymatic or cellular biocatalysts under a variety of operating conditions and to isolate products for further analysis
Discussion: A unit operation is a single conversion step in a process. It is of great merit to examine, for both academic and practical interest, the extent to which coal can be converted in a single step. By isolating a series of such steps, a complete bioprocess can be analyzed and optimized. Altered groupings of steps might lead to alterations in the product composition or to improved efficiencies. No research is currently under way with such a fundamental approach.

Initiative: Characterization of Bioreactor Parameters
Objective: To investigate kinetic mechanisms and carry out product analysis to permit the scaleup of bioreactor designs
Discussion: Some of the physical and chemical parameters that affect biological coal beneficiation and conversion processes have been defined through laboratory work on both aerobic and anaerobic cultures. More work is still needed in some systems. The real challenge will be to reproduce, and even improve upon, the results obtained in the laboratory in the use of long residence cultures in continuous, high-throughput, stirred, immobilized reactors, and to scale up these reactors. Some of this work is ongoing, as work on organisms is transferred from surface to submerged culture. No research in the biochemical engineering of these cultures is yet under way.

Initiative: Enhancement of Conversion Rates
Objective: To increase the contact surface area in biological coal conversion by decreasing particle size and to enhance reactivity by optimizing other variables, such as suspension density and viscosity
Discussion: The efficiency of heterogeneous reactions is, in broad terms, controlled by the degree to which the fluid and solid surfaces, or the reactant surfaces, are in contact. Ideally, adsorption of reactants onto, and desorption away from, the solid surface should be maximized, so that the overall reaction rate is controlled by the reaction kinetics. For maximum reactivity there is also an optimal concentration of fluid reactant molecules per unit area of solid surface, which is usually quantified in terms of bulk density. The effect of pulverization on the viscosity and density of the suspension as well as conversion rates for various particle-size fractions should be investigated.

Initiative: Control Systems for Biological Coal Conversion
Objective: To identify control methods, apparatus, and equipment for increasing the reliability of conversion and treatment process data
Discussion: In many areas of bioprocessing it is now possible to maintain control of the ongoing fermentation by means of on-line sampling and adjustment of reactor parameters. Since each fermentation, even though using common organisms and standard feed materials, is unique, this capability is of great use in maintaining a high specific production rate. Two types of instrumentation must be developed for these applications. First, new methods are needed to sample the process without disturbing the ongoing reactions. To accomplish this objective, one must investigate ways of determining and measuring those parameters that provide the best indication of the progress of the fermentation. These methods can include noninvasive methods to detect the health of the organisms or on-line analysis of the product stream. The second objective is to design and construct feed-

back mechanisms, primarily computer controlled, to alter the bioreactor conditions in response to the given analyses. A great deal of research in both of these areas is under way at universities, although none is specifically focused on the bioprocessing of coal. More information is needed about the coal organisms and their reactions as well as the products of these reactions.

Initiative: Optimization of Conditions and Reactors
Objective: To develop better biocatalysts and to design bioreactors with improved
            nutrient transport efficiencies as well as reduced equipment size and cost
Discussion: The efficiency of microorganisms, enzymes, or other biocatalysts in coal
conversion processes is highly dependent on physical and chemical conditions. Reactor designs must be based on optimal temperature, salinity, pH, and so on, and these conditions must be determined for each reaction medium. Coal particle size and rate of agitation may also exhibit strong influences on reaction rates. This research initiative focuses on the optimization of chemical and physical parameters for gas-producing, fluidized (agitated) systems and for fixed-bed systems in which liquid feed and products can percolate through crushed coal. In microbial systems, the attainment of steady-state conditions by gradual addition of catalysts, nutrients, buffers, or inoculum will be emphasized.

1.8 PROCESS DEVELOPMENT

This is the ultimate research stage in the proposed research plan for coal bioprocessing. During the course of this stage the work on smaller-scale processes and their optimization, which was accomplished during the process research stage, will be examined and the most promising processes selected for scaleup and demonstration of the new technology. Five processes have been selected for this report, within the category "processes," which seem to have potential at the current time. These processes are presented in Fig. 1.9. The actual dynamics and structure of these end processes will, of course, be determined by the research that is currently under way.

Although this is the final stage in the development of a commercial biogasification process, other supporting work will be necessary before an actual plant can be built. A

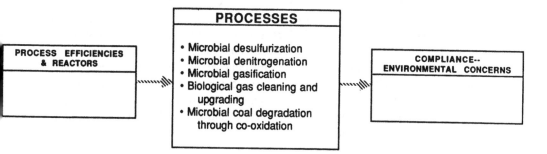

Figure 1.9 Research initiatives designed to produce working processes for the demonstration of new coal-processing technology. These initiatives correspond to individual processes with some expectation of accomplishment. Results of research at earlier stages will determine the actual system for demonstration.

large amount of this supporting work will entail the examination of the environmental implications of the new technology or of the specific plant configuration. This work is described in Section 1.9. There has been no prior work in process design and development that might be suitable for this program because supplementary research on coal biogasification has not progressed to the point where concepts or preliminary designs could be configured. The fact that none of this work is under way is not a source of concern at this time. Coal bioprocessing is a very young discipline. The overall effort is small and has focused primarily on understanding the fundamental behavior of the system. At the point where this understanding is sufficient to suggest demonstration projects, this fourth stage of the research plan will be implemented. It will no doubt be possible at that time to draw on bioprocessing experience in other areas, such as sewage treatment, pulping operations, and other biological industries, with respect to unit operations, materials handling, and process controls.

Initiative: Microbial Desulfurization
Objective: To identify bacteria that attack carbon-sulfur bonds specifically, converting the sulfur to oxygenated species that would be removed by subsequent separations processes
Discussion: The biological removal of toxic or process-limiting coal constituents can compete economically with conventional energy-intensive coal-cleaning methods. The mild process conditions required and the inexpensive feedstock materials, coupled with the ability of the organisms to regenerate the catalytic activity continually, are factors that provide economic incentive for the development of batch or continuous reactors capable of emulating laboratory and pilot bioprocesses for sulfur and metals removal. Work on the removal of inorganic sulfur using either *Thiobacillus ferrooxidans* or *Sulfolobus acidocaldarius* is the most advanced of these smaller-scale processes. Projections are currently under development on capital and operating costs of larger-scale reactions.

Initiative: Microbial Denitrogenation
Objective: To isolate microorganisms that degrade nitrogen-containing organic coal constituents, and to optimize the conditions under which such microorganisms will grow and produce the desired enzymes
Discussion: Nitrogen is another constituent of coal whose removal may prove more cost-efficient prior to combustion. Because nitrogen is often a limiting nutrient for the growth of microorganisms, it is likely that suitable biocatalysts can be developed to remove the organically bound nitrogen from coals. It may also be possible to engineer an organism to combine the capabilities of sulfur and nitrogen removable. Developing a commercial process to remove organically bound nitrogen from coals would require the isolation of cultures that specifically attack organic nitrogen compounds such as nitrogen heterocycles, subsequent strain characterization and improvement, and then the appropriate design and optimization of processes utilizing these organisms consistent with an overall coal conversion or combustion process.

Initiative: Microbial Gasification
Objective: To enhance the anaerobic fermentation of products derived from coal bioprocessing to obtain useful gaseous products such as hydrogen and methane
Discussion: A demonstrable large-scale microbial gasification process is the consummation of the research path discussed in this document. The objective of this work is to pro-

duce pipeline-quality gas or high-value gaseous products using raw coal as the starting material. A popular process concept involves prior degradation of the coal to provide a liquid material for subsequent production of methane. Parallels have been made to the pulp and paper industry, where lignocellulosic feedstocks are hydrolyzed prior to further processing. Supporting technologies for a coal biogasification process include coal beneficiation as well as gas cleaning and upgrading, both of which can be performed by microbial means. Necessary steps toward the development of a commercial biogasification process are the design of an effective multiphase and possible multistage reactor, as well as further characterization and optimization of potential processes.

Initiative: Biological Gas Cleaning and Upgrading
Objective: To develop a biological process for the economic removal of hydrogen sulfide and other contaminants from gasifier effluent gases and/or the upgrading of these gases to a product of higher value
Discussion: It will be possible to supplement existing gasification technology with biological process steps such as gas upgrading long before a complete microbial gasification plant is developed. This retrofit concept—that of substituting existing unit operations with biological reactors—will aid the development of a biogasification process by allowing the commercialization of intermediate steps in biogasification and by providing a smaller-scale means for energy process engineers to gain experience with bioreactors. Because microorganisms are able to convert gases such as carbon monoxide and carbon dioxide to compounds with a higher fuel value and are also tolerant of, or can remove, sulfur gases, microbial processes have a great potential for application in the areas of coal cleaning and upgrading. Smaller-scale processes are already available for gas upgrading. The addition of organisms that can metabolize hydrogen sulfide and carbonyl sulfide and subsequent coupling of these reactions to conventional gasification technology are required for the success of this initiative.

Initiative: Microbial Coal Degradation Through Co-oxidation
Objective: To evaluate co-oxidative biological processes for more efficient coal conversion
Discussion: As an example of co-oxidative processes, several bacteria are capable of oxidizing normal C_{10} to C_{20} alkanes, producing glycolipid surfactants that reduce the oil-water interfacial tension sufficiently to form low-viscosity oil-in-water emulsions. Such species have potential applications to pipeline coal transportation, coal processing, and coal process wastewater cleanup. A specific application to coal biogasification would be the development of a culture that can consume some of the alkyl side chains of coal while permeating and disrupting the coal matrix. In addition, extracellular extracts from these organisms might also be capable of permeating the coal matrix. The development of such a novel approach to biological coal conversion will require the characterization of a wide range of organisms for co-oxidative properties, examination of the products of such reactions, and the subsequent design and scaleup of suitable reactors.

1.9 COMPLIANCE: ENVIRONMENTAL CONCERNS

Compliance is an area of investigation that can determine the fate of a commercial process. Although this research is often neglected until brought to the developers' attention by regulatory requirements, the amount of research necessary and possible alterations in and disruptions of plant operations can be costly if they are not investigated early enough in the research process. These concerns are especially important for the use of bioprocesses because of human health concerns as well as noxious effects on the environment.

There has been no prior environmental work tailored to possible deleterious effects of coal biogasification because as yet there are no processes at even the pilot stage. However, a great deal of research has been performed on wastewater treatment of thermal coal gasifier wastes and similar activated sludge processes. These processes are similar to at least the methane-producing stage of coal biogasification and can be used as models for potential implications concerning methanogenesis. Until the various coal reactive microbial strains can be characterized and their biological behavior identified, it is difficult to determine any negative or positive effect on the environment from routine use or accidental discharge at other stages of biogasification.

In addition to acquisition of strains by isolation and selection, the use of genetic engineering techniques to improve process efficiencies may produce microorganisms that are resistant to natural constraints or that may during their lifetime transfer genetic characteristics to other species. Industries that have begun using these engineered organisms, such as the agricultural and pharmaceutical industries, are investigating potential environmental concerns. This research is at a preliminary level and relates to very specific organisms. The organisms currently used for coal conversion do not resemble these microbes, and environmental effects of their use or release must be investigated before a commercial plant is approved.

It is important that this research be implemented as soon as possible, as experience has shown that environmental and institutional problems often determine the pace of commercialization of new technologies. It would be possible for research in the five initiatives listed in the category "compliance–environmental concerns" to keep pace with the genetic research proposed in the category "genetic studies" at the biochemical research stage of the proposed research plan if there is early recognition of compliance requirements and if the appropriate planning schedules are developed. Potential areas of research are presented in Fig. 1.10.

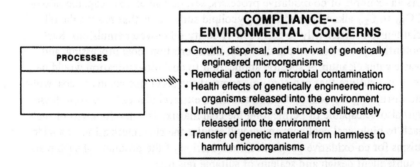

**COMPLIANCE--
ENVIRONMENTAL CONCERNS**

PROCESSES

• Growth, dispersal, and survival of genetically
  engineered microorganisms
• Remedial action for microbial contamination
• Health effects of genetically engineered microorganisms released into the environment
• Unintended effects of microbes deliberately
  released into the environment
• Transfer of genetic material from harmless to
  harmful microorganisms

Figure 1.10 Research initiatives devised to support compliance with environmental regulations related to coal handling and release of microorganisms during processing. Appropriate planning schedules must be developed so that this work can keep pace with organism development at the *biochemical research* stage of the research plan.

Initiative: Growth, Dispersal, and Survival of Genetically Engineered Microorganisms
Objective: To determine the mechanisms and assess the likelihood of microbe dissemination by insects or other animal vectors, by air and by water, in order to predict the survivability and growth of genetically engineered microorganisms
Discussion: The site of release of a nonindigenous organism is usually not the site where it can create a deleterious change. This often occurs because of the absence of suitable host animals or plants or the absence of suitable environmental conditions at the release site. Risk assessment thus requires information on the mechanisms and likelihood of dissemination of these organisms. The fields of epidemiology and pathology have developed good bases of knowledge on known pathogens, but little attention has been given to the organisms of current or likely future interest in biotechnology. There is also limited understanding of the traits contributing to successful dispersion of these microorganisms. In addition to dispersion, any organism—including engineered organisms—presents a problem only if it can grow and persist in the natural environment. It is not now possible to predict survival for either regulatory purposes or for the practical exploitation of microorganisms designed for biotechnological use. The current data base must be expanded for accurate prediction of which organisms will and will not proliferate and survive.

Initiative: Remedial Action for Microbial Contamination
Objective: To develop methods of containing and destroying genetically engineered organisms in the event that such organisms are accidentally released into the environment and found to be harmful
Discussion: A means must be found to contain and possibly destroy genetically engineered organisms in the unlikely event that despite research conclusions and testing before release, the organism is found to be harmful. Some precedents exist in the fields of human and veterinary medicine and plant pathology for means of remedial action, but the possible differences in the environmental behavior of genetically engineered microorganisms dictate that research be initiated for containment and remedial action. Attention should be given to adapting these precedents to altered organisms and the alternative conditions obtained in coal bioprocesses.

Initiative: Health Effects of Genetically Engineered Microorganisms Released into the Environment
Objective: To develop a base of information and expertise in public health as it pertains to the risk of communicable diseases deriving from genetically engineered microorganisms
Discussion: An investigation of the effects on public health, the risks of transmission of communicable diseases, and applications of relevant biomedical disciplines are needed to develop scientifically defensible risk assessments. As part of the development of this data base, knowledge must be obtained on the relationships between the microorganisms used for biotechnology purposes and human pathogens as well as toxin-producing organisms. Other important information concerns the transfer of genetic traits between organisms plus the dispersal and survival characteristics of implicated strains.

Initiative: Unintended Effects of Microbes Deliberately Released into the Environment
Objective: To determine the potential health and environmental effects of new
                 microbes other than the specific effects for which such organisms were in-
                 tended
Discussion: Microbes will be deliverately released to the environment as part of *in situ*
processes such as heap leaching. This deliberate release may not be expected to cause any
environmental damage. However, both the public sector and the scientific community are
concerned with the effects of microbes on natural communities and natural ecological
processes. EPA, for example, has had a research program aimed at measuring the effects
of chemicals on such communities and processes and has considered monitoring the influence of chemicals as part of its regulatory mission. Products of biotechnology need to
be evaluated in different ways from chemicals because these organisms are highly specific
for the plants, animals, and microorganisms that they may invade or harm, and the
generalizations so useful in toxicology may have little meaning for potential communicable agents. New methods of evaluation must be developed together with suitable testing
protocols, in order to generate the data necessary for informed decisions.

Initiative: Transfer of Genetic Material from Harmless to Harmful Microorganisms
Objective: To increase the understanding of gene transfer between microorganisms
                 under conditions that exist in nature
Discussion: Genetically engineered species that are harmful to the environment or to
human health may not themselves survive and multiply upon release. However, relevant
genetic information could be transferred *in situ* to other organisms. The latter organisms
might ultimately become harmful as a result of this transfer. Although enormous progress
has been made in recent years in increasing our understanding of gene transfer under highly artificial conditions, little information exists on gene transfer in nature. Suitable model
systems must be developed to allow laboratory testing of gene transfer, followed by
small-scale contained tests in natural settings. A greater understanding of the role of
vectors such as plasmids or bacteriophage in facilitating cross-species genetic transfer will
also support this work.

1.10 CONCLUSIONS

The foregoing summaries indicate that much of the research at the foundation stage is
under way or available from other research areas, such as coal science. Efforts should continue in all these areas to provide supporting information for later development of efficient processes as well as to enlarge the selection of organisms for future applications.

Research at the biochemical research stage is now under way and publications are
just beginning to appear in the open literature. This research is the key to all further developments and should be expanded. A much greater emphasis should be placed on determining biochemical pathways and mechanisms of action. Research is needed to elucidate specific enzymes responsible for transformations of coal and coal products. In addition, the focus can now shift from the collection of new organisms to modifications to
existing cultures through acclimation or through the use of genetic engineering techniques.

Existing investigations at the process research stage that are part of other tasks
should be continued as independent investigations. Efforts can be made at the present
time to apply new biochemical engineering techniques to coal bioconversion processes.
These include new reactor concepts, as well as new methods for controlling the progress
of the fermentation. Several existing laboratory efforts are now ready for transfer to this
stage.

It is not yet time to consider a concerted effort at the process development stage. Research results are too preliminary to predict the details of a commercializable process. However, it is not too early to begin consideration of the factors that will contribute to a successful commercial process. Flowsheets and benefit/cost predictions alone do not produce a working plant. Some consideration must be given to retrofit technology to bring these processes into existing plants, and to policy and regulatory issues that may exert a controlling influence in the future. Some discussion of these factors has occurred in relation to microbial coal processing and it is expected that these discussions will increase as progress is made at earlier stages of research.

The environmental impact stage, although listed last for the purposes of this research plan, should actually be implemented far in advance of process development. Much of the research described in this section is at the level of biochemical research and is a logical outgrowth of those initiatives. Answers must be found to many of the questions presented in Section 1.9 since it is unlikely that future regulations will be less strict than those currently in place. Some of these answers will be found in agricultural investigations or other environmental work, while some may be unique to coal processes.

ACKNOWLEDGMENTS

This work was supported by a contract with the Idaho National Engineering Laboratory. In addition, I would like to thank the following people, without whom this work would not have been accomplished: Douglas Blake, Shirley Frush, Peter Gray, Jaffer Mohiuddin, John Rezaiyan, Jagdish Tarpara, Dale Windsor of Sheladia Associates, Inc.; Young Yoo of Seoul National University; panel members: Patrick Dugan, William Finnerty, John Larsen, Nawin Mishra, Dharam Punwani, Edward Sybert; Michael McIlwain of the Idaho National Engineering Laboratory.

2

Biotechnology and Coal: A European Perspective

GORDON R. COUCH *IEA Coal Research, London, England*

2.1 INTRODUCTION

Traditional routes for coal conversion to high-grade gases and liquids usually involve coal cleaning, followed by treatment at high temperatures and pressures through a series of chemical processes and separation stages. These involve the close control of reaction conditions and high capital and operating costs, which make many processes uneconomical in current circumstances. The use of biotechnology could provide alternative process routes for coal conversion and/or could provide alternative methods for coal cleaning (e.g., sulfur removal) prior to use. In addition, its use might provide a method for exploiting the energy held in deep coal seams which cannot be extracted by conventional methods.

Biotechnology is a multidisciplinary field with its roots in the biological, chemical, and engineering sciences. The interrelationship is represented in Fig. 2.1. It arises from the combination of a number of specialist subjects, such as molecular genetics, microbial physiology, and biochemical engineering. It has been variously described, and is defined for the Organization for Economic Cooperation and Development (OECD) as "the application of scientific and engineering principles to the processing of materials by biological agents to provide goods and services" [14]. The subject is complex, due partly to its multidisciplinary base. It is inevitably affected by an innate conservatism, and there are difficulties associated with the mutual lack of understanding and communication between microbiologists and engineers [25]. Many published studies and much of the work undertaken to date, both in the United States and elsewhere, has been by monodisciplinary teams, frequently by microbiologists, and sometimes by hydrometallurgists specializing in ore leaching. Some of the studies suffer from poor coal characterization [9].

Coal does not appear to be an ideal material for biological attack. It is a highly variable and heterogeneous mixture. Some of the sulfur in coal is organically bound. A representative chemical structure for bituminous coal is shown in Fig. 2.2, modified to

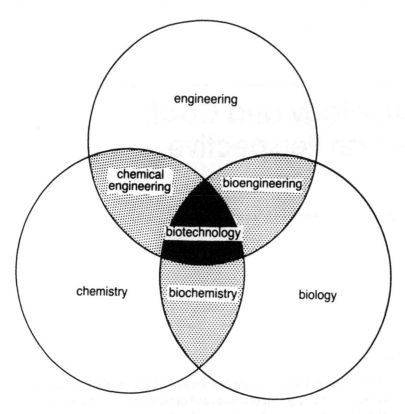

Fig. 2.1 Relation of biotechnology to scientific and engineering disciplines. (From Ref. 51.)

show the association with pyrite crystals. Given the background, it is perhaps remarkable that so much work has already been undertaken, and progress made. It is possible to discuss and assess the feasibility of a microbial desulfurization process and to report on microorganisms that bioliquefy both lignites and hard coals. The research and development work involves complex interactions, illustrated in Fig. 2.3. Most is laboratory based. There have been a few small-scale pilot-plant tests on desulfurization. Some work has been on model compounds or model mixtures rather than on coal itself. The interactions involve the mix of disciplines illustrated in Fig. 2.1, and this brings its own problems. The variability of coal (i.e., its nature and constitution) may be imperfectly understood by a microbiologist in a laboratory. Similarly, chemical engineers with wide practical experience may find it difficult to adapt their knowledge to the very precise nutritional and environmental requirements of particular microorganisms. Before exploitation, laboratory work should be replicated elsewhere, and in the course of some developments this has not proved to be easy. There can also be substantial problems in the scaleup of bioprocesses [44], and scaleup is regarded as a major obstacle to the successful commercialization of biotechnology-based products [45]. Some studies have suffered from an inadequate appreciation of the essential basis of other technologies and disciplines involved [9]. The Delft University work, in particular, has tried to take all the factors into account. Figure 2.3 represents some of the interactions involved that must be considered during the development of any possible process for large-scale use. In addition to the factors

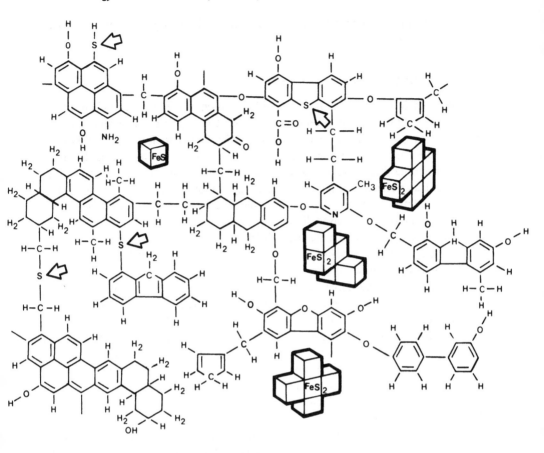

Fig. 2.2 Representative structure of bituminous coal showing sulfur linkages and the presence of iron pyrite FeS_2. (From Ref. 24.)

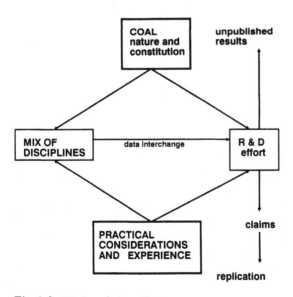

Fig. 2.3 Various interactions.

shown are economic considerations and a comparison with other possible process routes for achieving the same objectives.

2.2 THE NATURE OF BIOPROCESSES

A brief historical introduction and background discussion of the behavior and classification of microorganisms and the developments leading to the possibility of genetic engineering has been given to the IEA Coal Research review *Biotechnology and Coal* [19]. Considering the general nature of bioprocesses and their possible large-scale application, it is important to note that the process streams involved are commonly both aqueous and dilute. A particular bioreaction may take place only within a narrow band of reaction temperature and pH conditions, depending on the requirements of the particular microorganisms and/or enzymes involved. There may be competing or complementary reactions taking place, each of which may have different optimum conditions. The conditions used in a reaction vessel or series of reactors will be adjusted to obtain the best overall effect and may require precise process control.

For some bioprocesses, there is the need to maintain sterile conditions in parts (or all) of the process. This is necessary to ensure that once a reactor has been inoculated with a pure culture of the desired organism, or with a mixture of organisms, only those required to promote the particular reaction will multiply. Otherwise, undesired organisms may grow, resulting in the production of contaminating by-products. Even when non-sterile conditions are used, as with many desulfurization processes, the need to prevent the production of contaminating by-products is an important consideration. It is essential to use organisms that are highly competitive and to maintain the exact conditions under which they can operate.

Because the product stream from most fermentation processes is dilute and therefore large in volume, the capital cost of equipment, including that needed for dewatering and drying, is correspondingly high. Another characteristic of biological processes is that they are frequently inefficient in energy terms. Living creatures give out waste heat that gets into the product stream. The problem is that heat at 30 to 50°C is generally of little practical value. Cooling may be required to maintain the precise environmental conditions necessary for the growth of the organism. Work is going on to identify thermophilic organisms that grow at higher temperatures, and some bacteria are known to thrive above 100°C [13]. The diversity of bacteria living in hot-water areas, including sulfur bacteria, hydrogen-oxidizing bacteria, and elemental sulfur-respiring bacteria, is of particular interest. *Sulpholobus acidocaldarius* is a thermophilic organism with an optimal temperature range of around 70 to 80°C.

In principle, biotechnology can be used to bring about either a general breakdown of the coal molecule or to achieve the selective removal of particular components within its structure. Coal is basically carbonaceous material with mineral inclusions and it is such a heterogeneous substance that microorganisms may be able to attack, degrade, or use a number of different constituents. Bioprocesses might have significant advantages over the conventional methods of breakdown or cleaning currently in use, and they might open up the possibility of the economical production of new products. This is because the reaction conditions involved would be much "milder" than those required for the equivalent chemical transformations and because the production of specific products or the removal of certain impurities might be maximized. Simply because the coal molecule is large and the substance is such a heterogeneous mixture, coal does not appear to be an ideal material for biological attack.

The vast amounts of coal in the world exist because the carbon in the coal molecules was not degraded. This was due to the lack of oxygen and to imposed temperatures and pressures. Coal has biological origins and is largely insoluble in both water and many organic solvents, except under extreme conditions. The small size of the porous structure in most coals presents an obstacle to the ingress of microorganisms, and the outer surface is the main area available for attack. This is discussed in Section 2.3.2. In addition, there are substantial quantities of inorganic impurities present, in the form of quartz, clays, sulfur compounds, and others. There may be trace elements that will be toxic to various microorganisms [27].

Several routes for microbial coal modification are possible. Microorganisms may attack either the carbonaceous coal or the interspersed inorganic materials. One route is the depolymerization of the coal polymer and the breakage of various key links. This could provide the basis for liquefaction. A second route would be a reduction in oxygen content either through reduction of C=O to CH_2 or perhaps by decarboxylation to CO_2. This could improve the heating value. A third would be the removal of sulfur, nitrogen, or metals from the coal before combustion, which would reduce unwanted emissions [40].

In the 1960s considerable attention was paid to the possibility of using fine coal waste as a source of humus and to promote nitrogen fixation in the soil. Even the possibility of using coal fines as a basis for the production of microbiological food was considered [42]. None of these uses for coal fines has been developed on a commercial scale. In biological effluent plants treating the wastewaters from coal carbonization plants, microorganisms have been evolved that can break down process by-products such as phenols and higher-tar acids [41]. This suggests that certain structures in coal would be vulnerable to microbial attack. In addition, the phenomenon of methane formation in coal seams is well known and the feasibility of the biogasification of peat was demonstrated in the 1920.

Biotechnology has naturally tended to develop in the areas where it has been seen as most easily applicable, and consequently only a limited amount of work has been undertaken on coal. However, interest in possible applications has grown over the past few years and has been stimulated by the reported degradation of a North Dakota leonardite (an oxidized lignite) by the fungi *Polyporus (Trametes) versicolor* and *Poria monticola* [18], by the identification of microorganisms that attack bituminous coals [26], and by other work, particularly on sulfur removal.

The large complex molecular structures involved and their variability even in a coal of a particular rank mean that degradation to simple usable compounds is complicated, involving both reaction and separation. To find a microorganism or a series of microorganisms that will break down particular links or break into the ring structures is not likely to be an easy task. Chemical breakdown is also difficult, involving extreme conditions of temperature and pressure normally followed by a series of separation processes. There is considerable industrial experience with conventional chemical and pyrolitic processes starting with a coal feedstock, but these are costly, and bioprocesses might offer a more economical route to produce certain products. The development of bioprocesses might point the way for possible chemical conversions using milder and hence less costly conditions. The studies may also lead to new approaches for conventional chemistry as applied to coal.

2.3 COAL

2.3.1 Composition and Structure

Coal structure has been discussed by IEA Coal Research [20,21]. Essentially, the structure of mature coals consists of various arrangements or aromatic rings fused into small polycyclic clusters that are linked by aliphatic structures. Phenolic, hydroxyl, quinone, and methyl groups, among others, are attached to the rings [41]. The so-called Shinn model is regarded as a reasonably adequate interpretation of many of the structural parameters for hard coals, based on a U.S. bituminous coal (see Fig. 2.4). A comparison of the structure in coals of different ranks is presented in Table 2.1.

Much less work has been done to define the structure of the younger lignitic coals, but it is suggested that the first stage of the coalification reaction was the destruction of

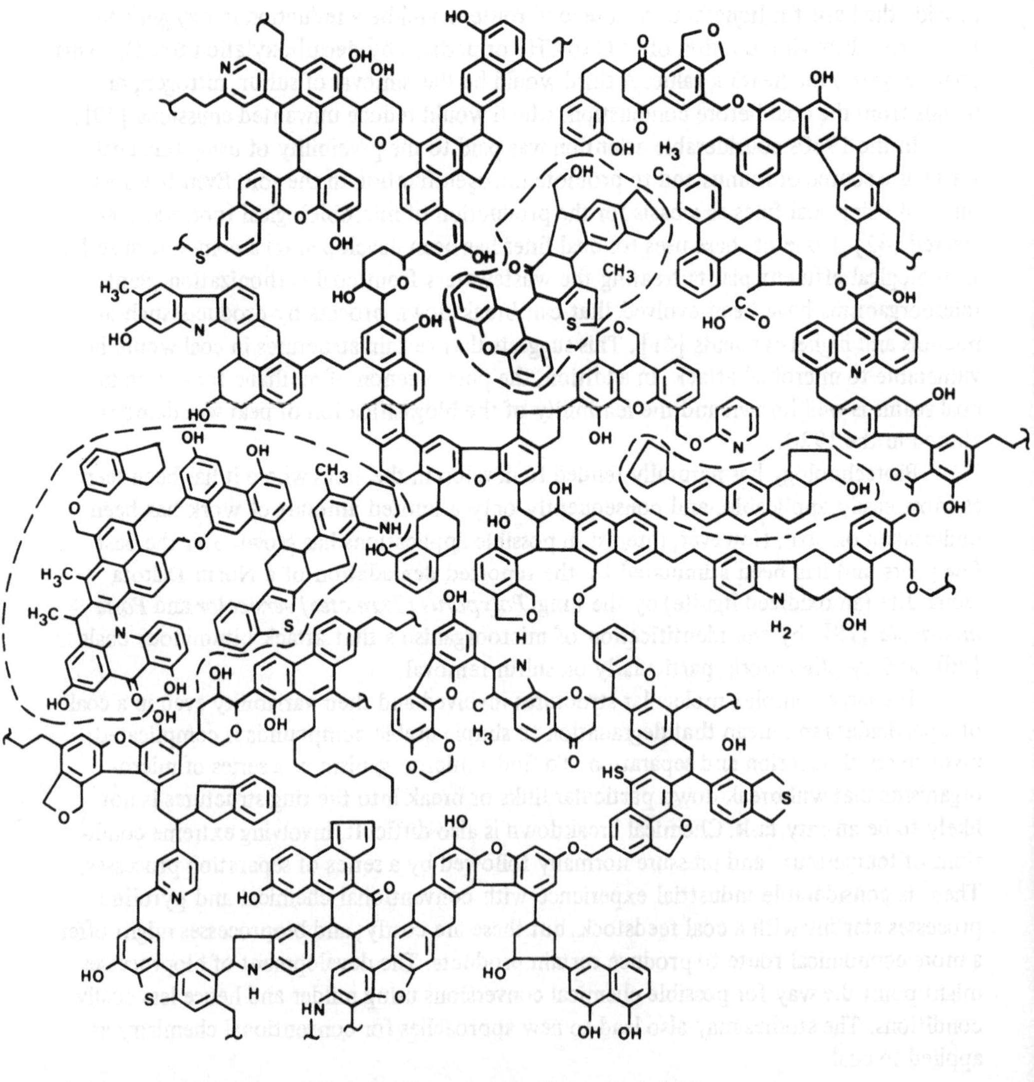

Fig. 2.4 Model of bituminous coal structure. (From Ref. 46.)

Table 2.1 Structural Comparison of Coals of Various Ranks

Coal rank	Carbon aromaticity (%)	Nature of monomers	Nature of cross-links
Lignite	30–50	Small, largely single-ring systems extensively substituted with O-functional groups ($-COOH$, $-OH$, $-OCH_3$); about 1 oxygen per 3 to 4 carbons	Many hydrogen bonds, probably some other cross-links; possibly salt bonds as in $COO-Ca-OOC$; few aliphatic cross-links; gel-like; water is an important structural component
Subbituminous	60	Still mostly single rings with some larger rings; O-groups on almost all rings ($-COOH$, $-OH$) less than on lignites; about 1 oxygen per 5 to 6 carbons	Mixture of hydrogen bonds and probably ethers; some aliphatic links though less important than ethers
Bituminous			
A	70	Mixture of ring systems, single ring still most common; about 1 oxygen per 9 carbons mainly $-OH$ functional groups	Mixture of aliphatic and ether cross-links
B		Significant increase in amount of larger rings; about 1 oxygen per 12 carbons, now almost entirely ring ether and $-OH$	Mostly aliphatic type, some scissile biphenyl types?
C	75–80	Degree of condensation of aromatics still greater; very few O groups; down to about 1 oxygen per 20 carbons	Nonreactive aliphatic bridges and biphenyl-type links?
Anthracite	95	Highly condensed aromatics, graphitic, commonly multiple rings; functional groups rare, only 1 oxygen to about 100 carbons	Almost entirely direct aromatic-aromatic links

Source: Ref. 46.

lignin to humic acids. Woody residues can intermingle with condensation polymers which are swollen by the water-soluble products from the degradation of cellulose and lignin [15,20]. A possible structure of brown coal, based on Rhenish material from West Germany, is shown in Fig. 2.5. Some of the variable parameters can be seen clearly in Table 2.1, where the nature of the monomers, the nature of the cross-links, and the degree of aromaticity are defined in coals of different rank. These data do not, however, highlight the variable presence of interspersed solids, such as noncombustible compounds forming ash, sulfur compounds, and trace metals. The ash content of a coal can be as high as 50% and the interspersed materials can have a significant effect on any coal usage process.

The presence of bacteriostats and bactericides in coal is an important consideration, affecting any possible bioprocess. Bacteriostats are selective inhibitors to the development of activity by microorganisms and are analogous to catalyst poisons. Bactericides have a

element	Rhenish brown coal (maf)
C	68.7
H	5.05
O	25.0
N	1.0
S (org)	0.25

typical composition

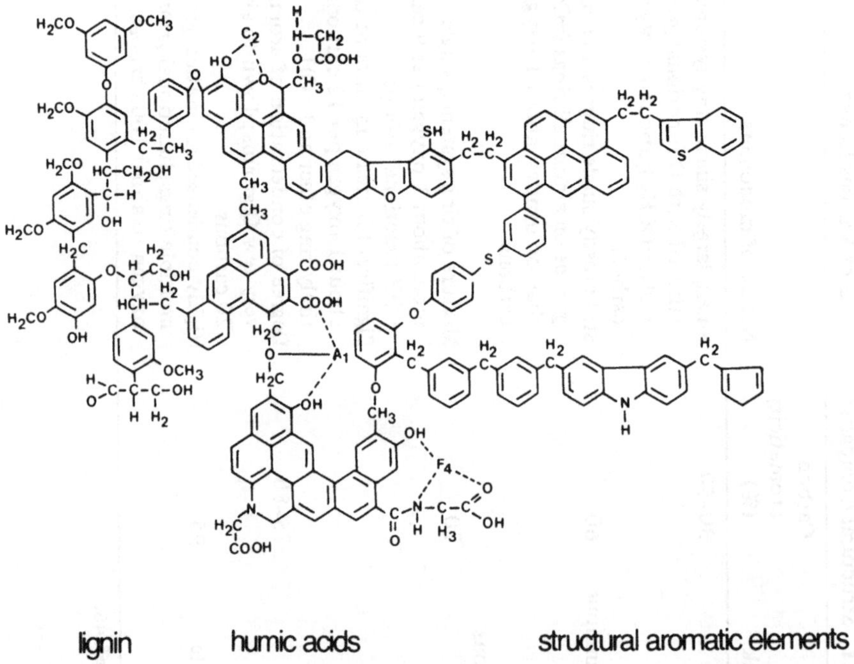

lignin   humic acids   structural aromatic elements

Fig. 2.5 Schematic composition of brown coal structure. (From Ref. 50.)

lethal effect on microorganisms either by interfering with a vital biochemical pathway or by destroying the molecular cell structure. They may either be present in the raw coal, or may be formed as a result of chemical reactions taking place during a particular process [32,43].

2.3.2 Porosity

In relation to the use of biotechnology, a knowledge of the porous structure of different coals is of relevance if the organisms involved are sufficiently small to penetrate part of the structure. It could mean that the surface area available for reaction is considerably increased and could indicate which coals would be more susceptible to bioattack. In considering the possibility of in situ coal conversion in deep seams by either conventional gasification or microbial action, a knowledge of the physical and chemical properties of the coal under the prevailing conditions is essential. This would include a knowledge of its porosity, which is illustrated in Fig. 2.6. The fact that microorganisms can develop in fine pores is reported in connection with work on the aerobic microbial conversion of phenols adsorbed on active carbon [35].

Coals have a complex and delicately structured pore system which gives them all the properties of viscoelastic colloids [3]. The structure is neither rigid nor fluid, but is deformable to the extent permitted by distortion of the cross-linkages from their equilibrium configurations. Consequently, coal porosity can be appreciably altered by adsorption of some of the fluids used for porosity measurement. Additionally, it seems certain that the interconnecting pores are far from cylindrical, and may consist of cracks varying from micrometer dimensions to apertures which are even closed to helium at room temperatures. The holes in the macromolecular structure form a branched system of very fine micropores with a large internal surface area. It is in this system that gases formed during coalification (CO_2, H_2O, and CH_4) are absorbed. Trapped gases and other organics can be released only at elevated temperatures. The micropore areas are surrounded by transport channels with diameters in the micrometer range suitable for the flow of liquids and gases in seams (see Fig. 2.6). They are responsible for the permeability of coal. The radius of these transport channels decreases with increasing rock pressure [35].

For the quantitative characterization of internal pore structure, estimates of the pore volume, surface area, and pore-size distribution are needed. Numerous techniques have been applied, but owing to the physical and chemical complexity of coals, it is sometimes difficult to select the most suitable methods for the purpose. The techniques and results have both been discussed thoroughly [30,37]. It is thought that there are three broad types of structure involved [31]:

1. An open structure characteristic of low-rank coals in the range up to about 85% carbon. These coals are highly porous.
2. The so-called liquid structure typical of bituminous coals in the range from around 85 to 91% carbon. Few pores are present.
3. The anthracitic structure common in higher-rank coals with a carbon content of over 91%, where there is less cross-linking and a porous structure is again present.

The difficulties involved in mercury porosimetry illustrate the problems involved in measuring and defining the pore structure. In principle, pore-size distributions can be determined by forcing mercury into coals at increasing pressure and measuring the volume of mercury penetration as a function of the pressure applied. At normal pressure,

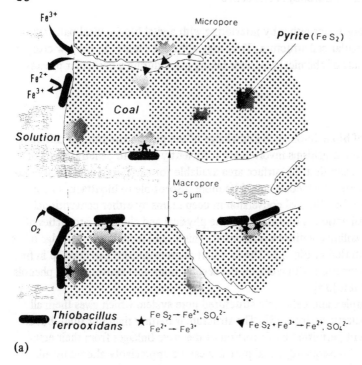

(a)

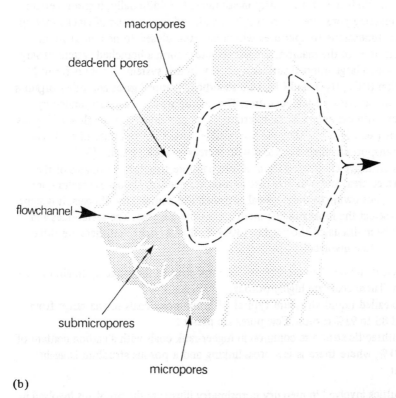

(b)

Fig. 2.6 (a) Bimodal pore structure of coal and oxidation of pyrite (from Ref. 33); (b) pore structure of hard coal (from Ref. 35).

mercury does not penetrate pores less than 10 μm (10,000 Å) in diameter. Porosimeters capable of operation at up to 400 bar are available, and at that pressure, pores of 36 Å diameter should be filled with mercury. However, there may be particle breakdown and deformation at high pressures, closed pores may be opened up, and the angle of contact and surface tension of mercury in small pores are not known [37].

Results are given in Table 2.2 for coals of various ranks using a 40 × 70 mesh cut. The higher proportion of large pores in the lignites is clear, together with the high proportion of very small pores in higher-rank coals. The variability in different coals of similar ranks is also demonstrated. There are some pores in all coals with a diameter of more than 300 Å. These can be as low as 10% of the total pore volume in anthracite, are generally around 20% (but can be as high as 50%) in bituminous coals, and can be in the range 60-90% in lignites. In lignites the pores can have diameters of about 0.1 μm (1000 Å).

The Size of Microorganisms

Individual cells are rarely greater than 1 mm in diameter, and some are only 1 nm (10 Å). The smallest microorganisms are the viruses, some of which can only be seen using an electron microscope, but these are not known to have any action on coal. Prokaryotes, including bacteria such as *Thiobacillus ferrooxidans*, have a relatively simple internal organization. The cells of many bacteria are regular cylinders, spheres, or spirals, but various distortions occur [39]. Considering the microorganisms on which much of the biodesulfurization work is based, size is dependent on the particular strain being used. To give an idea of the order of magnitude involved, *T. ferrooxidans* has a diameter of perhaps 0.2 μm (200 Å) and a length of 1 μm (10,000 Å). *Sulpholobus* is larger, with a diameter of perhaps 0.5 μm [11]. *T. ferrooxidans* is also referred to as being gram-negative rods some 0.5 by 1 to 2 μm [36]. The fungi associated with lignite degradation and bioliquefaction are eukaryotes with a more complex structure and are larger again than *Thiobacillus* and *Sulpholobus*. Many have branched filamentous hyphae. Although the organisms themselves would probably be unable to access even the large pores in lignite (say, 1000 Å diameter) it is possible that some of the hyphae might be able to grow into the coal. Hyphae can range in size from 0.1 to over 1.0 μm [38].

Does Coal Porosity Play a Part in Bioreactions?

Although some pores and transport channels are in principle accessible to some of the smaller organisms, the evidence to date is that the reaction rates are a function of the external surface area. This may be partly because most of the desulfurization work has been done on bituminous coals, where much of the pore structure is below 12 Å in diameter. *Thiobacillus* and *Sulpholobus* have diameters of about 2000 to 5000 Å, which means that they are too large to penetrate anything other than large cracks on the coal surface.

Work at Bergbau-Forschung on desulfurization does, however, indicate that pyrite particles can be removed from inside coal particles (see Fig. 2.7). This shows polished sections of coal particles before and after microbial desulfurization. The white pyrite crystals progressively disappear and dark holes are formed. This may indicate that microorganisms and their reaction products may be able to penetrate the coal matrix either through existing cracks or by generating new pores by removing other interspersed material. It is recognized that more basic research is required to explain the effect since the fissures leading to the pyrite are smaller than the bacterial cells. The porous structure of coal is illustrated in Fig. 2.6. The removal of pyrite from within the coal matrix accessible to molecules via diffusion in micropores can only be by dissolution due to chemical oxida-

Table 2.2 Gross Open Pore Distributions in Coals

Sample	Rank	% C (dry, ash free)	V_r^a (cm³/g)	V_1^b (cm³/g)	V_2^c (cm³/g)	V_3^d (cm³/g)	V_3 (%)	V_2 (%)	V_2 (%)
PSOC–80	Anthracite	90.8	0.076	0.009	0.010	0.057	75.0	13.1	11.9
PSOC–127	LV bituminous	89.5	0.052	0.014	0.000	0.038	73.0	Nil	27.0
PSOC–135	MV bituminous	88.3	0.042	0.016	0.000	0.026	61.9	Nil	38.1
PSOC–4	HVA bituminous	83.8	0.033	0.017	0.000	0.016	48.5	Nil	51.5
PSOC–105A	HVB bituminous	81.3	0.144	0.036	0.065	0.043	29.9	45.1	35.0
Rand	HVB bituminous	79.9	0.083	0.017	0.027	0.039	47.0	32.5	20.5
PSOC–26	HVC bituminous	77.2	0.158	0.031	0.061	0.066	41.8	38.6	19.6
PSOC–197	HVB bituminous	76.5	0.105	0.022	0.013	0.070	66.7	12.4	20.9
PSOC–190	HVC bituminous	75.5	0.232	0.040	0.122	0.070	30.2	52.6	17.2
PSOC–141	Lignite	71.7	0.114	0.088	0.004	0.022	19.3	3.5	77.2
PSOC–87	Lignite	71.2	0.105	0.062	0.000	0.043	40.9	Nil	59.1
PSOC–89	Lignite	63.3	0.073	0.064	0.000	0.009	12.3	Nil	87.7

Source: Ref. 28.

[a] Total open pore volume for pores accessible to helium as estimated from the helium and mercury densities.

[b] Pore volume contained in pores greater than 300 Å in diameter, as estimated from mercury porosimetry.

[c] Pore volume contained in pores in the diameter range 12 to 300 Å, as estimated from the adsorption branch of the N_2 isotherms.

[d] Pore volume contained in pores smaller in diameter than 12 Å, as estimated from $V_2 = V_2 - (V_2 + V_2)$.

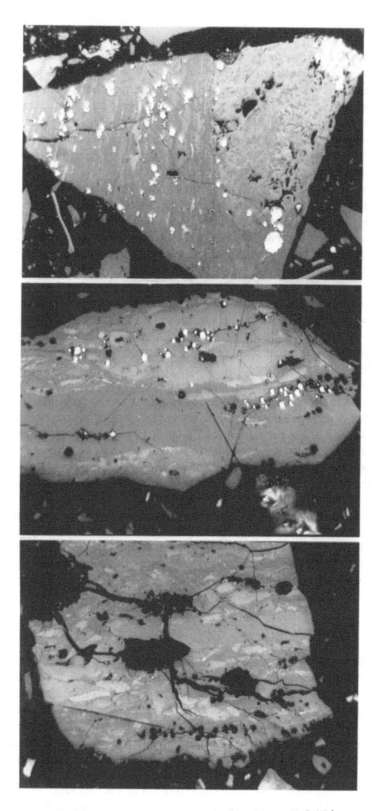

Fig. 2.7 Pyrite removal from inside coal particles. (From Ref. 35.)

tion with ferric iron. The ferrous iron produced must then be regenerated to ferric by the action of *Thiobacillus ferrooxidans* outside the micropore structure [33]. The whole question of pyrite accessibility via the pore structure is of far more importance when considering static percolation systems, commonly through much larger particles and lumps, than it is for slurry systems. The Bergbau-Forschung work also highlights the limitations of analytical techniques available. There are difficulties in determining exactly what sulfur is removed by microbial desulfurization. Elemental sulfur may either be present or be formed, and this is discussed by Beyer and others [7].

The required conditions for a bioreaction are a major consideration in assessing whether the porosity of coal might be a factor enhancing the rate of such reactions. There is potentially an increased surface area available if the organisms are small enough to penetrate the pore structure. However, bioactivity also requires the presence of the right nutrients and the absence of toxins. In the confined space of a pore where there may be little or no fluid flow, it would be more difficult to maintain the precise conditions needed. In addition, there could be a difficulty with the removal of product from the pore. The basic systems used for most of the biodesulfurization work involves the continuous input of growth substrates, agitation and mixing, and the removal of waste product, cells, and unused substrates. Reaction rates and growth rates can be controlled under these conditions, and this is the most commonly used system for large-scale production. If there is bioactivity in the pore structure of the coal, it could locally be a closed-growth system, which is more difficult both to define and control.

It should be noted that if there is to be any development toward the possibility of microbial extraction of deep coal reserves, this would almost certainly have to be based partly on the use of the inherent pore structure. The conditions involved are inhospitable in terms of temperature and pressure. In addition, such activity would be difficult to control. It is not regarded as being a likely direction of successful development work, but there is a strong incentive in terms of releasing the huge energy reserves in coal deposits currently too deep to mine.

A balanced view at this time would be that there is little evidence that coal porosity has any contribution or effect on the bioreactions observed to date. There are reasons to suppose that the dominant effects will always be at the surface of the coal particles, but the possibility of reaction inside pores should certainly not be discounted. There is significant European interest in investigating the phenomena further to assess the feasibility of deep-seam energy extraction by microbial means [35].

2.4 EUROPEAN CONTRIBUTION

The principal European effort in this work is summarized in Fig. 2.8. It has been looking particularly toward the implication of laboratory results in the design of an industrial-scale plant for desulfurization, and this is based largely on the work at Delft University. There has been recent work both at Bergbau-Forschung and Bochum in West Germany. There is also work at North Staffordshire Polytechnic, which has been looking at the possibility of a low-cost modification of an existing process to improve sulfur removal from fine coals. An important aspect of the research into microbial desulfurization is a different perspective on the question of the sulfur content of coals.

In the United States there is the common experience of the high-sulfur eastern bituminous coals, typified by those from Illinois, and the low-sulfur western low-rank coals typified by those from Montana and Dakota. The issue of sulfur removal is most

* At Delft Technical University,The Netherlands:
  extensive work on pyrite removal leading to a
  conceptual design and a feasibility study

* At Bergbau-Forschung, and at Bochum, FRG:
  on the kinetics of pyrite removal; sulphur removal
  from coal stacks has been considered

* At Cagliari, Italy:
  on organic sulphur removal

* At North Staffordshire Polytechnic, UK:
  on enhanced flotation and pyrite oxidation

* At Mid East Technical University, Turkey:
  on organic sulphur removal

There is also work on coal degradation at the Rhine
Frederick Wilhelm University, FRG, and there is interest
in the possibility of microbial action in deep coal seams
for methane production.

Fig. 2.8 Principal European work.

acute in association with eastern bituminous coals. In Europe and worldwide, the situation is more varied. Some valuable low-rank coals are high in sulfur content, and selected figures are included in Table 2.3. In the lower-rank coals the proportion of the sulfur that is organically bound is commonly higher. Thus there is more potential interest in organic sulfur removal, but this is a more difficult process than removing pyrite. There is also the interest of coal importers such as the Netherlands and Scandinavian countries, which will be dependent on what is available on the world market. This will include coals from Australia, Colombia, South Africa, the United States, and new contributors to the market, such as China. Traded coals generally contain relatively low sulfur contents, but local environmental requirements may still impose the need for sulfur removal, even from coals containing 1 to 1.5% sulfur.

Most of the European work has used *T. ferrooxidans*, whose sulfur-leaching abilities have been known for many years. It has looked in some depth at the reaction kinetics and the conditions for optimizing reaction rates for sulfur removal. U.S. work has looked in more detail at the use of other organisms, including *Sulpholobus acidocaldarius* and organisms of the *Pseudomonas* genus. There has been more work reported in the United States on organic sulfur removal, but this has concentrated on high-sulfur-content bituminous coals. The European, Turkish, and Indian work has been more limited but has included work on high-sulfur low-rank coals.

2.4.1 Delft University Study

The work at Delft has concentrated on considering the practical questions that must be addressed in taking a process from bench scale to industrial application. There is a substantial amount of published literature. In Table 2.4 some of the wide range of coals tested is listed. The work has centered on the use of *T. ferrooxidans* after a review of the

Table 2.3 Coal Analyses Illustrating Variability

	% S maf[a]	HHV (MJ/kg)
Typical U.S. Coals		
Illinois	3.3–4.1	32.9–33.6
Pennsylvania	1.2–3.9	34.5–36.4
Montana	0.5	32.8
North Dakota	0.7	28.7
Typical European Coals		
United Kingdom	0.7–0.9	33.0–36.4
Spain (brown coal)	2.1	21.2
Poland	1.0	33.8
West Germany	0.5–1.5	33.0–36.2
West Germany (brown coal)	0.5	26.4
Greece (lignite)	0.5–6.7	24.4–27.3
Turkey (brown coal)	5.1	23.7
Australia	0.4–0.6	31.4–31.7
Australia (brown coal)	0.5	28.3

[a]maf, moisture, ash free.

known sulfur-oxidizing bacteria [8]. The work on various coals established that good pyrite removal could be achieved from all of them, together with some removal of ash and heavy metals.

Kinetic studies indicated that pyritic oxidation is a first-order reaction for which a plug flow reactor where there is no backmixing is appropriate. However, because most of the bacterial biomass is attached to the solids, fresh incoming coal requires inoculation through intensive contact with coal particles already loaded with bacteria, and this re-

Table 2.4 Results of Leaching Experiments with Different Types of Coal[a]

Coal	Slurry density (%)	Duration (days)	Pyrite before leaching (%)	Pyrite after leaching (%)	Pyrite removal (%)
502GB28	10	14	1.03	0.13	88
502GB14	10	14	0.28	0.09	68
503BE33	10	14	0.13	0.04	69
504FR45	10	14	0.92	0.15	84
507DE30	10	14	0.21	0.06	71
508US19	10	14	0.26	0.04	85
509AU20	10	14	0.06	0.02	66
510US33	10	14	0.32	0.06	81
511US43	10	14	1.87	0.15	92

Source: Ref. 22.
[a]Coal samples from Australia, Belgium, France, the United Kingdom, the United States, and West Germany.

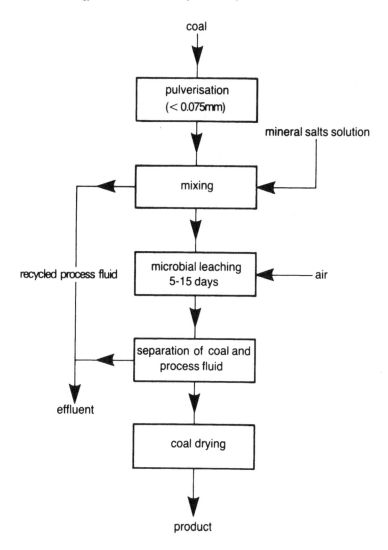

Fig. 2.9 Simplified flowsheet of a complete oxidation. (From Ref. 22.)

quires a mixed-flow reactor. To meet these conflicting requirements, a mixed-flow reactor followed by a plug flow reactor is proposed. A generalized plant flowsheet is given in Fig. 2.9. The studies then considered the practical implications of the long residence times required, and a trough-shaped reactor system similar to the Pachuca-tank reactor often used in hydrometallurgy was adopted (see Fig. 2.10). Air is jetted in at the bottom. There will be an upward current in the center and a downward stream in the outer parts. It is required that the first 10% of the trough shape have mixed flow and that the rest be plug flow. Studies have been made on laboratory-scale Pachuca reactors. Effluent treatment is considered and it is suggested that the only treatment required is the addition of lime to precipitate the ferric iron.

An economic assessment based on a design prepared by ESTS, Ijmuiden, indicated costs between $17 and $26 per metric ton for treating 100,000 metric tons per year (t/y) of a coal with 0.5% pyritic sulfur (see Fig. 2.11 and Table 2.5), and costs on the order of $10 per metric ion for a 1-Mt/y plant.

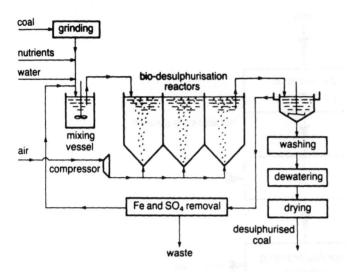

Fig. 2.10 Provisional design of biodesulfurization process based on Pachuca tanks. (From Ref. 34.)

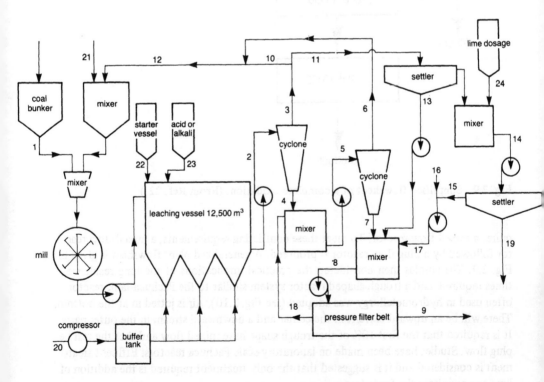

Fig. 2.11 Scheme for a 0.1-Mt/y installation. (From Ref. 10.)

Table 2.5 Composition of Streams in Fig. 2.11

Flow	Coal (kg/s)	Water (kg/s)	Inorganic sulfur (kg/s)	Description of the flow
1	3.2	0.2	0.016	Coal to be cleaned
2	3.6	10.5	0.053	
3	0.6	7.5	0.037	
4	3.0	3.0	0.016	
5	3.0	8.2	0.022	
6	0.0	5.2	0.013	
7	3.0	3.0	0.009	
8	3.2	6.7	0.0104	
9	3.2	1.5	0.0044	Cleaned product
10	0.4	5.1	0.024	
11	0.2	2.4	0.013	
12	0.4	10.3	0.037	
13	0.2	0.2	0.0014	
14	0.0	2.2	0.0116	
15	0.0	2.0	0.0	
16	0.0	1.5	0.0	Freshwater supply
17	0.0	3.5	0.0	
18	0.0	5.2	0.006	
19	0.0	0.2	0.0116	Waste stream (including gypsum, ferric iron, and heavy metals)

Flow 20, 25,000 m^3/h air; flow 21, nutrient; flow 22, starter culture; flow 23, acid or alkali for pH control; flow 24, lime

Source: Ref. 10.

The Dutch study suggests that the process is not really an alternative to stack gas cleaning for large utilities, but it could be applicable in certain other circumstances [10, 12]. In particular, it might be used in small plants, where the cost of stack cleaning would be too high; for waste streams high in pyrite content from coal preparation plants; and where coal-water mixtures are being prepared.

2.4.2 Other Work on Pyrite Oxidation

The other European work has been by Beier at Bochum; by Beyer, Ebner, and Klein at Bergbau-Forschung, West Germany; at Cagliari and Rome, Italy; and at the North Staffordshire Polytechnic, United Kindgom. The North Staffordshire work on enhanced flotation will be considered separately. No published papers on the recent Cagliari work have been identified, but some work in Rome on *T. ferrooxidans* has been reported. The Bochum work at the Advanced Technical College for Mining of the Westphalia Fund and that at Bergbau-Forschung parallels some of that at Delft University [4,5]. In the Bergbau-Forschung work, in tests on four different coals with sulfur contents ranging from 1.8 to 5.2%, 95% pyrite removal was achieved within 8 days under optimized conditions. The removal of pyrite granules from within coal particles has already been referred to, but it is recognized that the bacterial cells are larger than the pores in the coal [5]. This work used an air-lift reactor. Subsequently, some percolation tests were carried out involving coal with a grain size from 0.6 to 2.0 mm. In the static situation, a reduction in pyrite content from 2.3% to 0.6% was achieved in 60 days. Recently published papers look at the influence of pulp density and at bioreactor design [6] and at the development and application of a slurry reactor [33]. The emphasis in this work is to look at the influence of such parameters as superficial gas velocity, particle size, initial pyrite concentration, and slurry density, which are relevant to the design of an industrial process.

The Bergbau-Forschung work looked particularly at German coals that are difficult to clean in conventional processes. It was intended to proceed to consider the effects of scaleup, but reference is made to the "well-known difficulties of transferring microbiological findings to industrial process scale." Beier concluded from the work at Bochum that the biosuspension method for leaching/oxidizing the pyrite would be too costly, and that the only process worth pursuing technically was to percolate the microorganisms through a pile of fine coal. This might find application where piles of coal already arise as part of a handling or utilization process [4]. Results reported from Italy include two papers representing support work for biodesulfurization by *T. ferrooxidans*. In one, an attempt has been made to select mutant strains of the organism in order to get high oxidative efficiency at pH values as low as possible (to minimize precipitation effects) and with a high resistance to ferric ions [48]. In the other, some preliminary work on the characterization of plasmids in *T. ferrooxidans* prior to possible genetic engineering is reported [49].

2.4.3 North Staffordshire Work on Enhanced Flotation

A different approach to sulfur removal from coals has been pursued at the North Staffordshire Polytechnic in the United Kingdom. Essentially, this involves enhancing the separation of pyrite in a conventional flotation process using a range of bacteria, yeasts, and a suspension of fungal spores [47]. The early work was based on T. ferrooxidans [1]. Some of the results are given in Figs. 2.12 and 2.13. The effect appears to result from the selective modification of the surface properties of pyrite. A change in surface characteristics can be achieved with the oxidation of only a few molecular layers on the mineral surface, and only short reaction/residence times are involved.

The work is based on synthetic mixtures of coal and pyrite on a laboratory scale but has demonstrated the possibility of substantial improvement in pyrite separation in a flotation process. The technique has the advantage that it could be applied at low capital cost to an existing plant to improve the separation achieved. Among the agents that have been used to improve the separation during simulated froth flotation are:

Thiobacillus ferrooxidans
Pseudomonas maltophilia
Escherichia coli
Saccharomyces cerevisiae
Candida utilis
Spores of *Aspergillus niger*

Part of the impetus for seeking alternative agents to enhance the flotation separation process is that the use of *T. ferrooxidans* is seen as involving certain practical difficulties. Large cell yields would be needed, and the requirements for this have been considered. It was concluded that cultivation periods in excess of 500 hours in Pachuca reactors might be required [47]. In a study of three Turkish subbituminous coals, the effect of enhanced flotation was shown to be coal specific. In test work, one coal sample re-

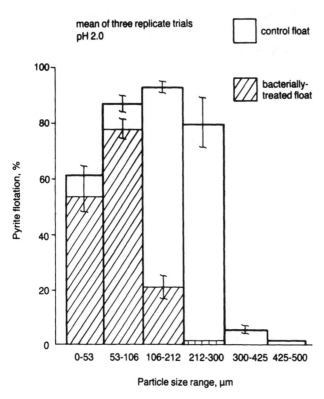

Fig. 2.12 Effect of particle size on pyrite flotation. (From Ref. 2.)

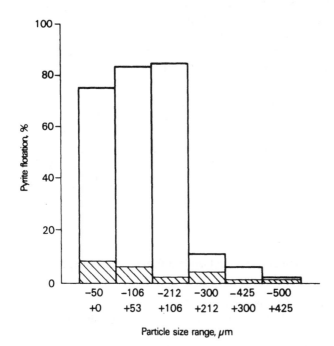

Fig. 2.13 Effect of increased bacterial dosage. (From Ref. 2.)

sponded well and 78% pyritic sulfur removal was obtained. In the other two coals, one gave poor flotation recovery by any method, and the other had very fine and reactive pyritic disseminated through it, making pyrite depression difficult. A conditioning time of 4 h was achieved [23].

2.4.4 Organic Sulfur Removal

Because of the inherent difficulty in removing organic sulfur from the coal structure, far less work has been done on this aspect. Published results from the Cagliari work in Italy have not been identified, but there has been some interesting work in Turkey related to high-sulfur low-rank coals. It has been reported that a strain of thermophilic *T. ferrooxidans* working above 50°C removed over 50% of the organic sulfur from some samples in a 10% lignite/water slurry [29]. In addition, up to 20% organic sulfur removal has been reported from India [16]. The U.S. work in this area has been discussed in the IEA Coal Research Review, *Biotechnology and Coal* [19].

The ability of microorganisms to remove organic sulfur from a coal is likely to be coal specific, because it depends on the form in which the sulfur is contained inside the structure. As yet, both the reported work and the results obtained have been scattered. It should be noted that there can be problems with organic sulfur analysis since this is commonly carried out by measuring total sulfur and subtracting the pyrite and sulfate sulfur. As it is a key subject, there is scope for replication of the work already reported from both Europe and the United States, and for attempting to establish which coals would be most susceptible to microbial attack for organic sulfur removal.

2.4.5 Fundamental Studies on Coal Degradation

In the United States, most of the work identified has been based on the possible degradation and bioliquefaction of lignite because it is in the younger coals that the original biological structures are better preserved. However, in West Germany, work has been undertaken which has established that there are fungi, yeast cultures, and bacteria that are enriched by attack on finely divided hard coal [26,27]. Out of a total of 1400 liquid and 1700 solid culture tests under variable conditions of pH, temperature, inoculum, coal type, and concentration, growth was observed in only 0.2% of the situations.

Such work is still in the very early stages of development and confirms the observations made about inhospitable nature of coal to most microorganisms. What is remarkable is that three mycelium sprouting fungi, two yeast cultures, and two bacterial cultures were identified as being enriched by the extraction of substances from the coal structure [26,27]. What may emerge from this work is new insight into the nature of coals and into their molecular structure, arising from analysis and study of the breakdown products of biological attack.

2.5 AREAS FOR FURTHER WORK

To establish a practical and cost-competitive microbial desulfurization process, the following questions need to be considered:

Can reaction rates be increased for pyrite removal?

Can organic sulfur removal be demonstrated as a practicable possibility in a range of coals?

Are the microorganisms sufficiently robust to be used on a large scale? Pilot- and eventually demonstration-plant work would be needed.

What are the technical requirements and costs of associated processes (culture preparation, nutrients, separation equipment, effluent treatment, etc.)?

In addition, the following need to be considered:

Larger-scale work on enhanced flotation, including work on various coals

Improved analytical techniques for organic sulfur measurement

Basic research is required to investigate the effect reported of pyrite removal from inside coal particles [5]. There is interest in the long-term possibility of microbial action in deep coal seams to release the stored energy, hopefully in the form of methane. This would involve a substantial and long-term research program, including:

An investigation into the properties of coals in deep seams

The development of methods to achieve increased permeability of the coal

The development of techniques for the in situ conversion of the coal and extraction of the products

2.6 PATENT ASPECTS

National patent laws vary considerably, particularly in the field of biotechnology, and to encourage technical development it would be helpful to harmonize patent protection standards. An OECD report on Biotechnology and Patent Protection was prepared stemming from a survey of the biotechnology industry which found that companies prefer to invest in countries that give strong and effective patent protection rights [17]. In the United States and Japan, the two countries that currently lead in industrial applications of biotechnology, natural organisms can be patented. This is not the case in some other OECD countries, including most European countries. OECD recommends that "discoveries" be allowed to be patented, whereas traditional patent law stipulates that only "inventions" qualify for patent protection. The distinction is important since the traditional approach only allows something new that has been thought up or invented to be the subject of a patent, not something that is found naturally in the environment.

The OECD proposal would enable researchers to patent many more microorganisms. It argues that the research and development effort needed to isolate, purify, and test the efficiency of a microorganism for a particular application justifies its patentability. The increasing use of biotechnology and the amount of research being undertaken has led the OECD to examine the entire field and to suggest a number of other modifications to patent procedure, including:

More effective protection to new genetically engineered plants

Faster processing of patent applications

Raising the effective life of a patent to 20 years to compensate for the increasing time taken in developing new products

Shifting the burden of proof in relation to patent infringement cases over microorganisms from the plaintiff to the defendant where there is a strong prima facie case of infringement

It is likely, however, that patent-related problems will remain. The scope of a patent claim may be unclear where similar organisms are perhaps produced by different means, when they are used for different purposes from those originally patented, or where different organisms perform the same functions. The precise guidelines laid down and the boundaries of particular patents are likely to be the subject of both discussion and development over the years as problems are encountered and tackled.

2.7 CONCLUSIONS

Although coal does not appear to be an ideal material for biological attack, a considerable amount of work has been undertaken. The interaction of widely differing scientific and technical disciplines in developing bioprocesses must be taken into account. A lack of understanding of the requirements of the necessary disciplines is a potential constraint to development. European effort has been directed toward the practical application of a bioprocess for pyrite removal and the development of plant designs. There has been work at Delft University, Bergbau-Forschung, Bochum, and North Staffordshire Polytechnic. Widely differing approaches have been adopted. In the Netherlands the emphasis has been on biooxidation in reactors. In West Germany this approach is being looked at together with the possibility of bioleaching in static heaps, and in the United Kindgom the main effort is on a change in surface properties giving rise to enhanced separation by flotation.

Biodesulfurization is not seen as being applicable to utility power station feedstocks unless and until organic sulfur can be removed. Only limited work is reported on organic sulfur removal. Work has been undertaken which demonstrates that some microorganisms attack bituminous coals. This may throw some light on coal molecular structure, and there is interest in the long-term possibility of the extraction of the energy in deep coal seams by microbial means. A great deal of research and development effort would be required to establish the feasibility of such a process.

REFERENCES

1. Atkins, A. S., Davis, A. J., Townsley, C. C., Bridgewood, E. W., and Pooley, F. D., Production of sulphur concentrates from the bio-flotation of high pyritic coals, in *International Conference, Sulphur 85*, London, Nov. 1985, pp. 83-104.
2. Atkins, A. S., Bridgewood, E. W., Davis, A. J., and Pooley, F. D., A study of the suppression of pyritic sulphur in coal froth flotation by *Thiobacillus ferrooxidans*, in *Coal Preparation*, Gordon and Breach, New York, 1987.
3. Bangham, D. H., (1943) *Ann. Rep. Chem. Soc., 40*, 29.
4. Beier, E., Removal of pyrite from coal using bacteria, in *Proceedings of the First International Conference on Processing and Utilisation of High Sulphur Coals*, Columbus, Ohio, Oct. 13-17, Elsevier, Amsterdam, 1985, pp. 653-672.
5. Beyer, M., Ebner, H. G., and Klein, J., Bacterial desulphurisation of German hard coal, in *6th International Symposium on Biohydrometallurgy*, Vancouver, British Columbia, Canada, Aug. 21-24 (R. W. Lawrence, R. M. R. Branion, and H. G. Ebner, eds.), Elsevier, Amsterdam, 1985, pp. 151-164.
6. Beyer, M., Ebner, H. G., and Klein, J., *J. Appl. Microbiol. Biotechnol., 24*, 342-346 (1986).
7. Beyer, M., Ebner, H. G., Assenmacher, H., and Frigge, J., *Fuel, 66*, Apr., 551-555 (1987).
8. Bos, P., and Kuenen, J. G., Microbiology of sulphur-oxidising bacteria, in *Microbial Corrosion* (A. D. Mercer, A. K. Tiller, and R. W. Wilson, eds.), The Metals Society, London, 1983, pp. 18-27.
9. Bos, P., Huber, T. F., Kos, C. H., Ras, C., and Kuenen, J. G., A Dutch feasibility study on microbial desulphurisation, in *6th International Symposium on Biohydrometallurgy*, Vancouver, British Columbia, Canada, Aug. 21-24 (R. W. Lawrence,

R. M. R. Branion, and H. G. Ebner, eds.), Elsevier, Amsterdam, 1985; also in *Microbial voorreiniging van steenkool*, Delft University of Technology, Delft, The Netherlands, 1986.

10. Bos, P., Huber, T. F., Kos, C. H., and Doddema, H. J., (1986) *Microbial Desulphurisation of Coal*, Department of Microbiology and Enzymology, Delft University of Technology, Delft, The Netherlands, 1986.

11. Bos, P., Delft Technical University, Delft, The Netherlands, private communication, 1987.

12. Bos, P., Microbial desulphurisation of coal: a feasible process? *Workshop on the Biological Treatment of Coals*, Washington, D.C., July 8-10, 1987.

13. Brock, T. D., Life at high temperatures, *Science, 230*, Oct. 11, pp. 132-138 (1985).

14. Bull, A. T., Holt, G., and Lilly, M. D., *Biotechnology; International Trends and Developments*, Organization for Economic Cooperation and Development, Paris, 1982.

15. Camier, R. J., Siemon, S. R., Battaerd, H. A. J., and Stanmore, B. R., The physical structure of brown coal, *Am. Chem. Soc. Div. Fuel Chem. Prepr., 24*(1), 203-209 (1987).

16. Chandra, D., Roy, P., Mishra, A. K., Chakrabarti, J. N., and Sengupta, B., Microbial removal of organic sulphur from coal, *Fuel, 58*, 549 (1979).

17. OECD initiates new biotech guidelines, *Chem. Eng. 93*(4), 16E-16F (1986).

18. Cohen, M. S., and Gabriele, P. D., Degradation of coal by the fungi *Polyporus versicolor* and *Poria monticola, Appl. Environ. Microbiol., 44*(1), 23-27 (1982).

19. Couch, G. R., *Biotechnology and Coal*, ICTIS/TR 38, IEA Coal Research Report, International Energy Agency, London, Mar. 1987.

20. Davidson, R. M., *Molecular Structure of Coal*, ICTIS/TR 08, IEA Coal Research Report, International Energy Agency, London, Jan. 1980.

21. Davidson, R. M., *Nuclear Magnetic Resonance Studies of Coal*, ICTIS/IS/TR32, IEA Coal Research Report, International Energy Agency, London, Jan. 1986.

22. Doddema, H. J., Huber, T. F., Kos, C. H., and Bos, P. Microbial coal desulphurisation, in *3rd European Coal Utilisation Conference*, Amsterdam, The Netherlands, Vol. 3, Oct. 1983, pp. 183-192.

23. Dogan, Z. M., Ozbayoglu, G., Hicyilmaz, C., Sarikaya, M., and Ozcengiz, G., in *6th International Symposium on Biohydrometallurgy*, Vancouver, British Columbia, Canada, Aug. 21-24 (R. W. Lawrence, R. M. R. Branion, and H. G. Ebner, eds.), Elsevier, Amsterdam, 1985, pp. 165-170.

24. Dugan, P. R., Microbial desulphurisation of coal and its increased monetary value, in *Workshop on Biotechnology for the Mining, Metal-Refining and Fossil Fuel Processing Industries*, Troy, N.Y., May 28-30, Wiley, New York, 1985, pp. 185-204.

25. Erlich, H. Z., and Holmes, D. S., Biotechnology for the mining, metal-refining and fossil fuel processing industries, *Biotechnology and Bioengineering Symposium 16*, Wiley, New York, 1986, p. 1.

26. Fakoussa, R., Kohle als Substrat für Mikroogranismen: Untersuchungen zur mikrobiellen Umsetzung nativer Steinkohle, Disseration, Rheinische Friedrich-Wilhelm Universität, Bonn, 1981.

27. Fakoussa, R., and Trüper, H. G., Coal as a microbial substrate under aerobic conditions, translation from *Kolloquium in der Bergbau-Forschung GmbH Biotechnologie im Steinkohlenbergbau*, Essen, West Germany, Jan. 1983, pp. 41-50.

28. Gan, H., Nandi, S. P., and Walker, P. L., Jr., *Fuel, 51*, 272-277 (1972).

29. Gockay, C. F., and Yurteri, R. N., Microbial desulphurisation of lignites by a thermophilic bacterium, *Fuel, 62*, Oct., 1223-1224 (1983).

30. Grimes, W. R., The physical structure of coal, in *Coal Science*, Vol. 1 (M. L. Gorbaty, J. W. Larsen, and I. Wender, eds.), Academic Press, New York, 1982, pp. 21-41.

31. Hirsch, P. B., *Proc. R. Soc. London, A226*, 143-169 (1954).

32. Hoffmann, M. R., Hiltunen, P., and Touvinen, O. H., Inhibition of ferrous ion oxidation by *Thiobacillus ferrooxidans* in the presence of oxyanions of sulphur and phosphorus, in *Proceedings of the First International Conference on Processing and Utilization of High Sulphur Coals*, Columbus, Ohio, Oct. 13–17, Elsevier, Amsterdam, 1985, pp. 683–688.

33. Höne, H. J., Beyer, M., Ebner, H. G., Klein, J., and Jüntgen, H., Microbial desulphurisation of coal: development and application of a slurry reactor, *Chem. Eng. Technol., 10*, 173–179 (1987).

34. Huber, T. F., Ras, C., and Kossen, N. W. F., Design and scale-up of a reactor for the microbial desulphurisation of coal: a kinetic model for bacterial growth and pyrite oxidation, in *3rd European Congress on Biotechnology*, Munich, Sept. 10–14, Verlag Chemie, Weinheim, West Germany, 1984, pp. 151–159.

35. Jüntgen, H., Research for future in situ conversion of coal, *Fuel, 66*, Apr. 443–453 (1987).

36. Kargi, F., Microbial methods for desulphurisation of coal, Tibtech., *Trends Biotechnol.*, Nov., 293–297 (1986).

37. Mahajan, O. P., and Walker, P. L., Porosity of coals and coal products, in *Analytical Methods for Coal and Coal Products*, Vol. 11 (C. Karr, Jr., ed.), Academic Press, New York, 1978, pp. 125–162.

38. McIlwain, M., Idaho National Engineering Laboratory, private communication, 1987.

39. Millis, N. F., The organisms of biotechnology, in *Comprehensive Biotechnology*, Vol. 1 (M. Moo-Young, ed.), Pergamon Press, Elmsford, N.Y., 1985, pp. 7–20.

40. Olson, G. J., Brinckman, F. R., and Iverson, W. P., (1986) *Processing Coal with Microorganisms*, EPRI Report AP-4472, Electric Power Research Institute, Palo Alto, Calif., Mar. 1986, p. 40.

41. Ponsford, A. P., Microbiological activity in relation to coal utilisation, part 1, *BCURA Mon. Bull., 30*(1), 1–18 (1966).

42. Ponsford, A. P., Microbiological activity in relation to coal utilisation, part 2, *BCURA Mon. Bull., 30*(2), 41–71 (1966).

43. Rogoff, M. H., and Wender, I., *Materials in Coal Inhibitory to the Growth of Microorganisms*, Report RI 6279, U.S. Department of the Interior, Bureau of Mines, p. 13.

44. Rosen, C., Biotechnology commercialization and scaling-up, *Chemical Economy and Engineering Review, 17*(7/8) (Jul/Aug), 1985, p. 5–11.

45. Shamel, R. E., Keil, M., Biotechnology opportunities and challenges: markets, engineering and regulation, *Chemical Economy and Engineering Review, 18*(7) (Jul/Aug) 1986, p. 5–8.

46. Shinn, The Structure of coal and its liquefaction products: a reactive model, in *International Conference on Coal Science*, Sydney, Australia, Oct. 28–31, 1985, pp. 738–741.

47. Townsley, C. C., and Atkins, A. S., Comparative coal fines desulphurisation using the iron-oxidising bacterium *Thiobacillus ferrooxidans* and yeast *Saccharomyces cerevisiae* during simulated froth flotation, *Process Biochem., 21*(6), 188–191 (1986).

48. Vian, M., Creo, C., Dalmastri, C., Gionni, A., Palazzolo, P., and Levi, G. *Thiobacillus ferrooxidans* selection in continuous culture, in *6th International Symposium on Biohydrometallurgy*, Vancouver, British Columbia, Canada, Aug. 21–24 (R. W. Lawrence, R. M. R. Branion, and H. G. Ebner, eds.), Elsevier, Amsterdam, 1985, pp. 395–401.

49. Visca, P., Valenti, P., and Orsi, N., Characterisation of plasmids in *Thiobacillus*

ferrooxidans strains, in *6th International Symposium on Biohydrometallurgy*, Vancouver, British Columbia, Canada, Aug. 21-24 (R. W. Lawrence, R. M. R. Branion, and H. G. Ebner, eds.), Elsevier, Amsterdam, 1985, pp. 429-441.

50. Wolfrum, Correlations between petrographic properties, chemical structure and technological behaviour of Rhenish brown coal, *Am. Chem. Soc. Div. Fuel Chem. Prepr., 28*(4), 11-55 (1983).

51. Yamada, Y., Preparation of engineering industry for biotechnology, *Chem. Econ. Eng. Rev., 18*(7), 14-20 (1986).

3

Economic Factors in the Bioprocessing of Coal

HERBERT J. HATCHER and LA MAR J. JOHNSON *Idaho National Engineering Laboratory, Idaho Falls, Idaho*

3.1 INTRODUCTION

The oil embargo of 1973-1974 clearly demonstrated the impact of national dependency on foreign oil imports. This event directed the attention of energy policymakers to the large reserves of domestic coal [1]. Between 1972 and 1982, coal's share of U.S. energy rose from a low of 17.3% to 22.1%. Still, due to 1987 oil demand, supplies, and prices, the future of coal is considered with skepticism. At the beginning of the twentieth century coal supplied 93% of U.S. energy requirements [1]. Today, oil supplies appear adequate, as do gas supplies. Since oil deregulation, rail transportation rates have risen sharply, raising coal costs also. Furthermore, environmental issues have become significant factors in coal economics, restricting usage in terms of quality and increasing coal processing costs.

The United States has the largest remaining coal reserves in the world. The locations of the major world reserves and their estimated amounts are shown in Table 3.1. These estimates should be regarded as projected values and are highly uncertain; however, they serve to indicate the vast supplies available and their source. Figure 3.1 projects types of energy sources up to the next century. Over the past 60 years, coal usage, as an energy supply, has varied little. If the estimates are correct, coal-derived energy will double during the next 15 years. Correspondingly, concern for the environmental impact of coal combustion and attention to coal as a resource with a number of facets of value will also increase.

Table 3.1 World Coal Reserves

Country	Coal production (10^9 metric tons)	Percent of world totals
USSR	600	25.8
China	500	21.8
United States	750	32.5
West Germany	140	6.2

Source: Ref. 2.

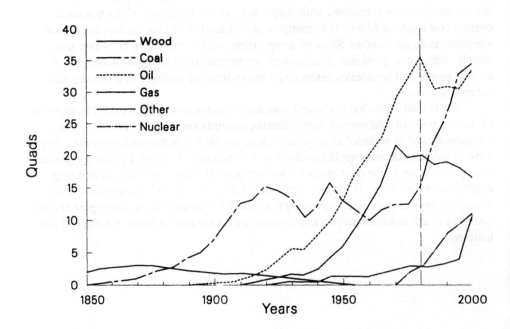

Fig. 3.1 Different types of energy sources for U.S. projections. (From Ref. 3.)

3.2 CHARACTERIZATION OF COALS

U.S. coal resources can be categorized as lignite, subbituminous, bituminous, and anthracite. Table 3.2 shows the significant properties and differences between these categories of coals. Formation of coal seems to take place through a continuous series of changes from living material (perhaps lignin-like) to lignite to bituminous to anthracite. The chemical-structural relationship can be illustrated in Fig. 3.2. In all cases the structures shown are gross simplifications. One can say that coal is a chemically dynamic substance whose structure is in constant but imperceptible flux, depending on conditions. The degree of change of chemical composition of coal is called the rank of that coal. Lignite is the lowest-rank and anthracite the highest-rank coal. Factors that lead to change from lower to higher rank are temperature, pressure, and time. The rate of chemical change doubles for a rise of 5 to 10°C, so that coals in lower seams are of higher rank than those in higher seams because of the temperature gradient of the earth's crust (increasing by about 1°C per 100 ft of depth). The variation of rank with depth is known as Hilt's law [6].

Identified coal reserves of the United States are shown in Table 3.3. The data provide important information as to the rank of deposits in various regions of the United States. Roughly three-fourths of the domestic supply of surface-minable coal is in the western half of the nation. Nearly equal amounts of underground reserves are found in the east and the west. All subbituminous and most lignite reserves are in the west. Surface mining is both cheaper and safer than underground mining. The Coal Mine Health and Safety Act of 1969 reduced productivity and increased the cost of underground mining. The Surface Mining Act of 1977 made environmental regulations related to coal production restrictive, thereby increasing the cost and limiting the expansion of surface mining.

The price of coal is determined by a complex variety of factors, including rank (related to depth of seam, which is variable), location, quantity purchased, contract arrangements, and quality. The quality of coal alone includes heating value, ash content, moisture content, and sulfur content [8]. As an example of the influence of location, the

Table 3.2 Approximate Values of Some Coal Properties in Different Rank Ranges

Component	Lignite	Subbituminous	Bituminous Volatile	Anthracite
Carbon (mineral-matter-free) (%)	65–72	72–76	76–90	93
Oxygen (%)	20	18	3–13	2
Oxygen, as COOH (%)	13–10	5–7	0	0
Oxygen, as OH (%)	15–10	12–10	1–9	0
Aromatic carbon atoms (% of total carbon)	50	65	70–90	90–95
Average number of benzene rings per layer	1–2	?	2–5	25?
Volatile matter (%)	40–50	35–50	10–46	10
Reflectance (vitrinite) (%)	0.2–0.3	0.3–0.4	0.5–1.8	4

Source: Ref. 4.

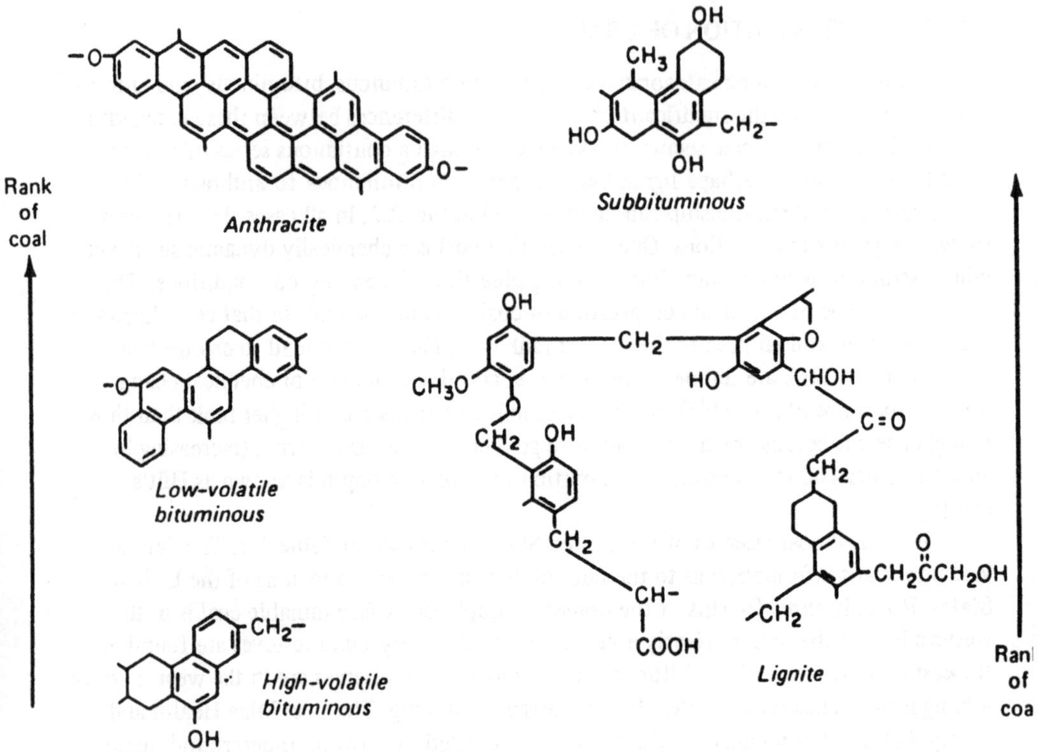

60 Hatcher and Johnson

Fig. 3.2 Relationship of chemical structures to ranks of coal in which structures appear to be present. (From Ref. 5.)

weighted-average value of production in western states in 1980 was $11.17 per ton ($0.55 per million Btu), while for Appalachian underground coal the value was $35.69 per ton ($1.55 per million Btu) [1]. Relatively little can be done to change most of the factors concerned with pricing so as to reduce coal costs. However, it is possible to treat coal so as to reduce ash and sulfur. Reduction of the content of ash and sulfur in coal increases fuel value by providing higher Btu per pound. Moreover, coal that presently cannot be used due to the high sulfur or toxic metal content could become usable.

The Clean Air Act amendments of 1970 provided for national air quality standards that set a limit on sulfur as sulfur dioxide of 0.03 ppm as an annual arithmetic mean concentration with an allowable 24-h maximum of 0.14 ppm, not to be exceeded more than once each year. In terms of performance standards, the rulings can restrict combustion of coal at large power plants to fuel that contains no more than 0.5 to 1.6% total sulfur [8]. The technology exists for removal of some forms of sulfur. However, due to the nature of coal, the problem is quite complex.

Table 3.4 gives the contents of three forms of sulfur in three different types of commonly used coal. Coals of the western United States tend to have a lower sulfur content (below 0.6%) than that of eastern coals (generally 2.0 to 3.5%) [10]. Moreover one finds western coals have a lower ash content than eastern coals (Table 3.5) in the United States [11]. In consequence of these variations, the nature of the treatment of most importance depends on the origin of the coal.

Table 3.3 Identified Coal Reserves of the United States (10^9 Metric Tons)[a]

State	Bituminous coal	Subbituminous coal	Lignite	Anthracite and semianthracite	Total
Alabama	12.1	—	1.8	—	13.9
Alaska	17.6	100.4	—	—	118
Arizona	19.2	—	—	—	19.2
Arkansas	1.5	—	0.4	0.4	2.3
Colorado	99	17.9	0.02	0.08	117
Georgia	0.02	—	—	—	0.02
Illinois	132	—	—	—	132
Indiana	29.8	—	—	—	29.8
Iowa	5.9	—	—	—	5.9
Kansas	17	—	—	—	17
Kentucky					
Eastern	25.6	—	—	—	25.6
Western	32.7	—	—	—	32.7
Maryland	1.1	—	—	—	1.1
Michigan	0.2	—	—	—	0.2
Missouri	28.3	—	—	—	28.3
Montana	2.1	160.4	102	—	264.5
New Mexico	9.8	45.9	—	0.004	55.7
North Carolina	0.1	—	—	—	0.1
North Dakota	0	—	318	—	318
Ohio	37.4	—	—	—	37.4
Oklahoma	6.4	—	—	—	6.4
Oregon	0.05	0.25	—	—	0.3
Pennsylvania	58	—	—	17.1	75.1
South Dakota	0	—	2	—	2
Tennessee	2.3	—	—	—	2.3
Texas	5.5	—	9.3	—	14.8
Utah	21	0.2	—	—	21.2
Virginia	8.3	—	—	0.3	8.6
Washington	1.7	3.8	0.1	0.005	5.6
West Virginia	91	—	—	—	91
Wyoming	11.5	111.7	—	—	123.2
Other states	0.5	0.03	0.05	—	0.5
Total	677.7	440.4	433.7	17.9	1569.7

Source: Ref. 7.
[a]Data as of January 1, 1974. Some figures shown in this table were rounded off to the nearest first or second decimal place and may therefore be slightly different from the original source. Dashes indicate that tonnage is too small to make a noticeable contribution and/or is included under other ranks.

Table 3.4 Sulfur Class (wT %)

Coal	Total	Pyritic	Sulfatic	Organic
Illinois	4.49	1.23	0.06	3.21
Kentucky	6.615	5.05	0.135	1.415
Texas lignite	1.20	0.40	—	0.80

Source: Ref.9.

Table 3.5 Percent Ash Content of U.S. Coals

Area	Range	Arithmetic mean	Geometric mean
Western United States	4.1–20	9.6	8.9
Eastern United States	6.1–25	12	12
Illinois basin	4.6–20	11	11

3.3 BIOLOGICAL INTERACTIONS WITH COAL

It is not the purpose of this chapter to discuss biological desulfurization or demineralization in detail, but some discussion of processes can aid in appreciating the advantages, or
disadvantages, of these processes. Most of the inorganic sulfur in coal is in the form of
pyrite (FeS_2). The mechanism of microbiological solubilization of pyrite is well known
(although details are debated). Removal of pyrite from coal would be analogous to removal of pyrite from mineral ores, such as copper. Some modification would be necessary, but as an example, consider pyrite removal from copper ore; approximately 20 to
25% of the total copper production in the United States is by microbial leaching of low-
grade copper ores and waste. The most common methods used for bacterial extraction of
mineral ores are dump and heap leaching. In dump leaching crushed ore is deposited on
an impermeable base and leach solutions are introduced, usually by spraying, flooding, or
injection. The liquid percolates through the ore and the pyrite is solubilized by the bacteria present in accordance with the following reactions:

$$FeS_2 + 3.5O_2 + H_2O \rightarrow FeSO_4 + H_2SO_4 \qquad\qquad (1)$$

$$2FeSO_4 + 0.5O_2 + H_2SO_4 \rightarrow Fe_2(SO_4)_3 + H_2O \qquad\qquad (2)$$

The product of reaction (2) is a strong oxidizing agent, and a strictly chemical reaction adds to the process of pyrite removal.

$$FeS_2 + Fe_2(SO_4)_3 \rightarrow 3FeSO_4 + 2S \qquad\qquad (3)$$

$$S + 1.5O_2 + H_2O \rightarrow H_2SO_4 \qquad\qquad (4)$$

The sulfur formed in reaction (3) is then oxidized by bacteria to sulfuric acid,
which promotes reaction (2). *Thiobacillus ferrooxidans* is a primary participant in reactions (1) and (2). Both *T. ferrooxidans* and *T. thiooxidans* are involved in reaction (4).
These bacteria are acidophilic gram-negative chemolithotrophic microorganisms. However, other acidophilic heterotrophic bacteria are also involved. Heap leaching is similar
to dump leaching, but the ore particle size is smaller and it is more concentrated. In this
case the ore is deposited in mounds on prepared drainage pads. The method might be applied to collection of metals from coal and coal ash as well as to desulfurization of coal.
Significant features of leaching processes are [12] : (1) low capital investment; (2) air,
soil, and water pollution controllable or not produced; (3) energy requirement limited to
pumping of solutions; and (4) easy to operate and adaptable to automation.

Using slurries of commercially pulverized coal and preconditioned culture from pyrite enrichment, extensive removal of pyritic sulfur has been demonstrated (Table 3.6).
The problem of organic sulfur removal from coal is much more complex than the removal
of inorganic sulfur. A variety of forms of organic sulfur are found in coal, including both
aliphatic and aromatic thiols and sulfides. Simple and complex thiophenes appear to be
major components of the organic fractions in many coals [14–16] . Furthermore, de-

Table 3.6 Treatment Percentage of Sulfur Before and After Biological Treatment

	Total S	Pyritic S	Organic S	SO_4^{2-}
Control coal blend				
45% 100–200 mesh	4.1	2.9	1.2	0.1
45% sub-200 mesh	5.4	4.2	1.0	0.2
After treatment				
100–200 (20% slurry)	1.8	0.1	1.6	0.1
Sub-200 (20% slurry)	2.0	0.5	1.3	0.2

Source: Ref. 13.

pending on the treatment of coal, the organic fraction can increase: for example, increased temperature, pressure, and exposure time. Many investigators are using dibenzothiophene and other types of model compounds to study the biodegradation of organic sulfur compounds in coal [15,16]. Results of these studies indicate that decomposition of thiophenic compounds by microorganisms is not difficult. However, microbial or enzymatic attack on the coal matrix with removal of sulfur from the integral structure does present steric problems. Nevertheless, some success at removal of organic sulfur has been reported [14].

Another aspect of coal bioprocessing with great potential for resource recovery as well as energy conservation and production is attracting considerable attention. It has been found that certain microorganisms are capable of degrading coal, forming a liquid product [17-20]. At this time only certain fungi and actinomycetes have been found to possess this capability. Current results indicate that products formed by solubilization are highly oxidized, highly aromatic, and of intermediate to high molecular weight [20]. The process appears to be extracellular. Following solubilization a granular residue remains which could have a high mineral content in an insoluble organic matrix. Pretreatment may consist of oxidation with nitric acid, hydrogen peroxide, ozone, or simply weathering (atmospheric exposure for a long period). So little is known regarding the nature of the products that one cannot speculate as to their ultimate value. Based on the types of products formed in nonbiological coal liquefaction, some products may have value as chemical feedstocks [21]. Under such circumstances, the value of coal as a natural resource may actually be upgraded.

A seldom mentioned but potentially significant aspect of coal beneficiation is recovery of metals. Considering the mechanism of bacterial attack on metallic sulfides, it would seem that any sulfide could be solubilized by leaching. Generalizing on the process mentioned previously, the reactions would be

$$2FeSO_4 + H_2SO_4 + 0.5O_2 \xrightarrow{\text{bacteria}} Fe_2(SO_4)_3 + H_2O \tag{5}$$

$$MS + Fe_2(SO_4)_3 \rightarrow MSO_4 + 2FeSO_4 + S \tag{6}$$

$$S + 1.5O_2 + H_2O \xrightarrow{\text{bacteria}} H_2SO_4 \tag{7}$$

In these reactions MS represents the particular metal sulfide present. The bacteria derive energy from the oxidations for synthesis of cell material. The suitability of 16 metal sulfides as energy sources for *T. ferrooxidans* has been examined [22]. Evaluation of various solid-state parameters indicated that no single property could account entirely for the suitability of the material for bacterial leaching. The most critical rate-determining

Table 3.7 Relationship Between Bacterial Counts
and Solubility Products for Various Metallic Sulfides

Sulfide	Solubility product	Approximate bacterial count
MnS	1.4×10^{-15}	5×10^8
FeS	3.7×10^{-19}	5×10^8
NiS	1.4×10^{-24}	5×10^7
CoS	3.0×10^{-26}	5×10^6
CdS	3.6×10^{-29}	5×10^6

factor for bacterial oxidation was the solubility products of the sulfides. Sulfides with high solubility products were good substrates for bacterial leaching, as indicated by the bacterial count (Table 3.7). The second-most-important character was the number of broken chemical bonds in the sulfide surface.

Enhanced leaching of cobalt in the presence of ferrous iron has been observed [23]. Bacterial growth was strongly inhibited by cobalt, but the inhibition was completely reversed by ferrous iron. A similar protective effect has been observed for ferrous iron against inhibition of bacterial growth by tin, nickel, zinc, silver, and mercuric ions. Improved leaching of sulfides using galvanically coupled systems has also been observed [24]. Rates of dissolution of copper and zinc increased tenfold for the coupled systems Cu FeS_2/FeS_2 and ZnS/FeS_2 as compared with electrically isolated systems. The implication of these studies is that bioleaching processes may be developed for a variety of metals, even when the metals are inhibitory for microbial growth.

Many metals are found in nature as oxides. Among these are industrially important metals such as aluminum and manganese. A limited amount of research has been conducted on the microbiological recovery of manganese. It has been reported that bacterial-mediated solubilization of solid manganese dioxide is a growth-related reducing reaction sustained by the presence of organic nutrients [25]. It has been reported that citric acid (a relatively common fungal product) can degrade aluminosilicate minerals. Methyl iodide, a marine algal product, solubilizes a variety of metals and metal sulfides [26]. Sugars present were converted to acids. Supplementation of media in which pure bacterial isolates were growing with soluble intracellular products from fresh- and saltwater algae improved soluble manganese yields by 24 and 100%, respectively.

3.4 ECONOMIC SIGNIFICANCE OF METALS IN COAL

Actually, incentive for conducting research on recovery of some of our most critically important metals from ores is lacking because recovery is economically unattractive or no domestic mines are in production. For example, chromium is a strategically important metal and the United States is a major chromium consumer, but there is no domestic chromium mine in production. The economy of the United States is, in large part, dependent on a rather small number of metals that are in short supply. Coal contains significant amounts of some of these metals. By means of methods similar to those described previously, a number of these metals could be recovered. In other cases, further research is necessary before such recovery would be economically feasible. Relevant facts

Table 3.8 Average Concentration of Certain Economically Important Elements in Coal and Recovery Significance in Terms of U.S. Demand, 1979

Element	U.S. average concentration (ppm)	Content in coal[a] (tons/yr)	U.S. production of metal (tons/yr)	U.S. demand for metal (tons/yr)
Chromium	15	12,000	0	50,000
Cobalt	7	5,600	416[b]	6,934
Gallium	7	5,600	3.3	10.4
Germanium	0.7	560	25.3	26.4
Manganese	100	80,000	31,000	1,250,000
Nickel	15	12,000	11,700	199,200
Titanium	800	640,000	231,000	618,000
Tungsten	2.5	2,000	3,317	11,850
Zinc	39	31,200	267,000	1,090,000

Source: Adapted from Refs. 27 and 29.
[a]Based on estimated 1979 production of 800 million tons per year (see Ref. 7).
[b]Based on 94% dependence of foreign sources.

regarding concentrations in coal and use of certain strategically important metals are shown in Table 3.8. Table 3.9 shows the principal uses and major sources for the metals shown on Table 3.8, and monetary values for some of the metals are listed in Table 3.10.

A variety of factors may have a significant impact on the economic recovery of metals from coal. Mixed communities of microorganisms may be more able to accumulate metals than some pure cultures [31]. Precipitation of sulfides in metal-rich effluents may take place through the activities of sulfate-reducing bacteria. As indicated previously, microbially assisted leaching is dependent on bacterial oxidation of iron and sulfide

Table 3.9 Uses and Major Sources of Certain Strategically Important Metals

Element	Principal industrial uses	Major sources
Chromium	Stainless steel alloys; manufacturing of refractory bricks, chemicals	South Africa, USSR
Cobalt	Steel alloys, nonferrous alloys, chemicals, and ceramics	Zaire, Zambia
Gallium	Fiber-optic diodes, lasers, computer memories	Switzerland, Canada
Germanium	Infrared optics, semiconducters, fiber-optic radiation detectors	Zaire, European countries
Manganese	Metal alloys, steel, dry cells, diodes	Australia, Gabon, South Africa
Nickel	Steel, nonferrous alloys chemicals	Australia, Botswana, South Africa
Titanium	Pigments, metal alloys	West Germany and United Kingdom for pigments, Japan and China for metal
Tungsten	Cutting and wear-resistant materials	Bolivia, Canada, China
Zinc	Various chemicals, paints, fabrics, lubricants, plastics, wood preservatives, soldering flux, batteries	Canada, Europe

Source: Refs. 29 and 30.

Table 3.10 Monetary Value of Selected Strategically Important Metals, 1980

Metal	Form	Value	Ref.
Chromium	Oxide	$1.90/lb	29
Cobalt	Chloride	$4.15/lb	
Gallium	Metal	$510/kg	28
Germanium	Metal	$558–834/kg	28
Manganese	Metal	$0.33¼/lb	29
Nickel	Metal	$3.45/lb	29
Titanium	Sponge	$4.85/lb	29
Tungsten	Paid	$12.85/lb	29
Zinc	Metal	$0.47/lb	29

minerals. The leaching of sulfides may be affected in dump, heap, and underground in situ leaching operations. In all cases, better understanding of the relationships between mixed cultures would certainly contribute toward improved performance. Current literature reports emphasize the significance of *T. thiooxidans* in the solubilization of metal sulfides. In cases where high sulfur-oxiding capability favors leaching, as for jarosite, a mixture of *T. ferrooxidans* and *T. thiooxidans* may favor leaching [30].

3.5 CURRENT STATUS OF APPLIED MICROBIAL DESULFURIZATION

It has been found that the rate of pyrite leaching is inversely proportional to particle size, as one would expect [31]. Analysis of coal after 110 days of column leaching gave the results shown in Table 3.11. The process resulted in removal of significant amounts of ash and inorganic sulfur (including pyrite), but little or no organic sulfur. Desulfurization of waste coal dumps using heap leaching is considered a feasible process, but obviously a method for removal of organic sulfur is needed.

Factors that have been reported to influence the rates of degradation of sulfide minerals include the morphology and electronic structure of the mineral surface, the surface area, temperature, pH, redox potential, oxygen partial pressure, and relative humidity [33]. Development of engineering technology based on optimal application of these factors is necessary for the most economical bioprocessing of coal. A recent feasibility study on microbial coal desulfurization leads to sober reflection on the process

Table 3.11 Analysis of Brogan Coal After 110 Days of Column Leaching at 226° to 230°C

	Percent initial coal	Percent leached coal	Percent removal
Total sulfur	6.93	4.05	41.5
Pyrite sulfur	4.24	1.60	62.3
Inorganic sulfur	1.10	0.29	73.6
Organic sulfur	1.60	2.16	−35.0
Ash	16.34	5.20	68.2

Source: Ref. 32.

[34] . A complete process design and cost analysis was conducted. Estimates indicated that a 1-million ton/year plant results in a desulfurizing price of approximately $10.00 per ton; organic sulfur was not removed.

Microbial desulfurization is a wet process; therefore, the process also provides a cleaning step in preparation of coal slurries for combustion. A 70% coal slurry has approximately the same bulk caloric value as dry coal [35] . Under supercritical conditions, viscosity of such a slurry approaches infinity and the slurry behaves as a solid. Water vapor present during combustion catalyzes and accelerates the burning of the carbon residue. Water provides hydroxyl radicals that react with carbon monoxide formed during burning, thereby decreasing the amount of carbon monoxide for reaction with oxygen and increasing the amount of oxygen that can burn the carbon.

Unquestionably, microbiological removal of pyritic sulfur from coal is economical and technically realistic, but adequate removal of organic sulfur can be seriously questioned. It is appropriately stated that organic sulfur is bound so firmly into the complex macromolecular structure of coal that removal is difficult to imagine [33] . Furthermore, accurate direct methods for the determination of organic sulfur compounds in coal are under development, but are not in common use. In consequence of these facts, microbial desulfurization of coal is not, at this time, an alternative to gas stack cleaning. As indicated previously, sulfur dioxide emission limitations require a high level of removal. Current gas cleaning processes are expensive, but they are accepted and used. For the present, microbial desulfurization of solid coal is best considered to be a cleaning step. Processing cost reduction by recovery of valuable materials in coal is a realistic goal.

The conclusions regarding desulfurization are only reinforced by the finding that adsorption of *T. ferrooxidans* is highly selective [36] . The adsorption appears to occur very rapidly, and predominantly at exposed pyrite inclusions, crystals, and along topographical defects in the structure. A two-stage adsorption process involving reversible and irreversible events may occur.

During the combustion of coal, sulfur escapes to the atmosphere as sulfur dioxide. In the United States, the amount of sulfur emitted in this manner exceeds the total amount of sulfur consumed nationally for industrial purposes. The emission is undesirable in two respects—sulfur dioxide is a serious acidic environmental pollutant and the loss of sulfur is a significant loss of a valuable resource. Problems begin initially at the mine. The presence of sulfur in coal causes the formation of acid. The acid formed causes corrosion of mining machinery. It has been reported that boiler superheater corrosion at high temperatures by sulfates has made necessary definition of a maximum operating temperature and consequently a limitation on the efficiency of stem power plants [37] . The cost of corrosion control is a major burden against the use of high-sulfur coal.

3.6 ECONOMIC FACTORS IN DESULFURIZATION

A variety of nonbiological technologies have been developed for desulfurization of coal. Technical details of these processes are not relevant to the purpose of this chapter, but the fact that all add significantly to the cost of coal is pertinent to this discussion. Three types of costs are associated with current methods for nonbiological coal desulfurization. These are reagent and waste disposal costs, energy-loss costs, and volume-related costs [38] . All are associated to a greater or lesser extent with desulfurization processes. The process used tends to be determined by the nature of the coal to be processed. Both precombustion and postcombustion methods are currently used. Microbial desulfurization of coal would be associated with precombustion treatment. As such, many of the sulfur-connected corrosion problems would be more easily controlled than at present. Removal of inorganic sulfur and metals would decrease ash, thereby increasing the heating value per unit weight. Biological desulfurization does not result in any fuel loss. Inclusion of a microbial treatment in conjunction with modified conventional desulfurization technology for removal of organic sulfur and finishing could be a realistic cost-saving process. It could also be a resource-conserving process.

Sulfur-containing products are used in the manufacture of petroleum, rubber, paper, sugar, pigments, fibers, and chemicals [39]. Up to 1984 fertilizer manufacture accounted for about 65% of international sulfur consumption. Agriculture was the primary determinant of sulfur use [40]. The situation has not changed significantly at this time. Major producers of elemental sulfur (brimstone) are the United States, the USSR, Poland, Mexico, and Iraq. It is also produced as a by-product from the processing of fossil fuels in Canada, the United States, the USSR, Saudi Arabia, France, and Japan [41]. Countries processing large amounts of natural gas, petroleum, and tar sands are good potential sources of sulfur as a by-product of fossil-fuel processing. At the present time over half the sulfur produced internationally is produced as a by-product of fossil-fuel processing. It is evident that as an element, sulfur is far from being difficult to obtain. Nevertheless, predictions based on present consumption trends indicate that there is a definite need for effective production and distribution of sulfur. Areas reporting sulfur deficiencies in 1983 are shown in Table 3.12. Other data of economic significance are shown in Table 3.13.

A number of new applications for sulfur have recently been developed. In terms of volume, the technologies that extend or replace asphalt as either sulfur-extended asphalt or Sulphlex are the largest [43]. Asphalt consumption in the United States for paving applications exceeds 20 million tons per year. That, in terms of sulfur used, represents about twice the current consumption of sulfur. It is almost three times the amount of sulfur that might be extracted from emissions of current coal-burning electric utilities and industrial boilers. Obviously, this is a market ready for sulfur that might be recovered from high-sulfur coal. Much of the asphalt used today is produced from petroleum. Approximately one-third of that petroleum is imported. Utilization of sulfur as a paving binder would lessen the importance of petroleum in asphalt production and decrease significantly the importance of imported petroleum.

Actually, the sulfur question is one of pricing and not of supply. A shortage of low-priced sulfur exists and there is a need for development of more uses of sulfur. A complex and abnormal supply-and-demand situation is creating the appearance of short supply and discourages research and development of new applications for sulfur. As of 1984 the United States produced only 20% of the world's supply of about 54 million tons. Internationally, additional sources of supply are under development. It has been reported that the Astrakhan sour gas deposits are being developed in the USSR. These are expected to yield far greater amounts of recovered sulfur than can be used in the USSR. In addition, new sources in Poland and Mexico are expected to increase the supply further.

U.S. government agencies have invested large sums of money in sulfur research and have found a number of new applications. These include sulfur foams, sulfur coatings, sulfur-aggregate concretes, sulfur-extended asphalt, and pavement binders. Other organizations involved in significant programs on sulfur utilization include, in Canada, Alberta Sulphur Research Ltd., McGill University, Ontario Research Foundation, and Shell Canada Ltd. In the United States research is, or has been, in progress at the Sulphur Institute, Southwest Research Institute, Amoco Petroleum Co., and Chevron Chemical Co. In

Table 3.12 Areas Reporting Sulfur Deficiencies in 1983

Africa	Louisiana	Venezuela
Cameroon	Maryland	Windward Islands
Central African Republic	Michigan	Asia
Chad	Minnesota	Burma
Congo	Mississippi	India
Ghana	Missouri	Indonesia
Guinea	Montana	Lebanon
Ivory Coast	Nebraska	Malaysia
Kenya	North Carolina	Sri Lanka
Malawi	North Dakota	Taiwan
Mali	Ohio	Thailand
Mozambique	Oklahoma	Australasia
Nigeria	Oregon	Australia
Senegal	Pennsylvania	Fiji
South Africa	South Carolina	New Guinea
Tanzania	South Dakota	New Zealand
Togo	Tennessee	Solomon Islands
Uganda	Texas	Europe
Upper Volta	Virginia	Belgium
Zaire	Washington	Bulgaria
Zambia	Wisconsin	Czechoslovakia
Zimbabwe	Wyoming	Denmark
Americas	Canada	Finland
United States	Alberta	France
Alabama	British Columbia	Iceland
Alaska	Manitoba	Ireland
Arkansas	Ontario	Italy
California	Saskatchewan	Netherlands
Colorado	Latin America	Norway
Florida	Argentina	Poland
Georgia	Brazil	Spain
Hawaii	Chile	Sweden
Idaho	Colombia	United Kingdom
Illinois	Costa Rica	USSR
Indiana	El Salvador	West Germany
Iowa	Honduras	Yugoslavia
Kansas	Puerto Rico	

Source: Ref. 43.

France, research at Societé Nationale Elf Aquitaine has contributed to sulfur utilization. The primary potential new uses of sulfur seem to be in structural materials, especially the previously mentioned sulfur asphalt compositions for road building, rigid sulfur foams for thermal insulation, and sulfur-based concretes for special applications where the properties of portland cement–based concrete were inadequate [42]. Trials have been conducted on the use of plasticized sulfur as a protective coating for conventional concrete surfaces exposed to corrosive conditions (as near certain chemical plants) and as a sealant for certain types of earthworks or dikes [43].

Table 3.13 Sulfur in All Forms, Production, Consumption, and Trade, 1982 (10^3 Tons Sulfur/Sulfur Equivalent)

Area	Production	Imports	Exports	Consumption
Western Europe	7,427	3,729	1,558	9,486
Africa	843	2,653	50	3,608
North America	16,960	1,728	7,073	12,242
Central America	2,118	93	1,064	1,119
South America	490	1,114	8	1,454
Asia	4,461	1,981	1,086	4,773
Oceania	195	631	–	911
Eastern Europe	6,654	1,386	4,016	3,735
USSR	9,132	1,163	177	10,144
China	2,414	346	–	2,981
World total	50,909	15,030	15,030	50,876

Source: Ref. 42.

3.7 CONCLUSIONS

Utilization of increasingly large amounts of fossil fuels will necessarily make available increasingly large amounts of sulfur. If costs of fossil fuels are to be minimized, the methods for sulfur removal will entail low-cost processes. Research by numerous investigators has now demonstrated that inorganic sulfur can be economically removed microbiologically. Removal of organic sulfur may require additional conventional sulfur-removal techniques.

Current studies on coal solubilization could have a major impact on removal of both inorganic and organic forms of sulfur as well as mineral content reduction and conservation. However, research in the solubilization area has not progressed sufficiently at this time to accurately foresee its influence. In any case, as regards sulfur, it is an element that has great potential for large-tonnage uses. Development of appropriate applications could both reduce coal-cleanup costs and provide useful products for industrial and civil development.

Minerals are potentially of considerable significance in regard to removal and recovery during precombustion processing of coal. Some of the metals are toxic, so their removal from coal is of importance in pollution control. Precombustion treatment of coal is particularly important because toxic metals are concentrated in ash and can subsequently be released in soil and water. They may also be released into the air during combustion or by wind from soil. The prevention of such mobilization by precombustion removal is a significant incentive for coal bioleaching studies [44].

Recovery of strategically valuable metals increases in importance as sources dwindle or become unreliable. Uses for metals such as germanium and gallium in highly sophisticated technical applications seem to be increasing. As has been indicated, coal is a potential source of a number of these metals. Availability of such critical materials contributes to international stability as well as to national security.

REFERENCES

1. Perry, H., Coal in the United States: a status report, *Science, 222*(4622), 377 (1983).
2. Valkovic, V., *Trace Elements in Coal*, Vol. 1, CRC Press, Boca Raton, Fla., 1983, p. 4.

3. Valkovic, V., *Trace Elements in Coal*, Vol. 1, CRC Press, Boca Raton, Fla., 1983,
    p. 5.
4. Valkovic, V., *Trace Elements in Coal*, Vol. 1, CRC Press, Boca Raton, Fla., 1983,
    p. 12.
5. Sternberg, H. W., Mechanism of coal liquefaction, in *Scientific Problems in Coal
    Utilization* (B. R. Cooper, ed.), Technical Information Department, U.S. Depart-
    ment of Energy, Washington, D.C., 1978, p. 53.
6. Francis, W., *Fuels and Fuel Technology*, Vol. 1, Pergamon Press, Elmsford, N.Y.,
    1965, p. 26.
7. Valkovic, V., *Trace Elements in Coal*, Vol. 1, CRC Press, Boca Raton, Fla., 1983,
    p. 6.
8. Dugan, P. R., Microbiological desulfurization of coal and its increased monetary
    value, in *Biotechnology and Bioengineering Symposium 16*, Wiley, New York, 1986,
    p. 185.
9. Attar, A., Evaluate sulfur in coal, *Hydrocarbon Process., 1*, 175 (1979).
10. Dugan, P. R., and Apel, W. A., Microbial desulfurization of coal, in *Metallurgical Ap-
    plications of Bacterial Leaching and Related Microbiological Phenomena* (L. E.
    Murr, A. E. Torma, and J. A. Brierley, eds.), Academic Press, New York, 1978,
    p. 223.
11. Valkovic, V., *Trace Elements in Coal*, Vol. 2, CRC Press, Boca Raton, Fla., 1965,
    p. 15.
12. Torma, E. E., and Bannegyi, I. G., Biotechnology in hydrometallurgical processes,
    Trends Biotechnol., 2, 13 (1984).
13. Dugan, P. R., Desulfurization of coal by mixed microbial cultures, in *Microbial
    Chemoautotrophy* (W. R. Strohl and O. H. Tuovinen, eds.), Ohio State University
    Press, Columbus, Ohio, 1984, p. 3.
14. Isbister, J. D., and Kobylinski, E. A., Microbial desulfurization of coal, in *Pro-
    ceedings of the First International Conference on Processing and Utilization of High
    Sulfur Coals*, Columbus, Ohio, Oct. 13–17, Elsevier, Amsterdam, 1985.
15. Hou, C. T., and Laskin, A. J., Microbial conversion of dibenzothiophene, *Dev. Ind.
    Microbiol., 17* (1976).
16. Monticello, D. J., Bakker, D., and Finnerty, W. R., Plasmid-mediated degradation
    of dibenzothiophene by *Pseudomonas* species, *Appl. Environ. Microbiol., 49*, 756
    (1985).
17. Krucher, R. V., Turovskii, A. A., Dzuhedzei, N. V., Pavlyuk, M. I., and Khmelnits-
    kaya, D. L., Cultivation of *Candida tropicalis* on coal substrates, *Mikrobiologiya, 46*,
    583 (1977).
18. Cohen, M. S., and Gabriele, P. D., Degradation of coal by the fungi *Polyporus versi-
    color* and *Poria monticola*, *Appl. Environ. Microbiol., 44*, 23 (1982).
19. Ward, B., Lignite-degrading fungi isolated from a weathered outcrop, *Sys. Appl.
    Microbiol., 6*, 236 (1985).
20. Scott, C. D., Strandberg, G. W., and Lewis, S. N., Microbial solubilization of coal,
    Biotechnol. Prog., 2, 131 (1986).
21. Gorin, E., Status of coal utilization technology, in *Scientific Problems of Coal Utili-
    zation* (B. R. Cooper, ed.), Technical Information Center, U.S. Department of
    Energy, Washington, D.C., 1978, p. 202.
22. Tributsch, H., and Bennet, J. C., Semiconductor-electrochemical aspects of bacterial
    leaching, Part 2, Survey of rate-controlling sulfide properties, *J. Chem. Technol.
    Biotechnol., 31*, 627 (1981).
23. Sugio, T., Domatsu, C., Tano, T., and Imai, K., Role of ferrous ions in synthetic
    cobaltous sulfide leaching of *Thiobacillus ferrooxidans*, *Appl. Environ. Microbiol.,
    78*, 461 (1984).
24. Mehta, A. P., and Murr, L. E., Kinetic study of sulfide leaching by galvanic interac-

tion between chalcopyrite, pyrite, and sphalerite in the presence of *Thiobacillus ferrooxidans* (30°C) and a thermophilic microorganism (55°C), *Biotechnol. Bioeng., 24*, 919 (1982).

25. Olson, G. J., and Kelly, R. M., Microbiological metal transformations: biotechnological applications and potential, *Biotechnol. Prog., 2*, 1 (1), (1986).

26. Mercz, T. J., and Madgwick, J. C., Enhancement of bacterial manganese leaching by microalgal growth products, *Proc. Australas. Inst. Min. Metall., 283*, 43 (1982).

27. Valkovic, V., *Trace Elements in Coal*, Vol. 1, CRC Press, Boca Raton, Fla., 1983, p. 59.

28. *Minerals Facts and Problems*, U.S. Department of the Interior, Washington, D.C., 1980.

29. *Minerals Yearbook*, Vol. 1, U.S. Department of the Interior, 1980.

30. *Chemical Marketing Reporter*, Nov. 17, 1986.

31. Norris, P. R., and Kelly, D. P., The use of mixed microbial cultures in metal recovery, in *Microbial Interactions and Communities*, Vol. 1, Academic Press, New York, 1982.

32. McReady, R. G. L., and Zentilli, M., A feasibility study on the reclamation of coal waste dumps by bacterial leaching, *CIM Bull.*, Apr. (1985).

33. Ralph, B. J., Biotechnology applied to raw minerals processing, in *Comprehensive Biotechnology*, Vol. 4 (M. Moo-Young, C. W. Robinson, and J. A. Howell, eds.), Pergamon Press, Elmsford, N.Y., 1986, p. 201.

34. Bos, P., Huber, T. F., Kos, C. H., Ras, C., and Kuenen, J. G., A Dutch feasibility study on microbial coal desulfurization, *6th International Symposium on Biohydrometallurgy*, Vancouver, British Columbia, Canada, Aug. 21–24, 1985.

35. Armson, R., Water makes coal a brighter fuel, *New Sci.*, Aug. 28 (1986).

36. Bagdigian, R. M., and Myerson, A. S., The adsorption of *Thiobacillus ferrooxidans* on coal surfaces, *Biotechnol. Bioeng., 28*, 467 (1986).

37. Engdahl, R. B., Impact of high sulfur on coal utilization, in *Processing and Utilization of High Sulfur Coals*, (Y. A. Attia, ed.), Vol. 9 of *Coal Science and Technology*, Elsevier, Amsterdam, 1985, p. 113.

38. Bullard, C. W., Technological options for coal desulfurization, in *Processing and Utilization of High Sulfur Coals* (Y. A. Attia, ed.), Vol. 9 of *Coal Science and Technology*, Elsevier, Amsterdam, 1985, p. 117.

39. Dugan, P. R., The relationship of biological sulfur cycling and coal industries to atmosphere acid and acid deposition, in *Processing and Utilization of High Sulfur Coals* (Y. A. Attia, ed.), Vol. 9 of *Coal Science and Technology*, Elsevier, Amsterdam, 1985, p. 101.

40. Detz, C. M., and Barvinchak, G., Microbial desulfurization of coal, *Min. Congr. J., 65*, 75 (1979).

41. Fike, H. L., Changing sulfur patterns, in *Processing and Utilization of High Sulfur Coals* (Y. A. Attia, ed.), Vol. 9 of *Coal Science and Technology*, Elsevier, Amsterdam, 1985, p. 769.

42. Sander, U. H. F., Fischer, H., Rothe, U., and Kola, R., in *Sulphur, Sulphur Dioxide, and Sulphuric Acid* (A. J. More, ed.), Verlag Chemie, Weinheim, West Germany, 1984.

43. Dale, J. M., Market prospects for coal-derived sulfur, in *Processing and Utilization of High Sulfur Coals* (Y. A. Attia, ed.), Vol. 9 of *Coal Science and Technology*, Elsevier, Amsterdam, 1985, p. 781.

44. Valkovic, V., *Trace Elements in Coal*, Vol. 2, CRC Press, Boca Raton, Fla., 1983, Chap. 2.

4

Pretreatment of Lignite

HANS E. GRETHLEIN *Michigan Biotechnology Institute, Lansing, Michigan*

To convert a heterogeneous, insoluble material into another product by microbiological means, it is necessary to break the insoluble material into soluble, usually smaller compounds that can pass through the microbial cell walls, where it serves as an energy source or nutrient in the metabolic cycles of the cell. For biogasification, a series of synergistic groups of organisms coexist in carrying out hydrolysis, acetogenesis, and methanogenesis.

The first step, the hydrolysis, is common to many biodegradations and has been studied in some detail for lignocellulose conversion to sugars by cellulase in a number of systems [1-3]. To some extent it is a useful model for the study of lignite biogasification. Often, the solubilization of the solid substrate is the rate-limiting step. The breakdown of the solid is facilitated by the presence of the proper extracellular enzyme. When the enzyme has access to the labile bond and the substrate has a reasonable number of labile bonds, the hydrolysis will result in producing sufficiently small molecules that are water soluble. Naturally, the breakdown products cannot be toxic to any of the many organisms that form this synergistic microbial system.

To overcome the enzyme inaccessibility problem, some type of pretreatment of the substrate is needed [4]. In the case of lignocellulose, for which there are many sources of cellulase, it is clear that the major requirement of a practical pretreatment is to increase the pore size of the microporous substrate to accommodate the relatively large micromolecules, the cellulase [5,6]. For example, native wood has about 0.5 mL of pore volume per gram of dry matter, which corresponds to about 500 m^2/g. Over 90% of these pores are less than 50 Å in width. Since the enzyme is estimated to be 51 Å in diameter, it is clear that the hydrolysis is, by and large, limited to the external bulk surface area of the lignocellulose. After fine grinding to 400 mesh, the particles still have less than 1 m^2/g of external area. Grinding to achieve significant surface area, comparable to 500 m^2/g, would mean a reduction to a particle size of 80 Å. Fortunately, more practical ways are available to increase the enzyme accessibility.

A thermal chemical pretreatment of lignocellulose can remove the easily hydrolyzable hemicellulose, which constitutes about 25% of the dry weight of wood and greatly increases the porosity of the solid residue. It is important that the removal of hemicellulose be done quickly to avoid subsequent lumic substance productions from the sugar formed by the thermal chemical hydrolysis of the hemicellulose [5,7]. How to carry out this pretreatment is discussed below.

The pretreatment is achieved in a continuous plug flow reactor as shown in Fig. 4.1. An aqueous slurry of substrate is suspended in the slurry tank with an agitator. The catalyst, acid such as H_2SO_4 or bases such as NaOH, is added to the slurry tank that is above the inlet of a Moyno Pump, a moving-cavity positive-displacement pump by Robbins and Myers (Springfield, Ohio). The purpose of the pump is to pressurize the slurry above the saturation pressure for the desired operating temperature in the reactor. Rapid heating of the slurry from room temperature to the desired pretreatment temperature, which is in the range 180 to 220°C for lignocellulose, is achieved by live steam injection into the pressurized slurry. The steam is generated in a self-contained electric steam generator. The flow of steam is controlled by a temperature controller and is proportional to the slurry flow. The reactor is a tube of 1/2-in. outside diameter which is well insulated. The outside wall temperature is monitored at several points along the length of the tube to ensure that, essentially, the reactor is isothermal. The residence time can be varied by adjusting the flow rate of the slurry and the length of the tube. The current design can operate from 4 to 20 s. A modified design with larger-diameter reactor tubes is being

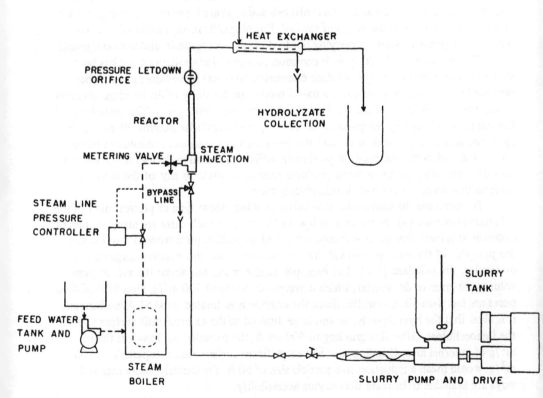

Fig. 4.1 Schematic diagram of reactor: live steam injection into slurry.

built to give up to 120 s of time. When the slurry passes through the orifice, the pressure drops to 1 atm. This causes an adiabatic flash, which cools and stops the pretreatment reactions. A heat exchanger condenses the flashed product and cools it to room temperature.

The reactor was designed originally to get the kinetics of cellulose hydrolysis with an acid catalyst [8]. Since overreaction can occur with a high-temperature hydrolysis, there is an optimum reaction time. This is shown in Fig. 4.2 for the acid hydrolysis of cellulose kraft paper. Note that at a given acid (catalyst) concentration, there is an optimum time for each temperature where the sugar (glucose) yield from the cellulose achieves a maximum value. As the temperature increases, the optimum yield increases with a reduction in time. This is typical of the case where the activation energy of the first reaction, the cellulose hydrolysis, is higher than the second reaction, the glucose decomposition.

This same reactor can be used to pretreat cellulose containing lignocellulosic materials for subsequent enzymatic hydrolysis. By using a lower temperature than one would for complete cellulose hydrolysis, it is possible to hydrolyze the hemicellulose selectively in the flow reactor and without overreacting the sugars formed from the hemicellulose [7]. In effect, a fraction of the lignocellulose is solubilized, leaving behind a biologically active solid residue of cellulose and lignin.

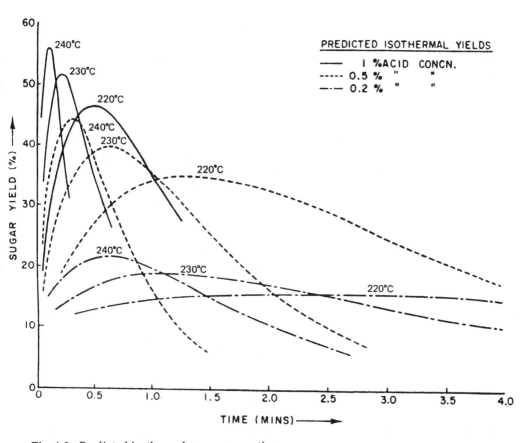

Fig. 4.2 Predicted isothermal sugars versus time.

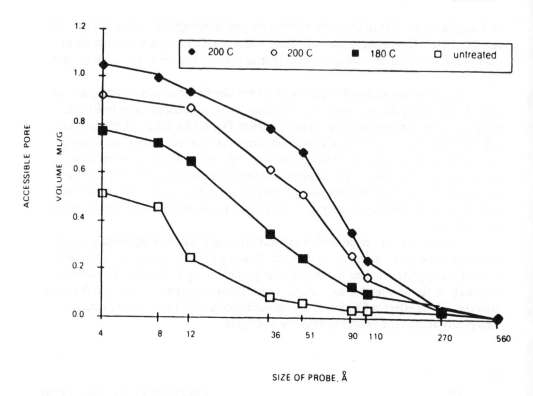

Fig. 4.3 Accessible pore volume for mixed hardwood and pretreated mixed hardwood in a plug flow reactor at 180, 200, and 220°C with 1% acid and 8 s. (From Ref. 6.)

Now let's look at the type of change in porosity the pretreatment of mixed hardwood can achieve in Fig. 4.3. Here the pore-size distribution of the wet solid residue is determined by the solute exclusion technique of Stone and Scallan [9] (it is essential that the solid not be dried after pretreatment, since this shrinks irreversibly all the micropores). The untreated mixed hardwood has 0.5 mL/g pore volume available to water (4 Å), of which about 0.07 mL/g is available to a 51-Å solute. Then, as the pretreatments at 180, 200, and 220°C are carried out with 1% H_2SO_4 for 8 s, the porosity of the residue is progressively increased. Note the significant increase in the region 50 to 100 Å over the original substrate. From the pore-volume distribution, it is possible to estimate the surface area accessible to a 51-Å enzyme [9]. The initial rate of enzymatic hydrolysis under a standard test condition of excess enzyme to substrate should be expected to be in proportion to the surface area available to the enzyme. This is what we found to a large degree in Fig. 4.4. Here not only are the different degrees of pretreated mixed hardwood included, but also a sample of poplar, white pine, and extracted pine. Note that the untreated samples are in the lower left corner of the figure at about 5% conversion in 2 h.

In the case of lignite, the structure can also be hydrolyzed at carbon-oxygen-carbon bonds such as point A in Fig. 4.5 with alkali as a catalyst. Since there is not the equivalent of an oxygen bond between every monomer as in a polysaccharide, the number of bonds that can be hydrolyzed in lignite are less than in cellulose. Thus a secondary reaction, such as an oxidation of carbon-carbon bonds, is necessary, as shown in Fig. 4.5 at point B.

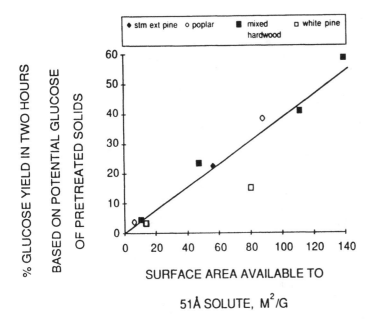

Fig. 4.4 Initial glucose yield as a function of surface area available to a 51-Å solute for various pretreated substrates by acid hydrolysis in a flow reactor. (From Ref. 6.)

While the composition of the soluble and residual products from pretreatment of lignite with alkali is unknown at this point of the investigation, it is apparent that the batch reactor, with its long heat-up and cool-down cycle, is not as effective as the flow reactor in producing soluble organic matter. For example, in Table 4.1 the first two rows show that the total dissolved volatile solids change from 5.9% to 3.9% as the batch reaction time with alkali increases from 120 min to 180 min. In Table 4.2 the TDVS is as high as 20% for the flow reactor for 60 s.

In the case of lignite, with so few oxygen bonds to hydrolyze, it was felt that the best pretreatment would be the one that produces the most soluble, bioconvertible organic matter. However, it remains to be shown what the reactivity of the residual lignite solids is to biogasification once suitable lignite consuming cultures are established.

In the meanwhile, the evaluation of the pretreatment for lignite has proceeded on the soluble fraction. Although alkaline hydrolysis gives only a modest increase in soluble matter, the combination of hydrogen peroxide (H_2O_2) and alkali are very potent. Again in comparing Tables 4.1 and 4.2, it is clear that the flow reactor gives more soluble material than the batch reactor when H_2O_2 is used with alkali. Figure 4.6 shows the increase in soluble volatile matter as a percent of the original total volatile matter in the lignite for various combinations of alkali and alkali with H_2O_2. The run, which gave over 77% solubilization, was a new batch of lignite with less ash content than the prior runs.

Defining the optimum pretreatment conditions is dependent on the biogasification culture used. To date the limited testing of biogasification of the pretreated lignite shows that a greater fraction of organic matter is converted to methane for the ultrafiltered soluble matter than for the plain filtered matter. The ultrafiltered matter is all less than 500 molecular weight.

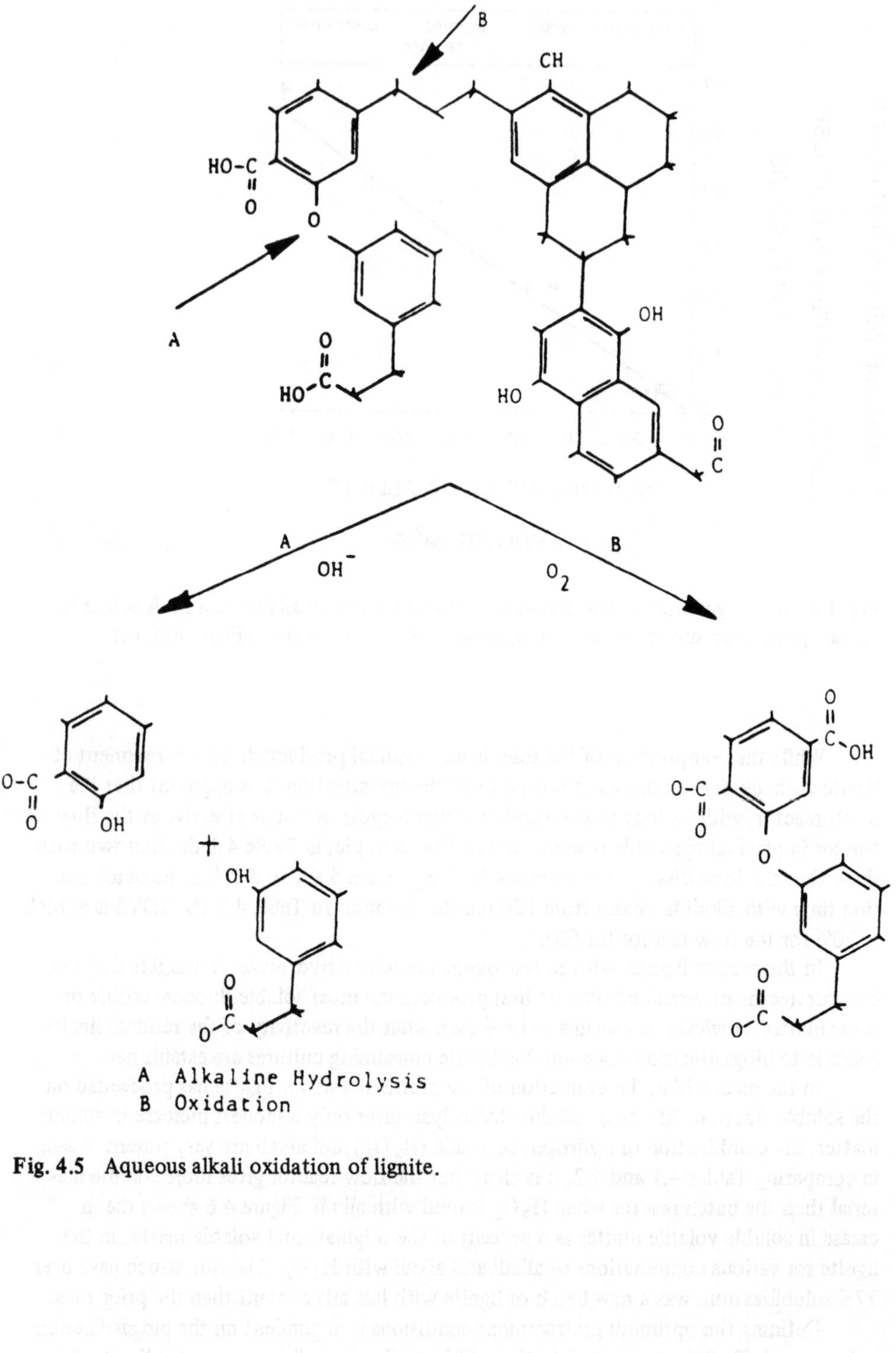

Fig. 4.5 Aqueous alkali oxidation of lignite.

Table 4.1 Summary of Batch Pretreatments

Reference number	Reaction temper-ature (°C)	Retention time (min)	Alkali concen-tration (% TS)	Other	TS (%)	TVS (% TS)	TDS (% TS)	TDVS (% TVS)
124279-A	250	120	10	N	9.92	68.55	7.97	5.88
124280-A	250	180	10	N	9.50	67.26	6.00	3.91
124282	250	30	10	5% EtOH	9.74	68.99	10.27	7.74
124288-A	250	30	10	1% H_2O_2, 5% EtOH	10.41	68.68	9.13	6.01
124280-B	250	30	10	1% H_2O_2	7.67	68.84	11.74	10.23
124283-B	250	30	10	1% H_2O_2	9.43	67.76	8.48	12.83
124283-A	250	15	10	1% H_2O_2	8.00	68.25	11.50	9.70
124283-C	250	60	10	1% H_2O_2	9.49	67.97	8.64	7.29
124283-D	250	90	10	1% H_2O_2	9.57	68.13	8.99	7.82
124288-B	285	30	10	1% H_2O_2	11.23	67.41	7.57	5.42

Table 4.2 Flow Reactor: Results of Extended Pretreatment-Solubilization (Filtration Method)

Retention time (s)	TS (%)	TVS (% TS)	TDS (% TS)	TDVS (% TVS)
	Alkali (10%) Pretreatment at 250°C			
0	9.72	56.85	2.26	1.28
20	9.72	56.85	6.37	5.92
40	9.72	56.85	13.43	17.98
60	9.72	56.85	26.77	20.00
80	9.72	56.85	26.92	19.57
100	9.72	56.85	27.45	16.13
	Alkali (10%) Pretreatment at 250°C with H_2O_2 (0.9%) Addition			
0	12.18	55.68	2.96	1.73
20	12.18	55.68	7.94	5.10
40	12.18	55.68	12.66	14.62
60	12.18	55.68	28.33	18.37
80	12.18	55.68	39.80	46.97
100	12.18	55.68	52.17	44.44

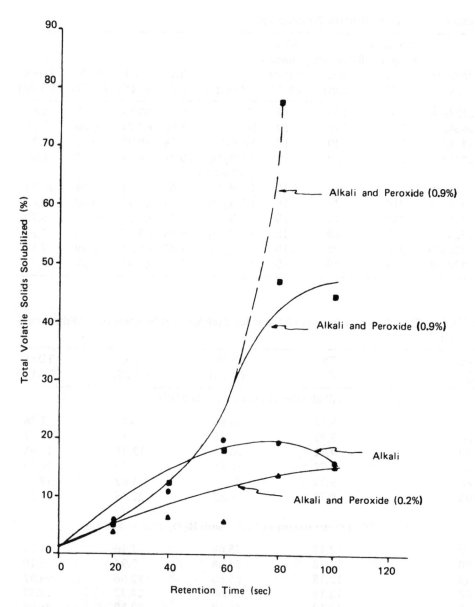

Fig. 4.6 Solubilization of volatile solids: continuous alkali pretreatment with and without peroxide addition (filtration method).

The effect of hydrolysis and oxidation, which is a free-radical decomposition, is initiated and accelerated by H_2O_2. The following initiation steps all occur:

$$H_2O_2 \rightarrow 2OH\cdot$$
$$H_2O_2 \rightarrow H\cdot + \cdot OOH$$

Both of these reactions are facilitated by some ions and hindered by other ions. The organic matter also cleaves by free-radical mechanisms such that

$$R{-}R' \rightarrow R\cdot + \cdot R'$$

is formed by a limited degree. However, with H_2O_2, there is an acceleration of free-radical compound formation during the propagation phase.

Some typical but not complete listings of free-radical propagation reactions are:

$$R-R' + \cdot OH \rightarrow R-OH + \cdot R' \quad \text{or} \quad R \cdot + R'-OH$$

$$R-R' + \cdot OOH \rightarrow R-OOH + \cdot R' \quad \text{or} \quad R \cdot + R'-OOH$$

$$R \cdot + R''H \rightarrow RH + R'' \cdot$$

$$R'' \cdot + ROOH \rightarrow RO \cdot + R''OH \quad \text{etc.}$$

Then there are termination steps, such as

$$R \cdot + H \cdot \rightarrow RH$$

$$R \cdot + R' \cdot \rightarrow R-R'$$

$$R \cdot + R'' \cdot \rightarrow R-R''$$

If $R'' \cdot$ is a high-molecular-weight radical, $R-R''$ is larger than the starting material. To know the optimum time and temperature of the pretreatment requires a detailed knowledge of all these species. This is not a likely prospect, so optimization has to be done empirically. However, it is clear that the longer times of the batch reactor give material that is less solubilized than the flow reactor. The final criterion has to be the degree of biogasification rather than soluble matter. This will naturally take into account what is least toxic and most accessible to the microbial system.

REFERENCES

1. Wood, T. M., and McCrae, S., in *Hydrolysis of Cellulose: Mechanisms of Enzymatic and Acid Catalysis* (R. D. Brown and L. Jusarek, eds.), *Advances in Chemistry Series, 181*, American Chemical Society, Washington, D.C., 1979, p. 81.
2. Mandels, M., in *Annual Reports on Fermentation Processes*, Vol. 5 (G. T. Tsao, ed.), Academic Press, New York, 1982, pp. 35–78.
3. Ladisch, M. R., Lin, K. W., Volock, M., and Tsao, G. T., Process considerations in the enzymatic hydrolysis of biomass, *Enzyme Microb. Technol., 5*, 82–102 (1983).
4. Grethlein, H. E., Pretreatment for enhanced hydrolysis of cellulosic biomass, *Adv. Biotechnol. Processes, 2*, 43–62 (1984).
5. Grethlein, H. E., The effect of pore size distribution on the rate of enzymatic hydrolysis of cellulosic biomass, *Biotechniques, 3*, 155–160 (1985).
6. Grethlein, H. E., and Converse, A. O., Understanding how pretreatment increases the rate of enzymatic hydrolysis of wood, *ACS Meeting*, Chicago, Sept. 1985.
7. Grethlein, H. E., Process for pretreating cellulosic substrates, U.S. patent, 4, 237, 226 1980.
8. McParland, J., Grethlein, H. E., and Converse, A. O., Kinetics of acid hydrolysis of corn stover, *Sol. Energy, 28*, 55–62 (1982).
9. Stone, J., and Scallan, A., Cellulose chemistry and technology, *Pulp Pap. Mag. Can., 69*, 343–358 (1968).

5

Bioreactor Study of the Microbial Degradation of Low-Rank Coals

VIJAY T. JOHN   *Tulane University, New Orleans, Louisiana*

MICHAEL D. DAHLBERG   *Pittsburgh Energy Technology Center, U. S. Department of Energy, Pittsburgh, Pennsylvania*

5.1 INTRODUCTION

The first evidence of the degradation of coals by fungi was given by Cohen and Gabriele [1], who reported that the basidiomycetes *Polyporus versicolor* and *Poria monticola* were capable of growing on and solubilizing lignite. Since then, other evidence of coal breakdown has appeared in the published literature [2-5]. Two important details emerged from previous research on coal degradation: (1) coals that are susceptible are low-rank coals such as lignite and, even more specifically, highly oxidized lignites such as leonardite, and (2) fungal species capable of degrading the lignin constituents of wood, the white-rot fungi, are also capable of degrading low-rank coals. The much greater susceptibility of highly oxidized coals to microbial attack was mentioned as early as 1961 by Rogoff et al. [6]; these authors speculated that coal degradation is essentially a surface phenomenon, that oxidation increases the surface area for enzymatic action, and that the additional hydroxyl and carboxyl groups generated through oxidation may provide sites for microbial attack. The degradative activity of the white-rot fungi should also be expected because low-rank coals such as lignite are generally believed to have their genesis in the lignin constituents of wood.

Much of the earlier work on coal degradation by fungal species was conducted using surface cultures grown over an agar layer. The cultures are then contacted with coal; the resulting liquid drops formed over a few days are then pipetted out for quantification and analysis. Submerged culture systems, wherein the fungus is cultured in a nutrient broth and contacted with coal in stationary or shake flask experiments, have also been reported as feasible in degradation [2]; however, a definitive analysis is still lacking.

In this chapter we report a preliminary study of the biodegradation of coal using a chemostat-type bioreactor operated in the batch mode. Thus the white-rot basidiomycete *Phanerochaete chrysosporium* was cultured in the chemostat and contacted with leonardite. The particular species was selected because extensive work has been carried

out on its growth and metabolism characteristics, and on factors that affect its ligninolytic activity [7-10]. Although details of the degradative action of *P. chrysosporium* are
too extensive to be reviewed, the relevant point is that ligninolytic activity is generated
when the fungus reverts to the idiophasic condition of secondary metabolism caused by
glucose, nitrogen, or sulfur deficiency in the growth medium. In this chapter we describe a method to analyze degradation subsequent to microbial-coal contact in the
chemostat bioreactor. Interpretations of the nature of coal degradation by *P. chrysosporium* are based primarily on Fourier transform intrared (FTIR) and ultraviolet (UV)-
visible (vis) spectrophotometric analyses.

5.2 MATERIALS AND METHODS

A defined culture medium was used [7]. All experiments were carried out in a 2-L fermentor (Bioflow from New Brunswick Scientific) operated in the batch mode using 1 L
of culture medium. Growth was initiated by adding an 8% (by volume) spore inoculum
standardized in all runs by an absorbance of 0.5 cm^{-1} at 650 nm (inoculum concentration of approximately 2.5×10^6 spores/mL). Air was bubbled through the reactor at a
flow rate of 1 L/min (1 v/v medium per minute), and the contents of the reactor were
maintained at 40°C and pH 4.2.

 Preliminary experiments to evaluate fungal growth levels were carried out by operating the bioreactor with a stirring speed of 500 rpm and maintaining the pH level by
adding KOH intermittently. At these conditions, the fungus grew as 1-mm pellets. A final
cell density of approximately 0.55 g dry weight per liter of solution was achieved before
the medium became nitrogen deficient. Addition of ammonium tartrate resulted in glucose deficiency and a final cell density of about 1 g dry weight per liter of solution. In the
next section we describe experiments where *P. chrysosporium* was cultured in the
presence of coal, and the results of such mycelia and coal contact.

5.3 COAL DEGRADATION EXPERIMENTS

To determine the effect of *P. chrysosporium* on low-rank coals, it is necessary to address
the following questions: (1) Does the microorganism really liquefy coal, and does it do so
through an oxidative process? (2) Can a simple scheme be devised to examine degradation
in a bioreactor? (3) Can the products of coal degradation be separated from other components in the bioreactor and identified through chemical characterizations? Three experimental studies were conducted to address these questions. Each study involved a 9-
to 11-day bioreactor run. A glossary of terms in Table 5.1 summarizes the procedures
used in the experiments; details are provided in the text. Thus, in the first study (bioreactor experiment I), 5 g of sterilized leonardite II (Knife River Coal Company, North
Dakota), of particle size <2 mm, was introduced into the chemostat together with spore
inoculum. The bioreactor operating conditions were as stated above, but the agitation was
reduced to a level just sufficient to prevent coal and fungus from settling to the bottom
of the reactor (<100 rpm). At this stirring speed, the growth of mycelia traps the coal
particles, and the mycelia and coal mass has a pulplike consistency, thus improving
mycelia and coal contact. Scanning electron micrographs of the mycelia and coal mass
showed extensive proliferation of mycelial filaments over the coal particles. Figure 5.1
illustrates mycelial protrusions at the edge of a coal particle at two different magnifi-
cations.

Table 5.1 Summary of Experiments

Bioreactor experiment I: The chemostat contents include leonardite, microorganism, and culture medium.

Bioreactor experiment II: The chemostat contents included leonardite and culture medium.

Bioreactor experiment III: The chemostat contents included leonardite extract, microorganism, and culture medium.

Leonardite extract (LE): The humic acid–type material obtained by carrying out the base dissolution and acid precipitation extraction procedure described in Fig. 5.2 with leonardite as the starting material.

Bioreactor extract (BE): The extract obtained through the same procedures of Fig. 5.2 but using the leonardite in contact with microorganism at the end of bioreactor experiment I as starting material.

Bioreactor-treated leonardite extract (BLE): The extracts again obtained using the same procedures of Fig. 5.2 but using leonardite extract in contact with microorganism at the end of bioreactor experiment III as starting material.

The prevalence of a constant pH (4.2) was taken as indicative of growth stoppage and the onset of glucose-deficiency-induced secondary metabolism. The volume of the fluid phase was kept constant at 1 L throughout the experiment by the addition of deionized water to compensate for evaporation losses. To determine if soluble products of biodegradation were being generated, samples of the fluid phase in the bioreactor were collected at different time intervals and centrifuged, and the percent transmittance scans of the supernatant were taken. Table 5.2 lists the time-dependent transmittance values at two wavelengths for bioreactor experiment I; note that the percent transmittance decreases to an asymptotically limiting value with time.

To rationalize the decrease in transmittance with time, one must review some relevant aspects of leonardite characteristics. Leonardite is naturally oxidized lignite; the oxidation leads to high-molecular-weight hydroxycarboxylic compounds termed humic acids. Extraction of humic acids from leonardite follows dissolution with base and subsequent precipitation with strong acid, as shown schematically in Fig. 5.2. The term "fulvic acids" refers to similar polycarboxylic acids with smaller molecular weights that are more soluble at low pH values. At the pH of 4.2 used in the bioreactor experiment, some of the humic/fulvic acids are solubilized in the culture medium, leading to an increase in the optical density. The question then is the relative amounts of solubilized material at pH 4.2 that are due to biochemical degradation of leonardite, and the amounts of solubilized fulvic/humic acids originally present in leonardite that simply diffuse out of the coal structure. The amount of material in the fluid phase at the end of the run was determined from the mass of the residue left after freeze-drying of a fluid sample. The mass determination yielded 425 mg of dissolved material per liter of fluid; however, this was an upper limit because components of the culture medium (primarily the polyacrylic acid buffer) were included in the mass determination. Clearly, this maximum amount of coal-based material in the fluid phase is a very small fraction of the total coal mass in the reactor (5 g).

To determine whether the decrease in transmittance of the fluid phase is due to microbial degradation of leonardite, a control experiment was conducted in the absence of fungal species (bioreactor experiment II). Thus 5 g of leonardite was introduced into

Fig. 5.1 Mycelial protrusions at the edge of a coal particle.

Table 5.2 UV-Visible Spectrophotometric Analysis of the Fluid Phase in the Bioreactor

Fluid phase sample	Time (days)	Percent transmittance	
		400 nm	700 nm
Bioreactor experiment I (leonardite + fungus)	1	32	96
	2	12	91
	4	7	86
	8	5	82
	10	5	80
Bioreactor experiment II (leonardite without fungus)	1	14	89
	5	2	76
	7	2	78
	9	3	79
Bioreactor experiment III (leonardite extract + fungus)	1	5	77
	3	3	74
	7	2	69
	11	1	65

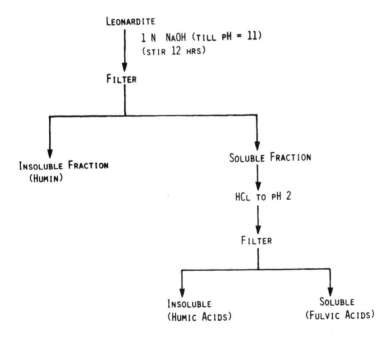

Fig. 5.2 Extraction of humic acids from leonardite.

the reactor; no spores were added, however, and the fluid phase contained only the diluted buffer component of the culture medium. Operating conditions were identical to those used in the previous experiment; samples were taken at discrete time intervals and centrifuged, and UV-vis scans of the supernatant were carried out. Table 5.2 again illustrates the results; note that the absorbance goes through a weak maximum at 5 days. This is simply because the equilibrium concentration of dissolved material at a pH of 4.2 is reached in a 5-day period; the removal of samples and the addition of deionized water to compensate for evaporation losses essentially dilute the fluid phase to the small extent, resulting in the observed minor decrease in absorbance after 5 days. Assuming similar chromophoric content with the fluid phase of bioreactor experiment I, the maximum concentration of dissolved material calculated from Beer's law and the transmittance at 600 nm was 556 mg/L of the fluid phase. A second observation is that the maximum optical density after 5 days is comparable to, and even slightly greater than, the maximum density in bioreactor experiment I. This is because in the absence of fungus, which tends to entrap the particles, the particles are subjected to more vigorous mixing and grinding, which results in a particle-size reduction and a greater diffusion rate of soluble material from the coal matrix. The comparable optical densities of the fluid phases in both bioreactor experiments make it clear, however, that the increase in transmittance cannot be attributed solely to microbial action on leonardite.

If there is to be significant biodegradation, the experimental observation that there are at most 570 mg of solubilized material for each 5 g of leonardite then implies that most of the degraded material must be insoluble at a pH of 4.2. Earlier research by Cohen and Gabriele [1] and Wilson and co-workers [4] on the characteristics of the droplets obtained from the coal and mycelia contact using surface cultures of the microorganism indicates that the material is very similar to humic acids. Since high-molecular-weight humic acids are insoluble in the fluid phase at pH 4.2, the solid phase in the bioreactor must be analyzed for evidence of degradation. Figure 5.3 illustrates schematically the possible constituents of the two-phase system in the bioreactor. The solid phase contains biomaterial (fungal mycelia), together with leonardite and any solid biodegraded product formed. To recover the biodegraded product, the contents of the reactor were first centrifuged to separate out the solid phase. The base extraction and acid precipitation procedure shown in Fig. 5.2 was then used on the solid phase. Filtration after treatment with NaOH separates out coal residues and fungal mycelia; the filtrate contains the humate-type material that is possibly produced by fungal metabolism on leonardite and that con-

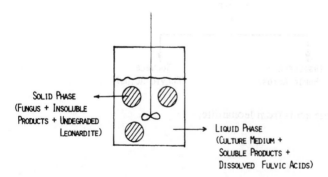

Fig. 5.3 Schematic of bioreactor contents.

stitutes a substantial fraction of leonardite, while the retentate contains cell debris and coal residues. The base-extracted material was then treated with HCl and refiltered, yielding a filtrate that was colorless, thus indicating complete precipitation of coal-based material left as the retentate. Cell lysis and release of intracellular material are bound to take place during the initial base treatment, leading to some contamination of the final coal-based material in the retentate. However, most of the cellular proteinaceous material probably remains in solution during the acid treatment step and is removed during the final filtration, where the retentate is the material of interest. In subsequent discussions, the acid-precipitated material from bioreactor experiment I is referred to as the "bioreactor extract" (BE).

An identical base extraction followed by acid precipitation was carried out for leonardite, resulting in extraction of the humic acid fraction of leonardite, amounting to 68% by weight. We refer to this extracted material as the "leonardite extract" (LE). Both the bioreactor extract and the leonardite extract were analyzed by FTIR and UV-vis spectroscopy, and elemental compositions of the two extracts and of the original leonardite were also determined. In obtaining the UV-vis spectra of the two extracts, the materials were solubilized with base and diluted to an approximate transmittance of 55% at 600 nm, after which a scan between 200 and 800 nm was carried out. The discretized results shown in the first two columns of Table 5.3 indicate that the transmittance values of the two extracts were almost identical over the entire wavelength range, thus implying no drastic differences in chromophoric contents.

Table 5.4 lists data on the elemental compositions of the extracts and of the original leonardite on a moisture-free basis. The extracts show reduced sulfur levels compared to leonardite. The bioreactor extract has an increased nitrogen content, probably owing to remnant-culture-medium nitrogen or protein. The C/O ratios in both extracts (1.88 and 1.47) are lower than the one in the original leonardite (2.08), as would be expected from material that contains primarily polycarboxylic compounds. The smaller ratio for the bioreactor extract may indicate further oxidation during fungal treatment, but this is not a clear conclusion.

Table 5.3 UV-Vis Scans on Alkali-Solubilized Extracts

Wavelength (nm)	Percent transmittance		
	Leonardite extract	Bioreactor extract	Bioreactor-treated leonardite extract
200	0	0	0
250	0	0	0
300	0	0	0
350	2	2	2
400	5	5	5
450	15	16	15
500	28	30	27
550	43	45	42
600	58	60	57
650	72	74	71
700	82	84	80
750	88	90	86
800	93	94	91

Table 5.4 Elemental Composition of Samples Subjected to FTIR (Percent of Dry Weight)

Element	Leonardite	Leondardite extract	Bioreactor extract	Bioreactor-treated leonardite extract
C	54.8	53.6	46.8	46.6
H	3.4	3.1	3.3	3.4
O	26.4	28.5	31.8	27.4
N	1.1	1.0	1.6	1.4
S	1.6	0.9	0.5	0.2
Ash	12.8	12.8	15.9	21.0

The FTIR scans on the two extracts are shown in Fig. 5.4b and c, together with a reference scan on untreated leonardite (Fig. 5.4a). A comparison of the spectra of the two extracts and leonardite shows that carboxylic group contributions (bands centered around 1720 and 1275 cm^{-1}) are present in the extract, as is to be expected of humic acid-type material. Note also that the spectra are almost identical.

Hence we concluded that over the 10-day period of fungal contact with leonardite, the microbial action on the substrate was negligible. To determine if this may be due to diffusional restrictions to enzyme penetration into the coal matrix, an experiment was carried out in which the humic acids content of leonardite (leonardite extract) was extracted and contacted with the microorganism, thus allowing easy access of the degradative enzymes to a potential substrate. Thus bioreactor experiment III was one wherein the original leonardite was replaced with the leonardite extract, and all other experimental procedures were identical to those followed in bioreactor experiment I. The time-dependent UV-vis scans of the supernatant are listed in Table 5.1; the scans do not indicate an asymptotic limit in transmittance values over the 11-day period of the run. The approximate concentration of the solubilized material after 11 days is 693 mg/L at a pH of 4.2.

After completion of the run, the solid phase (fungus + leonardite + extract + precipitated product) was centrifuged (7000 rpm, 15 min), and the base extraction and acid precipitation procedure of Fig. 5.2 was followed. The resulting material is termed the "bioreactor-treated leonardite extract" (BLE). The similarity of BLE to LE and BE is evident from UV-vis scans (Table 5.3) and elemental composition (Table 5.4). The C/O ratio of 1.7 is less than the ratios of leonardite (2.08) and the leonardite extract (1.88), and the low sulfur level (0.2 wt %) is probably due more to the extraction procedure than to biological removal. We feel the increase in ash content (21%) is due to biomaterial-based contamination.

The IR spectrum for the BLE sample is shown in Fig. 5.4d. Clearly, some change has taken place in the chemical constitution of the leonardite extract. The bands centered on 1720 cm^{-1} (C=O stretch in carboxylic groups) and at 1275 cm^{-1} (C–O stretch in carboxylic groups) that show up strongly in the IR spectra of the leonardite extract and

Fig. 5.4

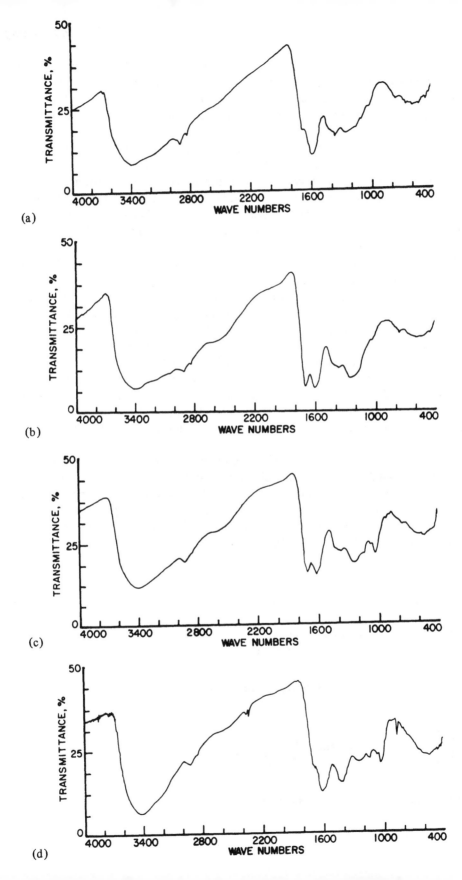

(a)

(b)

(c)

(d)

the bioreactor extract have been sharply reduced, indicating substantial decarboxylation. It would be expected that the loss of two oxygen atoms for every carbon, due to decarboxylation, would lead to an increase in the C/O ratio. We note, however, that the elemental analysis of the BLE sample in Table 5.4 does not reflect a pronounced change in this ratio.

No definite conclusions can be made from the other bands that show up in the bioreactor-treated leonardite extract. The loss of carboxylic groups indicates primarily the microbial action of P. chrysosporium on the leonardite extract. The question is: does decarboxylation occur during primary metabolism, wherein the microorganism utilizes carbon from the leonardite extract for growth, releasing CO_2 in respiration, or is it a result of secondary metabolic events, wherein degradative enzymes are synthesized? Carbon utilization by fungi follows typical growth and respiratory metabolic pathways [11] and is not discussed here. Degradative pathways for lignin may, however, provide information on the events taking place in secondary metabolism [12-14]. For example, the oxidative decarboxylation of vanillic acid, a lignin side-chain moiety, to methoxyhydroquinone has been reported to be catalyzed by P. chrysosporium (Fig. 5.5); such reactions may be prevalent in fungal action on the leonardite extract. However, one of the main modes of fungal attack on lignin is the oxidative opening of aromatic rings [12]; for example, methoxyhydroquinone is subsequently oxidized to maleyl acetate (Fig. 5.5), thus increasing the carboxylic content of the initial substrate. Experiments with increased durations of coal and fungal contact may reveal if such recarboxylation can occur. At this preliminary stage of our investigations, it is not yet understood if the decarboxylation observed is the consequence of microbial growth or of biodegradation during secondary metabolism.

Whereas the maximum degree of solubilization in the Bioflo experiments was less than 14% at pH 4.2, nearly complete solubilization of leonardite by P. chrysosporium mats on agar has been observed in our laboratory. The pH of the black drops produced generally ranges from 6 to 8. This difference may be because the degradation products were not solubilized at the low pH of the bioreactor. The influence of biosurfactants and ionic strength also needs to be investigated.

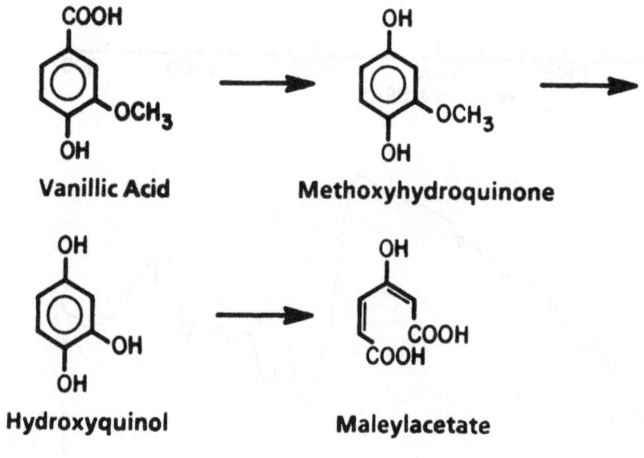

Fig. 5.5

5.4 CONCLUSIONS

Our experiments indicate that it is the oxidized humic acid–type material in leonardite that is susceptible to biodegradation. Leonardite is itself very slowly degraded, if at all, in submerged cultures, possibly owing to the difficulty of enzyme penetration into the coal matrix. However, if the humic acids are extracted from leonardite and subjected to microbial contact, such diffusional restrictions are removed and the substrate then becomes biodegradable. Further experiments with coal-swelling agents, such as γ-picoline or tetralin, may serve to indicate if reductions in diffusional limitations could improve the biodegradability of leonardite.

The base dissolution and acid precipitation process suggested offers a method to detect biodegradation in a complex system where separation of coal from microorganisms is difficult. The method developed to examine biodegradation is an indirect one; we examine changes in substrate characteristics through analysis of the acid-precipitated extracts. The combination of FTIR, UV-vis spectrophotometry, and elemental analysis can serve to clarify biodegradation. The FTIR technique in particular shows sensitivity to changes in the substrate chemical characteristics.

DISCLAIMER

Reference in this report to any specific commercial product, process, or service is to facilitate understanding and does not necessarily imply its endorsement or favoring by the U.S. Department of Energy.

ACKNOWLEDGMENTS

This work was supported by the Faculty Research Participation Program of the U.S. Department of Energy and was conducted at the Pittsburgh Energy Technology Center. The authors are grateful to Drs. B. Bockrath and R. Noceti for helpful discussions, to Mr. Schoffstall for laboratory assistance, and to Mr. B. Prozucek for the IR spectra.

REFERENCES

1. Cohen, M. S., and Gabriele, P. S., *Appl. Environ. Microbiol.*, *44*(1), 23 (1982).
2. Scott, C. D., Strandberg, G. W., and Lewis, S. N., *Biotechnol. Prog.*, *2*(3), 131 (1986).
3. Ward, H. B., *Syst. Appl. Microbiol.*, *6*, 236 (1985).
4. Wilson, B. W., Bean, R. M., Franz, J. A., Thomas, B. L., Cohen, M. S., Aronson, H., and Gray, E. T., *Energy Fuels*, *1*(1), 80 (1987).
5. Do Nascimento, H. C. G., Lee, K. I., Chou, S.-Y., Wang, W.-C., Chen, J. R., and Yen, T. F., *Process Biochem.*, *22*(1), 24 (1987).
6. Rogoff, M. H., Wender, I., and Anderson, R. B., *Information Circular 8095*, U.S. Bureau of Mines, Washington, D.C., 1962.
7. Jeffries, T. W., Choi, S., and Kirk, T. K., *Appl. Environ. Microbiol.*, *42*(2), 290 (1981).
8. Faison, B. D., and Kirk, T. K., *Appl. Environ. Microbiol.*, *49*(2), 299 (1985).
9. Ulmer, D. C., Leisola, M. S. A., Schmidt, B. H., and Fiechter, A., *Appl. Environ. Microbiol.*, *45*(6), 1795 (1983).
10. Janshekar, H., Brown, C., Haltmeier, Th., Leisola, M., and Fiechter, A., *Arch. Microbiol.*, *132*, 14 (1982).

11. Moore-Landeker, E., *Fundamentals of the Fungi*, Prentice-Hall, Englewood Cliffs, N.J., 1982.
12. Kirk, T. K., and Shimoda, M., in *Biosynthesis and Biodegradation of Wood Components* (T. Higuchi, ed.), Academic Press, New York, 1985.
13. Chen, C.-L., and Chang, H.-M., in *Biosynthesis and Biodegradation of Wood Components* (T. Higuchi, ed.), Academic Press, New York, 1985.
14. Higuchi, T., in *Biosynthesis and Biodegradation of Wood Components* (T. Higuchi, ed.), Academic Press, New York, 1985.

6

Microbiological Treatment of German Hard Coal

RENÉ M. FAKOUSSA *University of Bonn, Bonn, West Germany*

Research on the microbiology of hard coal and the idea of its utilization as a substrate for microorganisms is not new. The first experiments in this field, using brown coal primarily, were carried out at the beginning of this century [1-8]. Recently—due to the energy crisis and various environmental pollution problems—this idea has gained new currency and seems to be worthy of more basic research than has been the case [9]. In view of the increasing importance of hard coal as a raw material and energy source, it also seems helpful when developing new coal-processing technology to consider the eventual use of microorganisms in the degradation of hard coal. Research into microbial coal liquefaction has been stimulated in Germany especially by the technical problems associated with coal mining. In contrast to the United States and other countries, where mining is largely open-cast, hard coal seams in Germany are almost exclusively underground. There are extensive coal deposits throughout northern Germany. In the well-known coal mining area of the Ruhr valley, the deposits lie at depths of up to 1500 m (about 5000 ft) below the surface. The coal seams are tilted, so the depth increases to nearly 4000 m (about 13,000 ft) toward the northern coast. Even today there are great technical difficulties associated with mining coal at depths below 1500 m (about 5000 ft).

One utopian alternative to conventional mining techniques at depths greater than about 1500 m is the idea of using bacteria to solubilize the deposits. Due to the molecular complexity and physical structure of hard coal, there are various problems associated with microbiological degradation or even liquefaction of coal:

1. Hard coal is a solid material, having a macromolecular network structure. Therefore, the substrate is not present in the form of units that can be incorporated by microbial cells.
2. Hard coal is largely insoluble in water and organic solvents.
3. The lack of porosity in hard coal presents an obstacle to the penetration of microorganisms, so that the outer surface is almost the only area available for

attack. The pores of hard coal are too small to allow access to bacteria (about 10 to 100 Å).

4. Hard coal is not easily moistened with water; it is hydrophobic.
5. Hard coal is chemically very heterogeneous, as is also evident from the chemical diversity exhibited by the various types of hard coal.

Theoretically, there are four ways in which microorganisms may participate in the degradation of coal [9,9a] :

1. The incorporation of *whole* coal particles within the cell. In reality such a system is probably impossible.
2. Uptake of coal components that are soluble in the medium. In 1970, Rose and co-workers suggested in a patent the production of single-cell protein by using the water-soluble fraction from coal [10] . However, such a system has minimal economic importance due to the very small amount of water-soluble substances in the coal.
3. The production of cell-wall-bound extracellular enzymes for the degradation of coal in proximity to the cell. An example of such a system is one used by wood/cellulose-degrading fungi.
4. The most interesting mechanism: The degradation of coal by the release of extracellular enzymes or other cellular components.

The hard coal matrix is susceptible to enzymatic action if these areas can be made sterically accessible to microorganisms or enzymes. To date, the hard coal has been pretreated chemically (e.g., oxidized to carboxylic acids [11-13]) or only the extractable fraction has been used in previous microbiological studies (e.g., aliphatic compounds [10,14,15] . Other research groups used hard coal only as a "cosubstrate" [16-19] . Our goal was to use hard coal, which was not pretreated, an area not yet covered in any of the literature.

The first step was to achieve the greatest possible surface area (i.e., to grind the substrate as finely as possible). This was done by using a "counter jet mill" (i.e., fluid energy mill). The average particle size of the hard coal powder was 2.5 μm (see Fig. 6.1). Coal containing different amounts of volatile matter (15 to 35%) was used. The coals were characterized by their elemental composition (see Table 6.1) and maceral analysis (see Fig. 6.2).

Taking into consideration the mineral composition of hard coal, a nutrient solution was developed for the microorganisms. Metals that occur in dioxygenases and hydroxylases (Fe, Cu, Zn, Mn) were added in high concentrations. To satisfy the methodical preconditions for aerobic incubation of microbiological cultures with hard coal as a substrate, the nutrient suspension and solid culture media had to be prepared with coal powder as the main source of carbon: Ultrasound was used to disperse the coal in the nutrient suspensions. Solid culture media were prepared by pouring agar plates without coal and then adding a thin layer of hard coal with a glass spatula (Drigalski) [9,9a] . Dialyzed agar was used to exclude oligotrophic microorganisms. The cultures were incubated in a closed chamber to prevent contamination through airborne nutrients. The coal was also stored in airtight containers so that no foreign organic compounds could be absorbed.

In a comprehensive screening program, microorganisms from suitable locations were enriched. Burned regions of forest (0.5 to 20 years old) were used as screening locations, since charcoal is, with respect to its physical structure (pores, adsorptive characteristics),

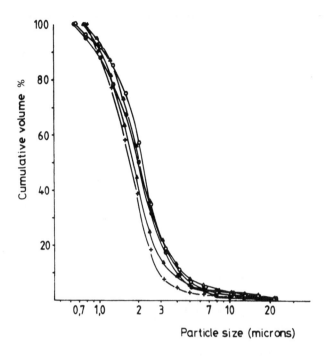

Fig. 6.1 Comparison of the grain-size distribution of the coal powder by Coulter Counter analysis. +, Coal with 15% volatile matter; △, coal with 20% volatile matter; ○, coal with 25% volatile matter; ▲, coal with 30% volatile matter; ●, coal with 35% volatile matter.

Table 6.1 Elementary Composition of the Coals Used (Weight Percent in Water-Free Coal)[a]

Elements/compounds	Type of coal				
	Ca. 15% volatile matter	20% volatile matter	25% volatile matter	30% volatile matter	Ca. 35% volatile matter
Total carbon	75.50	87.79	85.85	85.33	84.96
Carbonate CO_2	0.32	0.28	1.23	0.68	0.45
Total hydrogen	3.74	4.80	4.79	4.88	5.20
Total oxygen	2.01	2.71	3.92	4.13	4.13
Nitrogen	1.31	1.42	1.51	1.41	1.77
Total sulfur	7.10	0.87	0.63	0.99	0.71
Disulfide sulfur	5.55	0.24	0.06	0.22	0.18
Organic sulfur	1.55	0.63	0.57	0.77	0.53

[a]See Refs. 9 and 9a for description of the determination methods. The values for the ash composition are omitted.

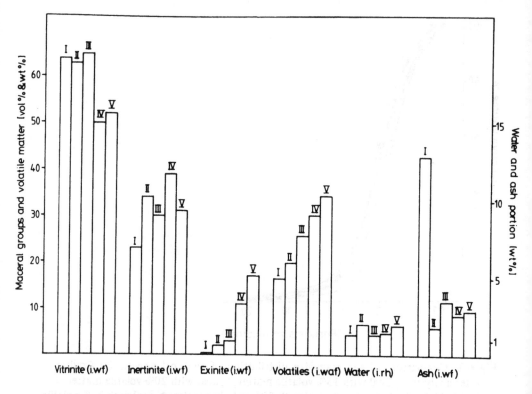

Fig. 6.2 Comparison of the most important characteristics of the types of coal used. I, Coal with 16% volatile matter; II, coal with 20.0% volatile matter; III, coal with 25.5% volatile matter; IV, coal with 30.1% volatile matter; V, coal with 34.4% volatile matter; i.rh, coal at the time of sampling; i.wf, water-free fuel; i.waf, in water- and ash-free fuel.

as well as in its elementary structure, related much more closely to hard coal than to oil (e.g., Refs. 20 and 21). A total of 1400 liquid cultures and 1700 solid cultures were tested. The pH values, temperature, inoculant, coal type and coal concentration, the incubation method, and substrate composition were varied [9]. Initially, three filamentous fungi were enriched (Figs. 6.3 and 6.4); two yeast (Fig. 6.5) and two bacterial strains were isolated. Growth was observed in only 0.2% of all cases, an exceptionally low rate. In the case of the fungi, attachment of coal particles at the cell wall could be observed (Fig. 6.4). This is possibly due to a hydrophobic interaction, in that bipolar and apolar molecules will dissolve the "bond" most readily. Embedding in mucilaginous substances does not occur.

Initially, pleomorphic bacteria that contained extremely hydrophobic cell walls were enriched. These were possibly organisms belonging to the mycobacteria or nocardia. The cell aggregations floating on the surface of the liquid could only be dispersed using detergents or oily substances. Spectroscopic investigations of the absorption in the ultraviolet (UV) region, which may be attributed to a soluble coal fraction, show that as the bacteria grow, the UV absorption decreases, indicating that the organisms are probably growing on the solubilized coal compounds.

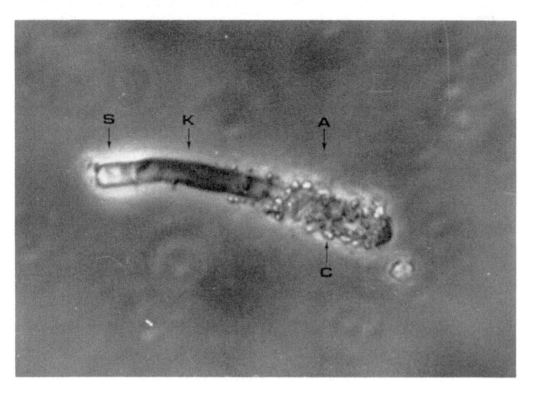

Fig. 6.3 Germinating hyphae of a fungus with sporal end (S). The K-zone of the sprouting tube has no affinity for coal particles. On the tip of the germ tube (A) newly formed cell walls start to adsorb coal (C) (magnification ca. 1600 ×).

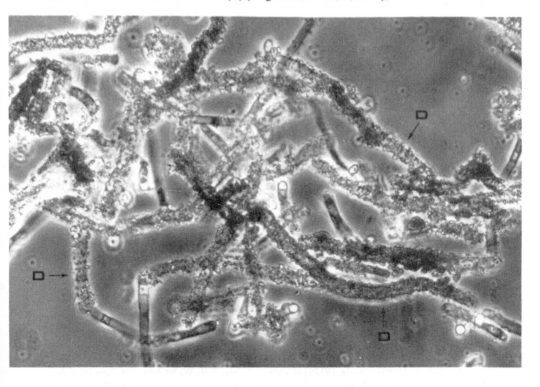

Fig. 6.4 Fungus culture in a coal suspension. In the end phase the hyphae are almost completely covered with adsorbed coal (D) (magnification ca. 1500 ×).

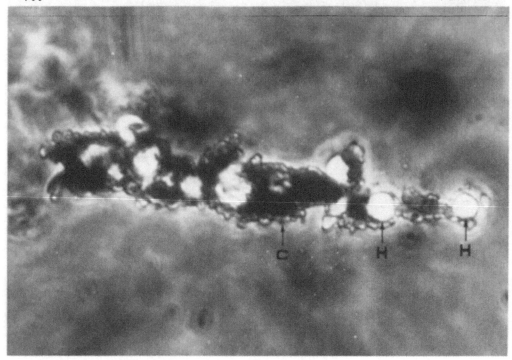

Fig. 6.5 Yeast culture in a coal suspension (magnification ca. 1600 ✕). H, yeast cells; C, coal particles.

Further screening concentrated on organisms capable of producing free extracellular material, which was active in altering the properties of the coal. A bacterium isolated and identified as *Pseudomonas fluorescens* showed some remarkable properties [9,9a] :

1. When growing in a mineral medium with coal, the protein yield was shown to increase in proportion to the amount of coal added to the culture (Fig. 6.6).
2. After some time (about 3 to 6 weeks, depending on the conditions) changes were evident in the culture. The culture medium turns brown, which may be due to the partial degradation of the coal. This browning of an aqueous coal suspension is clearly due to bacterial action, because it occurs within a few hours if the culture is grown in complex media and extracellular material is added to a coal suspension (Fig. 6.7).

The extent of these effects depended on the type of hard coal, the media composition, and the concentration of coal used. By using several controls, it could be established that the brown coloring of the culture supernatant is actually caused by substances dissolved out of the coal. They are neither of bacterial origin, nor are they very fine coal particles held in stable suspension by a bacterial surfactant.

3. This bacterial strain releases a detergent into the culture medium (Fig. 6.8).

The effect of the culture supernatant on the surface tension of water in comparison with sodium lauryl sulfate (SDS) is shown in Fig. 6.8. Water has a surface tension value of 72.6 mN/m, while the undiluted culture supernatant has a value of 25.5 mN/m,

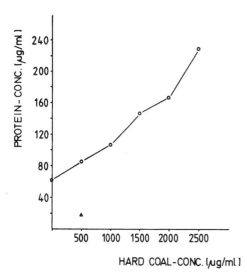

Fig. 6.6 Protein yields as a function of the hard coal concentration. ○, protein content minus the coal portion; △, sample poisoned with cyanide. The coal used contained 35% volatile matter, addition of normal carbon sources 0.008% = yeast extract and Casamino acids.

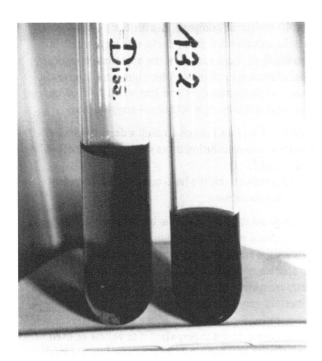

Fig. 6.7 Browning of aqueous coal suspension. Left, treatment with extracellular substances; right, supernatant of a culture medium after several weeks' incubation.

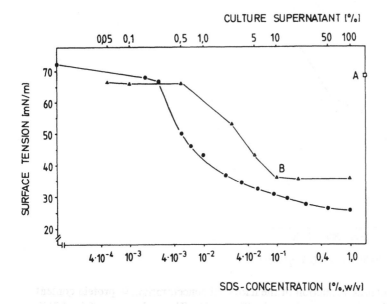

Fig. 6.8 Surface tension as a function of the concentration of surfactants. •, Dilute culture supernatant of a liquid culture; △, sodium lauryl sulfate solution (SDS, synthetic detergent); A, value of surface tension for uninoculated nutrient solution; B, critical micelle concentration for SDS.

corresponding to a reduction of over 40 mN/m in comparison with the uninoculated medium. Based on these values, it may be calculated that the surface energy, which is needed to wet coal, is reduced by two-thirds in the presence of the bacterial detergent.

As the strain excreted various substances (i.e., water-soluble pigments, detergents, and several enzymes, including lipases and proteinases), it was important to find out which of these had an effect on the coal. The following mechanisms are possible:

1. The bacterial surfactants lower the surface tension to such a degree that the coal may be extracted from the nutrient solution more easily. The extracted substances are absorbed and utilized.
2. The enzymes released by the bacteria attack the hard coal and the detergent acts only as a solvent or transport molecule.

Initially, the possibility was investigated as to whether the bacterial surface-active agent alone was responsible for the appearance of the brown substances—for example, by extracting the coal. To test this hypothesis, water and some synthetic detergent solutions were compared with regard to their efficiency in extracting components from coal. Rather surprisingly, it was found that pure water extracted coal to a higher degree than did some of the detergents used. This effect may be explained by the fact that the detergents stick to the coal and clog the pores of the coal. None of the detergents were able to extract coal to such an extent that the color of the supernatant was yellow or even brown. Consequently, the first possibility was excluded. Additional experimental work has shown that there is no obvious relationship between the surface tension of a hydrophilic liquid (or organic solvents) and the amount of material extracted from the coal (Table 6.2), which confirms our results. An indication that the second hypothesis seems

Table 6.2 Comparison of Surface Tension and Extraction Properties of Several Solvents

Solvent	Surface tension (dyn/cm)	Yield of coal extract (wt % waf)[a]
n-Hexane	18.4	0.0
Methanol	23	0.1
Ethanol	22	0.2
Acetone	23.5	1.7
Benzene	28.8	0.1
Pyridine	38	12.5
Diethyl ether	16.6	11.4

Source: Data from Ref. 22.
[a]waf, substance water- and ash-free.

to be correct is that up to 30% of the bacterial extracellular proteins was bound to the coal particles. Moreover, those types of hard coal that supported the best growth in culture showed the greatest affinity for extracellular proteins. Coke could bind only one-tenth of this amount. Control experiments with other proteins, such as bovine serum albumin, were negative.

Finally, it was shown that hard coal is changed through the extracellular bacterial substances: in addition to an increasing hydrophobicity of the coal particles, infrared (IR) spectroscopy showed that the transmission (in the wave number range 600 to 1500 cm^{-1}) increased, and at the same time the relative portion of aromatics was lowered (band range = 660 to 930, 1600 cm^{-1} wave number, Fig. 6.9). This leads to the supposition that the brown coloring (browning) of the culture supernatant is due to compounds from the coal. This is also indicated by the change of extractability of the coal. After treatment

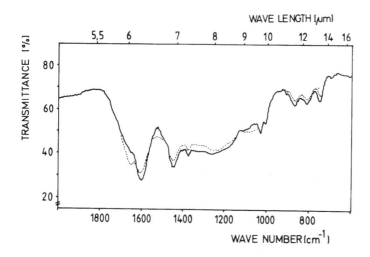

Fig. 6.9 Comparison of the infrared transmission of two samples of coal (rank of coal: 35% volatile matter). —, Coal after suspension in phosphate buffer; ---, coal after the action of a solution of extracellular, bacterial substances (duration of reaction ca. 20 h).

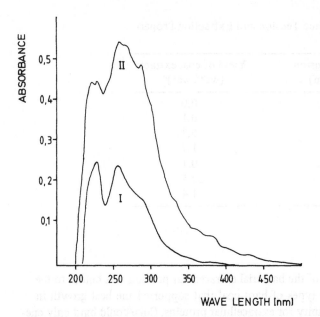

Fig. 6.10 Absorption spectra of two coal extracts. I, Coal after suspension in nutrient solution; II, coal after exposure to the action of extracellular bacterial substances. The coal sample contained 35% volatile matter.

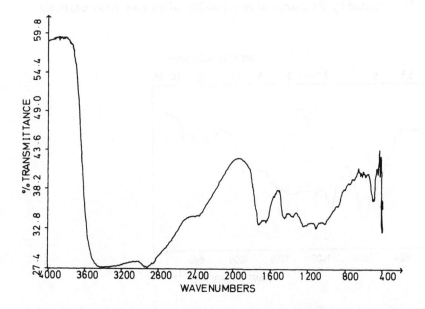

Fig. 6.11 Infrared spectrum of the isolated coal compounds released into the culture supernatant.

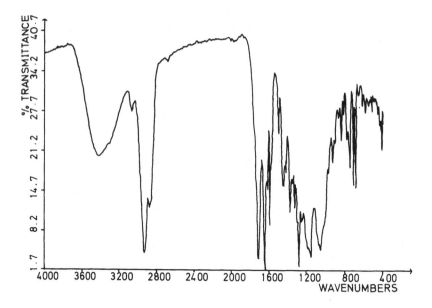

Fig. 6.12 Infrared spectrum of the methylated brown compounds of the coal released into the culture supernatant.

with microbial extracellular substances for about 20 h, the extractability with a solvent (hexadecane) was increased threefold (Fig. 6.10). The effects mentioned previously indicate that in addition to the external structure, the internal structure was also changed and cross-linkages were possibly broken.

The next step was to attempt to characterize the coal compounds released into the supernatant. The molecular weights as determined by ultrafiltration range between 50,000 and 100,000. The brown substances are hydrophilic, and their behavior on ion-exchange columns suggest that they contain a large number of ionic groups. The IR spectrum (Fig. 6.11) shows strong absorptions due to carboxyl groups and additional bands characteristic of substances derived from coal. These carboxyl groups were esterified with trimethylorthoformate [23]. The IR spectrum of the methyl esters is shown in Fig. 6.12. The presence of carboxyl (COOH) and hydroxyl (OH) groups is consistent with oxidative degradation of the coal. The results suggest that enzymatic action has partially altered the nature of the coal, perhaps creating a looser structure. More detailed investigations by solid-state ^{13}C NMR are currently under way.

CONCLUSIONS

Our experiments suggest that coals with a high content of volatile matter are attacked more easily by bacteria. A low inertinite content seems to be helpful. No further experiments were carried out to investigate the influence of the differences between the three maceral groups—inertinite, vitrinite, and exinite—due to the fact that no easy methods exist to separate them. Perhaps this may be possible in the future by selective degradation of one or more of these groups. The further extrapolation of such research makes it possible to consider the use of extracellular microbial substances for the decomposition of coal (e.g., splitting the aliphatic cross-linkages). Such a system would use little energy (normal temperatures and pressures!), in order to transform coal from a solid to another phase condition, which would be easier to process.

ACKNOWLEDGMENTS

This project was partially supported by the Max-Planck-Institut für Kohlenforschung, Mülheim/Ruhr. I would also like to thank Bergbauforschung, Essen (Prof. Kölling), and the Minister für Wirtschaft, Mittelstand und Technologie of Nordrhein-Westfalen for providing funds. I am especially thankful to Dr. Haenel from Max-Planck-Institut für Kohlenforschung for helping with the characterization of the "brown substances" and for helpful discussions. I am grateful to Prof. Trüper for his continuous and liberal support during my work. I thank Dr. B. J. Tindall for critical reading and for helping to prepare the manuscript. The elemental coal analysis was done by Ruhrkohle AG, Essen. The preparation of this manuscript was assisted by the use of translations made available by the Department of Energy, Pittsburgh Energy Technology Center (U.S.) and the International Energy Agency (U.K.).

REFERENCES

1. Potter, M. C., Bakterien als Agentien bei der Oxidation amorpher Kohle, *Zentralbl. Bakteriol. Parasitenk. II, 21*, 647–665 (1908).
2. Fuchs, S., Uber die Einwirkung von Bakterien auf Kohle, *Brennst. Chem., 8*, 324–326 (1927).
3. Lieske, R., Uber die in der Kohle lebenden Bakterien und einige ihrer Eigenschaften, *Gesammelte Abh. Kennt. Kohle, 9*, 27–29 (1928), Kaiser Wilhelm-Society, Germany.
4. Lieske, R., and Hofmann, E., Untersuchungen über die Mikrobiologie der Kohlen und ihrer natürlichen Lagerstätten. II. Die Mikroflora der Steinkohlengruben, *Brennst. Chem., 9*, 282–285 (1928).
5. Lieske, R., Biologie und Kohlenforschung, *Brennst. Chem., 10*, 437–438 (1929).
6. Lieske, R., Untersuchungen über die Verwendbarkeit von Kohlen als Düngemittel, *Brennst. Chem., 12*, 81–85 (1931).
7. Fischer, F., Biologie und Kohle, *Angew. Chem., 45*, 185–194 (1932).
8. Fischer, F., and Fuchs, W., Uber das Wachstum von Pilzen auf Kohle, *Brennst. Chem., 8*, 293–294 (1927).
9. Fakoussa, R. M., Kohle als Substrat für Mikroorganismen: Untersuchungen zur mikrobiellen Umsetzung nativer Steinkohle, Dissertation, Universität Bonn, 1981.
9a. Fakoussa, R. M., translation of the investigations of the microbial decomposition of untreated, hard coals, Coal as substrate for microorganisms: doctoral dissertation of R. Fakoussa, Bonn, 1981; prepared for U.S. Department of Energy, Pittsburgh Energy Technology Center; translated by the Language Center Pittsburgh under Burns and Roe Services Corporation, Pittsburgh, June 1987.
10. Rose, M. J., Carosella, J. M., Corrick, J. D., and Sutton, J. A., Method for producing protein by growth of microorganisms on a water extract of coal, U.S. Patent 3,540,983, 1970.
11. Fuchs, W., Fuchs, F., and Reid, J. J., Biological decomposition of hydroxy-carboxylic acids obtained from coal, *Fuel, 21*, 96–102 (1942).
12. Fischer, F., and Fuchs, W., Uber das Wachstum von Schimmelpilzen auf Kohle, *Brennst. Chem., 8*, 231–233 (1927).

13. Kucher, R. V., Turovskii, A. A., Dzumedzei, N. V., Bazarova, O. V., Pavlyuk, M. J., and Khmel'nitskaya, D. L., Cultivation of *Candida tropicalis* on coal-substrates, *Mikrobiologiya, 46*, 477–479 (1977).

14. Silverman, M. P., Gordon, J. N., and Wender, I., Food from coal-derived materials by microbial synthesis, *Nature, 211*, 735–736 (1966).

15. Brooks, J. D., and Smith, J. W., Microbiological oxidation of a coal tar fraction, *Aust. J. Chem., 19*, 1987–1989 (1966).

16. Darland, G., and Brock, T. D., A thermophilic, acidophilic *Mycoplasma* isolated from coal refuse pile, *Science, 170*, 1416–1418 (1970).

17. Johnson, G. E., Production of methane by bacterial action, U.S. Patent 3,640,846, Feb. 8, 1972.

18. McConville, T., and Maier, W. J., Use of powdered activated carbon to enhance methane production in sludge digestion, *Biotechnol. Bioeng. Symp., 8*, 345–359 (1978).

19. Spencer, R. R., Enhancement of methane production in the anaerobic digestion of sewage sludges. *Biotechnol. Bioeng. Symp., 8*, 257–268 (1978).

20. Neumüller, O. A., *Römpps Chemie-Lexikon*, Franckh'sche Verlagshandlung, Stuttgart, West Germany, 1972.

21. Winnacker, K., and Küchler, H., *Chemische Technologie*, Hanser, Munich, West Germany, 1970.

22. Marzec, A., Juzwa, M., Betly, K., and Sobkowiak, M., Bituminous coal extraction in terms of electron-donor- and -acceptor interactions in the solvent coal system, *Fuel Process. Technol., 2*, 35–44 (1979).

23. Cohen, H. and Mier, J. D., Esterification of carboxylic acids with triethyl orthoformate, *Chem. Ind.*, 349–350 (1965).

7

Biological Methane Production from Texas Lignite

ALFRED P. LEUSCHNER*, MARK J. LAQUIDARA, and ANNETTE S. MARTEL*
Dynatech Scientific, Inc., Cambridge, Massachusetts

7.1 INTRODUCTION

Conversion of Texas lignite to methane gas is being examined in this laboratory research program. The process under investigation is a two-stage process in which lignite is initially converted to a microbial substrate through thermochemical means and the resultant material is then biologically converted to methane and carbon dioxide gases. A schematic of the process is presented in Fig. 7.1.

The first step in this process requires that the mined lignite be crushed and ground to a powder. As a basis for this experimental program, a particle size of 0.7 mm (or 150 mesh) was selected. After crushing and grinding, water is added to the lignite, producing a 10% total volatile solids slurry (on a wet weight basis). The lignite slurry is then pretreated. The pretreatment unit used in this study is a plug from a reactor that was operated at short retention times. The lignite is treated at 250°C for 10 to 100 s. Alkali and hydrogen peroxide are added to the lignite slurry to improve pretreatment effectiveness. The objective of the pretreatment is to produce soluble, low-molecular-weight material from the lignite solids. After pretreatment the remaining, unreacted solids are removed from the lignite slurry. These solids are used as a low-grade fuel while the solubilized lignite solution undergoes fermentation.

This process is being evaluated on the basis of converting 20,000 tons of lignite per day. At a 10% solids slurry this represents a flow of approximately 23 million gallons of lignite slurry per day. Due to the relatively long retention times associated with anaerobic fermentation, the second stage in this process, and the flow rate of 23 million gallons per day, above ground, steel tanks were ruled out as being economically infeasible for this process. Thus a more economic method of producing large-volume reactors were required. In a report to Houston Lighting and Power Company prepared by Fenix & Scisson, Inc.

---

*Current affiliation: Remediation Technologies, Inc., Concord, Massachusetts

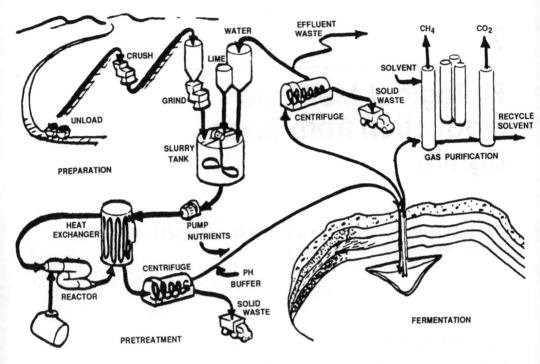

Fig. 7.1 Process schematic.

[2], two underground cavern designs were identified. The first design consisted of traditional room-and-pillar mining to produce a series of packed-bed digesters as shown in Fig. 7.2. The mined rock would be crushed to 1 to 2-in. stone and replaced into the digester. This stone would then form the support surface in the digester onto which the anaerobic microorganisms would attach.

The second reactor construction method utilizes solution mining of a salt dome to create a large volume in what is termed an inverted morning glory shape. A schematic of this reactor type is shown in Fig. 7.3. The anaerobic digester would be operated as a completely mixed reactor in this scenario. The most severe limitation in operating a digester of this type would be to prevent disolution of the cavern walls, which if allowed to occur, would cause the cavern to grow and eventually collapse. Fenix & Scisson, Inc. [2] estimated that if the reactor were operated at 75% of salt saturation growth of the reactor would be minimized and provide a 20-year reactor life. However, this restriction means that the anaerobic digestion process would have to be performed at salt concentrations far above known toxicity levels. To circumvent this problem it was decided to explore the possibilities of using anaerobic halophiles to perform this fermentation. Table 7.1 presents a list of the design criteria for the two reactors under consideration.

After digestion the liquid stream is subjected to liquid/solid separation. The solids resulting from this step are to be used as a low-grade fuel. The liquid requires disposal. The gas evolved from the process will contain methane and carbon dioxide. Gas purification is used to split the gas into its two components. The methane produced will be of sufficient quality to be injected into a gas pipeline. The carbon dioxide side stream will be sold for use in tertiary oil recovery.

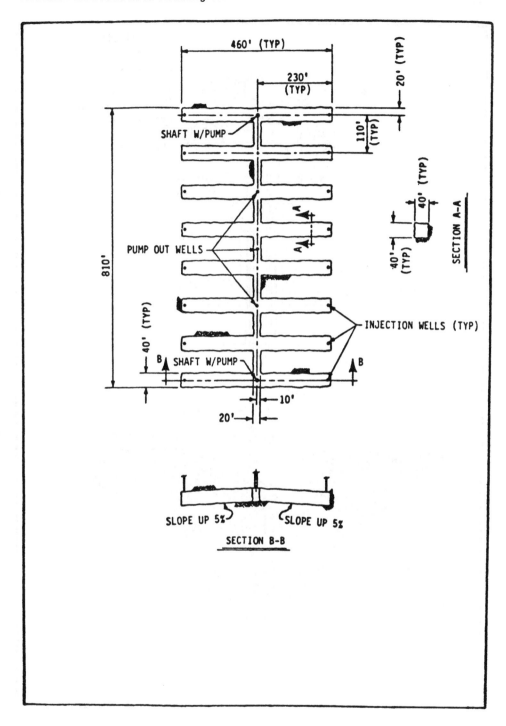

Fig. 7.2 Packed mine fermentor plan.

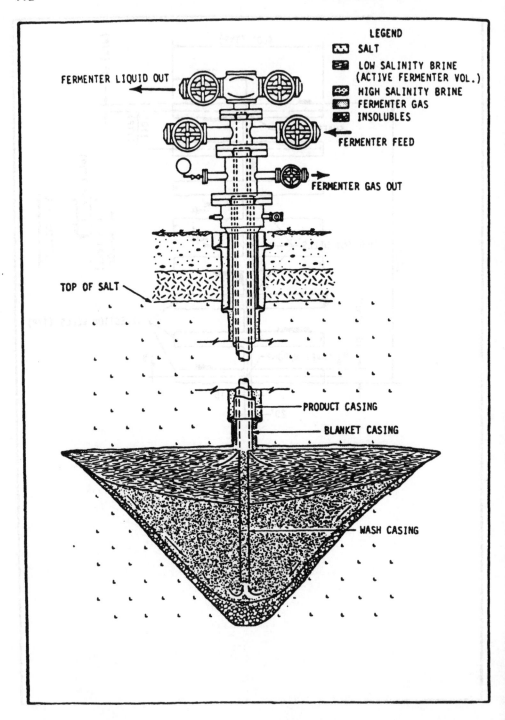

Fig. 7.3 Solution mine cavern fermentor, process flow.

Table 7.1 Design Criteria for Anaerobic Digesters

Parameter	Packed bed digester	Completely mixed digester
Reactor location	Rock cavern	Salt dome
Mining technique	Room and pillar	Solution
Reactor volume	2.3×10^7 gal	3.68×10^8 gal
Retention time	1 day	16 days
Temperature	35°C	35°C
Salt concentration	<1%	21%
Methane production	1.2×10^8 ft³/day	1.2×10^8 ft³/day

7.2 EXPERIMENTAL PROGRAM

The experimental program performed has examined the two major stages of this program: pretreatment and anaerobic fermentation. In this chapter we present a limited synopsis of the results from the pretreatment experiments; a detailed review of this portion of the program is presented in Chapter 4. Fermentation results are presented in the present chapter in greater detail.

7.3 PRETREATMENT

The objective of the pretreatment stage is to produce material capable of being fermented by anaerobic microorganisms. Thus solubilization of the lignite and subsequent breakdown of the molecules to low-molecular-weight simple aromatic compounds is what is desired. Solubilized lignite has been defined as that material which will pass through a 0.45-μm filter. Lower-molecular-weight cutoffs were determined by ultrafiltration of the solubilized material. Figure 7.4 presents the structural features of lignite. The matrix

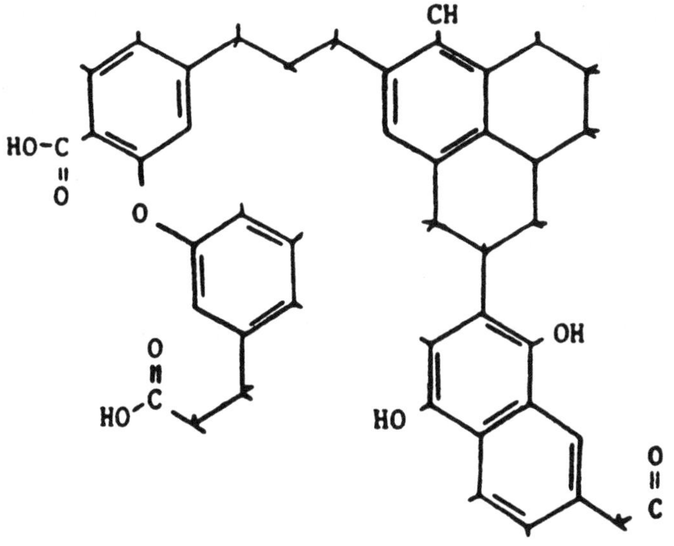

Fig. 7.4 Structural features of lignite. (Adapted with modification from Ref. 7.)

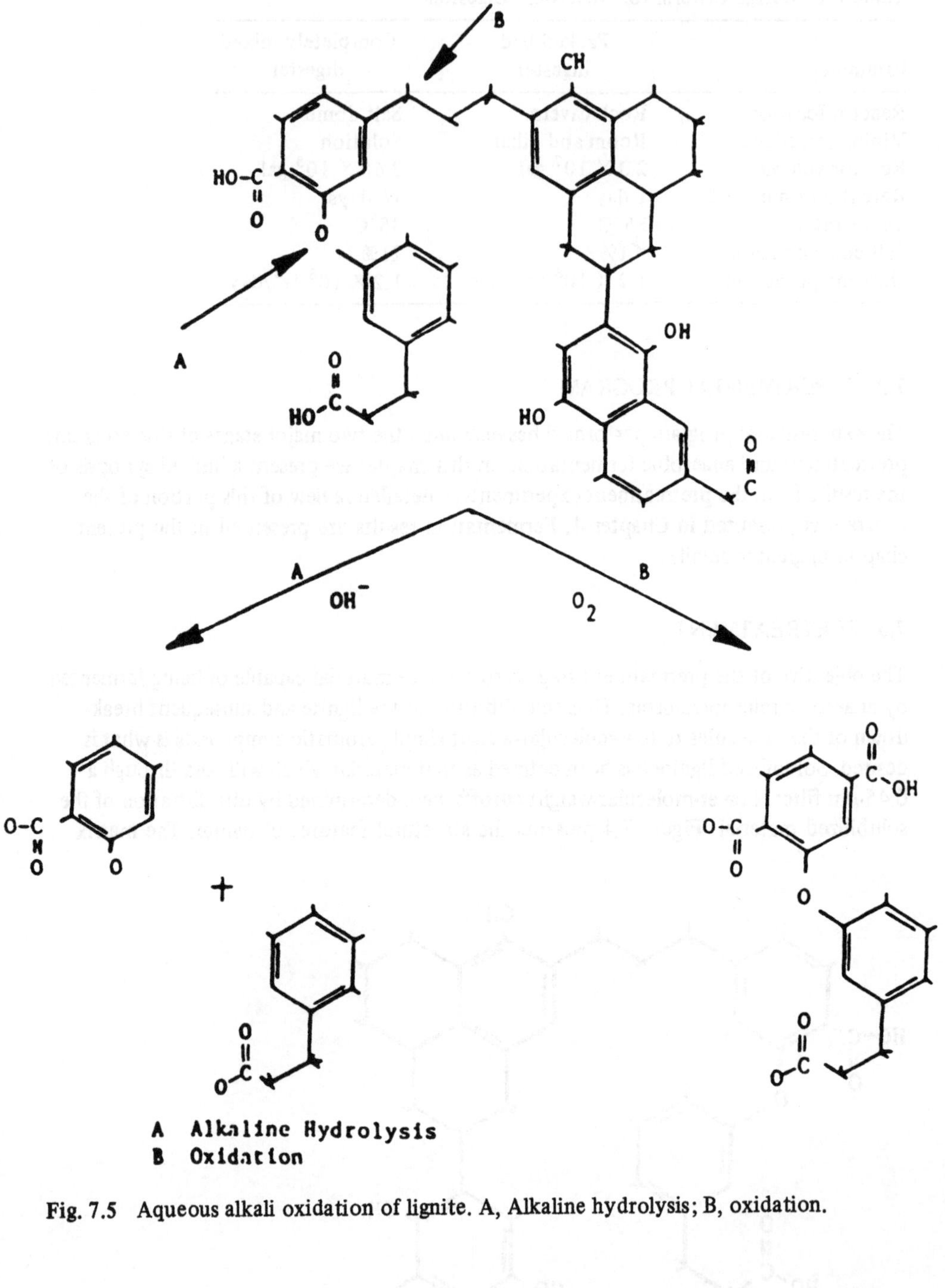

A   Alkaline Hydrolysis
B   Oxidation

Fig. 7.5 Aqueous alkali oxidation of lignite. A, Alkaline hydrolysis; B, oxidation.

shown represents an early stage in the coalification process where continuous deoxygenation results in structural elimination of water. Lignite is a low-ranked coal; thus a large portion of the matrix is composed of single-ring aromatic compounds joined in a network of carbon-carbon and carbon-oxygen bonds. It is the job of the pretreatment process to break these bonds and produce single-ring aromatic compounds. The process used to perform this is an aqueous alkali oxidation as depicted in Fig. 7.5. Alkali is used to break the carbon-oxygen bonds. Oxidation, in the form of hydrogen peroxide addition, is used to break carbon-carbon bonds. The utilization of this process thus reverses the coalification process. Both a batch pretreatment process and a continuous pretreatment process were used experimentally to explore this concept. The two reactors are shown in Figs. 7.6 and 7.7, respectively. A summary of the range of conditions at which pretreatments were performed is presented in Table 7.2.

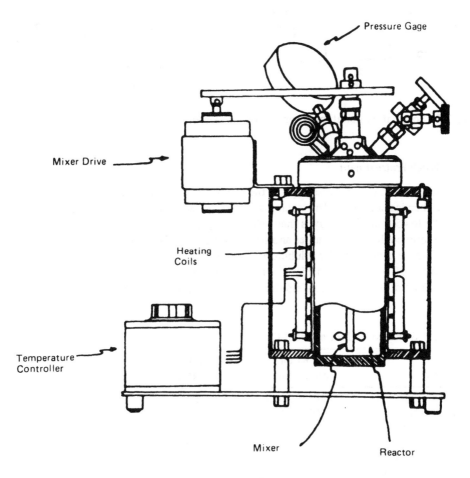

Fig. 7.6 Schematic of batch pretreatment reactor.

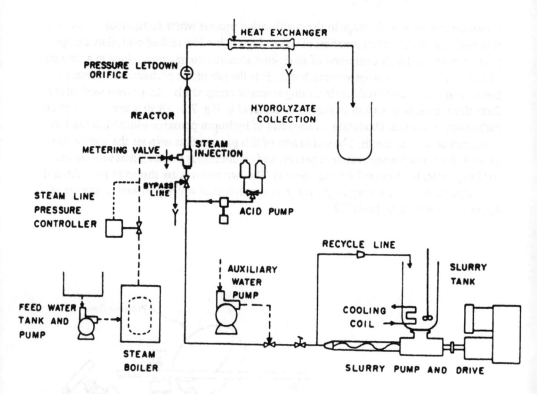

Fig. 7.7 Schematic of continuous pretreatment reactor.

7.3.1 Batch Pretreatment

The batch pretreatment process requires the reactor to be charged with the lignite slurry and any chemical additions. The reactor is then sealed and mixing is started. The heat-up time is 30 to 45 min. The slurry is then held at temperature for the prescribed length of time. Reactor cooling then requires another 30 to 60 min. This process has not been found effective in significantly solubilizing the organic fraction of lignite. Maximum solubilizations of 10 to 15% were achieved with this reactor. The most plausible explanation for the failure of this reactor to yield significant solubilization of lignite is the long reten-

Table 7.2 Summary of Pretreatment Conditions

Parameter	Batch pretreatment	Continuous pretreatment
Retention time	30–120 min	10–100 sec
Temperature	200–300°C	250°C
Alkali addition[a]	0–1%	0–1%
Peroxide addition[b]	0–5%	0–2.25%
Reactor heating time	30–45 min	0
Reactor cooling time	30–60 min	0

[a] As NaOH, expressed as a percent of total weight of slurry.
[b] H_2O_2, expressed as a percent of total weight of slurry.

tion times (and long heat-up and cool-down times) employed. It is hypothesized that initially, solubilization of the lignite occurred. This was followed by recondensation reactions, which converted solubilized material into particulates. The net result was low solubilization. To correct this problem, what was needed was a method to quickly heat and cool the lignite slurry and a reactor that would hold the slurry at temperature for short periods of time. The continuous pretreatment reactor served this purpose.

7.3.2 Continuous Pretreatments

All continuous pretreatments were performed at Dartmouth College under the direction of Dr. Hans Grethlein. The continuous reactor used was shown in Fig. 7.7. This reactor is designed to provide instantaneous heating of the incoming material through live steam injection. The reactor is a plug flow design; thus the retention time of the slurry in the reactor can be accurately controlled by varying the flow rate. The exit from the reactor is a blowdown orifice that provides instantaneous cooling of the pretreated material. Thus through these design features, specified retention times with no heat-up or cool-down times can be achieved.

To date, the most successful pretreatments have occurred using alkali (as NaOH) and hydrogen peroxide at retention times of approximately 80 s at 250°C. The present

Table 7.3 Summary of Continuous Pretreatment Data Utilizing Peroxide as an Oxidizing Agent

Reaction time (s)	NaOH addition[a]	H_2O_2 addition[a]	Total		Solubilized	
			TS (%)	TVS (% TS)	TDS (% TS)	TDVS (% TVS)
0	0.27	0	9.72	56.38	2.26	1.28
20	0.27	0	5.34	56.93	6.37	5.92
40	0.27	0	2.68	55.97	13.43	10.67
60	0.27	0	1.27	55.12	26.77	20.00
80	0.27	0	0.78	58.97	26.92	19.57
100	0.27	0	0.51	60.78	27.45	16.13
0	0.27	0	12.09	71.62	0.91	0.47
20	0.27	0.1	5.39	71.52	4.80	3.55
40	0.27	0.2	2.51	70.16	7.97	6.25
60	0.27	0.3	1.57	73.10	8.69	5.86
80	0.27	0.4	0.76	67.11	18.81	13.92
100	0.27	0.5	0.30	70.23	29.10	15.71
0	0.27	0	12.18	61.74	2.96	1.73
20	0.27	0.45	5.54	56.68	7.94	5.10
40	0.27	0.90	3.08	55.52	12.66	14.62
60	0.27	1.35	1.80	54.44	28.33	18.37
80	0.27	1.80	0.98	50.00	39.80	46.94
100	0.27	2.25	0.46	39.13	52.17	44.44
0	0.17	0	3.19	77.12	3.45	2.44
20	0.17	0.45	2.87	80.84	8.01	6.03
40	0.17	0.90	1.20	77.50	15.83	12.09
60	0.17	1.35	0.54	72.22	24.07	20.51
80	0.17	1.80	0.16	56.25	56.25	77.78

[a]Expressed as percent of total weight of slurry.

pretreatment reactor is incapable of achieving retention times longer than 20 s. Thus, to evaluate longer retention times, recirculation of the liquid through the reactor has been used. As an example, to achieve a 60-s retention time, the lignite slurry was passed through the reactor three times, each pass having a 20-s retention time. Data presented in Table 7.3 show the results of these analyses. The first set of data show alkali addition, without peroxide, at retention times of from 0 to 100 s. Solubilization of volatile solids increases to a maximum of 20% at a retention time of 60 to 80 s and then fall off to 16% at 100 s. The second set of data show the effect of hydrogen peroxide addition at a level of 0.1% added in each pass. No improvement in volatile solids solubilization is shown at this level of peroxide addition. The third set of data, 0.45% peroxide per pass through the

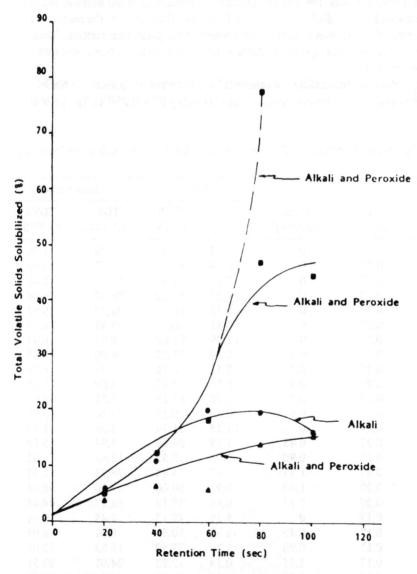

Fig. 7.8 Solubilization of volatile solids summary.

reactor, shows that solubilization increases significantly in this case, up to a maximum of 47% at 80 s. The final set of data are essentially a repeat of the previous set of data. Peroxide loadings were the same; however, a different lignite sample was used, one containing significantly less ash than the previous sample. Alkalinity requirements were less (determined by titrating the sample to a pH 10) in this case. Similar volatile solids solubilizations are shown. A summary of these data is shown in Fig. 7.8.

7.4 FERMENTATION

The second stage in this process is anaerobic fermentation of the pretreated lignite for the production of methane gas. The process, as described earlier, shows that traditional anaerobic digestion (through the use of an anaerobic packed bed) or digestion using anaerobic halophiles (in a salt cavern) would be required. Therefore, digestion of pretreated lignite in both of these regimes is being examined.

7.4.1 Anaerobic Fermentation

Anaerobic organisms from two sources—from a digester at a wastewater treatment plant and from the gut of higher termites—are being examined for conversion of lignite to methane gas in a nonhalophilic environment. The method used to examine the ability of sewage sludge anaerobes to convert pretreated lignite to methane gas has been to perform a series of batch digestions. Several pretreatment conditions have been examined as well as fractions of the pretreated material. Table 7.4 presents the starting conditions for the first phase of batch digestions. Nine batch digesters were set up using two fractions of pretreated lignite. Four digesters were fed pretreated lignite that was filtered to remove particulates. This material represents the feedstock that would be used in the full-scale process. A second set of four digesters were fed pretreated lignite that has been ultrafiltered. This fraction represents the low-molecular-weight (<500 MW) fraction of the pretreated lignite. The final reactor was a control, receiving only sewage sludge. These digestions were operated for 100 days at 37°C. Gas production from these reactors was monitored weekly, and at the end of 100 days the reactor contents were analyzed. Table 7.5 summarizes the data developed from this experiment and includes total solids (initial, final, removed), total volatile solids (initial, final, removed), total methane produced, total volatile solids removed based on methane production, and total volatile acids. In

Table 7.4 Starting Conditions for Batch Digesters

Reactor	TS (mg)			TVS (mg)		
	Lignite	Seed	Total	Lignite	Seed	Total
122140-F	61.2	285	346.2	23.4	161	184.4
122140-U	41.4	285	326.4	9.0	161	170.0
123021-BF	153.7	285	438.7	41.2	161	202.2
123021-BU	42.3	285	327.3	10.8	161	171.8
123014-BF	144.8	285	429.8	108.2	161	269.2
123014-BU	37.8	285	322.8	8.1	161	167.1
123017-BF	147.0	285	432.0	113.5	161	274.5
123017-BU	38.7	285	323.7	9.9	161	170.9
Control	0	285	285.0	0	161	161.0

Table 7.5 Data Summary for Phase I Batch Digestion

Reactor	TS (mg)			TVS (mg)			mL CH$_4$ prod.	TDS dest[a] (mg)	TVA (mg/Las Ac)	TVS conversion (%)		
	Initial	Final	Removed	Initial	Final	Removed				To gas	To TVS	Total
122140-F	346.2	260.0	86.2	184.4	130.0	54.4	20.4	40.8	26.8	22.1	6.3	28.4
122140-U	326.4	250.0	76.4	170.0	120.0	50.0	24.1	48.2	78.4	28.4	26.0	54.4
122021-BF	438.7	490.0	—	202.2	310.2	—	13.4	26.8	28.8	13.3	5.9	19.2
122021-BU	327.3	250.0	77.3	171.8	130.0	41.8	28.2	56.4	55.7	32.8	27.4	60.2
122014-BF	429.8	430.0	—	269.2	260.2	9.0	13.0	26.0	6.8	9.7	0.4	10.1
122014-BU	322.8	250.0	72.8	167.1	120.0	47.1	47.0	94.0	58.0	56.3	5.5	61.8
122014-BF	432.0	470.0	—	274.5	200.2	74.3	13.1	26.2	11.3	9.5	0.6	10.1
Control	285.0	210.0	75.0	161.0	100.0	61.10	20.5	41.0	8.3	25.5	0.9	26.4

[a]From methane production data.

addition, the last three columns of Table 7.5 present percent conversion of total volatile solids to methane gas, to volatile acids, and total conversion (gas plus acids). There is good agreement between the total volatile solids removed (as measured in initial and final samples) and the theoretical removals of total volatile solids based on methane production. Percent conversions of total volatile solids are shown for conversion to methane and conversion to volatile acids. Volatile acids are not produced in the pretreatment step but rather are a result of microbial breakdown of pretreatment products and are a direct precursor to methane production. A total conversion is also shown. These data indicate that those batch digesters fed ultrafiltered pretreated lignite achieved conversions of 50 to 60% whereas those receiving filtered pretreated lignite exhibited 10 to 30% conversions. Conversions of the filtered pretreated lignite in most cases did not reach the same degree of conversion as the control digester.

From batch digesters 122140-F, 122140-U, 122021-BF, and 122014-BU, four new batch digesters were started for phase II batch digestions. A summary of digester starting conditions is presented in Table 7.6. The seed for these reactors came from the phase I digesters that had been acclimated to the lignite. For this reason the percentage of pretreated lignite to seed was increased over what was used in phase I. The lignite used was subjected to alkali peroxide pretreatment. Digestion was again allowed to proceed for 100 days at 37°C. Gas production was monitored weekly and the reactor contents were analyzed upon completion. The data summary for this phase is shown in Table 7.7. Unfortunately, reactor 122140-F was lost. The other reactor receiving filtered pretreated lignite, 122021-BF, achieved a low conversion of only 12.6%. The other two digesters, which were fed ultrafiltered pretreated lignite, resulted in significant conversions of total volatile solids to methane (50% and 88%) and significant total conversions.

Several conclusions can be drawn from these data. It is apparent that conversion of that fraction of pretreated lignite less than 500 molecular weight occurs readily with these organisms once they have been acclimated to this substrate. Conversion of the entire soluble fraction does not occur to the same extent. In all cases where a sample of soluble lignite was fed to one batch reactor and the ultrafiltered lignite from that same sample was fed to a duplicate reactor, the total quantity of methane produced was greater for the digester receiving ultrafiltered material than for the digester receiving filtered material. It appears that some product has been formed during pretreatment which is inhibitory to the anaerobic process. It also appears that this inhibitory material is greater than 500 MW but is also soluble. Investigations are under way in an attempt to identify the constituent(s) causing this inhibition. Furthermore, these results point to the need to develop a pretreatment process capable of maximizing low-molecular-weight material. It is unclear from these data whether the higher-molecular-weight material would be bio-

Table 7.6 Starting Conditions for Phase II Batch Digesters

Reactor	TS (mg)			TVS (mg)		
	Lignite	Seed	Total	Lignite	Seed	Total
122140-F	87.5	91.0	178.5	52.5	45.5	98.0
122140-U	116.0	100.0	216.0	64.0	48.0	112.0
122021-BF	221.4	200.9	422.3	159.9	127.2	287.1
122014-BU	118.4	92.5	210.9	62.9	44.4	107.3

Table 7.7   Data Summary for Phase II Batch Digestion

Reactor	TS (mg)			TVS (mg)			mL CH$_4$ prod.	TDS dest[a] (mg)	TVA (mg/Las Ac)	TVS conversion (%)		
	Initial	Final	Removed	Initial	Final	Removed				To gas	To TVS	Total
122140-F	178.5	–	–	98.0	–	–	15.8	31.6	–	32.2	–	–
122140-U	216.0	194.0	22.0	112.0	130.2	–	28.2	56.4	28.6	50.4	31.8	82.2
122021-BF	422.3	320.0	102.3	287.1	225.3	61.8	13.2	26.4	6.9	9.2	3.4	12.6
122014-BU	210.9	167.0	43.9	107.3	113.9	–	42.6	95.2	22.9	88.7	22.0	110.7

[a]From methane production data.

degradable were it not for any inhibitory effects. It is clear that the low-molecular-weight material is biodegradable.

The second source of anaerobes being examined to convert lignite are from the gut of the higher termite *Nasutitermes*. It has been demonstrated that these termites are capable of lignin conversion [1]. However, this study did not determine whether this degradation was accomplished by the termite itself or by the gut microorganisms. Since lignite is "old" biomass lignin, it was decided to examine the ability of the organisms in the gut of *Nasutitermes* to metabolize lignite and convert it to methane gas.

Examination of this concept has been made exclusively in test tube fermentations to date. Figure 7.9 presents a summary of this experimental program to date, with quali-

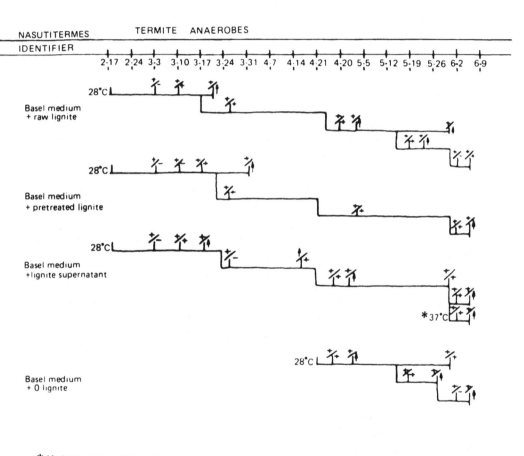

Fig. 7.9   Test tube fermentations in the gut of *Nasutitermes* to metabolize lignite and convert it to methane gas.

tative results on methane and carbon dioxide production. Basal medium, microbial inoculum plus raw lignite, pretreated lignite, and pretreated lignite supernatant were combined in test tubes. Each of these conditions was allowed to ferment, with gas analyses being performed. Each culture to date has gone through several transfers. The results show that each culture has exhibited methane production. A control set of test tubes, receiving no lignite, has also been established. This sample has also exhibited methane production. In the basal medium a limited amount of carbon is present. The concentration of methane evolved in the headspace gas is the least in the control test tubes. Equivalent and slightly higher methane concentrations are noted in the test tubes receiving raw lignite. Finally, the concentration of methane gas in the test tubes receiving pretreated lignite and pretreated lignite supernatant is significantly higher than all others. Although these results do not constitute proof that the lignite is being metabolized by these organisms, the results are encouraging. The next step in this experimental program will be to establish batch digesters inoculated with these organisms and various fractions of the pretreated lignite and raw lignite to assess quantitatively conversion of this material.

7.4.2 Halophilic Fermentation

The second form of anaerobic fermentation being examined is in high salt (NaCl) concentrations using anaerobic halophiles. There are two sources of organisms being used to perform this research from the Gulf of Mexico and the Great Salt Lake. This research has only been performed in test tubes. Anaerobic halophiles were obtained from the Gulf of Mexico at an offshore oil production platform. As part of the production operation, highly saline water from strata below the Gulf floor is pumped up to the platform, filtered, and then pumped into oil-bearing strata. This forces the oil out of this strata and up to the production platform. The filter eventually gets clogged by growth of anaerobic halophiles that originate in the water-bearing substrata. These organisms were the first source of anaerobic halophiles. Figure 7.10 presents a summary of the various analytical experiments evaluated to date with these organisms. Originally, four sets of test tubes were evaluated. The first contained haloanaerobium medium [5] in a N_2/CO_2 headspace. The second contained phosphate-buffered basal medium (PBBM) in a H_2/CO_2 headspace. The third set was identical to the second but with glucose added as an additional carbon source. The fourth set was again identical to the second, but with lignite ultrafiltrate added as an additional carbon source. In each case evidence of growth was initially observed; however, after several transfers, evidence of viable cultures stopped in all cases. We returned to the original sample and restarted a culture receiving lignite filtrate as a carbon source in a H_2/CO_2 headspace. Evidence of methane production occurred and this sample was transferred to culture tubes with a N_2/CO_2 headspace and an H_2/CO_2 headspace. In each case growth appears to have ceased.

Recently, four new cultures were established, again returning to the original sample as an inoculum source. Two samples are being grown with a H_2/CO_2 headspace, one receiving glucose and the other receiving lignite supernatant. The other set of two samples are being grown in a N_2/CO_2 headspace: one with glucose added, the other with lignite supernatant added. All four cultures presently are viable and exhibiting methane production. The aim now is to transfer these cultures successfully several times to show viability. Once this is achieved, the next step will be to introduce these cultures into batch digesters receiving pretreated lignite, for quantitative assessment of the decomposition of pretreated lignite.

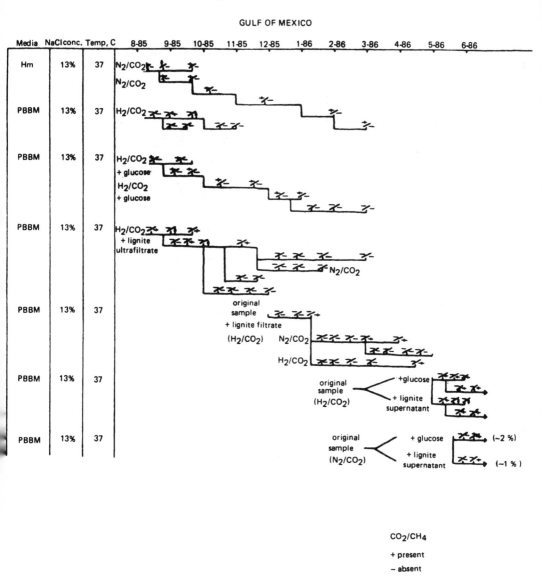

Fig. 7.10 Summary of various analytical experiments using anaerobic halophiles from the Gulf of Mexico.

The second set of anaerobic halphiles being evaluated in this study were obtained from sediment samples from the Great Salt Lake. The environment at the bottom of the Great Salt Lake is highly saline, with water samples containing 20 to 28% NaCl. Figure 7.11 presents the various analyses performed with these samples to date. The original sample was first transferred to a PBBM medium but showed no growth. The sample was also transferred to a haloanaerobium medium (Hm) and did exhibit growth. However, in successive transfers in this medium, growth was exhibited only in the presence of glucose, with no growth occurring when only pretreated lignite was present. This experiment was repeated with the same results.

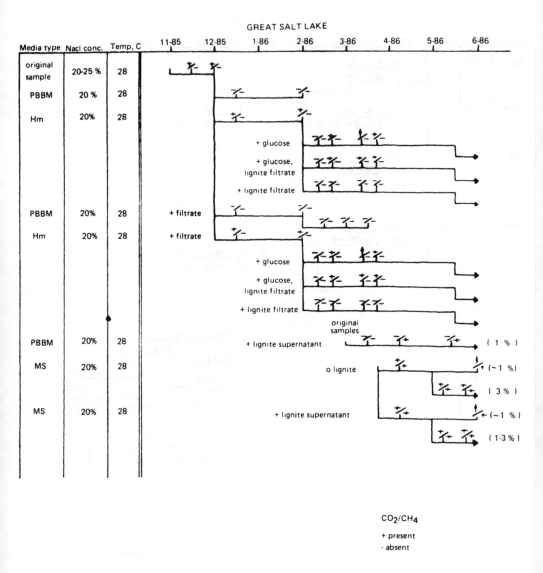

Fig. 7.11 Summary of various analytical experiments using anaerobic halophiles from the Great Salt Lake.

After these experiments, we returned to the original sample and inoculated it in a PBBM media with lignite supernatant. This sample has shown evidence of methane production. The original sample was also used as an inoculum in a minimal salt (MS) medium with and without lignite supernatant. Each of these samples has shown that methane production and headspace methane gas concentrations increased in the first transfer. The aim of this experiment will be to grow a sufficient volume of organisms to start batch digesters. The digesters will be operated on pretreated lignite to assess the ability of these organisms to convert this material and for quantitative assessment of their performance.

7.5 PROCESS ECONOMICS

The process under consideration was described at the beginning of this chapter. Preliminary estimates of capital, operation and maintenance, and methane production costs have been made. Original economic estimates were based on aqueous alkali pretreatment of the lignite followed by fermentation in a rock or salt cavern. Table 7.8 presents a list of assumptions and design criteria used in developing this analysis. The results of this analysis for two different reactor types are presented in Table 7.9. Capital costs for the two options range from approximately $225 to $240 million. Annual operation and maintenance costs are approximately $130 to $155 million. Two economic analysis were performed to determine unit gas cost. The standard analysis represents the selling price of the gas produced if the facility were examined as a processing operation. The break-even analysis represents the actual cost of producing the gas. As shown in Table 7.9, the unit gas costs are $2.70 to $2.80 per 10^6 Btu for the break-even analysis and $4.70 to $4.90 per 10^6 Btu for the standard analysis.

A sensitivity analysis was performed on the effect of pretreatment reactor size (varying the retention time from 10 to 120 s) and hydrogen peroxide addition (0 to 6% of the weight of the whole slurry) on the unit gas cost. Figure 7.12 shows the effect of varying the retention time in the pretreatment reactor on unit gas cost for the break-even and standard analyses. As shown, increasing the reactor size 12-fold (10 to 120 s retention time) has virtually no effect on the unit gas cost in all cases. The reason for this is that although this represents a significant increase in the size of the pretreatment reactor on a percentage basis, the actual increase in capital cost is only $82,000. This is such a low percentage of the overall capital cost of the process that it has no impact on the final unit gas cost.

Figure 7.13 presents the sensitivity analysis on hydrogen peroxide addition to the pretreatment process. In this case a dramatic effect on unit gas cost is shown. The use of hydrogen peroxide in this process is extremely costly, adding approximately $10 to $15 per 10^6 Btu for each percent of peroxide used. Peroxide usage effects both capital and operation and maintenance costs. Capital costs increase by approximately $38 million when going from a system without peroxide usage to a system using 6% peroxide. The capital cost increases are due to both the addition of a peroxide storage and handling system, and to the increased flow rates which result and cause increases in the size of all equipment. Similarly, operation and maintenance costs increase. Peroxide costs $6.45 per

Table 7.8 Process Conditions for Analysis of Biogas from Texas Lignite

---

20,000 Tons per day of coal (wet basis)
12.5% Volatile solids in coal slurry
10% Sodium carbonate on volatile solids
75% Conversion of volatile solids to fermentables
75% Conversion of fermentables to products
13.3% of products are methane
24.2% of products are carbon dioxide
50% Recycle
Rock cavern: 24-h retention time
Salt cavern: 16-day retention time
Pretreatment reactor: 30-s retention time

Table 7.9 Economic Analysis for Process Base Case

Case	Capital cost (%)	Annual O&M cost ($)	Methane cost ($/MMBtu) Standard analysis	Methane cost ($/MMBtu) Break-even analysis
Rock cavern plus mine conveyor	224×10^6	130×10^6	4.73	2.76
Salt cavern plus mine conveyor	239×10^6	155×10^6	4.90	2.79

Standard analysis:	*Break-even analysis:*
50% Financed	0% Financed
12% Interest	0% Interest
15-Year plant life	15-Year plant life
20% Return on investment	0 Return on investment
350 Days of operation per year	350 Days of operation per year
24 h of operation per day	24 h of operation per day

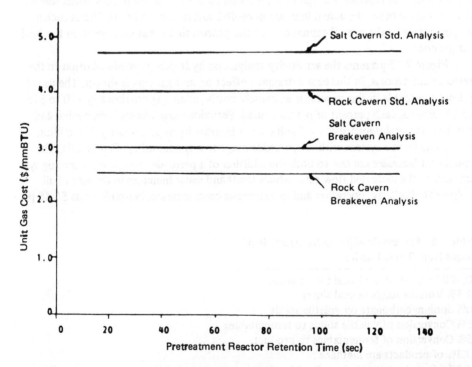

Fig. 7.12 Effect of varying the retention time in the pretreatment reactor on unit gas cost for the break-even and standard analyses.

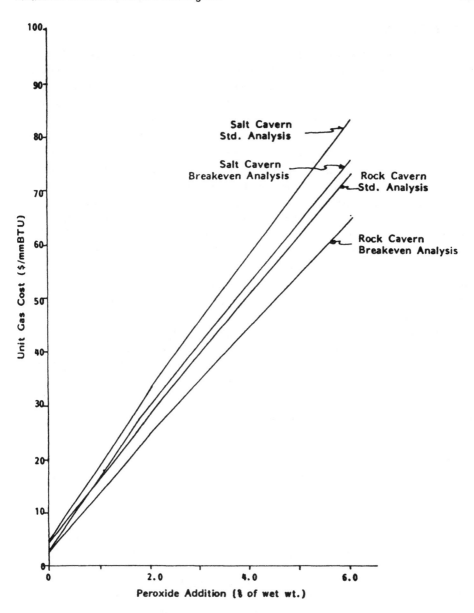

Fig. 7.13 Sensitivity analysis on hydrogen peroxide addition to the pretreatment process.

ton. The annual operation and maintenance budget increases from \$130 to \$155 million annually without peroxide addition to about \$3 billion with 6% peroxide usage. The experimental program has shown that the optimal dose of peroxide use achieved to date is 1.8%. Even at this level the unit gas cost increases to \$23 to \$25 per 10^6 Btu for the break-even analysis and \$27 to \$29 per 10^6 Btu for the standard analysis.

7.6 FUTURE WORK

Based on the experimental work accomplished to date and process economics, future experimental work will concentrate in several areas. Pretreatment of the lignite to produce a biodegradable substrate is the key to this process. Technical feasibility of producing soluble, low-molecular-weight lignite has been demonstrated. However, process economics have shown that peroxide usage at the levels presently employed is too expensive. Continued experimentation will be performed to minimize peroxide usage. In addition, other oxidizing agents will be explored. Process economics have also shown that increased retention time in the pretreatment unit does not significantly alter costs. Thus experimentation at longer retention times and lower pretreatment temperatures will be evaluated.

The anaerobic organisms acclimated to lignite slurries from a sewage treatment plan have been shown capable of producing methane gas in batch cultures. The next step will be to operate continuous digestion systems (both a CSTR digester and a packed-bed digester) to develop design information. Similarly, the organisms isolated from the gut of higher termites have clearly demonstrated their ability to produce methane from pretreated lignite in culture tubes. To analyze this batch quantitatively, digesters set up with these organisms as a seed will be utilized. In addition, continuous digestion will be evaluated to develop engineering design data.

The halophilic anaerobes being evaluated in this study have also demonstrated methane production from lignite. These organisms grow much more slowly than do non-halophilic anaerobes. Once a sufficient population of organisms has been established, both batch and continuous digestion of pretreated lignite with these bacteria will be examined. Finally, continuous updating of process economics will be evaluated based on results developed for the experimental program.

REFERENCES

1. Butler, J. H., and Buckerfield, J. C., Digestion of lignin by termites, *Soil Biol. Biochem., 11*, 507-513 (1979).
2. Fenix & Scisson, Inc., 1985, *Concepts for the Use of Underground Caverns as Anaerobic Bioreactors*, Tulsa, Okla.
3. Leuschner, A. P., Trantolo, D. J., Kern, E. E., and Wise, D. L., Biogasification of Texas lignite, presented at the *13th Biennial Lignite Symposium*, Bismark, N. Dak., 1985.
4. Sondreal, E., Willson, W., and Steinberg, V., *Fuel, 61*, 925-932 (1982).
5. Zeikus, J. G., Hegge, P. W., Thompson, T. E., Phelps, T. J., and Langworthy, T. A., Isolation and description of *Haloanaerobium pravalens* gen. nov. and sp.: an obligate anaerobic halophile common to Great Salt Lake sediments, *Curr. Microbiol., 9*, 225-234 (1983).

8

Biological Production of Ethanol from Coal Synthesis Gas

S. BARIK, S. PRIETO, S. B. HARRISON, E. C. CLAUSEN, and J. L. GADDY
University of Arkansas, Fayetteville, Arkansas

8.1 INTRODUCTION

Coal synthesis gas, a mixture of predominantly CO and H_2 with lesser amounts of CH_4, CO_2, and sulfur compounds, represents a valuable feedstock for the biological production of chemicals. Carbon monoxide may be converted to acetate through the action of anaerobic bacteria. The acetate may then serve as a biological precursor for other chemicals, such as butanol, ethanol, or higher-chain acids. Alternatively, the CO may be converted to CO_2, H_2, and other chemicals by blocking methane synthesis.

In this chapter we describe the biological utilization of synthesis gas to produce acetate and other chemicals. The reactions to produce acetate or CO_2 and H_2 from CO are well known in terms of organisms and reaction conditions. On the other hand, reactions to produce other chemicals from acetate and to produce other liquid hydrocarbons from coal gas are theoretically possible from an energy standpoint. The feasibility of these reactions needs to be determined and the best organisms identified and reaction conditions defined for these reactions.

8.1.1 Acetate Production from Coal Gas

Studies on bacterial CO metabolism, reported by many authors [18,23,30,31] demonstrate the formation of acetic acid, according to the reaction

$$4CO + 2H_2O \rightarrow CH_3COOH + 2CO_2 \qquad \Delta G^\circ = -16.15 \text{ kcal/mol} \qquad (1)$$

These bacteria include *Eubacterium limosum*, *Acetobacterium woodii*, *Butyribacterium methylotrophium*, *Clostridium thermoaceticum*, and *Peptostreptococcus productus*. Among these bacterial species, *P. productus* utilizes CO very rapidly, with a doubling time of less than 2 h and can grow with as much as 90% CO in the gas phase.

Carbon monoxide also can be oxidized to CO_2 by several anaerobic bacteria, particularly by the members of *Clostridium* [13,16,25,35,42] :

$$CO + H_2O \rightarrow CO_2 + H_2 \tag{2}$$

Methanobacterium thermoautotrophicum [11,43], and *Rhodopseudomonas gelatinosa* [12,48] are able to use CO as an energy source producing CO_2 and CH_4 or H_2, respectively. *Clostridium purinolytricum* can reduce CO_2 to formate. Formate is then converted to acetate [14]. Another *Clostridium* sp. can synthesize acetate from CO_2 during the fermentations of hypoxanthine [50].

Acetate is the key intermediate of C-1 catabolism by several anaerobic bacteria. Small amounts of other fatty acids, such as propionic, butyric, and valeric acids, have also been reported from CO_2 and H_2 by mixed bacterial populations [29].

In natural populations, CO_2, H_2, and acetate are usually converted to CH_4 by methane-producing or acetate-cleaving bacteria, found predominantly in anaerobic environments. Anaerobic bacteria belonging to *Methanosarcina* sp. and *Methanothrix* sp. cleave acetate to CH_4 and CO_2 [22,40] according to the reaction

$$CH_3COO^- + H_2O \rightarrow CH_4 + HCO_3^- \quad \Delta G° = -31 \text{ kJ/reaction} \tag{3}$$

All methane bacteria known to date can grow on H_2 and CO_2 to form CH_4 (for many references, see Thauer et al. [46]) according to the reaction

$$CO_2 + 4H_2 \rightarrow CH_4 + 2H_2O \quad \Delta G° = -131 \text{ kJ/mol} \tag{4}$$

Many anaerobic bacteria are also known to produce acetic acid from H_2 and CO_2 [3,32,37–39]. These bacterial isolates include *Acetobacterium* sp. [3], which produces a homoacetic fermentation by anaerobically oxidizing hydrogen and reducing CO_2 according to the reaction

$$4H_2 + 2HCO_3^- + H^+ \rightarrow CH_3COO^- + 4H_2O \quad \Delta G° = -25.6 \text{ kJ/reaction} \tag{5}$$

Acetobacterium woodii [32] and *Acetoanaerobium noterae* produced acetate from H_2 and CO_2 according to the reaction (5), but in addition to acetate, *A. noterae* produced some propionate and butyrate. Another chemolithotrophic bacteria, *Clostridium aceticum*, also produced acetate from H_2 and CO_2 [5]. However, *Clostridium acidiurici* produced acetate only from CO_2 using the glycine decarboxylase pathway [50].

P. productus, Eubacterium limosum, and *A. woodii* have been utilized in the University of Arkansas laboratories for the production of acetate as a methane intermediate from coal gas. *P. productus* has been shown to tolerate coal gas containing CO, H_2, CH_4, and CO_2 in producing quantitative yields of acetate from CO. Tolerance to the minor components of coal gas such as sulfur compounds, aromatics, and so on, is under investigation.

8.1.2 Utilization of Acetate as an Intermediate for Chemicals Production

Methanogenic enrichments utilizing acetate as a carbon source have been described by many authors (see, e.g., Refs. 2, 6, 7, 15, 19, 20, 24, 33, 36, 49, 51, and 52). In all of these cases, acetate was directly converted to CH_4 and CO_2. All of these papers dealt mainly with the bacterial enrichment, kinetics, yields, and pathways of acetate digestion to CH_4 by anaerobic bacteria.

Acetate assimilation by a sulfate reducer, *Desulfovibrio vulgaris*, was studied by Badziong et al. [1]. This bacterium growing on H_2 and sulfate, synthesized alanine, aspartate, glutamate, and ribose phosphate from [^{14}C] acetate and CO_2.

It should be mentioned here that acetylcoenzyme A (acetyl-CoA) is a key intermediate in the tricarboxylic acid cycle and the glyoxylate cycle. Thus acetyl-CoA can play an important role in producing many other chemicals. Acetate oxidation to CO_2 via the citric acid cycle was demonstrated by another anaerobic bacteria, *Desulfobacter postageti* [4,17]. The bacterium used sulfate as an electron acceptor during this process. The cell extracts of this bacterium were found to contain all the enzymes of the citric acid cycle.

Many strains of yeast (mostly *Candida* sp.) have been found to convert acetate to citric acid [44,45,47,54]. Acetic acid or calcium acetate was used as the substrate for citric acid fermentation by the yeast.

Organisms have been shown to utilize acetate in the production of higher-molecular-weight compounds, including propionic acid, butyric acid, and caproic acid. These organisms include species of *Clostridium, Butyribacterium*, and *Diplococcus* [29].

In addition, many other compounds theoretically can be produced from acetate, as indicated by the free energies of reaction. Table 8.1 presents a list of possible reactions utilizing acetate as the reactant with the associated free energies. Reactions with highly negative free energies have the most potential as being viable reactions. As noted, there is strong likelihood that ethanol, butyrate, butanol, valerate, and caproate may be produced from acetate.

Table 8.1 Production of Chemicals from Acetate

$2 \text{ Acetate}^- + 2H_2O \rightarrow \text{pyruvate} + HCO_3^- + 3H_2$
$\Delta G^\circ = +17.98 \text{ kcal/mol}$
$2 \text{ Acetate}^- + H^+ \rightarrow \text{Acetone} + HCO_3^-$
$\Delta G^\circ = +0.87 \text{ kcal/mol}$
$2 \text{ Acetate}^- + 2H^+ + 3H_2 \rightarrow 2,3\text{-butanediol} + 2H_2O$
$\Delta G^\circ = +2.0 \text{ kcal/mol}$
$2 \text{ Acetate}^- + H^+ + H_2 \rightarrow \text{isopropanol} + HCO_3^-$
$\Delta G^\circ = -4.06 \text{ kcal/mol}$
$\text{Acetate}^- + H^+ + 2H_2 \rightarrow C_2H_5OH + H_2O$
$\Delta G^\circ = -11.84 \text{ kcal/mol}$
$\text{Butyrate}^- + H^+ + 2H_2 \rightarrow \text{butanol} + H_2O$
$\Delta G^\circ = -3.9 \text{ kcal/mol}$
$\text{Acetate}^- + H_2O \rightarrow CH_4 + HCO_3^-$
$\Delta G^\circ = -7.4 \text{ kcal/mol}$
$2 \text{ Acetate}^- + H^+ + 2H_2 \rightarrow \text{butyrate}^- + 2H_2O*$
$\Delta G^\circ = -10.54 \text{ kcal/mol}$
$\text{Acetate}^- + \text{propionate}^- + H^+ + H_2 \rightarrow \text{valerate}^- + 2H_2O*$
$\Delta G^\circ = -21.1 \text{ kcal/mol}$
$\text{Acetate}^- + \text{butyrate} + H^+ + 2H_2 \rightarrow \text{caproate}^- + 2H_2O*$
$\Delta G^\circ = -21.1 \text{ kcal/mol}$

*Known and demonstrated.

Anaerobically, methanogens convert acetate to CH_4 and CO_2. Methane formation must therefore be blocked to force the microorganisms to make other feasible products. This may be achieved through changes in pH, use of methanogenic inhibitors, modification of nutrient constituents, change in partial pressure of hydrogen, and supplementation of other organic compounds. Culture enhancement and subsequent optimization can then be performed.

8.1.3 Production of Chemicals and Liquid Fuels from Synthesis Gas

Zeikus [55] and Wise [53] have reviewed the formation of liquid fuels and organic chemicals from biomass, including synthesis gas by anaerobic bacteria. However, both authors have pointed out the experimental difficulties in deriving liquid fuels from CO or coal gas. Levy et al. [29] and Daniels [10] presented theoretical considerations in producing various alcohols (methanol, ethanol, butanol, and propanol), ketones (acetone), and other chemicals (acetate, butyrate, and propionate). Both sets of authors suggested careful experimental designs while attempting to isolate bacterial cultures producing liquid fuels and chemicals from CO or coal gas. Although the authors showed many thermodynamically feasible products from CO or coal gas, they have cautioned the "biotechnologist" that obtaining such chemicals needs fruitful approaches and the selection of proper microorganisms.

On the basis of thermodynamic considerations, coal gas may be directly converted biologically to many low-molecular-weight compounds, including methanol, ethanol, n-propanol, n-butanol, propionic acid, and acetic acid [29,46]. Among all the possible anaerobic fermentations, the formation of methane is the most energetically favorable. As with acetate utilization, the production of several alcohols and organic acids from CO and H_2 or CO_2 and H_2 is theoretically feasible if the production of methane is successfully

Table 8.2 Theoretically Feasible Reactions Using CO, CO_2, and H_2

Compound	Net reactions	$\Delta G°$ (kcal/mol of product formed)[a]
Formic acid	$CO + H_2O \rightarrow HCOOH$[b]	+5.57
	$CO_2 + H_2 \rightarrow HCOOH$[b]	+10.35
Methanol	$CO + 2H_2 \rightarrow CH_3OH$	9.14
	$CO_2 + 3H_2 \rightarrow CH_3OH + H_2O$	−4.35
	Formate $+ H^+ + 2H_2 \rightarrow CH_3OH + H_2O$[c]	−5.2
Ethanol	$2CO + 4H_2 \rightarrow C_2H_5OH + H_2O$	−34.56
	$2CO_2 + 6H_2 \rightarrow C_2H_5OH + 3H_2O$	−24.99
Butanol	$4CO + 8H_2 \rightarrow CH_3(CH_2)_2CH_2OH + 3H_2O$	−80.0
	$4CO_2 + 12H_2 \rightarrow CH_3(CH_2)_2CH_2OH + 7H_2O$	−60.86
Acetic acid[c]	$2CO + 2H_2 \rightarrow CH_3COOH$	−22.73
	$2CO_2 + 4H_2 \rightarrow CH_3COOH + 2H_2O$	−13.15
Propionic acid	$3CO + 4H_2 \rightarrow CH_3CH_2COOH + H_2O$	−44.65
	$3CO_2 + 7H_2 \rightarrow CH_3CH_2COOH + 4H_2O$	−30.29
Acetone	$3CO + 5H_2 \rightarrow CH_3COCH_3 + 2H_2O$	−53.55
	$3CO_2 + 8H_2 \rightarrow CH_3COCH_3 + 5H_2O$	−39.20

[a]Calculated from Ref. 47.
[b]Known to be too transitory to accumulate in media.
[c]Known and demonstrated.

blocked. Table 8.2 presents examples of theoretical possibilities that can be obtained during CO_2 and H_2 or CO and H_2 catabolism.

As shown in the table, theoretically, many compounds can be produced, including alcohols, acids, and acetone. As with compound synthesis from acetate, organisms must be identified and isolated for these theoretical biological reactions. Careful consideration of product formation, the selection of microorganisms, and process optimization may lead to the industrial production of many important heterogeneous compounds.

8.1.4 Methane Oxidation by Bacteria

Methane utilization by many methane-oxidizing bacteria having methane monooxygenase enzyme activity has been demonstrated by many authors (see, e.g., Refs. 8, 9, 21, 26-28, 34, and 41). These bacteria include *Methylomonas methanica, Methanomonas methanooxidans*, and *Methylococcus capsulatus* and are generally classified under *Methylococcus, Methylomonas, Methylobacter, Methylosinus*, and *Methylocystis*.

Cell suspensions of *Pseudomonas (methylomonas) methanica* and *M. methanooxidans* produced methanol as a product during methane oxidation [21]. Patel et al. [34] isolated a facultative methane utilizing bacteria, *Methylobacterium* sp., which had an enzyme to convert many hydrocarbons, including alkanes (C_1-C_8) and substituted alkanes. This enzyme had high specific activity (93 nmol/min per milligram of protein) for converting methane to methanol. Methanotrophs oxidize methane to methanol, formaldehyde, formate, and finally to CO_2 [9]. All of these enzymes are now being characterized.

8.1.5 Conclusions

Theoretically, many chemicals can be produced by the action of microorganisms on synthesis gas, including methanol, ethanol, butanol, acetic acid, propionic acid, and acetone. Of these chemicals, only the organisms capable of producing acetate have been isolated and studied. Acetate may theoretically be used to produce ethanol, isopropanol, and higher alcohols. Acetate may be utilized by *Desulfovibrio vulgaris* to produce acetyl-CoA, and by *Candida* sp. to produce citric acid. Methane may be formed easily from either CO or acetate by the action of methane bacteria. Methanotrophs may be used to produce methanol or formaldehyde from methane. In general, little information is available in the literature for the biological conversion of synthesis gas to liquid fuels. Pure cultures may be utilized in a few instances, but natural sources of bacteria such as sewage sludge, soil, and so on, must be screened for CO or acetate conversion in most cases.

8.2 BACTERIAL SCREENING

Natural sources of bacteria, such as sewage sludge, animal wastes, and muds, serve as a final repository for a consortium of bacteria, capable of converting a wide variety of substrates to many different products. Almost all of the pure culture bacteria found in culture collections today originated from a mixed bacterial population isolated from such natural environments. The natural carbon flow in the mixed cultures in sewage sludge and animal rumen is the metabolism of substrate to methane. Thus methane inhibitors must be added to samples from these sources to stop methane production and to allow intermediates to form. Acclimation to CO, CO_2, and H_2 must be achieved over an extended period of time, during which the cultures are supplemented with basal salts, vitamins, minerals, and the gases.

8.2.1 Procedures

Several experimental studies were initiated in an attempt to produce alcohols from CO, acetate, and synthesis gas. Three natural sources of inocula were utilized: sewage sludge, chicken waste, and coal/coal mud samples. One milliliter of natural inocula sample was added to 9 mL of basal media containing acetate (80 nM), CO (52 to 55%, 2 atm), or synthesis gas as the primary carbon source. Monensin (1 μg/mL) and 2-bromoethane-sulfonic acid (BESA; 20 μM/mL) were used as methanogenic inhibitors. Anaerobic media were prepared using the Hungate method and were placed in serum tubes sealed with butyl rubber stoppers. Samples were inoculated in triplicate and were incubated at 37°C with the pH between 4 and 7. Gas samples were analyzed routinely for the consumption of synthesis gas. Liquid samples were removed and stored for the analysis of acids and alcohols.

The gas-phase composition was measured by gas-solid chromatography using a Porapak QS column. The liquid-phase composition was measured by gas-solid chromatography using separate procedures for alcohol and acid analysis.

A large number of these samples were inoculated at various conditions in an attempt to produce alcohols. All experiments were conducted in triplicate. After 3 to 4 weeks, the cultures were transferred with fresh media and gases, and the experiments repeated. Successive transfer of the best cultures allows the mixed population to evolve to utilize the desired substrate and produce the desired products.

8.2.2 Results: Inhibition of Methanogenesis

As mentioned previously, the natural tendency for most mixed bacterial cultures is to produce methane from organic substrates. In order to produce alcohols, which are intermediates in methane synthesis, inhibitors must be added to the cultures to block methane production. Therefore, the first step was to determine whether methanogenesis could be blocked with certain chemical inhibitors.

Table 8.3 Inhibition of Methanogenesis by an Enrichment Using Acetate as the Primary Energy Source

Substrate and inhibitors	pH	Methane (mol %)	
		Day 7	Day 14
Acetate only[a]	4	21.01	53.59
	5	36.87	52.49
	6	32.55	55.15
	7	42.80	60.59
Acetate + Monensin[b]	4	4.83	7.18
	5	5.11	5.29
	6	5.51	5.67
	7	5.11	5.32
Acetate + BESA[c]	4	0	0.34
	5	0.48	0.71
	6	0	0.45
	7	0.61	0.85

[a]8 mmol/mL, added anaerobically.
[b]1 μg/mL, filter sterilized.
[c]20 μM/mL, filter sterilized.

Table 8.4 Inhibition of Methanogenesis from CO Using Sewage Sludge as Primary Inoculum

Substrate and inhibitors	pH	Mole percent			
		Day 14		Day 35	
		CO	CH$_4$	CO	CH$_4$
CO[a]	4	51.7	0.6	47.3	0.25
	5	39.7	9.0	17.3	20.9
	6	41.9	7.7	21.5	16.4
	7	39.5	10.6	15.4	20.5
CO + Monensin[b]	4	53.7	0	46.3	0.47
	5	41.9	9.3	41.8	2.5
	6	41.9	4.9	47.1	6.1
	7	49.4	3.2	38.0	9.2
CO + BESA[c]	4	50.8	0.3	27.5	1.2
	5	46.7	0.8	23.8	3.2
	6	44.1	1.1	30.3	2.3
	7	37.8	2.7	5.1	4.8

[a]CO level was 52 to 55!, 2 atm on day 0.
[b]1 μg/mL filter sterilized.
[c]20 μM/mL, filter sterilized.

The inhibition of methanogenesis using CO and acetate as the substrates with a sewage sludge inoculum is shown in Tables 8.3 and 8.4. Without an inhibitor, acetate was metabolized to 61% methane in 14 days. With a BESA inhibitor, acetate produced less than 1% methane in 14 days. Similarly, CO produced 20% methane in 35 days without an inhibitor and less than 5% methane with a BESA inhibitor. Both monensin and BESA appear to be good inhibitors of methane production in mixed culture.

8.2.3 Results: Alcohol Production

The data in Table 8.5 were obtained for the samples taken 28 days after inoculation. The formation of acids was more pronounced than alcohol, particularly in the samples without BESA or monensin. When CO was used as the carbon source, methanol and ethanol were produced only at pH 7 without BESA or monensin, and no alcohols were detected at lower pH levels. Similarly, no alcohols were produced with BESA at pH 6 and 7. Methanol, ethanol, and butanol were the only alcohols produced with BESA at lower pH levels. With monensin, alcohols were produced at all pH levels.

When acetate was used as the carbon source, no alcohols were detected in samples without BESA and monensin, or with BESA at pH 6 and 7, or with monensin at pH 4 to 6. Propanol was formed in some samples, such as at pH 4 without BESA or monensin and with BESA at lower pH (pH 4 and 5) only. Methanol was formed in BESA-treated samples at pH 4 and 5. Both methanol and ethanol were detected at pH 7 treated with monensin. Although the acetate concentration was decreased, alcohols were not produced during the bacterial metabolism of acetate.

Similar results were obtained from the inocula with chicken waste. The coal mud samples were much less productive and were abandoned. From the preliminary screening data, mixed cultures were shown to produce alcohols when CO, synthesis gas, or acetate

Table 8.5 Production of Alcohols and Acids by Bacterial Enrichments Obtained from Sewage Sludge at Various Levels of pH and in the Presence of Methane Inhibitors[a]

pH	Inhibitors	+ BES, 1 µg/mL	+ Monensin, 20 µM/mL
		CO as Carbon Source[b]	
4	A, PR, 1B, NB	E, B, A, PR, 1B, NB	M, E, B, A, PR, 1B, NB
5	A, PR, 1B, NB	M, E, B, A, PR, 1B, NB	M, E, A, PR, NB
6	P, A, PR, 1B, NB	A, PR, 1B, NB	M, E, B, A, PR, 1B, NB
7	M, E, B, A, PR, 1B, NB	A, PR, 1B, NB	M, A, PR, 1B, NB
		Acetate as Carbon Source[c]	
4	P, A, PR, 1B, NB	M, P, A, PR, 1B, NB	A, PR, 1B, NB
5	A, PR, 1B, NB	M, P, A, PR, 1B, NB	A, PR, 1B, NB
6	A, PR, 1B, NB	A, PR, 1B, NB	A, PR, 1B, NB
7	A, PR, 1B, NB	A, PR, 1B, NB	M, E, A, PR, 1B, NB

[a]All samples were taken after 28 days of incubation. Samples were transferred at least five times before these analyses were done. M, methanol; E, ethanol; P, propanol; B, n-butanol; A, acetic acid; PR, propionic acid; 1B, isobutyric acid; NB, n-butyric acid.
[b]Gas phase contained 60% CO by volume (2 atm); balance 80:20.
[c]8 mM/mL concentration.

was used as sole carbon source. In general, alcohols were detected in the samples with lower pH and in the presence of methanogenic inhibitors such as BESA and monensin.

Since these analyses were made after 21 to 28 days of incubation, the exact time of alcohol production during bacterial metabolism must be determined. Alcohols could be produced early or late in the fermentation, or simultaneously with acid production. The next step in manipulating the mixed cultures for enriching specific microflora in producing alcohols is the quantification of the timing of alcohol production.

8.2.4 Results: Experiments to Quantify Alcohol Production

Selected experiments were carried out at pH levels from 4 to 7, monitoring gas concentration, alcohol concentration, and acid concentrations on a daily basis. The criteria used in selecting inocula for these experiments was based on previous results where sampling had taken place on a weekly to biweekly basis. These inocula consisted of both sewage sludge and chicken waste inocula at various pH levels and had showed either the best utilization of CO or synthesis gas, or the best production of alcohols of all samples at a given pH in the previous studies.

Five-milliliter aliquots of inocula were anaerobically transferred into 20 mL of liquid media in 65-mL stoppered bottles. The liquid media contained yeast extract, vitamins, basal salts, and either monensin or BESA as the methane inhibitor. The pH was adjusted to the desired level by the addition of acid or base as needed. The gas phase was initially at a pressure of 1 atm, and consisted of either CO in CO_2 and nitrogen (N_2) or a synthetic synthesis gas mixture of CO, H_2, CO_2, and CH_4. The gas phase with N_2 consisted of approximately 64 mol % CO, 24 mol % N_2, and 12 mol % CO_2. The synthesis gas phase consisted of approximately 15 mol % H_2, 73 mol % CO, 2 mol % CH_4, and 10 mol % CO_2.

The results of the CO and synthesis gas utilization studies are shown in Table 8.6 and Figs. 8.1 to 8.3. As noted, no data are shown at pH 4 since no CO or synthesis gas

Table 8.6 Synthesis Gas Utilization at pH 5: Chicken Waste Inoculum, BESA Inhibitor

	Gas-phase composition (mol %)					Alcohol production (g/L)				
Day	H_2	N_2	CO	CH_4	CO_2	MeOH	EtOH	PrOH	MePrOH	BuOH
0	14.6	0.9	73.1	2.0	9.0	0.004	0.001	–	–	–
1	14.4	1.0	71.3	2.0	11.1	0.013	0.011	–	–	0.002
2	14.2	1.5	70.3	1.9	11.9	0.010	0.016	0.001	–	–
3	13.4	0.9	68.8	2.8	14.0	–	0.015	0.002	0.008	–
4	14.1	0.8	69.1	2.2	14.5	0.017	0.017	0.006	–	–
5	20.0	1.4	63.7	1.7	13.0	0.010	0.026	0.004	0.004	–
6	13.6	1.3	68.8	1.9	15.8	0.009	0.032	0.007	0.002	0.002
7	13.5	1.1	67.8	2.1	15.6	0.012	0.045	0.010	0.010	0.019
8	13.9	1.3	64.9	2.1	17.3	–	0.070	0.015	–	0.008
9	13.4	1.3	64.1	1.9	19.2	0.012	0.078	0.021	–	0.008
10	13.0	1.9	61.5	2.1	21.4	0.013	0.099	0.024	–	0.004
11	12.6	1.6	58.1	1.9	25.7	0.016	0.105	0.017	–	–
12	12.8	2.3	54.1	2.1	28.7	0.011	0.067	0.019	0.003	–
13	11.9	1.3	50.8	2.7	33.2	0.013	0.117	0.013	–	–
14	10.8	9.6	38.6	2.3	38.6	–	–	–	–	–

was utilized at pH 5 and any slight alcohol production was attributed to the utilization of yeast extract. Total alcohol production as m moles of carbon is shown in the figures, and individual alcohol concentrations are noted in the table.

As noted in Fig. 8.1 at pH 5, the concentration of the gas phases decreased from 73% to 39% CO in 14 days. Very little H_2 or CH_4 was utilized. The CO_2 concentration increased from 9% to 30% during the experiment, indicating the production of CO_2 with alcohol and acid production. The predominant alcohol produced during the experiment was ethanol, reaching a concentration of 0.12 g/L after 13 days. Lesser amounts of methanol, propanol, methyl propanol, and *n*-butanol were also produced. A total alcohol yield of over 0.15 mmol of carbon equivalent in the form of alcohol was produced from 0.35 mmol of carbon equivalent in the form of carbon monoxide.

The experiments at pH 5 using a chicken waste inoculum were repeated after 14 days using the transfer method described previously. Figure 8.2 also shows the CO consumption and alcohol production during the experiment. As noted, CO conversion began much earlier after the transfer. Therefore, earlier transfer (14 days) benefits the evolution of the culture to produce alcohols.

The results for synthesis gas utilization at pH 7 with a chicken waste inoculum is shown in Fig. 8.3. The CO concentration fell from 63 mol % to 42 mol % in 11 days. However, only small amounts of alcohol were produced (0.03 mmol carbon equivalent maximum). Similar results were obtained at pH 6, with little alcohol production.

Media Background Studies

As mentioned previously, studies at pH 4 have shown that no CO or synthesis gas was utilized by the culture, while at the same time alcohols were being produced. Studies showed that the alcohols were apparently being produced from yeast extract or other media constituents. To see if the same problem was happening at other pH levels, the cultures at pH 5, 6, and 7 were inoculated into the liquid media without CO or synthesis gas. Nitrogen gas was used as the purge gas to ensure anaerobic conditions.

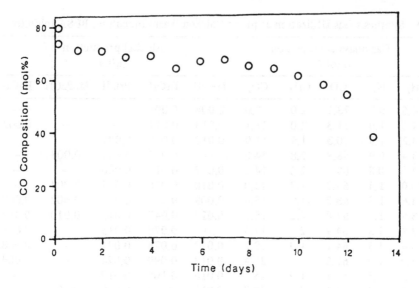

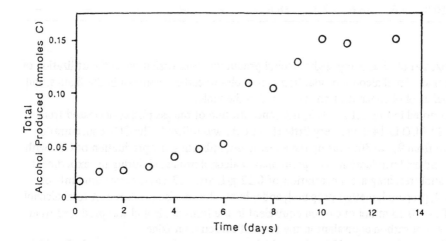

Fig. 8.1 Synthesis gas consumption and alcohol production at pH 5: chicken waste inoculum, BESA inhibitor.

The results of the experiments show that CO_2 was produced in the gas phase at all pH levels. Acid production was very high in each experiment, reaching levels of 0.5 to 2.0 g/L. Alcohol production, on the other hand, was essentially nonexistent, with levels of only 0.002 g/L at the end of the experiments. The higher initial alcohol concentrations were due to the transfer of some alcohol with the inoculum. Thus any alcohol production observed at pH 5, 6, and 7 must be from CO or synthesis gas. The high levels of acid present in the experimental studies may be due, in large part, to the conversion of media constitutents to acids. This result could be significant in the elimination of acid production from the cultures.

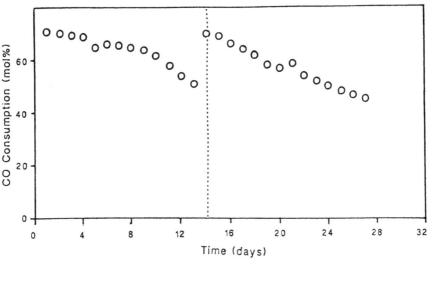

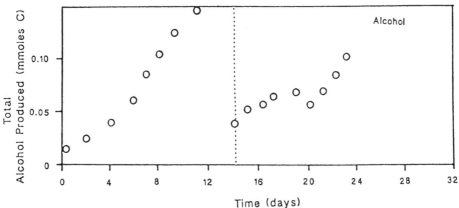

Fig. 8.2 Synthetic gas consumption and alcohol production at pH 5: chicken waste inoculum, BESA inhibitor.

Alcohol Production from CO at PH 5

Because of the encouraging results obtained previously at pH 5, several experiments were carried out with the pH 5 BESA inhibitor culture. In the first experiment, the culture was transferred after 14 days instead of the usual 28 to 30 days, feeding liquid media and synthesis gas. The utilization of CO and the production of alcohols was monitored as before. As shown in Fig. 8.1, CO utilization was more rapid after the 14-day transfer than in the prior run, falling to a level of less than 40% in 20 days. At the same time, the rate of alcohol production was similar. There are thus apparent benefits to more frequent transfer of the culture to aid in CO utilization.

A second experiment was carried out with the pH 5 BESA-inhibited culture in a shaker incubator. Previous experiments have been restricted to incubation without shaking. Since the culture has progressed to a point where CO-utilizing organisms are

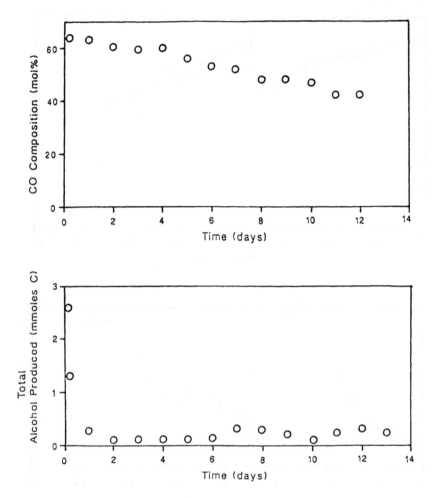

Fig. 8.3 Carbon monoxide consumption and alcohol production at pH 7: chicken waste inoculum, monensin inhibitor.

more dominant, shaking will be useful to provide better mass transfer to get CO into the liquid phase.

The bottles were inoculated with liquid media, BESA, culture, and synthesis gas and shaken gently to provide mass transfer. Gas-phase and liquid-phase compositions were monitored as before. Additional synthesis gas was added as needed, with no liquid media supplementations. The results of the experiment are shown in Fig. 8.4. As noted, the CO consumption was much faster with agitation, requiring only 2 days to utilize all the CO. Ethanol production reached 0.14 g/L in 2 days (comparable to 14 days without shaking) and 0.31 g/L in less than 3 days. At the same time, the levels of the other alcohols stayed nearly constant, with a maximum of 0.03 g/L of *n*-butanol produced in 3 days. A culture enriched for CO utilization, producing ethanol as a single produce, was indeed being developed, aided by agitation.

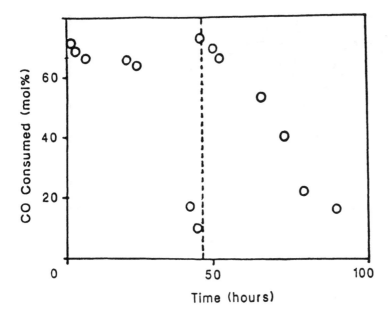

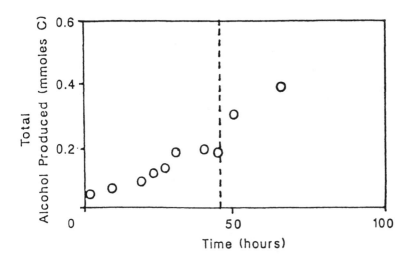

Fig. 8.4 Synthesis gas utilization and alcohol production at pH 5: chicken waste inoculum, BESA inhibitor, gentle agitation.

Further Studies with the pH 5 BESA Culture with Gentle Agitation

In an attempt to study further the pH 5 BESA culture with agitation, 5 mL of culture was inoculated into 50 mL of media. This inoculation level was far less than the 5 mL of culture in 20 mL of media used previously. A longer lag phase is thus expected prior to ethanol production. All other conditions were the same as reported earlier, including gentle agitation, pH 5, 0.5 mL of BESA inhibitor, and synthesis gas at 1 atm as the feed

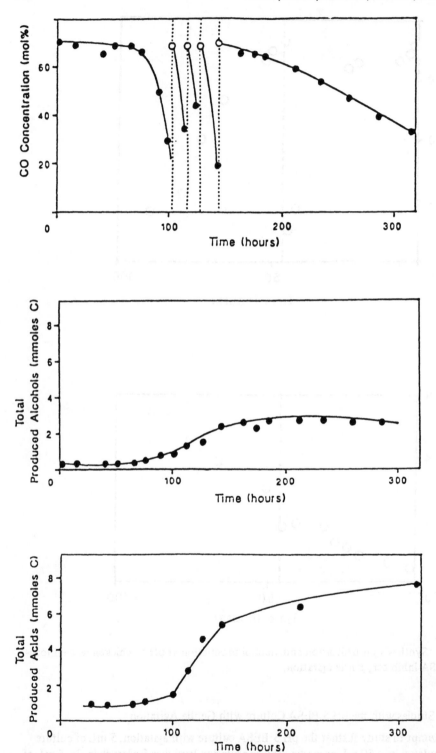

Fig. 8.5 CO conversion, alcohol and acid production from synthesis gas: pH 5, BESA culture, gentle agitation, 1 atm, 0.5 mL BESA added.

gas. The results of this experiment will serve as a baseline for future studies with the pH 5 BESA culture.

Figure 8.5 shows the results of this study. As noted, a long lag phase of about 76 h was required to develop cells prior to CO utilization and alcohol production. However, after a suitable number of cells had accumulated, 72% of the CO from synthesis gas was utilized in only 16 h. Many gas feedings were required in order to have sufficient CO for growth and production.

The maximum ethanol concentration reached in this study was about 1.25 g/L at 185 h after inoculation. The concentration of the other alcohols remained at essentially the initial background levels. The level of ethanol held constant at about the 1.25 g/L level from 185 h to the end of the experiment. Acid production continued to increase to the end of the experiment, reaching a level of 4.6 g/L. The predominant acid produced was acetic acid. The ratio of maximum acetic acid produced to maximum ethanol produced was 3.7.

Background studies utilizing liquid media and 0.5 mL of BESA at pH 5 were also carried out with gentle agitation to monitor the production of alcohols and acids from the media alone. Previous studies at pH 5 without agitation have shown that essentially no alcohols are produced without CO.

The results of the background studies show that no ethanol or other alcohol was produced beyond the initial background levels. Acetic acid production did occur, however, up to a level of 0.63 g/L. It is possible that a portion of the acid production during the batch experiments is from the liquid media instead of CO. Accurate material balance experiments are needed to help clarify this point.

Increased Pressure Studies

A third experiment was carried out to assess the effects of slightly increased pressure on the performance of the pH 5 BESA culture. The gas phase pressure was increased to 2 atm, holding the pH at 5, the BESA addition of 0.5 mL, and gentle agitation. Synthesis gas was used as the feed gas at 2 atm.

The results of the increased pressure experiment are shown in Fig. 8.6. An identical lag phase of 76 h was required prior to alcohol production, so that pressure did not affect acclimation. Again, after sufficient cells were available, CO utilization reached 68% in 15 h (essentially the same result as in Fig. 8.5). The maximum ethanol concentration reached 2.5 g/L, due to more CO being present at the slightly elevated pressure. The maximum acid concentration reached 2.9 g/L acetic acid. The ratio of maximum acetic acid concentration to maximum ethanol concentration was 1.2. Thus higher levels of ethanol are possible as well as lower ratios of acetic acid to ethanol. More will be said of this result later.

Inhibition Studies

Two studies were carried out to assess the benefits of methane inhibitors on acetate and alcohol production. In the first experiment the amount of BESA in the media was increased to 1 mL, in place of 0.5 mL in Fig. 8.5 and 8.6. In the second experiment, 0.5 mL of monensin was added to the media in place of BESA. All other conditions (pH 5, 1 atm synthesis gas feed, gentle agitation) for the pH 5 BESA culture were held constant.

The results of the increased BESA experiment are shown in Fig. 8.7. A longer lag phase of 115 h was required due to the increase in inhibitor concentration. After a sufficient number of cells had developed, up to 76% of the CO was utilized in 18.5 h, essentially matching the results of Figs. 8.5 and 8.6. Methane levels were essentially the

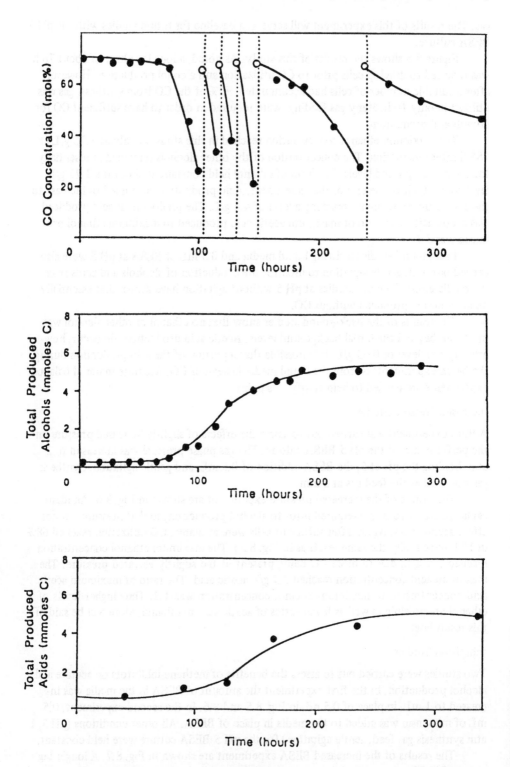

Fig. 8.6   CO conversion, alcohol and acids production from synthesis gas: pH 5, BESA culture, 2 atm, gentle agitation, 0.5 mL BESA added.

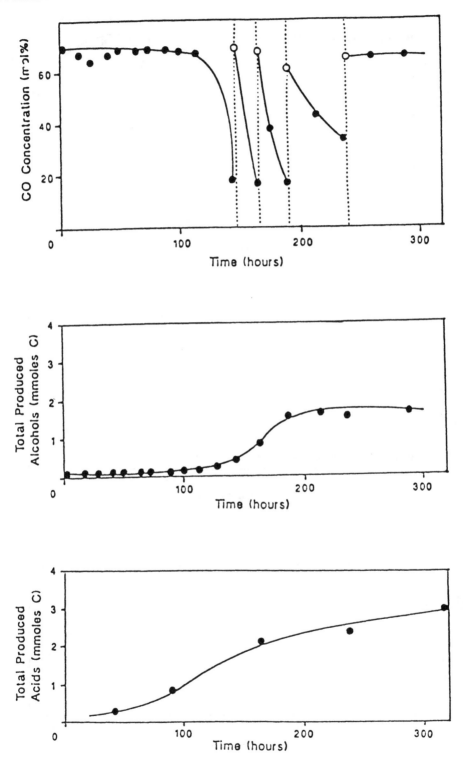

Fig. 8.7 CO conversion, alcohol and acid production from synthesis gas: pH 5, BESA culture, gentle agitation, 1 atm, 1 mL of BESA added.

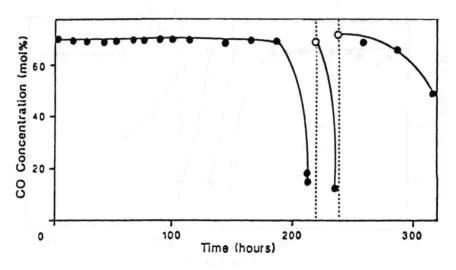

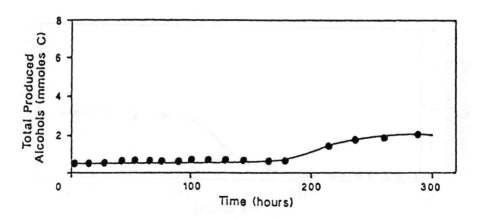

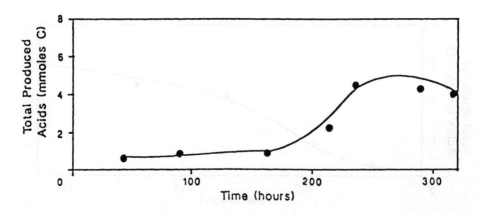

Fig. 8.8   CO conversion, alcohol and acid production from synthesis gas: pH 5, BESA culture, gentle agitation, 1 atm, 0.5 mL of monensin added.

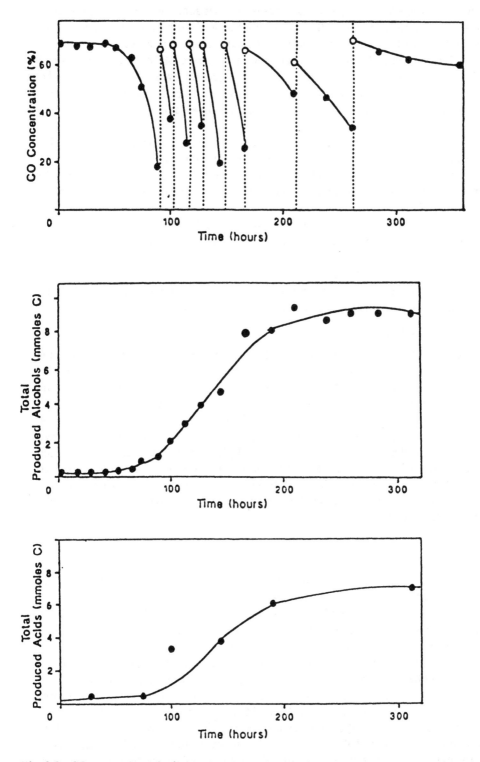

Fig. 8.9 CO conversion, alcohol and acids production from synthesis gas: pH 5, BESA culture, gentle agitation, no inhibitor.

same as in Figs. 8.5 and 8.6, indicating that the further addition of BESA has no beneficial effects in lowering the methane concentration in the gas phase below 1 to 3%. These
low levels of methane will be eliminated as the pure culture is attained.

The maximum ethanol concentration in the increased BESA experiment was 1.49
g/L. Again, the other alcohols remained at the background levels. The maximum acid concentration was 3.7 g/L acetic acid. The ratio of maximum acetic acid concentration to
maximum ethanol concentration was about 2.5, a slight ethanol increase over the result
presented in Fig. 8.5.

The results of the monensin inhibition experiment is shown in Fig. 8.8. In the
presence of monensin the lag phase exceeded 200 h. Monensin, known to be a more
active inhibitor than BESA, inhibited other reactions in addition to methane production.
After the culture was established, 82% of the CO was utilized in 18 h, a result similar to
the results of Figs. 8.5 to 8.7.

The maximum ethanol concentration was only 0.9 g/L acetic acid. The ratio of
the maximum acetic acid concentration to the maximum ethanol concentration was
about 3.

It is expected that these experimental results would have approached the results of
Figs. 8.5 to 8.7 if the experiments would have continued an additional 5 days. Still, there
was really no apparent benefit of adding monensin in place of BESA.

Removal of Inhibitor

An experiment was carried out without any inhibitors, since the level of methane in the
gas phase has remained low in all of the experimental studies. As in the other experiments, the pH was held at 5.0, gentle agitation was employed, and synthesis gas at 1 atm
was used as the feed gas. The results of the pH 5 BESA culture experiment without inhibitors is shown in Fig. 8.9. As in Figs. 8.5 and 8.6, the lag phase was 76 h. As with all
previous results, 72% of the CO from synthesis gas was utilized in 16.7 h. In this study,
the maximum ethanol concentration reached 4.3 g/L. The maximum acetic acid concentration reached 3.9 g/L yield, a ratio of the maximum acetic acid concentration to maximum ethanol concentration of only 0.9. It thus appears quite possible to increase the
ethanol concentration while decreasing the acid/alcohol ratio. The removal of the inhibitors apparently benefited alcohol production.

8.3 CONCLUSIONS

The following general comments and conclusions are offered. Further experimental work
will help clarify the results obtained in the experimental studies.

1. The only alcohol produced in the final screening studies is ethanol, and the only
acid produced is acetic acid. The small levels of other acids and alcohols are essentially
constant with time at the initial background levels.

2. Some acid (acetic acid) is produced from the liquid media only. Material
balances will help in determining the amount of acid produced from the media.

3. In observing the results obtained from the advanced screening studies, it appears
that the maximum acid level was 5 to 6 mmol C at the onset of constant alcohol production. This result indicates that acid may have been inhibitory to the mixed culture, thereby limiting the amount of alcohol that could be produced in these studies. Many possibilities exist for the reasons that the ratio of acid to alcohol decreased. If acid is removed
or eliminated from the media, alcohol levels could increase further.

4. A level of at least 4.3 g/L ethanol can be produced from the mixed culture. With gentle agitation, approximately 16 h is required to utilize 72% of the CO from synthesis gas at 1 to 2 atm total pressure.

5. No benefits are observed for alcohol production in the presence of BESA or monensin during the latter studies with the pH 5 BESA culture. Methane production has been essentially stopped, which was the primary purpose of the inhibitors.

8.4 FUTURE WORK

Immediate future work will concentrate on measuring cell concentrations in determining growth curves for the pH 5 BESA mixed culture. Also, carefully controlled and monitored experiments will be initiated to determine accurately cell, substrate, and product concentrations with time. These experiments will aid in following the stoichiometry of the reactions and will help complete the mass balance for the experiments.

Experiments have begun in attaining a pure culture to produce ethanol alone. The serial dilution method will be used for these purposes. As mentioned previously, microscopic examination has revealed a highly enriched mixed culture, well on the way toward attaining a pure culture.

These procedures, if successful, will yield cultures capable of producing alcohols in greater quantity than the concentrations obtained in the screening studies. Eventually, stirred tank and immobilized cell reactor studies will be initiated using pure or highly enriched mixed cultures to measure yields and reaction kinetics. The technical feasibility of biologically converting synthesis gas to ethanol can then be assessed.

REFERENCES

1. Badziong, W., Ditter, G., and Thauer, R. K., Acetate and carbon dioxide assimilation by *Desulfovibrio vulgaris* (Marburg), growing on hydrogen and sulfate as sole energy source, *Arch. Microbiol., 123,* 301–305 (1979).

2. Balba, M. T., and Nedwell, D. B., Microbial metabolism of acetate, propionate, and butyrate in anoxic sediment from the Colne Point Saltmarsh, Essex, U.K., *J. Gen. Microbiol., 128,* 1415–1422 (1982).

3. Balch, W. E., Schoberth, S., Tanner, R. S., and Wolfe, R. S., *Acetobacterium* new-genus of hydrogen oxidizing, carbon dioxide reducing anaerobic bacteria, *Int. J. Syst. Bacteriol., 27,* 335–361 (1977).

4. Brandis-Heep, A., Gebhardt, N. A., Thauer, R. K., Widdel, F., and Pfennig, N., Anaerobic acetate oxidation to carbon dioxide by *Desulfobacter postgatei.* 1. Demonstration of all enzymes required for the operation of the citric-acid cycle, *Arch. Microbiol., 136,* 222–229 (1983).

5. Braun, M., Mayer, F., and Gottschalk, G., *Clostridium aceticum* (Wierinsa), a microorganism producing acetate acid from molecular hydrogen and carbon dioxide, *Arch. Microbiol., 128,* 288–293 (1981).

6. Braun, M., Schoberth, S., and Gottschalk, G., Enumeration of bacteria forming acetate from hydrogen and carbon dioxide in anaerobic habitats, *Arch. Microbiol., 120,* 201–204 (1979).

7. Chynoweth, D. P., and Srivastrave, V. J., 1984. Biothermal conversion of biomass and wastes to methane, in *Biotechnology and Bioengineering Symposium 13* (C. D. Scott, ed.), *5th Symposium on Biotechnology for Fuels and Chemicals,* Gatlinburg, Tenn., May 10–13, 1983, Wiley, New York, 1984, pp. 539–556.

8. Colby, J., Stirling, D. I., and Dalton, H., The soluble methane manooxygenase of

Methylococcus capsulatus (Bath.): its ability to oxygenate *n*-alkanes, *n*-alkenes, ethers, and alicyclic, aromatic and heterocyclic compounds, *Biochem. J., 165*, 395–402 (1977).

9. Dalton, H., and Leak, D. J., 1985. Methane oxidation by microorganisms, in *Microbial Gas Metabolism* (R. K. Poole and C. S. Dow, eds.), Academic Press, New York, pp. 173–200.

10. Daniels, L., Comments on enzymatic synthesis or organic acids and alcohols from hydrogen gas, carbon dioxide and carbon monoxide, *Biotechnol. Bioeng., 24*, 2099–2101 (1982).

11. Daniels, L., Fuchs, G., Thauer, R. H., and Zeikus, J. G., Carbon monoxide oxidation by methanogenic bacteria, *J. Bacteriol., 132*, 118–126 (1977).

12. Dashekvicz, M. P., and Uffen, R. L., Identification of a carbon monoxide–metabolizing bacterium as a strain of *Rhodopseudomonas gelatinosa* (Mlisch) van Neil., *Int. J. Syst. Bacteriol., 29*, 145–148 (1978).

13. Diekert, G. B., and Thayer, R. K., Carbon monoxide oxidation by *Clostridium thermoaceticum* and *Clostridium formicoaceticum, J. Bacteriol., 136*, 597–606 (1982).

14. Duerre, P., and Andersen, J. R., Pathway of CO_2 reduction to acetate without a net energy requirement in *Clostridium purinolyticum, FEMS Microbiol. Lett., 15*, 51–56 (1982).

15. Fischer, J. R., Iannotti, E. L., Stahl, T., Garcia, A., III, and Harris, F. D., Bioconversion of animal manure into electricity and a liquid fuel, in *Biotechnology and Bioengineering Symposium 13* (C. D. Scott, ed.), *5th Symposium on Biotechnology for Fuels and Chemicals*, Gatlinburg, Tenn., May 10–13, 1983, Wiley, New York, 1984, pp. 527–528.

16. Fuchs, G., Schnitker, U., and Thayer, R. K., Carbon monoxide oxidation by growing cultures of *Clostridium pasteuranium, Eur. J. Biochem., 49*, 111–115 (1974).

17. Gebhardt, N. A., Linder, D., and Thauer, R. K., Anaerobic acetate oxidation to carbon dioxide by *Desulfobacter postgatei*. 2. Evidence from carbon-14 labeling studies of the operation of the citric-acid cycle, *Arch. Microbiol., 136*, 230–233 (1983).

18. Genthner, B. F. S., and Bryant, M. P., Growth of *Eubacterium limosum* with carbon monoxide as the energy source, *Appl. Environ. Microbiol., 43*, 70–77 (1982).

19. Godwin, S. J., Wase, D. A. J., and Forster, C. F., Use of upflow anaerobic sludge blanket reactor to treat acetate rich waste, *Process Biochem., 17*, 33–34, 45 (1982).

20. Henrich, R. A., Advances in bio gas to fuel conversion, *Biocycle, 24*, 28–31 (1983).

21. Higgins, I. J., and Quayle, J. R., Oxygenation of methane by methane-grown *Pseudomonas methanica* and *Methanomonas methanooxidans, Biochem. J., 118*, 201–208 (1970).

22. Huser, B. A., Wuhrmann, K., and Zehnder, A. J. B., *Methanothrix soehngenii* gen. nov. sp. nov., a new acetotrophic non-hydrogen-oxidizing methane bacterium, *Arch. Microbiol., 132*, 1–9 (1982).

23. Kerby, R., and Zeikus, J. G., Growth of *Clostridium thermoaceticum* on H_2/CO_2 or CO as energy source, *Curr. Microbiol., 8*, 27–30 (1983).

24. Khan, A. W., and Mes-Hartree, M., Metabolism of acetate and hydrogen by a mixed population of anaerobes capable of converting cellulose to methane, *J. Appl. Bacteriol., 50*, 283–288 (1981).

25. Kluyver, A. J., and Schnellen, C. G., On the fermentation of carbon monoxide by pure cultures of methane bacteria, *Arch. Biochem., 14*, 57–70 (1947).

26. Leadbetter, E. R., and Foster, J. W., Studies on some methane-utilizing bacteria, *Arch. Microbiol., 30*, 91–118 (1958).

27. Leadbetter, R. R., and Foster, J. W., Bacterial oxidation of gaseous alkanes, *Arch. Microbiol., 35*, 92–104 (1960).

28. Leak, D. J., Stanley, S. H., and Dalton, H., Implications of the nature of methane monooxygenase on carbon assimilation in methanotrophs, in *Microbial Gas Metabolism* (R. K. Poole and C. S. Dow, eds.), Academic Press, New York, 1985, pp. 201–208.

29. Levy, P. F., Barnard, G. W., Garcia-Martinez, D. V., Sanderson, J. E., and Wise, D. L., Organic acid production from carbon dioxide-hydrogen and carbon monoxide-hydrogen by mixed culture anaerobes, *Biotechnol. Bioeng., 23*, 2293–2306 (1981).

30. Lorowitz, W. H., and Bryant, M. P., *Peptostreptococcus productus* strain that grows rapidly with carbon monoxide as the energy source, *Appl. Environ. Microbiol., 47*, 961–964 (1984).

31. Lynd, L., Kerby, R., and Zeikus, J. G., Carbon monoxide metabolism of the methylotropic acidogen *Butyribacterium methylotrophicum, J. Bacteriol., 136*, 597–606 (1982).

32. Mayer, F., Lurz, R., and Schoberth, S., Electron microscopic investigation of the hydrogen oxidizing acetate forming anaerobic bacterium *Acetobacterium woodii, Arch. Microbiol., 115*, 207–214 (1977).

33. Mountford, D. O., and Asher, R. A., Changes in proportions of acetate and carbon dioxide used as methane precursors during the anaerobic digestion of bovine waste, *Appl. Environ. Microbiol., 35*, 648–654 (1978).

34. Patel, R. N., Hou, C. T., Laskin, A. L., and Felix, A., Microbial oxidation of hydrocarbons: properties of a soluble methane monooxygenase from a facultative methane-utilizing organism, *Methylobacterium* sp. strain CRL-26, *Appl. Environ. Microbiol., 44*, 1130–1137 (1982).

35. Postgate, J., Carbon monoxide as a basis for primitive life on other planets: a comment, *Nature, 266*, 978 (1970).

36. Powell, G. E., Hilton, M. G., Archer, D. B., and Kirsop, B. H., Kinetics of the methanogenic fermentation of acetate, *J. Chem. Technol. Biotechnol., 33B*, 209–215 (1983).

37. Sleat, R., Mah, R. A., and Robinson, R., Isolation and characterization of an anaerobic hydrogen utilizing acetate forming bacterium, 83rd Annual Meeting of the American Society for Microbiology, New Orleans, La., Mar. 6–11, *Abst. Annu. Meet. Am. Soc. Microbiol., 83*, 154 (1983).

38. Sleat, R., Mah, R. A., and Robinson, R., 1985. Acetoanaerobium noterae *gen. no., sp. nov.: An Anaerobic Bacterium That Forms Acetate from H_2 and CO_2*, School of Public Health, University of California, Los Angeles, 1985.

39. Sleat, R., Mah, R. A., and Robinson, R., *Acetoanaerobium noterae*, new-genus, new species, an anaerobic bacterium that forms acetate from hydrogen and carbon dioxide, *Int. J. Syst. Bacteriol., 35*, 10–15 (1985).

40. Smith, M. R., and Mah, R. A., Growth and methanogenesis by *Methanosarcina* strain 227 on acetate and methanol, *Appl. Environ. Microbiol., 36*, 870–879 (1978).

41. Stirling, D. I., Colby, J., and Dalton, H., A comparison of the substrate and electron donor specificities of the methane manooxygenases from three strains of methane-oxidizing bacteria, *Biochem. 7., 177*, 361–364 (1979).

42. Stephenson, M., *Bacterial Metabolism*, 3rd ed., Longmans, Green, London, 1949, pp. 95–96.

43. Stupperich, F., Hammel, K. F., Fuchs, G., and Thayer, R. K., Carbon monoxide fixation into the carboxyl group of acetyl coenzyme A during autotrophic growth of *Methanobacterium, FEBS Lett., 152*, 21 (1983).

44. Tabuchi, T., Tahara, Y., Tanaka, M., and Yanagiuchi, S., Organic acid fermentation by yeasts. IX. Preliminary experiments on the mechanism of citric acid fermentation in yeasts, *Nippon Nogei Kagaku Kaishi, 47*, 617–622 (1973).

45. Takayama, K., and Tomiyama, Citric acid and isocitric acid by fermentation, Ger. Patent 1,301,079, July 19, 1973.
46. Thauer, R. K., Jungermann, K., and Decker, K., Energy conservation in chemotrophic anaerobic bacteria, *Bacteriol. Rev., 41,* 100-180 (1977).
47. Uchio, R., Fermentative production of citric acid, Jpn. Patent 7,638,488, Mar. 3, 1976.
48. Uffen, R. L., Anaerobic growth of *Rhodopseudomonas* sp. in the dark with carbon monoxide as sole carbon and energy substrate, *Proc. Natl. Acad. Sci. USA, 73,* 3298-3302 (1976).
49. Valcke, D., and Verstraete, W., A practical method to estimate the acetoclastic methanogenic biomass in anaerobic sludges, *J. Water Pollut. Control Fed., 55,* 1191-1195 (1983).
50. Waber, L. J., and Wood, H. G., Mechanism of acetate synthesis from CO_2 by *Clostridium acidiurici, J. Bacteriol., 140,* 468-478 (1979).
51. Ward, D. M., Mah, R. A., and Kaplan, I. R., Methanogenesis from acetate: a nonmethanogenic bacterium from an anaerobic acetate enrichment, *Appl. Environ. Microbiol., 35,* 1185-1192 (1978).
52. Weber, H., Kulbe, K. D., Chmiel, H., and Troesch, W., Microbial acetate conversion to methane; kinetics, yields and pathways in a 2-step digestion process, *Appl. Microbiol. Biotechnol., 19,* 224-228 (1984).
53. Wise, D. L., Fuels and organic chemicals via anaerobic fermentation of residues and biomass, in *Biochemical and Photosynthetic Aspects of Energy Production* (A. San Pietro, ed.), Academic Press, New York, 1980, pp. 81-116.
54. Yoshinaga, F., Tsuchida, T., Nakase, T., and Okumura, S., Fermentative production of citric acid by *Candida,* Jpn. Patent 72,25,383, 1972.
55. Zeikus, J. G., Chemical and fuel production by anaerobic bacteria, in *Annual Review of Microbiology,* Vol. 34 (M. P. Starr, ed.), Annual Reviews, Palo Alto, Calif., 1980, pp. 423-464.

9

Anaerobic Degradation of Organic Compounds in Hypersaline Environments: Possibilities and Limitations

AHARON OREN   *The Institute of Life Sciences, The Hebrew University of Jerusalem, Jerusalem, Israel*

9.1 INTRODUCTION

Anaerobic digestion has long been a well-established procedure for the disposal of organic waste material. Theoretically, total conversion of most organic substrates, even relatively complex ones, to carbon dioxide and methane is possible under anaerobic conditions, the methane being an economically important by-product of the process. Total conversion of complex organic molecules to carbon dioxide and methane depends on a collaboration of a variety of microorganisms, performing processes such as depolymerization of complex polymeric substances; different fermentation processes leading to products such as hydrogen, carbon dioxide, ethanol, acetate, and other simple organic acids; oxidation of ethanol and short-chain organic acids to acetate with the production of hydrogen; and finally, hydrogen-consuming processes such as sulfate reduction and methanogenesis, being the terminal steps leading to complete degradation to carbon dioxide and methane.

Little is known as to the extent of anaerobic degradation of organic matter in hypersaline environments. The sediments of hypersaline water bodies are generally anaerobic, partly as a result of biological activity in the sediment and the overlaying water, and also because of the limited solubility of oxygen in hypersaline brines. The biology of anaerobic hypersaline environments has been studied relatively little, although it is curious to note that to the best of my knowledge, the first bacteria ever isolated from a hypersaline environment were anaerobes (although not halophilic ones): clostridia, causing tetanus and gas gangrene, isolated by Lortet from Dead Sea mud at the end of the nineteenth century [1].

As Dynatech R/D Company (Boston, Massachusetts) and the Houston Lighting and Power Company (Houston, Texas) expressed an interest in anaerobic biodegradation of organic materials in the presence of high salt concentrations, I have been studying the microbiology of anaerobic sediments of the Dead Sea as a model by isolating and characterizing the bacteria present and determining the processes performed by them. In this chapter I present some of the results and review our knowledge of the nature of bacteria able to grow in anaerobic hypersaline environments, and on their ecology. Part of the subject was reviewed earlier [2].

9.2 BACTERIAL TYPES INVOLVED IN ANAEROBIC BIODEGRADATION AND THEIR OCCURRENCE IN HYPERSALINE ENVIRONMENTS

Anaerobic degradation of complex organic substrates to carbon dioxide and methane is never achieved by one microorganism alone, but requires the cooperation of a variety of bacteria, performing different steps of the process, and as certain steps in the degradation can proceed only concomitantly with others, the different processes are closely interdependent [3]. Figure 9.1 presents a simplified scheme of the microbial conversions involved. In the first step, organic materials (if necessary after breakdown of polymeric substances into monomers) are fermented to simple products such as hydrogen, carbon dioxide, formate, ethanol, and simple organic acids. Part of these products can serve as substrates for further fermentations, as exemplified by the propionic acid fermentation of lactate, or the conversion of ethanol and acetate to butyrate, caproate, and hydrogen by *Clostridium kluyveri* (step 2). An extremely important, but for technical reasons relatively little studied process is that performed by proton-reducing acetogenic bacteria, converting simple alcohols and organic acids to acetate and hydrogen (step 3) in a process that is thermodynamically favorable only if the hydrogen pressure is maintained at extremely low values. Thus the organisms involved are obligate syntrophs, depending on other microorganisms (sulfate-reducing or methanogenic bacteria) that keep the hydrogen concentration sufficiently low. Hydrogen can serve as an energy source for different types of anaerobes: homoacetogenic bacteria (step 4, probably of minor importance in most ecosystems), sulfate-reducing bacteria (step 5), and methanogenic bacteria (step 6). The last two steps enable the final degradation of acetate to carbon dioxide, or to carbon dioxide and methane, respectively. Not all the processes above have been identified in anaerobic hypersaline environments, and our understanding of those that have been demonstrated is still incomplete. I will discuss our present knowledge on the anaerobic halophilic bacteria performing the various steps discussed above.

9.3 FERMENTATION PROCESSES BY HALOPHILIC BACTERIA

In the course of our studies on the microbiology of the Dead Sea we isolated a number of obligately anaerobic, moderately halophilic, fermentative bacteria. In the past a substantial part of the lake's water column was devoid of oxygen. Before 1979 the deep northern basin of the lake was "permanently" stratified [4], with an upper aerobic layer about 40 to 80 m deep, separated by a thermocline and pycnocline from the deeper brines. The lower water mass (down to the maximal depth of about 320 m) was anaerobic, and so were the bottom sediments. Anaerobic halophilic bacteria have been recovered from these sediments as early as 1943 [5], but unfortunately these isolates have not been preserved. In February 1979 an overturn of the lake's water column caused complete

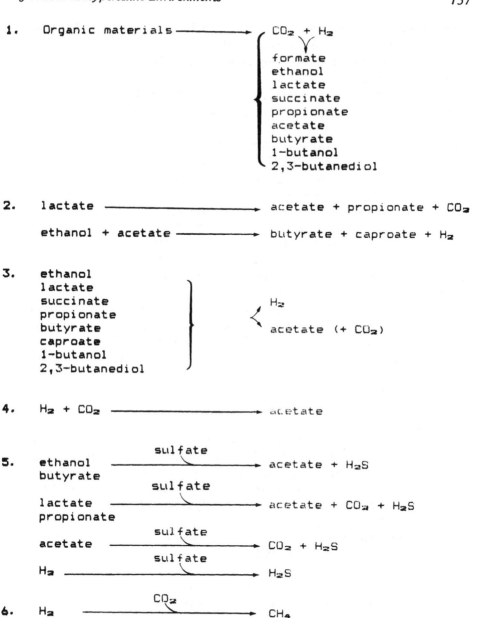

Fig. 9.1 Anaerobic food chain. (Modified from Ref. 3.)

mixing, and oxygen penetrated down to the bottom. However, even today the deeper layers of the bottom sediments remain anaerobic.

The following isolates of obligately anaerobic halophilic bacteria from the Dead Sea have been characterized:

1. *Halobacteroides halobius* [6], a very long, slender, rod-shaped, gram-negative bacterium (Fig. 9.2A), motile by means of peritrichous flagella. The type strain is ATCC 35273. The strain was isolated from an enrichment culture containing 80% Dead Sea water, pyruvate, and yeast extract. It grows with a doubling times as short as 55 min under optimal conditions (41°C, 1.5 M NaCl). The strain requires NaCl concentrations between 1.4 and 2.8 M, tolerates $MgCl_2$ concentrations of up to 1.5 M (in addition to 1.5 M NaCl), and grows best at 37 to 42°C.

2. *Clostridium lortetii* [7], a rod-shaped gram-negative bacterium, motile by means of peritrichous flagella and producing terminal endospores with attached gas vacuoles (Fig. 9.2B). The species was named for M. L. Lortet, who has isolated clostridia from Dead Sea sediments more than 90 years ago [1]. The type strain is ATCC 35059. The strain was isolated from an enrichment culture containing 80% Dead Sea water, lactate, and yeast extract, followed by dilutions in agar tubes. The strain requires NaCl concentrations between 1 and 2 M, and the optimum growth temperature is 37 to 45°C.

3. A strain, designated DY-1, being a rod-shaped, gram-negative bacterium, motile by means of peritrichous flagella (Fig. 9.2C), sometimes seen to produce terminal endospores, but without gas vacuoles (Fig. 9.2D). Minimal doubling times are as short as 40 min in media containing between 0.5 and 2 M NaCl and at 36 to 45°C (A. Oren, unpublished data, 1986).

4. Another obligately anaerobic, moderately halophilic bacterium was isolated by Zeikus and co-workers from bottom sediments of the south arm of the Great Salt Lake, Utah (Fig. 9.2E). The organism was described as *Haloanaerobium praevalens* [8], and the type strain has been deposited as DSM 2228. The bacterium may be identical to "*Bacteroides halosmophilus*" isolated earlier by Baumgartner [9] from salted anchovies and solar salt, but which has been lost. *H. praevalens* is gram negative and grows in NaCl concentrations between 0.4 M and saturation (optimally between 2.5 and 3.5 M) at temperatures between 25 and 45°C.

The four isolates of obligately anaerobic halophilic bacteria described above are all chemoorganotrophs, generating energy by means of fermentation. Minimal media supporting growth of *H. halobius* and strain DY-1 are simple: In addition to inorganic salts they should contain a suitable carbon and energy source (e.g., glucose), L-leucine, L-cysteine (as reducing agent), and vitamins. *H. halobius* requires biotin and *p*-aminobenzoate [6]. *C. lortetii* probably has complex nutritional requirements, and the minimal requirements of *H. praevalens* have not been reported.

H. halobius and strain DY-1 ferment carbohydrates: both utilize glucose, fructose, sucrose, and starch as the carbon and energy source [6]. *Halobacteroides halobius* uses, in addition, galactose and pyruvate. *C. lortetii* grows in a rich medium containing L-glutamate, yeast extract, casamino acids, and nutrient broth [7]; how many of the ingredients of this medium are actually required remains to be determined. The addition of glucose to this medium stimulates growth, and glucose is utilized, but only at the end of the exponential growth phase, probably after other, more favorable substrates have been depleted. The Great Salt Lake isolate *H. praevalens* ferments carbohydrates, peptides, amino acids, and pectin [8]. Fermentation products of the four isolates include acetate, ethanol,

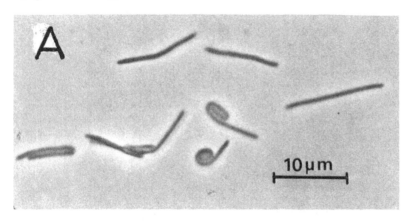

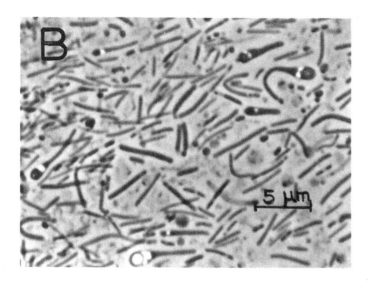

Fig. 9.2 Phase-contrast micrographs of anaerobic halophilic bacteria: (A) *Halobacteroides halobius*; (B) *Clostridium lortetii*; (C) strain DY-1 (young cells); (D) strain DY-1 (sporulating cell); (E) *Haloanaerobium praevalens*.

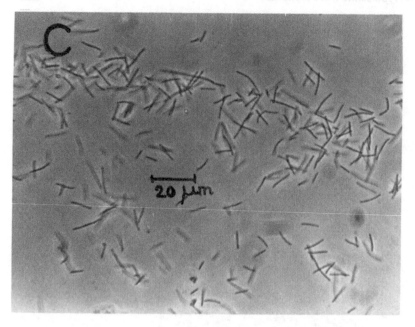

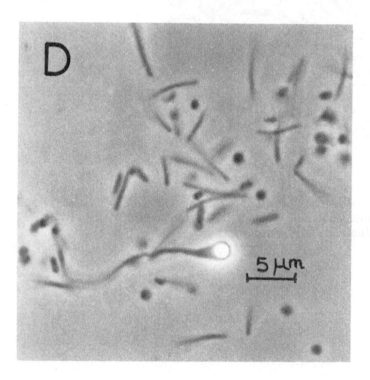

Fig. 9.2 (Continued).

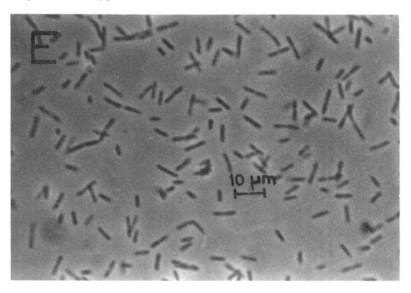

Fig. 9.2 (Continued).

hydrogen, carbon dioxide, and others (Table 9.1). *H. praevalens* also produces methyl mercaptan during degradation of methionine [8].

The four anaerobic halophilic bacteria described all stain gram negative; the gram-negative character of their cell envelope has been ascertained by means of electron microscopy in the case of *H. halobius* [6], *H. praevalens* [8], and *C. lortetii* [7]. The gram-negative character of the cell envelope of *C. lortetii* may preclude its classification in the genus *Clostridium*, and data on the structure of the 16S ribosomal RNA also do not support a close relationship of *C. lortetii* to the nonhalophilic clostridia (see below). The four bacteria share a low percentage of guanine plus cytosine in their DNA, between 27 and 31.5 mol %.

Comparative 16S oligonucleotide cataloging showed that *H. praevalens* and *H. halobius* obviously belong to the eubacterial kingdom, but do not show any clear relationship with any of the recognized subgroups within the eubacteria [10], except possibly with the spirochetes [11]. Moreover, the two halophilic anaerobes are related to each other [10], and a new family, the Haloanaerobiaceae, has been proposed to include these novel organisms. The isolates *C. lortetii* and DY-1 also belong to this group [2] (Fig. 9.3).

The possibility that facultatively anaerobic halophiles play a role in anaerobic degradation in hypersaline environments has not been documented but is theoretically feasible: The archaebacterial genus *Halobacterium*, although consisting of typically aerobic bacteria, contains a number of strains capable of anaerobic life, using different modes of energy generation in the absence of oxygen:

1. Light can serve as an energy source for anaerobic growth in bacteriorhodopsin-containing strains [12] (on condition that retinal is supplied, as its synthesis is oxygen dependent).

Table 9.1 Fermentation Products of Anaerobic Halophilic Bacteria[a]

	Acetate	Propionate	Isobutyrate	n-Butyrate	Isovalerate	Ethanol	Formate	CO$_2$	H$_2$
Haloanaerobium praevalens									
No glucose	100	15		38				15	35
Glucose added	100	36		152				180	200
Halobacteroides halobius	100					226		194	136
Clostridium lortetii	100	6	2	7	3			N.D.	N.D.
Strain DY-1	100			7		77	65	60	110

Source: Ref. 2; reproduced by permission.
[a] Amounts (in mmol) are given relative to acetate (=100). N.D., not determined.

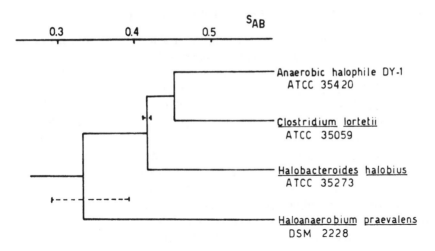

Fig. 9.3 Dendrogram showing the phylogenetic position of four isolates of anaerobic halophilic eubacteria, based on their 16S rRNA oligonucleotide similarity coefficients S_{AB}). (From Ref. 2; reproduced by permission.)

2. Nitrate can serve as an alternative electron acceptor in respiration in certain strains.
3. At least some strains can use arginine as an energy source for fermentation [12].
4. *Halobacterium vallismortis* has been described as a facultative anaerobe [13], but the nature of its energy-generation mechanism in the absence of oxygen has not been elucidated.

It is unknown whether these modes of anaerobic life in halobacteria have ecological importance. Similarly, we do not know if the moderately halophilic eubacterium *Vibrio costicola*, which is a facultative anaerobe, does grow anaerobically in nature. However, the ease with which strains resembling *V. costicola* could be isolated from anaerobic hypersaline sediments from the Dead Sea area (see below) suggests that this type of organism may also participate in anaerobic breakdown processes in the presence of high salt concentrations.

9.4 SUBSEQUENT STEPS IN ANAEROBIC DEGRADATION: THE PROPIONIC ACID FERMENTATION, PROTON-REDUCING ACETOGENS, AND HOMOACETOGENIC BACTERIA

To the best of my knowledge bacteria performing a propionic acid fermentation and proton-reducing acetogenic bacteria (steps 2 and 3 in Fig. 9.1) have never been isolated from a hypersaline environment. Preliminary experiments designed to enrich for methanogenic bacteria using Dead Sea sediments as inoculum, and using a mineral medium under a gas phase of hydrogen and carbon dioxide, yielded no methane, but slight growth was observed, and some acetate accumulated, suggesting the presence of homoacetogenic bacteria (step 4 in Fig. 9.1). However, further attempts to isolate these bacteria have failed.

9.5 SULFATE REDUCTION IN HYPERSALINE ENVIRONMENTS

Biological sulfate reduction occurs in many anaerobic hypersaline habitats, such as the Dead Sea [14-16], the Great Salt Lake [17], and salterns [18]. The anoxic lower water mass that existed in the Dead Sea before 1979 [4] contained between 0.23 and 0.56 mg of H_2S per liter [15]. The involvement of biological sulfate reduction in the formation of this sulfide was suggested by data on the isotopic composition of the sulfate and sulfide present in the sediments and in the lower water mass before the 1979 overturn of the lake: Sulfide in the lower water mass was relatively depleted of the heavy stable isotope ^{34}S ($\delta^{34}S$ = -19.6 to -21.7%), while the heavy isotope was abundant in the sulfate ($\delta^{34}S$ = +14.1 to +15.6%). In the interstitial waters of the reduced sediments these values were -17 and +16%, respectively [15]. Isotopic fractionations such as these are characteristic for bacterial dissimilatory sulfate reduction. Lerman [14] calculated the rate of sulfate reduction in Dead Sea sediments from the depth distribution of gypsum, and his estimations yielded very low rates, about 0.78 μmol sulfate/cm^3 per year. Direct measurements of sulfate reduction rates by injecting cores of anaerobic sediment from the shore of the lake with $^{35}SO_4{}^{2-}$ (according to Jørgensen and Fenchel [19]) yielded higher values: 10 to 25 nmol sulfate/cm^3 per day (at 16°C), 15 to 38 nmol/cm^3 per day at 26°C, and 42 to 136 nmol/cm^3 per day at 37°C. The rates were enhanced 80% by the addition of hydrogen or 40% by formate, but no stimulation was observed by the addition of lactate, acetate, or propionate [16]. All our attempts to isolate the organism(s) responsible for this sulfate reduction remained unsuccessful.

Somewhat higher sulfate reduction rates were reported by Zeikus [17] for sediments of the Great Salt Lake. In vitro rates at 30°C were 175 ± 22 nmol sulfate/cm^3 per day, and the rates were enhanced by the addition of lactate. Enrichment cultures for sulfate-reducing bacteria, using media with lactate as electron donor and salt concentrations of 18 to 23%, yielded curved rods resembling *Desulfovibrio*, and spore-forming rods. Still higher rates of sulfate reduction were measured in sediments of a saltern in California: rates between 0.6 and 0.1 μmol sulfate/g wet sediment per day were measured at salinities of 15 and 30% [18]. Sulfate and acetate were found to be present in these sediments above K_S concentrations, and therefore factors other than substrate availability, such as salinity, must have been limiting the rates.

Although the occurrence of dissimilatory sulfate reduction in hypersaline environments is thus well established, none of the organisms involved seems to have been isolated in pure culture and characterized. The most salt-tolerant isolate reported of which I am aware is a strain isolated by Trüper [20] from the transition zone near the *Atlantis II* hot brine in the Red Sea, a strain that tolerates 10 to 17% NaCl.

9.6 METHANOGENIC BACTERIA FROM HYPERSALINE SEDIMENTS

Only recently, halophilic methanogenic bacteria have been isolated in pure culture. Mathrani and Boone [21] isolated from a solar salt pond a methanogenic bacterium growing in salt concentrations from 1 to 3.5 M (optimum 2.1 M at 37°C). The strain grows only on methylamines and methanol. A strain of irregular cocci isolated from the Great Salt Lake [22] has a similar substrate specificity; it grows on methanol, methylamine, dimethylamine, and trimethylamine, but not on hydrogen and carbon dioxide, formate, or acetate. Optimal growth was observed at 1 to 2 M NaCl and 35°C (range of growth 0.5 to 3 M salt and 25 to 40°C). Another strain (*"Methanococcus halophilus"*) growing on methanol or methylamines was isolated by Zhilina [23] from a cyanobacterial mat from the hypersaline Hemelin Pool (Shark Bay, Australia). This isolate failed to grow on hydrogen and carbon dioxide or on formate, but after a long adaptation period it proved able to grow on acetate. It grew at NaCl concentrations between 1.5 and 15%, with an optimum at 7% at 26 to 36°C. Several other strains of methanogens were isolated from hypersaline lagoons in Crimea, USSR, growing in salinities of up to 30% and using methylamines as sole substrates for growth, being unable to metabolize hydrogen and carbon dioxide, acetate, or formate [24].

A methanogenic pleomorphic coccus able to grow on hydrogen and carbon dioxide or on formate has been isolated from a solar saltern [25]. No growth was observed on tri-methylamine, methanol, or acetate (although acetate appeared to be required for growth). It grew above 1.8 M salt (optimally at 2.7 M), and its optimal growth temperature was 35°C.

The sulfur springs on the shore of the Dead Sea near Ein Gedi (see below) may be a promising site to search for the presence of methanogenic bacteria. Gas bubbles rise from the pools formed by the springs, and methane is present in these bubbles. A single analysis of the gas performed by us showed a composition of approximately 88% nitrogen and 12% methane. No attempts have yet been made to assess the biogenic origin of this methane, to measure methanogenesis rates in situ, or to isolate methanogenic bacteria from the site.

9.7 DISTRIBUTION AND ECOLOGY OF ANAEROBIC HALOPHILIC BACTERIA

Although anaerobic sediments of hypersaline water bodies are often characterized by a very high organic matter content, and thus an abundant development of anaerobic halo-philes may be expected, surprisingly few data are available on the distribution of anaerobic halophilic bacteria and on their activities in situ. In anaerobic sediments from the bottom of the Dead Sea and on its shores, long rod-shaped bacteria of the type *H. halobius* are quite abundant, and by means of serial dilutions in growth medium we esti-mated numbers between 10^3 and 10^5 viable cells per gram of sediment [6]. On the abundance of *C. lortetii* and of sporulating anaerobes resembling strain DY-1, no quanti-tative data are available. Morphological types like those shown in Fig. 9.2D may be wide-spread in anaerobic sediments throughout the world; similar cells were observed in anaerobic mud from Australia, containing 21% NaCl (F. J. Post, personal communication, 1984).

The salinity of the Dead Sea water and sediments is much higher than the optimal values for the anaerobic bacteria isolated thus far from the Dead Sea, and this no doubt precludes high rates of degradation of organic matter in its sediments. This is evidenced by large amounts of accumulated organic carbon in the sediments (0.23 to 0.4% by weight). Even substrates that are easily degraded by many microorganisms, such as amino acids, were found in large concentrations (in oxidized sediments 60 to 120 mg/kg dry sediment, in reduced sediments 750 to 790 mg/kg, which is 8 to 12% of the total organic carbon, or 16 to 24% when humic and fulvic acids are not taken into account [26]).

H. praevalens is present in Great Salt Lake sediments in extremely large numbers: in sediments of the south arm of the lake more than 10^8 cells per milliliter were counted [8,17]. Using ^{14}C-labeled substrates, Zeikus [17] could demonstrate high rates of micro-bial decomposition of simple organic compounds in these sediments (e.g., in 11 days more than 20% of the label of [2-^{14}C]acetate, [U-^{14}C]glucose, and [3-^{14}C]lactate was found as $^{14}CO_2$). From these substrates no $^{14}CH_4$ was formed. The only substrates giving rise to the formation of methane were methyl mercaptan, methionine, and methanol [17,27]. No significant methanogenesis was observed from acetate or hydrogen and carbon dioxide, and counts of methanogenic bacteria, growing in a medium with methanol, hydrogen, and carbon dioxide, were as low as 10 cells/mL. In the Great Salt Lake hydrogen production is much less tightly coupled to its consumption than in freshwater environments: hydrogen is present in the sediments (salinity greater than 20%) in concentrations of up to 200 μM, while methane concentrations are as low as 4 μM [17]. In the Solar Lake (Sinai), methanogenesis was demonstrated in sediments of relatively low salinities (7 to 7.4%) [28], and up to 10^5 methanogenic bacteria per milliliter of sediment were counted. *Methanosarcina*-like organisms predominated, preferring methylamines as substrates.

9.8 SALT RELATIONSHIPS OF ANAEROBIC HALOPHILIC BACTERIA

The bacteria described above share a relatively high requirement for salt and tolerate high salt concentrations. However, it should be kept in mind that the range of salt concentrations tolerated differs from organism to organism, and this fact will have a profound impact on any attempt to use highly saline environments in biotechnological projects involving anaerobic breakdown of organic matter. This means that salt concentrations will have to be carefully controlled. All fermentative obligate anaerobic halophiles known thus far are moderate halophiles, which grow optimally at salt concentrations between 1 and 4 M NaCl, depending on the strain (Fig. 9.4). No anaerobes have yet been isolated that are able to grow well in saturated salt solutions (around 5.2 M NaCl), and thus under salt saturation no extensive degradation of organic matter can be expected.

We have studied the mechanisms enabling the anaerobic halophilic bacteria to withstand the high salt concentrations in the medium, in view of the fact that different groups of halophilic microorganisms possess different mechanisms to balance the cell contents osmotically with the external medium. For example, aerobic extremely halophilic archaebacteria (the genera *Halobacterium* and *Halococcus*) maintain high salt concentrations within the cells, a high sodium chloride concentration in the external medium being

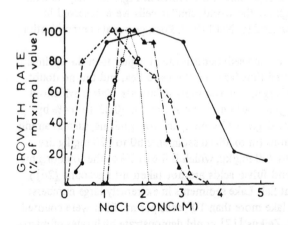

Fig. 9.4 Effect of NaCl concentration on the growth of *H. halobius* (▲), *C. lortetii* (○), *H. praevalens* (●), and strain DY-1 (△). (Data from Refs. 6 and 7 and our unpublished results.)

balanced by a high potassium chloride concentration inside. A different strategy of adaptation to high salt concentrations is found in halophilic and halotolerant eubacteria, halophilic cyanobacteria, and eukaryotic algae such as *Dunaliella*: In these organisms the enzymatic processes within the cells are inhibited by high salt concentrations, which precludes the existence of such high salt concentrations within the cells. Different organic "compatible solutes" have been shown to maintain an osmotic balance of the cell cytoplasm with the outside medium, while still permitting a high level of enzymatic activity. The obligately anaerobic, moderately halophilic eubacteria *H. praevalens* and *H. halobius* were found to contain high intracellular potassium concentrations (0.76 to 2.05 M, not well correlated with the external NaCl concentration), and high intracellular sodium concentrations (0.28 to 2.6 M, increasing with increasing extracellular sodium concentration). The sum of intracellular potassium and sodium concentrations approximated the total cation concentration of the medium, and internal chloride concentrations in *H. praevalens* equaled the external chloride concentration. No organic "compatible solutes" were found in significant concentrations [29].

Not only is the overall salt concentration important in determining the potential growth rates and activities of halophilic microorganisms, but the composition of the salt mixture is of no less importance. Hypersaline water bodies differ in the relative amounts of mono- and divalent cations. Thus the salt composition of the Great Salt Lake and of solar salt ponds reflects that of seawater, in which monovalent cations (especially sodium) dominate. Dead Sea water is extremely rich in divalent cations: around 1.8 M magnesium and 0.4 M calcium, in addition to 1.74 M sodium and 0.2 M potassium. Thus any organism living in the Dead Sea should be able to tolerate divalent cation concentrations that high. Studies of *Halobacterium* isolates from the Dead Sea (red aerobic chemoorganotrophic bacteria) showed that these not only tolerate high divalent cation concentrations, but even require them in much larger amounts than are required by comparable strains isolated from other hypersaline environments [30]. The anaerobic halophiles isolated thus far from Dead Sea sediments do not display such a marked requirement for divalent cations (*H. halobius*, *C. lortetii*, and strain DY-1 grow well at magnesium concentrations as low as 20 mM), and they are not extremely magnesium tolerant (*H. halobius* is inhibited above 1 M magnesium and does not grow above 1.5 M, *C. lortetii* grows optimally at 0.25 M magnesium and almost no growth was found at 0.75 M, DY-1 grows at magnesium concentrations of up to 0.9 M), which may explain low activities of these bacteria in situ in Dead Sea sediments.

In any case, it is obvious that degradation of organic matter in hypersaline environments, a process involving collaboration of a number of bacteria, will proceed efficiently only when the salt concentration and composition is in the range tolerated well by all participating organisms.

9.9 NEW ENRICHMENT EXPERIMENTS FOR ANAEROBIC HALOPHILIC BACTERIA FROM THE DEAD SEA AREA

9.9.1 Sampling Sites

Two sites on the shore of the Dead Sea were selected as sources of inocula for enrichments of anaerobic halophilic bacteria, and these are indicated in the map shown in Fig. 9.5. The first site chosen is a series of sulfur springs on the western shore of the lake a few kilometers south of Ein Gedi, a site known as Hamei Mazor. Figure 9.6 shows a picture of

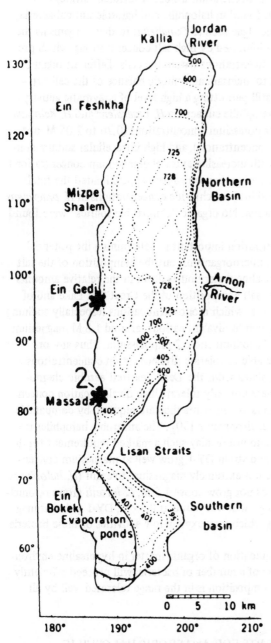

Fig. 9.5 Map of the Dead Sea, indicating the sampling sites: (1) sulfur springs near Ein Gedi, and (2) mud flats on the shore near Massada.

Fig. 9.6 Overview of the sulfur spring area on the shore of the Dead Sea near Ein Gedi.

the site. The water forms a number of pools on its way to the Dead Sea, and these are characterized by a rich microbial flora: mats of green cyanobacteria (*Phormidium*?) and red photosynthetic bacteria (*Chromatium*-like), while other areas are colored black (metal sulfides) or white (elemental sulfur). Ciliate protozoa, unicellular green algae (*Dunaliella*), and unicellular cyanobacteria are also present. Photosynthesis in the cyanobacterial mat is of the anoxygenic type, not sensitive to the inhibitor of oxygenic photosynthesis DCMU [31,32].

The water of the spring is highly saline: its specific gravity was around 1.11 to 1.115 g/cm^3, and it contained 189.2 g/L dry weight of salts (October 30, 1985). Analysis of the water (sampled April 17, 1986) showed the following composition: 1.34 M Na$^+$, 0.08 M K$^+$, 2.86 M Cl$^-$, and 0.72 M divalent cations (compare the analysis of Dead Sea surface water sampled on the same day: 1.707 M Na$^+$, 0.205 M K$^+$, 6.0 M Cl$^-$, and 2.044 M divalent cations. Na$^+$ and K$^+$ were determined by flame photometry, Cl$^-$ by titration with AgNO$_3$, and the divalent cation content was calculated by difference (note that the above-mentioned analysis of the spring water will account for only about 155 g of dissolved salts per liter).

The temperature of the water at the source was 39°C, its pH between 5.4 and 5.8, and it contained about 2.6 mM sulfide (as assayed by adding excess iodine and titrating with thiosulfate). No dissolved oxygen was found in the water sampled near the source, using a modified Winkler titration as described by Ingvorsen and Jørgensen [33].

The second site chosen as a source of inocula for anaerobic enrichments is an area on the western shore near Massada, just south of the pumping station of the Dead Sea

Fig. 9.7 Overview of the sampling site near Massada.

Works, Ltd. The site is characterized by a number of shallow pools of Dead Sea water, overlaying black mud (Fig. 9.7). This area was studied earlier by Yogev [16], who measured sulfate reduction rates and attempted to isolate halophilic sulfate-reducing bacteria. The obligately anaerobic halophilic spore-forming strain DY-1 (see above) was isolated from this site. No detailed analyses of the water were performed, but the salt concentration and composition should have been approximately identical to that of Dead Sea water.

9.9.2 Enrichment Culture Studies

Using black mud samples, in most cases from the sulfur spring site, as an inoculum, anaerobic enrichment cultures were set up using different organic compounds as the carbon and energy source. The basic medium used in most experiments contained 8.8, 14, or 15% NaCl as indicated, 0.37% KCl, 2% $MgCl_2 \cdot 6H_2O$, 0.74% $CaCl_2 \cdot 2H_2O$, 0.1 or 0.08% yeast extract, 0.0001% resazurin, 0.05% L-cysteine·HCl, and PIPES buffer to a final concentration of 25 mM, pH 7.0. Anaerobic techniques such as described by Balch et al. [34] were used throughout, and these included boiling of the media under nitrogen, use of reducing agents, and use of an anaerobic glove box to dispense media into tubes and to inoculate plates. The cysteine was added after boiling the medium under nitrogen, and PIPES and the different organic substrates used (see below) were added from separately autoclaved anaerobic concentrated solutions. Substrates were added generally to a final concentration of 0.5%, unless specified otherwise. In most experiments 10-mL portions of medium were used in 25-mL stoppered tubes under a gas phase of nitrogen. In other experiments the enrichment medium consisted of 50% Dead Sea water, enriched with 0.1% yeast extract and cysteine, resazurin, and organic substrates as above. Cultures were incubated in the dark at 28 or 37°C, and the turbidity developing was compared to that in control tubes that did not receive organic substrates except the small concentration of yeast extract and cysteine present in all tubes.

9.9.3 Studies on Isolated Strains

Organisms developing in the enrichment cultures were purified by steaking on agar plates, composed of the growth medium with the addition of 2% agar; plates were inoculated and incubated at room temperature in an anaerobic glove box. Alternatively, isolations were attempted aerobically, using a similar medium, but without resazurin and cysteine. Further work on isolated strains was performed as specified below.

9.9.4 Results

Enrichment Cultures

The first series of enrichment cultures using black mud from the sulfur springs near Ein Gedi was set up, using medium containing 8.8% NaCl, and tubes were incubated at 28°C. After 2 to 3 days of dense growth was observed of motile curved rod-shaped bacteria in media enriched with glucose, sucrose, starch, or lactose, after 4 days of incubation also on Na-L-glutamate, L-arginine, or casamino acids. On gelatin as substrate, growth remained slight; on 0.05% phenol, some growth was observed only after 6 days of incubation, which was probably due to growth at the expense of the yeast extract in the medium rather than the phenol. When the culture was transferred to a similar medium with 0.1 or 0.05% phenol or 0.5% Na-benzoate, no growth at all was observed in the presence of 0.1% phenol, and in the other tubes turbidity did not exceed that observed in the control tube without added substrate. Pure cultures were established of the bacteria developing on sucrose and on casamino acids (indicated below as strains SUC-1 and CAS-1).

Using the same inoculum as above, enrichments were set up using medium containing 50% (v/v) Dead Sea water, enriched with 0.1% yeast extract, cysteine, resazurin, and organic substrates as above. On glucose, sucrose, and starch, excellent growth was obtained of very long, slender, rod-shaped bacteria, known from our previous work as *H. halobius* [6]. On arginine or casamino acids, motile curved rods developed as above. Growth was slight on glutamate, lactose, and gelatin, and no growth at all was observed in the presence of 0.1% phenol.

Another series of enrichment cultures was set up with the same source of inoculum, using synthetic medium with 15% NaCl and 0.1% yeast extract, and tubes were incubated at 35°C. After 2 to 3 days of incubation, good growth of quite large motile rods was observed on glucose, sucrose, and starch. The dominating bacterium growing on starch was isolated (strain ST-1). Even after 5 to 6 days of incubation, no significant growth (compared to the control without substrate) was observed on lactose, casamino acids, glutamate, and L-arginine, and no growth at all in a medium supplemented with 0.02% phenol or with 0.5% Na-benzoate.

The last series of enrichments was made in synthetic medium with 14% NaCl, using as inoculum mud sampled from the Ein Gedi site. After 2 days of incubation at 37°C, profuse growth of slowly moving rod-shaped bacteria was observed on glucose, sucrose, raffinose, fructose, and arabinose. Growth slightly denser than in the control tube without added substrate was seen on casamino acids, tryptone, and D-ribose after 2 to 3 days, while no significant enhancement of growth was found with rhamnose, lactose, L-gluta-mate, L-arginine, 0.04% phenol, or 0.05% Na-benzoate. No growth at all was observed in the presence of 0.1% phenol or 0.5% Na-benzoate.

Studies with Isolated Strains

One isolate from the Massada salt flats, designated strain DY-1, was studied in detail. This strain is an obligate anaerobe, able to produce endospores. Its properties were described in part in the previous sections, and a full description of the organism will be published elsewhere.

The three strains isolated from the enrichment cultures described above (CAS-1, SUC-1, and ST-1) were characterized further. Strain CAS-1 appeared to be a facultative anaerobe, able to grow well at 28°C in aerobic medium containing 8.8% NaCl and other salts as used in the enrichment medium, 0.5% yeast extract, 10 mM PIPES, 0.5% sucrose, or casamino acids. The organism is gram negative, oxidase and catalase positive, does not produce indole, and makes no gas or nitrite from nitrate. The bacterium proved to be sensitive to vibriostatic agent 0/129. Starch is not digested, and acid, but no gas, is formed from mannose, glucose, sucrose, and mannitol but not from arabinose (the test medium contained 10% NaCl, 2% $MgCl_2 \cdot 6H_2O$, 0.74% $CaCl_2 \cdot 2H_2O$, 0.37% KCl, 0.1% beef extract, 1% peptone, 0.0018% phenol red, pH 7.4, in test tubes provided with Durham tubes). Strain CAS-1 grows well in NaCl concentrations from 1.25 to 16.25% (0.2 to 2.8 M), and only poorly at 3 M (at 25°C) (in addition to varying concentrations of NaCl, the medium contained 0.5% yeast extract, 1% $MgCl_2 \cdot 6H_2O$, 0.5% $CaCl_2 \cdot 2H_2O$, 0.25% KCl, pH 7, and 0.5% sucrose, added from a separately autoclaved concentrated solution). All the foregoing properties fit the species description of *V. costicola* [35].

Strain SUC-1 proved to be very similar to strain CAS-1 in all properties tested: It grows well in medium with casamino acids, and so does strain CAS-1 in medium with sucrose.

Strain ST-1 too proved to be a facultative anaerobe. While isolated anaerobically on plates containing 15% NaCl, 20 mM PIPES, 0.5% glucose, 2% $MgCl_2 \cdot 6H_2O$, 0.74% $CaCl_2 \cdot 2H_2O$, 0.37% KCl, 0.5% yeast extract, 0.05% cysteine·HCl, 0.0001% resazurin, and 2% agar, the isolate grew well aerobically on 15% NaCl-glucose plates similar to the medium above but without cysteine and resazurin. The bacterium is a motile rod, and is oxidase negative and catalase positive. No gas or acid is produced from glucose, sucrose, mannitol, or L-arabinose (using a medium containing 15% NaCl, 2% $MgCl_2 \cdot 6H_2O$, 0.74% $CaCl_2 \cdot 2H_2O$, 0.37% KCl, 0.1% beef extract, 1% peptone, 0.0018% phenol red, and 0.5% of the sugar tested). No further attempts were made to characterize the isolate.

9.10 CONCLUSIONS

It is becoming increasingly clear that even such extreme environments as the anaerobic sediments of hypersaline water bodies are inhabited by a variety of microorganisms. A number of fermentative obligate anaerobes have now been isolated and characterized. However, obviously not all types of anaerobic halophiles are known. The existence of halophilic methanogens and halophilic sulfate-reducing bacteria suggests that even at salinities approaching saturation, complete anaerobic degradation of organic compounds may be possible, with the participation of a variety of bacteria: polymer degraders, fermenters, sulfate reducers, and methanogens. In nature such a complete degradation is not always realized, as witnessed by the high content of organic matter in sediments of hypersaline water bodies. Hypersaline evaporitic environments have even been implied as the potential sites of hydrocarbon formation leading to the accumulation of oil [36].

The microbiology of hypersaline environments has not yet been studied sufficiently, and data on the distribution of the microorganisms thriving in them and their activities in situ are mostly lacking. There is no doubt that a wealth of novel and interesting microorganisms is waiting to be discovered in such habitats. Their isolation and characterization should complete our understanding of the ecology of hypersaline water bodies and enable estimates on the feasibility of the development of biotechnology processes based on anaerobic degradation in the presence of high salt concentrations.

ACKNOWLEDGMENTS

I thank E. Stackebrandt (Kiel), B. J. Paster, W. R. Weisburg, and C. R. Woese (Urbana), who all contributed to the data presented in this paper.

Support for the work described in this chapter was obtained under a grant from Houston Lighting and Power Company under a university participation program administered by Dynatech R/D Company, and a grant from the Israeli Ministry of Energy and Infrastructure.

REFERENCES

1. Lortet, M. L., Researches on the pathogenic microorganisms of the mud of the Dead Sea, *Palest. Expl. Fund., 48* (1892).
2. Oren, A., The ecology and taxonomy of anaerobic halophilic eubacteria, *FEMS Microbiol. Rev., 39,* 23 (1986).
3. Gottschalk, G., The anaerobic way of life of prokaryotes, in *The Prokaryotes: A Handbook on Habitats, Isolation, and Identification of Bacteria,* Vol. 2 (M. P. Starr, H. Stolp, H. G. Trüper, A. Balows, and H. G. Schlegel, eds.), Springer-Verlag, New York, 1981, p. 1415.
4. Steinhorn, I., Assaf, G., Gat, J. R., Nishry, A., Nissenbaum, A., Stiller, M., Beyth, M., Neev, D., Garber, R., Friedman, G. M., and Weiss, W., The Dead Sea: deepening of the mixolimnion signifies the overture to overturn of the water column, *Science, 206,* 55 (1979).
5. Elazari-Volcani, B., Bacteria in the bottom sediments of the Dead Sea, *Nature, 152,* 274 (1943).
6. Oren, A., Weisburg, W. G., Kessel, M., and Woese, C. R., *Halobacteroides halobius* gen. nov., sp. nov., a moderately halophilic anaerobic bacterium from the bottom sediments of the Dead Sea, *Syst. Appl. Microbiol., 5,* 58 (1984).
7. Oren, A., *Clostridium lortetii* sp. nov., a halophilic obligatory anaerobic bacterium producing endospores with attached gas vacuoles, *Arch. Microbiol., 136,* 42 (1983).
8. Zeikus, J. G., Hegge, P. W., Thompson, T. E., Phelps, T. J., and Langworthy, T. A., Isolation and description of *Haloanaerobium praevalens* gen. nov. and sp. nov., an obligately anaerobic halophile common to Great Salt Lake sediments, *Curr. Microbiol., 9,* 225 (1983).
9. Baumgartner, J. G., The salt limits and thermal stability of a new species of anaerobic halophile, *Food Res., 2,* 321 (1937).
10. Oren, A., Paster, B. J., and Woese, C. R., Haloanaerobiaceae: a new family of moderately halophilic, obligatory anaerobic bacteria, *Syst. Appl. Microbiol., 5,* 71 (1984).
11. Paster, B. J., Stackebrandt, E., Hespell, R. B., Hahn, C. M., and Woese, C. R., The phylogeny of the spirochetes, *Syst. Appl. Microbiol., 5,* 337 (1984).

12. Hartmann, R., Sickinger, H.-D., and Oesterhelt, D., Anaerobic growth of halobacteria, *Proc. Natl. Acad. Sci. USA, 77*, 3821 (1980).

13. Gonzalez, C., Gutierrez, C., and Ramirez, C., *Halobacterium vallismortis* sp. nov., an amylolytic and carbohydrate-metabolizing, extremely halophilic bacterium, *Can. J. Microbiol., 24*, 710 (1978).

14. Lerman, A., Model of chemical evolution of a chloride lake: the Dead Sea. *Geochim. Cosmochim. Acta, 31*, 2309 (1967).

15. Nissenbaum, A., and Kaplan, I. R., Sulfur and carbon isotopic evidence for biochemical processes in the Dead Sea ecosystem, in *Environmental Biogeochemistry*, Vol. 1 (J. O. Nriagu, ed.), Ann Arbor Science, Ann Arbor, Mich., 1976, p. 309.

16. Yogev, D., Sulfate reduction in anoxic sediments of the Dead Sea, M.Sc. thesis, The Hebrew University of Jerusalem, 1983 (in Hebrew).

17. Zeikus, J. G., Metabolic communication between biodegradative populations in nature, in *Microbes in Their Natural Environments*, Symposium 34 (J. H. Slater, R. Whittenbury, and J. W. T. Wimpenny, eds.), Society for General Microbiology, Cambridge University Press, Cambridge, 1983, p. 423.

18. Klug, M., Boston, P., Francois, R., Gyure, R., Javor, B., Tribble, G., and Vairavamurthy, A., Sulfur reduction in sediments of marine and evaporite environments, in *The Global Sulfur Cycle* (D. Sagan, ed.), NASA Technical Memorandum 87570, National Aeronautics and Space Administration, Washington, D.C., 1985, p. 128.

19. Jørgensen, B. B., and Fenchel, T., The sulfur cycle of a marine sediment model system, *Mar. Biol., 24*, 189 (1974).

20. Trüper, H. G., Bacterial sulfate reduction in the Red Sea hot brines, in *Hot Brines and Recent Heavy Metal Deposits in the Red Sea* (E. T. Degens and D. A. Ross, eds.), Springer-Verlag, New York, 1969, p. 263.

21. Mathrani, I. M., and Boone, D. R., Isolation and characterization of a moderately halophilic methanogen from a solar saltern, *Appl. Environ. Microbiol., 50*, 140 (1985).

22. Paterek, J. R., and Smith, P. H., Isolation and characterization of a halophilic methanogen from Great Salt Lake, *Appl. Environ. Microbiol., 50*, 877 (1985).

23. Zhilina, T. N., New obligate halophilic methane-producing bacterium, *Microbiology (Engl. Transl.), 52*, 290 (1983).

24. Zhilina, T. N., Methanogenic bacteria from hypersaline environments, *Syst. Appl. Microbiol., 7*, 216 (1986).

25. Yu, I. K., and Hungate, R. E., Isolation and characterization of an obligately halophilic methanogenic bacterium, presented at *Annual Meeting of the American Society for Microbiology*, New Orleans, La., 1983.

26. Nissenbaum, A., Baedecker, M. J., and Kaplan, I. R., Organic geochemistry of Dead Sea sediments, *Geochim. Cosmochim. Acta, 36*, 709 (1977).

27. Phelps, T., and Zeikus, J. G., Microbial ecology of anaerobic decomposition in Great Salt Lake, presented at *Annual Meeting of the American Society for Microbiology*, Miami Beach, Fla., 1980.

28. Giani, D., Giani, L., Cohen, Y., and Krumbein, W. C., Methanogenesis in the hypersaline Solar Lake (Sinai), *FEMS Microbiol. Lett., 25*, 219 (1985).

29. Oren, A., Intracellular salt concentrations of the anaerobic halophilic eubacteria *Haloanaerobium praevalens* and *Halobacteroides halobius, Can. J. Microbiol., 32*, 4 (1986).

30. Edgerton, M. E., and Brimblecombe, P., Thermodynamics of halobacterial environments, *Can. J. Microbiol., 27*, 899 (1981).

31. Cohen, Y., Padan, E., and Shilo, M., Facultative anoxygenic photosynthesis in the cyanobacterium *Oscillatoria limnetica, J. Bacteriol., 123*, 855 (1975).

32. Garlick, S., Oren, A., and Padan, E., Occurrence of facultative anoxygenic photo-synthesis among filamentous and unicellular cyanobacteria, *J. Bacteriol., 129,* 623 (1977).

33. Ingvorsen, K., and Jørgensen, B. B., Combined measurement of oxygen and sulfide in water samples, *Limnol. Oceanogr., 24,* 390 (1979).

34. Balch, W. E., Fox, G. E., Magrum, L. J., Woese, C. R., and Wolfe, R. S., Methano-gens: reevaluation of a unique biological group, *Microbiol. Rev., 43,* 260 (1979).

35. Shewan, J. M., and Véron, M., *Vibrio,* in *Bergey's Manual of Determinative Bac-teriology,* 8th ed. (R. E. Buchanan and N. E. Gibbons, eds.), Williams & Wilkins, Baltimore, 1974, p. 340.

36. Kirkland, D. W., and Evans, R., Source-rock potential of evaporitic environment, *Am. Assoc. Pet. Geol. Bull., 65,* 181 (1981).

This article was completed in September 1986; for reasons beyond the author's control it has not been possible to update it in accordance with recent developments in the field.

10

Chemical Methods of Enhancing the Biodegradation of Lignite

JOHN W. WANG[*], HENRIQUE C. G. DO NASCIMENTO[†], REBECCA S. HSU-CHOU[‡],
and TEH FU YEN *University of Southern California, Los Angeles, California*

10.1 INTRODUCTION

Coal is a major energy source, surpassed only by crude oil. Nevertheless, the use of coal is a real and potential threat to the environment since undesirable by-products generated by combustion are likely to be released into the atmosphere, water, and soil. In the interest of obtaining clean fuel, liquefaction and gasification of coal gained increasing attention in the 1960s and 1970s [1]. However, the massive use of solvents, high pressures, high temperatures, and expensive devices to withstand severe processing conditions make conventional liquefaction and gasification of coal economically unfeasible.

Since conventional processes have been proven to be costly, biological processes for coal liquefaction and gasification have become more and more attractive. In general, bio-liquefaction and biogasification of coal seem feasible under conditions that are less energy consuming, simpler, and cheaper than those of their conventional counterparts.

Microbial oxidation of coal has been strongly supported by early laboratory works in which biodegradation of coal was performed. Current studies have also shown that fungi partially degrade crushed lignite, yielding a black liquid that usually does not appear in cultures unless fungi and lignite are both present [2].

Despite early and recent efforts, bioliquefaction of coal is still incipient. In most environments and processing conditions, resistence of coal to biodegradation can be attributed to its stable molecular structure. Nonporous chemical structure and the presence of a limited number of functional groups are thought to be major factors contributing to the recalcitrant character of coals. Therefore, it seems desirable to use chemical methods

Current affiliations:
[*]Management Division, Department of Public Works, Los Angeles, California
[†]Engineering Division, South Coast Air Quality Management District, El Monte, California
[‡]Gtel Environmental Labs, Inc., Southwest Region, Torrence, California

to accelerate and to direct the breakdown of lignite into biodegradable compounds. This suggests that by providing both the necessary functional groups for the macromolecule of coals and changing their chemical structure, biodegradation will be optimized. The present work is an attempt to improve the biodegradable properties of lignite, since this low-rank coal is an abundant and promising substrate for producing useful chemicals through the application of biotechnology.

Taking into account the points raised here; the present research consists of three parts that were executed as a preliminary effort designed to optimize of the biodegradation of lignite. During the first stages of this work, various types of chemical oxidation methods were carried out to increase the yield of oxidized products from lignite and also possibly to improve the efficiency of biodegradation, since many of these products are known to be more acceptable than raw lignite to biological activity. The experimental oxidation methods were the following: (1) potassium permanganate, (2) sodium hypochlorite, (3) alkaline cupric, and (4) hydrogen peroxide-acetic acid.

After executing the foregoing methods, new ideas evolved and a new concept of incorporating biodegradable functional groups into the recalcitrant macromolecule of lignite was applied. Nitrogen enrichment of lignite was a preliminary attempt toward this new concept.

Another recent concept developed during the course of this research was chemical enhancement of lignite solubility into water. Modification of lignite by sodium metal and methanol was carried out to increase the conversion of water-soluble products derived from lignite. These products were thought to exhibit higher biodegradable characteristics than those of raw lignite.

Future experimental work toward the continuation and improvement of present research includes the study of the structure of metabolites generated through the biodegradation of lignitic substrates. Another area that deserves further development includes the isolation and identification of microbial species present in the soil, which yielded positive results for the biodegradation of methanol fraction of lignite and also for nitrogen-enriched lignite. The latter treatment needs to be redesigned to minimize such undesirable transformations in the structure of lignite that are generated in oxidation processes. Ideally, incorporation of desirable functional groups into the chemical structure of lignite is to be achieved with no loss of substrate and in such a way that the maximization of the process can be assessed.

10.2 OXIDATION

Biodegradation of low-rank coal such as lignite is not easy. Nevertheless, based on evidence obtained from chemical degradation of lignites and proposed features for its structure, it is speculated that biodegradation of lignite is feasible [3-5]. Since coal is believed to be of biological origin [6], it seems logical to think that certain portions of coal are biodegradable. If we can convert lignite into a form in which its original inhibitory factors can be removed, biodegradation would then be very possible.

Chemical oxidations have been used to study coal structure for a long time [4,6]. The relatively small compounds derived from lignite were found to be at least partially degradable. In order to investigate the biodegradable components in lignite matrix, various oxidation methods have been employed to study the substances derived from lignite.

10.2.1 Potassium Permanganate and Sodium Hypochlorite Oxidation [12]

Alkaline oxidation of lignitic coal is known to promote the breakdown of the chemical structure of lignite, producing high-molecular-weight organic acids [3,7]. These acids can be further oxidized to low-molecular-weight organic acids and carbon dioxide.

Although strong oxidation causes a significant loss of carbon as carbon dioxide, making it inefficient as a pre-process for biodegradation, potassium permanganate and sodium hypochlorite oxidation of lignite serve as a preliminary test having the following purposes: (1) to determine the concentrations of organic acids produced by oxidation in a solution of potassium permanganate or sodium hypochlorite, (2) to quantify the amount of lignitic carbon converted to both organic acids and carbon dioxide, and (3) to perform semiquantitative analysis of carbon dioxide.

Experimental Procedures

A lignite sample from Pennsylvania State University (PSOC-246) was ground and sieved. Lignite particles smaller than 100 mesh were collected for the oxidation experiments. Two types of experiments were carried out:

1. Gradual addition of potassium permanganate
2. Gradual addition of sodium hypochlorite

Experiments 1 and 2 were performed in a four-necked flask equipped with a pH electrode, a thermometer, a condenser connected to a KOH trap, and an additional oxidant funnel. They were carried out under similar experimental conditions. The procedure for these two experiments was as follows: 5 g of lignite sample was added to 500 mL of 10^{-2} M KOH solution, followed by addition of 3 drops of Tween-80 nonionic surfactant to wet the surfaces of lignite particles. This suspension was heated to 80 to 90°C. While this temperature range was maintained, known volumes of oxidant were gradually added to the alkaline suspension of lignite.

Oxidizing solutions were (1) 6 g of $KMnO_4$ in 100 mL of purified water, and (2) NaOCl solution with 5% minimum available chlorine for experiments 1 and 2, respectively.

Quantitative analyses of low-molecular-weight organic acids (oxalic acid, acetic acid, and formic acid) derived from previously described sodium hypochlorite and potassium permanganate oxidation were performed by a Dionex 2000i ion chromatograph. Semiquantitative analysis of CO_2 as CO_3^{2-} was carried out by trapping CO_2 in 1 M KOH solution, followed by ion chromatography.

To evaluate the conversion of lignite carbon to oxidized products, it was assumed that the organic content of lignite coal was 50%, of which 70% was carbon [10]. The following equation is used to determine the percent conversion of lignite carbon to organic acids:

$$L = \frac{A \times B}{C \times 0.7 \times 0.5}$$

where L is the percent conversion of lignite carbon to organic acids, A the organic acid concentration (mg/L), B the weight percent of carbon in a given organic acid, and C the concentration of lignite coal (mg/L).

Carboxylic acids derived from $KMnO_4$ oxidation of lignite were extracted and concentrated in organic solvent. The esterification of concentrated acid extracts followed the method published by Metcalfe and Schmitz [8]. After sample preparation, gas chromatographic analysis was conducted to determine carboxylic acids derived from oxidation of lignite.

Results and Discussion

To illustrate ion chromatography, initial and final chromatograms obtained from the experiment of the gradual addition of sodium hypochlorite are shown in Figs. 10.1 and 10.2. Oxidation of lignite by addition of sodium hypochlorite was found to yield the highest concentration of oxalic acid and formic acid (Table 10.1). Production of acetic acid was not detected by addition of sodium hypochlorite, whereas addition of potassium permanganate yielded concentrations of acetic acid greater than those for formic acid (Table 10.2). Production of oxalic acid by one slug of concentrated NaOCl (Table 10.3) was found to be lower than expected as compared to the other two experiments, where the concentration of lignite was more diluted. It was probably necessary to allow longer oxidation time in order to improve the acid production in the experiment using one slug of concentrated NaOCl.

Small concentrations of less than 100 mg/L of carbon dioxide were found in the CO_2 trap, whereas up to 2000 mg/L of carbon dioxide was found in some samples. This suggests that the gaseous CO_2 produced remained in the alkaline lignite suspension.

As shown in Table 10.4, it is possible to attain 10.0% conversion of the lignite carbon to oxalate carbon by sodium hypochlorite oxidation. In all experiments it was estimated that about 43% of the lignite carbon was converted to CO_2.

Despite the relatively small number of carboxylic acids identified, many unidentified carboxylic acids were recorded in the chromatogram obtained from the capillary gas

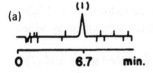

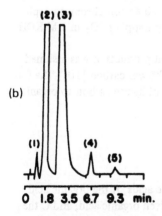

Fig. 10.1 Ion chromatography for oxalic acid; (a) initial [(1) sulfate (used as an internal standard)]; (b) final [(1) unidentified, (2) and (3) sodium hypochlorite oxidizing solution, (4) sulfate (used as an internal standard), (5) oxalic acid].

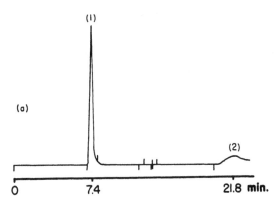

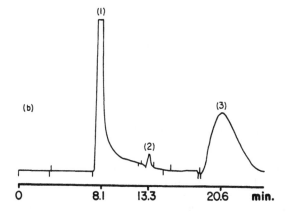

Fig. 10.2   Ion chromatography for formic acid, and carbon dioxide as carbonate; (a) initial [(1) unidentified, (2) carbon dioxide as carbonate] ; (b) final [(1) unidentified plus sodium hypochlorite oxidizing solution, (2) formic acid, (3) carbon dioxide as carbonate].

Table 10.1   Gradual Addition of NaOCl: Oxidation of Lignite at 80–90°C

Total time (h:min)	Total volume of NaOCl solution added (mL)	Oxalate (mg/L)	Formate (mg/L)	pH
0:00	0	30	0	8.6
2:30	100	162	59	11.0
3:18	150	244		11.1
5:00	250	357	125	11.1
5:42	325	472	162	11.1
6:00	425	577	227	11.1
6:30	525	715	305	11.9
8:00	680	943	478	11.7
8:40	845	1264	592	11.5
10:04	1007	1357	623	12.2
11:00	1213	1332	679	12.2

Source: Ref. 12.

Table 10.2 Gradual Addition of $KMnO_4$: Oxidation of Lignite at 80–90°C

Total time (h:min)	Total volume of $KMnO_4$ solution added (mL)	Oxalate (mg/L)	Formate (mg/L)	Acetate (mg/L)	pH
0:00	0	30	0	0	11.1
0:46	66	280	26	23	10.1
1:00	69.5	257	26	0	10.0
1:42	72.5	344	38	46	9.9
3:16	116	487	74	68	9.5
4:00	163.5	604	62	102	9.0
4:42	188.5	646	69	113	8.9
5:00	213.5	659	40	117	8.7
5:25	238.5	595	37	131	8.6
5:44	263.5	505	0	162	8.6

Source: Ref. 12.

Table 10.3 One Slug of Concentrated NaOCl: Oxidation of Lignite at 90–95°C

Lignite sample (g)	Total time (h:min)	Total volume of NaOCl solution added (mL)	Oxalate (mg/L)	Temperature range (°C)	pH
3	0:30	100	497	18–25	13.3
	1:30	100	–	90–95	13.3
	2:30	100	728	35–40	13.3
6	0:30	100	447	18–25	13.6
	1:30	100	–	90–95	13.6
	2:30	100	824	35–40	13.6

Source: Ref. 12.

Table 10.4 Conversion of Lignite Carbon to Organic Acids (% Efficiency)

Process[a]	Oxalate carbon	Formate carbon	Acetate carbon
A	5.1	0.5	1.9
B	10.5	5.0	–
C_1	1.9	–	–
C_2	1.1	–	–

Source: Ref. 12.
[a] A, gradual addition of KMn_4 –oxidation; B, gradual addition of NaOCl–oxidation; C_1, one slug of concentrated NaOCl–oxidation 3 g lignite; C_2, one slug of concentrated NaOCl–oxidation 6 g lignite.

Enhancing Lignite Biodegradation 183

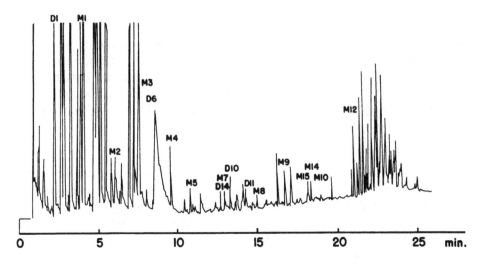

Fig. 10.3 Capillary gas chromatography for carboxylic acids. Table 10.5 lists identified species.

chromatography analysis as shown in Fig. 10.3. Table 10.5 lists the 12 monocarboxylic acids and five dicarboxylic acids that were identified.

Only one benzene carboxylic acid—terephthalic acid—was identified, suggesting that oxidizing conditions were severe enough to break aromatic rings. The variety and abundance of carboxylic acids found in the solution are attractive for biological work, since some microorganisms are known to be capable of growing on carboxylic acids. However, severe carbon loss in the form of carbon dioxide discouraged the application of strong oxidation as a chemical pretreatment for bioliquefaction of lignite.

Table 10.5 Analysis of Esterified Acid Fraction by Capillary Gas Chromatography

Monocarboxylic Acids			
M_1	C_6 Caproic	M_8	C_{14} Myristic
M_2	C_7 Heptanoic	M_9	C_{16} Palmitic
M_3	C_8 Caprylic	M_{10}	C_{18} Stearic
M_4	C_9 Pelargonic	M_{12}	C_{22} Behenic
M_5	C_{10} Capric	M_{14}	C_{16} Oleic
M_7	C_{12} Lauric	M_{15}	C_{18} Linoleic

Dicarboxylic Acids	
D_1	C_2 Oxalic
D_6	C_5 Glutaric
D_{10}	C_9 Azelic
D_{11}	C_{10} Sebacic
D_{14}	C_{10} Terephthalic

Source: Ref. 12.

10.2.2 Fractionation of Lignite and Alkaline Cupric Oxidation

Several investigators have reported that *Polyporus (Trameters) versicolor* is capable of growing on the surface of lignite [2,9-12]. Those findings suggest that lignite can serve as carbon source for biological growth. However, due to many discrepancies found among the results obtained by those investigators, further research is needed to investigate what is responsible for supporting biological growth on the surface of lignites. A fractionation procedure was developed for this investigation.

Fractionation of Lignite

A Beulah standard number 3 lignite from North Dakota was fractionated into three fractions: benzene-methanol fraction (A-1), aqueous alkaline fraction (A-2), and lignite residue (A-3). These three fractions were further subjected to biodegradation tests. Fourier transform infrared spectroscopy (FTIR) was employed to characterize functional groups present in these three fractions [19].

Biological experiments using *P. versicolor* for degrading the fractions obtained from the extraction of lignite and lignite residue showed some significant results. No biological growth was observed on both benzene-methanol fraction (A-1) and lignite residue (A-3). Conversely, the biological experiments using the diluted aqueous alkaline fraction (A-2)

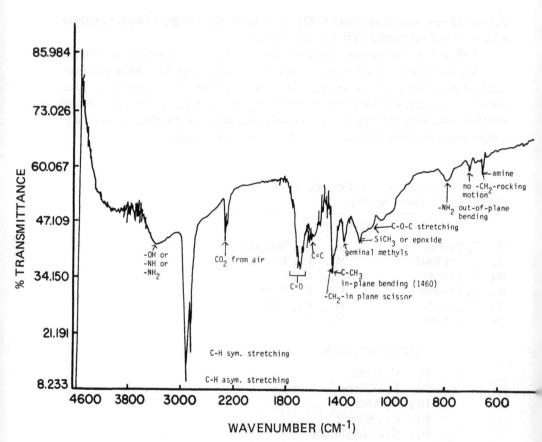

Fig. 10.4 FTIR spectrum of benzene-methanol fraction (A-1).

and raw lignite sample showed positive results after 14 days of incubation. These findings suggest that alkaline-soluble organic compounds trapped in the porous structure of lignite or weakly linked with macromolecular structure of lignite are responsible for sustaining fungal activity [20].

FTIR spectra are shown in Figs. 10.4 to 10.6. It was noted that strong absorptions corresponding to carboxylic acid groups were found in the IR spectrum of the alkaline fraction. It may be speculated that the presence of organic acids supports fungal growth. Nevertheless, it is unlikely that microorganisms can directly attack the core of the complex macromolecule of lignite without the aid of chemical pretreatment to break this complicated macromolecule down into certain subunits that would be more easily utilized as carbon source by microbes. Alkaline cupric oxidation was designed for this purpose.

Alkaline Cupric Oxidation [21]

In general, alkaline copper oxide is a relatively weak oxidant. According to the results of the cupric oxidation of model compounds, it was concluded that only weak linkages of aliphatic side chain in coal structure would be broken [6]. This mild oxidation might produce some acids that can be further degraded by microorganisms.

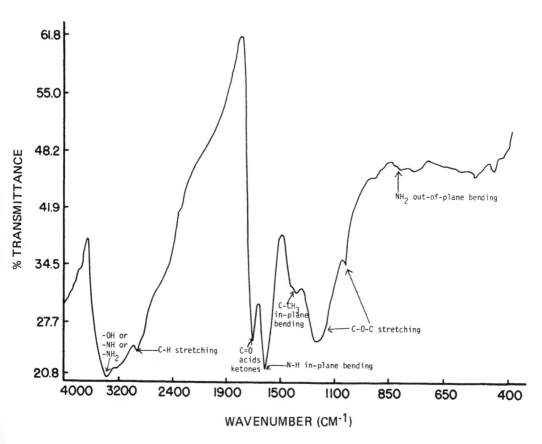

Fig. 10.5 FTIR spectrum of aqueous alkaline fraction (A-2).

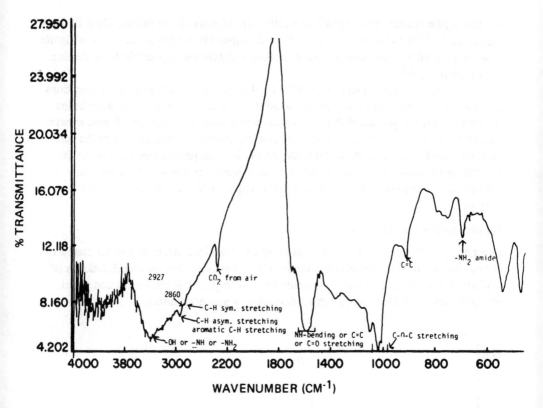

Fig. 10.6 FTIR spectrum of clean lignite residue (A-3).

Experimental Procedure. At first, moisture-free, less than 100 mesh Beulah Standard Number 3 lignite was fractionated. The fractionation procedure is illustrated by the scheme presented in Fig. 10.7. The fractionated lignite was then oxidized with alkaline cupric oxide in a stainless steel autoclave, under a pressure of 15 psi at 120°C for 6 h followed by refluxing for 3 h in open air.

The reaction mixture was separated into liquid and residue by centrifugation followed by filtration. The residue was believed to contain large quantities of copper salt, sodium salt, and a small amount of lignite residue.

The liquid was acidified approximately to pH 1. Organic acids were precipitated from the liquid phase and subsequently collected by centrifugation and filtration. The precipitated slush was extracted with methanol by means of Soxhlet for 40 h to separate methanol soluble material from nonsoluble residue.

The four fractions obtained (see Fig. 10.7) were tested further to calculate the recovery of each fraction. Since organic acids were expected to be present in the products of oxidation, and low-molecular-weight organic acids are likely to be extractable, ion chromatographic analysis of the methanol extractable fraction was executed to determine organic acids semiquantitatively.

Results and Discussions. The recoveries of organic material from each fraction were as follows:

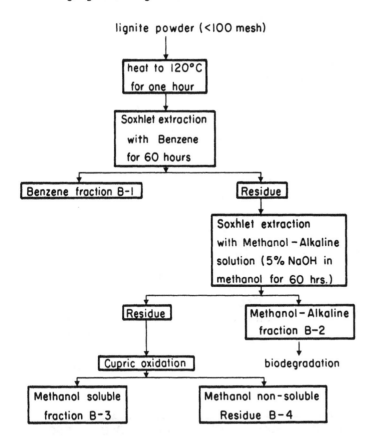

Fig. 10.7 Fractionation procedure.

B-1:	Benzene fraction	0.92%
B-2:	Methanol-alkaline fraction	0.94%
B-3:	Methanol extractable	30.35%
B-4:	Methanol nonextractable	13.99%
Ash		33.00%
Total		79.20%

Twenty percent loss was probably caused by CO_2 formation during oxidation. FTIR spectrum of methanol extractable fraction (B-3) showed strong absorption of OH stretching (3400 to 3200 cm^{-1}) and C=O stretching (1710 cm^{-1}) (see Fig. 10.8). As we know, these two absorption bands are characteristic of carboxylic acids. By comparing Fig. 10.8 with Fig. 10.5, it was found that methanol extractable fraction and alkaline fraction possess similar chemical features. Further biodegradation tests of methanol-soluble fraction yielded positive growth of soil bacteria. It seems that mild oxidation such as cupric oxidation can significantly enhance the biodegradability of certain portions of the macromolecule of lignite. It was also noted that IR spectra of methanol-extractable fraction obtained from cupric oxidation showed weak aromatic absorption around 2000 to 1800 cm^{-1}. IR spectra of alkaline fraction did not show aromatic properties. A reasonable explanation of these different results is that aromatic structures in lignite matrix were stripped out by means of cupric oxidation.

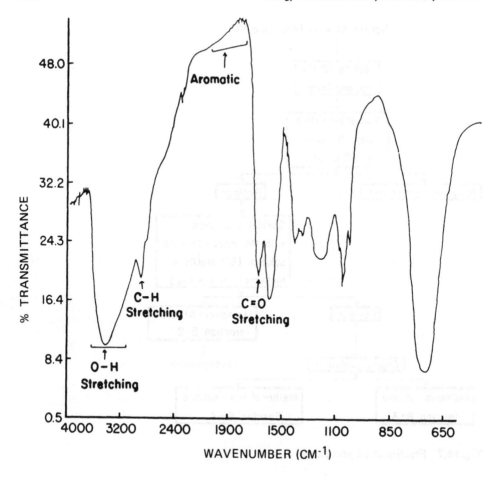

Fig. 10.8   FTIR spectrum of methanol-extractable portion (B-3).

Figure 10.9 shows the ion chromatography spectrum corresponding to lactic acid, formic acid, and acetic acid. The identification of previously mentioned organic acids was confirmed by spiking known compounds. The peak at the elution time of 6.5 ± 0.3 min (peak 1 in Fig. 10.9) is the only unidentified peak. It is believed that this peak is due to the presence of sulfate in the solution. However, it is also possible that some carboxylic acids having high ionic strength, such as benzenetetracarboxylic acid, were responsible for the generation of that peak. The concentrations of identified organic acids present in methanol extractable fraction derived from the cupric oxidation of lignite were as follows: 30 mg/L lactic acid, 60 mg/L formic acid, and 30 mg/L acetic acid.

10.2.3   Hydrogen Peroxide–Acetic Acid Oxidation [22]

Studies on chemical oxidation of lignite have suggested that strong oxidation results in massive carbon loss. Mild oxidation, such as cupric oxidation, significantly reduces carbon loss (20% carbon loss). Nevertheless, the yield of methanol-soluble fraction, the one of interest for its positive results regarding bacterial growth, has not been significant (about 30%).

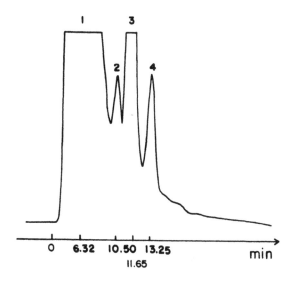

Fig. 10.9 Ion chromatography of the aqueous fraction of modified cupric oxidation; (1) sulfate and/or unidentified species, (2) lactic acid, (3) formic acid, (4) acetic acid.

To overcome this dilemma, it is necessary to use oxidizing agents that are strong enough to decompose lignite generating a high yield of methanol-soluble fraction, and at the same time capable of reducing carbon loss. It has been reported that an acetic acid–hydrogen peroxide mixture is able to oxidize humic acids without destroying phenolic groups [13]. Studier et al. [4] followed this method and obtained 80% conversion of lignite into methanol-soluble acids.

Following the procedure proposed by Schnitzer and Skinner [13] with slight modification, hydrogen peroxide–acetic acid oxidation was executed. A 76.4% conversion to methanol-soluble acids was obtained. In addition, 9.3% methanol-insoluble material were left, and 11.2% carbon was lost as CO_2.

Biodegradation tests of methanol-soluble acids obtained through $AcOH/H_2O_2$ oxidation was conducted semiquantitatively. Garden soil bacteria served as inoculum in this test.

Experimental Procedures

A volume of 200 mL of glacial acid was placed in a 500-mL three-necked round glass flask, equipped with reflux condenser, separatory funnel, and stirrer. It was submerged in a water bath that was maintained at $55 \pm 2°C$. Five grams of lignite powder and 20 mL of 30% H_2O_2 were added to the flask. H_2O_2 solution was added slowly drop by drop over a period of 3 h during which the suspension was stirred gently. Seven additional 30-mL aliquots of 30% H_2O_2 were added daily to the original suspension. The reaction was allowed to proceed for 8 days, and at the end of this period, the filtrate was separated from the residue. Vacuum distillation was used to obtain residue. The procedure for fractionation is presented schematically in Fig. 10.10. Methanol-soluble fraction was used as a substrate for biodegradation experiments utilizing garden soil microorganisms.

A soil column was used as bioreactor (20 mm ID; 10 cm in depth). The column was wrapped with heating tape and the temperature inside the column was kept at $30 \pm 3°C$.

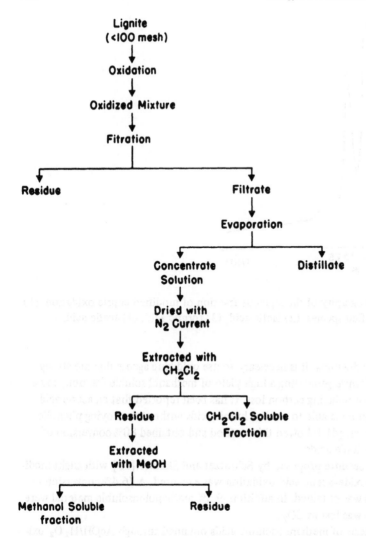

Fig. 10.10 Schematic separation for oxidized mixture after hydrogen peroxide–acetic acid oxidation.

Methanol-soluble acids derived from the oxidation of lignite were first diluted by mixing gummy organic acids with distilled water. The original total organic content (TOC) of this diluted solution was 124 mg/L. Two liters of diluted organic acids solution (TOC = 124 mg/L) was circulated through the soil column at a rate of about 2 days per cycle. Figure 10.11 illustrates the apparatus being used for this experiment. Samples were taken periodically from solution and analyzed. A TOC instrument was used to monitor the concentration of total organic carbon daily, and ion chromatography—Dionex 2000i—was used to monitor the concentration of biodegradable organic acids. Data obtained from those instruments were used to interpret the utilization of oxidized products by soil bacteria.

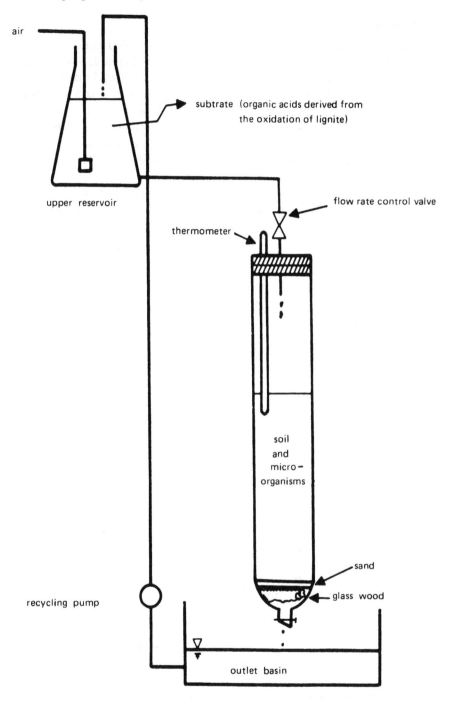

air

subtrate (organic acids derived from
the oxidation of lignite)

upper reservoir

flow rate control valve

thermometer

soil
and
micro—
organisms

sand

glass wood

recycling pump

outlet basin

Fig. 10.11 Apparatus of biodegradation conducted in a soil column.

Results and Discussion

Figure 10.12 shows TOC removal curves and Fig. 10.13 shows a biodegradation curve
based on the measurement of conductivity caused by organic acid in solution as analyzed
by ion chromatography. According to TOC removal curves, it was found that the lag
phase took about 2 days, and that the stagnant phase was reached after 6 days of incubation. The rate of TOC removal was calculated from Fig. 10.12 and then plotted against
TOC as shown in Fig. 10.14. The reaction rate increased with the decrease of TOB up
to 85 mg/L. After reaching the maximum rate, the rate of reaction started to decrease
and approached its stagnant point. A value of 41 mg/L TOC corresponds to nondegead-
able organic substances derived from the oxidation of lignite.

Based on the results found by ion chromatography (see Fig. 10.13), no lag phase
occurred and the reaction rate was constant for the first 5 days. As the concentration
of organic acids became more diluted (less than 10 ppm as acetic acid), the curve sud-

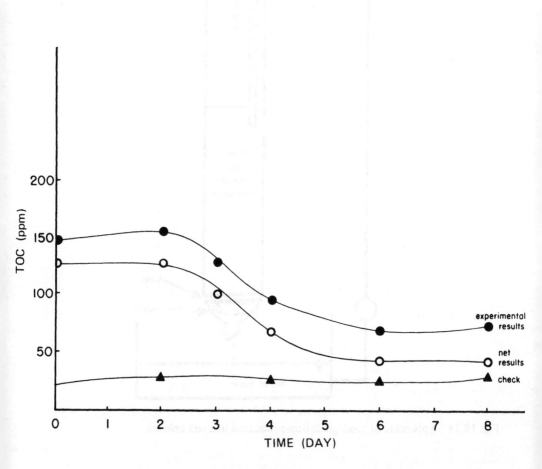

Fig. 10.12 TOC removal of organic matter derived from lignite by soil bacteria. (From
Ref. 23.)

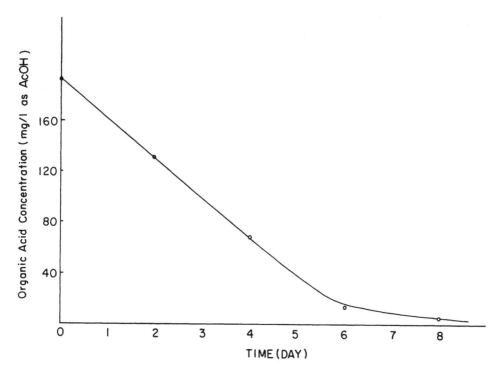

Fig. 10.13 Removal of biodegradable organic matter by soil bacteria. (From Ref. 23.)

denly changed its characteristic and reached the stagnant point at the eighth day of incubation.

The capability of soil microorganisms to degrade water-soluble organic acids derived from lignite was determined quantitatively. Our studies also revealed that a significant portion of oxidized product, about one-third, was relatively nonbiodegradable. This portion may be humic substances derived from lignite via a chemical oxidation process. By comparing the biodegradation curve based on inorganic content (IC) analysis with those based on TOC analysis, it was possible to conclude the following. (2) Some biodegradable organic acids were depleted very fast. However, these biodegradable organic acids did not contribute significantly to TOC. Therefore, a lag phase was observed in TOC analysis but not in IC analysis. (2) A stagnant phase was reached after 6 days of incubation. This was consistent for both TOC and IC analysis. (3) Significant removal of TOC occurred between the second and sixth days of incubation. (4) Before the fifth day of incubation, it was found that the reaction for biodegradable organic acids showed zero-order characteristics. This reaction rate was approximately 32 mg/L per day in terms of acetic acid (see Fig. 10.13). For TOC removal, the biodegradation rate reached its maximum point (38.9 mg/L day) on the fourth day of incubation (see Fig. 10.14). (5) A total of 41 mg/L TOC remained as nonbiodegeadable organic acids.

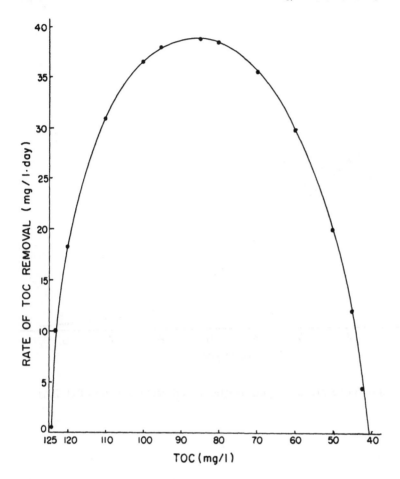

Fig. 10.14   Rate of TOC removal as a function of TOC concentration in medium.

10.3   NITROGEN-ENRICHMENT PROCESS TO ENHANCE BIODEGRADATION OF LIGNITE

In general, enzymes produced by microorganisms are able to decompose complicated nitrogen compounds such as proteins. Theoretically, proper C/N ratio can accelerate bio-degradation. Probably, the presence of amino groups in proteins makes it easily bio-degraded. Provided that lignite is impregnated with nitrogen, the rate of biological ac-tivity is expected to increase. Based on this recognition, incorporation of the amino group into the hydrocarbon skeleton of a recalcitrant organic matter such as lignite may stimu-late biodegradation.

Ammoniation of lignite is an example that illustrates the feasibility of nitrogen incorporation. It has been recognized for many years that fertilizers obtained from low-rank coals (e.g., lignites), when reacted with ammonia, would combine the properties of both humus and chemical nitrogen fertilizers [14]. It was reported that lignite treated with dilute nitric acid followed by extraction with ammonium hydroxide yields products containing over 10% nitrogen [15].

The evidence mentioned above suggests that nitrogen-enriched process can be utilized as pretreatment for biodegradation of lignite. The purpose of this pretreatment is to incorporate $-NH_2$ into the aromatic units of the macromolecular structure of lignite. By doing this, biodegradation of relatively nondegradable organic matter can be stimulated.

To incorporate desirable functional groups into the recalcitrant macromolecule of lignite, a nitrogen enrichment process was conducted through a three-step chemical process. The first step was chlorination of lignite by ferric chloride. The purpose of chlorination was to enlarge the porous properties of lignite [14]. Chlorine was removed by a subsequent nucleophilic substitution reaction, which was accomplished in the third step of this chemical process.

The second step of this process consisted of nitrating lignite by means of nitric acid. Nitration was performed to (1) increase the oxidation conditions of lignite, and (2) introduce nitro groups into aromatic rings, which in turn activate the chloro groups incorporated in the first step of the process.

Ammoniation, the third and final step, was carried out by treating lignite with ammonium hydroxide. This treatment was done to replace chloro groups and hydroxyl groups with amino groups.

As a result of this three-step process, nitrogen groups such as nitro and amino groups were incorporated into the macromolecule of lignite. Subsequently, the biodegradation of nitrogen-enriched lignite was conducted to assess the effect of this chemical process on biodegradation of lignite.

The present work is primarily qualitative rather than quantitative. It has the purpose of testing the possibility of optimizing biodegradation of lignite by chemically modifying the macromolecule of lignite through the incorporation of functional groups that would be compatible with ordinary microbial enzymatic systems.

10.3.1 Experimental Procedures

Raw lignite mixed with ferric chloride was placed in an autoclave. The reaction was performed at $200°C$ under a stream of N_2 gas. Reaction time was about 90 minutes. The coal sample reacted was stirred in distilled water for 8 h to remove ion salts and other soluble materials. The sample was then separated by filtration and oven dried.

Treated lignite was next subjected to mild oxidation with nitric acid (20% by volume) at $70°C$. Oxidation was carried out by making a slurry with lignite and distilled water (1:1 by weight) in a three-necked flask equipped with thermometer, stirrer, and graduated dropping funnel. The slurry was heated and when a temperature of $70°C$ was reached, dilute nitric acid (20% by volume) was added at a rate of 13 mL/min by means of a dripping funnel. The reaction lasted 2 h, and the temperature was kept at $70°C$. After that, the reaction mixture was filtered through a glass-fiber filter and the product was washed with 2 L of distilled water and subsequently dried for 4 h at $110°C$.

Dried nitrohumic acid was treated with 0.5 N ammonium hydroxide until pH 6 was reached. Following filtration, the residue was washed with 1 L of diluted HCl solution (0.005N HCl). The residue was then dried in an oven at $110°C$ for 2 h. The resulting product, nitrogen-enriched lignite, was then subjected to elemental analysis and microbial degradation. Elemental analysis of raw lignite, lignite subjected to nitration, and lignite subjected to ammoniation was conducted by Hoffman Laboratories Inc., Golden, Colorado.

Microbial degradation of nitrogen-enriched lignite was carried out by soil bacteria previously acclimated to oxidized products of lignite. The oxidation of nitrogen-enriched lignite by soil bacteria was performed in a 2-L flask at $30 ± 3°C$. Two liters of distilled water plus 10 mL of active suspension of bacterial cells were added to the flask. Bio-degradation of nitrogen-enriched lignite was monitored by measuring the concentration of nitrate and organic acids.

Ion chromatography was utilized to monitor the concentration of nitrate and bio-degradable organic acids. The concentration of biodegradable organic acids was expressed as milligrams per liter of acetic acid.

10.3.2 Results and Discussion

Table 10.6 shows the results of elemental analysis for raw lignite, lignite after nitration, and lignite after ammoniation. Based on the data listed in Table 10.6, it can be expected that the conversion of raw lignite to nitrogen-enriched lignite would take place to some extent, as suggested by the following steps: (1) chlorination and dehydration, (2) nitration and oxidation, and (3) dechlorination and ammoniation. It was also observed that the nitrogen content of nitrogen-enriched lignite was five times higher than that of the original raw lignite.

Figure 10.15 illustrates the behavior of organic acids and nitrate released into the medium as a result of microbial oxidation of nitrogen enriched lignite by soil bacteria during a period of 2 weeks. The increase of organic acids in the medium is proportional to that of nitrate between zero and the sixth day of incubation, as shown in Fig. 10.15. This observation fully demonstrated that biodegradation can be accelerated by impregnating nitrogen into the structure of lignite. It was also noted that immediately after reaching its maximum concentration at the sixth day of incubation, the concentration of organic acids being released to the medium decreased. Simultaneously, the concentration of nitrate also went down.

This pattern can be explained as, instead of attacking lignite, bacteria preferentially used organic acids and nitrate released into medium as their primary nutrients (see Fig. 10.15; sixth to tenth days of incubation). After depleting those dissolved nutrients, bacteria then attacked lignite. As a result of that, an increase in the concentration of organic acids and nitrate was observed. However, during this stage, the amount of organic acids released into the medium was much smaller than that of the previous stage. Perhaps it was due to microbial depletion of nutrients and the decrease in the release of nutrients from the lignite surface.

Table 10.6 Elemental Analysis of Three-Stage Lignites

	Percent raw lignite	Percent after nitration	Percent after ammoniation
Carbon	54.0	54.93	55.82
Hydrogen	3.2	2.0	2.04
Oxygen	23.72	23.06	23.40
Nitrogen	0.95	3.03	4.71
Chlorine	0.56	1.60	1.26
Ash	21.49	15.71	16.36
Weight loss after dryness	0.3	3.97	0.04

Source: Ref. 23.

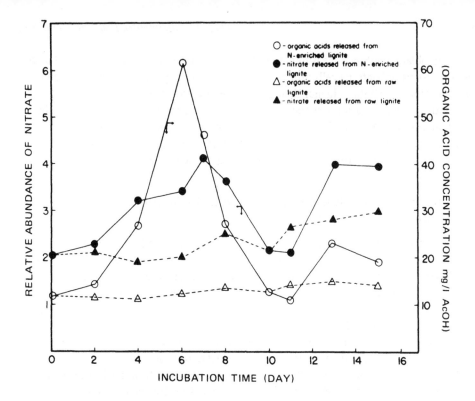

Fig. 10.15 Behavior of nitrate and organic acids released into the medium due to bacterial degradation of nitrogen-enriched lignite. (From Ref. 23.)

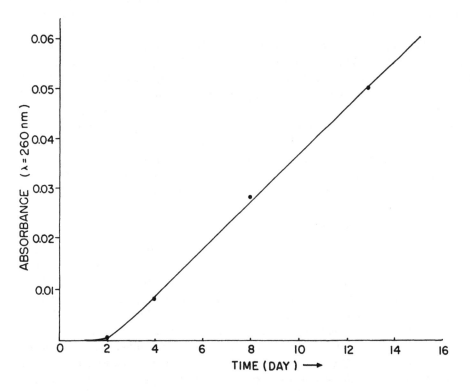

Fig. 10.16 Microbial growth as a function of incubation time measured by UV absorbance. (From Ref. 23.)

Microbial growth was measured directly through turbidity analysis (see Fig. 10.16). The total count of bacterial cells indicated that the bacterial population in the medium of the bioreactor was 10 times higher than that of the reference flask.

Nitrogen impregnation was found to be a promising pretreatment to optimize bio-degradation of lignite [27]. Nevertheless, the mechanisms of biodegradation need to be classified. To apply this novel pretreatment to lignite liquefaction processes, further studies associated with nitrogen impregnation of lignite need to be conducted.

10.4   CHEMICAL MODIFICATION TO ENHANCE THE SOLUBILITY OF LIGNITE IN WATER

Since the structure of raw coal is very complicated, microbes can hardly attack its structure. Our goal is to direct the process of chemical modification to enhance lignite solubility in water such that (1) certain weak bonds connecting adjacent ring structures can be broken, (2) specific functional groups exposed to the coal-water interface can be attacked, and/or (3) aromatic ring or heterocyclic ring structures exposed to the coal-water interface can be cleaved [29].

This process utilizes sodium metal and methanol to modify lignite. The weight increase observed in modified lignite is due to bond cleavage, sodium salt formation, and $-OCH_3$ addition. This observation is supported by Fourier transform nuclear magnetic resonance (FTNMR) and FTIR spectra. Possible mechanisms for the reaction of lignite with metal sodium and methanol were also proposed. The highly soluble modified lignite is more easily consumed by bacteria than the solid form of raw lignite. Modified lignite seems very promising for biodegradation processes. The biodegradation of the water-soluble fraction of modified lignite is used as the sole carbon source for biodegradation studies.

Oil-field soil bacteria RC-W, RC-P and fungus RC-F, which were isolated from oil-field soil of the Mobil Oil Lease Fault Block No. 5 M711E Well, were utilized to degrade the water-soluble fraction of modified lignite, which seems very promising for biodegradation processes. Yeast extracts could not only be a carbon source for biodegradation, but also an inducer, thus enhancing biodegradation. Two sets of experiments were designed to evaluate the effect and role of yeast extract. Experiment 1 uses lignite + minerals and lignite + minerals + yeast extract. The difference in biodegradation between two solutions can reveal the effect of yeast. Experiment 2 uses minerals + yeast extract without lignite as background and compared with lignite + minerals + yeast extract in order to obtain the net biodegradation from lignite under the influence of yeast extract [24].

10.4.1   Experimental Procedures

The modification process is shown in Fig. 10.17. Five grams of moisture-free, <100-mesh Beulah standard number 3 lignite was mixed with 100 mL of methanol. Five portions of 0.5 g of sodium metal were successively added into solution at an interval of 2 h. The reaction was further carried out for a total of 24 h at 65°C. Modified lignite was then filtered and oven dried at 100°C for 24 h. The weight of modified lignite was then measured. Two grams of dried modified lignite was mixed with 50 mL of distilled water and refluxed for 24 h at 80°C. The residual modified lignite was filtered and dried in an oven at 100°C for 24 h. The weight of residual modified lignite was measured to calculate the percentage of the water-soluble fraction (i.e., solubilized lignite). The pH of the filtrate was measured and saved for further biodegradation studies.

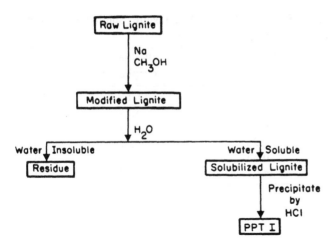

Fig. 10.17 Sodium metal/methanol modification process of lignite.

Elemental analysis of raw lignite, modified lignite, residue, and solubilized lignite that have precipitated by adding HCl were conducted by Hoffman Laboratories, Inc. A FTNMR spectrum of modified lignite dissolved in D_2O was taken. A FTIR scan of pellets of modified lignite mixed with KBr was also executed.

Table 10.7 shows the amount of minerals and yeast extract used. The water-soluble fraction of modified lignite was added into mineral solution or the minerals + yeast extract solutions in concentration of 1% in experiment 1 and 2% in experiment 2. All prepared substrate solutions were adjusted to pH 7 and then autoclaved before inoculation of bacteria RC-W, RC-P, and fungus RC-F. In experiment 1, solutions of 1% lignite + minerals and 1% lignite + minerals + yeast extract were inoculated with RC-W, RC-P, and RC-F and were incubated at 33°C for 10 days and 19 days. In experiment 2, solutions of minerals + yeast extract without lignite and with 2% lignite were inoculated with RC-W, RC-P, and RC-F, and were incubated at 33°C for 2 days and 7 days. All incubated solu-

Table 10.7 Amount of Minerals and Yeast Extract Used for Studies of Biodegradation of Water-Soluble Fraction of Modified Lignite

Component	Volume or weight
Yeast extract	1.0 g (ca. 0.1%)
K_2HPO_4	1.0 g
KH_2PO_4	0.5 g
NH_4Cl	0.5 g
Na_2SO_3	0.1 g
$CaCl_2$	0.1 g
$NaHCO_3$	0.1 g
$FeCL_3$	0.05 g
Tap water	100 mL
Distilled water	900 mL

Source: Ref. 23.

Table 10.8 Elemental Analysis of Raw Lignite, Modified Lignite, Solubilized Lignite Precipitate by HCl (Precipitate I), and Residue

	Percent raw lignite	Percent modified lignite	Percent precipitate I	Percent residue
C	50.22	39.89	56.59	48.39
H	3.29	2.86	3.81	3.35
O	25.74	26.88	32.71	21.23
S	2.43	1.95	1.31	0.62
N	0.74	0.55	0.95	0.69
Ash	17.58	28.07	4.63	26.35

Source: Ref. 28.

tions were autoclaved and were filtered through a 0.22-μm Millipore filter. The filtrates were analyzed by IC.

10.4.2 Results and Discussion

Table 10.8 shows the results of elemental analyses. The oxygen content increase in modified lignite is due to ester formation. Table 10.9 summarizes the results of quantitative data based on a sample of 100 g of raw lignite as starting material. Modified lignite showed a weight increase of 21.67% compared with the weight of raw lignite. This weight increase was due to sodium salt formation and methoxylation. Sodium salt contributed to the formation of ash and also to the weight increase of 16.58 g, which corresponded to 76.6% of the total weight increase of modified lignite. Methoxylation contributed to the organic portion, which indicated a weight increase of 5.09 g and consisted of 23.4% of the total weight increase of modified lignite.

The high solubility of modified lignite was probably due to hydrolysis of the sodium salt and ester, which yield 68% solubility on a moisture-ash-free basis.

FTNMR analysis of modified lignite was performed by NEOL FT-90 apparatus, and the resulting spectrum is shown in Fig. 10.18. The results of proton FTNMR are summarized in Table 10.10.

Table 10.9 Quantitative Analysis of Lignite, Intermediates, and Products (100 g of Raw Lignite as Starting Material)

	Weight (g)	Organics		Ash	
		%	g	%	g
Raw lignite	100.00	82.42	82.42	17.58	17.58
Modified lignite	121.67	71.93	87.51	28.07	34.16
Residue	37.91	73.65	27.92	26.35	9.99
Solubilized lignite (dewatered)	83.76	71.14	59.59	28.86	24.17
Precipitate I	50.42	95.37	48.09	4.63	2.33

Source: Ref. 28.

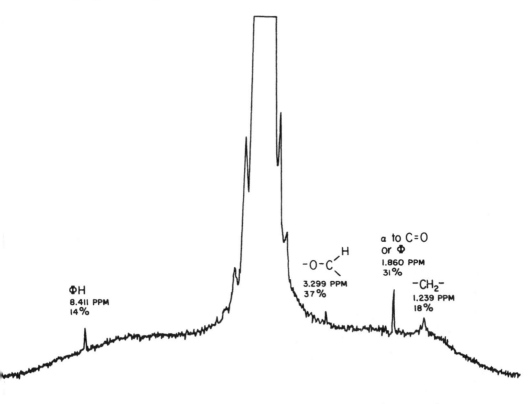

Fig. 10.18 NMR spectrum of modified lignite in D_2O.

Aromatic protons appeared at 8.411 ppm with an intensity of 14%. The oxygen in a $-O-C\overset{\diagup H}{\diagdown}$ proton could be derived from a hydroxyl group or ether linkage. The $-O-C\overset{\diagup H}{\diagdown}$ proton yielded an intensity of 37%. Due to $-OCH_3$ substitution, a proton α to C=O or an aromatic ring produced an intensity of 31%. C=O could be derived from ester, aldehyde, or ketone. A methylene group as $-CH_2-$ appeared at 1.239 ppm.

FTIR analysis of raw lignite and modified lignite was performed by Nicolet 5 DX apparatus. The spectra of raw lignite and modified lignite are shown in Figs. 10.19 and

Table 10.10 Proton FTNMR Spectrum of Modified Lignite in D_2O

Type of proton	ppm	Intensity (%)
ϕ H	8.411	14
$-O-C\overset{\diagup H}{\diagdown}$	3.299	37
α to C=O or ϕ	1.860	31
$-CH_2-$	1.239	18

Source: Ref. 24.

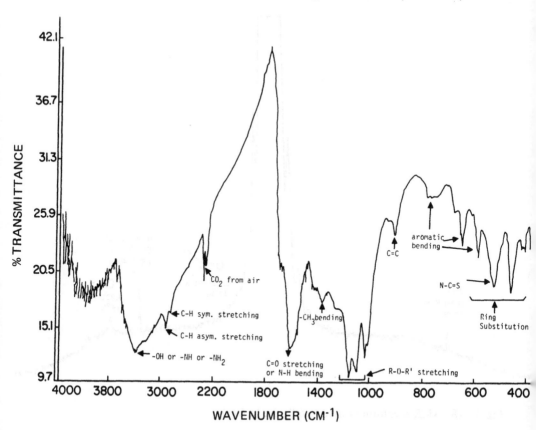

Fig. 10.19 FTIR spectrum of raw lignite.

10.20, respectively. Comparison of the two spectra showed that modified lignite has a stronger aromatic C–H stretch than does raw lignite. This is possibly due to hydrogen bond breakage, causing aromatic hydrogens in modified lignite to be less hindered than those of raw lignite. Modified lignite yielded stronger CH_3 bending due to $-OCH_3$ addition as well as weaker (than that in raw lignite) R–O–R stretching, due to breakage of the ether linkage.

Possible mechanisms for the reaction Na–CH_3OH/lignite are summarized in Fig. 10.21. When sodium metal is added to the CH_3OH-lignite solution, sodium will react with CH_3OH and form the active species CH_3O-Na^+ and hydrogen gas, as shown in equation (1) (see Fig. 10.21). There are several active sites present in the structure of lignite, such as $-COOH$ group, ether linkage, $-OH$ group, ketone linkage, and aldehyde group. These active sites are shown in Figs. 10.22 and 10.23. Figure 10.22 shows the lignite structure proposed by Sondreal [17]. Figure 10.23 shows a lignin structure that is similar to the lignite structure [18]. Some competition is expected to occur between active sites and based on the calculation of the maximum number of active sites, only 36% of the available active sites react with CH_3ON_a.

Figure 10.24 shows the effect of yeast extract using (1) fungus RC-F, (2) bacteria RC-W, and (3) bacteria RC-P by plotting oxalate concentration (ppb) versus incubated days. There is no oxalate in the lignite without yeast extract at zero days. The concentration of oxalate in the lignite with yeast extract is 322 ppb at zero days, which has been

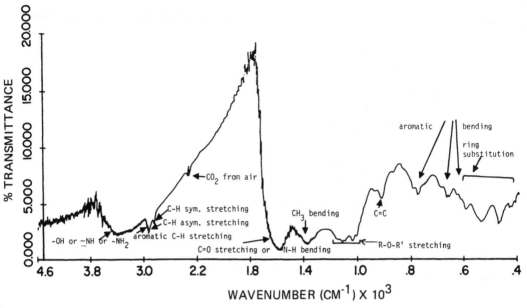

Fig. 10.20 FTIR spectrum of modified lignite.

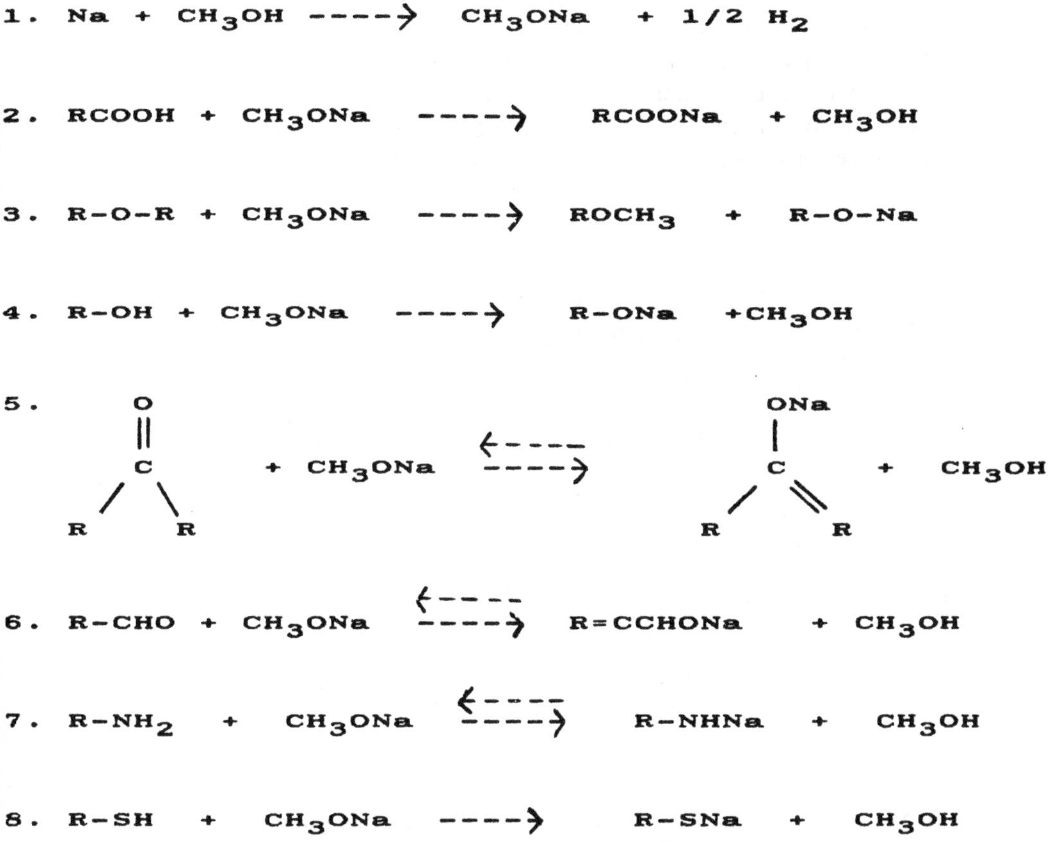

1. $Na + CH_3OH \dashrightarrow CH_3ONa + 1/2\ H_2$

2. $RCOOH + CH_3ONa \dashrightarrow RCOONa + CH_3OH$

3. $R\text{-}O\text{-}R + CH_3ONa \dashrightarrow ROCH_3 + R\text{-}O\text{-}Na$

4. $R\text{-}OH + CH_3ONa \dashrightarrow R\text{-}ONa + CH_3OH$

5.
$$
\underset{R \quad\quad R}{\overset{\overset{\displaystyle O}{\|}}{C}} + CH_3ONa \underset{\dashrightarrow}{\overset{\dashleftarrow}{}} \underset{R \quad\quad R}{\overset{\overset{\displaystyle ONa}{|}}{C}} + CH_3OH
$$

6. $R\text{-}CHO + CH_3ONa \overset{\dashleftarrow}{\underset{\dashrightarrow}{}} R\text{=}CCHONa + CH_3OH$

7. $R\text{-}NH_2 + CH_3ONa \overset{\dashleftarrow}{\underset{\dashrightarrow}{}} R\text{-}NHNa + CH_3OH$

8. $R\text{-}SH + CH_3ONa \dashrightarrow R\text{-}SNa + CH_3OH$

Fig. 10.21 Possible mechanism of Na/CH₃OH/lignite reaction. (From Ref. 25.)

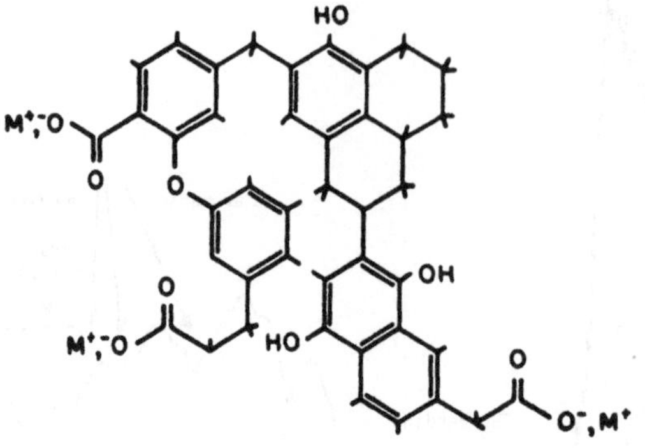

Fig. 10.22 Sondreal's partial lignite structure. $C_{42}O_{10}H_{35}$, MW = 699; C = 72, O = 23, H = 5 wt %; f_a = 0.67; O as $-COO^-$ = 60, O as $-OH$ = 30, percent O as $R-O-R$ = 10. (From Ref. 20.)

calibrated from curve II. We found that RC-F, RC-W, and RC-P can degrade the water-soluble fraction of modified lignite and produce oxalate as an end product without yeast extract. The differences among the three species are: (1) RC-F does not produce oxalate until 10 days of incubation; (2) RC-W produces oxalate at the beginning and stays stagnant after 10 days; and (3) RC-P produces oxalate at the initial stage and start decreasing after 10 days because of further degradation of oxalate.

From the curves of lignite with yeast extract (curve II), we can see that yeast extract can enhance the production of oxalate in three cases. We can conclude that yeast extracts do support biodegradation as a carbon source, so that the concentration of oxalate is higher than the solutions without yeast extract (curve I).

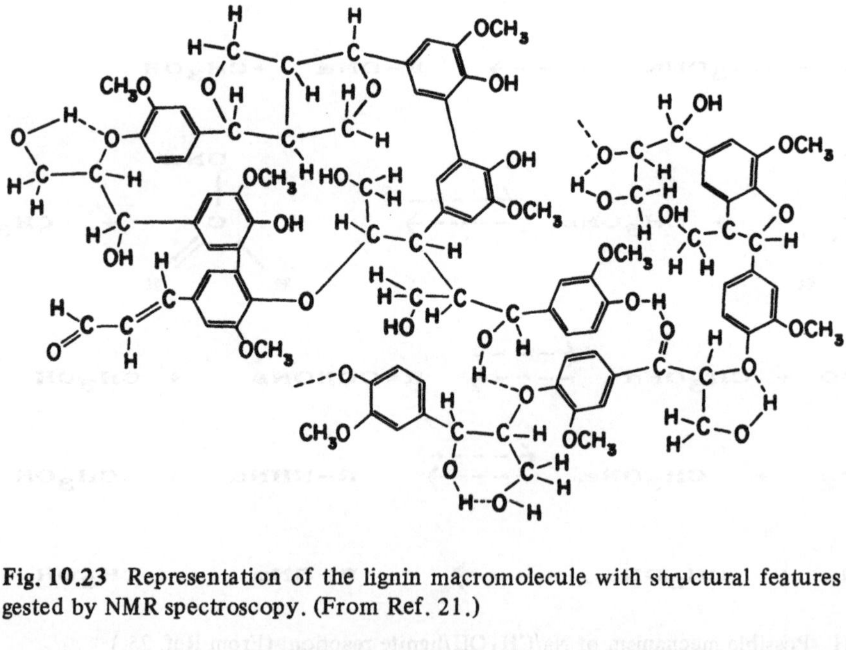

Fig. 10.23 Representation of the lignin macromolecule with structural features suggested by NMR spectroscopy. (From Ref. 21.)

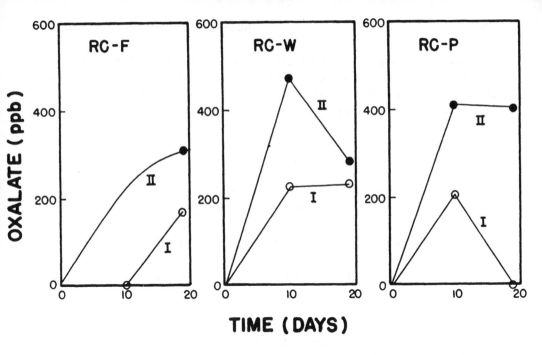

Fig. 10.24 The effect of yeast extract using fungus RC-F, bacteria RC-W, and bacteria RC-P. I, Water-soluble lignite plus minerals; II, water-soluble lignite plus minerals and yeast extract.

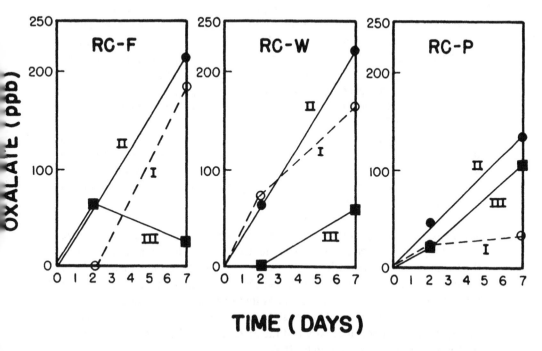

Fig. 10.25 Net biodegradation from lignite under influence of yeast extract using fungus RC-F, bacteria RC-W, and bacteria RC-P. I, Minerals plus yeast extract; II, minerals plus yeast extract and water-soluble lignite; III, net biodegradation from lignite (II – I).

Since we know that yeast extract can be a carbon source for biodegradation, experiment 2 used yeast extract as the sole carbon source which can be used as a background reference to deduce the effect of yeast extract as a carbon source. Thus the induced effect of yeast extract for biodegradation can be evaluated. Figure 10.25 shows the net biodegradation from lignite under the influence of yeast extract. Species RC-F, RC-W, and RC-P can utilize yeast extract as a sole carbon source and produce oxalate (Curve I). Species RC-F and RC-W can produce more oxalate than species RC-P. From Curve II we can see that more oxalate is produced in all three cases. When we subtract Curve I from Curve II we can obtain the net biodegradation from lignite, Curve III. Even though oxalate concentration from lignite increases in the three cases, the efficiencies are lower than the lignite without yeast extract except for species RC-F. So yeast extract can have a positive induced effect for fungus RC-F, and have a negative induced effect for bacteria RC-W and RC-P in a 7-day period. When two carbon sources, lignite and yeast extract, are coexistent, bacteria prefer to uptake yeast extract and produce oxalate [26].

10.5 CONCLUSION

Various oxidation processes of lignite were presented in this chapter. Potassium permanganate and sodium hypochlorite oxidation of lignite did not appear to be suitable treatments to enhance biodegradation of lignite due to the extreme oxidizing features of these two chemicals. This was confirmed by ion chromatography, which indicated that about 43% of the carbon structurally bonded to lignite was oxidized to carbon dioxide. High production of carbon dioxide was interpreted as a massive loss of potential substrate to sustain biological activity. Despite the undesirable results mentioned above, potassium permanganate and sodium hypochlorite oxidation of lignite produced many carboxylic acids, according to ion chromatography and gas chromatography analysis. The presence of carboxylic acids in the products of oxidation was an indication that biodegradation was possible, thus encouraging the search for more promising methods of treating lignite. The relationship between biodegradation and the occurrence of carboxylic acids was confirmed by a FTIR spectrum of the alkaline fraction of lignite, which supported the growth of white-rot fungi. These microbes were also capable of growing on the surface of raw lignite; however, no growth was observed on the residue of lignite after benzene-methanol extraction followed by alkaline extraction. FTIR of the latter showed a pronounced ester character. These findings suggest that alkaline solubles may be the biodegradable fraction of lignite matrix.

Alkaline copper oxide was employed as a weak oxidizing agent that would not produce massive amounts of carbon dioxide, and that would also specifically attack weak linkages of asphaltic side chain of the macromolecule of lignite. Comparing this oxidation method with previous ones, such as potassium permanganate and sodium hypochlorite, it was found that carbon losses were minimized (20% conversions of lignite carbon to CO_2). Another desirable feature of this method was that its methanol fraction supported the growth of a mixed culture of soil bacteria. Again, a FTIR spectrum demonstrated the relationship between microbial activity and the presence of carboxylic acids. The yield of methanol solubles was estimated to be 30%.

To improve the yield of the biodegradable fraction of methanol solubles, hydrogen peroxide–acetic acid oxidation was executed. The strength of this oxidizing agent combined with its specificity of not destroying phenolic compounds resulted in minimizing the carbon loss as carbon dioxide to 11.2% and maximizing the yield of methanol solubles to 76.4%. As was expected, the obtained methanol fraction was capable of sustaining the growth of a mixed culture of soil bacteria. Total organic acids and ion chromatography analysis demonstrated the utilization of products of oxidation present in the methanol-soluble fraction by soil bacteria. A mixed culture of bacteria acclimated to the methanol fraction of the products of hydrogen peroxide–acetic acid oxidation of lignite proved to be of value for biodegrading nitrogen-enriched lignite.

Incorporation of specific functional groups, such as an amino group in the recalcitrant macromolecule of lignite, to enhance biodegradation was a novel concept developed here. To test this concept, nitrogen enrichment of lignite through a series of chemical reactions was successfully produced in vitro. Semiquantitative data on the effect of nitrogen enrichment on the biodegradation of lignite were obtained [30].

Based on those results, modification of the chemical structure of lignite by introducing nitrogeneous groups in its macromolecular structure is a promising concept that deserves to be developed further for applications related to biotechnology of lignite.

Modification of the structure of lignite by means of sodium metal and methanol achieved its preliminary purpose, which was to produce a substantial conversion of lignite to water solubles, which was found to be 68%. This fraction is a promising substrate for biological activity, as suggested by FTNMR, FTIR, and elemental analysis.

The three species RC-F, RC-W, and RC-P can utilize the water-soluble fraction of modified lignite and produce oxalate. Species RC-W produced more oxalate than did species RC-P and RC-F. Yeast extract can act as a carbon source and as an inducer. For a 7-day period, yeast extract has a positive induced effect for fungus RC-F and a negative induced effect for bacteria RC-W and RC-P, due to the fact that bacteria prefer to uptake yeast extract as a carbon source when lignite and yeast extract coexist. Since oxalate is only a portion of the end products of biodegradation, a more detailed analysis of biodegraded solutions is needed before making further conclusions.

In short, the utilization of biotechnology to obtain useful products derived from fossil fuels is still in the development stage. Research in this field is coming up with novel approaches that are just beginning to show higher efficiency. Based on the results obtained by the research work presented in this chapter, the chemical structure of fossil fuels needs to be carefully modified to produce derivatives that are compatible with existing enzymatic systems.

As soon as this idea reaches more advanced stages, it will be possible to direct the modification of a given fossil fuel in such a way that the utilization of a given biological process will yield the expected and desired product. Genetic engineering could also play an important role in this approach.

ACKNOWLEDGMENTS

We acknowledge the financial support of this work from U.S. Department of Energy contract DEFG22-84PC70809. The authors would like to express grateful appreciation to Jau Ren Chen for his assistance in completing the final version of this chapter.

REFERENCES

1. Braunstein, H. M., Copenhaver, E. D., and Pfuderer, H. A., *Environmental, Health, and Control Aspects of Coal Conversion: An Information Overview*, prepared for Energy Research and Development Administration, 1977, Chap. 3.

2. Cohen, M. S., and Gabriele, P. D., Degradation of coal by the fungi *Polyporus* and *Ronia monticola, Appl. Environ. Microbiol., 44*(1), 23 (1982).

3. Young, D. K., and Yen, T. F., Oxidation of lignite into water-soluble organic acids, *Energy Sources, 3*, 49 (1976).

4. Studier, M. H., Hayatsu, R., and Winans, R. E., Analysis of organic compounds trapped in coal, and coal oxidation products, in *Analytical Methods for Coal and Coal Products*, Vol. 2 (C. Karr, Jr., ed.), Academic Press, New York, 1978, Chap. 21.

5. Dart, R. K., and Stretton, R. J., Biodegradation, in *Microbiological Aspects of Pollution Control*, 2nd ed., Elsevier, Amsterdam, 1980, p. 216.

6. Hayatsu, R., Winans, R. E., McBeth, R. G., Moore, L. P., and Studier, M. H., Structure characterization of coal: lignin-like polymer in coal, in *Coal Structure*, American Chemical Society, Washington, D.C., 1981, Chap. 9.

7. Elliot, M. A., *Chemistry of Coal Utilization*, Wiley, New York, 1981.

8. Metcalfe, L. D., and Schmitz, A. A., The rapid preparation of fatty acid esters for gas chromatographic analysis, *Anal. Chem., 33*, 363 (1961).

9. Ward, B., Lignite-degrading fungi isolated from a weathered outcrop system, *Appl. Microbiol., 6*, 236 (1985).

10. Korburger, J. A., Microbiology of coal: growth of bacteria in plain and oxidized coal slurries, *Proc. W. Va. Acad. Sci., 36*, 26 (1964).

11. Rogoff, M. H., *Microbiology of Coal*, Information Circular 8075, U.S. Bureau of Mines, Washington, D.C., 1962.

12. Yen, T. F., *Microbial Screening Test for Lignite Degradation, Quarterly Progress Report 1*, U.S. Department of Energy, Washington, D.C., Jan./Mar. 1985.

13. Schnitzer, M., and Skinner, S. I. M., The low temperature oxidation of humic substances, *Can. J. Chem., 52*, 1072 (1974).

14. Coca, J., Alvarez, R., and Ovledo, A. B., Production of a nitrogenous fertilizer by the oxidation-ammoniation of lignite, *Ind. Eng. Chem. Prod. Res. Dev., 23*(4), 620 (1984).

15. Schwartz, D., Asfeld, L., and Green, R., The chemical nature of the carboxyl groups of humic acids and conversion of humic acids to ammonium nitrohumates, *Fuel, 44*, 417 (1965).

16. Beall, H., Savage, L. A., and Curry, M., Reaction of coal with ferric chloride: effect on surface area and dependence on rank, *Fuel, 62*, 289 (1965).

17. Sondreal, E. A., Wilson, W. G., and Steinberg, V. I., Mechanisms leading to process improvements in lignite liquefaction using CO and H_2S, *Fuel, 61*, 925 (1982).

18. Ludwig, C. H., Nist, B. J., and McCarthy, J. L., Lignin. XIII. The high resolution nuclear magnetic resonance spectroscopy of protons in acelylated lignin, *J. Am. Chem. Soc., 86*, 1196 (1964).

19. Yen, T. F., do Nascimento, H. C. G., Chen, J. R., Lee, K. I., Hsu-Chou, R. S., and Wang, J. W., *Microbial Screening Test for Lignite Degradation, Quarterly Progress Report 2*, DOE contract DEFG22-84PC 70809, 1985.

20. do Nascimento, H. C. G., Lee, K. I., Chou, S. Y., Wang, W. C., Chen, J. R., and Yen, T. F., Growth of *polyporus versicolor* on lignite fractions, *Process Biochem., 22*(1), 24 (1987).

21. Yen, T. F., do Nascimento, H. C. G., Lee, K. I., Hsu-Chou, R. S., and Wang, J. W., *Microbial Screening Test for Lignite Degradation, Quarterly Progress Report 3*, DOE contract DEFG22-84PC 70809, 1985.

22. Yen, T. F., do Nascimento, H. C. G., Lee, K. I., Hsu-Chou, R. S., and Wang, J. W., *Microbial Screening Test for Lignite Degradation, Quarterly Progress Report 4*, DOE contract DEFG22-84PC 70809, 1985.

23. Yen, T. F., do Nascimento, H. C. G., Lee, K. I., Hsu-Chou, R. S., and Wang, J. W., *Microbial Screening Test for Lignite Degradation, Quarterly Progress Report 5*, DOE contract DEFG22-84PC 70809, 1985.

24. Yen, T. F., do Nascimento, H. C. G., Chen, K. C., Lee, K. I., and Hsu-Chou, R. S., *Microbial Screening Test for Lignite Degradation, Quarterly Progress Report 6*, DOE contract DEFG22-84PC 70809, 1985.

25. Hsu-Chou, R. S., and Yen, T. F., A novel chemical solubilization process for bene-ficiation of lignite, *Pacific Conference on Chemistry and Spectroscopy*, Irvine, Calif., 1987.

26. Yen, T. F., do Nascimento, H. C. G., Lee, K. I., Hsu-Chou, R. S., and Wang, J. W., *Microbial Screening Test for Lignite Degradation, Quarterly Progress Report 7*, DOE contract DEFG22-84PC 70809, 1985.

27. Yen, T. F., do Nascimento, H. C. G., Lee, K. I., Hsu-Chou, R. S., and Wang, J. W., *Microbial Screening Test for Lignite Degradation, Final Report*, DOE contract DEFG22-84PC 70809, 1988.

28. Yen, T. F., do Nascimento, H. C. G., Chen, K. C., Lee, K. I., Hsu-Chou, R. S., and Wang, J. W., *Microbial Screening Test for Lignite Degradation, Quarterly Progress Report 8*, DOE contract DEFG22-84PC 70809, 1985.

29. Hsu-Chou, R. S., do Nascimento, H. G. C., and Yen, T. F., Microbial action on coal, in *Sample Selection, Aging and Reactivity of Coal* (R. Klein and R. M. Wellock, eds.), p. 407–409, Wiley, New York, 1989.

30. Wang, J. W., do Nascimento, H. C. G., and Yen, T. F., Biodegradation of Nitrogen Enriched Lignite, *Reserves, Conservation, Recycling, 2*, 249–260 (1989).

11

Bioconversion of Solubilized Lignite: Pilot Reactor Studies

DON LAPIN and JACK V. MATSON *University of Houston, Houston, Texas*

11.1 INTRODUCTION

Anaerobic biological treatment (biomethanation) provides a means for converting waste or undesirable materials into energy in the form of methane gas, and into relatively harmless by-products. Anaerobic processes produce roughly 11,000,000 Btu of methane for each ton of chemical oxygen demand (COD) removed. In contrast, a typical aerobic biological process consumes 9,000,000 Btu of energy per ton of COD removed [11].

Energy production makes anaerobic treatment worthy of consideration for energy recovery from low-grade organic materials, which are marginally suited for conventional energy generation but which may be degradable by anaerobic organisms. One such material is lignite, a lower coal that contains as much as 50% ash and moisture by weight. Characteristics such as a tendency to smoulder in open railroad trucks make lignite an uneconomic material to transport to remotely located fossil power plants. Yet lignite represents slightly less than one-third of estimated coal reserves in the United States.

The molecular structure of lignite consists of large aromatic-ring polymers, joined together by carbon-carbon and ether linkages (Fig. 11.1). The polymer chains are too large to permit much biodegradation. However, hydrolysis of pulverized lignite breaks up the polymer and introduces more oxygen, producing much-lower-molecular-weight materials that are soluble in aqueous media, such as monocyclic species with multiple carboxyl ligands (benzene carboxylic acids). Recent experiments with the hydrolyzed material indicate that it is a suitable substrate for biomethanation.

This investigation focuses on reactor configuration and conditions for the anaerobic conversion of solubilized lignite feedstock into methane. Several anaerobic reactor (fermentor) types are applicable for the two site locations discussed by Leuschner et al. [13,14]: underground rock caverns and underground salt domes. Reactor costs are likely to influence process economics significantly; thus an evaluation of reactor design for the lignite conversion process is warranted.

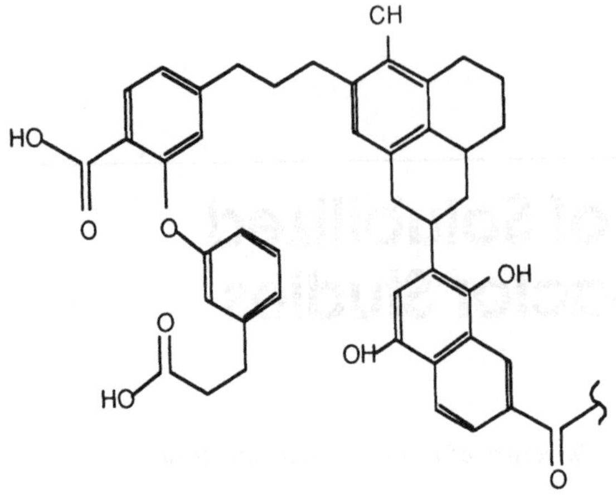

Fig. 11.1 Structure of lignite. (From Ref. 13.)

11.2 METHANOGENIC ASSOCIATIONS: REACTOR CONSIDERATIONS

More than do aerobic biological processes, methane-producing anaerobic fermentors require a symbiotic association of bacterial species. This association consists of at least three different bacterial types (Fig. 11.2) in four separate steps [3]. Hydrolyzing organisms convert complex organics to simple organics (sugars, amino acids, fatty acids) in the first step. For example, cellulose would be converted to sugar. Fermenting organisms, or "acid formers," convert the simple organics to organic acids, hydrogen, and carbon dioxide. $C3^+$ organic acids formed during fermentation are converted separately to acetate ("acetogenesis"). In the fourth step, methanogenic organisms reduce carbon dioxide to methane and water ("hydrogenotrophic" methanogens) and decarboxylate acetate to methane ("acetoclastic" methanogens).

Some 20 different species of methanogens are known; 15 or so reduce carbon dioxide, and two or three species decarboxylate acetate. One or two species are thought to perform both functions. This association of microorganisms requires that seed material used to start an anaerobic reactor be a mixed culture. The hydrolyzing organisms and the acid formers are facultative; they can function in aerobic or anaerobic environments. Methanogens are primitive organisms, obligate anaerobes that rely on the facultative hydrolyzers and fermentors to scavenge toxic oxygen from the reactor system. Because the highest concentration of active biomass occurs near the reactor inlet, residual dissolved oxygen disappears quickly after entering the reactor.

The slow growth rate of methanogens usually is the rate-limiting step in biomethanation systems. A slug dose of complex organic substrate can accelerate the faster-metabolizing acid formers, resulting in a net accumulation of acid that may "pickle" the reactor. Consequently, pH and volatile acid monitoring in a reactor are very important. Alkalinity must be added to buffer local pH variations. Since the acid formers tend to accumulate near the bottom (substrate inlet) of an upflow reactor, the pH and volatile acid profiles dip and bulge, respectively, at this location. For high organic loadings, effluent recycle may spread the conversion process more evenly through the reactor and dilute the influent.

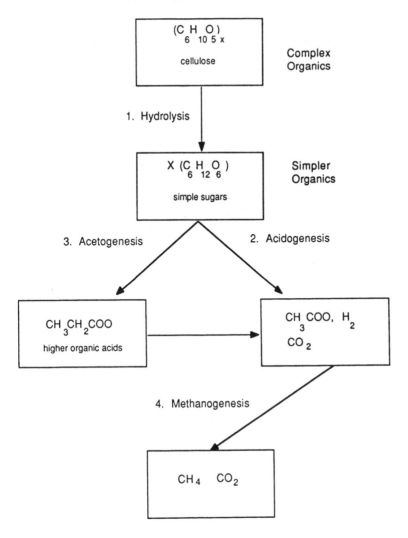

Fig. 11.2 Methanogenic associations in a fermentor, cellulose substrate. (Adapted from Ref. 3.)

Another effect of symbiosis relates to fermentor hydrogen concentrations. The standard free energy $[\Delta G^{\circ}(w)]$ of many hydrogen-producing reactions is positive (e.g., ethanol hydrolysis, benzoic acid fermentation, or propionate acidogenesis). Thus, at standard conditions, these reactions are energy *consuming*. To obtain energy for growth, the organisms effecting these conversions depend on the methanogens to consume hydrogen, reducing the partial pressure $[P(H_2)]$ to low levels. Figure 11.3 illustrates that $[10^{-6}$ atm $< P(H_2) < 10^{-4.2}$ atm for the ΔG of the acetogenesis/methanogenesis reaction couple to be negative (i.e., so that energy is produced and the reaction will proceed). With hydrogen concentration so low, little energy is available to the hydrogenotrophic methanogens, resulting in very low biomass yields for complete biomethanation of propionate and other organic acids [3,15]. Experimental studies have verified this observation (Fig.

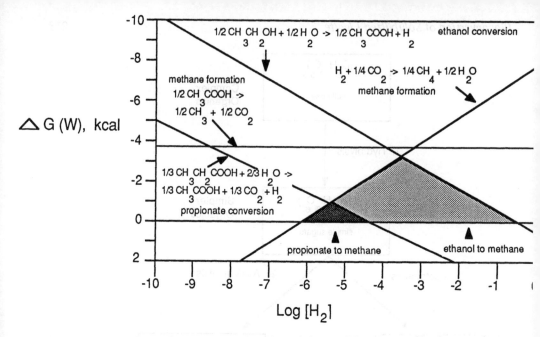

Fig. 11.3 Free energy of reaction as a function of the partial pressure of hydrogen; alcohol fermentation, acetogenesis, and methanogenesis. (From Ref. 15.)

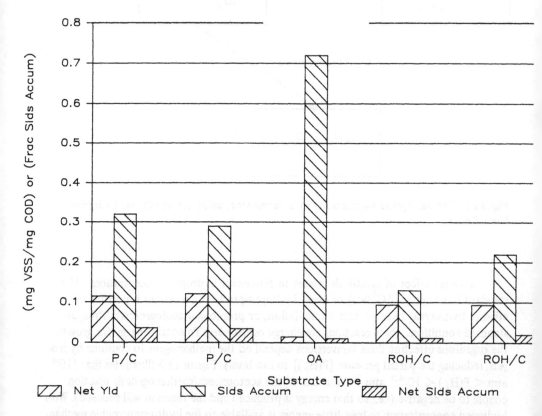

Fig. 11.4 Net biomass yields in anaerobic filters processing complex protein/carbohydrate substrate (P/C), organic acids substrate (OA), and alcohol/carbohydrate substrate (ROH/C). The net solids accumulation a product of both the net biomass yield and the solids retained (not washed out). Organic acid substrate produced the lowest net solids accumulation. (Adapted from Ref. 27.)

11.4). Methanogenic associations also feature an inherently low endogenous decay rate; the biomass diminishes very slowly, even in the absence of substrate. Accordingly, a reactor design that retains high biomass concentrations during flow interruptions, and which provides good mixing of substrate and solids when flow resumes, can be expected to perform well under load fluctuations and periods of dormancy.

11.3 CONVENTIONAL DIGESTER

The most common methane-producing anaerobic process, the digester, processes concentrated solids from primary and secondary waste treatment clarifiers, usually raw suspended matter and excess cell mass, respectively. Essentially a large, open reactor (Fig. 11.5), the digester's simple design makes it a candidate for solubilized lignite conversion in the salt dome location. The digester has its antecedents in late-nineteenth- and early-twentieth-century treatment of municipal waste sludges. Donald Cameron developed the septic tank (Fig. 11.6) to treat wastewater and solids together. Cameron recognized the

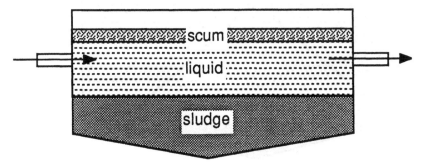

Unmixed Conventional Digester

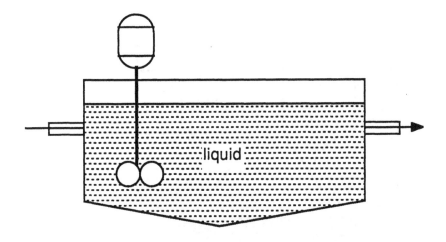

High Rate Digester (Mixing)

Fig. 11.5 Conventional anaerobic digesters. (From Ref. 15.)

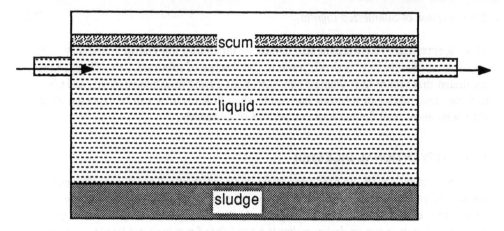

Septic Tank

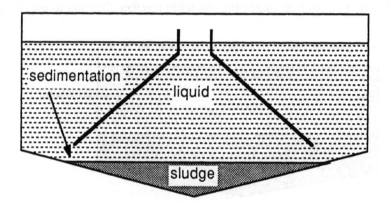

Travis Tank

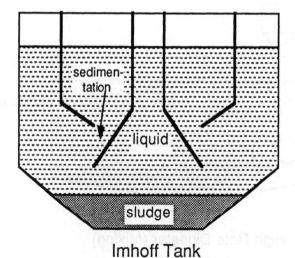

Imhoff Tank

Fig. 11.6 Early anaerobic treatment systems. (From Ref. 15.)

value of methane gas produced during sludge decomposition, and used some of the gas produced for heating and lighting at the Exeter, England, disposal works. Clark, Travis, and Imhoff improved on Cameron's design by permitting sludge to ferment by itself, separated from the wastewater flow. Heating the sludge in separate fermenting tanks improved methane yields, thereby permitting methane export for other uses. Mechanical mixing, introduced in the 1950s, produced better efficiencies by bringing bacteria and wastes closer together and by eliminating the surface scum layer, increasing effective capacity [15].

Despite its simplicity, the digester has shortcomings that make questionable its suitability for solubilized lignite conversion. Conventional digesters lack provision for retaining microorganisms in the system longer than the hydraulic residence time. Consequently, effective treatment depends on establishing conditions that optimize bacterial growth and maximize hydraulic residence time. Digesters are most effective for treating suspended matter that is highly concentrated in degradable organics over long detention times (20 days or longer). The process usually has been uneconomical when the substrate contains less than 10,000 mg/L of degradable organic material [25]. Heating is necessary to obtain high rates of gas production. Moreover, keeping what can be expected to be a large-volume underground digester well mixed with external pumps could be costly. With little solids retention capability, digesters respond poorly to fluctuating loads. Without effluent solids separation, poor effluent liquid quality results from high suspended solids concentration [2].

Even with effluent solids separation equipment, soluble substrate conversion may still be unsatisfactory. While lignite hydrolysis can produce high chemical oxygen demand (COD) concentration (consistent with digester treatment), filtration and ultrafiltration tests indicate that the biodegradable COD represents mostly soluble materials. Primarily soluble substrates promote a dispersed, difficult-to-settle biomass in a digester [25]. Thus, even when equipped with an effluent settler to recycle solids (as in the anaerobic contact process, see below), the digester may perform poorly due to high biomass washout.

11.4 UPFLOW ANAEROBIC FILTER

Anaerobic treatment in a conventional digester depends more on the microorganism or solids detention time in the reactor than on the substrate concentration or organic loading to the reactor [15]. Consequently, techniques to retain or recycle biological solids should improve conversion and volumetric efficiency in anaerobic systems. Young and McCarty incorporated this concept into a new reactor which they termed the "anaerobic filter" (Fig. 11.7); a packed or stationary media matrix within the reactor provides attachment sites for the biological solids. Flow of substrate enters the bottom of the filter and passes upward through the media matrix so that the filter is completely submerged. Methane gas produced by reaction trickles upward through the media and exits at the top. Anaerobic microorganisms attach to the media and accumulate in the void spaces of the media, so that the substrate contacts a large concentration of biological solids as it passes upward through the filter. The results are a high treatment efficiency at nominal temperatures using moderate-strength soluble substrates, and an effluent that is essentially free of biological solids [25,26].

Young and McCarty constructed three 183-cm-long, 15.2-cm-diameter (28.5 L empty volume) plexiglass columns. The media consisted of 25 to 38-mm quartzite stones,

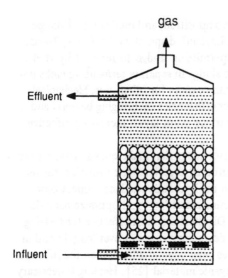

Fig. 11.7 Upflow anaerobic filter. (From Ref. 15.)

with a void fraction of 0.42. Synthetic substrates to the columns consisted of either protein-carbohydrate or volatile acids at 1500, 3000, or 6000 mg/L total COD. COD removal efficiency was high [typically above 90% with detention times of 18 h or more and organic loading up to 0.85 kg/m³ per day], with most of the removal taking place in the lower 0.5 m of filter height (Table 11.1). Methane production with protein-carbohydrate substrate was steady at approximately 22 L/day (0.72 L gas/L reactor volume per day), despite varied loadings. Lettinga et al. [12] later demonstrated anaerobic filter loadings of up to 10 kg COD/m³ per day (and shock loading of up to 17 kg COD/m³ per day) at 30°C in 23-L columns processing potato sap solutions. In Young and McCarty's experiments, conversion to methane was almost 100% for the volatile acids substrate; the total

Table 11.1 Anaerobic Filter Treatment of Volatile Acid Wastes[a]

Influent parameters				Effluent parameters			
CODi (mg/L)	θ (h)	Loading (kg/m3/day)	TSS (mg/L)	sCOD (mg/L)	tCOD (mg/L)	E (COD) (%)	Days of operation at steady state
1500	36	0.42	3	20	24	99.4	50
	18	0.85	3	135	139	90.5	36
	9	1.7	3	310	314	79.0	56
	4.5	3.4	4	470	476	68.4	40
3000	72	0.42	4	36	42	98.6	140
	36	0.85	7	230	240	92.0	22
6000	36	1.7	11	124	139	97.7	23
	18	3.4	16	772	794	86.9	35

Source: Ref. 25.
[a] θ, Hydraulic detention time; loading, kg as COD per cubic meter of reactor volume per day; TSS, total suspended solids; sCOD, soluble COD; tCOD, total (unfiltered) COD; E, efficiency of COD removal.

biological solids produced from the filter represented only 2.1% of the total COD removed by the filter.

11.4.1 Solids and Media Characteristics

The rolling action of rising gas bubbles in the anaerobic filters tended to form readily settling granular solids, which lay loosely in the interstices between stones. Solids did not readily become attached to the stones in the protein-carbohydrate filter or in the lower levels of the volatile acids filters. Later experiments by Young and Dahab [26] suggested that one-half to two-thirds of the total solids mass in anaerobic filters remained unattached in the interstitial void spaces. Upper-level solids were essentially all attached, and demonstrated lower activity than the loose solids at the lower levels. These observations suggest that the primary function of the media matrix is to promote solids retention through flocculation, granulation, and by preventing solids washout—rather than to promote retention through attachment. In particular, granulation contributes greatly to operational stability and is an essential factor in producing high removal efficiencies. The onset of suspended solids granulation coincided with rapid improvements in COD removal and methane production [30].

Experiments using different filter media tend to confirm the granulation/flocculation media functions. Poor filter performance using loose-fill Pall rings or perforated spheres suggested short circuiting (channeling) in the narrow voids around horizontal media surfaces. Narrow, horizontal passages also impeded solids transport, discouraging the granulation process and preventing effective wasting of solids, leading to plugging. In contrast, the open "honeycomb" passage design and cross-flow pattern of corrugated media facilitated solids transport and horizontal mixing while minimizing channeling; consequently, the performance improvement over the Pall rings or spheres was equivalent to a 50% increase in reactor volume [19,20,28,29].

The less active solids in the upper level of the filter may condition the solids that form at lower levels and move upward in the bed by gas bubbles or hydraulic displacement. (The filter bed would become plugged without this solids transport.) Conditioning reduces activity of these solids through decay and by promoting flocculation. For this reason, anaerobic filters shorter than 2 m may experience excessive solids washout and are not recommended [6,26]. At the same time, filter height beyond 2 m probably would not contribute materially to conversion. Thus anaerobic filter installation in a long, rectangular underground rock cavern probably would consist of a wide, shallow (2 m) bed with multiple bottom influent distributors.

11.4.2 Modeling and Flow Regimes

Anaerobic filters with clean media display almost ideal plug flow hydraulics. Solids accumulation and gas bubble production/evolution move the flow patterns more toward backmix [the continuously stirred tank reactor (CSTR)], but not completely since the solids would be evenly distributed in this case. Thus the flow regime in a mature filter is mixed, and conditions vary throughout the bed [27].

The mixed-flow regime makes mathematical modeling difficult. In fixed-film processes, diffusion limits passage of substrate from the bulk liquid into the biofilm, thereby controlling the rate of substrate uptake. The models of Williamson and McCarty [24] and Rittman and McCarty [17,18] incorporate diffusion (Fick's second law) and Monod kinetics into a single relationship to predict substrate gradient. Variations in biofilm thickness, surface area, and substrate concentration in different areas of the filter bed make this relationship difficult to apply to anaerobic filters [27], although it may be more suitable to modeling sludge blanket reactors [2] (discussed below).

Young developed the following relationships for upflow anaerobic filters, from his early experiments [26]:

Removal efficiency: $E = 100(1 - V\theta)$ (1)

Effluent quality: $S_e = \dfrac{\theta Q S_o}{aA}$ (2)

where E is the % COD removal, V the average upflow velocity through voids, θ the reactor coefficient, S_e the effluent concentration, Q the influent flow rate, S_o the influent concentration, a the media porosity, and A the reactor cross-sectional area. Hence efficiency of conversion decreases with increasing void velocity, and effluent quality decreases proportionately with flow rate.

11.4.3 Effect of Recycle

Consequently, effluent recirculation would be expected to reduce effluent quality by increasing velocity/flow, an expectation supported by testing [26]. However, effluent recirculation may reduce the concentration of some organics below toxic (inhibitory) levels, or may overcome local buffer concentration limitations by distributing biological reaction more evenly throughout filter height. The latter effect is particularly significant at higher organic loadings, where partial bed expansion by recirculation also may inhibit plugging by increasing void space of loose-fill media.

Cross et al. [5] and Suidan et al. [22] used recycle to fluidize upflow filters containing granular activated carbon media and processing coal gasification wastewater (synthetic for Suidan et al., Indianhead lignite for Cross et al.). To facilitate seeding and acclimation, both systems employed a loose-fill media packed bed prior to the fluidized bed filter (Fig. 11.8). The fluidized bed accounted for roughly 86% of the total methane production in the systems of Cross et al., which removed 86% COD, 93% phenol, and 99% cresols at a loading of 2.5 kg COD/m³ per day. In the study of Suidan et al., the first-stage, beryl-saddle-packed anaerobic filter removed very little COD, while the fluidized-bed filter accounted for 90% COD reductions. Suidan et al. concluded that the activated carbon filter was effective because of its ability to adsorb inhibitory organics, some of which might later desorb as the microorganisms acclimated to them. Cross et al. considered recirculation and fluidization important for minimizing gas entrainment, providing dilution and furnishing buffering capacity. Suidan et al. used fluidization for a 25% expansion, to reduce the risk of plugging.

11.4.4 Summary: Anaerobic Filter

The upflow anaerobic filter system has the following benefits:

1. High reliability, economic operation.
2. Higher treatment efficiencies at lower temperatures than the conventional digester are possible, due to long solids retention times and high solids concentrations.
3. Treatment of soluble substrates at moderate strength (1000 mg/L < COD < 10,000 mg/L).

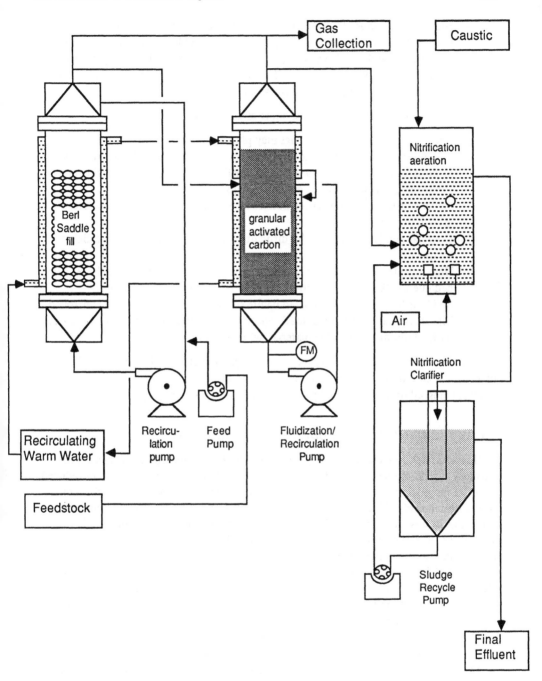

Fig. 11.8 Upflow anaerobic filter system of Suidan et al. (From Ref. 22.)

4. High stability under transient, intermittent loading. High solids retention, combined with an inherently low decay coefficient, enables the reactor to sustain long periods without flow/load. (A 4-month downtime, followed by a loading resumption at 2 kg COD/m³ per day at 3000 mg/L COD resulted in steady-state gas production in about 2 weeks [7].)

At the same time, the anaerobic filter has some disadvantages:

1. Excessive growth at high loadings may plug the filter bed.
2. Care must be exercised in selecting filter media, since types with narrow voids or horizontal surfaces may produce channeling and/or blockage, causing performance deteriorations and possibly reactor failure.
3. Media constitutes an unwanted additional expense, compared to a dispersed-growth process.
4. Granular solids, which may be difficult to form, are necessary for efficient treatment.
5. High organic loadings also may produce local pH fluctuations in the lower filter bed due to alkalinity exhaustion, possibly resulting in reactor failure.
6. Solids influx with the solubilized lignite substrate must be carefully controlled. The anaerobic filter has a limited capacity to accommodate suspended matter, particularly nondegradable material, without plugging the media.

Recirculation and/or bed expansion can mitigate the effects of high organic loading, but at the expense of mechanical complexity, pumping costs, increased washout, the potential for clogged underdrains, and possibly a more difficult seeding process.

11.5 UPFLOW ANAEROBIC SLUDGE BLANKET REACTOR

Another approach to increasing biological solids retention in anaerobic processes is to settle and recycle them back into the reactor. In 1955, Schroepfer and colleagues reduced detention times from 20 days or more to less than one day by using the anaerobic contact process on relatively dilute packinghouse wastes. The anaerobic analog of activated sludge, this process features an effluent settling tank that collects solids for recycle (Fig. 11.9). As noted earlier, however, primarily soluble substrate may cause much of the solids may remain dispersed, making them difficult to settle and recycle [25].

Separately, Stander collected and returned solids to laboratory reactors processing fermentation wastes in 1950; his experiments led to the "clarigester" reactor (Fig. 11.9), which uses a superincumbent settling tank to return solids to the digester [15]. Lettinga et al. [12] further developed this concept with the upflow anaerobic sludge blanket (UASB) reactor (Fig. 11.9), in which a well-flocculated sludge blanket or bed forms a stable phase, capable of withstanding mixing forces. A quiescent zone above the sludge bed enables sludge particles to flocculate and settle. A settler above the quiescent zone also captures sludge particles by entrapment in a secondary sludge blanket. The quiescent zone and/or settler limit solids washout to a level below the sludge bed growth rate [23].

Loading capability for the UASB appears to be higher than for the anaerobic filter, even when adjusting the filter capacity for volume occupied by the media. Lettinga et al. [12] indicate that 15 to 40-kg COD/m³ per day loadings and a 3 to 8-h detention times are possible. Low loadings (2 to 4 kg COD/m³ per day) may cause operational problems because occluded bubbles within the sludge bed may cause bed expansion, pulselike gas

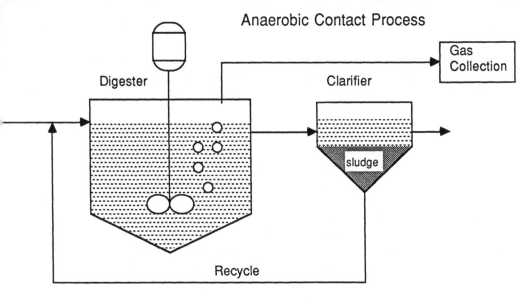

Anaerobic Contact Process

Gas Collection

Digester

Clarifier

sludge

Recycle

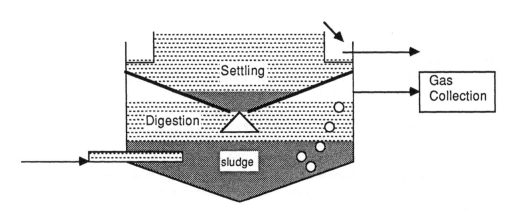

Anaerobic Clarigester

Settling

Gas Collection

Digestion

sludge

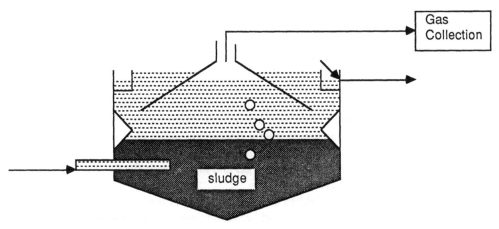

Upflow Anaerobic Sludge Blanket

Gas Collection

sludge

Fig. 11.9 Suspended-growth anaerobic processes designed to retain bacterial solids. (From Ref. 15.)

liberation, and local turnover (vertical mixing); at higher loadings the increased gas production causes the bed to resemble a boiling fluid, providing better agitation and thereby reducing channeling through canals and cracks in the sludge bed.

11.5.1 Modeling

The absence of media and the influence of rising gas bubbles causes UASB reactors to follow backmix hydraulics more than the anaerobic filter. Heertjes and van der Meer [10] defined three functional regions within their operational UASB reactor. The sludge bed contains the greatest accumulation of sludge particles, at the bottom of the reactor. Compactness of the sludge mass in this region hinders free flow of liquid, gas bubbles, and sludge particles. Above the sludge bed is the sludge blanket, which features very good mixing because of free-rising gas bubbles. The third region is the settler.

Heertjes and van der Meer conducted three Li^+ tracer tests on their UASB. Their hydraulics of their reactor corresponded to three backmix regions and one plug flow region (Fig. 11.10). The first backmix region (V_1) corresponded to the lower region of the sludge bed, near the inlet, while the second region (V_2) was near the interface between the sludge bed and the sludge blanket. The third backmix region was the sludge blanket (V_3), and the plug flow region was the settler (V_4). Short circuiting occurred between the inlet and the sludge blanket (V_3), and/or the second interfacial backmix region (V_2). The authors attributed the bypassing to density differences between influent and the sludge bed and/or to liquid transport in the wake of rising gas bubbles.

The amount of bypass and the size of the backmix regions varied with the amount of accumulated sludge in the reactor. Increasing sludge height promoted greater stabilization of the channels used by the bypassing streams, and resulted in a larger density difference between sludge and influent substrate. For these reasons, the authors recommended limiting sludge bed height to 2 to 3 m. With a minimum distance between the settler and the top of the sludge bed of 1 m, overall design height thus becomes 3 to 4 m. Increasing reactor volume (capacity) thus can be obtained only by increasing reactor cross section.

11.5.2 Summary: UASB

The UASB reactor offers some advantages over the upflow anaerobic filter for solubilized lignite treatment:

1. The UASB will not plug when handling substrates with a moderate concentration of undissolved solids. Short detention times (<24 h) make the UASB unsuitable for high suspended solids concentrations, however. Christensen et al. [4] recommend a substrate total suspended solids (TSS) limit of 500 mg/L. High concentrations of nondegradable solids dilute the bacterial mass, reducing efficiency. The presence of poorly flocculating solids may also necessitate mechanical (gas) mixing in the sludge bed.
2. The greater surface area between gas and liquid keeps floating solids from clogging gas ports [1].
3. Hydraulics of the sludge bed are closer to backmix than the anaerobic filter, leading to more efficient contact between substrate and active sludge. Backmix hydraulics also make the UASB less sensitive to low influent pH values, the presence of toxics, and to shock loadings.

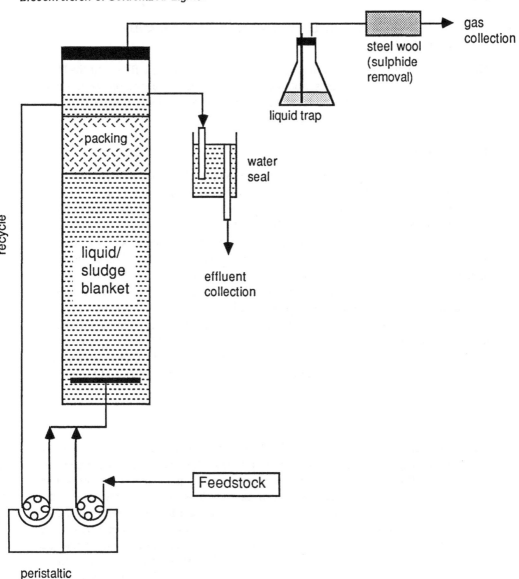

Fig. 11.10 Upflow blanket filter (UBF) apparatus of Guiot and van den Berg. (From Ref. 9.)

4. Loading rates can be much higher than are attainable with the anaerobic filter, even after accounting for the loss in volume to the filter media.

Like the anaerobic filter, however, the UASB requires granulated sludge for efficient operation. Granular sludge may be more difficult to develop in the UASB, however; sludge activity and settleability are very sensitive to startup procedures. Unlike the anaerobic filter, washout can be difficult to control, particularly during shock loadings, nutrient deficiencies, or with high concentrations of finely dispersed organic matter in the substrate. Careful settler design and control of sludge bed level are necessary.

11.6 HYBRID DESIGNS: THE UPFLOW BLANKET FILTER AND BAFFLED REACTOR

The previous discussion suggests that development of the anaerobic filter and the UASB have proceeded along similar lines; both designs seem to be converging on systems that provide large void spaces for granulated sludge development, yet also provide efficient solids/substrate mixing and minimize solids washout.

11.6.1 Upflow Blanket Filter

Guiot and van den Berg [9] discuss a reactor combining a fixed bed of plastic rings in the upper third of the vessel with a sludge blanket in the lower two-thirds (Fig. 11.11). Their 4.25-L test reactor, which they termed an "upflow blanket filter" (UBF), was capable of 93% COD removals at loadings of up to 26 kg COD/m^3 per day when processing sugar waste (sucrose) substrate. By contrast, a downflow anaerobic filter processing the same waste attained a maximum removal rate of 3 kg/m^3 per day. The UASB had a maximum methane production rate of 7 L/L per day. The authors noted no COD removals across the packing, not a surprising finding because of its location at the top of the reactor. They postulated that this fixed-bed region did enhance biomass retention by compacting the sludge and separating solids from the effluent. These functions resemble the characteristics of an efficient UASB settler. In a similar vein, Young [30] has suggested that anaerobic filter media should be placed only in the upper one-half to two-thirds of the reactor, to provide an open space for accumulation of anaerobic solids, and for initial contact between these solids and the influent wastes.

11.6.2 Baffled Reactor

Bachmann et al. [1] have performed a slightly different hybridization with the baffled reactor (Fig. 11.12), consisting of a series of upflow sludge blanket reactors sandwiched between downflow baffles. The baffled reactor does not require granular sludge, yet has the high void volume characteristic of UASB reactors. Substrate must flow under and over a series of vertical baffles as it passes from influent to effluent. Reactor solids rise and settle with gas production, but the baffled design makes horizontal motion relatively slow. Thus substrate may contact a large amount of biological solids, yet the effluent is relatively free of solids [2]. The authors employed effluent recycle (1) to reduce first-chamber organic acid production and the resulting low pH, and (2) to reduce first-chamber gelatinous bacteria growth, which caused short circuiting through channels in the sludge blanket.

A 6.3-L baffled reactor processing a complex protein-carbohydrate mixture (nutrient broth/glucose) attained 80 to 90% COD removal efficiencies at organic loadings up to 10.6 kg COD/m^3 per day at detention times ranging from 12 to 31 h. Maximum methane production was 3.7 L/L per day. This reactor performed similarly to a 0.4-L (void volume) anaerobic filter when both operated at 8 kgCOD/m^3 per day loading (Table 11.2), although the filter removed more COD and produced more methane. A 13-L baffled reactor processing nutrient broth-sucrose substrate removed 81 to 93% COD at loadings up to 11.4 kg COD/m^3 per day, and continued to show increases in both gas production and COD removal at loadings of 36.2 kg COD/m^3 per day, the highest examined. The authors observed that the reactor was capable of removing in excess of 24 kg COD/m^3 per day and producing more than 6 L/L per day of methane; they concluded that the maximum removal rate has yet to be determined.

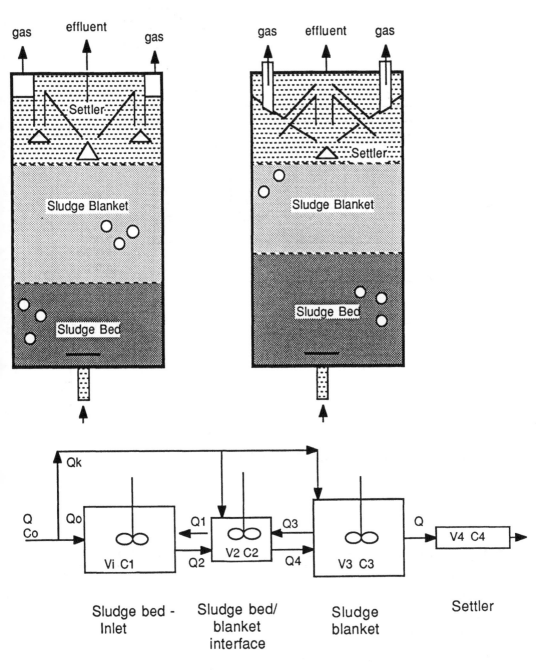

Fig. 11.11 Upflow anaerobic sludge blanket (UASB) reactor hydraulics. (Adapted from Ref. 10.)

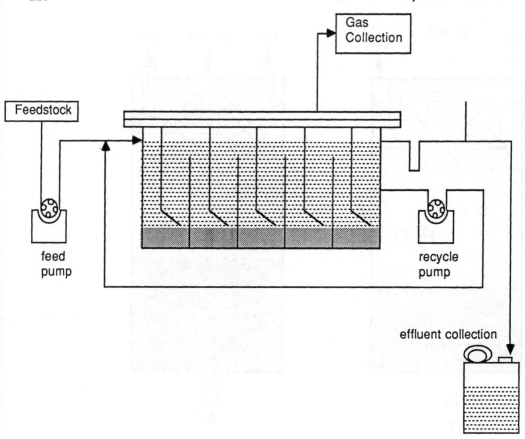

Fig. 11.12 Baffled anaerobic reactor. (From Ref. 1.)

Table 11.2 Comparative Performance for Baffled Reactor and Anaerobic Filter

Reactor type:	Baffled	Anaerobic filter
Parameter:	6.3-L	0.4-L
Influent COD (mg/L)	7600	8000
Organic loading (kg COD/m^3/day	8.3	8.0
Detention time (h)	22	24
COD removal efficiency (%)	82	92
Methane rate (L/L/day)	2.5	2.6
Methane in gas (%)	56	80
Effluent volatile acids (mg/L)	800	400

Source: Ref. 1.

The baffled reactor combines the stability and reliability of the anaerobic filter with the high void volume of the UASB reactor. It offers a simple design, which may be amenable to underground caverns. It requires no media, no special gas or sludge separation systems, and no granular sludge development. Its vertical baffle design minimizes bacterial washout from bed expansion. The absence of media minimizes clogging while maximizing void space for treatment efficiency. As with the UASB, however, short detention times limit solids-handling capability, and nondegradable solids can reduce treatment efficiency by diluting the microbial mass. Further, the narrow-rectangle geometry of a baffled reactor's passages may result in poor lateral mixing.

11.7 SOLUBILIZED LIGNITE TREATMENT: PILOT STUDY METHODOLOGY

11.7.1 Reactor Types

The pilot study program attempts to incorporate the most promising reactor designs and a range of realistic conditions into a treatability evaluation for the solubilized lignite. There are six reactors (Table 11.3, Fig. 11.13): three upflow anaerobic filters, two upflow sludge blankets (UASB), and one baffled reactor. One of the filter reactors has an external recycle pump, permitting media fluidization. A walk-in, climate/humidity-controlled room houses the reactors and related equipment.

The upflow filter reactors contain approximately 4 ft of crushed, sieved lignite particles, 6.3 to 9.5 mm (1/4 to 3/8 in. sieve sizes). This medium was a logical choice be-

Table 11.3 Pilot Study Reactor Configurations[a]

Parameter	Reactor 1	2	3	4	5	6
			Anaerobic filter			
Type	UASB	UASB				Baffled
Height	183	183	183	183	183	16.1
Width						13.7
ID	7.6	7.6	7.6	7.6	7.6	51.5
Media						
			0.63–0.95			
Size						
Height			113	118	121	
Subboard height			31	31	5.5	
Capacity						
Empty	8.01	8.01	8.01	8.01	8.01	10.37
With media	N/A	N/A	5.02	5.04	4.97	N/A
Operational[b]	7.71	7.56	4.85	4.86	4.74	7.34
Detention[c]	257	255	51	54	53	86
Mechanical mixing	Gas recycling	No.	No	No	Liquid recycling	Liquid recycling
Recycle	No	No	No	No	4–13	0–0.045
Separator	Conical	Conical	No	No	No	No

[a]Linear dimensions are centimeters, capacities are in liters, flows are in liters per meter, and detention time is in hours. The subboard represents height of the void space below the media support. All reactors are acrylic with rubber (Viton or Buna-N) O-ring seals and PVC, nylon, and stainless steel fittings.
[b]Including media, separator, and selected control level.
[c]Average values for the initial part of the study (Fig. 11.18).

Fig. 11.13 Pilot anaerobic reactors. From left: two upflow anaerobic sludge blanket (UASB) reactors, three anaerobic filters, and the baffled reactor. The leftmost UASB has provision for gas recycle, and the filter at extreme right has provision for liquid recycle; neither feature is presently in use.

cause of availability and cost. Its coarse surface may facilitate bacterial attachment. Draw-backs include its friability on handling and the possibility that the bacteria may degrade the media itself. As a loose-fill media, its void structure may be inferior to a good cross-flow design. (Loose-fill stone media reduced COD more effectively than did corrugated modular media for Dahab and Young [6], however; media selection is an uncertain process at present). Some optimization of media size probably is necessary; size selection for this study corresponds to Young's personal recommendation of 1/8 to 1/12 of reactor diameter. The two anaerobic filters without recycle contain roughly 1 ft (30 cm) of void space below the media support to permit sludge blanket development. In effect, the two filters are hybrid (UBF) designs.

The filter with recycle provision uses a progressive cavity pump to recycle effluent from the top of the reactor to the inlet area, below the media support. Figure 11.14 indicates that about 13% bed expansion is possible in a clean bed; although the pump is capable of more output, higher recycle rates cause the media to pulsate in an unstable manner. The two UASB reactors are essentially similar, except that reactor 1 has provision to recycle anaerobic gases for enhanced sludge bed mixing. As noted, gas mixing can improve sludge bed operation at low organic loading or in the presence of inert solids. The separator (settler) in these reactors collects rising gas bubbles in the center while diverting liquids and solids to the periphery (discussion below).

The baffled reactor is tabletop size and features five downflow/upflow chambers. Initially, a small variable-flow dc-motor peristaltic pump (Searle Buchler Instruments, Ft. Lee, New Jersey) provided both makeup and effluent recycle flows. An electronic

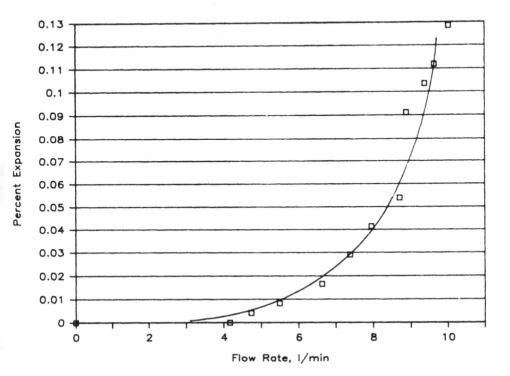

Fig. 11.14 Expanded bed reactor characteristics.

solenoid-operated diaphragm pump (Prominent Fluid Controls, Pittsburgh, Pennsylvania) has replaced the peristaltic pump on recycle. Each reactor contains a pH probe (located in the second chamber of the baffled reactor and 76 cm above the influent end of the others) for continuous monitoring of acid accumulation. The probes are inserted through an O-ring seal-and-valve apparatus to permit probe removal while the column is operational. An Orion Model 605 electrode switch (Orion Electronics, Cambridge, Massachusetts) permits one pH meter to monitor five electrodes at once. The pH meter is a data logging type, printing out pH/temperature/time values at programmed intervals (Omega Instrument Co., Stamford, Connecticut). For unknown reasons, the inserted pH probes consistently read below external measurements.

11.7.2 Process Flow

Figure 11.15 illustrates the overall flow scheme for the reactors. The preliminary scheme used a 57-L tank to prepare the feed mixture for the anaerobic reactors (substrate, plus vitamins, minerals, buffer, redox indicator, and sodium sulfide as a reducing agent; refer to Table 11.4 and the discussion below). Initial evaluations using a control substrate mix of sodium benzoate and D-glucose revealed that the full mixture was chemically unstable in the 57-L tank, however. Accordingly, the revised flow scheme uses in-line mixing of a concentrated substrate with dilution water (containing the other ingredients). Dilution water and concentrated substrate mix together in a stirred glass mixing vessel. The withdrawal of feed mixture from the mixing vessel causes the level to drop, causing an electronic float to activate electronic solenoid-driven diaphragm pumps (Liquid Metronics, Newton, Massachusetts) delivering the dilution water and substrate in the proper ratio to the mixing vessel.

Initially, electronic solenoid-driven diaphragm pumps (Prominent Fluid Controls, Pittsburgh, Pennsylvania) drew individually from the 57-L tank to supply makeup flow to

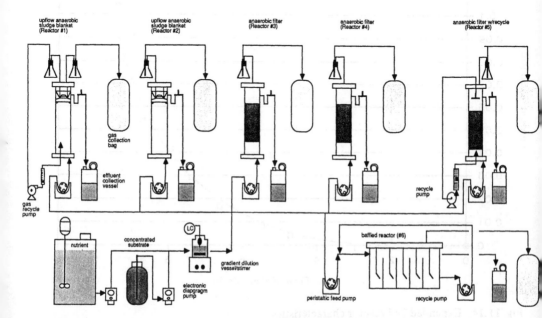

Fig. 11.15 Process flow diagram, pilot treatment system.

reactors 1 to 5. The pumps feature combination (stroke X frequency) turndown of approximately 1:200. Their performance was quite poor, unfortunately, perhaps due to their tendency to cavitate after collecting gas bubbles from the feed mixture in the diaphragm chamber. Multichannel peristaltic pumps (the Buchler pump cited previously and a Masterflex, Cole-Parmer, Chicago) eventually replaced all of the diaphragm makeup flow pumps. The Buchler has a turndown of roughly 1:10, while the Masterflex is a constant-speed pump; an external timer cycles the pump in 15-min intervals. A drip-chamber device upstream of the makeup flow pumps impedes bacterial growth back into the mixing vessel.

A feed mixture enters the bottom of each reactor (except the baffled reactor, in which fluid enters from one end) and exits through an inverted siphon at the top. Effluent drains by gravity into individual collection containers. Rising gas bubbles accumulate above the free liquid level. Exiting gas passes through liquid traps and then into Teflon or Tedlar gas bags. Periodic gas volume measurement requires transfer of the gas into a liquid displacement apparatus; a vacuum pump performs the transfer.

11.7.3 Separator Design

Separator design has proven to be particularly critical to effective operation of sludge blanket reactors 1 and 2. To reduce solids loss by gas entrainment, the separator must first divert gas bubbles away from the upflowing liquid and solids. A quiescent zone in the liquid path permits solids to settle into a secondary sludge blanket, which then filters the upflowing liquid of additional solids. The separator should provide a means for accumulating solids to fall back into the reactor.

Figure 11.16 illustrates the designs used thus far in this study. The separator in Fig. 11.16a channels liquid and solids into the center cone, after deflecting gas bubbles into the annulus. Hydraulic displacement drives effluent through an outlet at the top of the column. Entrained solids can settle near the bottom of the cone and return to the reactor. This design depends on an effective upper seal to keep the gas trapped in the annulus from escaping with the effluent. This separator was problematic in reactor 2 because of seal leakage, and because maintaining a steady gas-liquid interface in the annulus was difficult; liquid frequently plugged the gas product line, causing the interface to drop until gas entered the center cone.

The second separator design (Fig. 11.16b) uses an hourglass shape to deflect gas bubbles to a center hood for collection. Hydraulic displacement forces liquid to pass through a narrow gap between the center hood and the upper portion of the hourglass, out into the annulus for effluent collection. The gap permits solids to accumulate, forming the secondary sludge blanket. Because gas bubbles can penetrate the narrow clearance between the separator and the column walls, however, product gas accumulates in the liquid effluent space, dropping the gas-liquid interface until gas exits through the liquid effluent line. Further, smaller gas bubbles may be able to "hug" the shape of the hourglass in a sort of Coanda effect, ending up in the liquid section. A second gas collection line (from the liquid headspace) reduced the problem of gas accumulation by equalizing pressure between the center and annular sections of the separator.

Figure 11.6c illustrates a third separator design, now in use on reactor 2. Liquid and gas flows are similar to the second design (Fig. 11.16b), but the geometry is slighly different. A double-conical shape replaces the hourglass so that gas bubbles cannot follow the separator contours into the liquid section. The center section is open at the top, providing headspace equalization between center and annulus. The gap through which liquid must

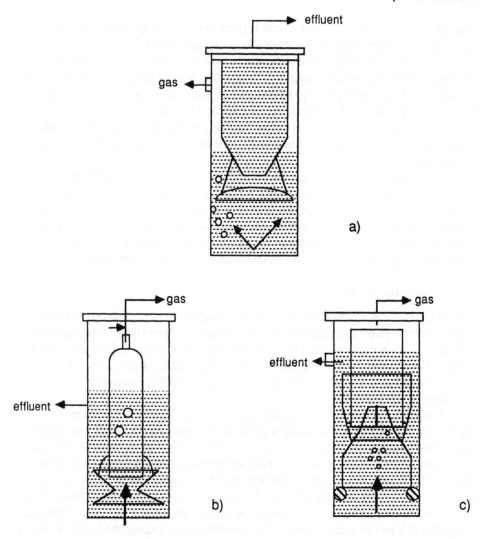

Fig. 11.16 Separator designs.

pass is much narrower, permitting more solids to accumulate. An oversized O-ring beneath the separator provides vertical support and prevents gas bubles from creeping up
the outside the separator.

Within hours of installation, this separator provided good solids separation and
liquid-gas interface control. The design may allow too much biomass solids to accumulate, however, with the result that gas generation from the secondary sludge blanket
disturbs effluent clarity. A modified version of this separator, currently under construction, features slots in the lower cone that facilitate sludge return to the lower part of the
reactor. For an in-depth discussion of separator design, the reader may refer to van der
Meer and de Vletter [23].

Figure 11.7 illustrates separator performance by showing effluent clarity as total
suspended solids (TSS, mg/L) in the two sludge blanket reactors 1 and 2. Effluent TSS
was much higher in reactor 2, with no separator, compared with reactor 1, which uses the

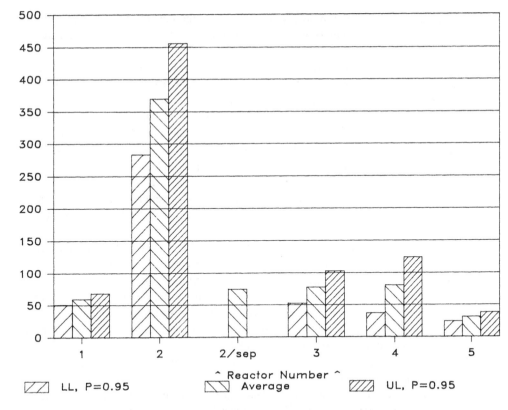

Fig. 11.17 Effluent quality (mg/L) in total suspended solids (TSS) from reactor 1 (separator, Fig. 11.16b); reactor 2 (no separator); reactor 2 (separator, Fig. 11.16c); reactors 3 to 5 (anaerobic filters). The UASB reactors (1 and 2) operated at roughly 250 h detention time, while the anaerobic filters operated at 50 h. Substrate is identical in all cases.

separator described in Fig. 11.16b. After installing the separator described in Fig. 11.16c, however, effluent from reactor 2 improved significantly. Both reactors operated at hydraulic detention times of roughly 250 h. For comparison, effluent from the anaerobic filters (reactors 3 to 5, with detention times around 50 h) also is shown.

11.7.4 Feed (Substrate) Composition

The temporary unavailability of pretreated solubilized lignite has prevented evaluation of anaerobic reactor performance with the real substrate material. Accordingly, a control substrate with D-glucose and sodium benzoate as carbon sources presently supplies the reactors. D-glucose, a simple carbohydrate, is a preferred substrate for the microorganisms to digest, while sodium benzoate, a ring compound, is less degradable and more likely to resemble portions of the solubilized lignite. Feeding pure benzoate to the cultures on startup resulted in inhibition—hence the mixture of carbohydrate and benzoate. Gradual phasing out of the D-glucose should be possible, since the author (and others, see Leuschner et al. [13]) has bottled cultures growing on pure benzoate. At this writing the substrate changed from a 50%/50% mix to an 80%/20% mix as % COD of benzoate/glucose.

Another lignite analog, sodium lignosulfonate, will be the subject of batch degradability testing; if successful, it could serve as a more accurate model substrate for the reactor tests.

Initially, the feed mixture going to the columns was a scaled-up version of the Anaerobic Toxicity Assay/Biomethanation Potential (ATA/BMP) medium of Owen et al. [16]. In addition to a carbon source, this medium consists of a major/trace ions mixture, a B-vitamin solution, a phosphate buffer, ferrous chloride, sodium sulfide (as a reducing agent), a redox indicator, and bicarbonate buffer. Originally intended to serve as a rich, complete medium for batch inoculations, the recipe required modification for large-scale preparation. These modifications eliminated boiling (sterilization) and pH adjustment with carbon dioxide sparging. Additionally, the major/trace ion levels were reduced by a factor of 5, on recommendation of Ed Bouwer (personal communication, 1986). A higher bicarbonate buffer dose was necessary to counteract the higher COD of the control substrate.

In this form, Owen's media was quite unstable. Within 48 h, the solution pH dropped precipitously. As pH approached the carbonic acid-bicarbonate equivalence point, the solution evolved large quantities of carbon dioxide. Gas evolution in the lines cavitated the feed supply pumps, causing flow to be unsteady. Discussion with McCarty (personal communication, 1986) revealed that Owen's media was unstable when exposed to the atmosphere; sulfide reducing agent forms sulfuric acid, and reactions with the trace heavy metals could take place. McCarty suggested using a simpler formula, excluding B vitamins and reducing agent, and adding trace metals either by separate (slug) dose or by using tap water to prepare the solution.

Table 11.4 compares the revised feed mixture with the original mixture. Ferric chloride solution replaces ferrous chloride suspension for convenience (reducing conditions in the reactors convert ferric ion to ferrous ion). The remaining ingredients are at much lower concentration, and trace metals (Ni, Co) are added separately by daily slug-dose injection. Combined with the process flow modifications described earlier, the revised mixture has been more stable, making steady flow possible through the anaerobic reactors. Table 11.5 lists general feed mixture and operating conditions, as recommended by Young [30].

11.7.5 Startup (Inoculation)

Reactor inoculation followed the two-stage method recommended by Young [30]. The first stage consists of a 1 to 2% (of reactor volume) inoculation with no makeup flow, with the reactor liquid-full of substrate. The mixture sits for 2 to 3 days to "condition" the reactor (i.e., to begin acclimation and remove traces of residual oxygen). The second stage consists of a heavier seed, about 5 to 10% of reactor volume for the anaerobic filters. Makeup flow starts after seeding, at a rate of roughly 0.1 lb COD/ft^3 (1.6 kg/m^3) per day and/or a 36 to 48-h detention time. Further seed may be added to accelerate startup after stable operation occurs. Care must be taken not to overseed the reactor, however, since an overseed will result in unrealistically good removal performance.

Seeding the anaerobic filters consisted of pressuring 50 mL of granulated sludge into the column bottoms, using 70% N_2/30% CO_2 gas. The seed material came from an idle 370-L reactor that had been processing carbohydrate substrate (courtesy of J. Young, University of Arkansas). The anaerobic filters were filled with substrate just prior to seeding. When the redox indicator turned from pink to clear, the second seed (0.5 L) was pressured into the column bottom.

Table 11.4 Original and Revised Feed Mixtures for Anaerobic Reactors 1 to 6

Organic constituent	Concentration as:		
	mg/L	sTOC	sCOD
Sodium benzoate	3000	1750	5000
D-Glucose	4688	1875	5000
Rezazurin (redox indicator)	0.9		

Inorganic constituent	Final concentration in mg/L (% of COD in parentheses)			
	Modified Owen's media		Revised formulation	
$(NH_4)_2HPO_4$	72.1	(0.72)	20.0	(0.2)
$CaCl_2 \cdot 2H_2O$	45.1	(0.45)	45.1	(0.45)
NH_4Cl	71.8	(0.72)	71.8	(0.72)
$MgCl_2 \cdot 6H_2O$	324.0	(3.24)	324.0	(3.24)
KCl	234.1	(2.34)	234.0	(2.34)
$MnCl_2 \cdot 4H_2O$	3.59	(0.04)		
$CoCl_2 \cdot 6H_2O$	5.40	(0.05)	0.92	(0.009)[a]
$NiCl_2 \cdot 6H_2O$			0.92	(0.009)[a]
H_3BO_3	1.03	(0.01)		
$CuCl_2 \cdot 2H_2O$	0.486	(0.005)		
$Na_2MoO_4 \cdot 2H_2O$	0.459	(0.005)		
$ZnCl_2$	0.378	(0.004)		
$FeCl_2 \cdot 4H_2O$	333.0	(3.33)		
$FeCl_3 \cdot 6H_2O$			167.3	(1.67)
$Na_2S \cdot 9H_2O$	450.0	(4.5)		
$NaHCO_3$	6667.0	(66.7)[b]	6667.0	(66.7)

Vitamin B Concentration (μg/L)

Biotin	18
Folic acid	18
Pyridoxine HCl	90
Riboflavin	45
Thiamine	45
Nicotinic acid	45
Pantothenic acid	45
B_{12}	0.9
p-Aminobenzoic acid	45
Thioctic acid	45

[a] Added as a daily injection (slug dose) to each reactor.
[b] Bicarbonate represents 50% of COD as alkalinity.

Table 11.5 Recommended Operating Conditions for Anaerobic Reactors

Item/element	Concentration/recommendation
Alkalinity	25% of COD minimum (startup)[a]
	10% of COD minimum (steady-state)
pH	6.5 minimum (6.8 ideal) in sludge zone
Temperature	25°C minimum, 38°C maximum (mesophilic)
	50°C minimum, 60°C maximum (thermophilic)
NH_3-N	10% of net solids yield, minimum, or 20 mg/L residual in solution
Phosphorus	15% of NH_3-N requirement, or 1 mg/L in solution
Sulfur	5% maximum H_2S in product gas, 10 mg/L total in solution, minimum
Ni, Co, Se	0.1 mg/L minimum in substrate stream

Source: Ref. 30.
[a]Young's personal recommendation for alkalinity at startup is 50% of COD.

Twice as much seed went into the two UASB reactors, for which startup is a more difficult and critical procedure. Lettinga et al. [12] recommend an initial flow condition of 0.1 to 0.2 g COD/g total solids (seed) per day for UASB reactors. Because the liter of seed added was roughly 20,000 mg/L TSS, the maximum recommended initial loading is then (1.0) (20) (0.2) = 4.0 g COD per day. At an influent COD of 10,000 mg/L, the required flow rate is (4/10) L/d or 0.278 mL/m, about 463 h of detention time in reactor 1.

The feed rate for reactors 1 and 2 has been about twice this recommended value due to low-flow metering limitations. Actual applied loading has been less than twice (below 8.0 g/day), however, because measured soluble TOC (sTOC) values are lower than calculated values. Figure 11.18b shows a plot of average applied organic loading, which for reactors 1 and 2 has been 0.329 and 0.336 g sTOC/L per day, respectively. Multiplying by 2.762 g COD/g TOC (50-50 mix of benzoate and glucose) and reactor volume (Table 11.3) yields 7.006 and 7.016 g COD/day.

11.7.6 Analytical Monitoring

Young's recommendations for anaerobic filter analytical monitoring appear in Table 11.6, along with the actual test frequencies. sTOC monitoring occurs daily; samples are 0.45 μm filtered and diluted for injection into a Dohrmann DC-80 (persulfate-assisted UV) carbon analyzer. The measured effluent samples are daily composites. pH samples are measured daily by withdrawing a sample from each reactor (76 cm from the bottom of the column reactors, the second chamber of the baffled reactor). pH and temperature monitoring occur continuously on one of the six reactors (rotating basis), but the pH values tend to read low, as noted previously. Organic acids are measured on a packed-column gas chromatograph with flame ionization detector (Perkin-Elmer Sigma 300), using the membrane-filtered effluent samples; the 2 mm (ID) × 1.83 m (L) gas column contains acid packing (AT-1100, Alltech, Inc., Chicago, Illinois). Effluent alkalinity measurements (Hach digital titrator) are run to pH 4.5. Volatile and total solids analyses of the composite effluent employ the Standard Methods glass-fiber filtration procedure [21]. Gas composition measurements are from the headspace of each reactor, analyzed using a Fisher gas partitioner (porous polymer and molecular sieve columns, thermal conductivity detector).

Table 11.6 Recommended/Actual Analytical Monitoring Procedures and Frequencies

Test	Test frequency		
	Startup	Long-term	Current (this study)
TOC or COD	6–8 h	1–2 per day	Daily
pH	6–8 h	1–2 per day	Daily
Organic acids	Daily	1–2 per week	2 per week
Alkalinity	2–3 per week	1 per week	1 per week
NH₃-N	1–2 per week	1 per week	[a]
Phosphorus	1–2 per week	1 per week	[a]
Effluent VSS/TSS	–	–	3 per week
Gas composition	–	–	1 per week
Gas quantity	–	–	3 per week
Micronutrients (Fe, Ni, Co, Mg, Ca)	1 per month	1 per month	[a]

Source: J. C. Young, personal recommendation, 1986.
[a] Analysis not performed yet.

11.7.7 Performance

Figure 11.18 contains four graphs summarizing reactor performance to date on the control substrate. In each graph, the center bar represents the average over the test period (10/22/86 to 11/22/86), while the left and right bars are an estimate of the 95% confidence interval around the average [$P(LL \leqslant$ average $\leqslant UL) = 0.95$]. With detention times around 255 h, the UASB reactors removed the most sTOC (about 70%) but generated the least gas (2.3 and 1.7 L/day for reactors 1 and 2, respectively). The anaerobic filters (reactors 3 to 5) removed less sTOC (23 to 33%, at 50 h of detention time) but produced more gas (3 to 4.5 L/Day). These low sTOC removal levels suggest that the anaerobic filters are degrading the glucose but not degrading the benzoate.

The baffled reactor (6) produced a similar amount of gas, 4 L/day, but at longer detention times (86 h) and with poorer sTOC removal (20%) than the anaerobic filters. The baffled reactor's biomass levels are small enough that liquid can bypass the sludge blanket in some of the compartments; combined with poorer lateral mixing, the bypass may contribute to higher sTOC in the effluent. Performance might have been slightly better in the period than indicated; the averages include a 12-h period of inadvertent shocking (concentrated substrate, no diluent/nutrient/buffer), after which TOC removals were negative for a few days. As biomass accumulates in the anaerobic reactors and the applied organic loading increases, the gas production rate can be expected to improve.

11.7.8 High-Salt Operation

Because of the possibility of using underground salt caverns as anaerobic reactors, saline conditions are of keen interest in this study. Unfortunately, the saline organisms available to the authors at this time (13% salt, Gulf coast sediments organisms) are not producing significant amounts of gas (Annette Martel, personal communication, 1986). Moreover, these cultures do not represent the full complement of organisms needed for a community (acid formers, acetogens, acetoclastic and hydrogenotrophic methanogens). Accordingly, for expedience the test program begins with freshwater conditions.

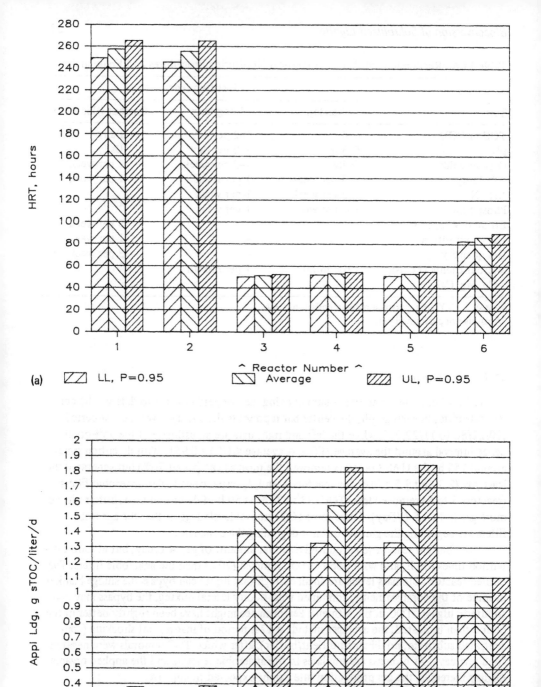

Fig. 11.18 Anaerobic reactor performance on 50-50 sodium benzoate/D-glucose mixture: (a) average hydraulic residence time; (b) average applied organic loading; (c) average percent sTOC removal; (d) average gas production (liters per day).

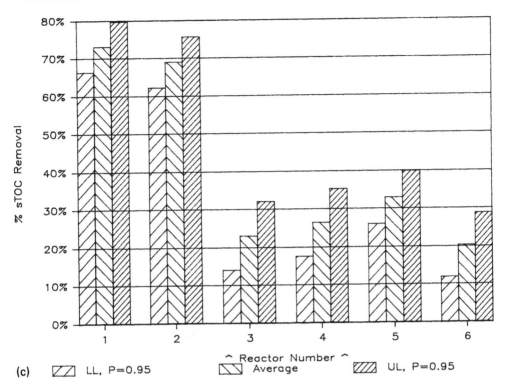

(c) ▨ LL, P=0.95   ▨ Average   ▨ UL, P=0.95

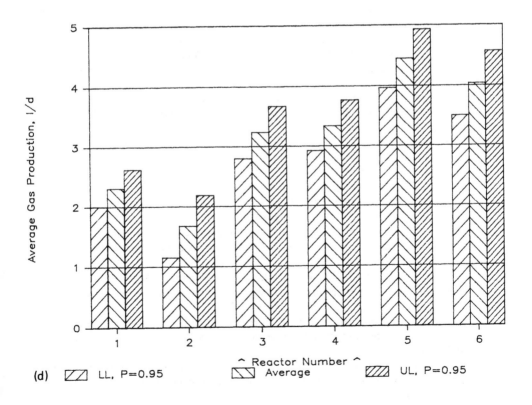

(d) ▨ LL, P=0.95   ▨ Average   ▨ UL, P=0.95

Acclimation of freshwater cultures to saline conditions may be possible. DeBaere et al. [8] were successful in slowly acclimating mixed freshwater cultures on an ethanol/ acetate substrate to high salt. Initial inhibition occurred around 65 g/L of NaCl; 95 g/L NaCl resulted in 50% reduction in production of gas and in substrate removal. At 120 g/L (12%) NaCl, the highest concentration tested, gas production declined to 15% of its fresh-water value (8 L/day) and TOC removal was less than 10% of its normal value. The 1-L polyurethane foam-filled reactors contained 99% *Methoanosarcina* as biomass. The maximum-tested 12% salt level is below the anticipated conditions inside a salt-dome fermentor, which does not lend much encouragement to the acclimation concept. The inoculation/salinity issue remains to be resolved.

11.8 CONCLUSIONS

High-rate anaerobic filter, sludge blanket, and hybrid reactors present the most inter-esting alternatives for biogasification of solubilized lignite. A combination of large void volume for good lateral mixing and retention mechanisms that maintain a high solids con-centration within the reactor provide a good environment for substrate conversion. Once under way, the pilot study program will provide useful comparisons between reactor types on operability and hydraulic/organic loading tolerance. This information, in turn, will lead to selection of the most appropriate reactor type for the full-scale system.

REFERENCES

1. Bachmann, A., Beard, V. L., and McCarty, P. (1984). *Comparison of Fixed-Film Reactors with a Modified Sludge Blanket Reactor*, Department of Civil Engineering, Stanford University, Stanford, Calif.
2. Bachmann, A., Beard, V. L., and McCarty, P. L., Performance characteristics of the anaerobic baffled reactor, *Water Res., 19*(1), 99–106 (1985).
3. Bouwer, E. J. (1984). Lectures on anaerobic fermentation at the University of Houston, Fall 1984.
4. Christensen, D. R., Gerick, J. A., and Eblen, J. E., Design and operation of an up-flow anaerobic sludge blanket reactor. *Water Pollut. Control Fed., 56*(9), 1059–1062 (1984).
5. Cross, W. H., Chian, E. S. K., Pohland, F. C., Harper, S., Kharkar, S., Cheng, S. S., and Lu, F., Anaerobic biological treatment of coal gasifier effluent, *Biotechnol. Bioeng. Symp., 12*, 349–363 (1982).
6. Dahab, M. F., and Young, J. C., Retention and distribution of biological solids in fixed-bed anaerobic filters, paper presented at the *First International Conference on Fixed-Film Biological Processes*, Kings Island, Ohio, Apr. 20–23, 1982.
7. Dahab, M. F., and Young, J. C., Response of anaerobic filters to transient loadings, paper presented at the *1983 National Conference on Environmental Engineering*, Boulder, Colo., July 6–8, 1983.
8. DeBaere, L. A., Devocht, M., Van Assche, P., and Verstraete, W., Influence of high NaCl and NH_4Cl salt levels on methanogenic associations, *Water Res., 18*(5), 543–548 (1984).
9. Guiot, S. R., and van den Berg, L., Performance of an upflow anaerobic reactor com-binding a sludge blanket and a filter treating sugar waste, *Biotechnol. Bioeng., 27*, 800–806 (1985).
10. Heertjes, P. M., and van der Meer, R. R., Dynamics of liquid flow in an up-flow reac-tor used for anaerobic treatment of wastewater, *Biotechnol. Bioeng., 20*, 1577–1594 (1978).

11. Hergenroeder, R., Parkin, G. F., and Speece, R. E., *Methane Fermentation of Coal Conversion Wastewater Constituents*, Environmental Studies Institute, Drexel University, Philadelphia, 1986.
12. Lettinga, G., van Velsen, A. F. M., Hobma, S. W., de Zeeuw, W., and Klapwijk, A., Use of the upflow sludge blanket (USB) reactor concept for biological wastewater treatment, especially for anaerobic treatment, *Biotechnol. Bioeng., 22*, 699–734 (1980).
13. Leuschner, A. L., Trantolo, D. J., Kern, E. E., and Wise, D. L., *Biogasification of Texas Lignite*, Dynatech R/D Company, Cambridge, Mass., and Houston Lighting and Power, Houston, Tex., 1985.
14. Leuschner, A. L., Laquidara, M. J., and Martel, A. S., Biological methane production from Texas lignite, presented at the *Biotechnology for Coal and Lignite Workshop*, The Woodlands, Houston, Tex., June 10–11, 1986.
15. McCarty, P. L., One hundred years of anaerobic treatment, paper presented at the *2nd International Conference on Anaerobic Digestion*, Travemünde, West Germany, Sept. 7, 1981.
16. Owen, W. F., Stuckey, D. C., Healy, J. B., Jr., Young, L. Y., and McCarty, P. L., Bioassay for monitoring biochemical methane potential and anaerobic toxicity, *Water Res., 13*, 485–492 (1979).
17. Rittman, B. E., and McCarty, P. L., Variable-order model of bacterial film kinetics, *J. Environ. Eng. Div. ASCE, 104* (EE5), Proc. Pap. 14067, 889–900 (1978).
18. Rittmann, B. E., and McCarty, P. L., Design of fixed-film processes with steady-state biofilm model, *Prog. Water Technol., 12*, 271–281 (1980).
19. Song, K.-H., and Young, J. C., Media design factors for fixed-film anaerobic filters, paper presented at the *58th Annual Conference of the Water Pollution Control Federation*, Kansas City, Mo., Oct. 7–10, 1985.
20. Song, K.-H., and Young, J. C., Media design factors for fixed-bed filters, *J. Water Pollut. Control Fed., 58* (2), 115–121 (1986).
21. *Standard Methods for the Examination of Water and Waste Water*, 16th ed., American Public Health Association, 1985.
22. Suidan, M. T., Siekerka, G. L., Kao, S.-W., Pfeffer, J. T., Anaerobic filters for the treatment of coal gasification wastewater, *Biotechnol. Bioeng., 25*, 1581–1596 (1983).
23. van der Meer, R. R., and de Vletter, R., Anaerobic treatment of wastewater: the gas-liquid-sludge separator, *J. Water Pollut. Control Fed., 54* (11), 1482–1492 (1982).
24. Williamson, K., and McCarty, P. L., A model of substrate utilization by bacterial films, *J. Water Pollut. Control Fed., 48*, 9–24 (1976).
25. Young, J. C., and McCarty, P. L., The anaerobic filter for waste treatment, *Proceedings of the 22nd Industrial Waste Conference*, Purdue University Engineering Extension Series, Vol. 129, 1967, pp. 559–574.
26. Young, J. C., and Dahab, M. F., Operational characteristics of anaerobic packed-bed reactors, *Biotechnol. Bioeng. Symp., 12*, 303–316 (1982).
27. Young, J. C., The anaerobic filter: past, present, and future, paper presented at the *3rd International Symposium on Anaerobic Digestion*, Boston, Aug. 14–20, 1983.
28. Young, J. C., and Dahab, M. F., Effect of media design on the performance of fixed-bed anaerobic reactors, *Water Sci. Technol., 15*, 369–383 (1983).
29. Young, J. C., and Song, K.-H., Factors affecting selection of media for anaerobic filters, paper presented at the *2nd International Conference on Fixed-Film Biological Processes*, Arlington, Va., July 10–12, 1984.
30. Young, J. C., Anaerobic filters for methane production from industrial wastes, paper presented at the *7th Miami International Conference on Alternative Energy Sources*, Miami Beach, Fla., Dec. 9–11, 1985.

12

Mechanism of Anaerobic Degradation of Aromatic Compounds in Texas Lignite by Bacteria

G. FUCHS *University of Ulm, Ulm, West Germany*

A. KRÖGER *J. W. Goethe-University, Frankfurt/Main, West Germany*

RUDOLF K. THAUER *University of Marburg, Marburg, West Germany*

12.1 INTRODUCTION

Aromatic compounds can be degraded biologically under aerobic or anaerobic conditions. While the mechanisms of aerobic degradation have been studied thoroughly [13], there are only few reports on the anaerobic pathways [1-12], which differ from the aerobic processes in that oxygenase reactions are not involved. Anaerobic bacteria are known that oxidize aromatic compounds with protons (H_2 production) [4,6,7,12], sulfate (H_2S production) [9], or nitrate (N_2 production) [1,3,5,10]. It is generally assumed that these bacteria use a common pathway of oxidation, irrespective of the electron acceptor. With protons as electron acceptor, the anaerobic oxidation of aromatic compounds is exergonic only if the partial pressure of H_2 is kept low. Therefore, these bacteria require the presence of H_2 utilizing anaerobes for growth (methanogens, acetogens, or sulfate reducers). These mixed cultures are not well suited for studying the biochemistry of the process. In contrast, with nitrate or sulfate, pure cultures growing on aromatic compounds can be obtained. We have isolated nitrate-reducing bacteria growing with monocyclic aromatic compounds. These are considered as model compounds of the constituents of pretreated Texas lignite.

12.2 AROMATIC COMPOUNDS IN TEXAS LIGNITE AS SUBSTRATES FOR ANAEROBIC BACTERIA

Raw and pretreated Texas lignite has been analyzed with respect to elemental composition and molecular weight (Table 12.1). The chemical summation formula of the treated lignite is approximately $C_7H_7O_2$. The lignite carbon has essentially aromatic character;

Table 12.1 Elemental Composition of Texas Lignite

Typical coal	Percent by weight				Residue	Atomic ratio			Chemical summation formula
	C	H	O	N		C	H	O	
Texas raw lignite[a]	45.2	3.7	17.1	1.0	32	3.75	3.7	1.07	$C_{1.0}H_{1.0}O_{0.28}$
Texas pretreated lignite[a]	52.5	3.9	23.1	0.9	19	4.38	3.9	1.44	$C_{1.0}H_{0.89}O_{0.33}$
Average lignite	50.7	4.4	12.9	1.0	32	4.22	4.4	0.81	$C_{1.0}H_{1.04}O_{0.19}$

[a]Dried at 2 mbar, 25°C.

the oxygen is mostly phenolic OH and carboxylic COO⁻. Therefore, benzene and hydroxylated and carboxylated benzenes can be considered as good model substances for thermodynamic calculations and experimental studies. The microbiological experience is that products of alkaline hydrolysis of lignite are fermented to a greater proportion the smaller the molecular weight. Therefore, the molecular weight of the water-soluble fraction of pretreated Texas lignite was determined by gel filtration on Sephadex G25; the molecular weight of the majority of the compounds was found to be in the order of that of cytochrome c (13,000). The handling of the samples was done at pH 10 and under aerobic conditions. At pH 7 most of the material was found to be insoluble in water. It is likely that low-molecular constituents of the pretreated sample had polymerized as a consequence of the exposure to air. To date, the oxygen-promoted polymerization of pretreated lignite has not been taken into account sufficiently. It is recommended that handling and storage of the pretreated lignite be done under standardized anaerobic and pH-controlled conditions in future studies.

12.3 ISOLATION OF BACTERIA CAPABLE OF ANAEROBIC GROWTH ON AROMATIC COMPOUNDS IN TEXAS LIGNITE

We have isolated several new bacteria that grow anaerobically with nitrate on benzoic acid and its derivatives, and on phenolic compounds, respectively. The properties of two representative strains are described here. The benzoate degrading strain KB 740 [5] grows optimally at 30°C and neutral pH on the compounds listed in Table 12.2. The com-

Table 12.2 Anaerobic Growth of the Benzoate Degrading Strain KB 740 on Various Benzoate Derivatives and Nitrate[a]

Aromatic compound	Substituent	Growth	Synthetase activity (%)
COOH, X (benzene)	H	+	100
	NH₂	+	100
	OH	−	2
	X = F	+	108
	Cl	−	12
	Br	−	2
	CH₃	−	18
COOH, X (benzene)	COOH	+	0
	X = NH₂	−	8
	OH	+	23
	Cl	−	2
COOH, X (benzene)	NH₂	−	2
	OH	+	9
	X = F	−	100
	Cl	−	3
	Br	−	0

[a]The specific activity of the enzyme(s) catalyzing the activation of the benzoate derivatives to the respective CoA-thioesters were measured in benzoate-grown cells. Full activity (100%) is equivalent to 0.55 μmol substrate/min per milligram of soluble bacterial protein. This strain was isolated by Braun and Gibson (see Ref. 5).

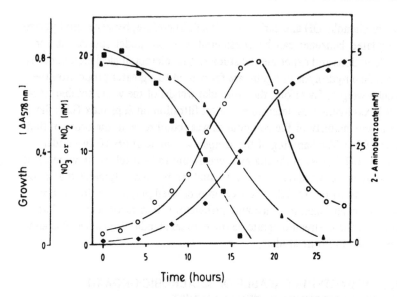

Fig. 12.1 Anaerobic growth of the benzoate degrading strain KB 740 on anthranilate plus nitrate in batch culture. A ΔA_{578nm} value of 1 corresponds to a cell concentration of 1.2 g cells (wet weight) per liter. △—△, 2-Aminobenzoate = anthranilate; ■—■, NO_3^-; ♦—♦, ΔA_{578nm}; ○—○, NO_2^-.

pounds are oxidized to CO_2. Concomitantly, nitrate is reduced to N_2, with nitrite as intermediate. The generation time with most of the substrates was approximately 5 h. With 5 mM substrate, the final cell concentration reached was 1 g wet weight per liter. A typical growth experiment is shown in Fig. 12.1. The phenol degrading strain S100 grows with a generation time of approximately 12 h at 30°C and neutral pH on the compounds listed in Table 12.3. CO_2 and N_2 were the products. A fed batch technique was used, since the phenols proved to be toxic at concentrations higher than 0.1 g/L. The final cell concentrations were approximately 1 g wet weight per liter.

12.4 BIOCHEMISTRY OF DEGRADATION OF SOME AROMATIC COMPOUNDS BY THE ISOLATED BACTERIA

The pathway of degradation has been investigated with one denitrifying *Pseudomonas* sp. and phenol as the substrate (Fig. 12.2). The first step appears to be carboxylation of phenol to 4-hydroxybenzoate, catalyzed by an inducible enzyme tentatively named phenol carboxylase (I). Cell extracts also contained an inducible 4-hydroxybenzoyl CoA synthetase (II), which activates the aromatic acid. This is followed by reductive dehydroxylation of the *para*-hydroxyl group, resulting in benzoyl CoA formation; this enzyme, 4-hydroxybenzoyl CoA reductase (dehydroxylating) (III), is also inducible. Finally, benzoyl CoA is reduced to a not-yet-defined CoA thioester of cyclohexanecarboxylate or cyclohexenecarboxylate; this reaction (IV) remains to be studied. We assume that the reduction product is further degraded via β-oxidation, essentially to acetylcoenzyme A, possibly via pimelyl-CoA as intermediate. The analogous degradation of cyclohexanol has been shown to proceed via the sequence of reactions shown in Fig. 12.3.

Table 3 Aromatic Compounds Used as Growth Substrate by the Phenol Degrading Strain S 100 in the Presence of Nitrate Under Anaerobic Conditions

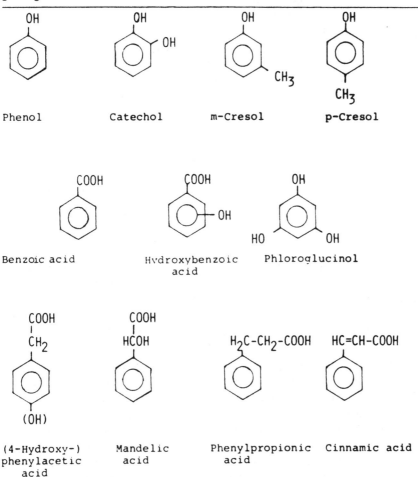

Phenol   Catechol   m-Cresol   p-Cresol

Benzoic acid   Hydroxybenzoic acid   Phloroglucinol

(4-Hydroxy-) phenylacetic acid   Mandelic acid   Phenylpropionic acid   Cinnamic acid

I   II   III   IV

Fig. 12.2 Suggested pathway of anaerobic phenol degradation by a denitrifying *Pseudomonas* sp. (From Ref. 14.)

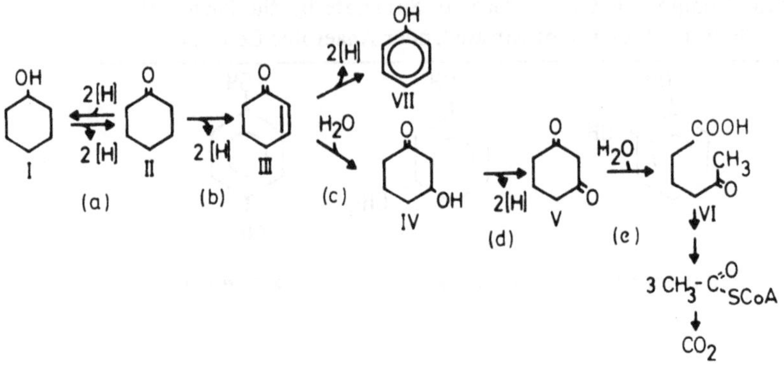

Fig. 12.3 Suggested pathway of anaerobic cyclohexanol degradation by a denitrifying *Pseudomonas* sp. I, Cyclohexanol; II, cyclohexanone; III, cyclohexenone; IV, 3-hydroxycyclohexanone; V, 1,3-cyclohexanedione; VI, 5-oxocaproic acid. (From Ref. 15.)

Studies on the pathways for degradation of other benzoate derivatives have been initiated. It appears that the enormous variety of aromatic compounds is channeled into a few central intermediates, such as phenol, benzoic acid, resorcinol, or phloroglucinol. These compounds are reductively metabolized. It is clear that the biochemistry and metabolic roules underlying anaerobic degradation of aromatic compounds are quite different from the aerobic metabolism (for literature, see Refs. 16 and 17).

Even a novel aerobic pathway involving the coenzyme A thioester of 2-aminobenzoic acid (anthranilic acid) has been disclosed in facultatively anaerobic, denitrifying bacteria. 2-Aminobenzoyl-CoA is formed during aerobic growth by a coenzyme A ligase and is metabolized in a reaction requiring 1 mol of oxygen and at least 2 mol of NADH per mole of 2-aminobenzoyl-CoA. The product of the reaction catalyzed by a novel flavoenzyme, 2-aminobenzoyl-CoA monooxygenase/reductase, is nonaromatic; preliminary data suggest 2-amino-5-oxocyclohex-1-ene carboxyl-CoA [18,19] (also, S. Ghisla, B. Langkau, K. Ziegler, R. Buder, G. Fuchs, unpublished results). The fate of 2-aminobenzoyl-CoA under anaerobic conditions is still at issue.

12.5 CONCLUSIONS

The results presented show that the anaerobic oxidation of aromatic compounds involves intermediary reduction steps, as proposed earlier. The enzymes involved catalyze reaction types which have not been encountered before in aerobic biological systems. This opens up new possibilities for the production of chemicals of commercial interest from aromatic compounds. The elucidation of the mechanism is also a prerequisite for a directed improvement of biogasification of Texas lignite.

REFERENCES

1. Bakker, G., Anaerobic degradation of aromatic compounds in the presence of nitrate, *FEMS Microbiol. Lett.*, *1*, 103–108 (1977).
2. Dutton, P. L., and Evans, W. C., The metabolism of aromatic compounds by *Rhodopseudomonas palustris*: a new, reductive, method of aromatic ring metabolism, *Biochem. J.*, *113*, 525–536 (1969).

3. Evans, W. C., Biochemistry of the bacterial catabolism of aromatic compounds in anaerobic environments, *Nature (London), 270*, 17–22 (1977).
4. Mountford, D. O., and Bryant, M. P., Isolation and characterization of an anerobic synthrophic benzoate-degrading bacterium from sewage sludge, *Arch. Microbiol., 133*, 249–256 (1982).
5. Schennen, U., Braun, K., and Knackmuss, H. J., Anaerobic degradation of 2-fluorobenzoate by benzoate degrading, denitrifying bacteria, *J. Bacteriol., 161*, 321–325 (1985).
6. Szewzyk, U., Szewzyk, R., and Schink, B., Methanogenic degradation of hydroquinone and catechol via reductive dehydroxylation to phenol, *FEMS Microbiol. Ecol., 31*, 79–87 (1985).
7. Taylor, B. F., Campbell, W. L., and Chinoy, I., Anaerobic degradation of the benzene nucleus by a facultatively anaerobic microorganism, *J. Bacteriol., 102*, 430–437 (1970).
8. Tschech, A., Anaerober Abbau von Phenolen und Benzoatderivaten, dissertation, Universität Konstanz, 1985.
9. Widdel, F., Anaerober Abbau von Fettsäuren und Benzoesäure durch neu isolierte Arten sulfat-reduzierender Bakterien, disseration, Universität Göttingen, 1980.
10. Williams, R. J., and Evans, W. C., The metabolism of benzoate by *Moraxella* species through anaerobic nitrate respiration: evidence for a reductive pathway, *Biochem. J. 148*, 1–10 (1975).
11. Young, L. Y., Anaerobic degradation of aromatic compounds, in *Microbial Degradation of Organic Compounds* (D. T. Gibson, ed.), Marcel Dekker, New York, 1984, pp. 487–523.
12. Young, L. Y., and Rivera, M. D., Methanogenic degradation of four phenolic compounds, *Water Res., 19*, 1325–1332 (1985).
13. Gibson, D. T. (ed.), *Microbial Degradation of Organic Compounds*, Marcel Dekker, New York, 1984.
14. Tschech, A., and Fuchs, G., Anaerobic degradation of phenol by pure cultures of newly isolated denitrifying pseudomonads, *Arch. Microbiol., 148*, 213–217 (1987).
15. Dangel, W., Tschech, A., and Fuchs, G., Anaerobic metabolism of cyclohexanol by denitrifying bacteria, *Arch. Microbiol., 150*, 358–362 (1988).
16. Evans, W. C., and Fuchs, G., Anaerobic degradation of aromatic compounds, *Annu. Rev. Microbiol., 42*, 289–317 (1988).
17. Berry, D. F., Francis, A. J., and Bollag, J.-M., Microbial metabolism of homocyclic and heterocyclic aromatic compounds under anaerobic conditions, *Microbiol. Rev., 51*, 43–59 (1987).
18. Ziegler, K., Braun, K., Böckler, A., and Fuchs, G., Studies on the anaerobic degradation of benzoic acid and 2-aminobenzoic acid by a denitrifying *Pseudomonas* strain, *Arch. Microbiol., 149*, 62–69 (1987).
19. Ziegler, K., Buder, R., Winter, J., and Fuchs, G., Activation of aromatic acids and aerobic 2-aminobenzoate metabolism in a denitrifying *Pseudomonas* strain, *Arch. Microbiol., 151*, 171–176 (1989).

13

Enzymatic Solubilization of Coal

MARTIN S. COHEN   *University of Hartford, West Hartford, Connecticut*

BARY W. WILSON and ROGER M. BEAN   *Battelle, Pacific Northwest Laboratories, Richland, Washington*

13.1 INTRODUCTION

Direct formation and recovery of a water-soluble product from microbial action on solid coal was first reported by Cohen and Gabriele [1]. Coal biosolubilization has become a subject of increasing interest as a possible approach to the utilization of low-rank coals. Although there are now a number of reports [2-7] and literature reviews [8,9] on this subject, little information is available on the chemical composition of either the liquefied products or on the enzymatic systems responsible for their formation.

Fakoussa [10] demonstrated that both fungal and bacterial strains could metabolize coals; however, he did not report on the recovery or chemical analysis of metabolic products. He was interested primarily in the alteration of coal structure caused by microbial metabolism as determined by increased extractability of the coal into various solvents after biological treatment. For the purposes of this discussion, it is important to distinguish between earlier work that yielded no microbial product and later work that yielded sufficient bioconverted material to allow chemical analyses.

The purpose of this chapter is to present the results of analytical chemistry and biochemistry experiments that have been applied to the characterization of biosolubilized products and enzymatic systems studied in our laboratories. These data will be evaluated in relation to present research goals and possibilities for future development of the biosolubilization phenomenon.

13.2 CURRENT VIEWS ON COAL STRUCTURE AS IT RELATES TO MICROBIAL SOLUBILIZATION

Coal rank is an important factor in determining the extent and rate of microbial solubilization. Experiments in several laboratories indicate that low-rank coals, such as leonardite and lignite, are more amenable to biosolubilization than are the higher-rank subbituminous and bituminous coals. Possible reasons for this difference in biosolubilization tendency among coal ranks include overall oxygen content, relative number of oxygen-containing linkages, differences in porosity, and differences in water content. Aromatic carbon content increases with coal rank while oxygen content decreases. Oxygen content in lignites and leonardites, as determined by elemental analyses, can exceed 30%. Subbituminous coals are generally below 23% oxygen and bituminous coals are generally between 3 and 14% oxygen. Porosity generally increases with decreasing coal rank. Figure 13.1 shows the main carbon environments in three ranks of coal. Note the larger number of oxygen linkages in the lignite model.

A recent report on coal structure [12] that we believe relates directly to biological solubilization focused on the determination of the oxygen linkages between aromatic moieties. Data from this work indicated a greater order in coal structure than was reported previously. Chung and Goldberg [12] demonstrated the importance of these oxygen linkages in holding the coal macromolecule together, by specific cleavage of ester and then ether linkages, using alcoholic NaOH at temperatures between 200 and 300°C.

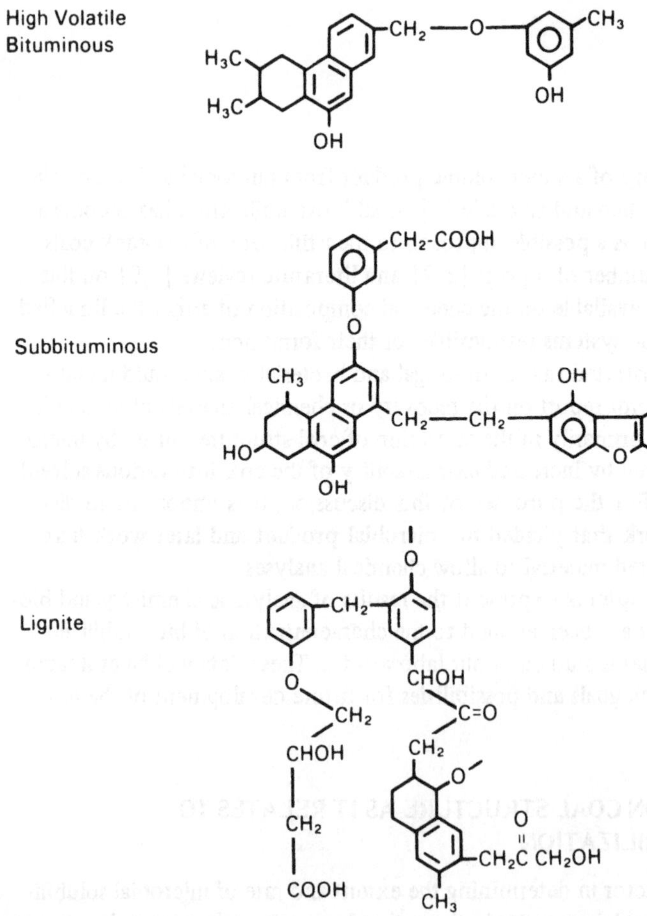

Fig. 13.1 Reported typical structures for lignite, subbituminous, and high-volatile bituminous coal types. (Adapted from Ref. 11.)

They reported that the relative abundance of oxygen linkages varies as a function of coal rank, as do the kinds of aromatic structures present. According to these authors, increased oxygen content in the lower-rank coals does not necessarily imply an increase in the number of oxygen-containing linking bonds. Oxygen also occurs in the form of hydroxyl groups or carboxylic acid functionalities that are not covalently bonded to other aromatic moieties. Although coal may not be as well ordered as reported by Chung and Goldberg [12], the order of relative biosolubility found for these coals correlates with their relative oxygen-linkage content.

Lignite coals occur mainly in near-surface deposits, with the most extensive deposits in the United States occurring in North Dakota and Texas. Lignites are reactive coals with moisture content, as mined, ranging up to 40% or higher. Chemically, lignites are distinguished by greater amounts of carboxylic acid functionalities than those of higher-rank coals. These lignites tend to be more porous in physical structure and are less aromatic. Elemental analysis of a typical North Dakota lignite (Beaulah South) is $C_{100}H_{62}O_{32}N_{1.5}$. Unlike many eastern bituminous coals, lignites are generally low in sulfur content.

Leonardite coal was used as a model coal for most of the results reported here. Leonardite is a highly oxidized form of lignite with a typical oxygen content of 28 to 29%, compared with 19 to 20% for other lignites. Leonardite occurs at shallow depths overlying or grading into the harder and more compact lignite. Leonardite is associated with virtually all lignite outcrops in North Dakota, where it is mined commercially. Lignite can be converted into leonardite by heating it to 150°C for 16 h in the presence of oxygen [13]. The elemental composition (based on 100 carbons) of the leonardite used in the experiments reported here is $C_{100}H_{78}O_{41}N_1$.

13.3 ORGANISMS INVOLVED IN MICROBIAL SOLUBILIZATION OF COAL

Both fungal and bacterial strains have been reported to solubilize coal. *Polyporus versicolor* (ATCC 12679) and *Poria monticola* (ATCC 11538) were the first to yield sufficient products for chemical analyses [1]. *Polyporus* is a homobasidiomycete fungus reported to digest lignin using a white-rot mechanism, whereas *Poria* is a homobasidiomycete fungus reported to degrade wood principally through digestion of cellulose and associated molecules. *Polyporus* proved to be much more effective in degrading the leonardite samples than was *Poria*. The results reported here are based principally on experiments with *Polyporus*.

The fungus was grown on Sabouraud maltose agar and in Sabouraud maltose broth cultures at 30°C with a relative humidity of 84% to 98% and pH 5.8. When sterilized pieces of leonardite coal were added to a continuous hyphal mat on Sabouraud maltose agar, the coal was degraded in a process we call microbial solubilization to a black, viscous liquid which we have termed the bioextract (Fig. 13.2). The process of solubilization was evident within 24 h and continued with the amount of bioextract increasing over several days until many of the coal pieces were completely liquefied. The bioextract was generally harvested from the fungal cultures after 5 days, freeze-dried, and stored desiccated at room temperature (Fig. 13.3). All data reported here are based on analyses of the freeze-dried bioextract. The neat bioextract appeared to contain no particulate matter when it was observed at 400X with a compound microscope, and could be passed through a 0.45-μm filter.

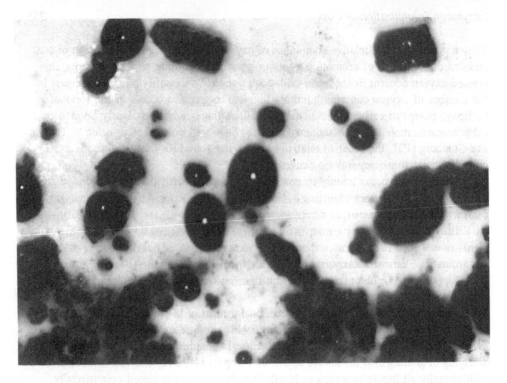

Fig. 13.2 Biosolubilized leonardite coal produced by action of the fungus *Polyporus versicolor*.

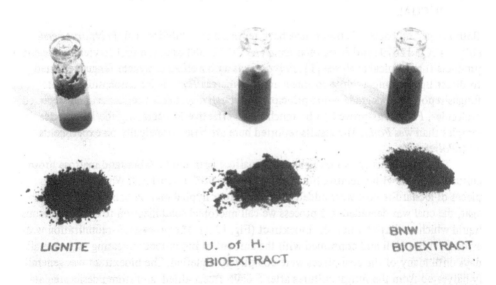

LIGNITE   U. of H.   BNW
                BIOEXTRACT   BIOEXTRACT

Fig. 13.3 Comparison of powdered leonardite coal with the freeze-dried bioextract produced by *Polyporus versicolor*.

Other microorganisms exhibiting lignolytic activity also appear capable of cleaving certain of the oxygen linkages occurring in low-rank coals. These include, but are not limited to, *Phanerochaete chrysosporium* [14], *Candida* sp. ML13 [6,7], *Penicillium waksmanii* [6], *Cunninghamella* sp. [6], and *Streptomyces viridosporus* (ATCC 39115).

Work in our laboratories using model compounds has shown differences between *P. versicolor* and *P. chrysosporium* in their ability to attack oxygen functionalities found in low-rank coals. *P. versicolor*, for example, demonstrated the ability to degrade both methoxybenzophenone and benzyl benzoate. *P. chrysosporium* degraded the benzyl benzoate (ester linkages) but displayed little ability to attack the methoxy benzophenone (diphenyl ketone linkages) [15]. The former compound contains aryl ester bonds analogous to those that occur in lignin. Both organisms formed carboxylic acids as products during degradation of one or both compounds. The higher solubilization rates observed for *P. versicolor*, compared with *P. chrysosporium*, may result from the ability of *Polyporus* to attack both types of oxygen-containing bonds in the coal.

13.3.1 Pretreatments of Coals

We have tested the effects on biosolubilization rates of pretreating several lignite, sub-bituminous, and bituminous coals with various chemical agents. Prior to oxidative pretreatment, the latter two coals were resistant to solubilization by *P. versicolor*. The nine coal samples tested were as follows:

> Beaulah-Zap lignite
> Montana Fort Union Bed Lignite
> Texas lignite
> Wyoming subbituminous B
> Illinois No. 6 bituminous [obtained from Pennsylvania State Coal Bank (PS)]
> Illinois No. 6 [obtained from Hydrocarbon Research Incorporated (HRI)]
> Texas lignite [obtained from Oak Ridge National Laboratory (ORNL)]
> Vermont lignite (ORNL)
> Mississippi lignite (ORNL)

Each of these nine coal types was treated with ammonium persulfate (0.2 M), hydrogen peroxide (30%), potassium permanganate (0.2 M), and nitric acid (8 M). The results are summarized in Table 13.1. Each of the coals was treated for 48 h followed by washing in distilled and deionized water until the washings were neutral, in the case of the acids, or until the washings were colorless. Coal pieces were dried overnight at 100°C, weighed, added to cultures of *Polyporus* grown in Sabouraud maltose agar cultures, and allowed to incubate with the fungus for 2 weeks. Untreated leonardite coal pieces were used as a standard against which the degradation of the other coals could be compared. Coal pieces were removed from the cultures, washed to remove liquefied coal and media, and then dried as described above. The amount of degradation was measured by weighing to determine the percent change in weight. There were five replicates of each treatment on each coal.

The pretreatment that produced the most subsequent degradation was nitric acid. Nitric acid (8 M) oxidized several of the coals to compounds that are susceptible to rapid solubilization by *Polyporus*. Solubilization of five of the nine coal samples (one sub-bituminous and four lignites) pretreated with nitric acid exceeded the solubilization of the untreated leonardite standard.

Table 13.1 Degradation of Nine Different Coals by *Polyporus versicolor* Following Chemical Coal Pretreatments

Coal type	Chemical pretreatment	Percent change in weight
	HNO$_3$	
Texas lignite		−80.10
North Dakota lignite		−89.40
Montana lignite		−77.70
Wyoming subbituminous		−84.80
Illinois No. 6 HRI		−11.60
Illinois No. 6 PSOC		−32.10
Texas lignite (ORNL)		−53.50
Vermont lignite (ORNL)		−72.40
Mississippi lignite (ORNL)		−34.10
	H$_2$O$_2$	
Texas lignite		−50.60
North Dakota lignite		−32.76
Montana lignite		−31.45
Wyoming subbituminous		−2.31
Illinois No. 6 HRI		−0.36
Illinois No. 6 PSOC		−12.65
Texas lignite (ORNL)		−21.70
Vermont lignite (ORNL)		−10.00
Mississippi lignite (ORNL)		−32.10
	Ammonium Persulfate	
Texas lignite		+8.15
North Dakota lignite		−1.23
Montana lignite		−5.06
Wyoming subbituminous		−4.78
Illinois No. 6 HRI		−0.06
Illinois No. 6 PSOC		−9.18
Texas lignite (ORNL)		−9.90
Vermont lignite (ORNL)		−26.0
Mississippi lignite (ORNL)		−7.20
	KMnO$_4$	
Texas lignite		+4.30
North Dakota lignite		−5.52
Montana lignite		−2.09
Wyoming subbituminous		+1.74
Illinois No. 6 HRI		+1.14
Illinois No. 6 PSOC		−1.50
Texas lignite (ORNL)		−1.90
Vermont lignite (ORNL)		−26.9
Mississippi lignite (ORNL)		−9.60
	Untreated Leonardite	
Leonardite		−67.52

The second most effective pretreatment of the coal samples was hydrogen peroxide. All coals lost weight with degradation of the lignite samples ranging from 31 to 50%, which compares favorably with the degradation of the untreated leonardite standard, which lost 68% in weight due to solubilization by the fungus. The Wyoming subbituminous and the Illinois No. 6 (HRI) bituminous samples showed very little degradation by the fungus, whereas the Illinois No. 6 (PS) lost more weight (13%) in these tests.

Ammonium persulfate was the third most effective pretreatment but was significantly less effective than nitric acid or hydrogen peroxide. One coal sample (Texas lignite) actually increased in weight during the tests. The other coal samples were degraded only slightly, with decreases in weight as high as 26% for the Vermont lignite, but ranging mostly between 1 and 19% for the other coal samples.

Potassium permanganate clearly was the least effective pretreatment tested. After this treatment, three of the coal samples increased in weight during incubation with *Polyporus*. These coals showed no clear pattern, consisting of one lignite, one subbituminous, and one bituminous coal. The Vermont lignite was degraded to the extent of 26.9% loss in weight. The other coals in which solubilization occurred were degraded only slightly with losses in weight ranging from 1.5 to 9.6%.

Nitric acid produced the most subsequent microbial solubilization of the coal samples; five of the samples were degraded more extensively than the untreated leonardite standard. Although ammonium persulfate apparently oxidized the coal samples, solubilization of the coals did not exceed 10% for most of the samples. Thus the coals were oxidized to structures that could not be degraded by the fungus. Degradation of treated samples did not exceed 10% except for the Vermont lignite; three of the coals showed no solubilization. *Polyporus* generally could not degrade the structures produced by pretreatment of coals with potassium permanganate. Hydrogen peroxide is a weak acid and can act as both an oxidizer and reducer. Although hydrogen peroxide pretreatment gave solubilization results that were second only to nitric acid pretreatment, it is not clear exactly how hydrogen peroxide altered the coals. Furthermore, the mechanism by which coals are changed by all of these pretreatments remains unclear.

13.4 CELL-FREE BIOSOLUBILIZATION OF COAL

It has been determined in our laboratories that cell-free filtrates of the medium in which *Polyporus* had grown have the ability to solubilize coal. Laccase activity was concentrated following the procedure of Wilson et al. [16]. Liquid minimal medium solutions were inoculated with a hyphal suspension of *Polyporus* and incubated at 25°C with agitation. After 7 days, the laccase inducer xylidine (2,5-dimethylanaline, 2×10^{-4} M) was added to the medium. At the end of the incubation period, the growth medium was filtered through three layers of cheesecloth to remove most of the fungal hyphae. Ultrafiltration through 10,000 MWCO filters (Amicon PM 10) reduced the volume to one-tenth of the original volume. The filtrate was then partially purified by column chromatography using a DEAE-52 ion-exchange column. Laccase activity was measured spectrophotometrically at 525 nm using syringaldazine as a substrate. One unit of the enzyme was defined as the amount of enzyme that resulted in conversion of 1.0 μmol of substrate per minute. The reaction mixture was composed of 1.0 mL of 0.2 M sodium phosphate buffer (pH 5.2), 0.15 mL of 0.1 mM syringaldazine in ethanol, 0.75 mL of distilled and deionized water, and 0.10 mL of enzyme preparation (the partially purified filtrate).

This enzyme preparation had the ability to solubilize leonardite coal. An undiluted and a 1:10 dilution of this preparation were added to 100 mg of leonardite and incubated for 6 days. Biosolubilization of leonardite was observed qualitatively as the color of the solution changed from a light straw color to brown, eventually becoming opaque (black). This color change was compared with appropriate controls, starting within 16 h of addition of enzyme solution and continuing over the following few days. Both the undiluted and the 1:10 dilution of the enzyme solution showed evidence of biosolubilization of the leonardite. The darkening of the enzyme solutions was determined quantitatively by measuring the absorbance of the solutions at 290 nm in a spectrophotometer. Figure 13.4 shows initial rates of leonardite solubilization by three concentrations of enzyme. The amount of solubilization was directly related to the activity of the enzyme solution in the

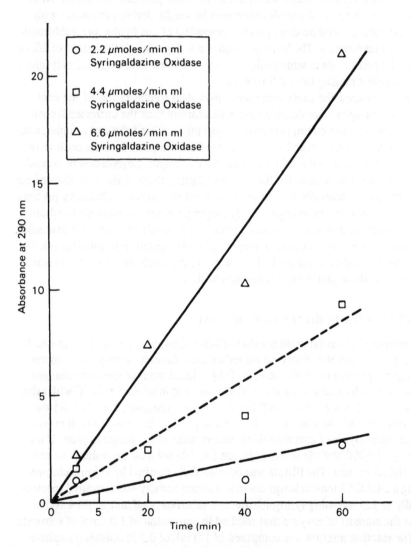

Fig. 13.4 Initial rates of leonardite biosolubilization by three cell-free enzyme preparations with different syringaldazine oxidase activities.

syringaldazine assay. We interpreted this to mean that the enzyme laccase, and possibly other enzymes contained in the solution, were involved in the biosolubilization of leonardite. It has been shown in our laboratories that most of the protein (as determined by absorbance at 280 nm), syringaldazine activity, and leonardite solubilization activity all eluted from the DEAE-52 ion-exchange column in the same fractions. This strongly suggests that one or more enzymes were responsible for the cell-free biosolubilization of leonardite.

13.5 ANALYTICAL CHEMISTRY OF THE BIOEXTRACT

Chemical analyses of the bioextract obtained from the action of *P. versicolor* on leonardite have recently been published [4] and the results are summarized here. The analyses discussed below were performed on the lyophilized bioextract. Concentration of solid material in the aqueous bioextract was typically 60 mg/mL. The material, after lyophilization, was a brown, flakelike solid with the ability to hold a static charge (Fig. 13.3).

13.5.1 Physical/Chemical Analyses

The solubility of the freeze-dried product in several solvents was determined by dissolving the particles to saturation over 2 h in solvents of varying polarities. Solutions were then evaporated to dryness and the solutes were weighed. Solvents of different polarity indexes dissolve different amounts of the bioextract (Fig. 13.5). The bioextract was more soluble in methanol and most soluble in water, where it formed in an opaque, black solution. Both the neat and the freeze-dried bioextracts were most soluble in water. The pH of the neat bioextract is about 7.2. Adjusting the solution to pH 4.5 by the addition of HCl, H_2SO_4, $HClO_4$, or $NaHSO_4$, caused some components of the bioextract to form a black flocculent to flaky precipitate that could be concentrated by centrifugation (Fig. 13.6). Further reduction of pH from 4.5 to 2.0 increased the amount of precipitate and the supernatant became clearer, as measured by a decrease in absorbance at 450 nm. When the pH of the solution was again raised to 4.5 by adding a base such as NaOH or $Na_2B_4O_7 \cdot 10H_2O$, the precipitate dissolved resulting in a black solution.

The number-average molecular weight was determined through vapor-pressure osmometry in aqueous solutions. Ultrafiltration was used to estimate molecular weight distributions of the freeze-dried bioextract. Aqueous solutions containing about 200 mg/L of organic carbon were slowly filtered with pressure through Diaflo ultrafilters with the use of a stirred cell. Organic carbon was determined in subsamples after filtration through >50,000-, >10,000-, >1000-, and >500-dalton filters with a coulometric carbon analyzer.

The molecular weight of solutes in the bioextract, as estimated by vapor-pressure osmometry, was 342 ± 18. This apparently low value is probably due to the contribution of low-molecular-weight counterions, such as sodium, potassium, and ammonium. Figure 13.7 shows the results of molecular weight measurements determined by ultrafiltration. These ranged from more than 50,000 daltons to less than 500 daltons.

Titrations with NaOH solutions were performed on the lyophilized bioextract after precipitation with 0.2 M HCl and lyophilizing again to remove the HCl. Two titrations were performed. For the first, the freeze-dried acid-precipitated material was titrated with 1.2500 mL of 0.1832 M (CO_2-free) NaOH in 5-μL increments added at 5 min per increment. Then an additional 2.0000 mL of titrant was added in 10-μL increments at 6 s per

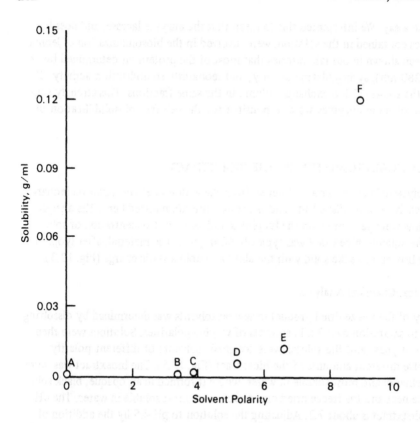

Fig. 13.5 Solubility of the freeze-dried bioextract in various solvents of different polarities (debye units). The solvents are hexane (A), methylene chloride (B), butanol (C), ethanol (D), methanol (E), and water (F).

increment. The second titration used 2.1500 mL of titrant added in 10-μL increments at 20 min per increment. The absence of distinct endpoints in the titrations (Fig. 13.8) indicated the presence of weak acids with a wide range of acid constants.

In the early stages of the titrations, some of the buffering capacity was due to the base-assisted dissolution of the solid. Once the solution became basic, there was evidence for a slow reaction of hydroxide ion with the coal in solution. The plot of the slower titration (curve A, Fig. 13.8) indicated that a large excess of hydroxide could be added to the solution and virtually all the hydroxide would slowly react with the coal. The data taken at 5 min per point (curve B, Fig. 13.8) showed more free hydroxide earlier in the titration (a higher pH), with an even more pronounced effect when the base was introduced more rapidly. The overall titration of the samples was a combination of two effects. Titration of relatively strong acids predominated in the lower pH regions of the titration, whereas in the higher-pH regions reaction of hydroxide ions with the coal predominated. The lack of a clean endpoint hampered estimating the number of acid/base sites in the molecules from the titration curves. Between 4 and 5 mol of base was needed to neutralize 1000 g of the coal sample. This translated to an equivalent weight of between 250 and 200 g per equivalent.

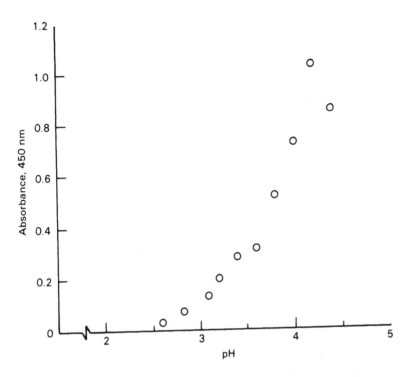

Fig. 13.6 Effect of pH on the solubility of the bioextract.

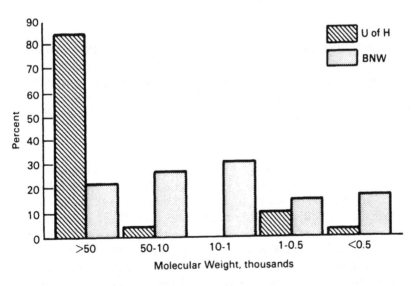

Fig. 13.7 Percent molecular weight distribution, determined by ultrafiltration, of chemical components in the bioextract as a function of pH.

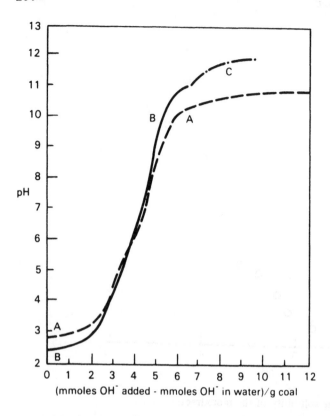

Fig. 13.8 Aqueous titrations of the acid-precipitated bioextract. Curve A, 20-min intervals between titration points; curve B, 5-min intervals between titration points; curve C, extension of curve B at 6-s intervals between titration points.

Elemental compositions, determined by combustion analysis, for the leonardite coal and freeze-dried bioextract are shown in Table 13.2. Analysis of ammonium ions on two different samples of product indicated that ammonium represented almost half (40.0% and 49.8%) of the product nitrogen. Table 13.3 shows the results from x-ray fluorescence determinations for trace metals in the bioextract and the feed leonardite.

13.5.2 Spectroscopic Analyses

Infrared (IR) and nuclear magnetic resonance (NMR) spectroscopy were employed to analyze the functional groups of the biosolubilized coal. The IR spectra of KBr pellets of the leonardite coal and the freeze-dried bioextract are shown in Fig. 13.9. The spectrum of the starting material is typical of that reported for lignite coals. The IR spectrum of the freeze-dried bioextract indicated that the aliphatic stretching band around 2950 cm^{-1} was detectably reduced in the product, compared with that of the starting coal. The 1700 to 1750-cm^{-1} shoulder, corresponding to ester or ketone functional groups, was also reduced in the products, perhaps because of further oxidation or ester hydrolysis. The broad IR absorption bands in the 3600 to 3000 cm^{-1} and 1650 cm^{-1} regions indicated the presence of OH^- and a carboxylate or quinone functionality in the products. A new absorbance peak, consistent with COO^- and/or NH_4^+, was also apparent at 1400 cm^{-1} in the spectrum of the products.

Table 13.2 Elemental Analyses of Starting Coal and Coal Products[a]

Element	Analysis (%)		
	Microbial product[b]	Leonardite[c]	Acid-precipitated product[d]
C	46.09 ± 0.11	54.95 ± 0.08	53.53
H	3.80 ± 0.03	3.60 ± 0.08	3.83
N	4.34 ± 0.15	0.80 ± 0.10	2.07
O	30.87 ± 0.11	30.24 ± 0.50	30.70
S	1.08 ± 0.01	1.06 ± 0.02	1.06
Ash	7.27 ± 0.28	8.32 ± 0.05	3.48

Source: Ref. 4.
[a]Values indicate range for duplicate determinations.
[b]Microbial product ($C_{10}H_{9.8}N_{0.8}O_{5.0}S_{0.1}$).
[c]Leonardite ($C_{10}H_{7.8}N_{0.1}O_{4.1}S_{0.1}$).
[d]Acid precipitate ($C_{10}H_{8.8}N_{0.3}O_{4.1}S_{0.1}$).

Table 13.3 X-ray Fluorescence Analyses of Starting Coal and Coal Products

	Analysis (%)								
	Na	K	Mg	Ca	Si	Al	Cl	Fe	S
Lignite	a	0.02		0.41	0.47	0.15	b	0.38	0.69
Biodegraded	2.77[c]	0.55	0.35[c]	0.47	ND	0.43	0.34	1.45	1.04
Acid precipitated		0.03		0.1	ND	0.30	2.3	1.41	1.0

Source: Ref. 4.
[a]Not determined.
[b]Not detected.
[c]By atomic absorption.

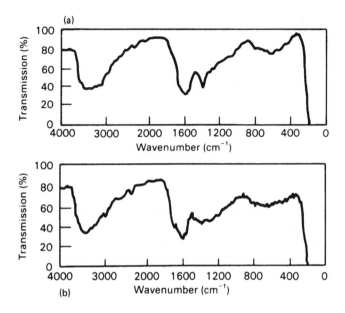

Fig. 13.9 Infrared spectra of leonardite and the bioextract as determined from KBr pellets: (a) infrared scan of the bioextract; (b) infrared scan of leonardite coal.

(a)

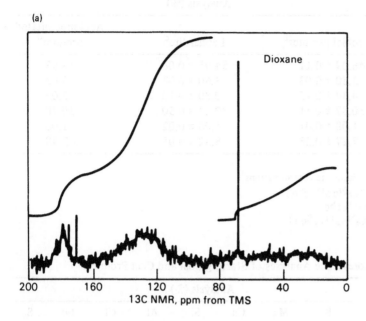

(b)

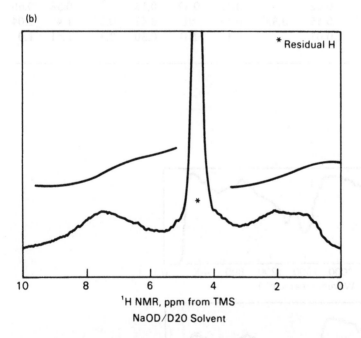

Fig. 13.10 (a) [1]H-Decoupled (nuclear Overhauser effect suppressed) ^{13}C nuclear magnetic resonance spectrum of the bioextract. Dioxane, 67.4 ppm, was used as an internal chemical shift and internal integration standard. (b) [1]H Nuclear magnetic resonance spectrum of the bioextract in NaOD/D_2O. The residual solvent OH⁻peak, indicated by an asterisk, is omitted. Methanol was added in a separate experiment as a chemical shift and internal integration standard.

The ^{13}C NMR spectrum of the freeze-dried acid-precipitated bioextract (Fig. 13.10a) detected 60% of the carbon found by elemental analysis. The remaining structure may have gone undetected because of its proximity to radical sites or because of chelation by carboxyl groups to iron. The NMR analysis indicated that the product consisted of 51% aromatic carbon, 20% carboxylate and/or quinone carbon, and 29% aliphatic carbon. The region of 175 to 185 ppm (Fig. 13.10a) corresponded to the chemical-shift range of ester, carboxylic acid, and carboxylate anions. Quinone carbonyl, at 184 to 189 ppm, could not be ruled out as a minor contributor to the observed band. However, ketone carbonyl (greater than or equal to 195 ppm) was not observed. Because the IR spectrum of the freeze-dried bioextract showed an insignificant amount of ester carbonyl, the band at 175 to 185 ppm can be attributed mainly to carboxylate carbon. The level of carboxylate carbon was much higher than that observed for lignites [17] and is consistent with the extensive oxidation found in leonardite. The proton NMR spectrum of the freeze-dried acid-precipitated bioextract (Fig. 13.10b) gave integral ratios that corresponded to 0.6% by weight aliphatic and 0.6% aromatic hydrogen, comparable with the direct chemical analysis of the freeze-dried bioextract (Table 13.2).

13.5.3 Calorimetric Measurements

Powdered leonardite coal samples and freeze-dried bioextract were pressed into pellets weighing approximately 1 g each and were burned in a nonadiabatic bomb calorimeter to determine their energy content. The average energy content of the freeze-dried leonardite coal sample was calculated to be 4.394 kcal/g (7378 Btu/lb). Average energy content for the bioextract was 4.208 kcal/g, which is equivalent to 95.7% of the original energy content of the leonardite on a gram-for-gram basis. However, 1 g of feed coal yielded less than 1 g of solid freeze-dried bioextract.

13.6 BIOLOGICAL AND ANALYTICAL ASSESSMENT OF POTENTIAL TOXICITY ASSOCIATED WITH THE BIOEXTRACT

An important consideration in evaluating coal-derived materials is the degree to which they may be potential mutagens or carcinogens. Therefore, we performed tests to determine if potentially carcinogenic polycyclic aromatic hydrocarbons (PAH) or their heterocyclic analogs were present in the bioextract [16]. In addition, a direct biological assay for mutagenic activity was performed.

Attempts to determine three- to six-ring PAH by gas chromatography (GC), GC mass spectrometry (GCMS), fast atom bombardment MS, and electron-impact MS yielded unsatisfactory results. These results indicated that only a small percentage of the total material in the bioextract was volatile enough and of low enough molecular weight to be analyzed successfully by these methods. However, we were not confident that the sample preparation techniques used would have extracted low levels of PAH from the polar solutes. We therefore performed direct supercritical pentane extraction/mass spectral analyses of the bioextract [18]. Material extracted under supercritical conditions of temperature and pressure was introduced directly into the chemical ionization source of the mass spectrometer. Results confirmed the absence of PAH or heterocycle PAC with three or more aromatic rings. Standard addition experiments indicated that the method was sensitive to less than 100 ppm for single components.

To compare the toxicologic potential of the bioextract with that of thermo-chemically produced coal liquids, we conducted microbial mutagenicity (Ames test) assays. As anticipated from the chemical analyses, which showed that the bioextract contained no detectable PAH, the Ames tests showed essentially no activity with *Salmonella typhimurium* TA98. The bioextract activity was not above background when compared with 10 revertants per microgram for a commercial fencing creosote and 43 revertants per microgram for SRC-II fuel oil blend.

13.7 STATE OF OXIDATION OF THE COAL, NATURE OF MICROBIAL ATTACK, AND STRUCTURE OF THE BIOSOLUBILIZED PRODUCT

As discussed above, microorganisms capable of degrading lignin are also capable of degrading coals. However, we have also determined that only certain coals are microbially degraded in their initial state and the process can be enhanced considerably by pretreatment with chemical oxidizing agents. Because of the high oxygen content of leonardite compared to lignite coals, oxygen addition is probably the critical oxidation process necessary for the biodegradation of coals. Our studies with different oxidizing agents have demonstrated that the way the oxygen is incorporated is also important in determining the extent of microbial attack. Hydrogen peroxide (30%) compares well with 8 M nitric acid in ability to render the coal susceptible to microbial attack, even though the latter is the more potent oxidant. Exposure of coal to air over long periods of time can similarly convert coal to a more biodegradable substrate. This fact supports the notion that specific oxygen-incorporation reactions, perhaps those arising through an initial peroxidation, are more useful in promoting microbial attack. Some coals may be better substrates for biosolubilization than others because they incorporate oxygen in ways that are more suitable for attack by fungi and other microorganisms. Chemical research studies of coal oxidation processes that correlate the mode of oxidation with microbiological biosusceptibility are needed to clarify these points. Further, such studies could be most useful in developing strains of microorganisms capable of coal degradation.

We have learned from our few experiments conducted with simple compounds that the organisms degrading leonardite coal exhibit some difference in selectivity of action. Thus far, *Polyporus*, which is a more versatile degrader than *Phanerochaete* of the model oxygenated aromatic compounds tested, is also the more potent coal biosolubilizer. Lignin bioconversion is thought to proceed through mechanisms involving susceptible ether and ester links as well as oxidative opening of substituted rings. Therefore, one could conclude that both of these mechanisms are operating in the case of coal biosolubilization, and that the best coal bioconverters can promote more than one type of oxidative hydrolytic reaction.

The chemical analyses of the biodegraded leonardite product reinforce the thought that coal biosolubilization proceeds through an oxidative hydrolysis mechanism. Both side-chain cleavage and ring-opening lead to carboxylic acid formation. We found carboxyl groups to be the most predominant oxygen-containing group in the bioextract produced by *Polyporus*.

Much of the chemical data obtained on biosolubilized coal are consistent with the material existing as a highly carboxylated macromolecular structure. From titration and osmometric molecular weight determinations, we conclude that bioconverted leonardite has one carboxyl group for every 250 to 350 daltons. At neutral pH, alkali metal ions and ammonium ions are the principal counterions to the carboxyl groups. This macromolecular salt is further solubilized by the presence of hydroxyl substituents. The molecular weight of the carbonaceous portion of the biodegraded product was demonstrated through ultrafiltration experiments to exceed 50,000 daltons. The apparent molecular weight distribution varies with pH, indicating that hydrogen bonding between macromolecules influences the values obtained. The high molecular weight of the coal product is shown by its progressive insolubility in solvents of decreasing polarity and its decreasing solubility in water (or any other solvent) at pH 4 or less.

The problem of accurately determining the molecular weight of biodegraded coal materials is a difficult one and worthy of considerable research effort. Changes in molecular weight distribution, if determined in a reliable and consistent manner, can give valuable clues to the mechanism of the biodegradation reactions. Until mole weights can be determined reliably, it will be very difficult to know the extent to which microbial attack continues after the coal is initially solubilized.

Based on the analytical data presented here as well as on previously published structures for lignite [11], we have drawn a model structure for the bioextract produced from the action of *P. versicolor* on leonardite (Fig. 13.11). The model reflects the approximate elemental composition in terms of C, H, O, N, and S, as well as the relative numbers of aromatic, aliphatic, and carboxyl carbons as determined by ^{13}C NMR. These data are consistent with the hydrogen environments as determined by proton NMR. This model also reflects the ammonium counterion concentration as determined by total ammonium analysis. We expect the model to be modified as additional structural information is obtained.

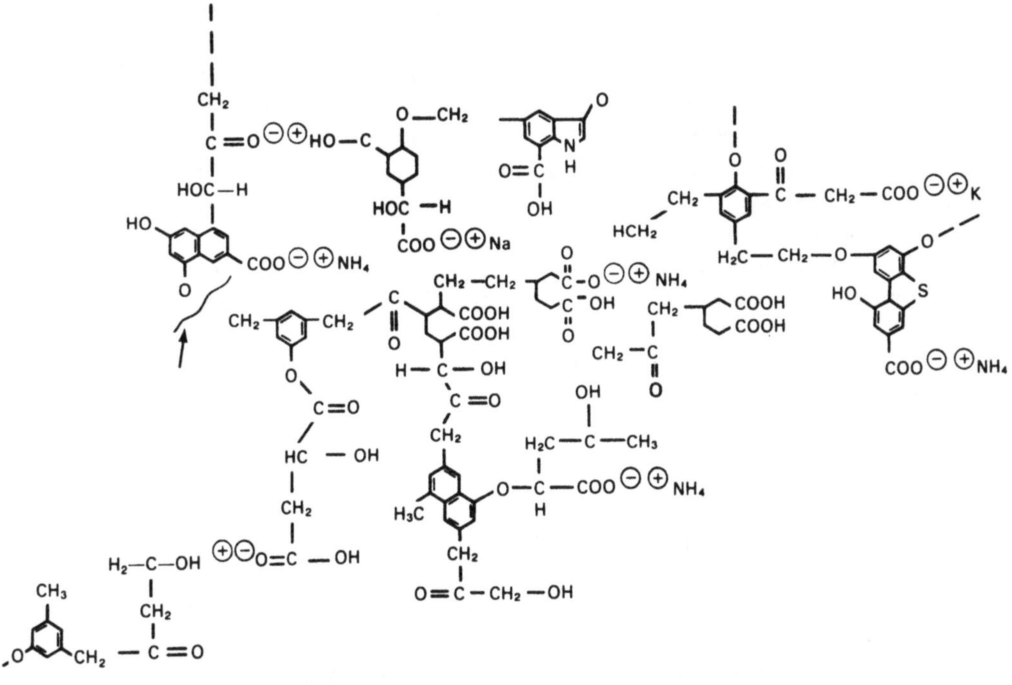

Fig. 13.11 Structural model for the bioextract produced by action of *Polyporus* on leonardite coal.

13.8 TECHNOLOGICAL APPLICATION OF A
BIOSOLUBILIZATION PROCESS

At present, microbial biosolubilization of coal is a relatively new process with several potential uses. Additional research will be needed to clarify the mechanisms of the chemical and biochemical processes involved in biosolubilization of coal. It might seem appropriate to limit this discussion to basic research studies designed to yield greater understanding of the principal reactions. However, consideration of practical applications at an early stage may provide additional stimulation for further improvements in technology.

Coal in liquid form presents unique possibilities for upgrading processes that are not available with solid coals. These processes may produce more valuable products or improve the fuel quality of the coal. A number of microbial processes can be suggested for upgrading coals. Some of the most important of these involve removal of sulfur, nitrogen, and ash, and production of clean gaseous fuels.

13.8.1 Removal of Pyritic and Organic Sulfur

The usefulness of much of the coal mined in the eastern United States is decreased by the presence of pyritic and organic sulfur. Direct burning of these coals in power plants results in sulfur dioxide emissions. Physical and chemical methods of sulfur removal are costly. Biological desulfurization of coal before combustion is a low-energy process that would have the potential of reducing operating costs and sulfur emissions into the atmosphere.

Some organisms reported to remove pyritic sulfur from coals are *Thiobacillus ferrooxidans, Sulpholobus acidocaldarius*, and "coal bug 1" developed by Atlantic Research Corporation [19]. Pyrite solubilization occurs at the surface of coal particles. The reaction rate has been reported to be dependent on coal particle size with effectiveness of pyrite removal increasing with smaller (<37 μm) particle size [9]. Biosolubilization of coals should greatly increase the amount of pyrite sites available for oxidation by microorganisms and therefore increase both the rate and extent of pyrite removal. Development of such a process would depend on isolating of microorganisms capable of solubilizing high-sulfur eastern coals or on discovery of pretreatments that would allow solubilization of these coals by known organisms.

The same microorganisms that can oxidize pyritic sulfur have been reported to remove organic sulfur from coal [9]. A thermophilic *Thiobacillus*-like strain (TH1) removed 50 to 57% of the organic sulfur contained in a sample of Turkish lignite [9]. If the lignite was first solubilized by microbial action, the rate and extent of organic sulfur removal from the coal should increase.

13.8.2 Removal of Nitrogen

The utility of coal as a fuel is reduced as the relative nitrogen content increases. Nitrogen shortens processing catalyst life and contributes to NO_x emissions upon combustion, creating one component of acid rain. Several microorganisms have been reported to metabolize nitrogen-containing heterocycles. An *Alcaligines* species, isolated from sewage sludge, has been reported to utilize indole as a sole carbon and nitrogen source [20]. Wang [21] reported that indole was degraded to methane and carbon dioxide by a methanogenic enrichment culture. *Bacillus* and *Nocardia* species isolated from soil metabolized pyridine and could use it as a sole source of carbon and nitrogen [22]. Bacteria isolated from sewage sludge have also been reported to use quinoline as a sole source of carbon, nitrogen, and energy [23]. The mechanism for microbial metabolism of nitrogen-containing heterocycles generally involves oxidative cleavage of the ring followed by removal of nitrogen with some loss of carbon. These heterocycles are generally located within the polymeric structure of the coal. Microbial solubilization of coal would expose virtually all of these structures to microbial action.

13.8.3 De-ashing of Coals

Another possible application of the biosolubilization process is de-ashing. As the elemental analysis presented here indicates, the silica content of the coal is removed from the coal upon solubilization. We can expect that dramatic reductions may also occur with other minerals, including pyrites, upon microbial dissolution. The analysis also shows that other minerals are not substantially reduced by the process, possibly because they are chelated or entrained. Further research is needed to determine the de-ashing potential for the microbial process.

13.8.4 Methane Production from Coals

A process for converting coal to aromatic hydrocarbons, methane, and CO_2 has been under investigation by Houston Lighting and Power Company [24]. In that process coal is first solubilized by caustic treatment followed by an anaerobic digestion step that produces methane and CO_2. The initial solubilization step takes place at 300°C and 55 atm to produce feed for the biodigester. A microbial process for producing a biosolubilized feed to the anaerobic reactor could result in a considerable savings in energy and chemicals. Microbial processes for direct production of methane from coals have not yet been developed [25]. A two-step all-microbiological process presents an attractice alternative.

13.9 CONCLUSIONS

Initial reports of Fakoussa [10] and Cohen and Gabriele [1] demonstrating the ability of certain fungi to depolymerize the coal macromolecule have been widely confirmed. Depolymerization and solubilization reactions in coal are now known to be catalyzed by extracellular enzyme systems. At least one of these systems, laccase, has been isolated and characterized. Work from other laboratories, however, suggests that other enzymes may also be capable of catalyzing these reactions in coals. For example, recent spot tests on more than 20 wild fungal strains capable of coal solubilization indicated that no more than four of these strains, including *P. versicolor*, were producing laccase [26]. On the other hand, earlier studies by Haskin and Obst indicated that laccase was common to many fungi [27].

Current data indicate that the coal rank and oxidation state are important factors in coal's susceptability to biosolubilization. Leonardite coal is solubilized most extensively by fungal action. Lignite and bituminous coals that have been oxidized by various pretreatments are solubilized more extensively than untreated samples of these coals.

Materials produced during biosolubilization are of high molecular weight, highly polar, and water soluble. The bioextract has higher oxygen content, less aliphatic carbon, and more carboxylate carbon than does the feed coal. Some of the nitrogen content of the bioextract is present as ammonium ions. The product is stabilized in aqueous solutions of neutral or basic pH because of solubilization through carboxylate anions; alkali metal and ammonium ions are the counterions. A model for the chemical structure of the solubilized coal products has been proposed.

Bioextract products tested to date have shown no genetic toxicity as determined by the Ames test. In addition, chemical analyses of the bioextract products gave uniformly negative results in tests for known toxic compounds associated with coal liquefaction processes.

Solubilization of different coal types may present unique possibilities for upgrading feed coals that are typically high in sulfur and nitrogen content. Solubilization of these coals would expose all of their sulfur and nitrogen sites to the solution. This process could be followed by sulfur and/or nitrogen removal by microorganisms. The combination of these two processes should allow removal of more sulfur and nitrogen from coals at much lower cost than is presently possible. Biosolubilization generates a unique feedstock for a variety of possible upgrading processes.

REFERENCES

1. Cohen, M. S., and Gabriele, P. D., Degradation of coal by the fungi *Polyporous versicolor* and *Poria monticola*, *Appl. Environ. Microbiol., 44*, 23 (1982).
2. Cohen, M. S., Aronson, H., and Gray, E. T., Jr., Degradation of lignite coal by *Polyporus versicolor*: initial characterization of products, in *Proceedings of the Direct Liquefaction Contractor's Review Meeting*, U.S. Department of Energy, Washington, D.C., 1986, p. IV48.
3. Cohen, M. S., Liquefaction of coal by *Polyporus versicolor*, in *Proceedings of the Biological Treatment of Coals Conference*, Idaho National Engineering Laboratory and U.S. Department of Energy, Washington, D.C., 1987.
4. Wilson, B. W., Bean, R. M., Franz, J. A., Thomas, B. L., Cohen, M. S., Aronson, H., and Gray, E. T., Jr., Microbial conversion of low-rank coal: characterization of biodegraded product, *Energy Fuels, 1*, 80 (1987).
5. Wilson, B. W., Pyne, J. A., Bean, R. M., Fredrickson, J. A., Stewart, D. L., Sass, E., Burnside, M., and Cohen, M. S., Microbial beneficiation of low-rank coals, in *Proceedings of the EPRI 11th Annual Conference on Clean Liquid and Solid Fuels*, Electric Power Research Institute, Palo Alto, Calif., 2, 47–62, 1988.
6. Ward, B., Lignite-degrading fungi isolated from a weathered outcrop, *Syst. Appl. Microbiol., 6*, 236 (1985).
7. Scott, C. D., and Strandberg, G. W., Microbial coal liquefaction, in *Proceedings of the Direct Liquefaction Contractor's Review Meeting*, U.S. Department of Energy, Washington, D. C., 1986, p. IV65.
8. Pyne, J. W., and Wilson, B. W., *Biologic Coal Beneficiation: Literature Review*, EPRI Report AP-4834, Electric Power Research Institute, Palo Alto, Calif., 1986.
9. Olson, G. J., and Brinkman, F. E., Bioprocessing of coal, *Fuel, 65*, 1638 (1986).
10. Fakoussa, R. M., Coal as a substrate for microorganisms: investigation with microbial conversion of national coals, Ph.D. thesis, Freidrich-Wilhelm Universität, Bonn, 1981.
11. Wender, I., Catalytic synthesis of chemicals from coal, *Am. Chem. Soc. Div. Fuel Chem. Prepr., 170*, 24(A) (1975).
12. Chung, K. E., and Goldberg, I. B., Chemical structural differences between two low-rank coals, in *Proceedings of the EPRI 11th Annual Conference on Clean Liquid and Solid Fuels*, Electric Power Research Institute, Palo Alto, Calif., 2, 83–121, 1988.
13. Fowkes, W. W., and Frost, C. M., *Leonardite: A Lignite Byproduct*, Report of Investigations 5611, U.S. Bureau of Mines, Washington, D.C., 1960.

14. Faison, B. D., and Kuster, T. A., Localization of ligninase protein in *Phanerochaete chrysosporium*, in *Proceedings of the 8th Symposium on Biotechnology for Fuels and Chemicals* (C. D. Scott, ed.), Wiley, New York, 1986, p. 261.
15. Campbell, J. A., Fredrickson, J. K., Stewart, D. L., Pyne, J. W., Wilson, B. W., Bean, R. M., and Cohen, M. S., Application of model compounds to the study of microbial degradation, *Abstracts of the ACS Meeting*, New Orleans, Aug. 30–Sept. 4, 1987.
16. Wilson, B. W., Lewis, E., Stewart, D. L., Li, S. M., Bean, R. M., Chess, E. K., Pyne, J. W., Cohen, M. S., and Aronson, H., Microbial processing of fuels, in *Proceedings of the Direct Liquefaction Contractor's Review Meeting*, U.S. Department of Energy, Washington, D.C., 1986, p. IV88.
17. Franz, J. A., and Camaioni, D. M., Study of deuterium transfer, isotope effects and structural distributions of products of reactions of coals in deuterated tetralin using ^{2}H and ^{13}C FT-n.m.r. and solid-state ^{13}C FT-n.m.r., *Fuel*, *63*, 990 (1984).
18. Smith, R. D., and Usdeth, H. R., New method for the direct analysis of supercritical fluid coal extraction and liquefaction, *Fuel*, *62*, 466 (1983).
19. Isbister, J. D., and Kobylinski, E. A., Microbial desulfurization of coal, in *Processing and Utilization of High Sulfur Coals* (Y. A. Attia, ed.), Elsevier, New York, 1985.
20. Claus, G., and Kutzner, H. V., Degradation of Indole by *Alcaligines* spec., *Syst. Appl. Microbiol.*, *4*, 169 (1983).
21. Wang, Y.-T., Suidan, M. T., and Pfeffer, J. T., Anaerobic biodegradation of indole to methane, *Appl. Environ. Microbiol.*, *48*, 1058 (1984).
22. Watson, G. K., and Cain, R. B., Microbial metabolism of the pyridine ring, *Biochem. J.*, *146*, 157 (1975).
23. Shukla, O. P., Microbial transformation of quinoline by a *Pseudomonas* sp., *Appl. Environ. Microbiol.*, *51*, 1332 (1986).
24. Leushner, A. P., Trantolo, D. J., Kern, E. E., and Wise, D. C., Biodegradation of Texas lignite, paper presented at the *Biennial Lignite Symposium on Technology and Use of Low-Rank Coals*, Bismark, N. Dak., May 1985.
25. Couch, G. R., *Biotechnology and Coal Use: A Review*, Electric Power Institute, Palo Alto, Calif., 1987.
26. Ward, B., Coal solubilizing fungi, in *Proceedings of the EPRI First Annual Workshop on Biologic Processing of Coal*, Electric Power Research Institute, Palo Alto, Calif., *5*, p. 13-16, 1988.
27. Haskin, J. M., and Obst, J. R., Syringaldazine: an effective reagent for detecting laccase and peroxidase in fungi, *Experientia*, *15*, 381 (1973).

14

Solubilization of Coal by Microbial Action

CHARLES D. SCOTT and SUSAN N. LEWIS *Oak Ridge National Laboratory, Oak Ridge, Tennessee*

14.1 INTRODUCTION

In the past, concepts for thermal/chemical coal liquefaction processes have been based on the use of relatively extreme operating conditions that result in high temperatures, high pressures, and corrosive environments. Because most biological interactions proceed in a rather mild environment, the potential use of microorganisms for coal processing appears to offer an attractive alternative. Microbial processes are under development for the removal of sulfur [1] from coal and for the removal of hazardous organic materials from coal-conversion wastewater [2]. Thus we know that microorganisms are active in the presence of coal and coal-derived substances.

Suggestions that microbial coal liquefaction might be possible had been made as early as the 1960s [3,4]. Since coal, especially the low-ranked type, has a structure somewhat similar to lignocellulosic material, it could be expected to undergo interaction with lignin-degrading microorganisms. Such interactions with both fungi and bacteria have now been verified by several research groups [5-8]. Various strains of fungi have been shown to solubilize lignite as well as subbituminous coals when used in surface cultures, and it was recently demonstrated that certain bacteria are capable of solubilizing coal in suspension cultures (G. W. Strandberg, personal communication, 1986). Oxidative pretreatment tends to enhance the microbial activity [9].

14.2 METHODS AND MATERIALS

The primary emphasis in the research on microbial coal solubilization has been on screening tests for a variety of organisms with several different types of coals. Both surface-culture and suspension-culture techniques were used, the former primarily for screening tests and the latter essentially in support of the study of various bioreactor configurations.

14.2.1 Microorganisms

A number of different types of fungi and bacteria have been shown to interact with coal and form a liquid product (Table 14.1). Of the several types of microorganisms, two strains of fungi were obtained from the American Type Culture Collection on the basis of their ability to degrade lignin; other fungi were isolated from coal samples or coal seams; and two different types of bacteria were selected due to their known interaction with lignin.

14.2.2 Types of Coal

Preliminary tests have been made using a variety of low-ranked coals, including a lenordite (a highly oxidized lignite), four lignites, and a subbituminous (Table 14.2). Each coal sample was size reduced (−20 + 40 mesh), sterilized at 120°C for 1 h, and dried at 95 to 100°C in air for 12 to 16 h.

14.2.3 Surface Cultures

Most of the surface-culture tests were carried out in a controlled environment at high humidity (>80% relative humidity) and at temperatures of 25 to 30°C in the presence of air on the surface of Sabouraud maltose agar [8]. In some cases, a mineral salt medium (modified Czapek-Dox medium) was used [10]. Approximately 30 mL of the sterile medium (autoclaved at 121°C for 45 min) was inoculated with the microorganisms in either a covered 0.2-L container or a petri dish.

Table 14.1   Some Microbial Species That Form a Liquid Product with Low-Ranked Coal

Microorganism	Source
Fungi	
Trametes versicolor ATCC 12679	American Type Culture Collection, Rockville, Md.
Poria placenta (monticola) ATCC 11538	American Type Culture Collection, Rockville, Md.
Penicillium waksmanii ML20	Isolated from Mississippi lignites by H. B. Ward, University of Mississippi[a]
Candida sp. ML13	Isolated from Mississippi lignites by H. B. Ward, University of Mississippi[a]
Aspergillus sp.	Isolated from an as-received lignite sample[a]
Paecilomyces sp.	Isolated from an as-received lignite sample[a]
Sporothrix sp.	Isolated from an as-received lignite sample[a]
Bacteria	
Streptomyces setonii 75Vi2	Obtained from D. L. Crawford, University of Idaho
Streptomyces viridosporous T7A	Obtained from D. L. Crawford, University of Idaho

Source: Reproduced in part from Ref. 8, with permission.
[a]Mycelia was removed from the coal surface with a sterile inoculation needle and cultured on sterile Sabouraud maltose agar. Spores from isolated colonies were transferred to sterile media for further purification.

Table 14.2 Coal Samples and Their Sources

Type of coal	Source
Mississippi lignite (probably type B)	Phillips Coal Co., via H. B. Ward, University of Mississippi
North Dakota lignite, type A	Coal Research Section, Pennsylvania State University
North Dakota leonardite	American Colloid Co., Skokie, Ill.
Texas lignite (probably type B)	Exxon Research and Development Co., Baytown, Tex.
Vermont lignite (probably type B)	Coal Research Section, Pennsylvania State University
Wyodak subbituminous	Amax Coal Co., Indianapolis, Ind.

Source: Reproduced in part from Ref. 8, with permission.

Since the fungi and bacteria used in these tests were filamentous organisms, a microbial mat was first allowed to develop on the agar surface. This action usually required 1 week to 10 days. Then the coal sample, in the form of small granules, was placed on the surface of the mat (Fig. 14.1). In a few tests, a membrane filter (cellulose acetate filter with 0.45-μm pores, Millipore Company, Bedford, Massachusetts) was placed on top of the microbial mat to prevent direct contact with the microorganisms. When quantitative tests were performed, the liquid product was collected periodically by pipette, the coal particle residues were washed and dried, and all dried materials were weighed.

14.2.4 Suspension Cultures

Several tests were carried out to observe the interaction between the organisms and coal particles in aqueous suspension. The organisms were cultured for several days in 250-mL shake flasks containing 150 mL of the nutrient medium, and then 5 to 10 g of sterilized and pulverized coal was added [8]. Both fungi and bacteria were used in these tests, but the bacteria seemed to be most effective (G. W. Strandberg, personal communication, 1986). Typically, after 2 days of contact, the suspended bacteria were separated from the coal by gentle agitation and decanting with several volumes of deionized, distilled water, and the dry weight of the residual coal was determined (drying at 16 to 18 h at 100°C). Alternatively, coal was added to the cell-free broth from a 7-day culture of the appropriate bacteria. The bacteria were subsequently removed by filtration using a 0.22-μm sterilizing filter (Millipore Corp., Bedford, Massachusetts). After 2 days of shaking at 30°C, the residual coal was recovered by either centrifugation (about 2000*g* for 10 min) or filtration (0.45-μm Millipore filter) and then washed extensively with deionized, distilled water; the weight was determined after drying.

14.2.5 Fluidized-Bed Contactor

A few tests were made with a small fluidized-bed bioreactor that was a 15-cm-long column in the form of an inverse cone with a diameter increasing from 1.25 cm at the entrance to 2.5 cm at the exit (Fig. 14.2). Particles of coal were suspended by pumping the solubilization fluid through the column; then the effluent was returned to the reservoir, where it was aerated prior to recycle.

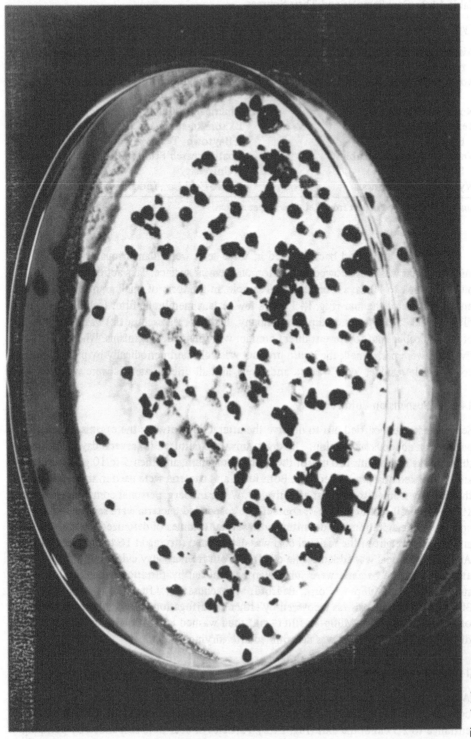

Fig. 14.1 Particles of North Dakota leonardite undergoing solubilization on a fungal mat of *Candida* sp.

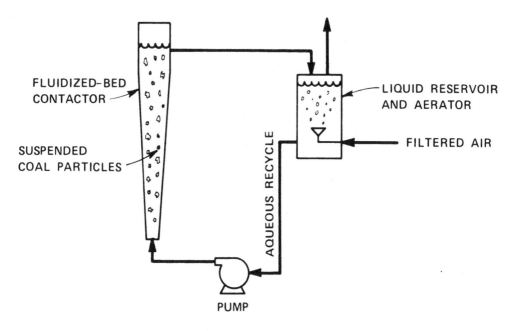

FLUIDIZED-BED CONTACTOR

SUSPENDED COAL PARTICLES

AQUEOUS RECYCLE

LIQUID RESERVOIR AND AERATOR

FILTERED AIR

PUMP

Fig. 14.2 Experimental system for studying microbial interactions with coal particles while being fluidized in a tapered reactor. (Reproduced, with permission, from Ref. 8.)

14.3 MICROBIAL INTERACTIONS WITH COAL

The coal solubilization tests, which have included both surface-culture and suspension-culture techniques, have been very preliminary in nature. However, the results obtained thus far clearly show that some types of coal can be substantially solubilized by various fungi and bacteria. Oxidative pretreatment of coal may result in much more reactive materials.

14.3.1 Solubilization by Surface Cultures

To date, most of the microbial solubilization tests have been qualitative in nature, utilizing surface cultures with the objective of evaluating a number of different organisms with several different types of coal. After coal particles were placed on the formed microbial mat, a period of 1 to 2 days was required before small liquid droplets began to form on the top surface of the particles.

Each of the microorganisms listed in Table 14.1 was capable of solubilizing, at least partially, one or more coal samples. However, *Candida* sp., *Penicillium waksmanii*, and *Sporothrix* sp. were the most generally effective fungi; and *Streptomyces setonii* was the most effective bacterium.

The formation of liquid products has been observed with six different types of coals and nine different types of microorganisms (Table 14.3). The color of these products ranged from water-clear to black, and the quantity varied from trace amounts (<0.05 mL per sample) to profuse amounts that encompassed the entire coal sample (as was found on some tests with the North Dakota leonardite).

Table 14.3 Liquid Products from Microbial Interactions with Various Types of Coal

Type of coal	Microorganism	Liquid product[a]
Mississippi lignite	*Aspergillus* sp., *Candida* sp. ML13, *P. waksmanii*, *T. versicolor* ATCC 12679	Clear, amber, and black; inconsistent and generally in moderate amounts
North Dakota lignite	*Aspergillus* sp., *Candida* sp. ML13, *Paecilomyces* sp., *P. waksmanii* ML20, *Sporothrix* sp.	Clear; moderate amounts
North Dakota leonardite	*Aspergillus* sp., *Candida* sp. ML13, *Paecilomyces* sp., *P. monticola* ATCC 11538, *Sporothrix* sp., *T. versicolor* ATCC 12679, *S. setonii* 75Vi2, *S. viridosporous* T7A	Brown to black; moderate to profuse amounts; some clear initially, turning brown to black after a few days
Texas lignite	*Aspergillus* sp., *Candida* sp. ML13, *Paecilomyces* sp., *Sporothrix* sp.	Clear; trace to moderate amounts
Vermont lignite	*Candida* sp. ML13, *Sporothrix* sp.	Brown to black; trace amounts
Wyodak subbituminous	*Paecilomyces* sp., *Aspergillus* sp., *Candida* sp., *Sporothrix* sp., *T. versicolor*	Clear; trace amounts
		Clear to black; trace amounts

Source: Reproduced in part from Ref. 8, with permission.

[a]Trace amounts, less than 10% of the coal surface covered with liquid; moderate amounts, 10 to 50% of the coal surface covered with liquid; profuse amounts, >50% of the coal surface covered with liquid.

Table 14.4 Liquefaction of North Dakota Leonardite with *Candida* sp. and *T. versicolor*

Run time (d)	Solubilization[a] (%)
Candida sp.	
3	0.7
5	2.5
7	8.2
10	11.6
14	84.0
14	83.0
14	92.8
T. versicolor	
14	55.5
14	75.5

Source: Reproduced in part from Ref. 8, with permission.
[a]Decrease in dry weight after drying overnight at 95°C following washing with water for 30 min.

With fungi, the solubilization appeared to occur most rapidly during the period of 10 to 14 days, at least under the conditions of our tests (Table 14.4). The degree of solubilization in three tests with *Candida* sp. for 14 days averaged 86.6%, which is approximately the level that would be expected if only the ash (14%) remained in the solid residue. *Trametes versicolor* was somewhat less effective in this type of test.

Two bacterial strains, *Streptomyces viridosporous* T7A and *S. setonii* 75Vi2, also solubilized coal in a manner similar to that observed with surface cultures of fungi. Of these, *S. setonii* was the most effective.

14.3.2 Suspension Cultures

Some tests were carried out to study the interaction between the microorganisms and coal particles in aqueous suspension. *Aspergillus* sp. (Fig. 14.3), *P. waksmanii*, and *T. versicolor*, the only fungi tested, were all observed to form microbial films on the coal. In each case, the microbial strains continued to increase the microbial film thickness with time until large masses of biomass particles, up to 3 to 4 mm in diameter and encompassing one or more coal particles, were formed. These results indicate that fungal attachment on lignite can easily be induced. Discoloration of the aqueous medium began within 1 day after addition of the coal particles, suggesting that solubilization had already started.

The bacterium *S. setonii* 75Vi2 was much more efficient than any of the fungi in suspension cultures. Within 3 to 4 h following the addition of coal (typically, North Dakota leonardite or nitric acid-treated Wyodak subbituminous coal) to a 7-day suspension culture of the *S. setonii*, solubilization became apparent as evidenced by discoloration of the broth. As shown in Table 14.5, up to about 80% solubilization of nitric acid-treated Wyodak coal was achieved in 2 days (G. W. Strandberg, personal communication, 1986).

The rapid onset of solubilization and the visible absence of direct association of the cells with the coal indicated that the liquefaction process was being effected by an extra-

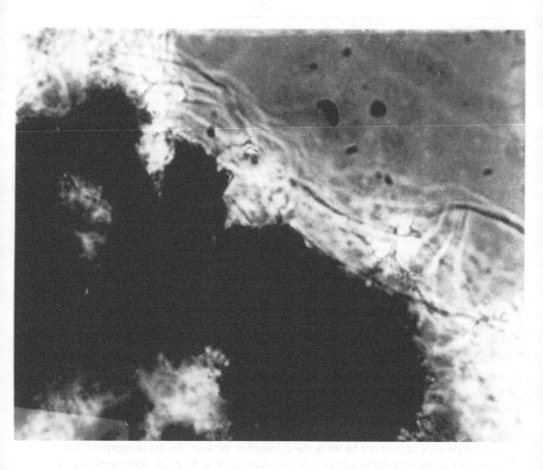

0   10   20

micrometers

Fig. 14.3 Fungal hyphae of *Aspergillus* sp. attached to particles of North Dakota leonardite in suspension culture. (Reproduced, with permission, from Ref. 8.)

Table 14.5 Solubilization of Coal by a Submerged Culture of *S. setonii* 75Vi2

Conditions	Coal[a]	Weight change[b] (%)
Sterile medium	Leonardite	−5
	Preoxidized subbituminous	+2
Submerged cultures[c]	Leonardite	−43
		−28
	Preoxidized subbituminous	−81

[a]North Dakota leonardite; Wyodak subbituminous coal pretreated by soaking for 48 h in 8 N HNO$_3$.
[b]Final dry weights of residual coal after incubation for 48 h with cultures.
[c]Coal was added to 7-day shake cultures (Sabouraud maltose medium, 30°C, 100 rpm, 2-in. stroke).

cellular component in the culture broth of *S. setonii* 75Vi2. This was readily demonstrated by removing the cells from a 7-day culture with a sterilizing filter before adding the coal. Extensive solubilization of both North Dakota leonardite and nitric acid–treated Wyodak coal occurred in the cell-free broth.

14.3.3 Preoxidation of Coal for Enhancement of Solubilization

The oxidation state of the coal seems to contribute to the microbial solubilization process. For example, the North Dakota leonardite, a highly oxidized lignite, was the best substrate for microbial interactions. Highly weathered Mississippi lignite obtained from outcroppings was also superior to the same coal found deep within the seam. This suggested that a highly oxidized coal would be the best feed material.

A series of tests showed that a mild oxidative chemical pretreatment would significantly enhance the reactivity. Sized particles of coal (20 to 40 mesh) were soaked in either 10, 20, or 30% (w/v) hydrogen peroxide or 5 to 8 M nitric acid at room temperature. Following this treatment, the samples were washed five to six times with approximately 75 to 100 mL of deionized distilled water and oven dried (95°C, 16 to 18 h). The third type of pretreatment consisted of exposing the coal particles to a 1 to 2% stream of ozone in oxygen.

Of those coals tested, only the North Dakota leonardite was susceptible to rapid and extensive solubilization as received. However, preoxidation of the other coals with one of the oxidizing agents has been shown to enhance significantly their susceptibility to solubilization (Tables 14.6 to 14.8). In general, the degree of solubilization increased with increasing reagent concentration and contact time. Pretreatment with 8 N HNO$_3$ for 48 h was subsequently used routinely in many tests (Fig. 14.4).

Table 14.6 Enhancement of Fungal Liquefaction of Coals by Preoxidation with Hydrogen Peroxide

		Liquefaction[a] (%)								
		Exposure time[b] (h)								
		24			48			72		
	Untreated	Hydrogen peroxide (% w/v)								
Coal	control	10	20	30	10	20	30	10	20	30
Mississippi lignite	<1	23	28	26	19	–	26	16	55	42
Texas lignite	<1	25	23	32	46	39	47	53	37	50
Vermont lignite	<1	16	18	29	16	41	30	25	18	26
Wyodak subbituminous	<1	34	18	15	56	19	15	15	13	15

Source: Reproduced from Ref. 9, with permission.
[a]Liquefaction tests were conducted with 7-day cultures of *Candida* sp. ML13 cultured on Sabouraud maltose agar at 30°C and 80% relative humidity. Liquefaction was monitored to cessation (14 to 21 days).
[b]Exposure to hydrogen peroxide at ambient temperature.

Table 14.7 Enhancement of Fungal Liquefaction of Coals by Preoxidation with Ozone

		Liquefaction[a] (%)			
	Untreated	Exposure time[b] (h)			
Coal	control	0.5	1.0	1.5	2.0
Mississippi lignite	<1	42	45	58	70
Texas lignite	<1	16	11	14	18
Vermont lignite	<1	24	26	32	38
Wyodak subbituminous	<1	16	7	9	10

Source: Reproduced from Ref. 9, with permission.
[a]See Table 14.6 for liquefaction test conditions.
[b]Coal exposed to stream of ozone (1 to 2 wt %) in dry oxygen flowing at the rate of about 200 mL/min.

Table 14.8 Enhancement of Fungal Liquefaction of Coals by Preoxidation with Nitric Acid

		Liquefaction[a] (%)					
		Exposure time[b] (h)					
		24		48		72	
	Untreated	HNO_3 (M)					
Coal	control	5	8	5	8	5	8
Mississippi lignite	<1	38	49	37	66	28	39
Texas lignite	<1	23	28	35	87	69	52
Vermont lignite	<1	79	48	41	37	58	55
Wyodak subbituminous	<1	23	38	24	34	71	73

Source: Reproduced from Ref. 9, with permission.
[a]See Table 6 for liquefaction test conditions.
[b]Exposure to nitric acid at ambient temperature.

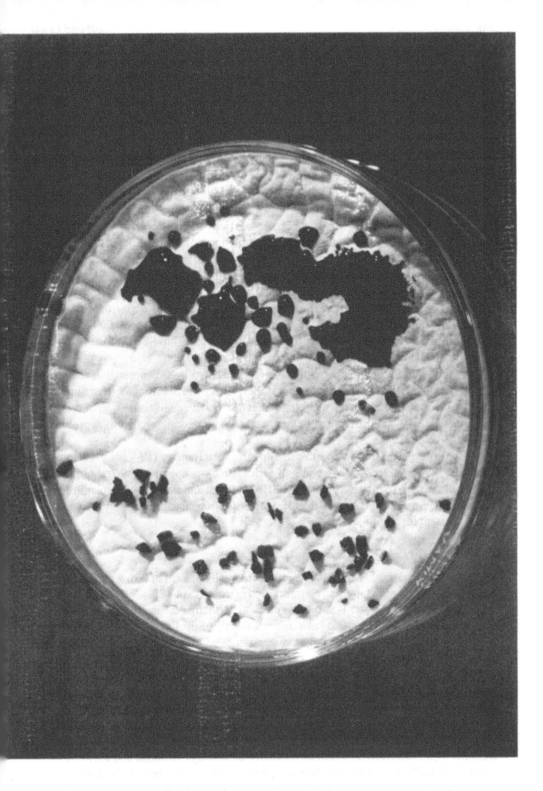

Fig. 14.4 Wyodak subbituminous coal particles undergoing solubilization on a fungal mat of *Candida* sp. The material on the left was untreated coal; the material on the right, which is almost completely solubilized, was pretreated with 8 N HNO$_3$ at 25°C for 48 h. (Reproduced, with permission, from Ref. 8.)

14.4 PRODUCT CHARACTERIZATION

Numerous analytical techniques have been used in an attempt to characterize the liquid product chemically. These include conventional methods for elemental analysis and physical properties, as well as various separations procedures and spectral analyses. At present, the results of this work will only allow a qualitative characterization; however, the material appears to be a complex mixture of moderate- to high-molecular-weight aromatic substances that are soluble in water [8].

14.4.1 Molecular Weight

The major fraction of the liquid product had a relatively high molecular weight, as confirmed by studies using Diaflo ultrafiltration membranes (Amicon Corp., Danvers, Massachusetts). Total organic carbon was used as an indicator of material retained during filtration on membranes of four different pore sizes (XM300, XM50, PM30, and PM10). The molecular weight ranges of all of the microbial products were similar. For example, it was found that with the liquid formed from the interaction of *P. waksmanii* with Mississippi lignite, 1.6% (on a dry weight basis) of the material was retained on a 300,000-MW membrane while 15.9% passed through a 30,000-MW membrane. Thus 82.5% of the material apparently had a molecular weight between 30,000 and 300,000.

A similar molecular weight range was also indicated by gel permeation chromatography on Bio-Gel P-200 (BioRad Laboratories, Richmond, California), where most of the material was eluted between a reference compound with a molecular weight of approximately 2,000,000 daltons and a compound with a molecular weight of approximately 50,000 daltons.

The same material was also subjected to mass spectrometry (MS) using low-ionization-energy (20-eV) electrons. The bulk of each sample was apparently charred in the capillary tube used to introduce the lyophilized material and thus was not volatilized. This charring was probably due to the low volatility of the high-molecular-weight substances present in the product.

14.4.2 Solubility of Liquefaction Product

The nonvolatile material in the aqueous product, usually present at 2 to 4 wt %, can be lyophilized and redissolved at concentrations greater than 18%. The solubility of the organic material in water is also pH dependent. Although the microbial process is usually initiated in a slightly acid environment, the pH of the liquid product is typically 7.7 to 7.8. Acidification to pH 1.0 by the addition of HCl results in the precipitation of about 90% of the dissolved matter. The precipitate was readily resolubilized by the addition of 0.1 N NaOH until a slightly basic pH was obtained. This behavior is somewhat similar to humic acids, although humic acids are known to be only sparingly soluble in water.

Qualitative studies on solubility were also performed with the black liquid product from the interaction of *Candida* ML13 with Mississippi lignite. The lyophilized solids from the liquid displayed only limited solubility in dimethyl sulfoxide and acetonitrile, apparent insolubility in pyridine, and complete solubility in water. This increase in solubility with successively more polar solvents suggests that the material is itself highly polar.

14.4.3 Chemical Speciation

The results of Fourier transform infrared spectroscopy (IR) and cross-polarization, magnetic-angle-spinning, dipolar-coupling solid-state ^{13}C NMR of the lyophilized product from *Candida* sp. interaction with North Dakota lignite gave a spectrum qualitatively similar to that of the lignite substrate. The product apparently has a high degree of aromaticity, with an increased number of carboxyl resonances and acid carbonyl or nitrogen groups. There is also an indication of an increase in tetrahedryl carbons attached to oxygen. The spectrum of the liquid product is consistent with the hypothesis that the product consists of a mixture of water-soluble hydroxy acids and solubilized coal from which the methylene component has been removed.

14.4.4 Separation of Constituents

Preliminary tests were made to separate the liquid product into individual constituents by using chromatography. In preparation for gas chromatography (GC), the material was first lyophilized and then extracted successively with organic solvents of increasing polarity (hexane, acetonitrile, and ethanol) for analysis. The organic-solubilized material was then introduced into a gas chromatograph equipped with a flame ionization detector and a capillary column (0.1 mm ID $\times$ 25 m) coated with SP-2100 (Supelco, Inc., Bellefonte, Pennsylvania). The temperature was increased from 60°C to 300°C at the rate of 8°C/min, and N_2 was used as the carrier gas at a flow rate of 1.25 mL/min.

Although much of the material was nonvolatile, as indicated by MS, a few peaks were detected in the extracts of the more highly polar solvents. Only a single chromatographic peak was found in the hexane-soluble fraction (lowest-polarity solvent), whereas three peaks were observed in each of the remaining solvents. Sample derivatization, which is employed to replace polar hydrogen with trimethylsilyl groups, tends to diminish the polarity and enhance the volatility of highly polar compounds; however, it failed to develop any new peaks for this material.

Some of the liquid products were subjected to liquid chromatography (LC) using a reversed-phase column (μBondapak C-18 in a 3.9 mm $\times$ 30 cm column, Waters Chromatography Division of Millipore Corporation, Milford, Massachusetts) and fluorescence and ultraviolet (UV)-absorbance detectors. The lyophilized product was dissolved in the starting eluant and then eluted at ambient temperature with a linear gradient of water to 95% ethanol (both solvents contained 1% v/v acetic acid) at a flow rate of 2 mL/min. A number of UV-absorbing and fluorescent chemical species, as yet unidentified, were separated (see Fig. 14.5). Although some similar chromatographic peaks were observed, *Paecilomyces* sp. appeared to produce a more complicated pattern than did *Candida* sp. As in the case of GC, there was evidence that much of the dissolved material did not exit the column.

14.4.5 Extracellular Microbial Processes

Extracellular processes are probably involved in the solubilization of coal by microorganisms; therefore, qualitative tests were made for proteinaceous material in the aqueous product. For example, the liquid product from the action of *Candida* sp. ML13 on North Dakota leonardite was contacted with Coomassie Brilliant Blue, a reagent that is known to bind with proteins, according to the method of Bradford [11]. This technique indicated that the protein content of the liquid was about 1.5 mg/mL.

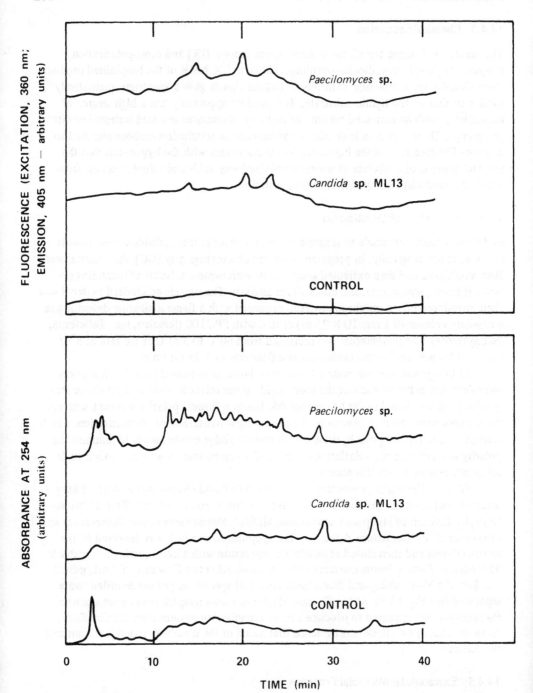

Fig. 14.5 Liquid chromatograms of aqueous products resulting from fungal action on lignite. The 0.39 mm × 30 cm column was packed with reversed-phase μBondapak C-18 and operated at 2.0 mL/min with a linear gradient of water to 95% ethanol with 1% acetic acid at ambient temperature. (Reproduced, with permission, from Ref. 8.)

Table 14.9 Comparison Between the Elemental Analyses of North Dakota Leonardite and the Liquid Products from Microbial Solubilization

Sample	Elemental analysis (wt %, dry basis)				
	C	O	H	N	S
North Dakota leonardite[a]	41.3	41.8	5.2	0.9	0.7
Liquid product from *Candida* sp. ML13	49.4	36.0	2.9	3.0	1.5
Liquid product from *P. waksmanii* ML20	49.4	38.7	3.4	3.9	1.1

Source: Reproduced from Ref. 8, with permission.
[a] Average of analyses for two different batches.

Sodium dodecyl sulfate polyacrylamide gel electrophoresis was also carried out on the same material according to the method of Weber and Osborne [12]. Staining of the gel with Coomassie Brilliant Blue following electrophoresis yielded a single band of material, corresponding to a molecular weight of about 60,000 daltons.

The fermentation broth from *S. setonii* was fractionated to isolate the active solubilization agent. The activity was determined by allowing each fraction to interact with Wyodak subbituminous coal that had been pretreated with nitric acid. Most of the activity was found to be in a fraction having a molecular weight greater than 1000 but less than 30,000. Although this material was not heat sensitive to temperatures below 100°C, the activity was usually less after exposure to a temperature of 120°C for greater than 30 min. This material may not be proteinaceous in nature.

14.4.6 Material Balance

Analyses of the reactive North Dakota lignite after removal of minerals (about 13 wt % of a quartzlike material was removed) and products resulting from fungal solubilization are shown in Table 14.9. Surprisingly, the increase in oxygen in the liquid product was marginal. The most significant change was in the nitrogen content, which was increased three- to fourfold in the liquid product as compared with the native coal. The increase in nitrogen content may be related to biological products.

14.4.7 Product Utility

Although the liquid product is still not characterized completely, its properties are sufficiently known to allow speculation on potential utility. Since the carbon content is approximately that of coal and the ash is not solubilized, the product could be a useful boiler feed if it is of sufficiently high concentration. The material is soluble in water; therefore, it would be an excellent feedstock for other bioconversion processes, such as the microbial conversion to alcohols or fuel gases. The product might also serve as suitable feedstock for aromatic chemicals or polymers.

14.5 BIOREACTOR CONCEPT

Two possible processing techniques are envisioned: (1) a gas-phase-continuous fermentation in a fixed-bed bioreactor, and (2) an aqueous suspension culture in a fluidized-bed bioreactor. The latter seems most likely at this time.

In a fluidized-bed bioreactor, small coal particles would be suspended (fluidized) in an upflowing aqueous stream that includes appropriate microorganisms or fluids derived from culturing microorganisms (Fig. 14.6). Air would be supplied to the bottom of the reactor to maintain adequate aeration in the reactor. As solubilization proceeds, the solid residues (small particulates of inorganics, etc.) would be swept out of the reactor and collected in a settling chamber. Fresh coal feed would be added at the top of the column either continuously or periodically. The liquid product would be collected as a side stream from the reactor liquid product, and the remainder of that liquid stream would be used for recycle to maintain fluidization within the reactor.

Some preliminary tests were made with a small, tapered, fluidized-bed contacting system (Fig. 14.2). The coal was fluidized by pumping an aqueous medium through the column; then the effluent was returned to the reservoir, where it was aerated prior to recycle. Typically, a suspension of fungi or the culture broth from a suspension of bacteria was used as the fluid phase. In each case there was evidence of solubilization. The *Candida* fungi attached rapidly to the coal particles, and solubilization occurred over a period of several days (Table 14.10). The fermentation broth from *S. setonii* 75Vi2 was

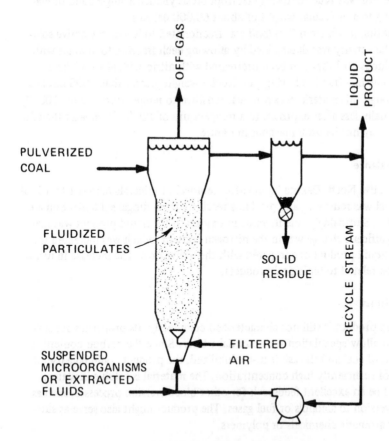

FLUIDIZED-BED CONTACTOR
(LIQUID PHASE CONTINUOUS)

Fig. 14.6 Concept for a fluidized-bed contactor for the microbial solubilization of coal.

Table 14.10 Solubilization of North Dakota Leonardite
Coal by *Candida* sp. ML13 in a Fluidized-Bed Bioreactor

Time	Degree of solubilization[a]	
(days)	Sterile system	*Candida* sp.
0	1	1
7	1.7	2.1
9	2.1	2.7
13	1.8	4.3

[a] Ratio of the absorbance at 292 nm of the fluidization medium at
the designated time compared with that at the start of the tests.

much more effective for carrying out a coal solubilization in the fluidized bed. Measurable amounts of coal were solubilized during the first day of exposure; and, as indicated above, more than 60% of the coal could be solubilized during the first 2 days of suspension culture.

14.6 CONCLUSIONS

Microbial solubilization of various low-ranked coals by both fungi and bacteria has been demonstrated. The most promising processing concept appears to be a fluidized-bed bioreactor in which active aqueous constituents derived from microorganisms effect the solubilization of suspended coal particles. The liquid product is composed of a mixture of aromatic compounds that are completely soluble in water. Experimental results obtained to date are very positive, but additional research will be required to define the system parameters completely and provide a basis for economic analysis.

ACKNOWLEDGMENT

Research supported by the Fossil Energy Advanced Research Technology Program, managed by the Pittsburgh Energy Technology Center, U.S. Department of Energy, under contract DE-AC05-840R21400 with Martin Marietta Energy Systems, Inc.

REFERENCES

1. Kargi, E., *Enzyme Microb. Technol., 4*, 13 (1982).
2. Donaldson, T. L., Strandberg, G. W., Hewitt, J. D., and Shields, G. S., *Environ. Prog., 3*, 248 (1984).
3. Korburger, J. A., *Proc. W. Va. Acad. Sci., 36*, 26 (1964).
4. Rogoff, M. H., Wender, I., and Anderson, R. B., *Microbiology of Coal*, Information Circular 8075, U.S. Bureau of Mines, Washington, D.C., 1962.
5. Fakoussa, R. M., and Trüper, H. G., Köhle als Microbielles Substrat unter aeroben Bedingungen, in *Biotechnologie im Steinkohlenbergbau*, Kolloquium in der Bergbau-Forschung GmbH (H. J. Rehm, ed.), Essen, West Germany, Jan. 20–21, 1983, p. 41.
6. Cohen, M. S., and Gabriele, P. O., *Appl. Environ. Microbiol., 44*, 23 (1982).
7. Ward, H. B., *Syst. Appl. Microbiol., 6*, 236 (1985).
8. Scott, C. D., Strandberg, G. W., and Lewis, S. N., *Biotechnol. Prog., 2*, 131 (1986).

9. Strandberg, G. W., and Lewis, S. N., *Appl. Biochem. Biotechnol.*, *18*, 355 (1988).
10. *Difco Manual: Dehydrated Culture Media and Reagents for Microbiology*, Difco Laboratories, Detroit, Mich., 1984, p. 257.
11. Bradford, M. M., *Anal. Biochem.*, *72*, 248 (1976).
12. Weber, K., and Osborne, M., *J. Biol. Chem.*, *244*, 4406 (1969).

15

Enzymatic Degradation of Texas Lignite

JUDITH K. MARQUIS *Arthur D. Little, Inc., Cambridge, Massachusetts*

BENEDICT J. GALLO *Boston University School of Medicine, Boston, Massachusetts*

Lignite is basically a waste coal that could, in fact, be a major energy source and chemical feedstock. In its raw form, lignite is difficult and costly to process into a usable and efficient source of energy. Some microbes, including fungi, bacteria, and protozoa, have been found to degrade lignin, the lignite precursor, to smaller polymeric and monomeric units. The most notable of these are the basidiomycete fungi, which include some of the agarics, and the white-rot fungi, such as *Phanerochaete chrysosporium*. One of a number of extracellular isozymic ligninases, diarylpropane oxygenase, has been extracted from *P. chrysosporium*, purified, and characterized [1]. The synthesis of diarylpropane oxygenase does not appear to be constitutive, and in this microbe, its synthesis is subject to both catabolite and nitrogen repression. Synthesis of diarylpropane oxygenase therefore requires the exhaustion of either the nitrogen or the carbon source (or both) in the growth medium. Media low in nitrogen and/or carbon sources can be used to culture *Phanerochaete*, but this results in limited biomass and low enzyme yields. Like the enzymatic degradation of cellulose, the enzymatic degradation of the more recalcitrant lignin should require the use of large quantities of lignase-type enzymes to actuate the degradation.

Phanerochaete chrysosporium is a white-rot fungus that is naturally found in forests and has the ability to degrade lignin in wood and lignocellulose pulps under certain conditions. The various ligninases were first isolated from this organism by Kirk and his coworkers (see discussion in Ref. 2). These investigators also characterized the enzymes with regard to their physical properties and their specific activities on lignin model compounds. They reported that ligninase is an extracellular enzyme and requires H_2O_2 for activity. Subsequently, Gold et al. [3] purified a ligninase or, more accurately, a family of ligninases, from this same organism and began to characterize the physicochemical properties of the enzyme proteins. They partially purified a peroxide-requiring diarylpropane oxygenase with a molecular weight of about 41,000 and containing a heme moiety.

Table 15.1 Fungal Strains Used

Trametes versicolor ATCC12679
Phanerochaete chrysosporium ME-446
Phanerochaete chrysosporium BKM F1767
Trichoderma reesi NG14

In 1982, Cohen and Gabriele [4] first reported that another fungus, *Trametes versicolor (Polyporus versicolor)*, could be grown on small pieces of unpretreated lignite. They concluded that the decay was based on the production of peroxidase enzymes, but the mechanism and biochemical pathways of lignite biodegradation were not specifically defined.

In our own laboratory, we have grown several strains of fungi in submerged culture and measured growth in terms of mycelial dry weight (Tables 15.1 and 15.2). Enzyme activities and extracellular protein production in the culture filtrate were measured but the data not shown. Enzyme activity was determined using the synthetic substrate veratryl alcohol, which is oxidized by ligninases to aldehydes (including anisaldehyde, benzaldehyde, and veratraldehyde) that can readily be identified by thin-layer chromatography. The production of veratraldehyde in culture filtrates incubated with veratryl alcohol was identified by thin-layer chromatography using a benzene/ethyl acetate solvent system and 2,4-dinitrophenol.

Renganathan et al. [5] demonstrated that the lignases are really a family of enzymes (i.e., enzymes that exist in multiple molecular forms), and that at least three of these forms are glycoproteins. Also, not all of these enzyme forms are peroxide dependent. It is particularly important to note that these multiple glycoproteins differ significantly in the kinetics of binding to lectins such as concanavalin A (carbohydrate-binding compounds) and may represent an unusual case where the carbohydrate portion of a glycoprotein has some modifying effect on the enzyme function.

The complexity of the enzyme system, particularly the interaction and interdependence of multiple enzymes and their different kinetic properties, has been the principal focus of our studies. In our preliminary studies, we achieved a 100-fold purification of ligninase activity by concanavalin A affinity chromatography, a technique that is not selective for the various molecular forms of the enzyme. We are continuing to pursue various methods for separating and purifying selected molecular forms.

Table 15.2 Estimate of Growth of Four Strains on 20 ml of Sabouraud G Medium

Strain	Mycelial dry weight (g)
Trametes versicolor ATCC12679	0.1640
Phanerochaete chrysosporium ME-446	0.1442
Phanerochaete chrysosporium BKM F1767	0.1331

Previous studies demonstrating the synthesis of fungal lignases have shown very low yields of total enzyme from stock strains and even from improved mutant strains. These low yields may be due to the poor biomass production resulting from the use of synthetic, yet well-supplemented media or due to the use of low quantities of one of the essential nutrients, such as nitrogen, carbon, sulfur, and phosphorus compounds, which, in normal medium amounts, repress the synthesis of ligninases in *P. chrysosporium*. We have also started experiments in modifying the growth medium for several of the above-mentioned fungi in order to maximize their ligninase synthesis so that their practical potential can be assessed. The media used in these studies are shown on Table 15.3. Our attempts to use a synthetic medium such as Vogel's inorganic basal medium with salts/supplements alone, or with small amounts of malt extract as an adjuvant basidiomycete growth stimulator, and with an ample amount of an easily metabolized monosaccharide or disaccharide, yielded poor growth and were overall unsatisfactory for use in enzyme kinetics. The addition of small amounts of organic nitrogen, vitamins, and Tween 80 (a nonionic detergent) greatly improved growth of the fungi but did not match the biomass yields generated when grown on the rich Sabouraud G organic medium.

In other growth studies fungi were grown in submerged, unshaken culture in a medium that contained pretreated (ball-milled) lignite as the main source of carbon. Table 15.4 shows the recalcitrance of pretreated (ball-milled) lignite to in vivo degradation by *T. versicolor* or *Trichoderma reesei* and the poor fungal growth that occurred. *T. versicolor* was used since in other ligninase synthesis studies we found that it produced more ligninase than either of the two strains of *P. chrysosporium*. Trichoderma reesei was included as a contrasting control microbe that has not been shown to synthesize any ligninase. We have conducted several preliminary regulatory studies for the purpose of increasing yields of diarylpropane oxygenase from *P. chrysosporium* BKM-F1767. In these

Table 15.3 Media Used

Vogels (Vogel S/S)	
Vogel salts/supplements concentrate	2%
Sabouraud G & GV*	
Neopeptone	1%
Maltose	2%
Malt extract	1%
Vogel salts/supplements	2%
Supplemented Vogels 1	
Malt extract	0.005%
Neopeptone	0.025%
Tween-80	0.25%
Vogel S/S	2.0%
Microelements	0.008%
Vitamin sets 1 and 2	0.005%
Substrate	
Supplemented Vogels 2	
Neopeptone	2%
Malt extract	0.5%
Vogel S/S	2.0%
Substrate	

Table 15.4 Estimate of Growth of Two Fungal Strains on Texas Lignite

	Estimate of mycelial dry weight of two 50-mL fungal cultures (g)	
Medium	Trametes versicolor ATCC12679	Trichoderma reesei NG14
Supplemented Vogel 1 without substrate (A)	0.0170	0.0298
Supplemented Vogel 1 and 1% maltose (B)	0.0482	0.0561
Supplemented Vogel 1 and 0.68% lignite (C)	0.7110	0.7162
B - A	0.0312	0.0263
C - A	0.6404	0.6864
C - A - 0.68% lignite	0.0162	0.0085

studies, a heavy mycelial pad was grown in a rich medium containing excess organic and inorganic nitrogen and an excess of the disaccharide maltose. After a period of time, the mycelial pad was partially washed free of the growing medium and the pad aseptically resuspended in various phosphate and tartrate buffers. The resuspended pads were then incubated for 2 days before final harvest of mycelium and culture filtrate. The undiluted, still rich, growing medium and the final buffer-based filtrates were then assayed for enzyme activity using the veratryl alcohol assay. Although the buffer-based filtrates still contain possibly repressive amounts of sugar, both the residual growing medium and the buffer-based filtrate did contain enzyme activity. These results indicate that the synthesis of this diarylpropane oxygenase enzyme is constitutive, and this constitutivity may not be repressed by high nitrogen or high maltose concentrations. We have also found that this constitutive ligninase is not peroxidase dependent in our assay system and may represent a new and previously unreported moiety of fungal diarylpropane oxygenase.

REFERENCES

1. Faison, B. D., and Kirk, T. K., Factors involved in the regulation of a ligninase activity in *Phanerochaete chrysosporium*, *Appl. Environ. Microbiol.*, *49*, 299-304 (1985).
2. Kirk, T. K., Tien, M., Croan, S., McDonach, T., and Farrell, R. L., Production of ligninase by *Phanerochaete chrysosporium*, personal communication from Dr. Kirk, 1986.
3. Gold, M. H., Kuwahara, M., Chiu, A. A., and Glenn, J. K., Purification and characterization of an extracellular H_2O_2-requiring diarylpropane oxygenase from the white rot basidiomycete, *Phanerochaete chrysosporium*, *Arch. Biochem. Biophys.*, *234*, 353-362 (1984).
4. Cohen, M. S., and Gabriele, P. D., Degradation of coal by the fungi *Polyporus versicolor* and *Poria monticola*, *Appl. Environ. Microbiol.*, *44*, 23-27 (1982).
5. Renganathan, V., Miki, K., and Gold, M. H., Multiple molecular forms of diarylpropane oxygenase, an H_2O_2-requiring, lignin-degrading enzyme from *Phanerochaete chrysosporium*, *Arch. Biochem. Biophys.*, *241*, 304-314 (1984).

16

Solubilization of Coal by *Streptomyces*

GERALD W. STRANDBERG *Oak Ridge National Laboratory, Oak Ridge, Tennessee*

16.1 INTRODUCTION

Although much is known about the microbial degradation and utilization of coal-derived organic compounds, relatively little information has been reported concerning the ability of microorganisms to attack the organic framework of intact coal. Rogoff et al. [1] reviewed the literature up to 1962. Although several bacteria and fungi had been found to be associated with coal, these authors concluded that no one had definitively demonstrated that unaltered coal could be attacked by microorganisms. However, they stated (prophetically) that members of the genus *Pseudomonas*, by virtue of their ability to attack polycyclic hydrocarbons; Actinomycetes, due to their role in humification; and several genera of fungi that could oxidize paraffins and humic acids were the logical organisms to consider in studying microbial coal oxidation. Members of all of the genera they listed have since been found to be capable of solubilizing coal.

In 1964, Korburger [2] found that *Escherichia freundii* and *Pseudomonas rathonis* grew on coal that had been pretreated with hydrogen peroxide, and that the extent of growth was dependent on the weight percent of coal in the growth medium. He took this as evidence that the coal was being attacked and utilized by the organisms.

There was apparently little further interest in this area until 1982, when it was demonstrated that certain lignin-degrading fungi could solubilize lignite coal [3]. Shortly thereafter, Fakoussa and Trüper [4] reported that based on spectroscopic evidence, a strain of *Pseudomonas fluorescens*, isolated from a mixed population cultured on a coal slurry, attacked and solubilized coal by means of an enzyme coupled with a surface-active agent that the organism produced.

Initial investigations at the Oak Ridge National Laboratory were directed at fungal coal solubilization [5,6]. However, the ability of fungi to solubilize coal appeared to be limited to intact mycelia cultured on a solid substrate [3,5,6]. This limitation, combined with the advantages of a submerged culture process, stimulated our interest in the ability of bacteria to solubilize coal. The perceived (at that time) relationship of lignin-degrading capability to coal-solubilizing capability led us to examine two lignin-degrading species of *Streptomyces*. The details of our studies with these organisms have been presented elsewhere [7,8]. In this report, certain aspects of what is known about coal solubilization by *S. setonii* are discussed and consideration is given to possible mechanistic aspects based on chemical coal-solubilizing processes.

16.2 COAL SOLUBILIZATION BY *STREPTOMYCES*

16.2.1 Surface Culture

Two species, *S. viridosporous* T7A and *S. setonii* 75Vi2 (obtained from D. L. Crawford, University of Idaho, Moscow, Idaho), solubilized coal in a manner similar to fungi (i.e., intact pieces of coal were solubilized when placed on the surface of agar-grown cultures) [7]. The rate of coal solubilization was more rapid (3 to 5 days) compared with fungi (14 days), but the extent of solubilization was not as great. The two bacterial species solubilized only 25 to 55% of the coal, while essentially complete solubilization could be achieved with fungi [5]. Furthermore, while fungi solubilized a naturally oxidized North Dakota lignite (American Colloid Co., Skokie, Illinois) to a greater degree than acid-pretreated coals (see Section 16.3.3), the opposite was true for the bacteria. We have examined only these two strains of *Streptomyces*. However, H. B. Ward (personal communication) has isolated several streptomycetes from the vicinity of coal deposits which solubilized coal when assayed in a manner similar to that indicated above.

16.2.2 Submerged Culture

An acid-pretreated Wyodak subbituminous coal (Amax Coal Co., Indianapolis, Indiana) was readily solubilized when added to shake-flask cultures of *S. setonii* 75Vi2 grown for 7 days prior to the addition of coal [7]. Submerged cultures of *S. viridosporous* T7A grown under identical conditions failed to solubilized coal. However, the composition of the growth medium for *S. setonii* has been shown to be of importance [8]; it may be that appropriate conditions were not used for *S. viridosporous*.

Although it has been presumed that enzymes, in particular ligninases, are responsible for coal solubilization by fungi, there is evidence, discussed in the next section, that coal solubilization by *S. setonii* is not enzyme initiated. In this regard, it would be of interest to know whether the strains of *Streptomyces* isolated by Ward can degrade lignin and solubilize coal when grown in submerged culture. Also, it would be appropriate to determine whether purified ligninase [9] has activity toward coal.

16.3 COAL SOLUBILIZATION BY AN EXTRACELLULAR PRODUCT FROM *STREPTOMYCES SETONII*

When acid-pretreated Wyodak coal was added to 7-day cultures of *S. setonii* (grown in the absence of coal), solubilization of the coal was evidenced by a blackening of the culture broth within a few hours [7]. This color change, along with the apparent lack of direct association (i.e., attachment) of the cells with the coal, suggested that coal-solubilizing activity was already present in the culture broth. The presence of extracellular coal-solubilizing activity was demonstrated by effecting coal solubilization with filter-sterilized (0.45 μm, Millipore Corp., Bedford, Massachusetts) culture broths (Table 16.1).

Table 16.1 Solubilization of Coal by a Cell-Free Culture Broth from S. setonii 75Vi2

Conditions[a]	Coal	Dry weight of coal (mg)		Coal solubilized[b] (%)
		Initial	Final	
Sterile medium	ND II	535	504	-6
	Wyodak-PT	509	518	+2
Cell-free growth broth[c]	NP II	550	423	-23
	Wyodak-PT	513	162	-69

[a]ND-II is a North Dakota lignite obtained from the American Colloid Corp., Skokie, Illinois, Wyodak-PT is a subbituminous coal obtained from the Amax Coal Co., Indianapolis, Indiana. It was pre-treated by soaking in 8 M nitric acid for 48 h (see Ref. 6).
[b]Weight of residual coal determined after 48-h incubation (30°C, 100 rpm, 2-in. stroke).
[c]Broth from 7-day culture was filter-sterilized using a 0.45-μm Millipore filter.

16.3.1 Characteristics of the Extracellular Coal-Solubilizing Component(s)

We have obtained only a limited amount of information about the characteristics of the extracellular coal-solubilizing component(s), as shown in Table 16.2. An alkaline pH is required for activity [8]. The combined extreme heat stability, relatively low molecular weight, and insensitivity to proteases indicate that the active component is not an enzyme. The characteristics noted and the similarity in action to other coal-solubilizing agents (see Section 16.3.2) suggest that this component is possibly an alkaline peptide or polyamine.

Pometto and Crawford [10] reported that the oxidative depolymerization and solubilization of lignin by S. viridosporous were maximal at an alkaline pH. This behavior was very similar to our findings with coal solubilization by S. setonii. They proposed that the alkalinity resulted from the action of the organism on nitrogenous substrates in the growth medium. Sabouraud maltose medium (Difco Corp., Detroit, Michigan), which was used for the studies of coal solubilization by Streptomyces, is rich in organic nitrogen compounds. To confirm further the similarity of our observations with those of Pometto and Crawford, we determined the utility of a variety of nitrogen-containing substrates to serve as a carbon and/or nitrogen source for growth and for the production of coal-solubilizing activity by S. setonii. As shown in Table 16.3, most of the substrates tested were suitable. However, the main points of interest are that (1) a carbohydrate (e.g., maltose) was not required; (2) glucose, for unknown reasons, suppressed the production of coal-solubilizing activity; and (3) the production of coal-solubilizing activity was always accompanied by an increase in the pH of the medium. On the other hand, an increase in the pH of the medium did not confirm the presence of coal-solubilizing activity.

Table 16.2 Characteristics of the Cell-Free Coal-Solubilizing Component(s)

High heat stability; 60 to 70% reactivity remained after 1 h at 121°C
Insensitive to acid and alkaline proteases
Molecular weight between 1000 and 10,000 daltons as determined by ultrafiltration studies
Maximal reactivity above pH 8
Probably not an enzyme
Possibly a basic polypeptide or polyamine

Table 6.3 Utility of Various Carbon and Nitrogen Sources for the Production of Extracellular Coal-Solubilizing Activity by *S. setonii* 75Vi2

Carbon/nitrogen for growth[a]	pH of culture		Coal solubilized[b] (%)
	Initial	Final (7 days)	
Neopeptone[c] (1%) plus maltose (4%)	5.8	8.9	67
Neopeptone[c] (1%) plus glucose (1%)	7.4	7.9	0
Neopeptone[c] (1%)	7.4	8.5	39
Yeast extract[c] (1%)	7.0	8.8	61
Amber-crude[d] (2%)	5.1	8.8	48
Amber-EHC[d] (1%)	5.8	6.0	0
Whey[e] (2%)	5.9	8.2	34
Casamino acids[c] (3%)	5.9	9.0	70
Glutamic acid (0.5%)	7.0	9.3	79
Valine (0.5%)	7.3	8.0	9
Asparagine (0.5%)	7.3	8.0	52

[a]The components of the medium were dissolved at the concentrations indicated in demineralized, distilled water and sterilized at 121°C, 20 min. A mineral supplement consisting of KCl, 0.05%; MgSO$_4$ ·7H$_2$O, 0.001%; and K$_2$HPO$_4$, 0.05% was included in the media containing only glutamic acid, valine, or asparagine.
[b]The extent of solubilization in sterile controls of each medium was <5%. The values shown are averages of duplicate determinations.
[c]Difco Laboratories, Detroit, Michigan.
[d]Amber Laboratories, Juneau, Wisconsin.
[e]Whole powdered cheese whey (partially demineralized), Flav-O-Rich, Inc., Knoxville, Tennessee.

16.3.2 Mechanistic Considerations

Our belief that the coal-solubilizing component is an alkaline peptide or polyamine stems from its general characteristics and its similarity in action on coal to that of alkaline buffers [8]. Additionally, Kersten et al. [11] found that nitrogen-containing solvents such as amides and amines exerted a similar solubilizing effect on nitric acid–treated coals. The mechanism of coal solubilization by these agents is unknown. There is evidence that the water-soluble product of fungal action on coal arises, in part, through a cleavage of methylene bridges between high-molecular-weight complexes of aromatic compounds, resulting in what might be termed a depolymerization of macromolecular structure of coal [5]. The products produced by the action of the bacterial extracellular components, buffers, and nitrogen-containing solvents have yet to be characterized.

16.3.3 Requirement for Pretreatment of Coal

As with growing cultures, the extracellular coal-solubilizing substance was found to be only partially active toward the naturally oxidized North Dakota lignite. Several other lignites (Texas, Mississippi, and Vermont) and Wyodak subbituminous coal were not solubilized by growing cultures or the extracellular substance without pretreatment with an oxidizing agent such as nitric acid [8]. As reported previously [6], while fungi can partially solubilize a variety of coals, pretreatment greatly enhances solubilization. The nature of the chemical/physical changes in coal which are produced by the oxidizing agents is not known. Chemical analysis showed no substantial increase in the oxygen content of the treated coals [6].

16.4 CONCLUSIONS

S. setonii solubilizes several coals, pretreated with an oxidizing agent, by means of an extracellular product produced in the absence of coal. The relatively low molecular weight, extreme heat stability, and insensitivity to proteases indicate that the extracellular coal-solubilizing substance is probably not an enzyme. The similarity in action of the substance to alkaline buffers and nitrogen-containing solvents suggests that it may be an alkaline peptide or polyamine. The identification of the coal-solubilizing substance generated by *S. setonii* and the characterization and comparison of the products of its action on coal to those of other chemical and biological (i.e., other bacteria, fungi) coal-solubilizing agents will be important to our understanding of the mechanism of coal solubilization.

ACKNOWLEDGMENTS

This research was supported by the Fossil Energy Advanced Research and Technology Program, managed by the Pittsburgh Energy Technology Center, U.S. Department of Energy, under Contract DE-AC05-84OR21400 with Martin Marietta Energy Systems, Inc. The author wishes to acknowledge the assistance of Susan N. Lewis, Chemical Technology Division, Oak Ridge National Laboratory, in conducting this research.

REFERENCES

1. Rogoff, M. H., Wender, I., and Anderson, R. B., *Microbiology of Coal*, Information Circular 8075, U.S. Bureau of Mines, Washington, D.C., 1962.
2. Korburger, J. A., Microbiology of coal: growth of bacteria in plain and oxidized coal slurries, *Proc. W. Va. Acad. Sci., 36*, 26 (1964).
3. Cohen, M. S., and Gabriele, P. D., Degradation of coal by the fungi *Polyporous versicolor* and *Poria monticola*, *Appl. Environ. Microbiol., 44*, 23 (1982).
4. Fakoussa, R. M., and Trüper, H. G., Köhle als Microbielles Substrat unter aeroben bedingungen, in *Biotechnologie im Steinkohlenbergbau*, Kolloquium in der Bergbau-Forschung GmbH (H. J. Rehm, ed.), Essen, West Germany, Jan. 20-21, 1983, p. 41.
5. Scott, C. D., Strandberg, G. W., and Lewis, S. N., Microbial solubilization of coal, *Biotechnol. Prog., 2*, 131 (1986).
6. Strandberg, G. W., and Lewis, S. N., A method to enhance the microbial liquefaction of lignite coals, *Biotechnol. Bioeng. Symp., 17*, 153 (1986).
7. Strandberg, G. W., and Lewis, S. N., Solubilization of coal by an extracellular product from *Streptomyces setonii* 75Vi2, *J. Ind. Microbiol., 1*, 371 (1987).
8. Strandberg, G. W., and Lewis, S. N., Factors affecting coal solubilization by the bacterium *Streptomyces setonii* 75Vi2 and by alkaline buffers, *Appl. Biochem. Biotechnol., 18*, 355 (1988).
9. Tien, M., and Kirk, T. K., Lignin-degrading enzyme from the hymenomycete *Phanerochaete chrysosporium* burds, *Science, 221*, 661 (1983).
10. Pometto, A. L., and Crawford, D. L., Effects of pH on lignin and cellulose degradation by *Streptomyces viridosporous*, *Appl. Environ. Microbiol., 52*, 246 (1986).
11. Kersten, M. H., Ramsey, J. W., and Brady, G. A., *Characteristics of Products Obtained by Oxidation of Anthracite with Concentrated Nitric Acid*, Report of Investigations 6535, U.S. Bureau of Mines, Washington, D.C., 1964.

17

Isolation and Application of Coal-Solubilizing Microorganisms

BAILEY WARD   *The University of Mississippi, University, Mississippi*

17.1 INTRODUCTION

At least as early as the first two decades of this century, there were investigators who were asking questions about microbial utilization of coals. Potter [1] and Schroeder [2] studied microbial interactions with coals and reached the conclusions that bacteria would utilize organic components of coal as nutrient substrates. Fischer and Fuchs [3] later tested a *Penicillium* sp. on a soft, brown coal and described evidence of growth on the coal. Rogoff et al. [4], in a rather comprehensive review of coal microbiology up until about 1960, described the growth of fungi (*Penicillium* sp. and *Trichoderma* sp.) on a North Dakota lignite. Best growth occurred on finely ground (–300 mesh) coal with added nitrogen source. An important conclusion of Rogoff et al. was that the degree of surface oxidation was the most important single factor affecting microbial utilization of coal carbon.

Chemically digested coals also have been tested for their ability to support microbial growth. Kucher et al. [5] used H_2O_2 or HNO_3 (concentrations and treatment methods not reported) on a high-grade coal to obtain aqueous extracts that supported the growth of a *Candida tropicalis* strain K41 given a supplemental growth medium. They noted that the degree of oxidation of the coal affected the concentration of growth-supporting, water-soluble substrates in the extract solutions.

Fakoussa and Trüper [6] reported on the interaction of a *Pseudomonas* sp. with hard German coal and observed spectroscopic evidence of changes in the coal structure. The involvement of surface-active agents produced by the bacteria was noted.

A report by Cohen and Gabriele [7] established the potential for applications of fungi for bioconversions of solid coal to solubilized products. These authors used the American Type Culture Collection strains of the wood-rot, basidiomycete fungi *Trametes (Polyporus) versicolor* ATCC 12679 and *Poria monticola* ATCC 11538 to achieve bioconversion of a highly oxidized (naturally weathered), North Dakota leonardite directly to water-soluble products. At about the same time, Ward [8] described the isolation from weathered lignite of several hyphomycete fungi that would grow on untreated, solid lignite and on ethanolic extracts of the same coal. Several of the hyphomycete isolates grew on raw Mississippi Claiborne lignite as a sole source of energy, carbon, and mineral nutrients [9]. Two of these fungal isolates would convert solid lignite directly to water-soluble products [10].

A note about terminology is in order at this point in the discussion. The generic term "biodegradation" refers to decomposition by biological means, and thus applies to many phenomena whereby complex materials are reduced to simpler components. "Bio-depolymerization" denotes a special case of biodegradation whereby polymers are reduced in size. "Bioconversion" is used to designate the event of a change in form of material, and the term does not necessarily imply biodegradation. The term "liquefaction" denotes a changing of a solid to a liquid; thus the use of "bioliquefaction" of coal would imply a direct conversion from solid to liquid coal, the latter which might or might not be hydrophilic.

The phenomenon that is the topic of this and several other chapters in this volume is known now only to be one whereby certain fractions of solid (mostly hydrophobic), oxidized coal are changed biologically to hydrophilic forms that readily enter into the aqueous phase of the culture medium. Thus the event is one of a bioconversion and the term "biosolubilization" is an appropriate one to use to refer to the special case of coal bioconversions resulting in water-soluble products. The more specific terms noted in the preceding paragraph might be reserved to apply to other special cases of coal bioconversions when they are known by experimental evidence to occur.

Recent efforts in coal bioconversion research have focused on pretreatments of biorecalcitrant coals to enhance microbial activity and on characterization of the solubilized product. Strandberg and Lewis [11] reported on enhanced bioconversions of several coals after oxidizing pretreatments with hydrogen peroxide, or nitric acid, or ozone. Scott et al. [12] described bioconversions of several coals and reported on chemical analyses of the solubilized products; chemically oxidized or naturally weathered samples of a lignite were more readily biosolubilized than were untreated or unweathered samples. Ward [13] described biosolubilization of several ranks of naturally weathered or chemically oxidized coals by a variety of fungi isolated from coal outcrops.

Sufficient evidence has been presented [14-16] to indicate that most coal biosolubilization is a phenomenon of base solubilization resulting from alkali production by developing or matured microbial cultures. A direct correlation between coal biosolubility and solubility in dilute alkaline buffer, and between coal oxygen content and either biosolubility or alkali solubility has been established [14]. Nevertheless, there also is evidence [17,18] for possible differences in mechanisms of coal biosolubilization for some strains of fungi. Attempts to implicate direct enzymatic activity in biosolubilization of a North Dakota leonardite by the wood-rot fungus *T. versicolor* (ATCC 12679) yielded some indications for enzyme involvement [19,20].

Nonoxidizing pretreatments with hydrochloric acid have been used [16,21] to enhance biosolubility of a variety of coals. The enhancement effects appeared to be associated with extraction of metal cations that had earlier (during coalification or as a consequence of weathering) formed insoluble complexes with oxidized coal functional groups. Hydrochloric acid extraction of apparently highly weathered (oxidized), but not readily biosoluble lignites, resulted in marked enhancement of biosolubilization [21].

This chapter describes general methods for isolating from natural sources a variety of coal-solubilizing strains of fungi and bacteria. Also presented are examples of application of methods for semiquantitative and quantitative biosolubility assays on a variety of coals, including some high-ranked coals pretreated to enhance biosolubility.

17.2 BIOLOGICAL PERSPECTIVE OF FOSSIL-FUEL BIOCONVERSIONS

Fossil evidence indicates that the prokaryotic bacteria have existed on earth for over 3 billion years. The eukaryotic fungi seem to have appeared some 2 billion years later. Most of the species of bacteria and all of the fungi derive their energy and carbon from organic compounds, and most of both kinds of microorganisms, given the opportunity, will metabolize organic substrates via aerobic processes. Microorganisms in general are notable opportunists that are capable of utilizing a wide variety of organic substrates, the presence of which often serves as the very stimulus for synthesis of specific enzyme catalysts. Extracellular enzymes facilitate the biodegradation of complex organic compounds, the breakdown products of which are transported across the microbial cell membrane into the intracellular milieu, where they are oxidized further to carbon dioxide and water via aerobic respiration, or to a variety of organic by-products via anaerobic processes. The natural form of the biologically available organic substrates is biomass.

Fossil fuels are derived from biomass, and represent complex organic carbon sources that have coexisted with bacteria and fungi for millions of years. Granted, the bulk of fossil-fuel deposits have not generally been accessible to microbial exploitation, but natural phenomena (geologic upheavals, erosion, seepage from subterranean deposits, and the like) have brought petroleum, methane, and coals to the surface, where they have been available to microorganisms for many years.

Methane can be metabolized by methylothropic bacteria, and petroleum fractions are known to be used by various bacteria (see the review by Atlas [22]). Studies on the processes of bioconversion of crude oil by microorganisms might have applications for coal-bioconversion research inasmuch as the oil represents a hydrophobic material composed of complex aromatic and aliphatic compounds derived through fossilization from biomass.

The eukaryotic alga *Prototheca zopfii* Kruger (earlier classified as a yeast-type fungus) is capable of using at least the alkane fractions of crude oil [23-25] and accumulating intracellularly certain complex aromatics [26]. Koenig and Ward [24] isolated and characterized a rapidly growing, crude oil-using variant (designated strain UMK-13) of the ATCC 30253 *P. zopfii* parent strain originally isolated from Chesapeake Bay by Walker et al. [27]. Metabolism of crude oil by strain UMK-13 began only after a rather sudden dispersion and emulsification of the oil. The emulsification occurred at a point when the cell population had reached an apparent critical density [16]. Microscopic examination of the suspension revealed that cells became attached to and partially embedded in the oil droplets prior to emulsification. After emulsification, individual oil droplets contained one or more cells. At this point, there occurred a dramatic increase in *n*-alkane utilization. Figure 17.1 illustrates the emulsification phenomenon for UMK-13 grown in an enriched organic nutrient medium on which a pool of mixed, pure crude-oil *n*-alkanes was floated. Although growth on crude oil proceeded at a slower pace than that on pure *n*-alkanes, the emulsification and utilization phenomena were similar. Figure 17.2 depicts *P. zopfii* cells in association with the hydrocarbon mixture just prior to and after

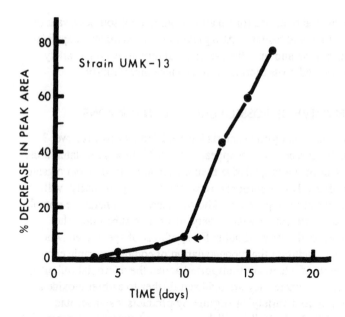

Fig. 17.1 Emulsification and utilization of a 1% v/v equal mixture of *n*-alkanes (hexadecane, tetradecane, tridecane, and octadecane) by *Protetheca zopfii* strain UMK-13 grown in Sabouraud dextrose broth, pH 5.0, 21°C. Arrow indicates point of emulsification. Analysis by gas chromatography of solvent extracts of growth suspension. Each data point represents pooled values of decrease in peak heights as reported by Koenig, D.W., Thesis, The University of Mississippi, May 1982.

emulsification. The emulsification phenomenon indicates the possibility of production of surfactants by *P. zopfii*. Because it is a liquid that can be emulsified by agents produced by microorganisms, and because it is a hydrophobic substrate to which organisms can become attached, petroleum represents a fossil fuel that can be attacked biologically.

The biodegradation of coals presents problems to microorganisms quite unlike those of petroleum. Not only is the coal substrate mostly a solid, but the organic matrix is made of a (presumedly) more biorecalcitrant structure than that of petroleum. Extracellular enzymes must contact their substrates before catalysis can occur, and under natural conditions such enzymes function in an aqueous environment. Moreover, if degradation products of remote substrates are to be utilized by the organism, they must be water soluble so as to reach the cell membrane to allow for uptake. Thus, for enzymatic catalysis of coal substrates to occur, one would expect that the aqueous phase of the biological milieu would extend to the level of the substrate. The very hydrophobic, solid, condensed nature of coals would indicate a quite limited degree of bioconversions, without chemical or physical modification of the coal structure. The importance of degree of coal oxidation on biosolubility has been established [14,16]. Yet, as will be shown in a following section, even some highly weathered lignites are more biorecalcitrant than are some weathered, higher-ranked coals. Other factors must be considered. Surface area and porosity would be expected to affect accessibility of substrates, as would amount and form of water. The production of surfactants by some microorganisms (e.g., as suggested earlier for *P. zopfii*, and by Fakoussa and Trüper [6] for a *Pseudomonas*) represents one possible mechanism whereby the coal hydrophobic matrix might become accessible to

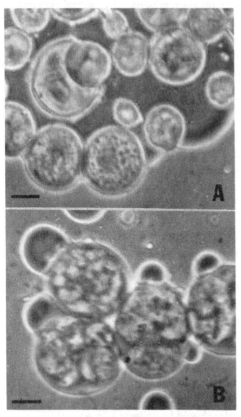

Fig. 17.2 Phase-contrast micrographs of *P. zopfii* strain UMK-13 from preemulsification (A) and postemulsification (B) of *n*-alkane mixture from Fig. 17.1 Scale = 6μm.

microbial attack. Secretion of metal chelators by some microorganisms might result in enhanced biosolubilization due to extraction of metal cations complexed with oxidized coal functional groups [16,21]. Studies on a variety of coals by a variety of microorganisms ultimately might reveal several mechanisms of microbial coal conversions.

17.3 ISOLATION AND HANDLING OF ORGANISMS

When our group first attempted (June 1982) to find coal-degrading microorganisms, we did not know of any reports describing direct biological conversions of coal substrates. The term "lignite" originated because of the coal's presumed derivation from lignin, the degradation of which by fungi already had been reported [28]. Lignite is also the lowest and softest rank of coals, having undergone less coalification than harder coals, and thus might be expected to be less biorecalcitrant than the latter. We reasoned that the extensive lignite deposits of the U.S. Gulf coast states might provide materials from which lignite-degrading organisms could be recovered.

Successful microbe hunters are those who search for their quarry where it is most likely to live and reproduce. Thus one would hope to obtain isolates of coal-degrading microorganisms from weathered coal samples collected from areas where environmental conditions favor growth and reproduction, and where soil and decaying vegetation (natural inocula) are mixed with the coals. Such sites are represented by natural outcrops occurring in creek beds where moisture is present and temperatures and desiccation are moderated by canopies of vegetation, the litter of which would host a variety of wood-decay microorganisms. Other favorable collecting sites are those where coals are exposed at old roadcuts, especially where water runoff and roadside vegetation mingle with the weathered coals. Natural outcrops or abandoned mines also provide good samples collected from weathered exposures or spoils.

When one wishes to isolate from natural habitats microorganisms that are capable of degrading biorecalcitrant materials, several factors should be considered. First, one must presume that conditions for growth often are less than optimal. Indeed, among microbial communities, the single factor of competition for nutrients might severely limit the population density of many species, yet reproductive structures (e.g., spores) and hardy vegetative structures would be present. Second, a given sample, especially of soil, should contain culturable spores or vegetative structures of diverse microbial forms representing a variety of fungi and bacteria. Third, if the objective is to obtain isolates capable of attacking specific (presumedly biorecalcitrant) organic materials such as crude oil or coals, one might expect that only a few (or none!) of the culturable organisms would exhibit the desired activity. Fourth, it is a fact that different microbial forms, even from closely related groups, exhibit markedly different growth requirements. For example, soil and wood-decay fungi in general tend to grow best at high relative humidity and at slightly acid pH, whereas many soil bacteria give best growth under alkaline conditions and will grow at low relative humidity. Thus choice of isolation and subsequent culture environment can be tailored to select for certain forms of microorganisms.

At the collection site, "biosamples" for isolation of microorganisms are collected from scrapings of highly weathered coals, coal and soil mixtures, and from decaying vegetation mixed with weathered coal. The biosamples should be collected "aseptically" to avoid subsequent isolation of organisms carried to the site by the collector. Twigs, stones, and the like, which already exist at the site, serve well as tools for manipulating the samples, which should be placed into sterile containers (e.g., Whirl-Packs, Nasco, United States) and sealed to retain moisture. Weathered coal samples for subsequent bioconversion tests are collected from surface exposures. Other samples are mined by handpick from several centimeters within the coal seam. The latter samples are referred to as being "semiweathered." Semiweathered samples are presumed to be less weathered than are surface-exposed samples of the same coal. When available, unweathered coals can be collected as they are exposed at the site of an active mine. Unweathered samples can be sealed under inert gas, immersed in deionized, gas-free water, or sealed and packed in dry ice to retard continued oxidation. In the laboratory, biosamples are used "full strength" and peppered onto isolation media or are prepared as serial dilutions in sterile water or nutrient broth, which are then streaked or spread onto agar medium and incubated.

There are several approaches that can be taken for recovery of potential coal degraders from natural samples. At one extreme, one can prepare purified agar substrates containing only sterile coal ("coal agar") pulverized and suspended throughout the agar medium; an inorganic pH buffer such as potassium phosphate can be added to stabilize pH at a desired value. Direct application of undiluted or diluted samples usually will yield several colonies per plate. Isolates so obtained are presumed to represent those capable of growing and reproducing under severely limited conditions for growth, and the presence of coal might select for those microorganisms that are capable of utilizing coal components.

However, the coal biosolubilization phenomenon, which is the subject of several chapters in this volume, appears to be one of co-metabolism, in which case best activity occurs when the organisms are grown on or in an enriched nutrient medium, with coal solubilization occurring rather as an opportunistic act by the microorganisms. Thus it would seem prudent to isolate as many microbial strains as possible by the use of full-strength media formulated to provide maximum organic and inorganic nutrients for the type of organisms being sought. Sabouraud dextrose or maltose 1 to 2% agar support the growth of many fungi; streptomycete bacteria grow well on YM agar, actinomycete isolation agar, and tryptic soy agar, all of which are available from Difco Laboratories in the United States. Antibiotics can be used to inhibit bacterial growth, whereas agents such as cycloheximide will inhibit the growth of fungi. When enriched media are used, dilutions of up to 10^{-4} of the original biosamples are usually required to yield plates with few enough colonies to permit selection and transfer.

A convenient method for isolating pinpoint, developing colonies or small clusters of spores from fungal hyphae is by use of a stereoscopic dissecting microscope housed in a laminar-flow transfer cabinet—plus a steady hand or a micromanipulator. Successive transfers of spores or isolated colonies from streak or spread-plate subcultures (along with antibiotics or antifungal agents as appropriate) are used to arrive at apparently axenic cultures.

Using various of these techniques over a period of about 2 years, we isolated from weathered coals several hundred individual cultures, each of which was screened for bioconversion activity on weathered lignites. Isolates that exhibited activity were studied further to establish uniqueness and to assign strain designations. Strain uniqueness was established by microscopic examination, macroscopic features of colony morphology and pigmentation, and growth habits on various nutrient media. Strain designation codes were assigned to reflect the source of each isolate (e.g., strain YML-1 was the first isolate derived from the York, Alabama site, Midway Lignite). About 40 apparently unique strains, among which several were bacteria, exhibited some degree of biosolubilization of at least one coal. Although some of the coal-solubilizing strains of fungi and bacteria were derived from subbituminous or bituminous coals, only data for lignite-derived strains of fungi are presented here.

17.4 SCREENING AND ASSAYS FOR COAL BIOCONVERSION ACTIVITY

Our first tests for coal bioconversions were designed to answer the question: Can microorganisms grow on raw coal as a sole source of carbon and energy? Biosamples were collected from a natural lignite (Claiborne Group) outcrop occurring in a creek bed located in the state of Mississippi. Test pieces of unexposed (semiweathered) coal, cut from within larger chunks, were inoculated with spores from about 40 fungal cultures isolated from the biosamples. Previously unexposed coal pieces (handled only with sterile, stainless-steel tools) were used to avoid growth due to the presence of allochthonous nutrient materials. Twelve of the tested isolates exhibited some degree of growth on the raw coal. Presoaking the coal in a dilute basal mineral solution enhanced growth markedly for some isolates. Table 17.1 lists the isolates, taxonomic identifies (where possible), and relative growth rates on the semiweathered Claiborne lignite (the ML strain designations for "Mississippi lignite" applied to the first isolates are retained here for consistency). Figure 17.3 depicts one of the isolates growing on the coal surface. The *Candida* sp. (strain ML-13) exhibited evidence of direct conversion of the lignite to a viscous, liquid product.

Table 17.1 Growth of Fungi on Mississippi Claiborne Lignite

Strain	Organism	Relative growth[a]
ML-2	Unidentified yeast	1
ML-5	Unidentified hyphomycete	2
ML-13	*Candida* sp.	5
ML-18	*Penicillium* sp.	4
ML-20	*Penicillium waksmani*	5
ML-24	*Mucor* sp.	2
ML-27	*Mucor lausannensis*	3
ML-30	Unidentified mycelial fungus	4
ML-31	*Penicillium* sp.	4
ML-33	*Paecilomyces* sp.	5
ML-34	*Penicillium* sp.	5
ML-36	*Aspergillus terreus*	4

[a]1, Least; 5, most extensive and rapid conspicuous colony development *on coal only.*

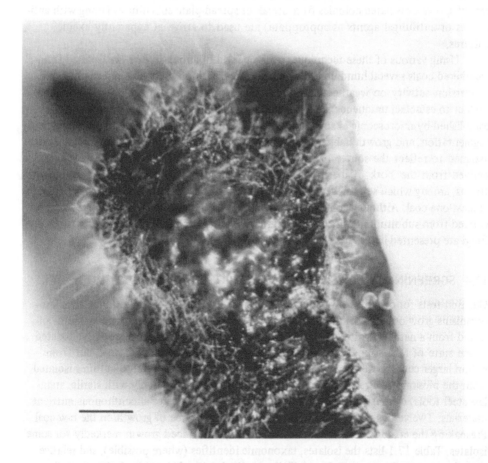

Fig. 17.3 Fungal strain ML-13 (*Candida* sp.) growing on untreated, semiweathered sterile (autoclaved) Mississippi Claiborne lignite with only purified water added to the coal. The glossy area in the center of the photograph represents a direct bioconversion of the coal to a viscous product. Scale = 2 mm.

Cohen and Gabriele [7] reported that *T. versicolor* and *P. monticola* were capable of growing on a highly oxidized North Dakota leonardite. We were not able to achieve growth of either *T. versicolor* or *P. monticola* on Claiborne lignite under our assay conditions. The leonardite is a more highly oxidized material than is the Claiborne semi-weathered lignite, and thus the latter probably is a more biorecalcitrant material.

In the case where the organisms grow on the coal as the sole source of carbon and energy, at least some of the organic matrix of the coal would be converted to biomass, thus reducing the energy value of the fossil fuel. The extent of energy loss would depend on the extent of growth, and we have observed the latter to be minimal compared to growth of the same organisms on enrichment media. Yet the evidence that an organism can attack the coal structure might indicate an application for bioconversion research and technology.

The work of Cohen and Gabriele [7] provided the basis for another approach to coal biodegradation assays. Another question to be asked is: Can an organism degrade or bioconvert a coal as a result of co-metabolism when the primary source of organic nutrients is an enriched medium? Cohen and Gabriele observed the formation of a liquid product from the North Dakota leonardite when *T. versicolor* or *P. monticola* were grown on an enrichment agar medium. Although we were not able to achieve the same results on Claiborne lignite by either of the basidiomycetes, two of our original 12 isolates, *P. waks-mani* (ML-20) and *Candida* sp. (ML-13 (Table 17.1), did produce liquid products [10], but only from a few of the many semiweathered coal pieces tested. Further tests revealed that surface-weathered samples of the same Claiborne lignite were more readily biosolu-bilized, yet there was still variability in degree of activity and percentages of coal pieces affected.

In the case of co-metabolism of coals, depolymerization of the coal matrix might not be accompanied by loss of organic carbon, or at least, the loss might be minimal. Loss in energy value would be in proportion to degree of oxidation of organic products. The co-metabolism approach, then, offers the potential for recovering most of the coal carbon in a variety of forms that might be useful for a variety of technological applications.

We have adopted variations of the co-metabolism assay to provide for measurements of coal bioconversion activity that will allow for comparisons among organisms and among different coals. When coal pieces are placed on the surface of a well-developed mat of a coal-solubilizing fungal culture, the solubilized product tends to accumulate as pools or drops on the fungal mat surface, although some product eventually diffuses through the mat into the agar below. This method allows for subjective evaluations of bioconversions and for collection of the solubilized product for subsequent analysis, but does not provide a convenient means by which rate and degree of activity can be expressed quantitatively. For fungi, we have seen that when the coal is placed on the nutrient agar prior to inoculation with organisms, developing mycelia cover the coal units, after which solubilized product diffuses into the agar around and beneath each coal unit. The diffusing product produces a colored zone that is clearly visible through the underside of the culture dish. In the case of some bacteria, the coal becomes solubilized without contact by the developing bacterial colony, a phenomenon indicative of an extracellular, coal-solubilizing agent (or agents) diffusing from the organism to the coal substrate. A similar phenomenon would probably be obscured by the rate and extent of fungal growth.

The methods described in this chapter were adopted to establish experimental approaches that would yield repeatable and reliable assays on coal bioconversions. The aim was to use methods that would minimize the considerable variability usually observed in results of biosolubilization assays. Only by use of standard assay techniques can one make meaningful comparisons of biosolubilization activity on different coals by different organisms and by different researchers.

We have noted considerable variability in biosolubilization activity even among small samples taken from one larger piece of coal. To minimize this variability, we attempted to reduce the heterogeneity of the samples by mixing particles prepared from large samples so as to arrive at test samples representative of the average whole. For the data reported herein, we used either solid coal pieces of about 0.2 cm^3 or aggregates of "sized" particles obtained by screening ground coal through 32 mesh (ASTM E-11 No. 35 sieve) then collected on 60 mesh (ASTM E-11 No. 60 sieve) to give a particle size range of 0.25 to 0.5 mm. We soaked all coals for 24 h in a dilute basal mineral enrichment medium (Table 17.2) to ensure that all samples presented a uniform inorganic supply to the microorganisms. Decanted coal preparations were sterilized by autoclaving in fresh mineral soak solution (sufficient to keep the coal wet) at 1.1 kg/cm^2 for 15 min, after which they were used at once or were sealed and stored at about $10°C$. Manipulations of sterile coals and microbial transfers were made using laminar-flow, biological safety cabinets.

Assays were carried out in the dark using Sabouraud dextrose agar (SDA) under about 70% relative humidity to reduce evaporation (and to optimize for growth of mycelial fungi). Higher relative humidity creates problems with unwanted growth of fungi on label tapes, cotton, and even glass surfaces, and causes interfering condensation inside culture vessels when these are removed from incubators into cooler environments for measurements. Controls for each assay should consist of coal-only and organism-only preparations on nutrient agar incubated with the experimentals. Some weathered or chemically oxidized coals produce a colored leachate when incubated on some nutrient agars. Leaching is most pronounced at $pH > 7$ and at temperatures above $30°C$, and varies with coal samples and nutrient medium. A further caveat: Some fungi and streptomycete bacteria produce dark, water-soluble products (in the absence of coal) that collect on the culture surface (fungi) or diffuse into the agar. These products often appear identical to solubilized coal.

Table 17.2   Mineral Supplement Solution[a]

Salt	Final concentration	
NH$_4$NO$_3$	2.50	(mg/L)
KH$_2$PO$_4$	1.75	
MgSO$_4$	0.75	
K$_2$HPO$_4$	0.75	
NaCl	0.25	
ZnSO$_4$	88.0	(μg/L)
FeCl$_3$	80.0	
CuSO$_4$	16.0	
MnCl$_2$	14.0	
MoO$_3$	7.0	
Co(NO$_3$)$_2$	5.0	

[a]Final pH adjusted to 5.5 for mycological media.

The diffusion of biosolubilized coal product into the agar around and beneath the coal units presents a means whereby quantitative measurements of activity can be made. One application allows for measurements of heterogeneity and visual estimates of degree and rate during initial screening for biosolubilization activity. Sized pieces or aggregates of sized coal are placed as uniformly spaced units on the nutrient agar surface, after which the preparations are inoculated with organism and incubated. Progress of solubilization is monitored visually and recorded as percentages of total coal units solubilized per lapsed time beyond which no additional activity occurs. It is a usual case for the more biorecalcitrant coals that some pieces do not exhibit any detectable solubilization. Table 17.3 presents data for percent solubilization of three different U.S. Gulf coast lignites by several active fungal strains derived from the same lignites. The coals were weathered samples of Mississippi Claiborne, Alabama Midway, and Mississippi Wilcox. The variability of susceptibility to microbial attack among the three lignites is apparent, as are differences in activity by any one organism on the different coals. Relative degree of activity can be expressed as visual estimates of amount of product formed [i.e., numerical values of 1 through 5 (with 5 representing maximal activity) can be assigned to indicate relative extent of biosolubilization].

More quantitative data can be generated by refinement of the method just described for initial screening assays. The zones of soluble product can be expressed as diameter or area of a circle that increases as a function of time of assay. We refer to this technique as the "diffusion zone assay." For standardization of assay conditions, it is desirable to control such factors as temperature and relative humidity, as well as concentration, depth, and uniformity of agar medium. For the data presented here, we used SDA (1.5% agar) medium poured to a depth of 5.5 mm in 100-mm-diameter culture vessels (Corning No. 3250). Condensation can be minimized by pouring the agar after it has cooled to near the point of solidification and by venting culture vessels, which should be cooled slowly. The sized coal units are placed as 3 to 4-mm-diameter aggregates evenly spaced on the surface of the nutrient agar medium. The agar surface should be free of liquid water from condensation, which hampers placement and maintenance of uniform coal units. Extent of separation of coal units from one another and from vessel boundaries

Table 17.3 Biosolubilization of Weathered Lignites by Lignite-Derived Fungi

	Coal[a]					
	Claiborne		Midway		Wilcox	
Strain	% Sol	Days	% Sol	Days	% Sol	Days
Ml-13	40	14	100	14	100	14
DML-12	46	11	27	10	100	6
NML-4	34	16	23	12	93	23
NML-10	27	17	31	16	80	11
RWL-5	35	23	33	20	90	21
RWL-18	24	22	20	21	40	24
YML-1	70	7	70	15	100	15
YML-21	33	12	33	12	100	11

[a]Prepared as ca. 5-mm³ pieces according to standard methods described in text; SDA medium, 25°C, 70–80% RH. % Sol, percent of total pieces solubilized of several replicate plates of at least 10 coal units; days, lapsed time beyond which no further solubilization occurred.

will affect the degree to which diffusion zones spread unhampered. A template circle drawn on paper to the size of the culture vessel and with locations and sizes of coal units marked will aid in uniform placement of coal units on all replicate culture vessels and by all personnel. As the diffusion zones increase in diameter, the outer margins become less distinct, thus limiting the precision of measurements after several days of biosolubilization activity. However, as noted later, different coals produce product zones of different color and intensity; some allow for longer time of assay and more precise measurements than do others. Increases in zone diameter with time yield a measurement of rate of solubilization. Replicate plates of several coal units each provide sufficient data for statistical treatment of results to establish validity of measurements. All data reported should be accompanied by details of all variables peculiar to the assays. The zone diameter (or area) measurements provide a convenient means by which effects on biosolubilization of such variables as temperature, pH, relative humidity, nutrient medium composition, and organism can be tested and expressed quantitatively.

Figures 17.4 and 17.5 represent data obtained from several diffusion zone assays of the three Gulf coast lignites. The organisms used were selected to represent a variety of taxonomic forms (Table 17.4), each of which exhibited relatively rapid solubilization of the weathered Mississippi Wilcox lignite. The data show that the rate and extent of activity differ among the coals and that different organisms act differently on different coals. The first data points represent the time of first appearance of solubilized product

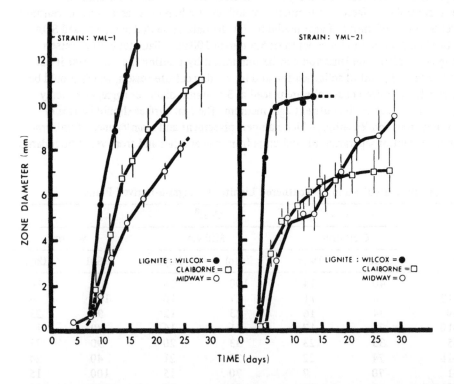

Fig. 17.4 Diffusion zone assays for comparison of biosolubilization of three different weathered lignites by two lignite-derived fungal strains. Conditions of assay: 25°C, 70 ± 5% relative humidity, 2% SDA medium (for details, see the text). Vertical bars represent 95% confidence levels for 18 coal unit replicates per point.

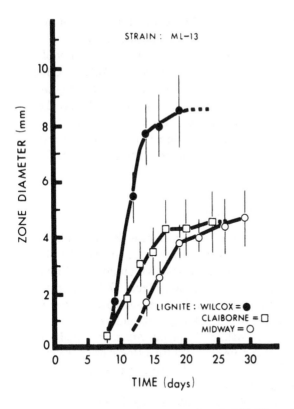

Fig. 17.5 Diffusion zone assays of strain ML-13; coals and conditions as for Fig. 17.4. Vertical bars represent 95% confidence levels for 18 coal units replicates per point.

after "time zero," which represents the time of inoculation and beginning of incubation. Increases in variability among zones with lapsed time are evident as increases in 95% confidence levels, indicated by the vertical bars at each mean data point.

As indicated in the figures by dashed lines extending from the last data points, when zone perimeters became diffuse or merged with one another, it was not possible to make accurate measurements, although in most cases, continued darkening of the agar indicated continued solubilization. There is the additional complication that some organisms themselves darken the agar to the point that solubilized coal product becomes obscure. Assays run for periods up to 20 days should provide adequate data for meaning-

Table 17.4 Selected Fungal Strains

Strain designation	Organism
ML-13	*Candida* sp.
DML-12	*Acremonium* sp.
RWL-40	Unidentified basidiomycete
YML-1	*Cunninghamella* sp.
YML-21	*Cunninghamella*-like
ATCC 12679	*Trametes (Polyporus) versicolor*

ful comparisons of rate and extent of solubilization of different coals by different organisms, and for coals subjected to different pretreatments to enhance activity.

Among the three Gulf coast lignites, weathering influences the susceptibility of two of the coals to bioconversion. Semiweathered and weathered Mississippi Wilcox lignites were solubilized about equally. The Alabama Midway weathered samples generally were somewhat more readily solubilized than were semiweathered samples, whereas the weathered Claiborne samples proved to be more readily attacked than were semiweathered samples. A fourth lignite, a highly weathered Colorado Kiowa coal collected from open ravines near the town of Kiowa, was the most biorecalcitrant of all weathered or semiweathered lignites tested.

17.5 BIOCONVERSIONS OF HIGH-RANK COALS

Given that rapid and extensive biosolubilization of some lignites is demonstrable, one then asks if harder coals can be attacked in a similar fashion. We chose two grades of Colorado subbituminous and two different samples of an Arkansas bituminous for biosolubilization assays using selected lignite-solubilizing strains (Table 17.4). The author collected Colorado subbituminous coals from the Green River region (abandoned strip mines near Steamboat Springs) and from the North Park region (abandoned shaft mine, natural outcrop, and active strip mine) near Walden, Colorado. Arkansas bituminous samples of the Lower Hartshorne Formation were collected from Horsehead Lake and near the city of Dardanelle. Except for the North Park active strip-mine samples, all coals were collected from weathered, surface exposures.

Initial screening assays revealed that only the weathered North Park coal was readily biosolubilized. Neither the weathered Green River nor the unweathered North Park samples exhibited any evidence of bioconversion when tested against several of our most active lignite degraders. The Arkansas bituminous coals showed only limited biosolubilization and only by a few of the fungal isolates.

Our observations that natural weathering renders some coals readily biosolubilized, or at least less biorecalcitrant, prompted us to test several methods of pretreatments of biorecalcitrant coals to enhance biodegradation. One aspect of weathering that seems to affect biodegradation is the oxidation state of the coal. We reasoned that oxidizing pretreatments might beneficiate biorecalcitrant coals. Others [12] have reported that pretreatment of some coals with strong (4 N or 8 N) nitric acid solutions enhanced bioactivity.

We used heat treatments, or pressurized oxygen atmosphere, or nitric acid treatments to enhance bioactivity on the coals. Heat treatment consisted of 120 h at 130°C in a flow-through, dry-air, convection oven. Oxygen treatment consisted of 25°C incubation of coals under 3 atm of 100% oxygen for 120 h. For acid treatments, we soaked coals for 48 h in two volumes of nitric acid (prepared at various strengths), after which they were washed with distilled water and brought to pH 5.5 before being autoclaved. All coal samples were sized (0.25 to 0.5 mm) and otherwise prepared as described for Figs. 17.4 and 17.5. We also pretreated the Colorado Kiowa lignite, which was about as resistant to microbial attack as were the weathered bituminous coals.

Table 17.5 contains representative data for effects of pretreatments on susceptibility to biological attack of Colorado lignite (Kiowa) and subbituminous or Arkansas bituminous coals. The unweathered North Park coal was not beneficiated by any pretreatment used, yet the same coal in weathered form (but otherwise untreated) was readily bio-

Table 17.5 Pretreatments of Biorecalcitrant Colorado Subbituminous and Lignite Coals[a]

| Strain | Kiowa lignite, weathered and treated[b] | | | | | Northpark subbituminous | | Green River subbituminous |
	U	H	O_2	0.1 N	1.0 N	Untreated and weathered	Treated[c] and unweathered	Treated[d] : 8.0 N, H_2NO_3
ACL-13	0	0	1	3	1	5	0	5
DML-12	0	0	1	2	1	5	0	—
RWL-40	1	1	1	3	0	5	0	3
YML-1	0	1	0	2	2	5	0	—
YML-21	0	1	0	4	3	5	0	5

[a]Comparative values of degree of solubilization; "5" denotes the maximum for condition of diffusion zones merging or complete coloration of agar to point where individual zones could not be discerned.
[b]Treatment code; U, untreated; H, heat-treated; N, nitric acid concentration.
[c]Treatments as for Kiowa lignite, no 8 N treatment.
[d]Treatments as for Kiowa lignite, plus 8 N treatment; dash indicates no data.

Table 17.6 Biosolubilization of Untreated or Treated Weathered Arkansas Bituminous Coals[a]

Strain	Dardanelle					Horse Head				
	U	H	O_2	0.1 N	1.0 N	U	H	O_2	0.1 N	1.0 N
ACL-13	0	0	0	0	5	0	0	4	0	0
DML-12	0	1	4	1	5	1	0	1	1	2
YML-1	0	4	4	2	5	1	1	2	1	1
YML-21	1	2	1	5	5	1	1	1	2	3
RWL-40	0	0	1	2	5	1	1	0	0	0

[a]Codes, conditions, and value assignment as for Table 17.5.

solubilized. The Green River coal required pretreatment with 8N nitric acid before biological activity occurred. The weathered but relatively biorecalcitrant Colorado Kiowa lignite yielded best activity after 0.1 N acid pretreatment; we cannot account for the decreased activity for 1.0 N acid pretreatment.

Table 17.6 presents results of tests on several pretreatments of weathered Arkansas bituminous coals. For the Dardanelle samples, 1.0N acid pretreatment yielded best re-

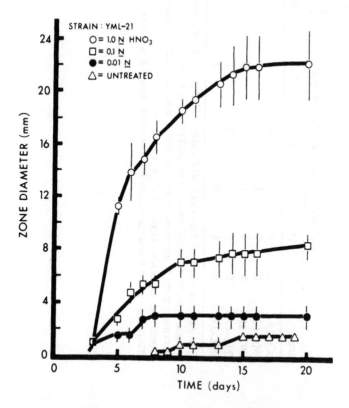

Fig. 17.6 Effects of mild nitric acid pretreatments on biosolubilization of weathered Arkansas bituminous coal by strain YML-21. Diffusion zone assays as for Fig. 17.1 but at 30°C. Vertical bars represent the SE for three replicates.

sults for all strains tested. Other pretreatments gave variable results for different organisms.

The data for Tables 17.5 and 17.6 were derived from experiments run at 25°C. At higher temperatures (e.g., 30°C), and for some organisms, the rate of biosolubilization increased and even untreated Dardanelle or Horse Head exhibited increased activity, especially for strain YML-21. Figure 17.6 illustrates the activity of YML-21 on acid-treated Dardanelle coal at 30°C and shows a clear acid-concentration effect. Comparison of the data of Fig. 17.4 with those of Fig. 17.6 reveals that the 1.0N acid treatment of Dardanelle bituminous resulted in greater total biosolubilization activity than that observed for naturally weathered Mississippi Wilcox lignite. We note here that the biosolubilized product from the acid-treated Dardanelle produced distinct zone margins that allowed for measurements beyond those possible for Mississippi Wilcox. Our observations in general indicate similar differences in appearances of products produced from different coals, and by different organisms on any one coal.

17.6 DISCUSSION

Biodegradation and bioconversion of fossil fuels appear to be common events in nature. Several types of petroleum-degrading microorganisms have been isolated from oil-polluted habitats [22,23]. Fungi and bacteria capable of coal bioconversions also are abundant in areas of weathered coal exposures. Coal-solubilizing microorganisms can be recovered from coals of various ranks and types, with a diversity of taxonomic microbial groups represented. In the laboratory, the coal biosolubilization phenomenon appears to progress in the same manner for almost all organisms, resulting in the production of dark, water-soluble coal products. Superior coal solubilizers probably are those that produce relatively large amounts of alkaline materials, or perhaps those that somehow (e.g., via metals chelation or production of surfactants) render the coal structure more readily solubilized by alkaline or other materials produced by the organisms. Solubilized coal products would be directly accessible to enzymatic attack. In nature, solubilized coal probably serve as a carbon and energy source for any number of microbial types, which ultimately could completely degrade and metabolize the organic materials. Coal solubilizers might have a selective advantage in environments enriched with the weathered coals.

The assay methods described in this chapter provide a means by which comparative, quantitative tests can be made directly on growing cultures of bacteria or fungi. The diffusion zone assay allows for time-course assays without sacrifice or disturbance of the cultures. The assay method can be used to identify superior coal-solubilizing microorganisms and to make comparative tests of coal biosolubility.

ACKNOWLEDGMENTS

Portions of the work described herein were supported by the Mississippi Mineral Research Institute (Grant 833S) and the U.S. Department of Energy, Grant DE-FG22-85PC80913. However, any opinions, findings, conclusions, or recommendations expressed herein are those of the author and do not necessarily reflect the views of DOE.

REFERENCES

1. Potter, M. C., Bacteria as agents in the oxidation of amorphous carbon, *Proc. R. Soc. London, Ser. B, 80*, 239–259 (1908).

320   *Ward*

2. Schroeder, H., The bacterial content of coal, *Zentralbl. Bakteriol. Parasitenk., 41*, 460–469 (1914).
3. Fischer, F., and Fuchs, W., Uber das Wachstum von Pilzen auf Kohl, *Brennst. Chem., 8*, 293–295 (1927).
4. Rogoff, M. H., Wender, I., and Anderson, R. B., *Microbiology of Coal*, Information Circular 8075, U.S. Bureau of Mines, Washington, D.C., 1962.
5. Kucher, R. V., Turovskii, A. A., Dzumedzei, N. V., Bazarova, O. V., Pavlyuk, M. I., and Khmel'nitskaya, D. L., Cultivation of *Candida tropicalis* on coal substrates, *Mikrobiologiya, 46*(3), 583–585 (1977).
6. Fakoussa, R. M., and Trüper, H. G., Köhle als Microbielles Substrat unter aeroben Bedingungen, in *Biotechnologie im Steinkohlenbergbau*, Kolloquium in der Bergbau-Forschung GmbH (H. J. Rehm, ed.), Essen, West Germany, Jan. 20–21, 1983, pp. 41–49.
7. Cohen, M. S., and Gabriele, P. D., Degradation of coal by the fungi *Polyporus versicolor* and *Poria monticola, Appl. Environ. Microbiol., 44*, 23–27 (1982).
8. Ward, H. B., Coal-degrading fungi isolated from a Mississippi lignite outcrop, *Proceedings of Symposium on Biological and Chemical Removal of Sulfur and Trace Elements in Coal and Lignite*, Department of Natural Resources, State of Louisiana, 1982, pp. 149–163.
9. Ward, B., Lignite-degrading fungi isolated from a weathered outcrop, *Syst. Appl. Microbiol., 6*, 236–238 (1985).
10. Ward, B., Apparent bioliquefaction of lignite by fungi and their growth on lignite components, *Proceedings, Bioenergy 84 World Conference and Exhibition* (E. Egneus and A. Ellegard, eds.), *Biomass Conv., 3* (1985).
11. Strandberg, G. W., and Lewis, S. N., A method to enhance the microbial liquefaction of lignite coals, *Biotechnol. Bioeng. Symp., 17*, 153–158 (1987).
12. Scott, C. D., Strandberg, G. W., and Lewis, S. N., Microbial solubilization of coal, *Biotechnol. Prog., 2*(3), 131–139 (1986).
13. Ward, B., *Biodegradation and Bioconversion of Coals by Fungi*, Pittsburgh Energy Technology Center document DOE/PC/80913-8, U.S. Department of Commerce, Springfield, VA, 1988.
14. Quigley, D. R., Ward, B., Crawford, D. L., Hatcher, H. J., and Dugan, P. R., Evidence that microbially produced alkaline materials are involved in coal biosolubilization, *Appl. Biochem. Biotechnol., 21*, 753–763 (1989).
15. Strandberg, G. W., and Lewis, S. N., Factors affecting coal solubilization by the bacterium *Streptomyces setonii* 75Vi2 and alkaline buffers, *Appl. Biochem. Biotechnol., 18*, 355–361 (1988).
16. Ward, B., Quigley, D. R., and Dugan, P. R., Relationships between lignite biosolubility or alkali solubility and natural weathering, *Proceedings, IGT Symposium on Gas, Oil, and Coal Biotechnology*, New Orleans, La., Dec. 5–7, 1988, in press.
17. Yeh, G. J. C., Ward, B., Quigley, D. R., Crawford, D. L., and Meuzelaar, H. L. C., Numerical comparisons between the pyrolysis mass spectra of twelve U.S. coals and their relative solubility in microbial cultures or alkaline buffer, *Amer. Chem. Soc. Div. Fuel Chem. Prepr., 33*(4), 612–622 (1988).
18. Ward, B., and Sanders, A., Solubilization of coals by fungi, in *Proceedings: 1989 Symposium on Biological Processing of Coal and Coal-Derived Substances*, Electric Power Research Institute, Palo Alto, Ca., 3157–3165, 1989.
19. Cohen, M. S., Bowers, W. C., Aronson, H., and Grey, E. T., Cell-free solubilization of coal by *Polyporus versicolor, Appl. Environ. Microbiol., 53*, 2840–2843 (1987).
20. Pyne, J. W., Stewart, D. L., Fredrickson, J., and Wilson, B. W., Solubilization of leonardite by an extracellular fraction from *Coriolus versicolor, Appl. Environ. Microbiol., 53*, 2844–2848 (1987).

21. Quigley, D. R., Breckenridge, C. R., Dugan, P. R., and Ward, B., Effects of multivalent cations on low-rank coal solubility in alkaline solution or microbial cultures, *J. Energy and Fuels, 3*, 571–574 (1989).
22. Atlas, R. M., Microbial degradation of petroleum hydrocarbons: an environmental perspective, *Microbiol. Rev., 45*, 180–209 (1981).
23. Walker, J. D., Colwell, R. R., and Petrakis, L., Degradation of petroleum by an alga, *Prototheca zopfii, Appl. Microbiol., 30*, 79–81 (1975).
24. Koenig, D. W., and Ward, H. B., *Prototheca zopfii* Krüger strain UMK-13 growth on acetate or *n*-alkanes, *Applied Environ. Microbiol., 45*, 333–336 (1983).
25. Koenig, D. W., and Ward, B., Growth of *Prototheca zopfii* Krüger on crude-oil as a function of pH, temperature, and salinity, *Syst. Appl. Microbiol., 5*, 119–123 (1984).
26. Pore, R. S., Removal of hydrocarbon pollutants from water by *Prototheca zopfii*, Water Research Institute, West Virginia University, Morgantown, W. Va., 1980.
27. Walker, J. D., Colwell, R. R., Vaituzis, Z., and Meyer, S. A., A petroleum-degrading achlorophylous alga, *Prototheca zopfii, Nature, 254*, 423–424 (1975).
28. Crawford, D. L., Lignocellulose decomposition by selected *Streptomyces* strains, *Appl. Environ. Microbiol., 35*(6), 1941–1045 (1978).

18

Focus of Research in Coal Bioprocessing

LEAH REED   *Sheladia Associates, Inc., Rockville, Maryland*

18.1 INTRODUCTION

Microbial processing of coal and minerals has been suggested by several authors for many potentially economically favorable reasons. Among these are low capital and operating costs, less-energy-intensive operation, and the ability, in the case of sulfur removal from coal, to remove finely disseminated compounds that would be overlooked by mechanical cleaning processes. So far, the rates of transformation are too low to be economically attractive, but each new set of results brings the field closer to practical applications.

Because of the many advances in the field of biotechnology, these techniques are finding application in a number of novel areas. One of these "new" areas is that of coal bioprocessing. However, "new" is a relative term because scientists were examining the possibility of microbial action on coal as early as the 1950s. However, in the intervening years, the focus of this coal research has changed. In 1947, Colmer and Hinkle were the first to report the isolation of *Thiobacillus ferrooxidans* from coal mine drainage [8]. From that period until the mid-1960s the interest was in bacterial leaching of metals or the production of acid in mine drainage. With the advent of the energy problems of the 1970s the focus changed to examining the use of these microorganisms for energy production. The current focus is still on energy production, but ranges over all aspects from enhanced oil recovery to the detoxification of wastewaters.

The major division in coal biotechnology is between the use of microorganisms for coal cleaning or coal beneficiation and their use in transforming coal into a liquid or gaseous product. A wide range of microorganisms is used, from simple autotrophic bacteria to the filamentous fungi. It is primarily the autotrophic bacteria that can reduce the sulfur and mineral content of coal, while other bacteria and fungi are responsible for liquefaction and gasification.

18.2 INORGANIC SULFUR REMOVAL

Due to public concern over the environmental issue of acid rain, legislators and people in the utility and manufacturing sectors have sought ways to decrease sulfur emissions to the atmosphere. The relationship between the use of high-sulfur fuels and increased acid precipitation is well documented. This fact has led to increased interest in the beneficiation of fossil fuels such as coal by removing sulfur prior to combustion. Microbial processes are being investigated because of their specific catalytic properties, potential high yields, low process temperatures and pressures, and tolerance to trace levels of poisons.

Because coal cleaning relates directly to the early body of work on the role of bacteria in acid mine drainage, this area of microbial coal interaction has seen the most progress. Several organisms have been isolated that can remove either the organic or the inorganic sulfur from coal; an organism can rarely remove both. In addition, process schemes have been developed and economic assessments made to determine the competitiveness of these techniques with chemical and mechanical coal cleaning processes.

18.2.1 Mechanism

In the case of inorganic sulfur, the mechanisms for the process have been defined. Both indirect contact and direct contact mechanisms are operative. In the indirect contact mechanism, the organisms oxidize the iron in pyrite to ferric ions which then catalyze the further oxidation of pyrite through the following scheme:

$$14Fe^{2+} + (7/2)O_2 + 14H^+ \rightarrow 14Fe^{3+} + 7H_2O$$
$$FeS_2 + 14Fe^{3+} + 8H_2O \rightarrow 15Fe^{2+} + 2SO_4{}^{2-} + 16H^+$$

The ferric ion is therefore the primary oxidant, with bacteria responsible for regenerating the active species [15].

The direct contact mechanism requires attachment of the microorganisms to the pyrite. The ferric ion is not involved, as evidenced by bacterially assisted removal of the sulfur in iron-free sulfides. At low cell densities, most of the cells are attached to the pyrite, so sulfur oxidation predominates. There is a transfer to oxidation of iron when the pyrite is depleted. At higher cell densities and higher concentration of ferrous ion, the rate of oxidation of iron (ferrous ion to ferric ion) is faster than the oxidation of sulfur (pyrite to ferrous sulfate), because a large number of organisms are left unattached.

18.2.2 Organisms

The organisms capable of these transformations are sulfur-oxidizing acidophilic autotrophs, which use carbon dioxide as their source of carbon. These organisms include *Thiobacillus ferrooxidans, Thiobacillus thiooxidans, Sulfolobus acidocaldarius,* as well as other thermophilic *Thiobacillus*-like organisms. Mixed cultures have also been used successfully in the laboratory for pyrite removal. Mixed cultures that have been tested include *T. ferrooxidans* and *T. thiooxidans,* and *Leptospirillum ferrooxidans* and *T. thiooxidans,* as well as unknown cultures isolated from acid mine waters.

T. ferrooxidans will oxidize ferrous ion or reduced sulfur as an energy source. Carbon dioxide provides the primary carbon source, although the organism can adapt to heterotrophic growth on glucose. The adapted organisms contain alphaketoglutarate dehydrogenase and reduced NAD oxidase. The iron-oxidizing enzymes also seem to be inducible. *T. ferrooxidans* is an aerobic acidophilic bacterium found in acid mine waters. Its optimum temperature is in the range 25 to 35°C; optimum pH in the range of 2 to 3; redox potential <400 mV. Ammonia, phosphate, and other minerals are also necessary in the culture medium. The organism is found on the surface of particles in coal refuse piles, but not in the internal pores. Using various coals and slurry concentrations, Sproull et al. found that *T. ferrooxidans* was able to remove 80% of the sulfur and 45% of the initial iron content [17]. In addition, they observed the leaching of several other minerals due to the acidic conditions, leading to a 15 to 27% reduction in ash content and a 50 to 450°C increase in the ash fusion temperature of the treated coals. Hoffman observed a 90% removal of pyrite in 8 to 12 days [9]. *T. thiooxidans* will oxidize sulfur but not iron. It does not attack pyrite, although it is useful in co-culture with *T. ferrooxidans*.

Dugan and Apel found that mixed enrichment cultures of acidophilic microorganisms isolated from acid mine waste were capable of removing pyritic sulfur from 20% slurries of commercial grade pulverized coal [8]. For a coal containing 3.1% pyrite initially, 97% of the pyrite was removed in 5 days. *T. ferrooxidans* or *T. thiooxidans* alone did not have this rate of sulfur removal. They were able to culture these organisms directly from the coal or pyrite slurry with no inoculation or addition of nutrient medium.

S. acidocaldarius was isolated from an acidic hot spring in Yellowstone National Park. It is a thermophile and acidophile, preferring temperatures of 50 to 80°C with an optimum at 70 to 75°C, and a pH range of 0.9 to 5.8 with an optimum at 2 to 3. It is a chemoautotrophic organism requiring reduced sulfur as an energy source, but it is capable of growth on a limited number of simple organic molecules, including yeast extract. *Sulfolobus* has been reported to remove both inorganic and organic sulfur from coal. Currently, several researchers are independently investigating the removal of inorganic and organic sulfur compounds using this organism. Kargi and Robinson found that 96% of the inorganic sulfur was removed from a 5% slurry [11]. Chen and Skidmore demonstrated a time of 3.1 days for 90% removal of inorganic sulfur, much faster than for *T. ferrooxidans* [4]. Their highest bench-scale reaction rate was 10.21 mg S/L per hour, as compared to the 4.5 mg S/L per hour reported by Kargi and Robinson. Some of this difference may be due to a difference in starting coals and particle size. *Sulfolobus brierleyi* is a larger member of the *Sulfobus* species, which will also oxidize pyrite. This organism has requirements similar to those of *S. acidocaldarius* but cannot utilize yeast extract. It is neither as efficient nor as rapid at oxidizing sulfur as is *S. acidocaldarius*.

18.2.3 Process Variables

The effects of various process variables have been studied in many laboratories for both *T. ferrooxidans* and *S. acidocaldarius*. A microbial coal process concerns mass transfers in sold, liquid, as well as gas phase. The microbes, suspended in a liquid phase, are growing on the surface of coal particles (solid phase). Nutrients such as carbon dioxide and oxygen (gas phase) must diffuse through the liquid medium and through the layer of products formed by the microorganisms. Other important process variables include particle size, slurry density, agitation rate, temperature, pH, and nutrient concentrations.

Particle Size

The size of the particle used in a leaching process is critical because the organisms work on the surface of the particles. The reactions follow Michaelis-Menten kinetics; that is, at low surface area, the rate is dependent on surface area, whereas at high surface area, it is dependent on cell concentration. Therefore, to optimize the reaction, one must reduce the particle size or increase the pulp density of coal in the reactor. The optimum size will depend on the pyrite content and distribution as well as the economics of grinding. There is a lower limit to the optimum particle size, at which point recovery becomes a problem.

In experiments with *T. ferrooxidans*, Sproull et al. found that decreasing the particle size uniformly increased the removal of sulfur [17]. Murphy et al. [14] and Chen and Skidmore [4] drew similar conclusions from work on *S. acidocaldarius*. They used coal ground to 65, 150, and 200 mesh. Sulfate production (from oxidation of pyrite) followed first-order kinetics with respect to pyrite content with 90% removal at 3.1 days for the 200-mesh samples. Because of the different rates at different particle sizes, the rate must actually be dependent on surface area as well as concentration. To support this assumption, Hoffman observed a linear rate of sulfur removal with respect to the total external surface area using *T. ferrooxidans* [9].

Surface area is important because of bacterial attachment to the coal particles during the reactions. In experiments using *S. acidocaldarius*, Chen and Skidmore demonstrated attachment of the microorganism to the coal particle [3,4]. They investigated the rate and extent of this attachment using activated carbon, pyrite, and several different coals ground to 270 to 325 mesh. In all cases, attachment occurred within the first minute of incubation. The attachment to activated carbon, pyrite, and one of the coals was irreversible. The irreversible attachment followed a Langmuir isotherm, although theoretically one might not apply this curve to the process. The available pyrite surface areas on the various substrates did not correlate with cell adsorption. The authors comment that surface area, surface charge, and hydrophobicity might all play a role in this phenomenon.

DiSpirito et al. have reported sorption of the organism *T. ferroxidans* to various fine-grained particulates [5]. Sorption rates to materials such as fluorapatite, glass beads, and pyrite had relatively high initial rates. More than 25% were sorbed within the first 5 min of exposure, with a steady state achieved in 1 h. Slower initial rates were observed with flowers of sulfur, with a concomitant increase in time required for achievement of a steady state. This rate was increased by preincubating the material in filter-sterilized spent ferrous iron media for 24 h. At increasing concentrations of particulate material, more cells were sorbed but not proportionally. Multiple-cell aggregates were not observed in any experiment. Ultraviolet-irradiated cells or those with the lipopolysaccharide layer removed had similar initial rates but lower overall sorption to the particulates. These cells failed to grow on ferrous iron.

Slurry (Pulp) Density

Greater than 50 to 60% slurries are needed in order to improve the process economics of microbial desulfurization. Large cell concentrations are also needed at this high coal loading. However, slurry density has a profound effect on mass transfer rates within the reaction mixture, and high concentrations may actually reduce the rate of desired reactions. Higher costs may be incurred due to increased agitation and aeration requirements, as well as the need to remove metabolic heat. The last requirement leads many researchers to propose the use of thermophiles, because less process cooling will be required.

Kargi and Robinson compared the amount of sulfur removed by *S. acidocaldarius* from an untreated coal slurry at several concentrations [11]. At a 5% loading, 50% of the total (96% of the inorganic) sulfur was removed, while at a 10% loading, 40% of the total sulfur was removed. At densities of 15 to 30%, particle agglomeration was evident along with reduced solubility and transfer of gaseous nutrients. All of these factors may reduce the rate of sulfur oxidation at higher concentrations.

Chen and Skidmore as well as Murphy et al. examined slurry concentrations of 5, 20, and 40% using the same microorganism [4,14]. They found higher overall rates of sulfur removal at higher concentrations, but when these rates were normalized to the amount of remaining sulfur, the rates were actually lower at higher concentrations. The rate constant for the 5% slurry was an order of magnitude larger than those for the 20% or 40% slurry. They note that other researchers have reported that the rate normalized to available sulfur remains constant up to a 15% slurry concentration.

Using *T. ferrooxidans*, Detz and Barvinchek noticed very reduced rates of sulfur oxidation if slurry concentrations were greater than 20% by weight [8]. Sproull et al. have documented the same effect [17]. The probable explanation is that higher coal concentrations lead to reduced mass transfer, loss of uniform particle distributions due to poor mixing, and agglomeration.

Agitation/Aeration Rate

Agitation distributes nutrients as well as causing increased contact between the cells and the substrate. Cultures can be mechanically agitated or aeration can be arranged to cause agitation. All researchers found that better mixing and increased oxygen and carbon dioxide transfer increased the rate of desulfurization. An increase in aeration rate was found to have a beneficial effect on the reaction rate up to 4 SCFH (standard cubic feet per hour). Some of this effect may have been due to the increased agitation. A much more detailed analysis of factors, such as reactor shape and the physical properties of the slurry, must be done before aeration and mixing can be defined for a scaled-up process.

Temperature

Temperature is an important process variable due to the economics of a process as well as the chemical and biological requirements. All microorganisms have temperature optima—temperature ranges at which they function most efficiently. In addition, because some reactions, especially for inorganic sulfur removal, are chemical rather than biological in nature, increasing the temperature can increase the reaction rate. Many researchers have recommended the use of thermophiles because at higher temperatures, there is less chance of contamination from unwanted organisms. In an economic sense, there is a necessary balance between the necessity for removing metabolic heat in culturing mesophiles and the possibility of adding process heat for the maintenance of thermophiles.

The *Thiobacilli* are primarily mesophiles, growing at temperatures of 25 to 35°C, while *Sulfolobus* is a thermophile, with a temperature range of 50 to 80°C. Within these broad temperature ranges, investigators have determined optima for sulfur removal. For example, Sproull et al. noted a 20% reduction in sulfur removal when the temperature was raised from 27°C to 37°C using *T. ferrooxidans* [17]. Chen and Skidmore investigated the temperature optimum for sulfur removal using *Sulfolobus* between 67 and 83°C [4]. A maximum rate of 10.21 mg/L per hour was observed at 72°C.

Acidity

All of the organisms mentioned so far as acidophiles, so that the optimum pH is in the range of 2 to 3. Within that range, large differences in reaction rate can be found. Dugan and Apel found that their mixed enrichment cultures were most effective in a pH range of 2 to 2.5 [8]. There was a rapid release of sulfate as the pH dropped below 2.5, with the most reaction observed between pH 2.3 and 2.8. *Sulfolobus* will grow in the range pH 0.9

to 5.8, with an optimum for sulfur removal of pH 2 to 3. A narrower optimum range has not been determined for this organism.

Another factor to consider in determining the optimum pH for these reactions is precipitation of nutrients or products. Aside from removing nutrients from solution, precipitation blocks reaction sites on the coal surface and adds sulfur (as sulfate) to the solid coal. Kargi has reported that several iron salts—sulfate, hydroxide, and phosphate—will precipitate onto the coal particles [10]. The reaction rate may then be limited by the diffusion of nutrients through the layer of precipitated products. He observed considerably less deposition in the pH range 1.5 to 1.7.

Nutrient Concentrations

The concentration of nutrients—those chemicals the microorganisms require for growth—can have a profound effect on the rate and extent of a particular conversion. It may seem natural that providing the organisms with an abundant supply of nutrients would enhance any process, but this is often not true. Adding nutrients will curtail the progress of transformation by switching the cells back into a growing phase, especially for reactions that normally take place during the stationary phase, when the cells are crowded and nutrients are scarce. In the case of inorganic desulfurization, the concentrations of certain key nutrients do affect the rate of reaction.

Because these organisms are primarily autotrophs, the concentration of carbon dioxide, the principal carbon source, is of importance. Air, which is very low in carbon dioxide, is normally used as the source. The rate of reaction should increase as the partial pressure increases. Kargi found that for a concentrated suspension of *T. ferrooxidans*, using an external supply of carbon dioxide along with air under good agitation increased the rate appreciably [10]. Sproull et al., using the same organism, observed no effect with changes in carbon dioxide concentration [17]. Detz and Barvinchek also noted that carbon dioxide at normal partial pressure was sufficient [8]. Dugan and Apel found that a 5% carbon dioxide/95% oxygen supplement increases the lag time because the larger amount of carbon dioxide buffers the pH [8]. There was a slight increase in sulfur removal by pure and mixed cultures. Bartel and Skidmore, using *S. acidocaldarius*, observed a small increase with carbon dioxide at a 0.007 mole fraction, but no increase at higher concentrations [2]. They also found that when the oxygen concentration was high, changes in the carbon dioxide concentration had no effect.

Kargi and Robinson tested the effect of adding yeast extract and peptone to cultures of *S. acidocaldarius* [11]. In all cases, sulfur removal decreased. In contrast, Chen and Skidmore found that adding yeast extract increased the growth without affecting sulfur oxidation [4]. They also found that at low salt concentration, the rate and extent of desulfurization increased with salt concentration. At higher salt concentrations, desulfurization was limited and actually decreased, possibly due to precipitation. Lacey and Lawson have shown that the precipitation of basic ferric sulfate hinders the access of the bacteria to reaction sites [12]. Precipitation also adds solid sulfate to the coal, increasing its sulfur content. Dugan and Apel examined the sulfate release from coal or pyrite by *T. ferrooxidans* or acid mine waste enrichment cultures using several media [8]. The 9K basal salts medium produced a larger release of sulfur, but the lag time remained the same. No increase was seen with the addition of trace minerals or phosphate. Adding ammonium sulfate increased both the rate of sulfur removal and the extent. When Sproull et al. increased their nutrient medium fivefold, 15% less removal of sulfur by *T. ferrooxidans* occurred [17].

Coal Type and Condition

Although microorganisms increase the rate of iron and sulfur removal for all coals tested, the type of coal, the initial sulfur content, and the weathering of the coal are all important to the reaction rate for desulfurization. Kargi and Robinson have shown that the rate of sulfur removal depends on the sulfur content [11]. In experiments with coal of 11% sulfur content, the rate was 13 mg S/L per hour, while at a sulfur content of 4%, the rate was only 4.5 mg S/L per hour. These translate into surface reaction rates of 4.1 X 10^{-4} mg S/cm and 1.4 X 10^{-4} mg S/cm every 2 h. In field studies, Dugan and Apel found the largest $^{14}CO_2$ uptake in 2- to 3-year-old coal refuse [8]. Very little reaction occurred in fresh coal or coal that had weathered for 40 years. The reactions took place in the top 8 to 10 cm.

Flotation

Attia and Elzeky have described a nonleaching method for pyrite removal with *T. ferrooxidans*, which is based on froth flotation [1]. They have investigated the effects of several process variables on coal and pyrite flotation, including concentration of frothing agent, pH, and bacterial preconditioning. For a 5% slurry, coal recovery increased from 28 to 92% as the amount of frothing reagent (MIBC) was increased from 0.74 to 9.12 lb/ton, with the optimum at 7.88 lb/ton. The recovery of pyrite increased from 12% to 53% as the frothing reagent increased from 0.64 to 8.41 lb/ton. There is a differential in flotation, which the microorganisms increase. Recovery of clean coal decreased with increasing pH, again with some differential between coal and pyrite—56 and 28% recovery at pH 8, respectively, versus 88 and 86% recovery at pH 2. Neither the nutrient medium nor bacterial conditioning caused any loss of recovery of the coal. However, a 50% reduction in pyrite flotation was achieved with 60 min of conditioning with a culture that had been grown for 6 weeks on a pyrite substrate.

18.2.4 Process Schemes

The first commercial-scale experiment for the removal of inorganic sulfur from coal was performed by Capes in 1973 [8]. He used 10% weathered coal as the inoculum added to run-of-min coal. The sulfur content was reduced from 6.1% to 2.7%. Most discussions of a microbial process for desulfurization conclude that the process is still too slow to accommodate an economically sized reactor volume. Several workers are investigating ways to increase the efficiency of such a process.

In a process scheme suggested by Kargi, the coal is pretreated with acid to increase the susceptibility of the surface for microbial attack [10]. Some ash is also removed in this treatment. A two-stage reactor is recommended so that organic and inorganic sulfur can be removed. The coal can be desulfurized in pipelines, reducing the negative time effect and decreasing the capital costs. However, the disadvantages of this scheme include lack of environmental control, deposition of products, and corrosion of the pipes.

Sproull et al. have proposed the use of a lagoon to leach inorganic sulfur by the action of *T. ferrooxidans* [17]. Their process includes mixing, leaching, clarifying, and dewatering steps, along with recycle of the sludge. An economic analysis for a process that removes 90% of the inorganic sulfur shows costs on the order of $16 per ton, including pulverizing and drying. This process also removes a portion of the ash, increasing the ash fusion temperature of the product.

Vaseen has described the design of a scaled-up processing plant for sulfur removal using *T. ferrooxidans* [18]. A plug flow vessel is proposed that would be rotated slowly. Features of this vessel would include internal ridges to aid in suspending and mixing the slurry and manifolds carrying heated air and slurry in, and surplus air and desulfurized slurry out. A commercial-scale operation would consist of a primary bioreactor with log phase bacterial growth from recycle of 5 to 20% of the effluent slurry, plus two or more secondary bioreactors for stationary and death phases of the cycle. An example was given of a modular plant with a capacity of 100,000 tons per year. Drum rotation would be 1 to 20 rpm with an 80% liquid depth. A greater length-to-diameter ratio would accommodate processes requiring a longer reaction time. Retention time would be 1.93 days in the primary bioreactor and 6.92 days in the secondary reactors.

Chen and Skidmore have investigated a semicontinuous as well as a pilot-scale process for inorganic sulfur removal using the organism *S. acidocaldarius* [4]. In semicontinuous operation, a 5 to 15% slurry was run until sulfur oxidation slowed, at which point a volume of the slurry was replaced with a fresh sample. The sample error was great in these experiments, but at higher solids loadings there was a noticeably lower extent of sulfur removal, even at longer residence times. The pilot-scale reactions were run in 55-gallon drums. Solids loadings of 5 to 45% were tested with temperature control and some agitation. The top, middle, and bottom of the reactors were sampled. The mixing and mass transfer were not adequate; at 5% the rate was 5.14 mg S/L per hour, which is much lower than bench-scale results.

Murphy et al. envision a process scheme based on the use of *S. acidocaldarus* at pH 2 [13]. Microbes would be inoculated into a semibatch reactor to release sulfur from the coal as soluble sulfate ions. A water cleanup and recycle system would be used to remove excess sulfur and heavy metals. The acidic vapors from the off-gas would be condensed and mineral sulfates and residual acids would be removed by a scrubber. The coal would be washed, filtered, and dried.

Another process for pyrite removal has been described by Isbister, related to work on froth flotation by Attia and Elzeby [1]. This is not a bioleaching process. *T. ferrooxidans* was used to alter the surface chemical and flotation behavior of pyrite, which was then removed mechanically by use of a flotation column. During an incubation step, *T. ferrooxidans* attached to the pyrite particles in the coal. Air and water containing a frothing agent were forced into the flotation column. The incubated coal could then be fed in at different heights for processing. The cleaned coal was removed from the top of the column while the ash and pyrite sank to the bottom because of their increased affinity for water. The tailings with the attached microorganisms were recycled to reinoculate the coal. The entire process took 20 min, with good recovery of the coal. At low pH both pyrite and ash removal were enhanced [1].

18.2.5 Economics

Dugan has examined the value added to coal by sulfur removal as a factor in the competitiveness of microbial processes [6,7]. A decreased ash content, for example, results in a higher Btu value per ton. There is also an increase in value due to lower environmental cleanup requirements necessary with a lower-sulfur coal. Dugan has attempted to quantify this increase in value using the differential in rates paid by Ohio utilities for lower-sulfur coals [6,7]. These rates are normalized on a Btu content basis for consistency. At least as far back as 1978, utilities in Ohio have been paying a premium for low-sulfur coal.

When total sulfur was in the range of 1 to 3%, a premium of approximately $13 per ton per 1% reduction in sulfur was paid, while in excess of $26 per ton per 1% reduction in sulfur was paid when the total sulfur was less than 1% [6]. Using a larger data set, Dugan found that a $25 to $31 increase in value corresponded to a decrease in sulfur from 4.6 to 0.6% [7].

In comparison, several authors are quoted on the process costs for microbial desulfurization [7]. Detz and Barvinchak calculated $12 per ton, which would be approximately $19.36 today. Isbister and Kobalinsky reported costs between $21 and $25 per ton. And Bos and colleagues at the Delft School in the Netherlands have estimated between $5 to $10 per ton (converted from Dutch guilders) processing costs. A comparison of these data—low-sulfur premium prices and microbial processing costs—demonstrates that the costs of the microbial processes under development are quite competitive with prices for as-mined low-sulfur coal.

18.3 ORGANIC SULFUR REMOVAL

Much less work has been done on the removal of organic sulfur from coal. The forms of organic sulfur in coal are much more complex, of unknown structure, and the sulfur is bound to the carbon and is therefore part of the coal matrix. A few organisms have been identified that can oxidize the organic sulfur forms in coal. Most of these have been isolated from soil, especially oil-contaminated soil. For these experiments, dibenzothiophene has been used as a model compound representing the most refractory form of organic sulfur. It is an example of a sulfur-containing polycyclic aromatic hydrocarbon. Other representative compounds are thiophene, benzothiophene, benzonaphthothiophene, and so on.

18.3.1 Organisms

Isbister has isolated an aerobic organism identified as *Pseudomonas* from soil, and through mutagenesis has selected a strain that can remove sulfur from dibenzothiophene and leave the carbon structure intact [21,22]. This organism only removed organic sulfur. There was no response with pyrite. The mechanism of action does not involve breakage of the carbon skeleton of the molecule. As elucidated in ^{14}C and ^{35}S studies, the aromatic carbons bound to the sulfur are hydroxylated, with subsequent release of the sulfur as sulfate leaving a dihydroxybiphenyl. The enzyme has been found in the cell envelope and can be induced with benzoate. The organism does not have to be alive for desulfurization to occur. Approximately 34% of the organic sulfur was removed from the coal by this organism.

Monticello et al. have demonstrated that in cultures of *Pseudomonas* isolated from an oil-contaminated soil sample, the ability to degrade dibenzothiophene is a plasmid-mediated process [24]. Three cultures were isolated that had the ability to oxidize dibenzothiophene to characteristic water-soluble products. The strains were identified as *Pseudomonas alcaligenes*, *Pseudomonas stutzeri*, and *Pseudomonas putida*. None of the organisms could grow on dibenzothiophene as a sole carbon, sulfur, or energy source. In addition, the products caused inhibition of dibenzothiophene oxidation and also of cell growth. Dibenzothiophene oxidation could be induced by naphthalene or salicylate or to a lesser extent by dibenzothiophene itself, and was inhibited by the presence of succinate. Conversion by a cell-free extract was also demonstrated.

Hatcher reported the oxidation of dibenzothiophene by unidentified micro-organisms isolated from oil refinery soil [20]. The oxidation seemed to follow the accepted scheme with red, orange, and yellow water-soluble products formed from the reactions. The rate of dibenzothiophene degradation reached a maximum on the first day of incubation and then remained constant to slightly increasing for the next 9 days. This behavior paralleled the growth curve for the organisms, which also reached a maximum rate on the first day. In the absence of dibenzothiophene, the growth decreased. Growth was stable in the presence of the compound, showing no toxic effects. A possibility of product inhibition, such as demonstrated by Monticello et al., was observed [24].

As mentioned in the inorganic section of this chapter, the organism *S. acidocaldarius* is also capable of oxidizing dibenzothiophene to colored water-soluble compounds. *S. acidocaldarius* is a thermophilic archaebacterium that requires sulfur for energy if it is growing as an autotroph. Krawiec, who has worked with this organism for dibenzothiophene oxidation, noted that such structures as thiophenes are not normal to bacteria, and that the oxidation may be a form of detoxification [23]. He has found that after 10 days of incubating the organism with dibenzothiophene, virtually none of the compound remains. The oxidation seems to be associated with the late exponential phase of the bacteria. No colored compounds were seen in his submerged culture work.

18.3.2 Process Variables

Isbister reported on the effects of various process variables on the removal of organic sulfur by CB-1, a *Pseudomonad*. [21]. She examined the effects of coal type, rapid mixing, wet milling versus dry milling, inoculum size, retention time, and reactor design. In addition, Isbister and Kobylinski have reported on the effects of microbial processing on coal characteristics, such as particle size, settling, and Btu value [22].

Several different coals were examined. For Illinois No. 6, which has 2.1% organic sulfur, 24% was removed by treatment with CB-1. For a Consolidation coal, 25% of the organic sulfur was removed. The best results were achieved with a Homer City underflow—57% removal. The latter coal was of a finer mesh size and was stored under water, which prevented oxidation. Both of these factors are relevant to the higher yield. The organisms must attach to the particles, just as for inorganic sulfur removal, so particle size affects the reaction. For Illinois No. 6 ground to -60 mesh, organic sulfur removal was 17%, while at -270 mesh the removal was 32%. In addition, oxidized coal cannot be used for this particular organism.

Rapid mixing was investigated as a way to escalate attachment of the organisms to the coal. There seemed to be no effect on the reaction. For inorganic sulfur removal, attachment occurred less than 1 min after contact. It is likely that a similar process operates here. Because CB-1 cannot tolerate oxidized coal, the effects of wet versus dry milling were investigated. The assumption was that dry milling might promote oxidation of the coal. However, there was no difference in sulfur removal between the two grinding methods.

A dose study was undertaken to determine the optimum inoculation size for processing. The inoculum size was varied from 1.8×10^9 to 1×10^{10} for Homer City coal, and from 2.9×10^9 to 4.4×10^{10} for Kentucky No. 9, both ground to -60 mesh. For the former coal, the critical dose is between 1.8 and 4×10^9, while for the latter, 1×10^{10} is the critical inoculum size.

Because reducing the retention time allows for a smaller (and therefore more economical) reactor size, the effects of retention time on the extent of desulfurization were investigated. The initial 48-h retention time was reduced to 24 h and then to 12 h. A final time of 9 h was selected, based on experiments down to a 3-h retention time.

The final parameter investigated in these studies was reactor design [21]. These were very preliminary experiments with columns, a stirred reactor, a rotating drum, and a drais mill. Attempts were made to increase the solids loading; up to 38% was used. All reactors showed some desulfurization, although none to the extent of the original experiments. This subject is still under investigation.

The particle size of the coal was altered by microbial processing [22]. Electron micrographs indicated a smoothing of the particle edges, which might have been due to microbes adhering to the particles. The increase in particle size could have been caused by hydration at the sulfur site during oxidation. The treated coal had a lower settling rate than the raw coal, although the largest particles settled out immediately in both cases. Hydroxylation at the former sulfur site is proposed as an explanation for this behavior because this would alter the surface properties of the coal. A change in additive formulation for coal-water fuel use may be necessary with the treated coal. The Btu contents of the treated and untreated samples were comparable. Even though the addition of oxygen might in theory reduce the value, the microorganisms attached to the coal contribute to the carbon content.

18.3.3 Process Schemes

Because much less is known about the structures of organic sulfur in coal, the organisms, and removal mechanisms, very little work has been done leading to process schemes. Most of the work concentrates on organism collection and optimization. In the bench-scale processing unit described by Isbister, the organisms are fermented separately and then metered into the coal reactor [21,22]. An ultrafiltration system was used for media recycle and to concentrate the organisms before they enter the reactor. The cleaned coal comes out as the overflow from the reactor, while the sulfur (as sulfate) is contained in the water phase. The capacity of this unit is 10 to 30 lb of coal per day.

A semipilot plant was also constructed by the same group. The treatment section is capable of handling 2500 lb of coal per day. Again, the organisms are cultured separately and then added to the coal after it is suspended in water. A successful test run was made for 6 weeks of continuous operation. Economic estimates for this process including flotation removal of pyrite are on the order of $20 per ton of coal.

18.4 LIQUEFACTION

The use of microorganisms to liquefy coal is under investigation as another method for producing upgraded liquid fuels from coal. The research is preliminary and primarily concerns the use of lower-rank coals such as lignite and leonardite. Liquefying the coal—or solubilizing it as some researchers have designated the process—provides an ash and inorganic sulfur-free material for which fuel and nonfuel applications have been proposed.

18.4.1 Mechanism

The microbial liquefaction work is based on the resemblance of coals, especially the lower-rank lignite and leonardite, to lignin, a naturally occurring aromatic component of woody materials. Lignin and cellulose are the two major constituents of biomass, with lignin the more difficult to degrade. Several fungi and a few bacteria possess the ability to degrade lignin. A large amount of work has been done on the mechanisms of lignin oxidation by white rot and other fungi.

Polyporus versicolor decomposes lignin by digesting polyphenylpropane polymers and associated aromatic structures. The enzymes polyphenol oxidase and peroxidase are key to these transformations, as well as the enzyme laccase (benzendiol: oxygen oxidoreductase). *Poria monticola*, a brown-rot fungus, digests polysaccharides using betaglucosidases, possibly with the contribution of nonenzymatic free-radical reactions. *Phanerochaete chrysosporium*, another lignin-active fungus, is known to produce ligninases during secondary metabolism (i.e., when the culture has stopped growing).

Laccase is a phenol oxidase known to be associated with lignin degradation. It catalyzes the oxidation of a phenolic compound to a quinone. The molecular weight of this enzyme is roughly 60,000 daltons. Peroxidase is another enzyme suspected to be a ligninase, which converts a phenolic compound to a quinone in the presence of hydrogen peroxide. The enzyme catalase is present in many bacterial systems to detoxify hydrogen peroxide and may be associated with these systems.

18.4.2 Organisms

In 1982, Cohen and Gabriele first reported the degradation of lignite by the filamentous fungi, *P. versicolor* and *P. monticola* [26] . The basic assumption underlying this work is that lignite still bears a strong chemical resemblance to lignin, the natural substrate for these oxidation enzymes. The liquefaction of lignite, like the degradation of lignin, is an oxidative process that can also occur under anaerobic conditions. As might be expected from the different mechanisms, the products from *Poria* and *Polyporus* degradation of lignite are different. Cohen and Gabriele also found that only half of their cultures produced liquid from lignite, so some adaption of the organisms may be necessary to degrade the "modified" substrate [26] . Several other research teams are now investigating the subject of coal liquefaction using microorganisms. To date, there has been a great deal of cooperation among the groups, with both cultures and coal samples being shared. A larger number of fungal cultures have been discovered that liquefy lignite, as well as a few bacterial species.

P. versicolor and *P. monticola* are basidomycete fungi that can grow in minimal liquid or solid media when supplemented with crushed lignite. No growth occurs in these media in the absence of coal. These organisms can also grow directly on crushed lignite, although hyphal growth is less than on lignite agar. Growth is determined as the production and extension of hyphae. *P. monticola* is less productive than *P. versicolor* on coal. The cultures can be maintained on Sabouraud maltose agar or broth in the absence of coal.

During the growth, the solid coal is degraded to a black liquid product, some of which diffuses into and darkens the agar. On the solid lignite, the vegetative hyphae grew up onto the coal and covered it, following the grain pattern of the lignite. When the cultures were grown on Sabouraud maltose agar for several days, and a polypropylene grid was placed between the culture and sterile coal, the coal was also liquefied. These experiments demonstrate the presence of an extracellular enzyme or factor of some sort.

Other organisms have since been used to transform coal into a liquid product. As with the proceeding work, most of this work involved surface culture, where the fungi were grown on an agar surface and the sterilized coal was then placed on the fungal mat for reaction. Droplets ranging from colorless to black formed on the surface of the coal in a large number of cases.

Scott and Strandberg have investigated organisms such as *Candida, Pennicillium waxsmanii, Streptomyces, P. chrysosporium, P. versicolor*, and *Aspergillus* [28,29]. An additional three productive cultures were isolated from Texas lignite. These proved to be *Aspergillus, Paecilomyces*, and either *Trichoderma* or *Sporotrichum* species. Of these, the *P. waxmanii* was the most prolific producer of liquid from coal.

Dahlberg has been involved with surface culture screening of fungi for the liquifaction of various coal types [27]. He has worked with *Geosmithia argillacea, P. chrysosporium, P. versicolor*, and *Candida*. Culturing methods were the same as other researchers, with a fungal mat established first and then coal placed on it for liquefaction. Yellow drops were formed initially, which turned black and coalesced after further incubation.

In addition to the fungi, Scott and Strandberg have isolated bacteria that can liquefy coal [28,29]. Two species of *Streptomyces, S. setonii* and *S. viridosporous*, produced a liquid product from coal. The liquefaction proceeded more rapidly than that with fungi, beginning within the first 2 days, but was not as extensive. Bacteria are easier to culture, so they have a greater potential for use in large-scale fermenters.

18.4.3 Submerged Cutlure Studies

All the screening studies mentioned previously used surface culture methods; the organisms were grown on an agar surface with coal subsequently added. In considering scaling this work up to a pilot or commercial process scale, most researchers project that a submerged culture—growing the organisms and processing the coal in suspension—will be needed to provide direct contact between the organisms and the substrate.

Several groups have investigated the use of fungi in submerged culture with pulverized coal. Wilson et al. experimented with a column for liquefaction [30]. Coal was liquefied on the column, to 8 or 10%, but there was a great deal of plugging. They also used a submerged culture without substrate to produce the enzyme laccase for characterization. Scott and Strandberg experimented with shake flasks and small fluidized beds [28,29]. They observed ball-like growths formed around coal particles by *P. waxmanii* and *T. versicolor* in nutrient supplemented media. *Candida*, however, was observed as individual cells in direct proximity to the surface of particles. It is presumed that these morphologies affect the rate of solubilization.

Candida was used on a small fluidized bed that had total recycle of the medium. Cheese whey was added to this system as a polyelectrolyte, which increased attachment of the microorganisms to the coal. The reaction with fungi in submerged culture did not proceed as well as on the surface culture. In *Streptomyces* (bacterial) cultures, however, liquefaction began roughly 2 days after inoculation and at twice the rate of the fungal system. As in the fungal systems, it is clear that an extracellular process was involved, because the cell-free broth also degraded the coal.

18.4.4 Enzyme Studies

Because the active species in coal solubilization are extracellular products, researchers have tested, grown, and isolated products from the fermentation broths. Some screening has also been done to demonstrate production of known ligninases from strains that also liquefy coal. Dahlberg has examined five cultures for the production of laccase, peroxidase, and catalase [27]. The first two are directly associated with lignin oxidation. Of these, only *T. versicolor* tested positive for laccase, and *Candida* for peroxidase. *T. versicolor, P. chrysosporium*, and *Candida* all produced catalase.

Wilson et al. have used a small-scale reactor to produce laccase from *P. versicolor* [30]. The enzyme production was induced by xylidine, a known laccase inducer, and then harvested from the culture. In tests of the crude enzyme preparation, a fairly good conversion of coal to soluble product ensued as evidenced by darkening of the reaction solution. Electrophoresis of the crude preparation showed that several enzymes were present. It is possible that the other enzymes also contribute to the coal degradation. Scott and Strandberg have also isolated an enzyme from their fungus cultures with a molecular weight of 60,000 daltons, as measured by polyacrilamide gel electrophoresis [29]. A likely identification of this enzyme is also laccase.

Streptomyces also produces an extracellular factor that liquefies coal, as evidenced by reactions using the cell-free broth. The reaction is apparent within 2 to 3 h after addition of the coal. The active species does not seem to be a protein because the molecular weight is between 1000 and 10,000 daltons (by ultrafiltration) and the activity is extremely stable to heating. After 10 min at 120°C, no change was observed, and only 30 to 40% was lost after an hour at this temperature. The activity was finally destroyed after 3 h of heating. There is also no loss of activity after protease treatment. Because the medium grows alkaline during the production of this factor and maximum activity is seen at pH 8 to 10, Scott and Strandberg have suggested that the active specie may be a basic organic compound [29].

18.4.5 Coal Type and Condition

All researchers have documented the fact that the type of coal, its size, as well as its storage and pretreatment conditions, affect the rate and extent of solubilization. Several groups have investigated these parameters, using a range of coals from leonardite to bituminous under various pretreatment conditions. Scott and Strandberg observed that larger pieces of coal are harder to liquify than smaller ones [28,29]. They used four cultures—*Candida, Penicillium, Polyporus,* and *Paecilomyces*—in screening experiments with different kinds of coal. Product inhibition may be a possible explanation for these results because coal extracts are known to be toxic to many organisms. In Dahlberg's experiments with *Candida,* 2 to 7 days were required for drops to form on -60 mesh, while for larger chunks of leonardite, 14 to 19 days were required. Similar experiments with York lignite spread as a thin film of particles on the fungal mat gave drops that formed at random over the film. When this coal was placed in piles, a ring of liquid was produced around the perimeter. No drops formed on chunks of this coal. *Geosmithia,* which was isolated from North Dakota lignite, produced drops on ground coal within 2 to 5 days; 2 to 7 days were required for chunks, with 4 to 9 days necessary for obvious liquefaction. On ground coal, 2 to 15 days were required for drops to form using *Phanerochaete*; 7 to 13 days for chunks, with 19 to 27 days required for obvious liquefaction.

The lowest-rank coals, the lignite and the leonardite, are the best substrates for liquefaction, with subbituminous and bituminous coals hardly liquefying. Dahlberg has noted that the reactivity of the coal was proportional to the oxygen content, which is higher in lower-ranked coals [27]. To determine the effect of oxidation, most workers have investigated pretreatment of the coal with oxidizing agents ranging from air to permanganate. Cohen has treated several different coals—North Dakota lignite, Montana lignite, Wyoming subbituminous, Texas lignite, and Illinois No. 6—with air, peroxide, ammonium persulfate, and potassium permanganate [25]. In all cases, pretreatment increased degradation by the fungi. Peroxide treatment gave the best response in these experiments. There was little degradation by the fungi when the lignite was not at least exposed to air.

Scott and Strandberg also observed the best liquefaction by fungi on well-weathered coal [28,29]. A black liquid was produced from Mississippi lignite, while a colorless liquid resulted from incubation of Texas and North Dakota lignites. Of the oxidative pretreatments—air, ozone, peroxide, and nitric acid—the latter proved the most effective. For example, with no pretreatment, Wyodak coal remained unchanged after incubation with fungus. After a 48-h pretreatment with 8 N nitric acid, the coal was well solubilized by the fungi. There was no appreciable increase in oxygen content after treatment.

Dahlberg has also investigated the effects of pretreatment on liquefaction of coal [27]. He used 1 N and 2 N nitric acid for different treatment periods with and without surfactant. He also examined the effect of water washing. The water washing had no effect on the liquefaction rate. With one normal nitric acid and Tween 80, good results were achieved overnight with chunks but not with ground lignite. The same treatment with 2 N nitric acid produced good results for almost all the coals.

18.4.6 Product Characterization

In their original paper on the degradation of coal by *P. versicolor* and *P. monticola*, Cohen and Gabriele interpreted the differences in the infrared analyses of lignite and of their liquid product to be a modification of the aromatic rings similar to that seen with pyrolysis of coal [26]. Since that time, a host of other analytical techniques have been applied to the analysis of the liquid product from microbial liquefaction.

There have been two approaches to collecting the liquid product. For those researchers employing surface culture, the liquid was removed from the plate via a Pasteur pipette. Others have developed small-scale reactors, although this material is known to be contaminated with the microbial growth medium. In addition, it has proven difficult to remove the remaining unsolubilized coal particles from the biomass in suspension.

Cohen has taken the liquid material from the agar surface by pipette and lyophilized it [26]. The lyophilized extract was then subjected to a series of tests, including solubility in a range of organic solvents. Of the solvents tested—hexane, methylene chloride, butanol, ethanol, methanol, and water—the extracts were soluble only in water. There was very limited solubility in methanol. Other researchers have also noted the limited organic solubility of this material. The material can be precipitated with acid and redissolved with base. Solubility increased monotonically as the pH increased. At any point, the soluble material could be filtered through a 0.2-μm filter. No solids were observed after centrifugation or filtration of the original material and no particles were seen under 400X magnification at a concentration of 60 mg solids/mL. Scott and Strandberg have demonstrated solubility in water up to 20% by weight [29]. The bulk of their material was soluble only in an aqueous medium. The energy content of the extract as measured by calorimetry was 4.39 kcal/g lignite or 96% of the energy content of the original coal.

The elemental analysis data on the bioextract provided by Wilson et al. show that, in essence, water has been added to the lignite with loss of some sulfur and metals [30]. Their x-ray fluorescence data confirm the reduction in silicon, and x-ray diffraction data demonstrate the loss of pyrite. The elemental analysis data of Scott and Strandberg demonstrates an increase in carbon and decrease in hydrogen, based on the dry weight of the solid material, coupled with a decrease in oxygen and nitrogen [29]. There was some increase in sulfur, which may be due to sulfur-containing amino acids from the micro-organisms.

Cohen has also titrated the liquid product, following the uptake of hydroxide versus change in pH [25]. The titration progressed, exhibiting no endpoint for the freeze-dried product. Near the beginning of the titration, a series of weak acids were present. Precipitating the bioextract with acid altered its response to this procedure. A series of stronger acids appeared first, followed by the weaker acids, then a plateau was observed. The speed of addition of hydroxide was significant, showing that the base was reacting with the coal. The equivalent molecular weight from the titration experiments was 200. Osmometry measurements carried out by Wilson et al. gave a molecular weight of 350 [30]. They consider these weights comparable, and point out that the nature of the counterions present would change the apparent molecular weight.

Both Wilson et al. and Scott and Strandberg determined the molecular weight by ultrafiltration [28-30]. Wilson et al. determined a molecular weight concentration greater than 50,000 for Cohen's unprecipitated material, while that of his own group was between 1000 and 10,000 [30]. The former material when precipitated still gave a concentration above 50,000, but at pH 10.4 the majority of the weight was in the range 10,000 to 50,000. Either there is a strong association at low pH, or bonds are broken by the addition of hydroxide. Scott and Strandberg found a molecular weight distribution from 30,000 to 300,000 by ultrafiltration [28,29]. Gel permeation chromatography also gave results in this molecular weight range—80% in the range 10,000 to 100,000, with 50% greater than 30,000.

The behavior on polyacrylamide gel electrophoresis (PAGE) of the bioextract collected from agar plates was examined by Cohen [25]. The neat extract, which had a pH of 7.5, remained primarily at the top of the gel, although two bands were observed elsewhere on the gel. If the extract was precipitated and redissolved at pH 9, more of the material traveled down the gel. At pH 13, the charge on the particles became more negative, so more material spread out on the gel. At pH 4, the precipitated extract remained at the origin. There did not seem to be more discrete bands at the higher pH, just movement of a large bulk of material.

Nuclear magnetic resonance (NMR) spectroscopy has been used as a measure of the aromaticity and the functional groups present in the bioextract. Scott and Strandberg reported a solid-state ^{13}C NMR analysis of the lyophilized product that was similar to the original lignite [29]. There was a high degree of aromaticity, including an increase of carboxyl and acidic carbonyl groups. A loss of methylene carbon was seen along with an increase in tetrahedral carbon attached to oxygen. Dahlberg has also noted similarities between the bioextract and the original coal, except for increased carbonyl functions in the product [27].

Wilson et al. used both ^{13}C and proton NMR for anlayses of both the neat and precipitated bioextracts isolated by Cohen [30]. Approximately half of the carbon and half of the carboxylate oxygen was detected by ^{13}C NMR. By proton NMR, the aromatic/HDO/acetone/aliphatic ratio in the product was 11:61:12:13, which corresponded to 0.5% aromatic, 2.6% hydroxyl, and 0.5% aliphatic hydrogen by weight. A total of 95% of the hydrogen was detected. For the acid precipitated material, proton NMR gave aromatic/methanol/aliphatic ratios of 10:3.3:9.5, which corresponded to 0.6% aliphatic and 0.6% aromatic hydrogen. The results from the precipitated and unprecipitated bioextract are therefore comparable. Proton NMR of material from the bioreactor had a higher aromatic content relative to the plate-derived bioextract. The aliphatic portion of the spectrum was contaminated by signals from the nutrient medium and was therefore unusable. Infrared spectroscopy (IR) is another measure of functional groups in a solid or liquid material. The analysis performed by Wilson et al. showed the appearance of carboxylate in the bioextract, coupled with loss or masking of the CH_2 stretch band.

All of the foregoing analyses have convinced researchers that they are dealing with a complex mixture of compounds, containing a high degree of aromaticity and oxygen functionalities. Scott and Strandberg have proposed structural features such as an aromatic nucleus with a large number of phenolic groups, primary amines, and carbonyls which might be polycondensed aromatic rings with hydroxyl substituents [29]. Wilson et al. have proposed a structure similar to lignin, but taking into account the results from NMR, IR, elemental analysis, and so on [30]. There are few fused aromatic rings and a large amount of carboxylation, consistent with the analytical data.

18.4.7 Process Schemes

Because research is currently involved with screening studies and elucidating the mechanisms of these reactions, little work is yet described on bioprocess schemes. The exceptions are the work of Wilson et al. in harvesting the enzyme laccase, and bench-scale process work by Scott and Strandberg [28-30]. As mentioned in Section 18.4.3, Scott and Strandberg have experimented with several reactor types, using both fungi and bacteria [28,29]. Their fixed-bed reactor resembles surface culture but uses a slowly moving bed. There is a continuous addition of particulates, nutrient solution, and organisms, together with aeration of the bed.

In an alternative scheme, the fluidized bed, particulates are introduced as the bed is degraded by the organisms. As the particles decrease in size, the ash filters out. The coal liquid product is a side stream from the liquid recycle. The final product of either of these processes is a feed material with a high carbon and energy content, which has potential use for the production of petrochemicals as well as for fuel.

18.5 GASIFICATION

Very little work is under way for direct coal gasification using microorganisms. However, a large body of research in related areas is available for process modeling and supplementary information. These areas include peat and biomass fermentation, anaerobic sewage treatment, and oxidation of simple and complex aromatic compounds. There are also numerous publications on methane-producing bacteria, although these anaerobic organisms are not as well known as some of the aerobic bacteria.

Two approaches have been developed for the production of methane from coal. Both require chemical processing of the substrate prior to fermentation. In the first approach, the coal is gasified using conventional coal gasification technology. The product gas is then upgraded to almost pure methane through anaerobic fermentation. Some of this work is ready for development to a pilot-plant scale. The second approach is modeled on processes for municipal waste, biomass, and peat gasification. The coal is oxidized with aqueous alkali for depolymerization, then fermented anaerobically to methane. Various stages of this conversion scheme are under investigation. The alkaline hydrolysis state has been used commercially by the pulping and paper industry for many years. That process is currently being adapted to lignite hydrolysis. Other stages are currently at the research or bench-scale level of development.

18.5.1 Mechanism

The production of methane involves three stages of degradation. First, a complex substrate such as coal, biomass, or peat—all of which can be considered polymers—must be broken down to smaller organic compounds. Then these compounds are fermented to acetate or carbon dioxide, the two substrates for methanogenesis. Finally, the acetate and carbon dioxide are converted to methane. A different population of microbes is required for each stage.

A complex mixture of fermentative bacteria carry out the first stage of degradation. Most of these organisms are obligate anaerobes, poisoned by the presence of oxygen, although some are facultative. They do not oxidize fatty acids, nor do they use amino acids or peptides as their major energy sources. From the complex substrate, they produce organic acids, alcohols, carbon dioxide, and hydrogen. The second stage is dominated by the acetogenic bacteria. Few of these species have been isolated and studied. These organisms ferment fatty acids, alcohols, and aromatic acids to acetate, hydrogen, and carbon dioxide. The partial pressure of hydrogen is critical at this stage because an increase will prevent formation of the desired products.

The methanogens take either carbon dioxide and hydrogen or acetate and produce methane. This reduction of carbon dioxide is necessary to the energy metabolism of the organism, as carbon dioxide serves as the main electron acceptor for the organism. These are the only organisms that can convert acetate and hydrogen to methane without light or electron acceptors other than carbon dioxide. The absolute requirement for hydrogen makes these organisms ideal for co-culture with the acetogenic bacteria. The usual ratio between the acetogenic bacteria and methanogens is between 1:26 and 1:256 [35].

18.5.2 Organisms

A common approach to selecting organisms for gasifying coal is to use an initial inoculum from some natural process. Untyped cultures have been prepared from sewage sludge, chicken waste, coal pile runoff, and chemical waste. Pure cultures have also been used for conversion to acetate or carbon dioxide, as well as methanogenesis. Several authors have also suggested nonenzymatic processes, such as alkali oxidation, for the initial depolymerization of the coal.

Gaddy has used both mixed and pure cultures of anaerobic organisms for the conversion of coal gas to methane [32-34]. Mixed cultures were enriched from a sewage sludge inoculum by progressively higher concentrations of the coal gas. The organisms were grown under anaerobic conditions in a chemostat at 37°C and constant pH. The coal gas consists primarily of carbon monoxide, with some carbon dioxide, hydrogen, and sulfur gases as well. Gaddy has acclimated his cultures to grow on up to 90% carbon monoxide as a sole carbon source. At least 3 months of acclimation was necessary because carbon monoxide is normally toxic to these microorganisms.

Clausen has used natural inocula from sewage sludge, chicken waste, coal pile runoff, and chemical waste to investigate the conversion of coal gas to liquid fuels [31]. Chicken waste was the least successful of the inocula. These cultures were grown using anaerobic techniques at pH 5 to 7, 37°C, with carbon monoxide, carbon dioxide, and acetate serving as the primary carbon sources and basal salts added as nutrients. A 10% inoculum was used with 50% transfer weekly. It was necessary to block methane synthesis with bromoethane sulfonic acid or monensin to observe products other than methane. Less than 1% methane is produced over the course of a 35-day incubation with either of these inhibitors. Small amounts of methanol, ethanol, and butanol were produced. Gaddy has also investigated the use of pure cultures for the conversion of coal gas to methane [32-34]. Two sets of organisms were examined in series and co-culture—organisms that converted carbon monoxide to carbon dioxide or acetate, and organisms that produced methane from these intermediate substrates.

The organisms *Rhodospirillum rubrum* and *Rhodopseudomonas gelatinosa* produce carbon dioxide and hydrogen from carbon monoxide and water. Two cultures of the former and one of the latter were examined for viability, growth rate, and conversion rate. The *R. gelatinosa* was unsatisfactory because it grew slowly and growth was inhibited by a concentration of 50% carbon monoxide. One of the species of *R. rubrum* performed well and was used for further studies.

Eubacterium limosum, *Acetobacterium woodii*, *Butyribacterium methylotrophium*, *Clostridium thermoaceticum*, and *Peptostreptococcus productus* are all capable of converting carbon monoxide and water to acetate and carbon dioxide. These organisms were also examined for viability, growth rate, and conversion rate. The *A. woodii* and the *E. limosum* did not respond well to the carbon monoxide. Of the remaining organisms, the *P. productus* was the most successful, with a 2-h doubling time and growth on 80% carbon monoxide. Quantitative yields of acetate were observed.

Methanogenic bacteria are very different from most other species of bacteria. The evolutionary divergence of the archaebacteria is the most ancient known. The major differences are the lack of a peptidoglycan cell wall, the presence of polyisoprenoid ether-linked lipids in the cell membranes, and immunologically unique transfer and ribosomal RNA [35]. They are slow growers, requiring 1 to 9 days for doubling time, depending on species and growth conditions. Normal polysaccharide reserves are absent, so cell death occurs when substrates are diminished, up to 90% cell loss in 2 days [37]. Their optimum temperatures are in the range 35 to 40°C, although those isolated from hot springs grow at much higher temperatures. The optimum pH is in the range 6 to 8. Methanogens are strict anaerobes, requiring the redox potential of the medium to be below -300 mV. This, as well as other factors, such as a tendency to flocculate, can make them difficult organisms to isolate and culture in pure form.

All methane bacteria can convert carbon dioxide and hydrogen to methane. The number that can convert acetate is much more limited. Both *Methanosarcino* sp. and *Methanothrix soehngenii* have this capability. These organisms can be isolated from rumen or from sewage sludge, where 80% of the methane is derived from acetate. Optimum temperatures for these bacteria are in the range of 30 to 37°C.

For their co-culture work, Gaddy and Clausen have used *P. productus*, which produces acetate, and a *Methanothrix* enrichment culture [32-34]. The conversion of carbon monoxide was almost the same as with the pure culture of *P. productus*, with 88% of the stoichiometric maximum converted. A co-culture with *Methanosarcina* was not successful because the organism is sensitive to the presence of carbon monoxide and washed out of the culture. Neither *Methanobacterium* nor *Methanospirillum* was used successfully in co-culture [34].

Wise has proposed a process using large underground rock or salt caverns for large-scale fermentation of hydrolyzed lignite [39]. Sewage sludge could be used as an inoculum for the rock cavern fermentation, but the salt caverns will require alternate organisms that can tolerate the high salt concentrations that will leach into the medium. Several of his collaborators are engaged in isolating halophiles from natural sources such as the Dead Sea or the deep brine wells used for pumping out offshore oil wells. Some of these organisms are found at concentrations of salt as high as 18%. Another group of halophiles isolated from deep thermal vents will metabolize aromatic compounds, which are suggested products of the initial hydrolysis step.

18.5.3 Process Variables

Gaddy and Clausen have investigated a number of process variables in relation to the conversion of coal gas to methane [32-34]. All processes were anaerobic at 37°C, with the pH, ammonia nitrogen, cell density, and ATP concentration monitored. The gas composition and flow rate were measured at the entrance and exit of the reactor. The variables investigated included concentration of feed gas, gas retention time, agitation rate, continuously stirred versus immobilized cell reactor, pressure, and sulfur content of the feed gas.

The concentration of the feed gas was varied as part of the acclimation process. The mixed culture was acclimated first in batch culture and then in the continuous-flow apparatus. Carbon monoxide is toxic to many organisms, so 3 months of gradually increasing gas concentration were required. As methane production ensued, the flow of gas was increased. During this period, the culture was buffered to prevent pH changes due to cell death. After 9 months of acclimation, 96 to 97% of the substrate was converted in a gas retention time of 8 to 10 h. In 2 h, only 20% was converted. By one year of acclimation, 95% conversion occurred in 2 h with the mixed culture. The gas retention time was varied at a constant agitation rate of 500 rpm.

The agitation rate was observed to be the most critical variable. These experiments were performed in a continuously stirred tank reactor (CSTR) with the mixed culture at atmospheric pressure while the conversion of carbon monoxide was being monitored. At 500 or 600 rpm, there was 99% conversion of the carbon monoxide and 90% conversion of the hydrogen. The conversion of carbon monoxide dropped to 70% at 400 rpm, and 38% at 200 rpm. The authors conclude that a high rate of agitation is necessary for good mass transfer between the insoluble substrates (carbon monoxide and hydrogen gas) and the organisms. At 600 rpm, a closed mass balance is observed for this process, producing a mixture of 40% methane and 60% carbon dioxide.

To obviate the need for high agitation rates, experiments were performed using a fixed-film reactor [33,34]. The bacteria were attached to an inert support either chemically or physically to achieve high cell density, which reduces the reactor size as well as increasing the throughput rate. The immobilized cells were then placed in a plug flow column with no agitation. The rate and extent of the reaction is dependent on the cell mass. A very high conversion by immobilized *P. productus* was observed in a gas retention time of 0.8 h, which improved with acclimation time. This is comparable to the 0.73-h retention time observed for this organism in the CSTR. A mixed methane bacteria culture was also immobilized for the conversion of acetate to methane. The 2-h retention time observed corresponds to a very small reactor because of the small liquid feedstream required. These retention times may decrease as the cell mass increases. Optimization of other process variables—such as pH, temperature, nutrient level, and inoculum size for this reactor—should be the next set of parameters for investigation.

Reactions were also performed in a Parr reactor at varying elevated pressures [32-34]. The cultures were acclimated slowly so as not to impair the cultures. A 5-psi increase in pressure was followed by several weeks acclimation period until the culture again reached a steady state. Total conversion occurred at the same retention time as at atmospheric pressure. The rate of conversion was proportional to pressure up to 75 psi. At 75 psi, the effective retention time was 24 min, while at 150 psi it was 12 min. At higher pressure, the reactor size would also be proportionately smaller.

The last parameter investigated was the effect of sulfur gases on organism toxicity [33,34]. Because it is expected that up to 2% hydrogen sulfide or carbonyl sulfide will be present in the effluent from a coal gasifier, it was necessary to determine the effect of these compounds on the cultures. For the *P. productus* culture, no effect was seen up to a concentration of 2% hydrogen sulfide. At 5%, a pronounced negative effect was observed, with culture death occurring at 10%. For carbonyl sulfide, no effect was seen at 1%, with some inhibition above 2% and culture death at 5%. This organism does seem to acclimate to the presence of sulfur gases, however. The mixed culture showed no effect from 1% of either sulfur compound, with slight impairment at 2%.

18.5.4 Process Schemes

As mentioned at the beginning of this section, there are no processes for the biogasification of coal. However, several authors have developed potential process schemes involving chemical as well as biological treatment of the substrate. These schemes are concerned primarily with the biogasification of lignite for several reasons. First, lignite has a high moisture content, which makes it unsuitable or expensive for use in more conventional processes. For a biogasification process the moisture content is an advantage, because aqueous systems are necessary for the microorganisms. Second, lignite is considered a young coal, which has not undergone a great deal of polymerization and oxygen loss. It is therefore much closer in structure to the original biological components than a higher-rank coal.

The process scheme envisioned by Wise includes an aqueous alkali pretreatment of the lignite, followed by anaerobic fermentation of the hydrolysate in underground caverns [39]. The resulting gas would then be purified to remove the carbon dioxide and any trace sulfur gases. This lignite process was modeled on a similar process developed for the biogasification of peat. Fermentation might also be used for the production of intermediates for organic synthesis or for the production of liquid fuels. The large amount of carbon dioxide resulting from this process has the potential for use in tertiary oil recovery.

As described by Leuschner et al., in the alkali oxidation process the complex aromatic ring structure of the lignite would be broken up to yield single ring aromatics and other organic acids [36]. Primarily carbon-oxygen (ether) bonds are broken, although some aliphatic carbon-carbon bonds are also broken. A high total carbon conversion can result from this process at temperatures less than 300°C and a pressure of 55 atm. For example, Wise reported a 77% solubilization of lignite at 250°C; conversions for peat and subbituminous coal were 96 and 30%, respectively [39]. This kind of process has been used in large-scale pulping operations for many years. The first work in this area with coal began around the time of World War I and has continued sporadically through the 1950s and 1960s in various countries.

The proposed fermentation would take place in large underground caverns, which could be maintained at 60°C. The digester size in such a cavern is 40 ft by 40 ft by 230 ft. In a parallel pilot process developed for peat fermentation, 20 model aromatic compounds were studied for methane production. All of the compounds could be fermented. Fermentation pathways have also been investigated from the standpoint of sewage digestion. Clark and Fina measured yields up to 90% of theoretical for the fermentation of benzoic acid [36]. As long as benzoic acid was added to the reaction mixture, methane and carbon dioxide were produced. In other studies, bioflavinoids have been used as models of plant-derived aromatic compounds that might be related to coal hydrolysis products. All were degradable by rumen bacteria.

The New Zealand Energy Research and Development Committee (NZERDC) has reviewed the research on the oxidation of peat, biomass, and aromatic xenobiotics [38]. Their purpose was to assess the feasibility of designing a lignite gasification process to convert Southland lignite to methane or liquid fuels. Thermochemical gasification was considered uneconomical.

The feasibility of peat biogas formation was demonstrated in the 1920s, although conversion rates and methane yields were low. Wet or dry feed was usable and the process could be coupled to wet harvest of the feed. Large or small peat deposits were suitable. The refractory material was more accessible when the particle size was decreased. A 270% increase in gas production was observed. Pretreatment increased the yields by converting polymers to water-soluble substrates for fermentation. Enzymatic pretreatments, such as with the fungus *P. chrysosporium*, catalyzed the cleavage of carbon-carbon bonds in the alkyl side chains of lignin, also causing hydrolysis. The same enzymes are probably used by this fungus in liquefying lignite, as discussed in a previous section of this chapter. Peat or its pretreatment products could be fermented by a batch or continuous process to methane using a sludge inoculum with a retention time of 15 to 60 days. Most of this work is at the bench scale. Little pilot- or full-scale experimentation has occurred.

No work was discovered by these authors on the biogasification of lignite itself. However, some chemical pretreatments have been reported to produce high yields of hydrocarbons, which can be used in refining processes. In addition, lignite can serve as a substrate for the aerobic growth of fungi. The fermentability of all kinds of aromatic and phenolic compounds has been used as support for the design of fuel-producing lignite bioprocesses.

Chynoweth has described a process for the in situ hydrolysis of peat that might be applicable to use with lignite deposits that have a permeable structure and high moisture content [38]. Aqueous alkali is injected into the deposit and the hydrolyzed products then pass to an active fermentation zone within the deposit. Products of this fermentation are then harvested for methane production in a conventional bioreactor. No mining is required for this process, with concomitant savings. There are also savings in waste disposal because the spent liquor can be returned to the site. When the active zone becomes depleted, the injection point can be moved, allowing a productive lifetime of 15 to 17 years for a peat bog of 3000 ha. Feasibility would be site specific, depending on geological and hydrological features.

The NZERDC has also considered the problem of waste disposal from a large-scale lignite bioprocess. They suggest that for lignite mined in a conventional manner, the inert residue could be dewatered and returned to the mine. Flash methanolysis is also suggested for use with the residue for the production of chemical feedstocks, such as ethylene, benzene, and carbon monoxide. An effluent treating process would also be necessary for separating and recycling biomass.

Preliminary economic analysis of a dilute acid hydrolysis process, coupled with an ethanol fermentation and biogasification reactor, concludes that the capital costs of the pretreatment reactor far outweigh the costs of the latter processes. Where the feedstock is already solubilized, methane can be produced for approximately $5 per gigajoule. Adding gathering (mining) and pretreatment costs raises the cost by a factor of 2.5.

REFERENCES

1. Attia, Y. A., and Elzeky, M. A., Biosurface modification in the separation from coal by froth flotation, in *Processing and Utilization of High Sulfur Coals* (Y. Attia, ed.), Elsevier, New York, 1985, pp. 673–698.
2. Bartel, P., and Skidmore, D., Gas composition effects on microbial coal desulfurization with a thermophilic bacterium: *Sulfolobus acidocaldarius*, unpublished manuscript.
3. Chen, C. Y., and Skidmore, D., Adsorption of *Sulfolobus acidocaldarius* cells on coal particles, unpublished manuscript.
4. Chen, C. Y., and Skidmore, D., Microbial coal desulfurization with thermophilic microorganisms, unpublished manuscript.
5. DiSpirito, A. A., Dugan, P. R., and Tuovinen, O. H., Sorption of *Thiobacillus ferroxidans* to particulate material, *Biotechnol. Bioeng.*, 25(4), 1163–1168 (1983).
6. Dugan, P. R., The value added to coal by microbial sulfur removal, in *Processing and Utilization of High Sulfur Coals* (Y. Attia, ed.), Elsevier, New York, 1985, pp. 717–726.
7. Dugan, P. R., Economics of coal desulfurization, in *Proceedings of the Biological Treatment of Coals Workshop*, June 23–25, 1986, pp. 65–82.
8. Dugan, P. R., and Apel, W. A., Microbial desulfurization of coal, in *Metallurgical Applications of Bacterial Leaching and Related Microbiological Phenomena* (L. E. Mull, A. E. Torma, and J. A. Brierley, eds.), 1978, pp. 223–250.
9. Hoffman, M. R., et al., Kinetics of the removal of iron pyrite from coal by microbial catalysis, *Appl. Environ. Microbiol.*, 42(2), 259–271 (1981).
10. Kargi, F., Microbial coal desulfurization, *Enzyme Microb. Technol.*, 4(1), 13–19 (1982).
11. Kargi, F., and Robinson, J. M., Removal of sulfur compounds from coal by the thermophilic organism *Sulfolobus acidocaldarius*, *Appl. Environ. Microbiol.*, 44(4), 878–883 (1982).
12. Lacey, D. T., and Lawson, F., Kinetics of the liquid phase oxidation of acid ferrous sulfate by the bacterium *Thiobacillus ferroxidans*, *Biotechnol. Bioeng.*, 12, 29–50 (1970).
13. Murphy, J., et al., Coal desulfurization by microbial processing, in *Processing and Utilization of High Sulfur Coals* (Y. Attia, ed.), Elsevier, New York, 1985, pp. 643–652.
14. Murphy, J., et al., Microbial desulfurization of coal by thermophilic bacteria, unpublished manuscript.
15. Silverman, M. P., Mechanism of bacterial pyrite oxidation, *J. Bacteriol.*, 94(4), 1046–1051 (1967).
16. Silverman, M. P., and Lundgren, D. G., Studies on the chemoautotrophic iron bacterium *Ferrobacillus ferroxidans*, *J. Bacteriol.*, 77(5), 642–647 (1959).
17. Sproull, R. D., et al., Enhancement of coal quality by microbial demineralization and desulfurization, in *Proceedings of the Biological Treatment of Coals Workshop*, June 23–25, 1986, pp. 83–94.
18. Vasen, V. A., Commercial microbial desulfurization of coal, in *Processing and Utilization of High Sulfur Coals* (Y. Attia, ed.), Elsevier, New York, 1985, pp. 699–715.
19. Wakao, N., et al., Bacterial pyrite oxidation. II. The effect of various organic substances on release of iron from pyrite by *Thiobacillus ferroxidans*, *J. Appl. Microbiol.*, 29(3), 177–185 (1983).
20. Hatcher, H., Biological coal desulfurization, in *Proceedings of the Biological Treatment of Coals Workshop*, June 23–25, 1986, pp. 38–51.
21. Isbister, J., Biological removal of organic sulfur from coal, in *Proceedings of the Biological Treatment of Coals Workshop*, June 23–25, 1986, pp. 18–37.

22. Isbister, J. D., and Kobylinski, E. A., Microbial desulfurization of coal, in *Processing and Utilization of High Sulfur Coals* (Y. Attia, ed.), Elsevier, New York, 1985, pp. 627–641.

23. Krawiec, S., Some experimental observations, criteria, and speculations about bacterial desulfurization of organic components in coal, in *Proceedings of the Biological Treatment of Coals Workshop*, June 23–25, 1986, pp. 52–64.

24. Monticello, D. J., Bakker, D., and Finnerty, W. R., Plasmid-mediated degradation of dibenzothiophene by *Pseudomonas* species, *Appl. Environ. Microbiol., 49*(4), 756–760 (1985).

25. Cohen, M., Solubilization of coal by *Polyporus versicolor*, in *Proceedings of the Biological Treatment of Coals Workshop*, June 23–25, 1986, pp. 95–113.

26. Cohen, M. S., and Gabriele, P. D., Degradation of coal by the fungi *Polyporus versicolor* and *Poria monticola*, *Appl. Environ. Microbiol., 44*(1), 23–27 (1982).

27. Dahlberg, M., Some factors influencing the bioliquefaction of lignite, in *Proceedings of the Biological Treatment of Coals Workshop*, June 23–25, 1986, pp. 172–193.

28. Scott, C. D., and Strandberg, G. W., Microbial coal liquefaction, in *Proceedings of the Biological Treatment of Coals Workshop*, June 23–25, 1986, pp. 128–140.

29. Scott, C. D., and Strandberg, G. W., *Microbial Coal Liquefaction, Quarterly Reports*, U.S. Department of Energy, Washington, D.C., 1985–1986.

30. Wilson, B. W., et al., Microbial beneficiation of low rank coals, in *Proceedings of the Biological Treatment of Coals Workshop*, June 23–25, 1986, pp. 114–127.

31. Clausen, E., Production of liquids from coal synthesis gas, in *Proceedings of the Biological Treatment of Coals Workshop*, June 23–25, 1986, pp. 214–225.

32. Gaddy, J., Biological conversion of coal synthesis gas, in *Proceedings of the 5th Annual Gasification Contractors Meeting*, June 25–27, 1985, pp. 260–267.

33. Gaddy, J., Production of methane from coals synthesis gas, in *Proceedings of the Biological Treatment of Coals Workshop*, June 23–25, 1986, pp. 194–213.

34. Gaddy, J., Biological conversion of coal synthesis gas, in *Proceedings of the 6th Annual Gasification Contractors Meeting*, June 24–25, 1986 (M. R. Ghate and L. A. Baker-Jarr, eds.), pp. 121–132.

35. Kirsop, B. H., Methanogenesis, *CRC Crit. Rev. Biotechnol., 1* (2), 109–159 (1984).

36. Leuschner, A. P., et al., Biogasification of Texas lignite, unpublished manuscript.

37. McInerney, M. J., and Bryant, M. P., Review of methane fermentation fundamentals, in *Fuel Gas Production from Biomass* (D. L. Wise, ed.), CRC Press, Boca Raton, Fla., 1981, pp. 19–46.

38. New Zealand Energy Research and Development Committee, *Biological Production of Methane from Southland Lignite: A Literature Survey*, National Technical Information Service, Springfield, Va., 1985.

39. Wise, D. L., Methane production from coal derived materials, in *Proceedings of the Biological Treatment of Coals Workshop*, June 23–25, 1986, pp. 226–232.

19

Lignite Refinery Economic Study

ERNEST E. KERN and KATHLEEN GRIBSCHAW *Houston Lighting and Power Company, Houston, Texas*

19.1 INTRODUCTION

The purpose of this feasibility study was the design and economic analysis of a lignite refinery and a biogasification process. Heat and material balances for the process were performed on the IBM Personal Computer with the aid of ChemCad, a software package purchased from COADE. Equipment sizes were then calculated and the capital cost for each process was determined by utilizing a computer program from McGraw-Hill, called PRICE.

Both processes were studied under two different operating conditions, resulting in an analysis of four cases. Although none of the cases analyzed produced revenues sufficient to achieve a 15% return on investment at prevailing economic conditions and level of technical development, one of the lignite refinery cases produced a positive cash flow of $32 million per year. This process will become more feasible in the future as product prices and conversion rates increase and efficiency improvements are developed. Therefore, the lignite refinery is favored over biogasification alone. Figures 19.1 to 19.4 show the conceptual design of the lignite refinery.

Lignite is in abundance in Texas, but since it is composed of nearly 50% moisture and ash, it is not adaptable for the same uses as bituminous coal. Therefore, Houston Lighting and Power Company has decided to find new uses for this natural resource. A lignite refinery concept, shown in Fig. 19.5, was introduced, whereby lignite is mixed with sodium carbonate and water to produce a slurry. This slurry is sent to a reactor that breaks down the lignite to produce water-soluble organic compounds. A second reactor then converts 50% of these acids to aromatic compounds. The stream is then split. The aromatics are sent to distillation for individual product recovery and the acids are sent to a bioreactor, which is an underground cavern filled with microorganisms. These "bugs" feed on the acids produced in the first reactor and convert them to carbon dioxide and methane. Another version of this process called biogasification is shown in Fig. 19.6. In

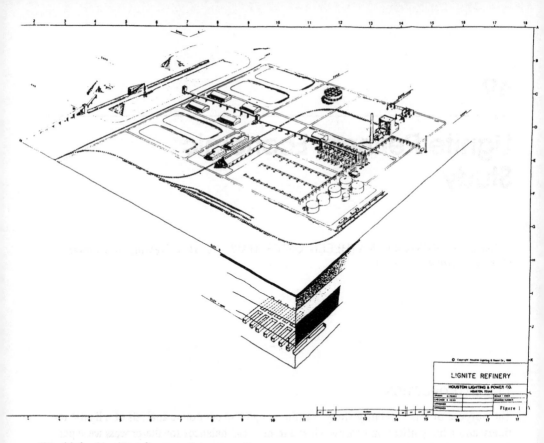

Fig. 19.1 Lignite refinery artist conception.

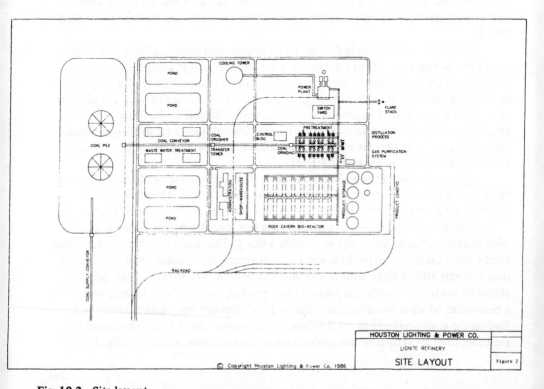

HOUSTON LIGHTING & POWER CO.

LIGNITE REFINERY

SITE LAYOUT | Figure 2

Fig. 19.2 Site layout.

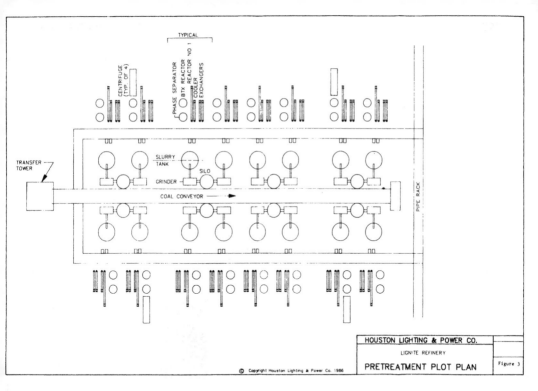

Fig. 19.3 Pretreatment plot plan.

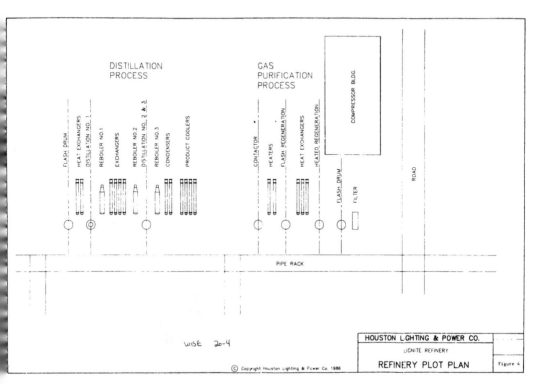

Fig. 19.4 Refinery plot plan.

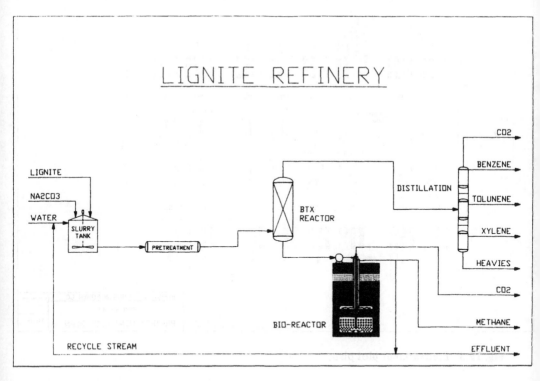

Fig. 19.5 Lignite refinery concept.

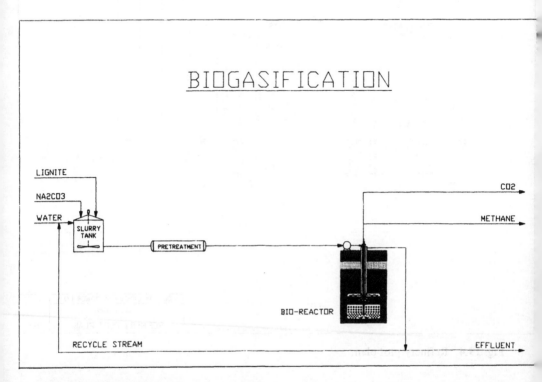

Fig. 19.6 Biogasification.

this process the second reactor and subsequent distillation phase has been eliminated. All the acids produced in the first reactor are sent to the bioreactor for conversion to gases. These two processes have both been analyzed under two different operating conditions giving a total of four cases. A summary of these cases follows.

	Conversion of VS to acids	Ratio of VS to H_2O	Ratio of VS to NA_2CO_3
Case 1: lignite refinery	50%	1:10	10:2
Case 2: biogasification			
Case 3: lignite refinery	75%	1:7	10:1
Case 4: biogasification			
VS = Volatile Solids			

19.2 LIGNITE CHEMISTRY

Lignite is ranked lower than bituminous coal because it is in an earlier stage of coalification. As seen in Fig. 19.7, the chemical structure of lignite is composed of aromatic clusters and oxygen groups. These oxygen groups are more prevalent in lignite than in higher-ranked coals, and it is the bonds between the carbon and oxygen molecules that break down under alkaline hydrolysis, thereby producing water-soluble organic acids. This transformation occurs in the first reactor. Conversion of 50% of the acids to carbon dioxide and immiscible aromatics occurs in the second reactor. The remaining acids are fermented by the microorganisms in the bioreactor and converted to carbon dioxide and methane.

CHEMICAL STRUCTURE OF LIGNITE

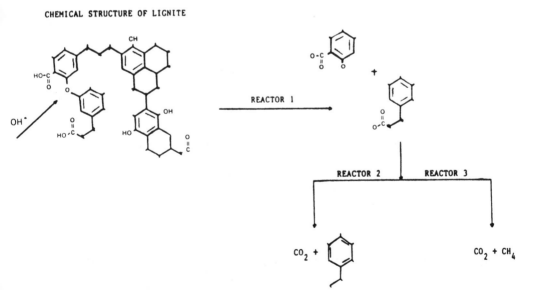

Fig. 19.7 Chemical structure of lignite.

19.3   LIGNITE COMPOSITION

Lignite is composed of volatile solids, moisture, and ash. Seven different core samples were analyzed for these components, and their results can be found in Fig. 19.8. In this chapter, volatile solids refer to the volatile matter and the fixed carbon found in the sample. A more detailed analysis of the volatile solids, showing the percentage of carbon, hydrogen, nitrogen, sulfur, and oxygen is also presented. Since the composition can vary dramatically, the process was designed using the average composition of the seven samples.

Based on a lignite feed rate of 20,000 tons/day or 1,667,000 lb/h, it follows that the feed contained 865,200 lb/h volatile solids, 527,600 lb/h water, and 274,000 lb/h ash. This initial calculation sets the basis for the entire design, since the amount of volatile solids present in the process determines the amount of products produced.

CORE SAMPLE	1	2	3	4	5	6	7	AVG.
PROXIMATE								
MOISTURE	31.94	32.98	33.63	30.80	29.02	32.48	30.67	31.65
ASH	10.54	11.76	12.14	20.07	21.95	16.23	22.48	16.45
VOLATILE	29.41	32.41	26.78	25.27	26.02	27.64	25.81	27.62
FIXED C	28.11	22.85	27.45	23.86	23.01	23.65	21.04	24.28
	100.00	100.00	100.00	100.00	100.00	100.00	100.00	100.00
SULFUR	1.39	1.88	2.70	1.40	0.82	1.50	0.76	1.49

AVG. VOLATILE SOLIDS = AVG. VOLATILES + AVG. FIXED C
= 27.62 + 24.28
= 51.90

BREAKDOWN OF VOLATILE SOLIDS

CORE SAMPLE	1	2	3	4	5	6	7	AVG.
ULTIMATE								
C	41.52	36.71	38.75	35.22	33.81	36.39	34.49	36.70
H	3.23	2.98	2.95	2.48	2.77	2.75	2.77	2.85
N	0.91	0.76	0.85	0.65	0.65	0.81	0.62	0.75
S	1.39	1.88	2.70	1.40	0.82	1.50	0.76	1.49
O	10.47	12.93	8.98	9.38	10.98	9.84	8.21	10.11
TOTAL	57.52	55.26	54.23	49.13	49.03	51.29	46.85	51.90

LIGNITE FEED	VOLATILE SOLIDS =	51.90% =	865,200 LB/HR
	ASH	= 16.45% =	274,200 LB/HR
	MOISTURE	= 31.65% =	527,600 LB/HR
	TOTAL	= 100.00% =	1,667,000 LB/HR

Fig. 19.8   Lignite composition.

For simplicity, the amount of sulfur present in the lignite feed has not been shown throughout the process, but appears leaving the bioreactor flowsheet in the form of H_2S. The H_2S could be converted to elemental sulfur or sulfuric acid and then sold.

19.4 CASE STUDIES

19.4.1 Case 1

The process flowsheets have been divided into three main areas. First, the front-end flowsheet (Fig. 19.9) includes the equipment from coal handling to phase separation. The distillation flowsheet (Fig. 19.10) shows the carbon dioxide and aromatics stream through flashing, distillation, and product separation. The bioreactor flowsheet (Fig. 19.11) follows the stream containing acids, water, and dissolved solids through preparation conversion, product recovery, and recycle.

Corresponding stream information, including composition, temperature, and pressure data can be found in Fig. 19.12. To obtain these values, heat and material balances were calculated using the IBM Personal Computer with the aid of ChemCad, a software package purchased from COADE. The programs generated from this analysis are on file with the Houston Lighting and Power Company.

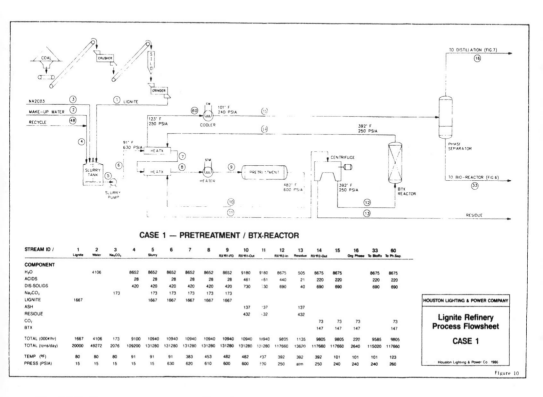

Fig. 19.9 Pretreatment process flowsheet.

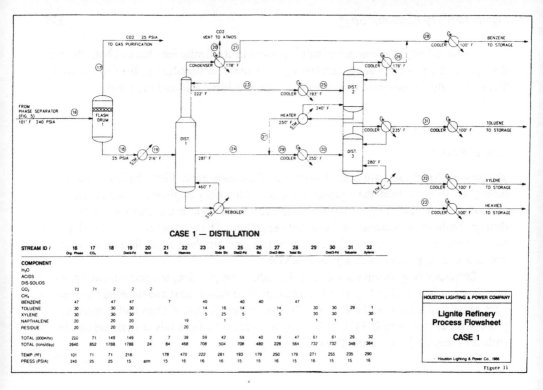

Fig. 19.10 Distillation process flowsheet.

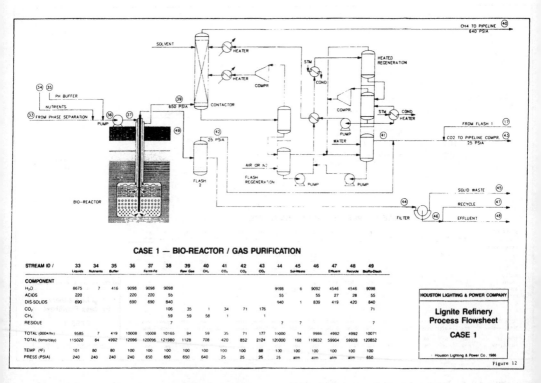

Fig. 19.11 Bioreactor process flowsheet.

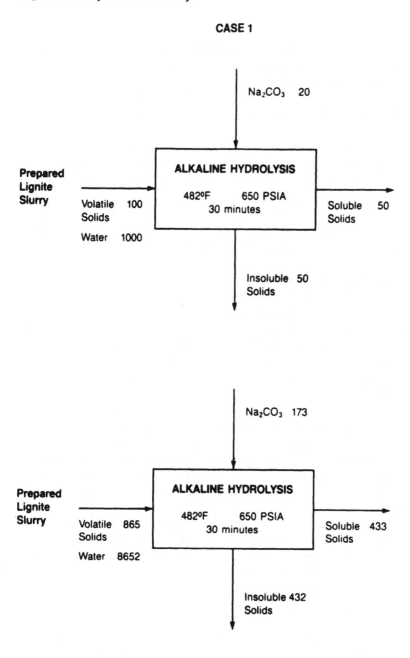

(Flows shown in thousand pounds per hour.)

Fig. 19.12 Alkaline hydrolysis flow ratios.

Process Design

Front-End Flowsheet. As stated previously, lignite contains nearly 50% moisture and ash. Because of this it is not economical to transport the lignite by rail before processing, the refinery should be situated next to the lignite mining operation. A conveyor system will transport the lignite from the mine to the refinery site, where it will be stored in silos. The lignite will be crushed and ground before utilization as a process feedstock.

In the first stage of the process, water and sodium carbonate are mixed with lignite in a slurry tank. A specific ratio of these feedstocks is necessary. The top of Fig. 9 shows the relative amounts of the three components used to form the slurry. The diagram was based on research performed by Dynatech R/D Company under contract to Houston Lighting and Power Company. Since the average amount of volatile solids in the 20,000-ton/day lignite feed was calculated to be 865,200 lb/h (see Fig. 19.8), the ratios in Fig. 19.12 were proportionally converted to the inlet flows. These flows can be seen in the bottom portion of Fig. 19.12.

After these three streams are mixed, the slurry is sent through a pump that increases the pressure of the stream from 15 psia to 630 psia. The stream is then sent through a series of three heat exchangers. In the first exchanger, the stream is heated from 91°F to 383°F by the outlet stream from the second reactor. The stream is further heated by the stream from reactor 1 to 453°F. A final temperature of 482°F is obtained from a steam heat exchanger or steam injection. A 10-psi drop in pressure occurs in each exchanger, causing the outlet stream from the final heat exchanger to enter the first reactor at 482°F and 600 psia.

During alkaline hydrolysis, which occurs in the first reactor, the lignite bonds are broken and converted to ash, residue, dissolved solids, water, and organic acids. The conversion was based on information from Dynatech. Referring again to the bottom portion of Fig. 19.12, the 432,000 lb/h of insoluble solids is considered residue, and the 433,000 lb/h of soluble solids is considered organic acids. This amounts to a 50% conversion of volatile solids to organic acids. (Although various acids may be produced, the design is based on benzoic acid.)

As stated at the bottom of Fig. 19.8, the lignite feed contains 527,600 lb/h of water and 274,200 lb/h of ash. The outlet stream of the reactor also shows an increase of 310,000 lb/h dis-solids. This increase occurs from the 173,000 lb/h of sodium carbonate added previously and half the ash in the lignite feed.

The outlet stream from the first reactor is then cooled to 437°F through heat exchange with the incoming stream. Next the flow is sent to a centrifuge, which removes the ash and residue from the stream. Approximately 5% of the water, acids, and dis-solids are also removed during this step. The resulting residue contains 50% moisture and amounts to 13,620 tons/day. Instead of spending money to dispose of this waste, it could be burned to produce steam and/or electricity for the process. Any excess energy could then be sold. At this point, no credit or disposal costs have been assumed for the solid residue.

The remainder of the stream now at 250 psia proceeds to the second reactor. In this reactor 50% of the acids are converted, one-third of the converted acids producing carbon dioxide and the other two-thirds producing various aromatic compounds. (Because of the chemical structure of lignite, no straight-chain hydrocarbons are assumed to be present in the system.)

After the stream is cooled to 101°F, it is sent to a phase separator. Since the aromatics are immiscible in water, they are easily separated into two streams. The water and the remaining acids proceed to the bioreactor while the aromatics and carbon dioxide are sent to distillation. (The solubility of hydrocarbons in water, and vice versa, is assumed to be negligible.)

Distillation Flowsheet. Although a wide range of aromatics are produced in the second reactor, a few chemicals have been selected for ease in making the calculations. Benzene, toluene, and xylene will represent the light fraction of the stream, and naphthalene and chrysene will be representative of the heavy fraction. The exact percentages of these chemicals are not known, but values have been estimated. One must realize that the optimum design and operating conditions for the distillation column varies with the components that need to be separated. Also, since the product prices for the lighter aromatics are significantly higher than for the heavy aromatics, the product sales will be greatly influenced by the ratio of lights to heavies. (For example, the selling price of benzene was assumed to be $1.48/gal, while the selling price of No. 2 fuel oil is $0.67/gal.) Without actual laboratory data, an estimation of the aromatic stream composition is all that is possible. This estimation will provide the basic design criteria until further data can be gathered.

Before proceeding to the distillation columns, the stream containing carbon dioxide and aromatics is flashed to 25 psia to remove most of the carbon dioxide, which is then compressed to 1000 psia and sent to a pipeline for distribution. The remainder of the stream is heated to 216°F (the bubble point of the stream) and enters the first distillation column.

The column contains 40 stages and has a reflux ratio of 2.5. Four streams are taken from this column: a distillate, a bottom, and two side streams. The remaining carbon dioxide is vented at the condenser, leaving an essentially pure benzene distillate. The bottom stream, which is virtually 100% heavies, is sent to storage and will be sold for boiler fuel. A mixture of benzene, toluene, xylene, and a small amount of naphthalene is found in the two side streams.

The upper side stream leaves the seventeenth stage of the column at 222°F and is cooled to its bubble point of 193°F before entering the second distillation column, which has 30 stages and a reflux ratio of 2.0. This column separates the stream into a top and bottom flow. The top stream containing 100% benzene is merged with the other benzene stream from the top of the first column and is then sent to storage. The bottom flow leaving the second column is joined with the lower side stream from the twenty-fifth stage of the first column. This combined stream is cooled to 255°F and sent to the third distillation column.

Utilizing 30 stages and a reflux ratio of 2.5, this column splits the incoming flow, producing a top toluene stream and a xylene stream coming from the bottom of the column. Both streams are then sent to separate storage facilities. (In this design, no attempt was made to separate the *ortho-*, *meta-*, and *para*-xylene isomers. The mixed isomers will be sold as a solvent or fuel additive.)

Bioreactor Flowsheet. The microorganisms in the bioreactor require certain nutritional requirements to convert the pretreated lignite to carbon dioxide and methane. Dynatech has determined the necessary ratio of carbon to nitrogen to phosphorus to be 100:5:1 by weight. A deficiency exists for the solubilized lignite in both the nitrogen and phosphorus levels. This deficiency is corrected by the addition of nutrients. Urea, which contains 46% nitrogen by weight, is added along with hydrogen phosphate, which contains 52% phosphorus by weight. Hydrogen chloride, which contains 64% water by weight, is also added to give the solution the proper pH level. (The nutrients and buffers appear in the stream information and computer programs as water.)

The upper half of Fig. 19.13 shows the ratio of additives, and carbon dioxide and methane produced based on 50 parts soluble solids. Since half the acids were converted to aromatic chemicals, 220,000 lb/h of soluble solids remains to enter the bioreactor. The flows of the other streams are proportionally converted to the actual flows and can be seen in the lower portion of Fig. 19.13. This combined stream is injected into the bioreactor, which operates at 650 psia and 100°F.

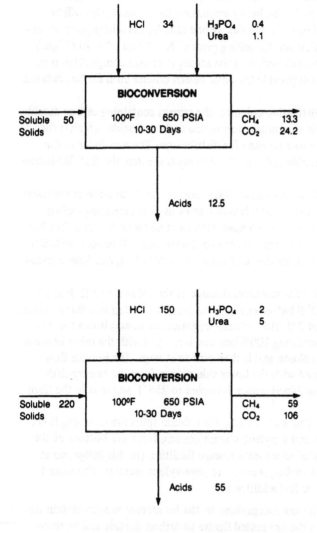

(Flows shown in thousand pounds per hour.)

Fig. 19.13 Bioconversion flow ratios.

The idea of situating the bioreactor in an underground rock cavern was developed because of the high cost of placing the reactors above ground. A side view of the cavern can be seen in Fig. 19.14. The inlet stream is pumped down into the reactor at the ends of the carvern, while the gas and remaining liquid stream are recovered from the cavern's center. Figure 19.11 shows that the cavern resembles a series of parallel hallways, 40 ft wide, 40 ft high, and 460 ft long. A gradual upward slope is found between the injection walls and the pump-out wells. (Figures 19.14 and 19.15 have been taken from a report performed by Fenix & Scisson for Houston Lighting and Power regarding underground caverns.)

Crushed rock fills the cavern to serve as the substrate for the growth of the microorganisms. During the process, these microorganisms convert 75% of the acids remaining in the solubilized lignite to carbon dioxide and methane. Figure 19.13 shows the specific amounts of these gases. The gas stream is then recovered from the bioreactor and purified. Hydrogen sulfide is also separated at this point and could be converted to sulfuric acid or elemental sulfur for future sale.

Since some of the carbon dioxide produced in the bioreactor is trapped in the liquid stream, this stream is flashed to release the gas and the CO_2 is compressed to 1500 psia and sent to a pipeline for distribution. The remaining liquid stream is sent to a rotary filter, resulting in a 168-ton/day solid waste stream composed of 50% sludge and 50% moisture. This stream is then combined with the residue stream from the front end of the process. The combined stream could be used as fuel in a fluidized-bed combustor to produce steam and/or electricity for the process; any excess energy could be sold. After the filter, the remaining liquid stream is split, with 50% of the flow going to an effluent pond for cleanup and disposal while the other 50% is recycled to the beginning

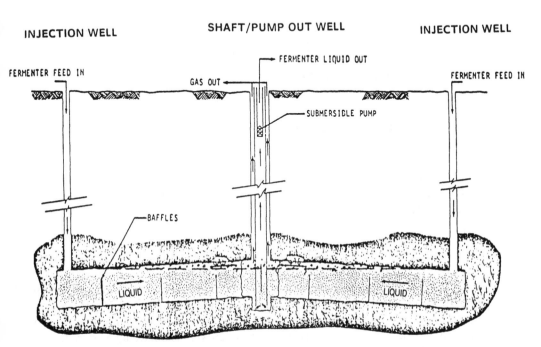

Fig. 19.14 Packed mine fermentor process flow.

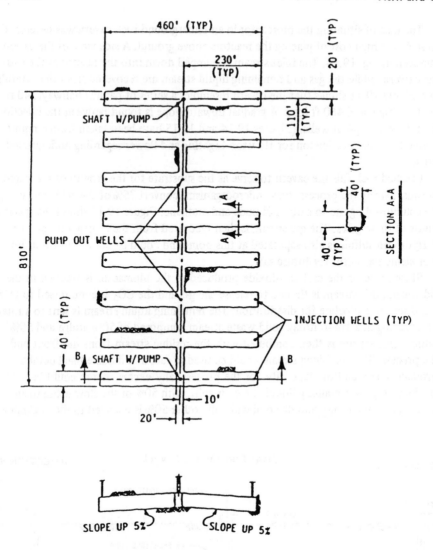

Fig. 19.15 Packed mine fermentor plan.

of the process. At this point, makeup water is added to the recycle stream to raise the water flow rate to the required 8,652,000 lb/h (see Fig. 19.12). In this particular design, 4,106,000 lb/h makeup water is necessary for the process.

Plant Cost Estimate

A software program from McGraw-Hill called PRICE was utilized to estimate the capital required to construct the lignite refinery described previously. Equipment for the process was sized by using the information gained from the heat and material balances performed by ChemCad. Supplier quotes were obtained for the carbon dioxide and methane compressors. The Selexol process was utilized as the method of gas purification, and the capital investment and operating costs for this step were estimated by Norton, Inc. The cost for the underground cavern was obtained from a report written by Fenix and Scisson for Houston Lighting and Power. Additional values were obtained from Dynatech.

Two contingencies are added to the battery limits capital cost for the plant. A 25% process contingency is used since the extent of development of the process data is only in the calculation phase. If the technology develops enough to obtain pilot-plant data, this process contingency could be reduced to 10%. In addition, a 25% project contingency is added, since the level of detail in the engineering design is only at the development flow-sheet stage. When the detail of design is increased to include piping drawings, the project contingency can be dropped to 10%.

The PRICE program utilized the Marshall and Swift average equipment cost index for the petroleum industry for the first quarter of 1986. The final result of this program is a total estimated project cost of $262 million. This figure refers to all the on-site costs for the plant. Off-site costs, which include such items as roads, buildings, laboratories, and utility access, amount to $92 million or 35% of the total on-site costs. The sum of the on-sites and off-sites gives a total erected cost of $354 million.

19.4.2 Case 2

Process Design

In Case 2, the second reactor has been eliminated. There is no conversion of acids to aromatics, and consequently, the distillation step has been removed. In this process, all the acids produced in the first reactor are sent to the underground bioreactor for conversion to carbon dioxide and methane, which are the only two products produced from this process. The process flowsheets and the corresponding stream information for this case can be seen in Figs. 19.16 to 19.18.

The process conditions for the second case remain the same, with a 50% conversion to volatile solids to acids, a 1:10 ratio of volatile solids to water, and a 10:2 ratio of volatile solids to sodium carbonate. The description of the process is similar to case I; therefore, the details will not be repeated here.

Figure 19.18 shows the relative amounts of the three feedstocks used to form the slurry. Since there is no conversion of acids to aromatics, the 469,000 lb/h of acids sent to the bioreactor in this case is approximately double that shown in case 1. A corresponding increase in nutrients and buffers is necessary, resulting in a larger production of carbon dioxide and methane. The actual values of these flows are shown in Fig. 19.19.

Plant Cost Estimate

The capital necessary to construct this biogasification plant was estimated using the PRICE program. The capital cost of the rock cavern and gas purification has increased proportionally with the increased flow rates of the acids and gases. Equipment sizes for the process were calculated using the new flow rates and heat duties resulting from the ChemCad program. All equipment related to the distillation phase of the process have been eliminated in this case. The final result of the PRICE program is an on-site cost of $230 million. With an addition of $80 million in off-site costs, the total erected cost of the plant is $310 million.

19.4.3 Case 3

Process Design

Process flowsheets and corresponding stream information for the third case can be seen in Figs. 19.20 to 19.22. This case is also a lignite refinery similar to that shown in case 1,

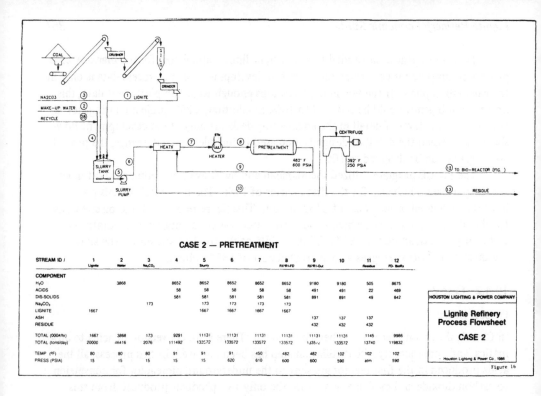

CASE 2 — PRETREATMENT

STREAM ID /	1	2	3	4	5	6	7	8	9	10	11	12
	Lignite	Water	Na_2CO_3		Slurry			RX-#1-FD	RX-#1-Out		Residue	FD-BioRx
COMPONENT												
H_2O		3868		8652	8652	8652	8652	8652	9180	9180	505	8675
ACIDS				58	58	58	58	58	491	491	22	469
DIS-SOLIDS				581	581	581	581	581	891	891	49	842
Na_2CO_3			173	173	173	173	173	173				
LIGNITE	1667			1667	1667	1667	1667	1667				
ASH									137	137	137	
RESIDUE									432	432	432	
TOTAL (000#/hr)	1667	3868	173	9291	11131	11131	11131	11131	11131	11131	1145	9986
TOTAL (tons/day)	20000	46416	2076	111492	133572	133572	133572	133572	1335?2	133572	13740	119832
TEMP (°F)	80	80	80	91	91	91	450	482	482	102	102	102
PRESS (PSIA)	15	15	15	15	15	620	610	600	600	590	atm	590

HOUSTON LIGHTING & POWER COMPANY

Lignite Refinery Process Flowsheet

CASE 2

Houston Lighting & Power Co. 1986

Figure 16

Fig. 19.16 Pretreatment process flowsheet.

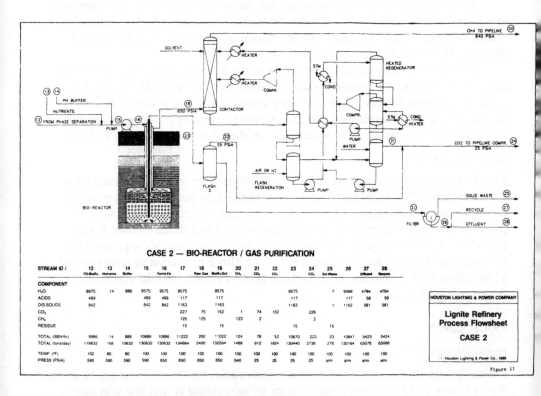

CASE 2 — BIO-REACTOR / GAS PURIFICATION

STREAM ID /	12	13	14	15	16	17	18	19	20	21	22	23	24	25	26	27	28
	FD-BioRx	Nutrients	Buffer		Ferm-Fd		Raw Gas	BioRx-Out	CH_4	CO_2	CO_2		CO_2	Sol Waste		Effluent	Recycle
COMPONENT																	
H_2O	8675	14	886	9575	9575	9575		9575				9575		7	9568	4784	4784
ACIDS	469			469	469	117		117				117			117	58	59
DIS-SOLIDS	842			842	842	1163		1163				1163		1	1162	581	581
CO_2						227	75	152	1	74	152		226				
CH_4						125	125		123	2		2					
RESIDUE						15		15				15		15			
TOTAL (000#/hr)	9986	14	886	10886	10886	11222	200	11022	124	76	152	10670	220	23	10847	5423	5424
TOTAL (tons/day)	119832	168	10632	130632	130632	134664	2400	132264	1488	912	1824	130440	2736	276	130164	65076	65088
TEMP (°F)	102	80	80	100	100	100	100	100	100	100	100	100	100	100	100	100	100
PRESS (PSIA)	590	590	590	590	650	650	650	650	640	25	25	25	25	atm	atm	atm	atm

HOUSTON LIGHTING & POWER COMPANY

Lignite Refinery Process Flowsheet

CASE 2

Houston Lighting & Power Co. 1986

Figure 17

Fig. 19.17 Bioreactor process flowsheet.

CASE 2

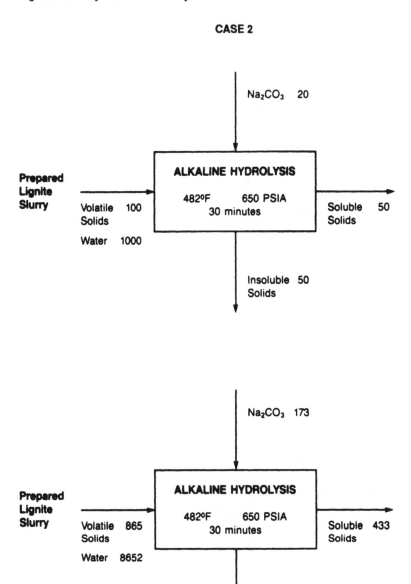

(Flows shown in thousand pounds per hour.)

Fig. 19.18 Alkaline hydrolysis flow ratios.

CASE 2

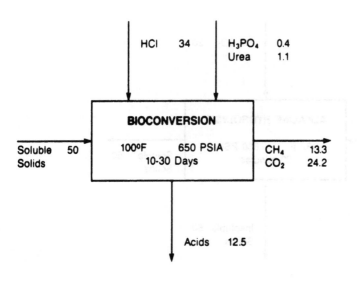

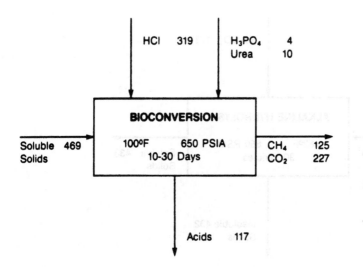

(Flows shown in thousand pounds per hour.)

Fig.19.19 Bioconversion flow ratios.

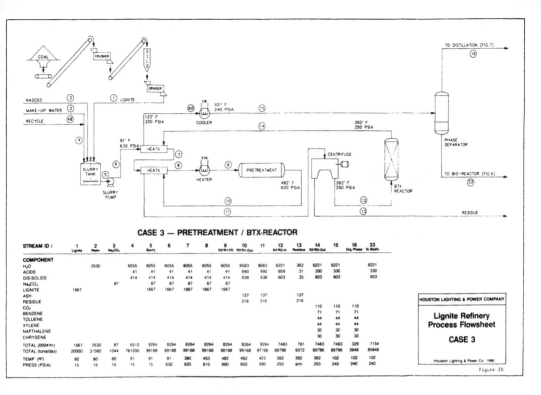

Fig. 19.20 Pretreatment process flowsheet.

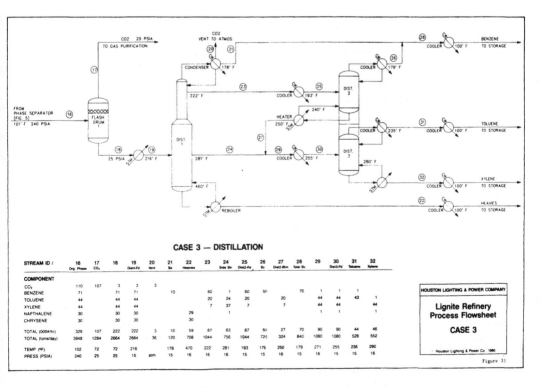

Fig. 19.21 Distillation process flowsheet.

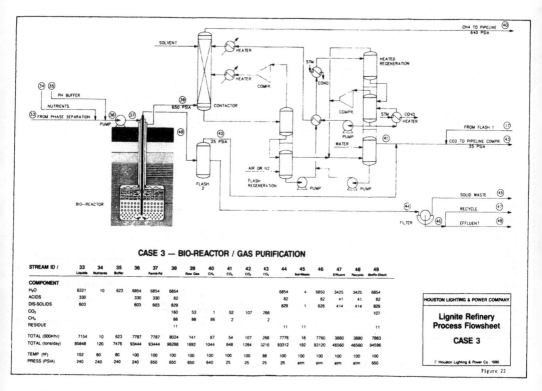

CASE 3 — BIO-REACTOR / GAS PURIFICATION

STREAM ID /	33 Liquids	34 Nutrients	35 Buffer	36	37 Ferm'd-Fd	38	39 Raw Gas	40 CH4	41 CO2	42 CO2	43 CO2	44	45 Sol-Waste	46	47 Effluent	48 Recycle	49 BioRe-Disch
COMPONENT																	
H2O	6221	10	623	6854	6854	6854						6854	4	6850	3425	3425	6854
ACIDS	330			330	330	82						82		82	41	41	82
DIS-SOLIDS	603			603	603	829						829	1	828	414	414	829
CO2							160	53	1	52	107	266					107
CH4							88	88	86	2	2						
RESIDUE						11						11	11				11
TOTAL (000#/hr)	7154	10	623	7787	7787	8024	141	87	54	107	268	7776	16	7760	3880	3880	7883
TOTAL (tons/day)	85848	120	7476	93444	93444	96288	1692	1044	648	1284	3216	93312	192	93120	46560	46560	94596
TEMP (°F)	102	80	80	100	100	100	100	100	100	100	88	100	100	100	100	100	100
PRESS (PSIA)	240	240	240	240	650	650	650	640	25	25	25	25	atm	atm	atm	atm	650

HOUSTON LIGHTING & POWER COMPANY

Lignite Refinery Process Flowsheet

CASE 3

© Houston Lighting & Power Co. 1986

Figure 22

Fig. 19.22 Bioreactor process flowsheet.

but the process conditions have changed. Conversion of volatile solids to acids in the first reactor has increased from 50% to 75%. There are 12.5% volatile solids in the lignite slurry, which corresponds to a 1:7 ratio of volatile solids to water. In addition, the sodium carbonate has also been reduced to half the amount previously needed in the slurry. Figure 19.22 shows the pertinent information on the alkaline hydrolysis and bioconversion phases of the process.

Plant Cost Estimate

When the water flow rate was reduced by 30%, the equipment was downsized, and this resulted in a lower capital cost for the plant. The on-site cost determined by the PRICE program was $206 million. When an off-site cost of $72 million is added, the total erected cost for the refinery becomes $278 million.

19.4.4 Case 4

Process Design

Figures 19.23 and 19.24 show the process flowsheets and corresponding stream information for class 4. This case, which is biogasification alone, has flowsheets similar to case 2, but the process conditions are identical to those stated in case 3. To reiterate, the conversion of volatile solids to acids has increased to 75%, the water flow rate has been reduced by 30%, and the amount of sodium carbonate needed to produce the slurry has been cut in half. The relative amounts of the three feedstocks used to form the slurry and

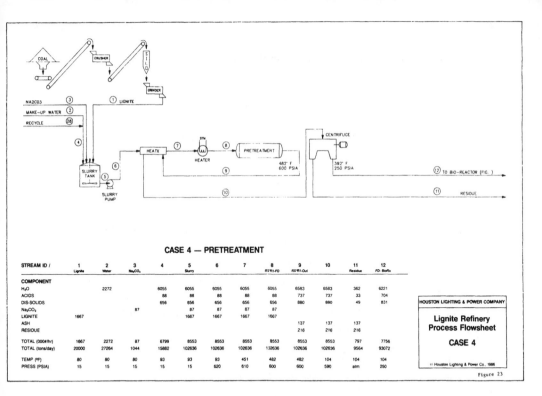

Fig. 19.23 Pretreatment process flowsheet.

CASE 4 — PRETREATMENT

STREAM ID /	1 Lignite	2 Water	3 Na₂CO₃	4	5 Slurry	6	7	8 RX'R1-FD	9 RX'R1-Out	10	11 Residue	12 FD-BioRx
COMPONENT												
H₂O		2272		6055	6055	6055	6055	6055	6583	6583	362	6221
ACIDS				88	88	88	88	88	737	737	33	704
DIS-SOLIDS				656	656	656	656	656	880	880	49	831
Na₂CO₃			87		87	87	87	87				
LIGNITE	1667				1667	1667	1667	1667				
ASH									137	137	137	
RESIDUE									216	216	216	
TOTAL (000#/hr)	1667	2272	87	6799	8553	8553	8553	8553	8553	8553	797	7756
TOTAL (tons/day)	20000	27264	1044	15882	102636	102636	102636	102636	102636	102636	9564	93072
TEMP (°F)	80	80	80	93	93	93	451	482	482	104	104	104
PRESS (PSIA)	15	15	15	15	15	620	610	600	600	590	atm	250

HOUSTON LIGHTING & POWER COMPANY

Lignite Refinery Process Flowsheet

CASE 4

(c) Houston Lighting & Power Co., 1986

Figure 23

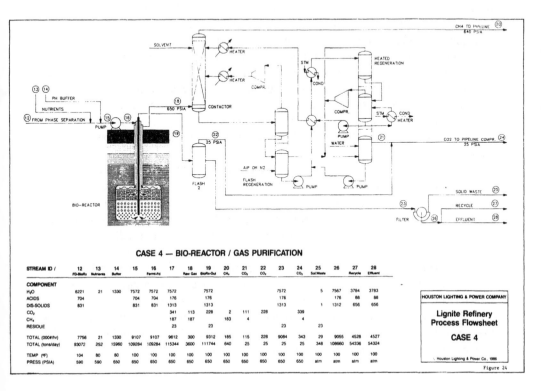

Fig. 19.24 Bioreactor process flowsheet.

CASE 4 — BIO-REACTOR / GAS PURIFICATION

STREAM ID /	12 FD-BioRx	13 Nutrients	14 Buffer	15 Ferm-Fd	16	17	18 Raw Gas	19 BioRx-Out	20 CH₄	21 CO₂	22 CO₂	23	24 CO₂	25 Sol.Waste	26	27 Recycle	28 Effluent
COMPONENT																	
H₂O	6221	21	1330	7572	7572	7572		7572				7572		5	7567	3784	3783
ACIDS	704			704	704	176		176				176			176	88	88
DIS-SOLIDS	831			831	831	1313		1313				1313		1	1312	656	656
CO₂						341	113	228	2	111	228		339				
CH₄						187	187		183	4			4				
RESIDUE								23				23		23			
TOTAL (000#/hr)	7756	21	1330	9107	9107	9612	300	9312	185	115	228	9084	343	29	9055	4528	4527
TOTAL (tons/day)	93072	252	15960	109284	109284	115344	3600	111744	640	25	25	25	348	108660	54336	54324	
TEMP (°F)	104	80	80	100	100	100	100	100	100	100	100	100	100	100	100	100	100
PRESS (PSIA)	590	590	650	650	650	650	650	650	650	650	650	650	650	atm	atm	atm	atm

HOUSTON LIGHTING & POWER COMPANY

Lignite Refinery Process Flowsheet

CASE 4

Houston Lighting & Power Co., 1986

Figure 24

the nutrient and buffer flow rates needed in the bioconversion can be seen in Figs. 19.25 and 19.26.

Plant Cost Estimate

The PRICE program computed an on-site cost of $222 million. (This is below the $230 million of on-site costs computed for case 2, due to the reduced water flow rate resulting from smaller equipment sizes.) When an off-site cost of $78 million is added, the total erected cost of this plant becomes $300 million.

19.4.5 Comparison of Cases

Raw material and product flow rates resulting from the Chem Cad programs are summarized in Fig. 19.27. All four cases are based on a lignite fuel rate of 20,000 tons/day or 1.67 million lb/h. The sodium carbonate was reduced by half when the conversion of volatile solids to acids was increased from 50% to 75%. The amount of nutrients and buffers necessary for bioconversion changes proportionally with the flow rate of acids to the underground reactor. In addition, the amount of makeup water needed for the process is greatly reduced in the 75% conversion cases.

Carbon dioxide and methane are the only products produced in the biogasification process. In the lignite refinery, aromatic chemicals are produced in addition to these two gases. The product flow rates increase by 50% in the last two cases because of the increased conversion of volatile solids to acids.

After these flow rates were obtained, an economic analysis of each case was performed using the values listed in Fig. 19.28. This analysis resulted in the tabulation of the yearly operating costs and revenues found in Fig. 19.29. Since a marketing survey was not conducted, it is assumed that all products produced in each case can be sold at the prevailing market price. A summary of the economic analysis, which includes total erected cost, total operating cost, and total revenues, can be found in Fig. 19.30.

It should be noted that total revenues exceed total operating costs in only one case: the lignite refinery at 75% conversion. In this case, a positive operating cash flow of $32 million per year is obtained. The plant was depreciated over 10 years using a salvage value of 5% of the total erected cost. A present-value calculation was used to determine the yearly cash flow necessary to give a 15% rate of return, which in case 3 amounted to $81 million.

One method of increasing the cash flow to the $81 million necessary for a 15% ROI is to increase the product revenues by 13.4%, to $414 million. This can be accomplished by raising the product prices or by increasing conversion, so a larger amount of products can be produced from the same amount of lignite feed. Conversely, a 14.7% reduction in operating costs to $284 million would also lead to a cash flow fo $81 million per year.

All of the comparisons discussed so far have been based on cases utilizing a rock cavern fermenter. In addition to these data, total erected costs have been recalculated for the four cases if a salt cavern is used in place of a rock cavern. The results, which are shown at the bottom of Fig. 19.30 have been obtained by changing the cavern cost in the PRICE program. The capital cost needed in each case is larger for the salt cavern than for the rock cavern. Therefore, if one has a choice of plant locations, it would be beneficial to construct the plant utilizing a rock cavern fermenter.

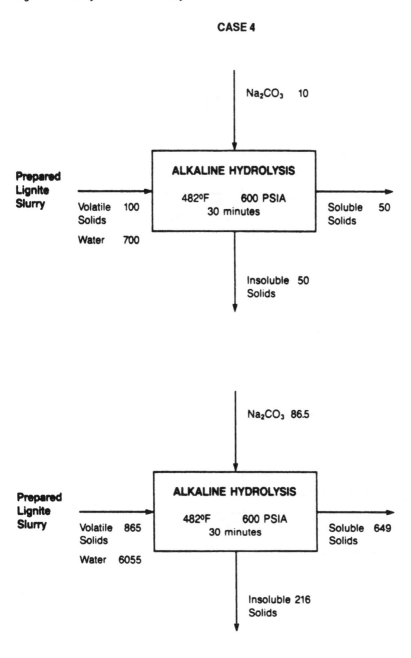

(Flows shown in thousand pounds per hour.)

Fig. 19.25 Alkaline hydrolysis flow ratios.

CASE 4

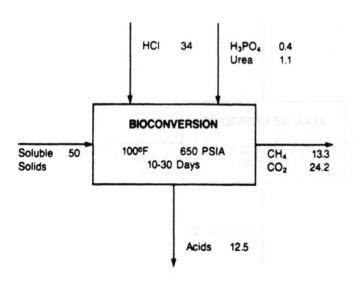

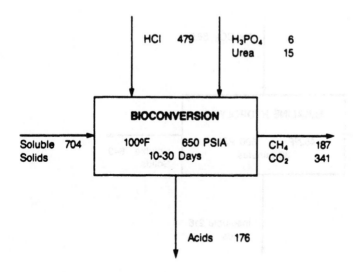

(Flows shown in thousand pounds per hour.)

Fig. 19.26 Bioconversion flow ratios.

	Case 1 Refinery (50%)	Case 2 Bio (50%)	Case 3 Refinery (75%)	Case 4 Bio (75%)
RAW MATERIALS				
Lignite	1667	1667	1667	1667
Na_2CO_3	173	173	86	86
Nutrients	7	14	10	21
Buffer	416	886	623	1330
Make-up Water	4106	3868	2630	2272
PRODUCTS				
Carbon Dioxide	176	226	226	339
Methane	58	123	86	183
Benzene	44	—	66	—
Toluene	29	—	44	—
Xylene	30	—	45	—
Heavies	38	—	58	—

(Flows shown in thousand pounds per hour.)

Fig. 19.27   Raw materials and product flowrates.

COST OF RAW MATERIALS
Lignite	—	$ 12/ton
Na_2CO_3	—	$ 69/ton
HCl	—	$ 57/ton
Urea	—	$200/ton
H_3PO_4	—	$ 3.1/ton

REVENUES FROM PRODUCTS
Methane	—	$3.00/million BTU
Carbon Dioxide	—	$ 25/ton
Benzene	—	$1.48/gal
Toluene	—	$1.37/gal
Xylene	—	$1.40/gal
Heavies	—	$ .67/gal

COST OF UTILITIES
Cooling Water	—	$ .03/1000 gal
Process Water	—	$ .30/100 gal
Steam	—	$2.50/million BTU
Electricity	—	$ .05/KWH

Fig. 19.28   Values of economic analysis.

CHANGING RAW MATERIAL COSTS

Raw Material Cost	Revenues Needed For 15% ROI	Percentage Increase In Current Revenues
currently $249	$414	13.4%
up 5% to $261	$427	17.0%
up 10% to $274	$439	20.2%
up 20% to $299	$465	27.4%
up 30% to $324	$490	34.2%

CHANGING LIGNITE COSTS

Lignite cost	Revenues Needed 15% ROI	Percentage Increase In Current Revenues
$12/ton	$414	13.4%
$15/ton	$433	18.6%
$18/ton	$453	24.1%
$21/ton	$473	29.6%
$24/ton	$492	34.8%

* Revenues shown in $million/yr

** Current Revenue = $365 million/yr

Fig. 19.29   Changing raw material and lignite costs.

	Case 1 Refinery (50%)	Case 2 Bio (50%)	Case 3 Refinery (75%)	Case 4 Bio (75%)
Onsites (rock)	$ 262	$ 230	$ 206	$ 222
Offsites	92	80	72	78
Total Erected Costs	354	310	278	300
Raw Materials	224	334	249	414
Utilities	50	24	56	19
Fixed Costs	35	31	28	30
Total Operating Costs	309	389	333	463
Total Revenues	243	91	365	136
Current Yearly Cash Flow	− 66	− 298	32	− 327
Cash Flow Needed for 15% ROI	104	91	81	88
Onsites (salt)	309	281	243	288
Offsites	108	98	85	101
Total Erected Cost	417	379	328	389

(Figures shown in $ millions per year.)

CONCLUSIONS

- Current state of technology does not meet hurdle return rates.
- Need to focus future efforts on increasing conversion while improving efficiency.
- Conversion of acids to aromatics favored over biogasification alone.

Fig. 19.30   Summary of economic analysis.

19.4.6   Extended Analysis of Case 3

As stated previously, the lignite refinery with 75% conversion of volatile solids to acids is the only case with a positive yearly cash flow. Since this case represents the most feasible design of the four cases analyzed, a closer look is warranted. A sensitivity analysis was performed to determine what effect changes in raw material costs and lignite costs in particular would have on product revenues. Figure 19.29 shows the result of this analysis.

Total raw material costs are currently $249 million. Product revenues of $414 million are necessary to achieve a 15% ROI. This corresponds to a 13.4% increase over the current revenues of $365 million. These raw material costs were then increased in five stages to 130% of the original value, while all other costs were held constant. The revenues needed for a 15% return under these conditions were then calculated.

The analysis is then repeated to determine the effect of changing lignite prices on product revenues. Lignite costs have been increased from the current price of $12 per ton to $24 per ton, while all other costs were held constant. The results of this analysis can be seen at the bottom of Fig. 19.29.

As stated previously, no credit or disposal cost has been assumed for the residue generated in the process. In case 3, 781,402 lb/h of residue is obtained from stream 11 by using a centrifuge. An additional 15,837 lb/h of solid waste is generated from stream 44 by using a filter. The streams are then combined, resulting in a total residue flow-rate of 797,239 lb/h.

Although an exact composition for the stream was not available at this stage of the analysis, it is estimated that the stream is composed of 46% water, 22% ash, and 32% volatile solids. If a heating value of 3000 Btu/lb is assumed, steam could be generated at a rate of 2.4 million lb/hr by burning this residue. Pyropower has given a capital cost estimate of $45/lb per hour of steam, which results in a capital investment of $107.6 million. Revenues generated from this steam production amount fo $47 million per year.

If this steam generation is incorporated into the economic analysis, the total erected cost rises to $386 million and the total revenues increase to $412 million per year. The revenues necessary for a 15% return on investment are $457 million. This corresponds to an increase of 11% over the current revenues of $365 million.

If a sample of the residue could be obtained from Dynatech, and a complete analysis run on the sample, a more accurate economic analysis could be obtained. If the heating value of the stream is determined to vary substantially from the estimated 3000 Btu/lb, the economics of the process would be altered dramatically. This analysis would also determine if a support fuel is needed for the process; lignite could be used as the support fuel.

19.5 CONCLUSIONS AND RECOMMENDATIONS

The results of this chapter clearly show that conversion of acids to aromatics in the lignite refinery process is favored over biogasification alone. Since the current state of technology and present natural gas prices do not meet hurdle return rates, there is a need to focus future efforts while improving the efficiency of the overall process.

The data presented in this report were based on the most accurate information available at the present time. Further research should be conducted by Dynatech to gain a better estimation of the products produced in the three reactors and the relative proportions of these products. In addition, a market survey should be conducted to confirm that all the products produced in the system can be sold.

20

Methanation of Synthesis Gas Using Biological Processes

S. BARIK, J. L. VEGA, E. R. JOHNSON, E. C. CLAUSEN, and J. L. GADDY
University of Arkansas, Fayetteville, Arkansas

20.1 INTRODUCTION

The United States has massive reserves of coal. In fact, coal is by far the largest fossil energy source in the United States, equivalent to about 750 billion barrels of oil, or a 300-year supply at the current rate of consumption [32]. Subbituminous and lignite coals account for about 60% of this reserve, but contain less than 40% fixed carbon and have an energy content of 5500 to 8000 Btu/lb [18]. The easiest method of utilizing coal is by direct combustion with air to recover the energy content. However, the sulfur and ash content of much of this coal is so high that environmental regulations prevent its burning, or the energy content is so low that the transportation of the coal to power plants is economically prohibitive. Nevertheless, coal will be the major source of both fuel and chemicals for the future and these problems must be resolved.

Coal may be converted to a more convenient energy source and a valuable chemical feedstock by gasification and removing undesirable components. In most gasification processes, the coal is hydrogasified by adding steam and energy at 300 to 1000 psig and at temperatures as high as 2700°F. A gas consisting of more than 50% hydrogen and carbon monoxide is usually produced, having a heating value of about 450 Btu/SCF. A typical range of synthesis gas compositions from several gasification processes is shown in Table 20.1. In addition to CO and H_2, small amounts of CO_2, CH_4, H_2S, and COS are also produced.

To upgrade synthesis gas, the low-Btu gas typically undergoes a catalytic shift reaction where carbon monoxide and water are converted to carbon dioxide and hydrogen. Following purification to remove excess carbon dioxide and hydrogen sulfide, the hydrogen and carbon dioxide are catalytically reacted to methane and water. The water is then removed to give a gas that is 95 to 98% methane, with an energy content of 980 to 1035 Btu/SCF.

Table 20.1 Typical Composition of Pyrolysis Gas[a]

Component	Percent of total
Hydrogen	25–35
Carbon monoxide	43–65
Carbon dioxide	1–20
Methane	0–7
Sulfur as H_2S or COS	0–1

[a] Average of data taken from Westfield Lurgi gasifier, Shell Kopper gasifier, and Cool Water Gasification Project (*Energy Progress*, 1982).

The temperatures required in most methanation processes are 570 to 650°C at pressures of 40 to 250 psig [24]. The Exxon process employs gasification and methanation in a single step at 1290°F [13,22]. Most catalysts used are nickel based; however, the Exxon process utilizes a potassium-based catalyst [13,22]. Poisons for these catalysts include chlorine and sulfur [21,29]. Free carbon, nitrogen, and oxygen are also potential nickel-based catalyst poisons [27]. Liquid fuels such as alcohols and petrochemicals also may be produced from coal synthesis gas. Examples of these processes include the Mobil MTG Process and the Advanced Coal Liquefaction processes [26]. Other processes use the coal directly to produce a liquid fuel, which can be used to produce such derivatives as ethylene, propylene, and BTX [30]. These processes require elevated temperatures and pressures, often as high as 1400°F and 1500 psig [26]. Because high temperatures and pressures are required in the catalytic gasification and liquefaction process, losses in thermal efficiency and high energy costs occur. Thus, processes employing lower temperatures and pressures for even a portion of the overall process scheme would be advantageous.

20.1.1 Microbial Coal Conversion

Microorganisms may be used to convert coal to chemicals either directly or indirectly by their action on synthesis gas. Microbial processes offer certain advantages over chemical conversions. Microorganisms exist and carry out conversions at ambient temperatures and pressures, which should result in substantial energy and equipment savings. Also, yields from microbial conversions are often as high as 95%, since the microorganism utilizes only a small fraction of the substrate for energy and growth. Under proper conditions, microbial conversions are quite specific, generally converting a substrate into a single product, with perhaps a few by-products. These advantages are offset by slower reaction rates and special reactor considerations, such as sterility, nutrient provision, and so on.

Direct Coal Conversion

Direct microbial conversion involves the selection of a microorganism to produce liquid and gaseous fuels by direct biological action on coal, perhaps in situ. Although the prospects for success are limited, there is tremendous potential for this type of coal conversion process. Indicative of the problems associated with direct microbial action on coal is the apparent toxicity of liquefied coal products and wastewater streams from the coal conversion processes [12,17,19,25,28]. Also, it has been extremely difficult to get active flora to flourish on the ground cover composed of coal mine spoil or tailings [9,16].

Significant attention has been devoted to the microbial conversion of the inorganic (pyrite) fraction of sulfur in coal. Microorganisms such as *Thiobacillus ferrooxidans, T. thiooxidans,* and *Sulfobus acidocaldarius* have been found to remove as much as 96% of the pyritic sulfur in a 5-day period [1,23,3]. Many species of fungi and a few bacterial strains have been found to degrade lignin. These include *Coriolus versicolor, Sporotrichum pulverulentium, Phanerochaete chrysosporium, Aspergillus fumigatus, Metulius tremellosus,* and some species of *Acetiomycetes* and *Eubacteria* [2,7,8,11].

Some lignolytic fungi have been shown to be capable of degrading lignite coal. *Polyporous versicolor* and *Poria monticola* have been grown on lignite agar medium. Basic salts also are required and a liquid exudate containing various chemical products was obtained [6]. Also, research at the University of Mississippi has isolated fungi from a natural lignite outcrop. These isolates have been shown to utilize lignite as a sole source of organic carbon and minerals. Some of these fungi have been identified as *Mucor* sp., *Penicillium* sp., *Paecilomyces* sp., and *Candida* sp. The isolation of these organisms for direct conversion of lignite is quite promising. However, as expected, the rates of conversion are extremely slow and the development of a commercial process will be difficult.

Indirect Coal Conversion

A more promising biological approach, perhaps, is the conversion of raw synthesis gas by microorganisms capable of producing alcohols, acids, aldehydes, and so on, from CO, CO_2, H_2, and H_2O. Therefore, a two-step process is required: conventional coal gasification, followed by biological liquefaction. Some of these biological reactions are known to occur, especially those that produce organic acids. Others that produce alcohols from acids or syngases appear promising and possible, but have not been examined. The known reactions use anaerobic organisms and give reasonable rates of reaction. These microorganisms usually produce a single product, so that purification systems are simplified. Ordinary temperatures and pressures are used so that substantial capital and energy savings are possible. A cheaper source of synthesis gas can be used, since the problem of sulfur poisoning of the catalyst can be eliminated, because sulfur tolerance of microorganisms can often be developed or sulfur can be removed biologically, if necessary. Thus indirect biological conversion offers a technically feasible and novel process for producing liquid hydrocarbons and/or alcohol fuels from crude synthesis gases. This approach may prove to be economically superior to chemical conversion processes.

20.1.2 Microbiology

The primary reactions in the biological conversion of synthesis gas to methane are the formation of methane precursors (acetate or H_2 and CO_2) and the biomethanation reactions. The organisms *Rhodopseudomonas gelatinosa* and *Rhodospirillum rubrum* utilize CO to produce CO_2 and H_2 by the water gas shift reaction:

$$CO + H_2O \rightarrow CO_2 + H_2 \tag{1}$$

The organisms *Peptostreptococcus productus, Acetobacterium woodii,* and *Eubacterium limosum* produce acetate by the reaction

$$4CO + 2H_2O \rightarrow CH_3COOH + 2CO_2 \tag{2}$$

Methanosarcina sp. utilizes the CO_2 and H_2 to produce CH_4 by the reaction

$$CO_2 + 4H_2 \rightarrow CH_4 + 2H_2O \tag{3}$$

Methanothrix soehngenii utilizes acetate to produce CH_4 by the reaction

$$CH_3COOH \rightarrow CO_2 + CH_4 \qquad\qquad\qquad (4)$$

It may also be possible to convert CO directly into methane. *Methanobacterium thermoautotrophicum* [10] has been reported to produce methane according to

$$4CO + 2H_2O \rightarrow CH_4 + 3CO_2 \qquad\qquad\qquad (5)$$

The growth of this organism was reported to be very slow and was inhibited by high concentrations of CO. It has also been reported that other methanogenic bacteria can convert CO to CH_4 according to the equation [14,15,33,34]

$$CO + 3H_2 \rightarrow CH_4 + H_2O \qquad\qquad\qquad (6)$$

All of the organisms above can be isolated from natural sources. Sewage sludge from anaerobic treatment plants and animal rumen are the sources from which most anaerobic bacteria have been isolated. Sewage sludge usually contains a consortium of various methanogenic bacteria. In anaerobic digestion processes, 80% of the methane is produced from acetate by equations (2) and (4). Some of the organisms above are maintained in pure culture collections. Therefore, coal gas conversions might be carried out with pure cultures or with mixed cultures, isolated from natural sources, and acclimated to the toxic environment of carbon monoxide.

All of these organisms function at optimal temperatures of 30 to 37°C and ordinary pressures, although higher pressures have been used successfully [5]. Processes to produce methane from coal gas using the microorganisms above would be very simple. The synthesis gas is introduced into a reactor containing a culture of the organisms where the CO, CO_2, and H_2 are converted into CH_4, followed by separation of the excess CO_2, to produce pipeline-quality gas. The culture might consist of a mixture of microorganisms to carry out all the reactions above, a pure culture of a single organism to carry out reaction (4) or (5), or a co-culture of two bacteria to carry out the series of reactions (1) and (3) or (2) and (4).

Since biological reactions are much slower than chemical reactions, retention time and reactor volume are important considerations in developing an efficient and economical process. Cultures that result in reaction kinetics that give high yields and complete conversion are also important to conserve raw materials and avoid recycle.

20.1.3 Purpose

The purpose of this chapter is to demonstrate the technical feasibility of biological synthesis gas conversion to methane in both mixed and pure cultures. Mixed culture, co-culture, and series experiments are studied to determine the most efficient culture to produce methane from synthesis gas. Various reactor schemes, including fixed-film and stirred-tank reaction vessels, are investigated to define the system that maximizes cellular density and reaction kinetics.

20.2 EXPERIMENTAL RESULTS

20.2.1 Mixed Culture Studies

Studies have been conducted to determine the feasibility of developing an acclimated mixed culture to produce methane from synthesis gas. A sewage sludge inoculum was utilized. Mixed-flow reactors were used. Atmospheric pressure studies were conducted in New Brunswick Model C30 chemostats with agitation rates up to 1000 rpm. High-pressure studies were conducted in Parr 822 HC-17 magnetic drive agitated reactors with pressure limits to 600 psi. Synthetic coal synthesis gas (45% CO, 30% H_2, 15% CO_2, and 10% CH_4) was introduced continuously through spargers in the mixing vortex. Liquid volumes of 600 mL to 1 L were maintained with periodic supply of nutrients for micro-organism growth. The reactors were contained in an incubated room at 37°C. Procedures to ensure strict anaerobic conditions were used.

Development of the culture was achieved by acclimation to synthesis gas first in batch and then in continuous flow. Culture acclimation was a slow procedure requiring approximately 3 months of gradual gas introduction into the sludge culture. Since CO is toxic to many organisms, the liquid culture was buffered to pH 7.3 to prevent pH increase during the death of noncontributing organisms in the culture. As the culture began to produce methane, the flow rate of synthesis gas was increased, and a viable cell mass began to emerge. As the culture continued to acclimate, data collection began.

At steady state, the gas composition and flow rates were monitored at the entrance and exit of the reactor. The pH, ammonia nitrogen, cell density, and ATP concentrations were also monitored.

The results of single-stage biomethanation studies carried out at 500 rpm and variable gas retention times (based on liquid volume) are shown in Table 20.2. These data were collected after approximately 9 months of acclimation, and show the carbon monoxide and hydrogen conversions as a function of retention time in the chemostat. The early culture required an 8- to 10-h retention time to yield 96 to 97% conversions of carbon monoxide and hydrogen. At a 2-h retention time, the conversion was only about 20%.

The steady-state results of Table 20.3 were obtained after a year of continuous operation and acclimation of the mixed culture. These results, shown at a gas retention time of 1.94 h and variable agitation rate, indicate that a 90% conversion of CO and H_2 can occur at a 2-h retention time and an agitation rate of 500 rpm. It should be noted that a 2-h retention time is quite good for a biological reactor; ethanol fermentation, for example, requires 30 to 40 h. Nevertheless, a 2-h retention time translates into very large

Table 20.2 Single-Stage Conversion Experiments at 500 rpm (Early Studies)

Retention time (h)	CO conversion (%)	H_2 conversion (%)
16.7	93.9	93.7
11.1	94.5	95.4
8.3	96.7	87.9
6.7	87.5	90.2
5.6	88.0	87.2
4.8	84.1	84.7
4.2	84.9	81.9
3.7	45.7	8.5
2.8	34.1	14.1
1.9	23.2	17.7

Table 20.3 Mixed-Culture Single-Stage Conversion of CO and H_2

| Agitation (rpm) | Gas retention time (h) | OD at 580 nm | Conversion (%) | |
			CO	H_2
200	1.94	2.25	0	36.6
300	1.94	2.35	40.48	62.28
400	1.94	2.30	85.16	85.78
500	1.94	2.55	89.80	90.52
600	1.94	2.15	96.28	93.03

reactors for coal gas conversion. While these reactors are simple and inexpensive, future efforts should concentrate on improving reaction rates to reduce retention time.

The major reason for the improvement in the culture with time was the enrichment of the culture. Organisms necessary for CO and H_2 conversion became more dominant in the mixed culture, and thus the population of the essential CO and H_2 utilizing organisms increased. Therefore, with time the mixed culture evvolved toward a pure culture of organisms capable of converting CO and H_2 to methane. No efforts have been made at this time to identify the organisms present in this culture.

As noted in Table 20.3, the conversions are improved with agitation rate, requiring 600 rpm for nearly complete conversion. The product gas composition under these conditions provides a closed mass balance, with a concentration of almost 40% methane and 60% CO_2. At low agitation rates, the biological reactions are mass transfer limited. Solubilities of the reacting gases are very low and transport of the reactants from the gas into the liquid phase and to the solid phase is quite slow. Therefore, high agitation rates are necessary to enhance the conversion.

Under these conditions, high pressure would be expected to speed the transfer and overall reaction rate. Experiments have been conducted at higher pressures to determine the benefits. The results of these experiments at 75 psig show that total conversion can be achieved in the same retention time as at atmospheric pressure. Therefore, reaction rates are proportional to pressure, at least to the limits tested (75 psig). Experiments up to 150 psig are presently being successfully initiated. Operation at 150 psig has the direct effect of reducing gas retention time from 2 h to the equivalent of 12 min, compared to atmospheric pressure operation.

20.2.2 Pure Culture Studies

Faster reaction rates and shorter retention times are sought for these reactions. Methods to improve reaction rates in biological reactions include increasing the cell density and utilizing microorganisms that have faster reaction kinetics in metabolizing substrate to products. Therefore, other microorganism systems are being studied to further improve the reaction kinetics.

Methane Precursor Screening Studies

Pure cultures of CO converting bacteria were screened for their ability to convert CO to methane intermediates. CO converters were chosen since the natural sources, such as sewage sludge, are deficient in these bacteria, thus requiring long acclimation times. Methane bacteria are well defined and were not included in these screening studies. Conversion to both acetate and CO_2 and H_2 were considered since pure culture methane bacteria are able to use both products as substrates. The reactions for converting CO to methane intermediates were given in equations (1) and (2).

The anaerobic techniques for the preparation and use of media for these pure cultures were essentially those of Hungate [20] as modified by Bryant [4] and Balch and Wolfe [3]. The basal salts media used to grow the bacterial cultures are listed in Tables 20.4 to 20.7. For the preparation of the media, basal salts, yeast extract, and other constituents in appropriate amounts were taken in a round-bottomed flask and were boiled for 2 to 3 min under a N_2/CO_2 (80:20) atmosphere. $NaHCO_3$ (0.35 g per 100 mL) was added when the media was cooled to about 55°C. Media, in appropriate amounts, was then pipetted anaerobically into serum stoppered tubes (18 × 150 mm) or serum bottles (ca. 120 or 160 mL) and crimp sealed. The tubes and bottles were then autoclaved at 121°C for 20 min.

Sodium sulfide (2.5%) was used to reduce the media prior to use. The media was always inoculated (5 to 10% inoculum) with a log phase culture. A CO or coal gas mixture was added using either a sterile syringe fitted with a one-way valve or a gas manifold in connection with a vacuum pump. The final pressure for the tubes or the bottles was 2 atm. The gas manifold was used for adjusting pressure (15 to 50 psi) in the bottles inoculated with the organisms during pressure studies. Different percentages of CO (20 to 90%) or coal gas were also introduced using this manifold.

Control tubes without CO or coal gas were run simultaneously with each experiment. The cultures of *P. productus* were incubated (flat) in the dark at 37°C, and *A. woodii* was incubated at 30°C. In some experiments, *P. productus* tubes were agitated at

Table 20.4 Composition of Media for the Growth of *Rhodopseudomonas gelatinosa* and *Rhodospirillum rubrum*, Strain I

	Amount per 100 mL
Mineral 1[a]	50 mL
EDTA (0.425%)	0.1 mL
Pfenning's trace metal[b]	0.1 mL
B vitamins[c]	0.5 mL
Trypticase	0.1 g
Resazurin (0.1%)	0.1 mL
Sodium bicarbonate	0.35 g
Sodium sulfide (2.5%)	2.0 mL
Gas phase, N_2:CO_2	60%:40%

[a]Contained per liter, K_2HPO_4, 9.0 g; KH_2PO_4, 9.0 g; $(NH_4)_2SO_4$, 9.0 g; NaCl, 18.0 g; $MgSO_4 \cdot 7H_2O$, 3.6 g; $CaCl_2 \cdot 2H_2O$, 2.4 g; and NH_4Cl, 0.2 g.
[b]Contained per liter: $ZnSO_4 \cdot 7H_2O$, 0.1 g; $MnCl_2 \cdot 4H_2O$, 0.03 g; H_3BO_3, 0.3 g; $CoCl_2 \cdot 6H_2O$, 0.2 g; $CaCl_2 \cdot H_2O$, 0.01 g; $NiCl_2 \cdot 6H_2O$, 0.02 g; $Na_2MoO_4 \cdot 2H_2O$, 0.03 g, and $FeCl_2 \cdot 4H_2O$, 1.5 g.
[c]Contained per 100 mL: B_6 (pyridoxine) HCl, 1.9 mg; B_1 (thiamine) HCl, 0.95 mg; nicotinic acid, 0.95 mg; p-aminobenzoic acid, 0195 mg; biotin, 0.38 mg; folic acid, 0.38 mg; B_{12} crystalline, 0.16 mg; B_2 (riboflavin), 0.95 mg; lipoic acid, 0.95 mg; and pentothenic acid, 0.95 mg.

Table 20.5 Composition of Media for the Growth of *Acetobacterium woodii*[a]

	Amount per liter
Yeast extract	1.0 g
NH_4Cl	1.0 g
$MgSO_4 \cdot 7H_2O$	0.1 g
KH_2PO_4	0.4 g
K_2HPO_4	0.4 g
Wolfe's vitamin[b]	10.0 mL
Wolfe's mineral[c]	10.0 mL
Fructose (25 mL of 20%)[d]	5.0 g
Resazurin (0.1%)	0.1 mL
Sodium sulfide (2.5%)	20.0 mL
$NaHCO_3$	3.0 g
Gas phase, $N_2:CO_2$	80%:20%

[a]Adjust pH to 6.7.
[b]Contained per liter: folic acid, 2.0 mg; pyriodoxine hydrochloride, 10.0 mg; riboflavin, 5.0 mg; biotin, 2.0 mg; thiamine, 5.0 mg; nicotinic acid, 5.0 mg; pentothenic acid, 5.0 mg; vitamin B_{12}, 0.1 mg; PABA, 5.0 mg; and thioctic acid, 5.0 mg.
[c]Contained per liter, nitriloacetic acid, 1.5 g; $MgSO_4$, 3.0 g; $MnSO_4$, 0.5 g; NaCl, 1.0 g; $FeSO_4$, 0.1 g; $CaCl_2$, 0.1 g; $CoCl_2$, 0.1 g; $ZnSO_4$ 0.1 g; $CuSO_4$, 0.01 g; $Alk(SO_4)_2$, 0.01 g; H_3BO_3, 0.01 g; and Na_2MoO_4, 0.01 g.
[d]Filter sterilized and added to sterilized media.

Table 20.6 Composition of Media for the Growth of *Rhodospirillum rubrum*, Strain II[a]

	Amount per liter
Malic acid	2.5 g
Yeast extract	1.0 g
$(NH_4)_2SO_4$	1.25 g
$MgSO_4 \cdot 7H_2O$	0.2 g
$CaCl_2 \cdot 2H_2O$	0.07 g
EDTA (0.425%)	1.0 mL
KH_2PO_4	0.6 g
K_2HPO_4	0.9 g
Pfennig's mineral[b]	1.0 mL
B vitamins[c]	7.5 mL
Resazurin (0.1%)	1.0 mL
Sodium sulfide (2.5%)	20.0 mL
Gas phase, $N_2:CO_2$	80%:20%

[a]Adjust pH to 7.0.
[b]Contained per liter: K_2HPO_4, 9.0 g; KH_2PO_4, 9.0 g; $(NH_4)_2SO_4$, 9.0 g; NaCl, 18.0 g; $MgSO_4 \cdot 7H_2O$, 3.6 g; $CaCl_2 \cdot 2H_2O$, 2.4 g; and NH_4Cl, 10.0 g.
[c]Contained per 100 ml: B_6 (pyriodoxine) HCl, 1.9 mg; B_1 (thiamine) HCl, 0.95 mg; nicotinic acid, 0.95 mg; *p*-aminobenzoic acid, 0.95 mg; biotin, 0.38 mg; folic acid, 0.38; B_{12} crystalline, 0.16 mg; B_2 (riboflavin), 0.95 g; lipoic acid, 0.95 mg; and pentothenic acid, 0.95 mg.

Table 20.7 Composition of Media for the Growth of *Peptostreptococcus productus*

	Amount per 100 mL
Pfenning's mineral[a]	5.0 mL
Pfenning's trace metal[b]	0.1 mL
B vitamins[c]	0.5 mL
Resazurin (0.1%)	0.1 mL
Yeast extract	0.2 g
Sodium bicarbonate	0.35 g
Sodium sulfide (2.5%)	2.0 mL
Gas phase, $N_2:CO_2$	80%:20%

[a]Contained per liter: KH_2PO_4, 10.0 g; $MgCl_2 \cdot 6H_2O$, 6.6 g; NaCl, 8.0 g; NH_4Cl, 8.0 g; and $CaCl_2 \cdot 2H_2O$, 1.0 g.
[b]Contained per liter: $ZnSO_4 \cdot 7H_2O$, 1.0 g; $MnCl_2 \cdot 4H_2O$, 0.03 g; H_3BO_3, 0.3 g; $CoCl_2 \cdot 6H_2O$, 0.1 g; $CaCl_2 \cdot H_2O$, 0.01 g; $NiCl_2 \cdot 6H_2O$, 0.02 g; $Na_2MoO_4 \cdot 2H_2O$, 0.03 g; and $FeCl_2 \cdot 4H_2O$, 1.5 g.
[c]Contained per liter: biotin, 20 mg; folic acid, 20 mg; pyridoxal hydrochloride, 10 mg; lipoic acid, 60 mg; riboflavin, 50 mg; thiamine hydrochloride, 50 mg; calcium-D-pentothenate, 50 mg; cyanocobalamin, 50 mg; PABA, 50 mg; nicotonic acid, 50 mg; and Na_2SeO_3, 10 mg.

200 rpm using a temperature controlled New Brunswick incubator shaker (Model G-25). All incubations of *R. rubrum* (strains I and II) are *R. gelatinosa* were performed under tungsten light at 30°C.

Several pure culture bacteria that convert CO into CO_2/H_2 or acetate were studied, comparing viability, growth rates, and conversion rates. The results of this screening study are shown in Table 20.8. On the basis of prolonged screening and optimization studies, two anaerobic bacteria were selected as having significant promise. *Rhodospirillum rubrum*, strain II, utilizes CO according to equation (1) to produce H_2. *Peptostreptococcus productus* utilizes CO according to equation (2) to produce acetate. Both of these organisms have been found to be capable of utilizing up to 90% CO in producing methane intermediates. Neither H_2S nor COS, in concentrations up to 2%, affect the rate of production, although bacterial growth is inhibited by higher concentrations of COS. These organisms are available for use in combination with methane bacteria in series reactors and as a co-culture in a single reactor to attempt to further reduce reaction times.

Reactor Experiments Involving Pure Culture Organisms

Various reaction schemes were employed, including series operation and co-culturing, for the pure cultures of CO-utilizing bacteria in combination with methane-producing bacteria. Both stirred-tank and fixed-film reactors were utilized. These feasibility studies are continuing and preliminary results are available.

Series Operation. One possible method to improve on the results obtained with the mixed culture study is to employ a series of two pure culture reactors. In the first reactor, a pure culture is used to convert CO to a methane precursor. The second reactor contains a pure culture of methane bacteria and is used to convert the precursor to methane. This arrangement could provide an improvement over single-stage mixed culture conversion, since only the organisms needed for the conversions are present, and noncon-

Table 20.8 Survey of Shift Conversion Organisms

Organism	Maximum optical density[a]	Conditions of growth	Doubling time (h)	Time required to utilize CO (days)	Maximum CO tolerance (%)	Major products
Peptostreptococcus productus	0.83 (50)[b]	37°C, dark	4–6	4 (50)[b]	90	Acetate + CO_2
Rhodopseudomonas gelantinosa	0.16 (20)	20°C, dark	30	12 (20)	30	$H_2 + CO_2$
Rhodospirillum rubrum (I)	0.46 (50)	30°C, tungsten light	35	11 (50)	60	$H_2 + CO_2$
Rhodospirillum rubrum (II)	1.0 (50)	30°C, tungsten light	24	10 (50)	60	$H_2 + CO_2$
Acetobacterium woodii	0.50 (25)	37°C, dark	28	7 (25)	Tested up to 40	Acetate + CO_2

[a] Initial O.D. subtracted.
[b] Growth with %CO and %CO utilization.

tributing organisms are excluded from the culture. Also, in a series operation, each stage can operate at the optimum requirements of pH, nutrients concentration, and so on, for that microorganism.

Two reactor types, continuously stirred tank reactor (CSTR) and fixed film, have been used in the conversion of CO to acetate in the first reactor of the series operation. First, the anaerobic bacteria *Peptostreptococcus productus* was fed coal bas at various gas retention times in a chemostat reactor, as described previously. A summary of the results of these experiments is shown in Table 20.9. As noted, 90% conversion of CO can be obtained at a gas retention time of 0.73 h, more than double the maximum gas flow rate in the mixed culture studies. This translates into a reactor one-half the size of the mixed culture reactor.

Peptostreptococcus productus was also employed in a fixed-film reactor in an effort to reduce the operating costs for agitation. This reactor is characterized by the attachment of bacteria to an inert support to yield a very high cell density. This type of reactor has been shown to yield productivities (reaction rates) as high as 20 times the values of batch and stirred-tank vessels when producing ethanol by fermentation. Also, because of the high cell densities, reactor size is significantly reduced for a given retention time and conversion. Since a plug flow column reactor is employed, no agitation is required. Primarily results indicate that a 50-min retention time can be utilized to give high conversions of CO to acetate. The performance of the reactor is a function of cell mass and performance continues to improve with acclimation time.

A fixed-film reactor is also being utilized for acetate conversion to methane. In the reactor, a mixed methane bacterial culture attached to an inert support is being fed acetate at pH 5. No buffering of the feed is required to maintain the pH at 7.0 in the reactor.

Table 20.9 Conversion of Coal Gas by *P. productus* in a Bioflo Reactor[a]

Agitation (rpm)	Gas retention time (h)[b]	CO conversion (%)
250	1.56	13.7
	0.73	12.4
300	1.56	32.3
	0.73	23.3
400	1.56	69.8
	0.73	54.7
500	1.56	87.1
	0.73	60.2
600	1.56	88.0
	0.73	80.6
700	1.56	88.0
	0.73	80.6
800	1.56	90.2
	0.73	86.9
850	1.56	91.7
	0.73	87.3

[a]Feed gas composition: 65% CO, 22% H_2, 11% CO_2, 2% CH_4.
[b]A minimum of three to four gas retention times were required to reach optimum conversion.

Table 20.10 Conversion of Acetate to Methane by *Methanothrix* in a Fixed-Film Reactor

Retention time (h)	Acetate conversion (%)	Methane concentration in the gas phase (mol %)
12	100	50
6	100	50

Table 20.10 presents the data for the conversion of acetate to methane in the second-stage reactor using *Methanothrix* sp. A retention time of either 12 h or 6 h gives total acetate conversion with a methane concentration of 50%. The retention time probably can be reduced as the cell mass builds up in this reactor. However, the size of this second-stage reactor is not large since it converts only the small liquid stream from the gas phase reactor. A 6-h retention time translates into a reactor of about one-thirtieth the size of the gas phase reactor. Therefore, the retention time in this reactor is less critical, and 6 h is acceptable.

The potential of series operation in reducing retention time, and thus reactor volume, has been demonstrated without expensive agitation. These reactions, however, still need to be optimized for pH, temperature, nutrients level, inoculum, reactor type, and so on.

Co-culture Operation. An obvious economically beneficial alternative to series operation is to utilize two bacteria together in a co-culture to produce methane. In this manner, a CO-utilizing bacteria can produce a methane precursor, and a methane bacteria can utilize the precursor to produce methane in the same reactor. Co-culturing, if successful, can eliminate a reaction vessel, while maintaining reaction conditions near optimal for each individual bacteria.

A successful co-culture was employed using *P. productus* with an enrichment of *Methanothrix* sp. The results of continuous conversion studies as a function of agitation rate in a CSTR are shown in Table 20.11. The conversion of CO parallels the performance of *P. productus*, which was nearly identical to the pure culture *P. productus* results of Table 20.9. Methane formation was 88% of the stoichiometic maxium.

The co-culturing process was not always successful. When *Peptostreptococcus productus* and *Methanosarcina* sp. were co-cultured in continuous reactors, the *Methano-*

Table 20.11 Coal Gas Conversion by *P. productus* in Co-culture with *Methanothrix* sp. in a CSTR

Agitation (rpm)	Retention time (h)	Conversion (%)	
		CO	H^2
250	1.45	21.36	21.82
300	1.45	38.74	54.60
400	1.45	72.94	72.19
500	1.45	79.40	82.47
600	1.45	91.10	85.10

sarcina did not grow, but instead washed out of the reactor in a short period of time. *P. productus* was unaffected and continued to produce acetate. *Methanosarcina* is sensitive to the presence of CO and did not grow even at low CO concentrations. Similar attempts at co-culturing *P. productus* with *Methanobacterium* and *Methanospirillum* were also unsuccessful.

Summary/Significance. Pure cultures of *P. productus* and *R. rubrum* have been utilized to produce methane precursors from CO. *P. productus* can achieve near complete conversion of CO to acetate in a CSTR in a retention time of 0.7 h, less than half that required for the mixed culture. In a fixed-film reactor, without agitation, total conversion can be achieved in 0.8 h. A small second-state reactor then produces methane from acetate. A co-culture of *P. productus* and *Methanothrix* sp. has been developed that gives 88% of theoretical methane conversion from CO and H_2 in 1.5 h in a CSTR.

20.2.3 Sulfur Toxicity Studies

Coal synthesis gas may contain varying quantities of sulfur, present as H_2S or COS, depending on the sulfur content of the original coal. Generally, a total sulfur content of 2% or less is expected. For example, typical gas compositions from a Texaco gasifier, operating with a high-sulfur coal, include 1.2 and 0.5% H_2S and 0.07 and 0.05% COS for oxygen blown and air blown, respectively. These sulfur compounds could affect conversion performance, due to potential toxicity to the bacteria employed.

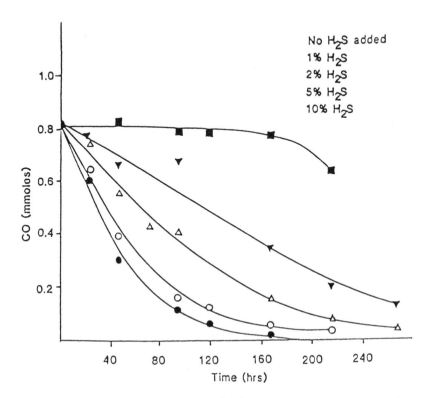

Fig. 20.1 Effect of H_2S concentration on the metabolism of CO by *P. productus*.

Batch sulfur toxicity experiments (COS, H_2S) were carried out on *P. productus*, *Rhodospirillum rubrum*, and the mixed culture to assess the capability of the organisms to grow and produce products in the presence of sulfur gases. H_2S and COS concentrations were varied from 0 to 10%, while maintaining all other fermentation conditions in the culture constant.

The effect of hydrogen sulfide on *P. productus* at 0, 1, 2, 5, and 10% concentrations is shown in Fig. 20.1. As noted, there is no appreciable effect on CO metabolism up to 2% H_2S. At 5% H_2S, a noticeable decrease in rate is apparent. At 10% the culture becomes inactive. However, *P. productus* does appear to be acclimating to the hydrogen sulfide after a period of time.

Figure 20.2 shows the results of COS addition to *P. productus*. Again, no effect on conversion was seen at a 1% COS level. However, a large decrease in CO conversion was noted when 2% or greater COS was present.

The same plots are given in Fig. 20.3 and 20.4 for H_2S and COS added to the mixed single-stage bacterial culture. As was noted with *P. productus*, no significant reduction in CO conversion was noted at concentrations below 2% H_2S or COS.

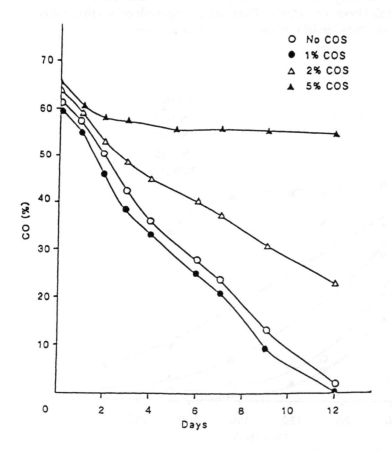

Fig. 20.2   Conversion of CO by *P. productus* when grown in the presence of COS.

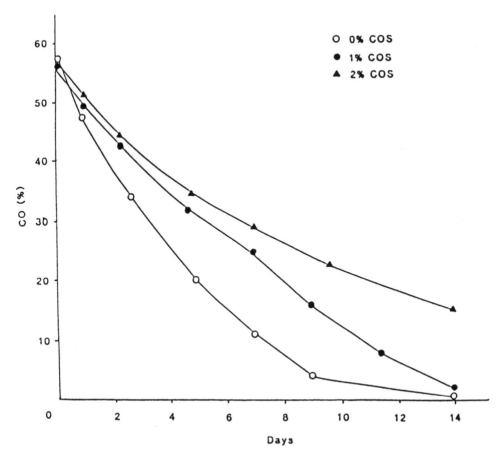

Fig. 20.3 Bacterial conversion of CO in the presence of H_2S when using the mixed culture.

Summary/Significance. The results presented in Figs. 20.1 to 20.4 are quite significant. At typical levels of total sulfur (2% or less), no significant decreases in conversion by either the mixed culture or *P. productus* occurs. Therefore, sulfur, in low concentrations, is not inhibitory to these biocatalysts and does not have to be removed prior to the biomethanation process.

20.3 CONCLUSIONS

In conclusion, the feasibility of biological reactions in converting synthesis gas to methane has been demonstrated in both mixed and pure cultures. Complete conversion of coal gas to methane has been achieved in a retention time of 2 h with a mixed culture, and an equivalent retention time of 24 min at higher pressure. The use of higher pressures will probably further reduce retention times. For the pure culture, with *P. productus* and *Methanothrix* sp., either in series or in co-culture, the retention times are 45 min to 1½ h. These cultures have not been run at higher pressures. Sulfur concentrations up to 2% do not appreciably affect the performance of these cultures. Future efforts will concentrate

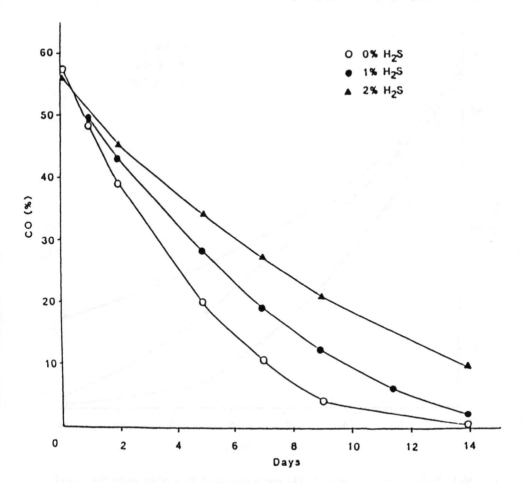

Fig. 20.4 Bacterial conversion of CO in the presence of COS when using the mixed culture.

on the optimization of the cultures and the type of reactor to further minimize the retention time. Investigations of the mass transfer and kinetic relationships in this heterogeneous system are necessary to understand and manipulate the parameters to maximize the reaction rate.

REFERENCES

1. Alpec, J., Bioengineers are off to the mines, *High Technol.* (1984).
2. Ander, P., and Erickson, K. C., Selective degradation of wood components, *Physiol. Plant., 41*, 239–248 (1974).
3. Balch, W. E., and Wolfe, R. S., New approach to the cultivation of methanogenic bacteria: 2-mercaptoethane sulfonic acid (HS-COM)-dependent growth of *Methanobacterium ruminantium* in a pressurized atmosphere, *Appl. Environ. Microbiol., 32*, 781–791 (1976).
4. Bryant, M. D., Commentary on the Hungate technique for culture for anaerobic bacteria, *Am. J. Clin. Nutr., 25*, 1324–1328 (1972).

5. Clausen, E. C., and Gaddy, J. L., *Biological Conversion of Coal Synthesis Gas*, Annual Report, U.S. Department of Energy, Morgantown Energy Technology Center, Morgantown, W. Va., May 1985.
6. Cohen, M. S., and Gabriele, P. D., Degradation of coal by the fungi *Polyporous versicolor* and *Poria monticola*, *Appl. Environ. Microbiol.*, *44*, 23 (1982).
7. Crawford, D. L., *Conversion of Lignocellulose by* Actinomycete *Microorganisms*, final report of NTIS B-293015, U.S. Department of Commerce, Washington, D.C., 1979.
8. Crawford, R. L., *Assessment of Bacteria for Lignocellulosic Transformations*, NTIS PB 80-108681, U.S. Department of Commerce, Washington, D.C., 1979.
9. Daft, M. J., and Hacskaylo, E., Arbuscular mycrorrhizas in the anthracite and bituminous coal wastes of Pennsylvania, *J. Appl. Ecol.*, *13*, 523 (1976).
10. Daniels, L., Fuchs, G., Thauer, R. K., and Zeikus, J. G., Carbon monoxide oxidation by methanogenic bacteria, *J. Bacteriol.*, *132*, 118-126 (1977).
11. Drew, S. W., Glasser, W. G., and Hall, P. L., *Enzymatic Transformation of Lignin: Final Report of Research Project, July 1, 1976–June 30, 1979*, 1979.
12. Eastmond, D. A., Muehle, C. M., Proce, R. L., Hutchens, C. A., Booth, G. M., and Lee, M. L., Acute toxicity, bioconcentration, and elimination of a coal liquid in freshwater organisms, 1983, p. 451.
13. Euker, C. A., Jr., and Wesselhott, R. D., Coal gasification in a process development unit, *Energy Process, 1*, 1-4 (1981).
14. Fisher, F., Leiske, R., and Winzer, K., Umestzung in des Kohlenoxyds, *Biochem. Z.*, *236*, 247-267 (1931).
15. Fisher, F., Leiske, R., and Winzer, K., Uber die Bildung von Essigsäure bei der biologischen Umsetzung von Kohlenoxyd und Kohlensäure mit Wasserstoff zu Methan, *Biochem. Z.*, *245*, 2-12 (1932).
16. Fresquez, P. R., and Lindemann, W. C., Soil and rhizosphere microorganisms in amended coal mine spills, *Soil Sci. Soc. Am. J.*, *46*, 751 (1982).
17. Giddings, J. M., *Four-Hour Algal Bioassays for Assessing the Toxicity of Coal-Derived Materials*, EPA-600/9-81-018, U.S. Environmental Protection Agency Office of Research and Development, Washington, D.C., 1981.
18. Hammond, A. L., Metc, D., and Maugh, T. H., III, *Energy and the Future*, American Association for the Advancement of Science, Washington, D.C., 1973.
19. Hill, J. O., Giere, M. S., Pickrell, J. A., Hahn, F. F., and Dahl, A. R., in vitro and in vivo toxicity of potential organometallic compounds associated with coal conversion processes, *Annu. Rep. Inhalation Res. Inst.*, 406 (1979).
20. Hungate, R. E., The anaerobic mesophilic cellulolytic bacteria, *Bacteria. Bacteriol.*, *14*, 1-44 (1950).
21. Jockel, H., and Triebskorn, B. E., GASYNTHAN process for SNG, *Hydrocarbon Process.*, 93 (1973).
22. Kaplan, L. J., Methane from coal aided by use of a potassium catalyst, *Chem. Eng.* (1982).
23. Kargi, F., and Robinson, J. M., The removal of sulfur compounds from coal by thermophilic organism *Sulfolobus acidocaldarius*, *Appl. Environ. Microbiol.*, *44*, 878-883 (1982).
24. Koh, W. K., What's happening with coal-gasification technology? *Hydrocarbon Process.*, 85 (1982).
25. Lee, D. D., Scott, C. D., and Hancher, C. W., Fluidized-bed bioreactor for coal-conversion effluents, *J. Water Pollut. Control Fed.*, *51*, 974 (1979).
26. Mills, G. A., Synfuels from coal progress in the USA, *Energy Prog.*, *2*, 57 (1982).
27. Moeller, F. W., Roberts, H., and Britz, B., Methanation of coal gas for SNG, *Hydrocarbon Process.*, 69 (1974).

28. Pfaender, F. K., Singer, P. C., Lamb, J. C., III, and Goodman, R., Effects of bio-
    logical treatment on the potential impact of coal conversion wastewaters, *Dept.
    Energy Symp. Ser., 54*, 541 (1981).
29. Richardson, T., SNG catalyst technology, *Hydrocarbon Process.*, 91 (1973).
30. Ryan, D. G., and Aczel, T., EDS coal liquifaction products as petrochemical feed-
    stocks, *Energy Prog., 2*, 87 (1982).
31. Silverman, M. P., Rogoff, M. H., and Wendes, I., Removal of pyritic sulfur from coal
    by bacterial action, *Fuel, 42*, 113-124 (1963).
32. Speeks, R., and Klusmann, A., *Energy Prog., 2*, 2 (1982).
33. Stephenson, M., and Strickland, L. H., The bacterial formation of methane by the
    reduction of one-carbon compounds by molecular hydrogen, *Biochem. J.*, 1417-
    1527 (1933).
34. Wolfe, R. S., Microbial formation of methane, *Adv. Microb. Physiol., 6*, 107-146
    (1971).

21

Anaerobic Metabolism of Aromatic Compounds

JOHN D. HADDOCK and JAMES G. FERRY   *Virginia Polytechnic Institute and State University, Blacksburg, Virginia*

21.1 INTRODUCTION

Aromatic compounds constitute a class of organic molecules that are of major importance in the global storage and cycling of carbon. Petroleum, peat and coal deposits, lignin, plant phenolics, and proteins contain a large aromatic fraction. Human activities have increased the levels of natural and synthetic aromatic compounds released into the environment through the exploitation of natural resources and widespread application of pesticides. While microbial degradation or alteration of aromatic compounds by aerobic microorganisms is well known and major degradative pathways have been elucidated (e.g., Refs. 1–3), anoxic degradative pathways are poorly understood. This is partially the result of difficulties in working with anaerobic procedures, but also because the anaerobic metabolism of aromatic compounds was previously thought to be a minor degradative route in nature. However, as early as 1934, Tarvin and Buswell [4] demonstrated that a variety of aromatic compounds were degraded under the strictly anaerobic conditions of methanogenic fermentations. Since the first demonstration of anaerobic aromatic degradation in pure cultures [5], an increasing number of bacteria of diverse physiological types have been shown to degrade a wide variety of aromatic compounds. These organisms generally fall into three physiological groups: (1) photosynthetic bacteria; (2) anaerobic respirers, including dissimilatory sulfate and nitrate reducers; and (3) nonrespiring anaerobes. The latter group includes fermentative organisms, obligate proton-reducing acetogens, and those involved in one-carbon metabolism. This review includes published studies concerning anaerobic degradation of monoaromatic compounds by mixed as well as pure cultures. The reader is also referred to the reviews of Evans [6], Sleat and Robinson [7], Young [8], Berry et al. [9], and Evans and Fuchs [10]. Scheline [11] reviewed anaerobic metabolism in the gastrointestinal tract from a pharmacological point of view and includes some information on reactions involving aromatic compounds. Anaerobic degradation of aromatic amino acids has been reviewed by Barker [12].

21.2 PHOTOMETABOLISM

Although Tarvin and Buswell [4] first demonstrated the anaerobic degradation of several aromatic compounds to methane and CO_2 in mixed cultures derived from sewage sludge, little information was gained on the organisms or the biochemical pathways involved. The first demonstration of anaerobic aromatic degradation by a microorganism in pure culture was reported by Scher and Proctor [13]. A *Rhodopseudomonas* sp. was isolated that could utilize benzoate as an organic electron donor. Several other strains of *Rhodopseudomonas* spp. and *Rhodospirillum rubrum* grew anaerobically on benzoate, protocatechuate, and catechol in the presence of light [5]. They proposed a degradation pathway analogous to aerobic pathways in which oxygen is incorporated into the aromatic ring prior to ring cleavage. However, Dutton and Evans [14] proposed that *R. palustris* photo-metabolized aromatic compounds by a different route. They found that aerobic *p*-hydroxybenzoate degradation involved oxygenases, while addition of air to cells growing anaerobically on benzoate stopped benzoate degradation. Also, catechol, a key intermediate of the aerobic pathway, was not degraded anaerobically. Hegeman [15] showed that enzymes involved in aerobic metabolism of *p*-hydroxybenzoate by *R. palustris* were induced by aerobic conditions and the presence of the substrate. In contrast, approximately 100-fold lower enzyme levels were found for cells grown anaerobically on this substrate. Evidence that anaerobic degradation involved a different mechanism was provided by the radiotracer studies of Dutton and Evans [16,17]. They found that the aromatic ring was reduced before being cleaved, that oxygenases were absent, and that anaerobic benzoate degradation was inducible. They proposed enzyme-catalyzed reactions similar to those involved in the β-oxidation of fatty acids. Guyer and Hegeman [18] proposed a similar pathway based on evidence obtained with mutants of *R. palustris* and radioisotopes. The failure of initial attempts to show benzoate degradation in cell extracts [17] was subsequently shown by Dutton and Evans [19,20] to involve the presence of fatty acids released during cell rupture, and that carboxyl groups were inhibitory. It was suggested that the inhibitory effects may have been a result of competition for cofactors involved in fatty acid metabolism. Coenzyme A involvement in benzoate degradation to pimelate was shown in *R. palustris* cell extract when it was discovered that benzoate conversion to benzoyl-CoA required ATP and Mg^{2+} in phosphate buffer [21]. In addition, conversion of cyclohex-1-enecarboxylate, an intermediate of reductive benzoate degradation to pimelate, required NAD. Hutber and Ribbons [22] showed that the coenzyme A derivative of cyclohexane carboxylate was also formed by *R. palustris* and that the enzymes responsible for cyclohexanecarboxyl-CoA and benzoyl-CoA were induced when these compounds were growth substrates under anaerobic as well as aerobic conditions. All of the major enzymes involved in fatty acid β-oxidation were demonstrated in cell extracts. Harwood and Gibson [23] studied the uptake of benzoate by *R. palustris* and found that the intracellular formation of benzoyl-CoA was likely the first step involved in degradation and that it was responsible for maintaining a concentration gradient that allowed cells to take up benzoate from the medium at concentrations below 1 μM. Geissler et al. [24] purified a soluble benzoate-coenzyme A ligase from this organism. The enzyme required Mg^{2+}-ATP for activity and was insensitive to oxygen. Fluorobenzoates were also efficient substrates for the enzyme. Uptake of 4-hydroxybenzoate by this organism has also been studied, and apparently a different ligase enzyme is involved with this substrate [25]. Harwood and Gibson [26] have shown that several different strains of *R. palustris* were able to grow anaerobically as well as aerobically on a wide range of substituted aromatic acids. Some compounds were degraded only under anaerobic but not aerobic conditions, and vice versa. Experiments with mutants suggested the presence of two pathways of anaerobic metabolism that passed through benzoate or 4-hydroxybenzoate, respectively.

21.3 RESPIRATORY DEGRADATION

21.3.1 Denitrification

Several aromatic compounds have been shown to be degraded anaerobically by denitrifying organisms that utilize nitrate as the terminal electron acceptor. These organisms are facultative anaerobes and often also have the ability to degrade the same aromatic compounds by aerobic pathways that are distinct from the anaerobic pathways.

Oshima [27] first described two soil organisms resembling *Pseudomonas* spp. that were able to anaerobically degrade protocatechuate (3,4-dihydroxybenzoic acid) with nitrate in the medium. However, degradation occurred only when the two organisms were co-cultured. Taylor et al. [28] isolated *Pseudomonas* strain PN-1 from soil using *p*-hydroxybenzoate with nitrate as the terminal electron acceptor. This organism was also able to degrade benzoate anaerobically. Benzoate, *p*- and *m*-hydroxybenzoate, and protocatechuate were degraded aerobically without nitrate. Aerobic degradation involved the enzyme protocatechuic acid-4,5-oxygenase, which had a specific activity about 50-fold greater in extracts of aerobically compared to anaerobically grown cells. Therefore, they concluded that the aerobic and anaerobic pathways of aromatic degradation were different. Taylor and Heeb [29] added support to this hypothesis using ^{14}C-labeled benzoate. Protocatechuate and catechol, key intermediates of aerobic degradation pathways [2] failed to trap any radioactivity when *Pseudomonas* PN-1 was grown anaerobically on [^{14}C]benzoate. *Pseudomonas* PN-1 was shown to anaerobically degrade several methoxylated aromatic compounds to the corresponding hydroxylated derivatives [30]. Dehydroxylation then produced benzoic acid, which was proposed as a central intermediate for the anaerobic degradation pathways of some aromatic compounds. Phenylpropane lignin derivatives were only degraded anaerobically, while most of the tested methoxylated aromatics that lacked a three-carbon side chain were demethylated both aerobically and anaerobically. Blake and Hegeman [31] reclassified this organism as *Alcaligenes xylosoxidans* subspecies *denitrificans* PN1. They found a small 17.4-kilobase plasmid that carried the genes necessary for anaerobic benzoate degradation in this organism. The plasmid was transmissible to strains of *P. stutzeri* and *P. aeruginosa* giving them the ability to grow anaerobically on benzoate. Williams and Evans [32,33] described a *Moraxella* sp. capable of anaerobic nitrate respiration with benzoate, phloroglucinol (1,3,5-trihydroxybenzene), and several phenylpropane derivatives. Cyclohexane-carboxylate (CHCA), 2-hydroxy CHCA, and adipic acid were identified as degradation products leading them to suggest a degradative pathway similar to the reductive pathway utilized by photosynthetic microorganisms [17]. Phenol has also been used as a substrate for the isolation of denitrifiers [34]. Three gram-negative, flagellated rods were obtained, but growth was slow in pure culture. Enrichment cultures grew faster and were able to degrade a variety of aromatic compounds, including benzoate, 3,4-dihydroxybenzoate, *m*- and *p*-hydroxybenzoate, and *o*-, *m*-, and *p*-cresol. Catechol oxygenase levels were low and [^{14}C]phenol was used to show some incorporation of substrate into cell material. The list of aromatic compounds degraded by nitrate-respiring organisms includes the ubiquitous pollutants known as phthalates [35,36]. Aftring et al. [35] described the anaerobic dissimilation of o-, m-, and p-phthalate by denitrifying mixed bacterial cultures. Aftring and Taylor [36] described the isolation of a *Bacillus* sp. that required nitrate to anaerobically degrade o-phthalate to CO_2. Benzoate was also degraded and was suggested to be an intermediate via decarboxylation of phthalate. Taylor and Ribbons [37] showed that o-phthalic acid was decarboxylated to benzoate anaerobically in the presence of nitrate. Nozawa and Maruyama [38] isolated a *Pseudomas* sp. that degraded o-, m-, and p-phthalate as well as benzoate, cyclohex-1-ene-carboxylate, and cyclohex-3-ene-carboxylate under denitrifying conditions.

para-Cresol (4-methylphenol) was anaerobically metabolized by a syntrophic association of two denitrifying organisms [39]. Two gram-negative, facultative anaerobes were isolated from polluted river sediment. Isolate PC-07 oxidized p-cresol to p-hydroxybenzoate, which was oxidized by PB-04 to undetermined ring-fission products. Degradation by both organisms was nitrate dependent and was substrate specific since neither organism degraded the other aromatic compound. *para*-Hydroxybenzyl alcohol and p-hydroxybenzaldehyde were detected as intermediates of p-cresol degradation by PC-07 [40]. *para*-Ethylphenol, but not p-propylphenol, toluene, or the ortho and meta isomers of cresol, was degraded by this organism. The aromatic hydrocarbons, m-xylene (1,3-dimethylbenzene) and toluene, were degraded primarily to CO_2 with nitrate reduction in anaerobic laboratory aquifer columns [41]. Toluene, benzaldehyde, benzoate, m-toluylaldehyde, m-toluate, m-cresol, and p-hydroxybenzoate were degraded in columns previously adapted to m-xylene under denitrifying conditions [42].

Anaerobic degradation of aromatic compounds with ring substituents other than hydroxyl, methoxyl, and methyl groups has also been shown for dissimilatory nitrate reducers. Braun and Gibson [43] isolated three strains of *Pseudomonas* spp. capable of anaerobic growth on anthranilate (2-aminobenzoate) with nitrate as the electron acceptor. Taylor et al. [44] showed that anaerobic cell suspensions of *Pseudomonas* PN-1 grown on p-hydroxybenzoate could metabolize o- and p-fluorobenzoate with release of the fluoride ion into the medium. However, this organism would not grow at the expense of m-fluorobenzoate. Schennen et al. [45] showed that the anthranilate-degrading *Pseudomonas* spp. described by Braun and Gibson [43] could grow on 2-fluorobenzoate and that an inducible enzyme, benzoyl-coenzyme A synthetase, was involved in the initial degradation reaction. Ziegler et al. [46] showed the presence of a Mg^{2+}-ATP-dependent aryl-CoA synthetase of broad substrate specificity in *Pseudomonas* strain KB 740⁻ grown on benzoate under denitrifying conditions. Reduction of benzoyl-CoA was not detected, however 2-aminobenzoyl-CoA reduction was detected in a partially purified preparation from extracts of cells grown anaerobically on 2-aminobenzoate. NADH as well as NADPH served as reductant for the reaction.

21.3.2 Sulfate Reduction

An anaerobic organism has been isolated that utilizes sulfate as the terminal electron acceptor and benzoate as the electron donor and carbon source [47]. This organism, a new species of sulfate reducer, was named *Desulfonema magnum* and was found to utilize the aromatic compounds 4-hydroxybenzoate, hippurate, phenylacetate, and 3-phenylpropionate, as well as C_1–C_{10} fatty acids, succinate, and fumarate [48]. The compounds 2- and 3-hydroxybenzoate and cyclohexane carboxylate were not utilized. Another sulfate reducer, *Desulfococcus niacini*, which degrades nicotinate to CO_2 and ammonia, was isolated from marine sediments [49]. It also degraded 3-phenylpropionate to CO_2 and to benzoate, which was not further metabolized. *Desulfobacterium indolicum* and *Desulfobacterium phenolicum* were isolated from marine sediments after enrichment with indole and phenol, respectively [50,51]. *D. indolicum* degrades indole, anthranilic acid, and quinoline to CO_2, with sulfate as the electron acceptor. *D. phenolicum* also metabolizes indole and anthranilic acid as well as phenol, *p*-cresol, benzoate, 2- and 4-hydroxybenzoate, phenylalanine, phenylacetate, and 4-hydroxyphenylacetate. *Desulfobacterium catecholicum* was isolated from a catechol-degrading enrichment culture derived from an estuarine sediment [52]. This strict anaerobe grew slowly with catechol, resorcinol, hydroquinone, benzoate, 4-hydroxybenzoate, protocatechuate, 2-aminobenzoate, phloroglucinol, and pyrogallol as aromatic substrates and either sulfate or nitrate as electron acceptors. It also grew on 2-hydroxybenzoate with nitrate but not sulfate as the electron acceptor. Nitrate was reduced to ammonium.

21.4 NONRESPIRATORY DEGRADATION

Few anaerobes capable of the oxidization of aromatic compounds in the absence of light or exogenous electron acceptors have been isolated. In some cases this is attributable to the slow growth rate of these organisms because of the limited amount of energy available from incomplete oxidation of the substrates and a dependence on other organisms for the removal of fermentation end products. Organisms that degrade trihydroxylated aromatic compounds in pure culture have been described, while those that degrade aromatic compounds with two or fewer hydroxyl groups have only been obtained in defined mixed cultures. A recent report has shown that certain facultative anaerobes belonging to the Enterobacteriaceae are able to degrade ferulic acid slowly while growing anaerobically in pure culture [53]. These organisms appear to represent exceptions to the generalization that only trihydroxylated aromatic compounds are cleaved by pure cultures under fermentative conditions. Pure cultures of anaerobes that metabolize only ring substituents of aromatic compounds without degrading the ring itself have been isolated. As members of microbial food webs these organisms play an important role in mineralization through the production of aromatic substrates necessary for the growth of anaerobes that can degrade the ring.

21.4.1 Pure Cultures

The first organism isolated in pure culture with the capacity to ferment an aromatic compound was a *Clostridium* sp. that anaerobically degraded the nitrogen-containing heterocyclic aromatic compounds nicotinic acid and nicotinamide [54]. The isolate from river mud fermented nicotinic acid to acetate, propionate, ammonia, and CO_2. The first step in the fermentation was a reversible hydroxylation resulting in the formation of 6-hydroxynicotinate [55]. The organism was lost, however, before further studies could be made. Pastan et al. [56] reisolated a similar organism that was subsequently named *Clostridium barkeri* [57]. Tsai et al. [58] proposed a pathway for nicotinic acid degradation that involved partial ring reduction to 1,4,5,6-tetrahydro-6-oxonicotinic acid. The organism had high concentrations of a vitamin B_{12} derivative subsequently shown to be involved in a rearrangement reaction after ring cleavage [59]. Purification and characterization of the hydroxylase that catalyzed nicotinic acid oxidation to 6-hydroxynicotinic acid showed a requirement for NADP as the electron acceptor [60]. Nicotinic acid hydroxylase activity in cell extract of *C. barkeri* was increased by growth on a medium supplemented with sodium selenite. Maximal activity occurred when the selenium concentration in the medium was 10^{-7} M [61]. Holcenberg and Tsai [62] partially purified the next enzyme in the pathway that catalyzed the reduction of 6-hydroxynicotinic acid to 6-oxo-1,4,5,6-tetrahydronicotinic acid. Since this enzyme required reduced ferredoxin for activity and would not substitute reduced NADP, it appeared that the oxidation and subsequent reduction reactions were not directly coupled through NADP(H).

The first organisms capable of fermenting a nonheterocyclic aromatic compound, phloroglucinol (1,3,5-trihydroxybenzene), were isolated from the rumen and were identified as strains of *Streptococcus bovis* and a *Coprococcus* sp. [63]. The *Coprococcus* sp. produced 2 mol each of acetate and CO_2 per mole of phloroglucinol fermented [64]. As with other anaerobes discussed so far, degradation involved ring reduction. Patel et al. partially purified and characterized phloroglucinol reductase, which catalyzed the reduction of phloroglucinol to dihydrophloroglucinol, with NADPH as the electron donor [65]. *Rhodopseudomonas gelatinosa* [21] and *Pelobacter acidigallici* [66,67] also have NADPH-dependent phloroglucinol reductase activity. *P. acidigallici*, isolated from salt and freshwater sediments, fermented gallate (3,4,5-trihydroxybenzoate), 2,4,6-trihydroxybenzene, pyrogallol (3,4,5-trihydroxybenzene), and phloroglucinol to acetate and CO_2 [66]. This organism cooperated with *Acetobacterium woodii* and *Desulfobacter postgatei* in defined co-culture as the second member of a food chain that mineralized 3,4,5-trimethoxybenzoate [68]. *A. woodii* O-demethoxylated the substrate to gallate, which was fermented by *P. acidigallici* to acetate and CO_2. The acetate was then oxidized to CO_2 by the sulfate reducer *D. postgatei*. Samain et al. [67] have proposed a unified pathway for degradation of trihydroxylated aromatics through (1) decarboxylation of the acidic constituents, if present; (2) rearrangement of hydroxyl groups to ring positions 1, 3, and 5 to form phloroglucinol; (3) ring reduction to dihydrophloroglucinol; and (4) degradation to acetate and CO_2. *Eubacterium oxidoreducens*, a recent rumen isolate, ferments gallate, pyrogallol, phloroglucinol, and crotonate to acetate, butyrate, and CO_2 [69]. A unique requirement for growth on gallate was the addition of formate or H_2, which were oxidized in approximate stoichiometric amounts with the aromatic substrate. NADP-dependent formate dehydrogenase (FDH) and NADPH-dependent phloroglucinol reductase (PR) activities were detected in cell extracts and a scheme involving transfer of electrons from formate through NADP(H) to phloroglucinol was proposed [70].

We have mass-cultured *E. oxidoreducens* in 10-L fermentors on 50 mM gallate plus 50 mM formate and have characterized the FDH and PR in cell extract (Table 21.1 and Figs. 21.1 to 21.3). As with many formate dehydrogenases, enzyme activity in cell extract from *E. oxidoreducens* was extremely sensitive to air but was protected from inactivation by sodium azide. Protection by azide should facilitate the purification and study of this enzyme. FDH activity, as assayed by reduction of methyl viologen (MV), was recovered in the soluble fraction after centrifugation of cell extract at 105,500 g for 1.5 h in anaerobic phosphate buffer containing 20% sucrose. No significant activity was detected in the membrane fraction. We have purified the phloroglucinol reductase from *E. oxidoreducens* to electrophoretic homogeneity [72b]. The enzyme is an α_2 homodimer with a native molecular weight of 78,000 daltons. No metals or cofactors were found, suggesting that the reaction mechanism may involve a direct hydride transfer from NADPH to phloroglucinol. Enzyme activity was insensitive to exposure to aerobic conditions and only NADPH and phloroglucinol were substrates for the enzyme.

Table 21.1 Kinetic Constants for Formate Dehydrogenase and Phloroglucinol Reductase in Cell Extract of *Eubacterium oxidoreducens*

Enzyme	Substrate	Range of concentrations used (mM)	Km (mM)	V_{max} (U/mg)[a]
Formate	Formate	0.1–10	0.29	1.43[b]
dehydrogenase	NADP	0.0135–0.157	0.14	3.9[c]
	MV[d]	0.192–38	1.4	1.94[b]
Phloroglucinol	Phloroglucinol	0.02–1.9	0.23	2.65[e]
reductase	NADPH	0.0184–0.185	0.05	6.45[e]

[a]Units/mg protein: 1 unit is μmol per minute.
[b]Units methyl viologen reduced.
[c]Units NADP reduced.
[d]Methyl viologen.
[e]Units NADPH oxidized.

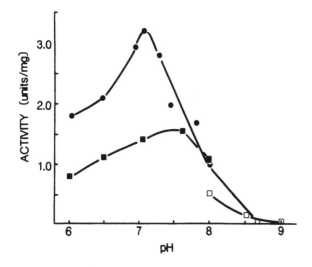

Fig. 21.1 Effect of pH on phloroglucinol reductase and formate dehydrogenase activity in cell extract of *Eubacterium oxidoreducens*. Cells were grown anaerobically in 10-L fermentors on 50 mM gallate and 50 mM formate in media described previously [69]. Cell extract was prepared anaerobically with a French pressure cell as described previously [148] but with the following modifications: 2 g (wet weight) liquid N_2 frozen cell paste were thawed in 4 mL of anaerobic 75 mM potassium phosphate buffer, pH 7.5, that contained 10 mM sodium azide and 2 mM 2-mercaptoethanol. Centrifugation of lysate was at 32,000 g for 20 min. Cell extract was diluted 1/40 in anaerobic 50 mM potassium phosphate buffer, pH 7.5 prior to being assayed, and 5 to 6 μg protein was added per assay. Phloroglucinol reductase assays were performed in 50 mM potassium phosphate (●) or 50 mM sodium borate (○) buffer as described previously [65] except the assay mixture contained in μmoles: NADPH, 0.1; phloroglucinol, 1.0. One unit of activity is μmoles NADPH oxidized per minute. Anoxic formate dehydrogenase assays were performed in 50 mM potassium phosphate (■) or 50 mM sodium borate (□) buffer as described previously [129] except the anaerobic assay mixture (0.52 mL), contained in μmoles: 2-mercaptoethanol, 10; NADP, 0.11; sodium formate, 10. Activity was measured by recording the increase in absorbance at 340 nm. The extinction coefficient used was 6.2 mM^{-1} cm^{-1}. One unit of activity is μmoles NADP reduced per minute.

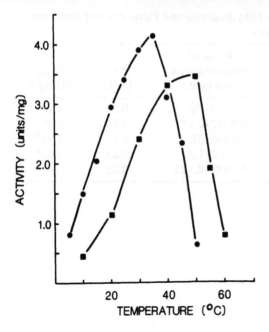

Fig. 21.2 Effect of temperature on phloroglucinol reductase and formate dehydrogenase activity in cell extract of *Eubacterium oxidoreducens*. Assays were performed as described in Fig. 21.1 in 50 mM potassium phosphate buffer, pH 7.5. ●, Phloroglucinol reductase; ■, formate dehydrogenase.

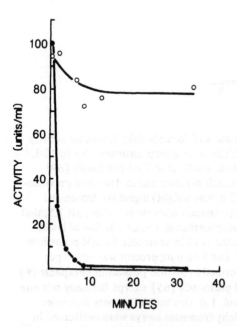

Fig. 21.3 Effect of air on formate dehydrogenase activity in *Eubacterium oxidoreducens* cell extract with and without sodium azide. Activity was assayed in undiluted cell extract as described in Fig. 21.1 with 20 mM methyl viologen in place of NADP as the electron acceptor, and the increase in absorbance at 603 nm was recorded. An extinction coefficient of 11.3 mm^{-1} cm^{-1} was used. ○, with 6.7 mM sodium azide; ●, without sodium azide.

Krumholz et al. [71] proposed a complete pathway for gallate degradation in this organism and demonstrated the presence of a number of intermediates and enzyme activities. Of significance was the demonstration of the ring-cleaving enzyme, dihydrophloroglucinol hydrolase, and detection of 3-hydroxy-5-oxohexanoate as the product. Enzymes involved in β-oxidation and substrate-level phosphorylation were also detected. All of the enzymes assayed appeared to be constitutive since crotonate grown cells had enzyme activities comparable to those in gallate plus formate grown cells. Pyrogallolphloroglucinol isomerase, the second enzyme in the proposed pathway of gallate metabolism in *E. oxidoreducens* was partially purified and characterized [72a]. The enzyme had a native molecular weight of approximately 230,000 daltons and showed a single major band corresponding to 92,000 daltons on denaturing polyacrylamide gels. The enzyme contained iron, molybdenum, and acid-labile sulfide, and the reaction mechanism appeared to involve hydroxylation of pyrogallol to benzenetetrole followed by dehydroxylation to phloroglucinol.

Attempts to isolate an organism that fermented aromatic compounds with fewer than three hydroxyl groups attached to the ring were unsuccessful [73-75]. An organism from phenylacetate enrichments of anaerobic digestor sludge was isolated on succinate but required co-culture with a *Wolinella* sp. to degrade several aromatic compounds including mono- and dihydroxy aromatics [76]. Thermodynamic considerations offer an explanation for these results [73]. Equations (1) and (2) show that the reactions for fermentation of benzene or benzoate to acetate and hydrogen are thermodynamically unfavorable under standard conditions [77].

$$C_6H_6 + 6H_2O \rightarrow 3CH_3COO^- + 3H^+ + 3H_2 \quad \Delta G^{o\prime} = +72.2 \text{ kJ/mol} \tag{1}$$

$$C_6H_5CO_2^- + 7H_2O \rightarrow 3CH_3COO^- + 3H^+ + 3H_2 + HCO_3^- \quad \Delta G^{o\prime} = +70.6 \text{ kJ/mol} \tag{2}$$

However, Ferry and Wolfe [73] hypothesized that removal of benzoate fermentation products by methanogens allowed benzoate to be degraded with a favorable change in free energy. Equation (3) [74] shows that removal of hydrogen results in a favorably negative change in free energy.

$$C_6H_5CO_2^- + 4.75H_2O \rightarrow 3CH_3CO_2^- + 3H^+ + 0.25HCO_3^- + 0.75CH_4 \quad \Delta G^{o\prime} = \tag{3}$$
$$-31.1 \text{ kJ/mol}$$

Equations (4) to (6) [77] show the effect that the number of ring hydroxyl groups has on the change in free energy for the fermentation.

$$C_6H_5OH + 5H_2O \rightarrow 3CH_3COO^- + 3H^+ + 2H_2 \quad \Delta G^{o\prime} = +6.55 \text{ kJ/mol} \tag{4}$$

$$C_6H_4(OH)_2 + 4H_2O \rightarrow 3CH_3COO^- + 3H^+ + H_2 \quad \Delta G^{o\prime} = -78.1 \text{ kJ/mol} \tag{5}$$

$$C_6H_3(OH)_3 + 3H_2O \rightarrow 3CH_3COO^- + 3H^+ \quad \Delta G^{o\prime} = -158 \text{ kJ/mol} \tag{6}$$

An increase in hydroxyl group number decreases the amount of hydrogen produced and increases the thermodynamic favorability of the reactions. Hydroxyl groups may destabilize the aromatic nucleus through inductive effects on the π electrons of the ring [74,78]. Theoretically, an organism capable of fermenting dihydroxy aromatics such as catechol, resorcinol, or hydroquinone should be able to grow in pure culture since the free-energy

change is negative [equation (5)] . However, to date, only defined co-cultures of mono-
and dihydroxy aromatic degraders with hydrogen-utilizing organisms have been described [74,76]. Equation (6) shows a relatively large negative $\Delta G^{\circ\prime}$ for trihydroxybenzenes. As previously mentioned, organisms have been isolated in pure culture that can
ferment trihydroxybenzenoids [63,66,69].

      In addition to the influence of the number of ring hydroxyl groups on the thermo-
dynamics of aromatic fermentation, their relative position on the ring also affects the
degradation pathway [74,75,78]. Since β-oxidation appears to be the mechanism by
which aromatic ring carbon is metabolized anaerobically [6] hydroxyl groups at ring
positions that would interfere with this process are usually removed, while those at positions where β-oxidation would occur are conserved [75]. The reductive removal of
hydroxyl groups may provide a small amount of energy for an aromatic degrader or an
organism specializing in such a reaction; however, a source of reducing equivalents would
then be necessary [78]. These authors have calculated that such a reaction would be
thermodynamically feasible with electrons at the redox potential of the pyruvate/lactate
couple but not at the level of the fumarate/succinate couple. Rearrangement of ring
hydroxyl groups to positions favoring β-oxidation occurs during the fermentation of gallic
acid and pyrogallol by *P. acidigallici* [56,67] and *E. oxidoreducens* [69].

21.4.2 Syntrophic Cultures

The first benzoate-degrading organism described from a methanogenic consortium was
obtained in co-culture with a hydrogen-utilizing *Desulfovibrio* sp. [79]. The benzoate degrader *Syntrophus buswellii* [80] also grew with the methanogen *Methanobacterium
formicicum* in the absence of sulfate. *S. buswellii* would not grow in pure culture on benzoate or on any other substrate tested. Since degradation of benzoate to acetate, CO_2
and hydrogen is thermodynamically unfavorable [73], removal of H_2 by the hydrogenotrophs *Desulfovibrio* sp. or *M. formicicum* was thought to be necessary for benzoate
degradation by *S. buswellii*. Two additional syntrophic strains, P-2 and PA-1, have been
isolated in co-cultures from phenylacetate or phenol-degrading consortia [76]. *Wolinella
succinogenes* was used as the hydrogenotroph, which reduces fumarate to succinate [81].
Strain P-2 in co-culture with *W. succinogenes* degraded benzoate, phenylacetate, hydro-
cinnamate, and phenol anaerobically. Strain PA-1 in co-culture with *W. succinogenes* degraded the same aromatics and also resorcinol (1,3-dihydroxybenzene), 2-aminobenzoate,
ferulate (4-hydroxy-3-methoxycinnamate), gallate, pyrogallol, 4-aminophenol, phloroglucinol, and catechol. Contrary to other described aromatic fermenters, which degrade
only one or a few compounds, strains PA-1 and P-2 had an unusually broad substrate
range for aromatic compounds. Shelton and Tiedje [82] isolated a syntroph that fermented benzoate in co-culture with a sulfate reducer and a methanogen. The benzoate degrader resembled *S. buswellii* and cooperated with other organisms in the degradation of
3-chlorobenzoate by a methanogenic consortium. An anaerobic microbial food web
capable of 3-chlorobenzoate utilization was reconstituted from a pure culture of a dechlorinating organism and a defined co-culture of a benzoate fermentor and a hydrogen-
utilizing methanogen [83]. One-third of the hydrogen produced during benzoate fermentation to acetate and CO_2 was utilized by the dechlorinating organism for reductive dechlorination of 3-chlorobenzoate. The methanogen utilized the remainder of the hydrogen produced to reduce CO_2 to methane. Tschech and Schink [74] isolated two strictly
anaerobic gram-positive, spore-forming, motile syntrophs from freshwater sediments. The

isolates, which were placed in the genus *Clostridium*, degraded resorcinol and 2,4- and 2,6-dihydroxybenzoate in co-culture with *Campylobacter* spp., an association that did not involve interspecies hydrogen transfer. The nature of the dependance of degradation on the *Campylobacter* sp. could not be determined. However, additional isolates from marine sediments were obtained in association with sulfate reducers and methanogens. Benzoate, 2-hydroxybenzoate, and 3-hydroxybenzoate were syntrophically degraded by gram-negative anaerobes isolated in co-culture with *D. vulgaris* [75].

Syntrophic associations involving interspecies hydrogen transfer are not restricted to the degradation of aromatic compounds. Obligately syntrophic organisms have been isolated and described for the anaerobic degradation of propionate, butyrate, and volatile fatty acids up to C-7 [82,84–86]. As with many aromatic compounds, fermentation of fatty acids with the production of H_2 as the electron sink is thermodynamically unfavorable. Syntrophic association with hydrogenotrophs therefore lowers the H_2 partial pressure sufficiently so that the reactions become favorable. Fatty acid-fermenting syntrophs probably utilize the β-oxidation pathway. *Syntrophomonas wolfei* [87] has been shown to possess all of the necessary enzymes required for β-oxidation [88]. Syntrophic organisms occupy an important niche in many anaerobic systems by completing the mineralization of less energetically favorable end products produced by fermentation of complex organic compounds. These organisms usually have slow growth rates in the laboratory, so that aromatic compounds with two or fewer hydroxyl groups may not be degraded in an anaerobic habitat like the rumen, which has a relatively short turnover time [69].

21.4.3 Metabolism of Ring Methoxyl Groups

Anaerobes have been isolated that remove the methyl group from aromatic compounds such as syringic acid (4-hydroxy-3,5-dimethoxybenzoate) and ferulic acid. O-Demethylation results in the formation of hydroxylated aromatics [89], discussed previously as being important in determining subsequent degradation pathways. The methoxyl groups may be metabolized by dissimilation to acetate through the folate-dependent Wood pathway [90]. Bache and Pfennig [89] were the first to show anaerobic O-demethylation of aromatic compounds for strains of *Acetobacterium woodii* isolated from freshwater sediment and sewage sludge enrichments with vanillate (4-hydroxy-3-methoxybenzoate), syringate, and 3,4,5-trimethoxycinnamate as substrates. Ten different methoxylated aromatics were stoichiometrically metabolized to their corresponding hydroxylated derivatives and acetate; however, the aromatic ring was not attacked. Growth yields of *A. woodii* on methoxylated aromatics were increased by reduction of the double bond in the acrylate side chain of caffeate. A corresponding decrease in acetate formation occurred [89,91]. Frazer and Young [92] isolated an organism that O-demethylated syringate, vanillate, and ferulate. Radiotracer studies with vanillate showed that the O-methyl group carbon was converted to CO_2 and acetate in a 2:1 ratio [93]. As with *A. woodii*, reduction of the double bond in the side chain of ferulate resulted in decreased acetate formation, suggesting the diversion of electrons of methoxyl group oxidation from acetate production to reduction of the acrylate side chain of ferulate. Reduction of the side chain, in place of acetate formation, may result in a greater overall negative free-energy change [91]. A facultative anaerobic *Enterobacter* sp. was isolated from ferulate enrichment cultures that transformed ferulate aerobically and anaerobically [94]. Anaerobically, O-demethylation, reductive dehydroxylation, and reduction of the acrylate side chain were observed, but the aromatic ring itself was not attacked. The organism contained two plasmids that did not appear to be involved in coding for enzymes involved in ferulate metabolism [95]. *Clostridium pfennigii*, isolated from steer rumen fluid, metabolized methoxylated aromatics to their hydroxylated derivatives and butyrate [96]. In addition, pyruvate or carbon monoxide could be utilized as substrates. Mountfort and Asher [97] have isolated an anaerobe that metabolized the methoxyl group of several methoxylated aromatics but did not reduce the side-chain double bond of phenylacrylate derivatives. Rather, butyrate and formate were produced, in addition to acetate. DeWeerd et al. [98] concluded that there was no relationship between the enzyme systems for O-demethylation of 3-methoxybenzoate and reductive dechlorination of 3-chlorobenzoate by DCB-1, an anaerobic dechlorinating organism.

21.5 METHANOGENIC FERMENTATION OF AROMATICS

The first demonstration that aromatic compounds were degraded under anaerobic conditions was shown by Tarvin and Buswell [4] when it was shown that several aromatic compounds were completely converted to methane and CO_2 by anaerobic enrichment cultures. Clark and Fina [99] confirmed that benzoic acid could be degraded to methane and CO_2 by anaerobic enrichments. However, the substitution of catechol and protocatechuate in place of benzoate stopped gas production, suggesting that these key intermediates of aerobic degradation pathways [2] were not involved. The low recovery of $^{14}CH_4$ added as [^{14}C] bicarbonate, suggested that benzoate was the source of most of the methane carbon. Benzoate labeled at carbons 1 and 7 (ring and carboxyl carbons, respectively), when added to rumen fluid or sewage sludge-derived enrichment cultures, converted the carboxyl carbon to CO_2 while the ring carbon was metabolized to methane [100]. Nottingham and Hungate [101] also used ring-labeled [^{14}C] benzoate to show anaerobic degradation of benzoate to methane and CO_2 by a sewage sludge-derived mixed culture. The fate of carbon-4 of benzoate in sewage sludge enrichments was shown by Fina et al. [102] to favor production of CO_2 over CH_4 by a factor of about 4.5 to 1. Propionate was detected as an intermediate with the carboxyl group containing the C-4 of benzoate. Acetate and butyrate were the only labeled volatile fatty acids detected when [^{14}C] benzoate was added to methanogenic enrichment cultures [103]. Ferry and Wolfe [73] examined the microorganisms involved in methanogenic benzoate degradation. They showed that at least two populations of organisms were necessary for benzoate degradation. One group was involved in degradation of benzoate to H_2, formate, acetate, and CO_2, which were subsequently utilized as substrates by methanogenic bacteria. Based on the thermodynamics of the reactions involved, they concluded that a symbiotic relationship existed in which product removal by the methanogens was necessary for benzoate degradation to be energetically favorable. Grbic-Galic and Young [104] showed that the inhibition of methanogenesis with bromoethanesulfonic acid (BESA) still allowed aromatic degradation, but the concentration of metabolic intermediates increased. Studies on benzoate metabolism in lake sediment samples showed an almost immediate degradation of benzoate to methane, but that enrichment cultures required a long lag phase that was reduced by prior adaptation to fatty acid degradation [105]. In other studies acetate shortened the lag phase for benzoate degradation presumably by stimulating methanogens, which remove end products of benzoate fermentation [99]. Cyclohexane carboxylate (CHCA) and 2-hydroxy CHCA, intermediates of the proposed reductive pathway of anaerobic aromatic degradation [6], were degraded without a lag in the benzoate-degrading consortia described by Ferry and Wolfe [73], while cyclohex-1-ene-carboxylate, another proposed intermediate, did not support methanogenesis. However, this compound was detected in extracts of sheep rumen liquor cultures that produced methane from benzoate [106]. The absence of an uptake mechanism for some intermediate compounds may preclude their use as exogenously supplied substrates. Keith et al. [107] found that benzoate-degrading sewage sludge enrichments utilized cyclohex-1-ene-carboxylate. Grbic-Galic and Young [104] proposed a reductive pathway that linked the degradation of benzoate and ferulate through common intermediates. Several fatty acids, including seven-carbon mono- and dicarboxylic acids, as well as straight- and branched-chain volatile fatty acids, were detected in their benzoate and ferulate enrichment cultures. Although acetate-utilizing methanogens have not been isolated from any methanogenic benzoate enrichments, organisms with a morphology similar to acetate-degrading *Methanothrix* sp. have been observed [73,105].

Benzoate has served as an effective model compound for investigating anaerobic aromatic metabolism, but other aromatic compounds have also been shown to be degraded to CH_4 and CO_2. Enrichment cultures derived from several different habitats were shown to degrade aromatic amino acids to CH_4 and CO_2 [108]. Horowitz et al. [109a] compared the anaerobic aromatic biodegradation potential of lake sediment and sewage sludge. Various substituted benzoate and benzene derivatives were degraded under methanogenic conditions. However, analines were generally resistant to degradation. Indole, a potential intermediate of the anaerobic degradation of tryptophan [108] was degraded in methanogenic cultures established with inoculum from an anaerobic activated carbon filter used to treat water containing indole, quinoline, and methylquinoline [109b]. Healy et al. [110] showed that a methanogenic consortium degraded the aromatic lignin derivative ferulate by pathways that converged with proposed benzoate degradation pathways. Healy and Young [111] obtained enrichment cultures capable of producing CH_4 and CO_2 from catechol and phenol. Healy and Young [112] also found that 11 different aromatic compounds, several of which were lignin monomers, supported methanogenesis. There was a substantial lag phase in gas production after first exposure to the compounds, but upon subsequent feeding, adaptation of the microbial community shortened or eliminated the delay. Some cultures were simultaneously adapted to structurally similar compounds that had not been supplied as substrates originally. Dwyer et al. [113] studied the kinetics of methanogenic phenol degradation by a bacterial consortium immobilized in thin spaghetti-like strands of agar. Immobilization reduced the maximum degradation rate and apparent Km, while providing protection from otherwise inhibitory substrate concentrations. Balba et al. [114] examined degradation of methoxy-benzoates and protocatechuic acid (3,4-dihydroxybenzoate) in methanogenic consortia enriched with benzoate. O-Demethylation occurred readily, but the resulting phenolics required an adaptation period prior to degradation to methane. Knoll and Winter [115] found that the presence of high levels of H_2 and CO_2 in a methanogenic sewage sludge-derived consortium reduced the rate of phenol degradation and that benzoate was formed from phenol and CO_2. Decarboxylation of *p*-hydroxybenzoate to phenol and proto-catechuic acid to catechol was observed. Methanogenesis from methoxylated aromatics involved fermentation of methoxyl groups to acetate by organisms similar to *A. woodii* [77]. The resulting hydroxylated aromatic was further degraded to acetate and was not dependent on product removal by methanogens if three hydroxyl groups were present on the ring. If, however, catechol was formed, the ring was not degraded. The reason for the latter result was not clear since the calculated free-energy change for conversion of catechol to methane and CO_2 is favorable. Other investigators have found catechol to be degradable in methanogenic enrichments [111,112,114]. Hydroquinone (1,4-dihydroxybenzene) and catechol (with hydroxyl groups para and ortho to one another, respectively) were degraded in methanogenic cultures, probably after dehydroxylation to phenol [78]. In contrast, resorcinol and resorcylates with meta hydroxyl groups were not degraded by the enrichments. Tschech and Schink [74] found that enrichments supplied with resorcinol or resorcylate isomers degraded these compounds to methane. Freshwater isolates were obtained from the enrichments in defined co-culture with a *Campylobacter* sp. that degraded 2,4- and 2,6-dihydroxybenzoate to acetate and butyrate. Hydrogen-utilizing methanogens or sulfate reducers would not replace *Campylobacter*, and its role remained unexplained. Marine isolates degraded the two aromatics to acetate and H_2 and required co-culture with methanogens to scavenge H_2. It was suggested that hydroxyl groups in positions meta to one another, as in resorcinol and resorcylates, are in favorable positions for degradation to acetate and butyrate, while ortho- and para-positioned dihydroxylated aromatics, such as catechol and hydroquinone, must be dehydroxylated to phenol prior to degradation. Reduction of the aromatic ring may not be necessary since the hydroxyl groups could stabilize a dione tautomer, which could then serve as a ring cleavage substrate [74]. Young and Rivera [116] proposed phenol as the central intermediate for methanogenic metabolism of substituted phenols, since phenol was detected after BESA inhibition of methanogenesis of mixed anaerobic enrichment cultures that degraded *p*-cresol, hydroquinone, and phloroglucinol.

Field and Lettinga [117] investigated the toxicity of gallotannic acid to methanogenesis during the anaerobic digestion of volatile fatty acids by granulated sludge. Gallotannic acid, a polyester of gallic acid, severely decreased rates of methane production relative to unamended controls and those with additions of gallic acid and pyrogallol. However, gallotannic acid as well as gallic acid and pyrogallol were rapidly degraded by the sludge.

Aromatic compounds with ring substituents other than, or in addition to, carboxyl and hydroxyl groups have been shown to be susceptible to degradation in anaerobic habitats. Sewage sludge- and sediment-derived enrichment cultures reductively dehalogenated aromatic compounds [118]. Complete mineralization to methane and CO_2 occurred for some compounds that required complete dehalogenation before the aromatic ring could be attacked. A lag time in dehalogenation occurred for chloro-, bromo-, and iodobenzoates in lake sediment upon initial exposure; however, the lag was eliminated or greatly reduced after acclimation to the substrate [119]. Number, ring position, and type of substituent groups all affected degradation. The kinetics of dechlorination of various chlorobenzoates in anaerobic lake sediments and methanogenic enrichment cultures was shown to follow Michaelis-Menten kinetics for the first chlorine atom. If, however, a second chlorine atom was present, its removal was competitively inhibited by the parent substrate. This effect was found for both 3,5-dichlorobenzoate and 4-amino-3,5-dichlorobenzoate [120]. Boyd and Shelton [121] compared the degradation of isomers of mono- and dichlorophenols in acclimated and unacclimated sewage digestor sludge. Cross-acclimation studies suggested that different populations of microorganisms were responsible for degradation of 2- and 3-chlorophenol. While both populations separately degraded 4-chlorophenol, a mixture of the two appeared to be responsible for 2,4- and 3,4-dichlorophenol degradation. Shelton and Tiedje [82] isolated several bacteria of different trophic levels from enrichment cultures growing on 3-chlorobenzoate. Degradation of the substrate apparently involved a symbiotic relationship between a dechlorinating organism, a benzoate degrader, and hydrogen oxidizers. An organism capable of reductively de-chlorinating 3-chlorobenzoate to benzoate was isolated and found to grow on a rumen fluid medium. An assessment of the anaerobic degradation of substituted phenols showed that ease of degradation was dependent on the type and position of the substituent group [122]. *para*-Chlorophenol and *o*-cresol (2-methyl phenol) were the most resistant to degradation. Methoxylated phenols were demethylated to the corresponding hydroxyl-ated phenol. Nitrophenols were also degraded. Pentachlorophenol, a widely used preserv-ative and pesticide, was degraded to methane and CO_2 after sequential reductive removal of chlorine by two to three microbial populations. Pentabromophenol was similarly de-graded [123]. Gibson and Suflita [124] examined four anaerobic habitats for the ability to degrade benzoate, phenol, and chlorinated aromatic pesticides. While benzoate and phenol were degraded anaerobically at all sites, degradation of various chlorinated aro-matics varied among the habitats. The presence of sulfate inhibited dechlorination. Smo-lenski and Suflita [125] found that *o*-, *m*-, and *p*-cresol degradation in anaerobic alluvial sand aquifer material was greater under sulfate-reducing conditions than under methano-genic conditions. The aromatic hydrocarbons toluene and benzene were degraded in methanogenic consortia via ring hydroxylation as shown by incorporation of ^{18}O-labeled water [126]. A major route for toluene degradation in the cultures appeared to involve *p*-cresol and benzoic acid as intermediates while benzene degradation involved phenol as an intermediate [127].

21.6 ANAEROBIC LIGNIN DEGRADATION

Lignin is a heterogeneous high-molecular-weight aromatic polymer in which the basic phenylpropane subunits are randomly linked by intercarbon and ether bonds between aromatic nuclei and side chains [128]. Lignin, the second most abundant terrestrially produced organic polymer [129] is the greatest source of the benzene ring in the environ-ment [130]. Lignin is resistant to microbial attack and until recently was not considered to be degraded under anaerobic conditions [131–133]. Vanillic acid, a monoaromatic lignin derivative, and an aromatic dimer with an ether linkage were degraded in methano-genic lake sediments, but high-molecular-weight synthetic lignin was not [134]. This led Zeikus et al. [134] to conclude that the high-molecular-weight property was the major impediment to anaerobic lignin degradation. Dehydrodivanillin, a diaromatic lignin derivative, was shown to be degraded anaerobically by a mixed culture of rumen or-ganisms [135] but contrary to most studies, little gas production was reported. Colberg and Young [136,137] have shown an inverse correlation between increasing lignin molecular weight and its susceptibility to anaerobic degradation. Solubilized lignin frag-ments, "oligolignols," with a molecular weight of approximately 600 were shown to be partially degraded to volatile fatty acids, monoaromatic compounds, methane, and CO_2 in anaerobic enrichment cultures [138]. Inhibition of methanogenesis with BESA re-sulted in increased conversion of substrate and accumulation of intermediate degradation products relative to uninhibited cultures. Apparently, interspecies hydrogen transfer [139] of reducing equivalents to methanogens is not required for anaerobic depolymeri-zation of lignin oligomers to monoaromatic subunits. Recently, Benner et al. [140,141] and Benner and Hodson [142] using sensitive radioisotope techniques, have shown slow but significant degradation of high-molecular-weight lignin derived from a variety of sources; degradation occurred in anaerobic freshwater and marine sediments. Coniferyl alcohol, an aromatic lignin precursor, was completely degraded to CO_2 and CH_4, with ferulic acid appearing as an intermediate [143]. Ferulic acid was shown by Healy et al. [110] to be degraded methanogenically through a pathway proposed to merge with the benzoate degradation pathway proposed by Evans [6]. Methanogenic degradation has been demonstrated for a large number of monoaromatic lignin derivatives [77,104,112]. Limited attack on ring substituents of hydroxycinnamic acids in anaerobic enrichment cultures was shown by Nali et al. [144]. Reactions involing reductive removal of p-hydroxyl groups and reduction of the side-chain double bond were indicated by the degradation products. Ohmiya et al. [145] isolated an organism identified as *Wolinella succinogenes* that utilized ferulic acid as an electron acceptor when grown on a yeast extract medium. Disruption of methanogenesis by BESA resulted in the accumulation of a number of intermediate ferulate degradation products that were not detected in uninhibited cultures [146]. Removal of methanogenesis as an electron sink for anaerobic aromatic degradation appeared to cause a shift from usual degradation pathways. Balba and Evans [147] showed that the C_3-phenyl compounds phenylpropionate and cinnamic acid were degraded without a lag to methane and CO_2 by benzoate-adapted cultures, while phenylacetate degradation required an adaptation period. They suggested that β-oxidation of phenylpropionate yielded benzoate, which was readily degraded by the benzoate-adapted microbial population, while β-oxidation of phenylacetate was blocked since the carbon beta to the carboxyl group was part of the aromatic ring.

21.7 CONCLUSIONS

Despite the fact that aromatic rings have increased stability from the negative resonance energy through delocalization of π electrons in the ring, numerous microorganisms utilize aromatic compounds as sources of carbon, energy, and even electron acceptors. In aerobic environments aromatic compounds are more readily degraded by the involvement of oxygenase enzymes and with oxygen as the terminal electron acceptor. However, in anaerobic environments microorganisms have evolved diverse strategies to unlock the potential of aromatics to support growth. Ring reduction, co-metabolism of substrates, and interspecies hydrogen transfer result in the anaerobic metabolism of a wide range of substituted aromatic compounds. The oxidation level of ring carbons and the presence or absence of terminal electron acceptors appear to be major determining factors governing the mechanisms involved in degradation of aromatic compounds in anoxic habitats. Consequently, there is great metabolic diversity among anaerobes which is highlighted by the fact that many of the described organisms are often new species.

The activity of anaerobes is important in carbon cycling in anoxic habitats and for the removal of toxic compounds released into the environment. The potential for exploitation of their specialized metabolic reactions has not been overlooked; however, the lack of basic research concerning the physiology, biochemistry, and genetics of anaerobic aromatic degradation pathways currently limits their potential usefulness.

ACKNOWLEDGMENTS

We appreciate the support of Houston Lighting and Power and the Biobased Materials Center of the Virginia Center for Innovative Technology for our research, and the efforts of Susan B. Irons, who typed the manuscript.

REFERENCES

1. Evans, W. C., The microbial degradation of aromatic compounds, *J. Gen. Microbiol.*, *32*, 177-184 (1963).
2. Gibson, D. T., Microbial degradation of aromatic compounds, *Science, 161*, 1093-1097 (1968).
3. Dagley, S., Catabolism of aromatic compounds by microorganisms, *Adv. Microb. Physiol.*, *6*, 1-46 (1971).
4. Tarvin, D., and Buswell, A. M., The methane fermentation of organic acids and carbohydrates, *J. Am. Chem. Soc., 56*, 1751-1755 (1934).
5. Proctor, M. H., and Scher, S., Decomposition of benzoate by a photosynthetic bacterium, *Proc. Biochem. Soc., 76*, 33 (1960).
6. Evans, W. C., Biochemistry of the bacterial catabolism of aromatic compounds in anaerobic environments, *Nature, 270*, 17-22 (1977).
7. Sleat, R., and Robinson, J. P., The bacteriology of anaerobic degradation of aromatic compounds, *J. Appl. Bacteriol., 57*, 381-394 (1984).
8. Young, L. Y., Anaerobic degradation of aromatic compounds, in *Microbial Degradation of Organic Compounds* (D. T. Gibson, ed.), Marcel Dekker, New York, pp. 487-523 (1984).
9. Berry, D. F., Francis, A. J., and Bollag, J. M., Microbial metabolism of homocyclic and heterocyclic aromatic compounds under anaerobic conditions, *Microbiol. Rev., 51*, 43-59 (1987).
10. Evans, W. C., and Fuchs, G., Anaerobic degradation of aromatic compounds, *Annu. Rev. Microbiol., 42*, 289-317.
11. Scheline, R. R., Metabolism of foreign compounds by gastrointestinal microorganisms, *Pharmacol. Rev., 25*, 451-523 (1973).
12. Barker, H. A., Amino acid degradation by anaerobic bacteria, *Annu. Rev. Biochem., 50*, 23-40 (1981).
13. Scher, S., and Proctor, M. H., Studies with photosynthetic bacteria: anaerobic oxidation of aromatic compounds, in *Symposium on Comparative Biochemistry of Photoreactive Systems* (M. B. Allen, ed.), Academic Press, New York, 1960, pp. 387-394.
14. Dutton, P. L., and Evans, W. C., Dissimilation of aromatic substrates by *Rhodopseudomonas palustris, Biochem. J., 104*, 30-31 (1967).
15. Hegeman, G. D., The metabolism of p-hydroxybenzoate by *Rhodopseudomonas palustris* and its regulation, *Arch. Mikrobiol., 59*, 143-148 (1967).
16. Dutton, P. L., and Evans, W. C., The photometabolism of benzoic acid by *Rhodopseudomonas palustris*: a new pathway of aromatic ring metabolism, *Biochem. J., 109*, 5p-6p (1968).
17. Dutton, P. L., and Evans, W. C., The metabolism of aromatic compounds by *Rhodopseudomonas palustris, Biochem. J., 113*, 525-536 (1969).
18. Guyer, M., and Hegeman, G., Evidence for a reductive pathway for the anaerobic metabolism of benzoate, *J. Bacteriol., 99*, 906-907 (1969).
19. Dutton, P. L., and Evans, W. C., Inhibition of the photometabolism of aromatic acids in *Rhodopseudomonas palustris* by a lipid component of the organism, *Biochem. J., 107*, 28p-29p (1968).
20. Dutton, P. L., and Evans, W. C., Inhibition of aromatic photometabolism in *Rhodopseudomonas palustris* by fatty acids, *Arch. Biochem. Biophys., 136*, 228-232 (1970).
21. Whittle, P. J., Lunt, D. O., and Evans, W. C., Anaerobic photometabolism of aromatic compounds by *Rhodopseudomonas* sp., *Biochem. Soc. Trans., 4*, 490-491 (1976).

22. Hutber, G. N., and Ribbons, D. W., Involvement of coenzyme A esters in the metabolism of benzoate and cyclohexanecarboxylate by *Rhodopseudomonas palustris*, *J. Gen. Microbiol.*, *129*, 2413-2420 (1983).

23. Harwood, C. S., and Gibson, J., Uptake of benzoate by *Rhodopseudomonas palustris* grown anaerobically in light, *J. Bacteriol.*, *165*, 504-509 (1986).

24. Geissler, J. F., Harwood, C. S., and Gibson, J., Purification and properties of benzoate-coenzyme A ligase, a *Rhodopseudomonas palustris* enzyme involved in the anaerobic degradation of benzoate, *J. Bacteriol.*, *170*, 1709-1714 (1988).

25. Merkel, S. M., Eberhard, A. E., Gibson, J., and Harwood, C. S., Involvement of coenzyme A thioesters in anaerobic metabolism of 4-hydroxybenzoate by *Rhodopseudomonas palustris*, *J. Bacteriol.*, *171*, 1-7 (1989).

26. Harwood, C. S., and Gibson, J., Anaerobic and aerobic metabolism of diverse aromatic compounds by the photosynthetic bacterium *Rhodopseudomonas palustris*, *Appl. Environ. Microbiol.*, *54*, 712-717 (1988).

27. Oshima, T., On the anaerobic metabolism of aromatic compounds in the presence of nitrate by soil microorganisms, *Z. Allg. Mikrobiol.*, *5*, 386-394 (1965).

28. Taylor, B. F., Campbell, W. L., and Chinoy, I., Anaerobic degradation of the benzene nucleus by a facultatively anaerobic microorganism, *J. Bacteriol.*, *102*, 430-437 (1970).

29. Taylor, B. F., and Heeb, M. J., The anaerobic degradation of aromatic compounds by a denitrifying bacterium, *Arch. Microbiol.*, *83*, 165-171 (1972).

30. Taylor, B. F., Aerobic and anaerobic catabolism of vanillic acid and some other methoxy-aromatic compounds by *Pseudomonas* sp. strain PN-1, *Appl. Environ. Microbiol.*, *46*, 1286-1292 (1983).

31. Blake, C. K., and Hegeman, G. D., Plasmid pCB1 carries genes for anaerobic benzoate catabolism in *Alcaligenes xylosoxidans* subsp. *denitrificans* PN-1, *J. Bacteriol.*, *169*, 4878-4883 (1987).

32. Williams, R. J., and Evans, W. C., Anaerobic metabolism of aromatic substrates by certain micro-organisms, *Biochem. Soc. Trans.*, *1*, 186-187 (1973).

33. Williams, R. J., and Evans, W. C., The metabolism of benzoate by *Moraxella* species through anaerobic nitrate respiration, *Biochem. J.*, *148*, 1-10 (1975).

34. Bakker, G., Anaerobic degradation of aromatic compounds in the presence of nitrate, *FEMS Microbiol. Lett.*, *1*, 103-108 (1977).

35. Aftring, R. P., Chalker, B. E., and Taylor, B. F., Degradation of phthalic acids by denitrifying, mixed cultures of bacteria, *Appl. Environ. Microbiol.*, *41*, 1177-1183 (1981).

36. Aftring, R. P., and Taylor, B. F., Aerobic and anaerobic catabolism of phthalic acid by a nitrate-respiring bacterium, *Arch. Microbiol.*, *130*, 101-104 (1981).

37. Taylor, B. F., and Ribbons, D. W., Bacterial decarboxylation of *o*-phthalic acids, *Appl. Environ. Microbiol.*, *46*, 1276-1281 (1983).

38. Nozawa, T., and Maruyama, Y., Dentrification by a soil bacterium with phthalate and other aromatic compounds as substrates, *J. Bacteriol.*, *170*, 2501-2505 (1988).

39. Bossert, I. D., Rivera, M. D., and Young, L. Y., *p*-Cresol biodegradation under denitrifying conditions: isolation of a bacterial coculture, *FEMS Microbiol. Ecol.*, *38*, 313-319 (1986).

40. Bossert, I. D., and Young, L. Y., Anaerobic oxidation of *p*-cresol by a denitrifying bacterium, *Appl. Environ. Microbiol.*, *52*, 1117-1122 (1986).

41. Zeyer, J., Kuhn, E. P., and Schwarzenbach, R. P., Rapid microbial mineralization of toluene and 1,3-dimethylbenzene in the absence of molecular oxygen, *Appl. Environ. Microbiol.*, *52*, 944-947 (1986).

42. Kuhn, E. P., Zeyer, J., Eicher, P., and Schwarzenbach, R. P., Anaerobic degradation of alkylated benzenes in denitrifying laboratory aquifer columns, *Appl. Environ. Microbiol.*, *54*, 490-496 (1988).

43. Braun, K., and Gibson, D. T., Anaerobic degradation of 2-aminobenzoate (anthranilic acid) by denitrifying bacteria, *Appl. Environ. Microbiol.*, *48*, 102–107 (1984).

44. Taylor, B. F., Hearn, W. L., and Pincus, S., Metabolism of monofluoro- and monochlorobenzoates by a denitrifying bacterium, *Arch. Microbiol.*, *122*, 301–306 (1979).

45. Schennen, W., Braun, K., and Knackmuss, H.-J., Anaerobic degradation of 2-fluorobenzoate by benzoate-degrading, denitrifying bacteria, *J. Bacteriol.*, *161*, 321–325 (1985).

46. Ziegler, K., Braun, K., Böckler, A., and Fuchs, G., Studies on the anaerobic degradation of benzoic acid and 2-aminobenzoic acid by a denitrifying *Pseudomonas* strain, *Arch. Microbiol.*, *149*, 62–69 (1987).

47. Widdel, F., Methods for enrichment and pure culture isolation of filamentous gliding sulfate-reducing bacteria, *Arch. Microbiol.*, *134*, 282–285 (1983).

48. Widdel, F., Kohring, G. W., and Mayer, F., Studies on dissimilatory sulfate-reducing bacteria that decompose fatty acids. III. Characterization of the filamentous gliding *Desulfonema limicola* gen. nov., sp. nov. and *Desulfonema magnum* sp. nov., *Arch. Microbiol.*, *134*, 286–294 (1983).

49. Imhoff-Stuckle, D., and Pfennig, N., Isolation and characterization of a nicotinic acid-degrading sulfate-reducing bacterium, *Desulfococcus niacini* sp. nov., *Arch. Microbiol.*, *136*, 194–198 (1984).

50. Bak, F., and Widdel, F., Anaerobic degradation of indolic compounds by sulfate-reducing enrichment cultures, and description of *Desulfobacterium indolicum* gen. nov., sp. nov., *Arch. Microbiol.*, *146*, 170–176 (1986).

51. Bak, F., and Widdel, F., Anaerobic degradation of phenol derivatives by *Desulfobacterium phenolicum* sp. nov., *Arch. Microbiol.*, *146*, 177–180 (1986).

52. Szewzyk, R., and Pfennig, N., Complete oxidation of catechol by the strictly anaerobic sulfate-reducing *Desulfobacterium catecholicum* sp. nov., *Arch. Microbiol.*, *147*, 163–168 (1987).

53. Grbic-Galic, D., *O*-demethylation, dehydroxylation, ring-reduction and cleavage of aromatic substrates by Enterobacteriaceae under anaerobic conditions, *J. Appl. Bacteriol.*, *61*, 491–497 (1986).

54. Harary, I., Bacterial fermentation of nicotinic acid, *J. Biol. Chem.*, *227*, 815–822 (1957).

55. Harary, I., Bacterial fermentation of nicotinic acid. II. Anaerobic reversible hydroxylation of nicotinic acid to 6-hydroxynicotinic acid, *J. Biol. Chem.*, *227*, 823–831 (1957).

56. Pastan, I., Tsai, L., and Stadtman, E. R., Nicotinic acid metabolism, *J. Biol. Chem.*, *239*, 902–906 (1964).

57. Stadtman, E. R., Stadtman, T. C., Pastan, I. R. A., and Smith, L. D. S., *Clostridium barkeri* sp. n., *J. Bacteriol.*, *110*, 758–760 (1972).

58. Tsai, L. I., Pastan, E., and Stadtman, E. R., Nicotinic acid metabolism. II. The isolation and characterization of intermediates in the fermentation of nicotinic acid, *J. Biol. Chem.*, *241*, 1807–1813 (1966).

59. Kung, H. F., Cederbaum, S., Tsai, L., and Stadtman, T. C. Nicotinic acid metabolism. V. A cobamide coenzyme-dependent conversion of alpha-methyleneglutaric acid to dimethylmaleic acid, *Proc. Natl. Acad. Sci. USA*, *65*, 978–984 (1970).

60. Holcenberg, J. S., and Stadtman, E. R., Nicotinic acid metabolism. III. Purification and properties of nicotinic acid hydroxylase, *J. Biol. Chem.*, *244*, 1194–1203 (1969).

61. Imhoff, D., and Andreesen, J. R., Nicotinic acid hydroxylase from *Clostridium barkeri*: selenium-dependent formation of active enzyme, *FEMS Microbiol. Lett.*, *5*, 155–158 (1979).

62. Holcenberg, J. S., and Tsai, L., Nicotinic acid metabolism. IV. Ferredoxin-dependent reduction of 6-hydroxynicotinic acid to 6-oxo-1,4,5,6-tetrahydronicotinic acid, *J. Biol. Chem.*, *244*, 1204-1211 (1969).

63. Tsai, C. G., and Jones, G. A., Isolation and identification of rumen bacteria capable of anaerobic phloroglucinol degradation, *Can. J. Microbiol.*, *21*, 794-801 (1975).

64. Tsai, C. G., Gates, D. M., Ingledew, W. M., and Jones, G. A., Products of anaerobic phloroglucinol degradation by *Coprococcus* sp. Pe₁5[1,2], *Can. J. Microbiol.*, *22*, 159-164 (1976).

65. Patel, T. R., Jure, K. G., and Jones, G. A., Catabolism of phloroglucinol by the rumen anaerobe *Coprococcus*, *Appl. Environ. Microbiol.*, *42*, 1010-1017 (1981).

66. Schink, B., and Pfennig, N., Fermentation of trihydroxybenzenes by *Pelobacter acidigallici* gen. nov. sp. nov., a new strictly anaerobic, non-sporeforming bacterium, *Arch. Microbiol.*, *133*, 195-201 (1982).

67. Samain, E., Albagnac, G., and Dubourguier, H. C., Initial steps of catabolism of trihydroxybenzenes in *Pelobacter acidigallici*, *Arch. Microbiol.*, *144*, 242-244 (1986).

68. Kreikenbohm, R., and Pfennig, N., Anaerobic degradation of 3,4,5-trimethoxybenzoate by a defined mixed culture of *Acetobacterium woodii*, *Pelobacter acidigallici*, and *Desulfobacter postgatei*, *Microb. Ecol.*, *31*, 29-38 (1985).

69. Krumholz, L. R., and Bryant, M. P., *Eubacterium oxidoreducens* sp. nov. requiring H₂ or formate to degrade gallate, pyrogallol, phloroglucinol and quercetin, *Arch. Microbiol.*, *144*, 8-14 (1986).

70. Krumholz, L. R., Crawford, R. L., Hemling, M. E., and Bryant, M. P., A rumen bacterium degrading quercetin and trihydroxybenzenoids with concurrent use of formate or H₂, in *Plant Flavinoids in Biology and Medicine: Biochemical, Pharmacological, and Structure-Activity Relationships*, Alan R. Liss, New York, 1986, pp. 211-214.

71. Krumholz, L. R., Crawford, R. L., Hemling, M. E., and Bryant, M. P., Metabolism of gallate and phloroglucinol in *Eubacterium oxidoreducens* via 3-hydroxy-5-oxohexanoate, *J. Bacteriol.*, *169*, 1886-1890 (1987).

72a. Krumholz, L. R., and Bryant, M. P., Characterization of the pyrogallol-phloroglucinol isomerase of *Eubacterium oxidoreducens*, *J. Bacteriol.*, *170*, 2472-2479 (1988).

72b. Haddock, J. D., and Ferry, J. G., 1989. Purification and properties of phloroglucinol reductase from *Eubacterium oxidoreducens* G-41, *J. Biol. Chem.*, *264*, 4423-4427.

73. Ferry, J. G., and Wolfe, R. S., Anaerobic degradation of benzoate to methane by a microbial consortium, *Arch. Microbiol.*, *107*, 33-40 (1976).

74. Tschech, A., and Schink, B., Fermentative degradation of resorcinol and resorcylic acids, *Arch. Microbiol.*, *143*, 52-59 (1985).

75. Tschech, A., and Schink, B., Fermentative degradation of monohydroxybenzoates by defined syntrophic cocultures, *Arch. Microbiol.*, *145*, 396-402 (1986).

76. Barik, S., Brulla, W. H., and Bryant, M. P., PA-1, a versatile anaerobe obtained in pure culture, catabolizes benzenoids and other compounds in syntrophy with hydrogenotrophs, and P-2 plus *Wolinella* sp. degrades benzenoids, *Appl. Environ. Microbiol.*, *50*, 304-310 (1985).

77. Kaiser, J.-P., and Hanselmann, K. W., Fermentative metabolism of substituted monoaromatic compounds by a bacterial community from anaerobic sediments, *Arch. Microbiol.*, *133*, 185-194 (1982).

78. Szewzyk, U., Szewzyk, R., and Schink, B., Methanogenic degradation of hydroquinone and catechol via reductive dehydroxylation to phenol, *FEMS Microbiol. Ecol.*, *31*, 79-87 (1985).

79. Mountfort, D. O., and Bryant, M. P., Isolation and characterization of an anaerobic syntrophic benzoate-degrading bacterium from sewage sludge, *Arch. Microbiol.*, *133*, 249-256 (1982).

80. Mountfort, D. O., Brulla, W. J., Krumholz, L. R., and Bryant, M. P., *Syntrophus buswellii* gen. nov., sp. nov.: a benzoate catabolizer from methanogenic ecosystems, *Int. J. Syst. Bacteriol., 34*, 216–217 (1984).

81. Wolin, M. J., Wolin, E. A., and Jacobs, N. J., Cytochrome-producing anaerobic vibrio, *Vibrio succinogenes*, sp. n., *J. Bacteriol., 81*, 911–917 (1961).

82. Shelton, D. R., and Tiedje, J. M., Isolation and partial characterization of bacteria in an anaerobic consortium that mineralizes 3-chlorobenzoic acid, *Appl. Environ. Microbiol., 48*, 840–848 (1984).

83. Dolfing, J., and Tiedje, J. M., Hydrogen cycling in a three-tiered food web growing on the methanogenic conversion of 3-chlorobenzoate, *FEMS Microbiol. Ecol., 38*, 293–298 (1986).

84. Boone, D. R., and Bryant, M. P., Propionate-degrading bacterium, *Syntrophobacter wolinii* sp. nov. gen. nov., from methanogenic ecosystems, *Appl. Environ. Microbiol., 40*, 626–632 (1980).

85. McInerney, M. J., Bryant, M. P., and Pfennig, N., Anaerobic bacterium that degrades fatty acids in syntrophic association with methanogens, *Arch. Microbiol., 122*, 129–135 (1979).

86. Henson, J. M., and Smith, P. H., Isolation of a butyrate-utilizing bacterium in co-culture with *Methanobacterium thermoautotrophicum* from a thermophilic digester, *Appl. Environ. Microbiol., 49*, 1461–1466 (1985).

87. McInerney, M. J., Bryant, M. P., Hespell, R. B., and Costerton, J. W., *Syntrophomonas wolfei* gen. nov. sp. nov., an anaerobic, syntrophic, fatty acid–oxidizing bacterium, *Appl. Environ. Microbiol., 41*, 1029–1039 (1981).

88. Wofford, N. Q., Beaty, P. S., and McInerney, M. J., Preparation of cell-free extracts and the enzymes involved in fatty acid metabolism in *Syntrophomonas wolfei, J. Bacteriol., 167*, 179–185 (1986).

89. Bache, R., and Pfennig, N., Selective isolation of *Acetobacterium woodii* on methoxylated aromatic acids and determination of growth yields, *Arch. Microbiol., 130*, 255–261 (1981).

90. Ragsdale, S. W., and Wood, H. G., Acetate biosynthesis by acetogenic bacteria, *J. Biol. Chem., 260*, 3970–3977 (1985).

91. Tschech, A., and Pfennig, N., Growth yield increase linked to caffeate reduction in *Acetobacterium woodii, Arch. Microbiol., 137*, 163–167 (1984).

92. Frazer, A. C., and Young, L. Y., A gram-negative anaerobic bacterium that utilizes O-methyl substituents of aromatic acids, *Appl. Environ. Microbiol., 49*, 1345–1347 (1985).

93. Frazer, A. C., and Young, L. Y., Anaerobic Cl metabolism of the O-methyl-^{14}C-labeled substituent of vanillate, *Appl. Environ. Microbiol., 51*, 84–87 (1986).

94. Grbic-Galic, D., Fermentative and oxidative transformation of ferulate by a facultatively anaerobic bacterium isolated from sewage sludge, *Appl. Environ. Microbiol., 50*, 1052–1057 (1985).

95. Grbic-Galic, D., and LaPat-Polasko, L., *Enterobacter cloacae* DG-6: a strain that transforms methoxylated aromatics under aerobic and anaerobic conditions, *Curr. Microbiol., 12*, 321–324 (1985).

96. Krumholz, L. R., and Bryant, M. P., *Clostridium pfennigii* sp. nov. uses methoxyl groups of monobenzenoids and produces butyrate, *Int. J. Syst. Bacteriol., 35*, 454–456 (1985).

97. Mountfort, D. O., and Asher, R. A., Isolation from a methanogenic ferulate degrading consortium of an anaerobe that converts methoxyl groups of aromatic acids to volatile fatty acids, *Arch. Microbiol., 144*, 55–61 (1986).

98. DeWeerd, K. A., Suflita, J. M., Linkfield, T., Tiedje, J. M., and Prtichard, P. H., The relationship between reductive dehalogination and other aryl substituent removal reactions catalyzed by anaerobes, *FEMS Microbiol. Ecol., 38*, 331–339 (1986).

99. Clark, F. M., and Fina, L. R., The anaerobic decomposition of benzoic acid during methane fermentation, *Arch. Biochem. Biophys., 36*, 26–32 (1952).
100. Fina, L. R., and Fiskin, A. M., The anaerobic decomposition of benzoic acid during methane fermentation. II. Fate of carbons one and seven, *Arch. Biochem. Biophys., 91*, 163–165 (1960).
101. Nottingham, P. M., and Hungate, R. E., Methanogenic fermentation of benzoate, *J. Bacteriol., 98*, 1170–1172 (1969).
102. Fina, L. R., Bridges, R. L., Coblentz, T. H., and Roberts, F. F., The anaerobic decomposition of benzoic acid during methane fermentation. III. The fate of carbon four and the identification of propanoic acid, *Arch. Microbiol., 118*, 169–172 (1978).
103. Shlomi, E. R., Lankhorst, A., and Prins, R. A., Methanogenic fermentation of benzoate in an enrichment culture, *Microb. Ecol., 4*, 249–261 (1978).
104. Grbic-Galic, D., and Young, L. Y., Methane fermentation of ferulate and benzoate: anaerobic degradation pathways, *Appl. Environ. Microbiol., 50*, 292–297 (1985).
105. Sleat, R., and Robinson, J. P., Methanogenic degradation of sodium benzoate in profundal sediments from a small eutrophic lake, *J. Gen. Microbiol., 129*, 141–152 (1983).
106. Balba, M. T., and Evans, W. C., The methanogenic fermentation of aromatic substrates, *Biochem. Soc. Trans., 5*, 302–304 (1977).
107. Keith, C. L., Bridges, R. L., Fina, L. R., Iverson, K. L., and Cloran, J. A., The anaerobic decomposition of benzoic acid during methane fermentation. IV. Dearomatization of the ring and volatile fatty acids formed on ring rupture, *Arch. Microbiol., 118*, 173–176 (1978).
108. Balba, M. T., and Evans, W. C., Methanogenic fermentation of the naturally occurring aromatic amino acids by a microbial consortium, *Biochem. Soc. Trans., 8*, 625–627 (1980).
109a. Horowitz, A., Shelton, D. R., Cornell, C. P., and Tiedje, J. M., *Dev. Ind. Microbiol., 23*, 435–444 (1982).
109b. Wang, Y.-T., Suidan, M. T., and Pfeffer, J. T., Anaerobic biodegradation of indole to methane, *Appl. Environ. Microbiol., 48*, 1058–1060 (1984).
110. Healy, J. B., Jr., Young, L. Y., and Reinhard, M., Methanogenic decomposition of ferulic acid, a model lignin derivative, *Appl. Environ. Microbiol., 39*, 436–444 (1980).
111. Healy, J. B., Jr., and Young, L. Y., Catechol and phenol degradation by a methanogenic population of bacteria, *Appl. Environ. Microbiol., 35*, 216–218 (1978).
112. Healy, J. B., Jr., and Young, L. Y., Anaerobic biodegradation of eleven aromatic compounds to methane, *Appl. Environ. Microbiol., 38*, 84–89 (1979).
113. Dwyer, D. F., Krumme, M. L., Boyd, S. A., and Tiedje, J. M., Kinetics of phenol biodegradation by an immobilized methanogenic consortium, *Appl. Environ. Microbiol., 52*, 345–351 (1986).
114. Balba, M. T., Clarke, N. A., and Evans, W. C., The methanogenic fermentation of plant phenolics, *Biochem. Soc. Trans., 7*, 1115–1116 (1979).
115. Knoll, G., and Winter, J., Anaerobic degradation of phenol in sewage sludge, *Appl. Microbiol. Biotechnol., 25*, 384–391 (1987).
116. Young, L. Y., and Rivera, M. D., Methanogenic degradation of four phenolic compounds, *Water Res., 19*, 1325–1332 (1985).
117. Field, J. A., and Lettinga, G., The methanogenic toxicity and anaerobic degradability of a hydrolyzable tannin, *Water Res., 21*, 367–374 (1987).
118. Suflita, J. M., Horowitz, A., Shelton, D. R., and Tiedje, J. M., Dehalogenation: a

novel pathway for the anaerobic biodegradation of haloaromatic compounds, *Science, 218,* 1115-1117 (1982).

119. Horowitz, A., Suflita, J. M., and Tiedje, J. M., Reductive dehalogenations of halobenzoates by anaerobic lake sediment microorganisms, *Appl. Environ. Microbiol., 45,* 1459-1465 (1983).

120. Suflita, J. M., Robinson, J. A., and Tiedje, J. M., Kinetics of microbial dehalogenation of haloaromatic substrates in methanogenic environments, *Appl. Environ. Microbiol., 45,* 1466-1473 (1983).

121. Boyd, S. A., and Shelton, D. R., Anaerobic biodegradation of chlorophenols in fresh and acclimated sludge, *Appl. Environ. Microbiol., 47,* 272-277 (1984).

122. Boyd, S. A., Shelton, D. R., Berry, D., and Tiedje, J. M., Anaerobic biodegradation of phenolic compounds in digested sludge, *Appl. Environ. Microbiol., 46,* 50-54 (1983).

123. Miksell, M. D., and Boyd, S. A., Complete reductive dechlorination and mineralization of pentachlorophenol by anaerobic microorganisms, *Appl. Environ. Microbiol., 52,* 861-865 (1986).

124. Gibson, S. A., and Suflita, J. M., Extrapolation of biodegradation results to groundwater aquifers: reductive dehalogenation of aromatic compounds, *Appl. Environ. Microbiol., 52,* 681-688 (1986).

125. Smolenski, W. J., and Suflita, J. M., Biodegradation of cresol isomers in anoxic aquifers, *Appl. Environ. Microbiol., 53,* 710-716 (1987).

126. Vogel, T. M., and Grbic-Galic, D., Incorporation of oxygen from water into toluene and benzene during anaerobic fermentative transformation, *Appl. Environ. Microbiol., 52,* 200-202 (1986).

127. Grbic-Galic, D., and Vogel, T. M., Transformation of toluene and benzene by mixed methanogenic cultures, *Appl. Environ. Microbiol., 53,* 254-260 (1987).

128. Wenzyl, H. F. J., *The Chemical Technology of Wood,* Academic Press, New York, 1970.

129. Higuchi, T., Biodegradation of lignin: biochemistry and potential applications, *Experientia, 38,* 159-166 (1982).

130. Cain, R. B., The uptake and catabolism of lignin-related aromatic compounds and their regulation in microorganisms, in *Lignin Biodegradation: Microbiology, Chemistry, and Potential Applications* (K. T. Kirk, T. Higuchi, and H.-M. Chang, eds.), CRC Press, Boca Raton, Fla., 1980.

131. Hackett, W. F., Connors, W. J., Kirk, T. K., and Zeikus, J. G., Microbial decomposition of synthetic ^{14}C-labeled lignins in nature: lignin biodegradation in a variety of natural materials, *Appl. Environ. Microbiol., 33,* 43-51 (1977).

132. Zeikus, G., Fate of lignin and related aromatic substrates in anaerobic environments, in *Lignin Biodegradation: Microbiology, Chemistry, and Potential Applications* (K. T. Kirk, T. Higuchi, and H.-M. Chang, eds.), CRC Press, Boca Raton, Fla., 1980.

133. Odier, E., and Monties, B., Absence of microbial mineralization of lignin in anaerobic enrichment cultures, *Appl. Environ. Microbiol., 46,* 661-665 (1983).

134. Zeikus, J. G., Wellstein, A. L., and Kirk, T. K., Molecular basis for the biodegradative recalcitrance of lignin in anaerobic environments, *FEMS Microbiol. Lett., 15,* 193-197 (1982).

135. Chen, W., Ohmiya, K., Shimizu, S., and Kawakami, H., Degradation of dehydrodivanillin by anaerobic bacteria from cow rumen fluid, *Appl. Environ. Microbiol., 49,* 211-216 (1985).

136. Colberg, P. J., and Young, L. Y., Biodegradation of lignin-derived molecules under anaerobic conditions, *Can. J. Microbiol., 28,* 886-889 (1982).

137. Colberg, P. J., and Young, L. Y., Aromatic and volatile acid intermediates observed during anaerobic metabolism of lignin-derived oligomers, *Appl. Environ. Microbiol., 49,* 350–358 (1985).
138. Colberg, P. J., and Young, L. Y., Anaerobic degradation of soluble fractions of [^{14}C-lignin]lignocellulose, *Appl. Environ. Microbiol., 49,* 345–349 (1985).
139. Iannotti, E. L., Kafkewitz, D., Wolin, M. J., and Bryant, M. P., Glucose fermentation products of *Ruminococcus albus* grown in continuous culture with *Vibrio succinogenes*: changes caused by interspecies transfer of H$_2$, *J. Bacteriol., 114,* 1231–1240 (1973).
140. Benner, R., Maccubbin, A. E., and Hodson, R. E., Anaerobic biodegradation of the lignin and polysaccharide components of lignocellulose and synthetic lignin by sediment microflora, *Appl. Environ. Microbiol., 47,* 998–1004 (1984).
141. Benner, R., Maccubbin, A. E., and Hodson, R. E., Preparation, characterization, and microbial degradation of specifically radiolabeled [^{14}C]lignocelluloses from marine and freshwater macrophytes, *Appl. Environ. Microbiol., 47,* 381–389 (1984).
142. Benner, R., and Hodson, R. E. Thermophilic anaerobic biodegradation of [^{14}C]-lignin, [^{14}C]cellulose, and [^{14}C]lignocellulose preparations, *Appl. Environ. Microbiol., 50,* 971–976 (1985).
143. Grbic-Galic, D., Anaerobic degradation of coniferyl alcohol by methanogenic consortia, *Appl. Environ. Microbiol., 46,* 1442–1446 (1983).
144. Nali, M., Rindone, B., Tollari, S., Andreoni, V., and Treccani, V., Anaerobic microbial conversion of three hydroxycinnamic acids, *Experientia, 41,* 1351–1353 (1985).
145. Ohmiya, K., Takeuchi, M., Chen, W., Shimizu, S., and Kawakami, H., Anaerobic reduction of ferulic acid to dihydroferulic acid by *Wolinella succinogenes* from cow rumen, *Appl. Microbiol. Biotechnol., 23,* 274–279 (1986).
146. Grbic-Galic, D., Anaerobic production and transformation of aromatic hydrocarbons and substituted phenols by ferulic acid–degrading BESA-inhibited methanogenic consortia, *FEMS Microbiol. Ecol., 38,* 161–169 (1986).
147. Balba, M. T., and Evans, W. C., The methanogenic fermentation of ω-phenylalkane carboxylic acids, *Biochem. Soc. Trans., 7,* 403–405 (1979).
148. Schauer, N. L., Ferry, J. G., Properties of formate dehydrogenase in *Methanobacterium formicicum, J. Bacteriol., 150,* 1–7 (1982).

22

Isolation of Extremely Thermophilic, Fermentative Archaebacteria from Deep-Sea Geothermal Sediments

HOLGER W. JANNASCH *Woods Hole Oceanographic Institution, Woods Hole, Massachusetts*

22.1 INTRODUCTION

There are two roads to obtain bacterial strains potentially useful for biotechnically important transformations: genetic engineering and the classical enrichment culture approach. The latter implies taking advantage of nature's genetic engineering and has—since the work of Winogradsky and Beijerinck—provided us with all the information on the presently known diversity of microbial metabolism. Chances for finding new metabolic types arise with the discovery of new microbial habitats. If characteristic environmental conditions suggest certain chemical or biochemical conversions, the existence of microorganisms involved in these transformations might be possible. Enrichment or selective growth experiments are the techniques of choice for obtaining pure cultures of such organisms, either by providing favorable energetics for the presumed transformation or by making use of certain predicted nutritional requirements for growth.

Oceanographic studies of tectonic spreading centers along the East Pacific Rise resulted in such discoveries of hitherto unknown microbial habitats [1,2]. The most striking biological phenomenon, the chemolitho-autotrophic production of organic carbon for the subsistence of copious populations of deep-sea invertebrates in the immediate vicinity of hydrothermal vents at depths of 1800 to 3700 m, led to the study of aerobic processes, primarily the bacterial oxidation of reduced sulfur compounds and methane. Methanogenic archaebacteria were also found in anoxic pockets near "black smoker" vents [3], but the absence of sediments at most of the lava-bottom spreading centers and the high temperature (350°C) of the anoxic and highly reduced hydrothermal fluid made extensive anaerobic microbial growth unlikely.

This situation changed with the discovery of the Guaymas Basin vent site in the Gulf of California. First observations suggested an anoxic, organic-rich, high-temperature sediment: in short, a novel habitat for the potential existence of new types of anaerobic microorganisms. Recently, extensive core collections made it possible to embark on a well-designed enrichment and isolation study.

Work during our first cruise to the Guaymas Basin vent site (*Atlantis II/Alvin* Cruise 120/28) concerned (1) aerobic microbial transformations of reduced sulfur compounds with the ensuing chemoautotrophic reduction of carbon dioxide to organic carbon, and (2) an extensive core collection with the aim of laboratory studies on enrichments and isolations of a variety of anaerobic chemoautotrophic as well as heterotrophic bacteria. Emphasis for part of this work, specifically discussed in this paper, is directed toward heterotrophic processes that might occur at temperature ranges between 70 and 110°C (i.e., the presently known upper temperature limit of microbial activities), including occasional tests for growth at temperatures up to 140°C.

The general aim of this study is to compare the mode of microbial attack on complex polymers such as hydrocarbons, lignite, and lignite hydrolysates, and to search for biotechnologically interesting isolates. At the time of this review for the Houston workshop, the work is in its initial stage. Many relevant observations made by other cruise participants are still unpublished and referred to as such.

22.2 GUAYMAS BASIN HYDROTHERMAL VENT SITE

In contrast to the offshore "warm" and "hot" hydrothermal vents of the East Pacific Ocean floor spreading centers, their extension into the Gulf of California (Fig. 22.1) results in a very special situation. Processes accompanying this tectonic activity "result in high heat flow (locally exceeding 1.2 watts m^{-2}) and dike and sill intrusions into the overlaying unconsolidated sediments [4,5]. The sediments in the basin accumulate at a rate of more than 1 m/1000 yrs and have covered the rift floors to a depth of up to 400 m" [6]. Thus the primary vents are overlayed by young and rapidly depositing sediments (Fig. 22.2) of terrigenous origin [7,8].

Furthermore, a pelagic input of diatomaceous detritus adds a substantial organic component to the deposits. The sediments are composed of 30 to 50% diatoms, 30 to 45% detrital clay, 10 to 15% calcareous nannofossils, 4 to 15% feldspar, 3 to 10% quartz, and 2 to 5% organic carbon [9]. In accordance with some expected and some unexpected chemical reactions between hydrothermal fluid and overlying sediment, the ion composition of Guaymas Basin vent emissions (Table 22.1) shows distinct differences from those of other vent sites where the hydrothermal solutions exit directly from basalt [10-12]. The East Pacific Rise 21°N site, about 100 miles south of the Baja California peninsula (Fig. 22.1), represents a typical example of the latter. "Black smokers" have been discovered at this site for the first time. It was unexpected that hydrothermal solutions at the Guaymas Basin site retain indeed enough metals and sulfides after their passage through the sediments to produce black smokers of considerable size. However, as compared to sediment-free vent sites, the Guaymas Basin emissions are less acid and therefore contain lower concentrations of metals and sulfur (Table 22.1). The relatively high pH and alkalinity lead to the formation of dissolved $CaCO_3$ as well as to an increased degradation of organic matter. The latter represents the source of the relatively high concentrations of ammonia that characterize the Guaymas Basin emissions.

A deep seawater circulation (Fig. 22.2A) through the oceanic crust was predicted early [13] as being the source of hydrothermal fluid emissions at oceanic spreading centers as discussed in a summarizing paper by Sleep [14]. At the Guaymas Basin vent site these emissions reach the ocean floor at a depth of 2000 m and range from typical

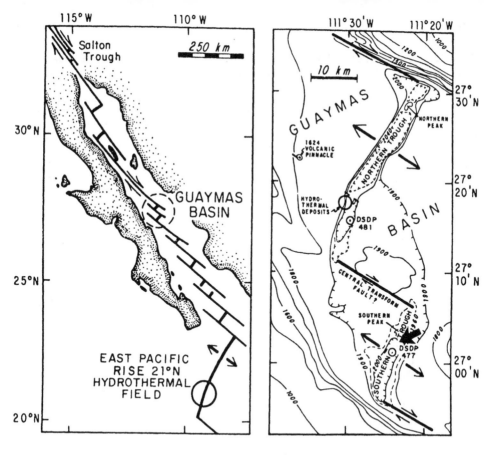

Fig. 22.1 Locations of the Guaymas Basin and 21°N East Pacific Rise vent sites and topography of Guaymas Basin; large arrow: Angel Rock area, depth 2003 m. (Modified from Ref. 7.)

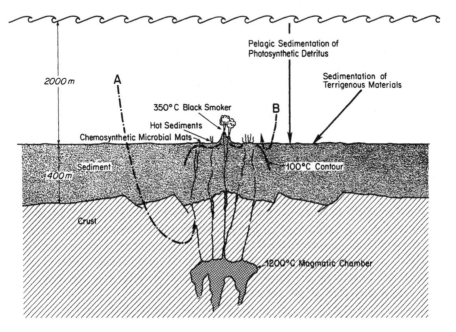

Fig. 22.2 Scheme of the sediment-covered Guaymas Basin vent site. A and B, seawater circulating systems (see the text).

Table 22.1 Composition of Hydrothermal Emissions of a Typical Vent at the Guaymas Basin (2000 m) and the 21°N (2000 m) Ocean Spreading Centers As Compared to Ambient Seawater

Element[a]		Guaymas Basin	21°N	Seawater
Fe	(μM)	56	1664	0.001
Mn	(μM)	139	960	0.001
Co	(nM)	5	213	0.03
Cu	(μM)	1	35	0.007
Zn	(μM)	4.2	106	0.01
Ag	(nM)	230	38	0.02
Pb	(nM)	265	308	0.01
Na	(mM)	489	432	464
K	(mM)	48.5	28	9.79
Li	(μM)	1054	891	26
Mg	(mM)	29	0	52.7
Ca	(μM)	29	15.6	10.2
Sr	(μM)	202	81	87
Ba	(μM)	12	7	0.14
Cl	(mM)	601	489	541
SO_4	(mM)	0	0	27.9
Si	(mM)	12.9	17.6	0.16
Al	(μM)	0.9	5.2	0.01
NH_3	(mM)	15.6	0	0
H_2S	(mM)	5.82	7.3	0
pH		5.9	3.4	7.9
Alk. mEq		10.6	-0.4	0.16

Source: Data from Ref. 10.
[a] $mM = 10^{-3}$ mol/kg; $\mu M = 10^{-6}$ mol/kg; $nM = 10^{-9}$ mol/kg.

black smokers with exit temperatures of 150 to 355°C (averaging at 315°C) to peripheral areas of hot pore water seeps. The latter are believed to be caused by a countercurrent shallow water circulation (Fig. 22.2B). A variable temperature structure of the surface sediments appears to be an important factor in inorganic chemical reactions, but even more so in organic and most likely biological transformations.

22.3 ORGANIC CONSTITUENTS OF GUAYMAS BASIN SEDIMENTS

As commonly found in anoxic sediments, several fatty acids, especially acetate and propionate, were reported from the Guaymas Basin vent site (C. S. Martens, unpublished). Other analytical work (R. B. Gagosian, unpublished) deals with the search for "biological markers." Most groups of organisms (e.g., diatoms, dinoflagellates, crustaceans) leave behind, after decomposition, certain characteristic chemical fingerprints which are used as source indicators of population quantity, periodicity, and so on. In such studies considerable amounts of five classes of lipids have been found: steroids, fatty alcohols, hydrocarbons, wax esters, and glycerol ethers.

In the context of the present discussion, the ether-linked lipids are of special importance for the occurrence of archaebacteria. Commonly marked by tetraether linkages, two types of archaebacteria might be expected to occur at vent environments: the thermo-acidophilic Sulfolobales and methanogenic bacteria. Of the two vent isolates so far obtained and studied in detail [3,33], one belongs to the genus *Methanococcus* and contains no tetraether lipids, but a newly described macrocyclic diether [15].

The most abundant organic constituents of Guaymas Basin sediment are hydrocarbons. It is generally assumed that thermal alterations of deposited organic carbon resulted in a number of petroleum components, especially primary olefins, kerogen, and gasoline-type hydrocarbons [7,16,17]. On upward migration to the seafloor, the heavier hydrocarbons condensed and were deposited with sulfides. Two dredge samples contained complex bituminous materials, both consisting of different amounts of sulfur, aliphatic aromatic and napthenic hydrocarbons, and asphaltic compounds [8]. From these samples 15 individual volatile hydrocarbons have been obtained. In addition, 26 polynuclear aromatic compounds were listed.

In a more recent analytical study from certain sediment core material (D. Bazylinski, unpublished), the paradox has been observed that the less biodegradable aromatic hydrocarbons appear to be depleted, while readily degradable alkanes are abundant at the sediment surface. A preliminary explanation contends that high-temperature alterations in rising hydrothermal solutions diminish aromatic constituents in lower sediment layers, leading to a recondensation of alkanes at low temperatures when reaching the sediment surface.

22.4 TEMPERATURE PROFILES OF GUAYMAS BASIN SEDIMENT

The extreme heterogeneity of the top sediment layers within relatively limited spaces became apparent when temperature profiles were taken. The particular set of data presented in Fig. 22.3 was obtained from five locations about 5 to 10 m apart, all within the so-called "Angel Rock" area. A small black smoker with an exit temperature of 240°C was situated within this area. It appears that one group of measurements, designated by the *Alvin* Dive numbers 1614a, 1614c, and 1615, showed very similar profiles in the upper 30 to 40 cm of sediment, but considerable irregularities below that depth. Profiles 1606 and 1614b demonstrated quite regular temperature increases with depth, but at different levels. Other profiles taken during this cruise (unpublished cruise report by J. F. Grassle) are principally similar. The major key to the observed heterogeneity appears to be explained by preliminary measurements (F. L. Sayles, unpublished) indicating an advective upward, but also downward flow of pore water in the top sediment layers. This study was done by tracer injection in the same area where the temperature profiles were measured.

22.5 EXTREMELY THERMOPHILIC BACTERIA

Initiated by the work of Brock [18-20] and Zeikus [21], the existence of extremely thermophilic microorganisms capable of growing above 80°C has been established through the work of Zillig's laboratory [22-24]. As an offspring of this group, Stetter extended this research to include the sulfur-reducing methanogens [25] and set the present highest optimum growth temperature of an isolated strain (*Pyrodictium*) at

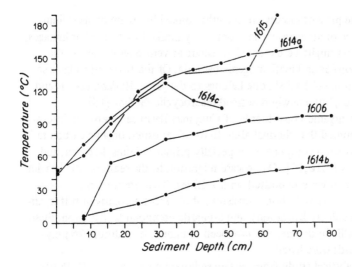

Fig. 22.3 Temperature profiles of a top sediment layer at five locations within the Angel Rock area (see Fig. 22.1) at the Guaymas Basin vent site; depth 2003 m (see the text).

105°C [26-28]. These authors also address the systematic and phylogenetic relationships of the extremely thermophilic bacteria [24,29] which belong with a few exceptions to the archaebacterial kingdom.

The present state of the art in the knowledge of the extremely thermophilic archaebacteria distinguishes the acidophilic aerobic sulfide and sulfur oxidizers, comprised in the order Sulfolobales and Thermoproteales, from the anaerobic methanogens falling into the orders Methanobacteriales, Methanococcales, and Methanomicrobiales. (No thermophilic representatives of the orders Halobacteriales and Thermoplasmales have yet been found.) A most unusual finding concerns two species of the genus *Sulfolobus* (*S. brierly* and *S. ambivalens*) which grow either anaerobically by reducing, or aerobically by oxidizing elemental sulfur [30,31].

While the source material for these studies originated from terrestrial or shallow marine springs [32], so far only two extreme thermophiles were isolated from deep-sea hydrothermal vents [3,33]. A new species of a methanogenic bacterium of the genus *Methanococcus* was isolated from a sample collected at the 21°N East Pacific Rise vent site. This organism has a maximum growth rate of 0.036 h^{-1} (doubling time of 28 min) at 86°C [3]. Another isolate from the same sample site represents a heterotrophic, sulfur-respiring archaebacterium with optimal and maximal growth temperatures of 92 and 98°C, respectively [33].

Microbial activity of growth at about 100°C was indicated in experiments by Baross et al. [34] using black smoker samples as inoculum to a number of different media. From data of another experiment it was concluded that a consortium of similar bacteria grew at temperatures of at least 250°C [35,36]. In the light of some critical evaluations of the data [29,32,37,38] and missing confirmation, the issue is still unresolved. Based on the information presented in these papers, the upper temperature limit for growth is believed to be below 150°C. Ahern and Klibanov [39] specifically discuss the time factor involved in the thermal inactivation of enzymes.

22.6 ISOLATION OF EXTREME THERMOPHILES FROM HYDROTHERMAL VENTS: ONGOING STUDIES

Our results of testing for microbial activity in emitted black smoker water (ATP determinations, microscopic counts, DNA production from labeled adenine, incubations in various media at 95°C) were negative in all those samples (taken with the Edmond-Walden sampler [12]) that contained no magnesium (i.e., were uncontaminated with ambient seawater). Exit temperatures below 350°C as well as a definite magnesium content indicated intrusion of ambient seawater and were often accompanied by the detection of microbial activity; the rate of DNA synthesis in such samples was highest at 90°C [40]. These data, as well as the above mentioned isolation of extremely thermophilic bacteria [3,33], suggest a widespread occurrence of high-temperature adapted organisms in the immediate vicinity of hot vents.

Since we are dealing with oxic/anoxic interfaces in the top sediment layers at the Guaymas Basin vent site, we also have to consider the existence of sulfate-reducing thermophiles in our studies. Widdel [41] has revolutionized this area of microbiology by isolating and describing not only six new genera of sulfate reducing bacteria from freshwater sediments, but also demonstrating a much wider substrate specificity of these organisms than formerly known, including acetate and hydrogen as electron donors. Only part of this work has been published [41-46]. Thermophilic sulfate- and thiosulfate-reducing bacteria have been isolated and described by Schink and Zeikus [47] and Zeikus et al. [48].

Work on thermophilic bacteria receives a strong impetus from biotechnological goals. The production of proteins, the fermentation of starch and cellulose [49-51], and possible conversion of lignites are the predominant areas of emphasis.

Our interest in the search for organisms conducting such transformations deepened when the existence of hot and nutrient-rich sediments of the Guaymas Basin vent site were discovered (see above) and promised a much larger variety of metabolic types of bacteria to be present than at the black smoker habitat. Our enrichment media and procedures follow in principle the anticipated types of metabolism listed in Table 22.2. Incubation temperatures up to 100°C are done in covered water baths. We are using pressurized incubations at temperatures between 100 and 140°C and pressures between 1 and 250 atm. Six pressure reactors, containing sample tubes, are heated in silicone oil, monitored by thermocouple readout, and controlled by adjustable voltage. Pressure is applied hydrostatically before temperatures exceed 100°C and vapor pressure effects are compensated for by a gas buffer system (accumulation vessels) individually joined to each

Table 22.2 Metabolic Types of Extremely Thermophilic Bacteria

Type of metabolism	Energy-producing reaction
Lithoautotrophy	
S° respiration	$H_2 + S^{\circ} \rightarrow H_2S$
$SO_4{}^{2-}$ respiration	$4H_2 + SO_4{}^{2-} \rightarrow S + 4H_2O$
Methanogenesis	$4H_2 + CO_2 \rightarrow CH_4 + 2H_2O$
S° oxidation	$2S^{\circ} + 3O_2 + 2H_2O \rightarrow 2H_2SO_4$
S^{2-} oxidation	$H_2S + 2O_2 \rightarrow H_2SO_4$
Heterotrophy	
S° respiration	$(org.)-H + S^{\circ} \rightarrow H_2S$
$SO_4{}^{2-}$ respiration	$(org.)-H + SO_4{}^{2-} \rightarrow S^{2-} + H_2O$
Fermentation	$(org.)-H \rightarrow -COOH, CO_2$
O_2 respiration	$(org.)-H + O_2 \rightarrow 2H_2O$

pressure reactor. These incubators allow us to screen a total of six independent combinations of elevated temperature and pressure at one time.

To become familiar with the type of organisms described by the Zillig/Stetter group, prior to our first cruise to this site, we obtained sediment samples from a submarine volcanic spring near Pozzuoli, Italy. We were able to isolate two different strains that were both new to the present systematic scheme of thermophilic archaebacteria. Our strain NS–C [52] grows optimally at 88°C above a pH of 7 and reduces elemental sulfur to H_2S. It is a strictly anaerobic heterotroph which is insensitive to free oxygen, however, at a lower temperature range where no growth occurs (i.e., below 50°C). It cannot be grouped into the "thermoacidophiles" [27] and resembles morphologically and by its G + C content *Sulfolobus* [53], although it is not autotrophic or aerobic. By the absence of muramic acid and the type of antibiotic sensitivity found, it appears to belong to the archaebacteria.

The second isolate, NS–E [54], is unique by its eubacterial characteristics despite certain similarities to extremely thermophilic archaebacteria: namely, the reduction of elemental sulfur, neutrophily, heterotrophy, and oxygen sensitivity. It differs from strain NS–C by its more pronounced H_2S sensitivity (i.e., high cell concentrations can be obtained only if H_2S is constantly stripped from the culture to levels of less than 1 mM). The organism appears morphologically to be similar to Stetter's [32] strain MSB8, but he describes it as fermentative and did not observe sulfur reduction.

During our work with these sulfur-reducing isolates, we observed a certain abiological portion of this process. While in cultures incubated above 80°C the reduction of elemental sulfur is exponential, it follows a linear pattern in the absence of the bacteria. The biological sulfur reduction in these organisms does not seem to be essential for growth but increases the cell yield up to fourfold [55].

We are presently working with two extremely thermophilic strains isolated from Guaymas Basin sediment collected during the 1985 cruise. They are heterotrophic and sulfur-reducing archaebacteria containing phytanyl diethers and dibiphytanyl diglycerol tetraethers in different proportions (T. A. Langworthy, unpublished). They are distinct by their intact polar lipid pattern and by the fact that one is motile; the other one is not. According to their nucleic acid composition, archaebacterial lipid pattern, sensitivity to antibiotics, cell morphology, and certain physiological characteristics, they are tentatively grouped into the genus *Desulfurococcus*.

A larger number of similar isolates have been recently obtained and are now being studied. They can briefly be characterized as being generally coccoid (single or paired), growing optimally at temperatures of 85 to 90°C and maximally at 90 to 96°C. The doubling time at optimum growth temperature is around 0.5 h. The isolates are not acidophilic and do not grow below pH 5.5. Growth rates of isolates with no absolute requirement for sulfur are reduced in its absence to about 20% and the cell yield to 3 to 9%. In growing cultures sulfide reaches levels of 25 mM or more. Growth can still proceed at 100 mM sulfide. Below the minimum growth temperature of about 50°C the isolates are tolerant toward free oxygen. A systematic and comparative physiological study of these isolates is in progress.

ACKNOWLEDGMENTS

Work reported in this overview and not specifically acknowledged in the text was done by C. O. Wirsen and S. J. Molyneaux in my laboratory. The research was supported by NSF Grant OCE86-00581 and by ONR N00014-88-K-0386. Contribution No. 3687 from the Woods Hole Oceanographic Institution.

NOTE IN PROOF

In the last paragraph of this chapter we mentioned a number of novel isolates of hyperthermophilic archaebacteria. After the submission of the manuscript in January 1987, detailed information on these isolates has appeared in Refs. 56 to 58.

REFERENCES

1. Jannasch, H. W., and Wirsen, C. O., Chemosynthetic primary production at East Pacific Ocean floor spreading centers, *BioScience, 29*, 492 (1979).
2. Jannasch, H. W., and Mottl, M. J., Geo-microbiology of deep sea hydrothermal vents, *Science, 228*, 717 (1985).
3. Jones, W. J., Leigh, J. A., Mayer, F., Woese, C. R., and Wolfe, R. S., *Methanococcus jannaschii* sp. nov., an extremely thermophilic methanogen from a submarine hydrothermal vent, *Arch. Microbiol., 136*, 254 (1983).
4. Williams, D. L., Becker, K., Lawver, L. A., and Von Herzen, R. P., Heat flow at the spreading center of the Guaymas Basin, Gulf of California, *J. Geophys. Res., 84*, 6757 (1979).
5. Einsele, G. J., Gieskes, J. M., Curray, J., Moore, D. G., Aguayo, J., Aubry, M. P., Fornari, D. J., Guerrero, J. C., Kastner, M., Kelts, K., Lyle, M., Matoba, Y., Molina-Cruz, A., Niemitz, J., Rueda, J., Saunders, A. D., Schrader, H., Simoneit, B. R. T., and Vacquier, V., Intrusion of basaltic sills into highly porous sediments, and resulting hydrothermal activity, *Nature, 283*, 441 (1980).
6. Simoneit, B. R. T., Hydrothermal petroleum: composition and utility as a biogenic carbon source, *Bull. Biol. Soc. Wash., 6*, 49 (1985).
7. Lonsdale, P. J., Bischoff, J. L., Burns, V. M., Kasner, M., and Sweeney, R. E., A high-temperature hydrothermal deposit on the seabed at a Gulf of California spreading center, *Earth Planet. Sci. Lett., 49*, 8 (1980).
8. Simoneit, B. R. T., and Lonsdale, P. F., Hydrothermal petroleum in mineralized mounds at the seabed of Guaymas Basin, *Nature, 295*, 198 (1982).
9. van Andel, T. H., Recent marine sediments of the Gulf of California, in *Marine Geology of the Gulf of California* (Tj. H. van Andel and G. Shor, eds.), *Am. Assoc. Pet. Geol. Mem., 3*, 216 (1964).

10. Edmond, J. M., and Von Damm, K. L., Chemistry of ridge crest hot springs, *Bull. Biol. Soc. Wash.*, *6*, 43 (1985).

11. Edmond, J. M., Von Damm, K. L., McDuff, R. E., and Measures, C. I., Chemistry of hot springs on the East Pacific Rise and their effluent dispersal, *Nature*, *297*, 187 (1982).

12. Von Damm, K. L., Chemistry of submarine hydrothermal solutions at 21°N, East Pacific Rise and Guaymas Basin, Gulf of California, Ph.D. dissertation, Massachusetts Institute of Technology/Woods Hole Oceanographic Institution, Joint Program in Oceanography, 1983.

13. Lister, C. R. B., Qualitative models of spreading-center processes, including hydrothermal penetration, *Tectonophysics*, *37*, 203 (1977).

14. Sleep, N. H., Hydrothermal convection at ridge axes, in *Hydrothermal Processes at Sea Floor Spreading Centers* (P. A. Rona, K. Boström, L. Laubier, and K. L. Smith, Jr., eds.), Plenum Press, New York, 1983, p. 71.

15. Comita, P. B., and Gagosian, R. B., Membrane lipid from deep-sea hydrothermal vent methanogen: a new macrocyclic glycerol diether, *Science*, *222*, 1329 (1984).

16. Simoneit, B. R. T., Hydrothermal effects on organic matter: high vs. low temperature components, *Org. Geochem.*, *6*, 857 (1984).

17. Simoneit, B. R. T., Philip, R. P., Jenden, P. D., and Galimov, E. M., Organic geochemistry of deep sea drilling project sediments from the Gulf of California: hydrothermal effects on unconsolidated diatom ooze, *Org. Geochem.*, *7*, 173 (1984).

18. Brock, T. D., *Thermophilic Microorganisms and Life at High Temperatures*, Springer-Verlag, New York, 1978.

19. Brock, T. D., Brock, M. L., Bott, T. L., and Edwards, M. R., Microbial life at 90°C: the sulfur bacteria of Boulder Spring, *J. Bacteriol.*, *107*, 303 (1971).

20. Tansey, M. R., and Brock, T. D., Microbial life at high temperatures: ecological aspects, in *Microbial Life in Extreme Environments* (D. J. Kushner, ed.), Academic Press, London, 1978, p. 158.

21. Zeikus, J. G., Thermophilic bacteria: ecology, physiology and technology, *Enzyme Microbiol. Technol.*, *1*, 243 (1979).

22. Zillig, W., Stetter, K. O., Wunderl, S., Schulz, W., Priess, H., and Scholz, I., The *Sulfolobus "Caldariella"* group: taxonomy on the basis of the structure of DNA-dependent RNA polymerase, *Arch. Microbiol.*, *125*, 259 (1980).

23. Zillig, W., Tu, J., and Holz, I., Thermoproteales: a third order of thermoacidophilic archaebacteria, *Nature*, *293*, 85 (1981).

24. Zillig, W., Schnabel, R., Tu, J., and Stetter, K. O., The phylogeny of archaebacteria, including novel anaerobic thermoacidophiles in the light of RNA polymerase structure, *Naturwissenschaften*, *69*, 197 (1982).

25. Stetter, K. O., and Gaag, G., Reduction of molecular sulphur by methanogenic bacteria, *Nature*, *305*, 309 (1983).

26. Stetter, K. O., Ultrathin mycelia-forming organisms from submarine volcanic areas having an optimum growth temperature of 105°C, *Nature*, *300*, 258 (1982).

27. Fischer, F., Zillig, W., Stetter, K. O., and Schreiber, G., Chemolitho-autotrophic metabolism of anaerobic extremely thermophilic archaebacteria, *Nature*, *301*, 511 (1983).

28. Stetter, K. O., König, H., and Stackbrandt, E., *Pyrodictium* gen. nov., a new genus of submarine disc-shaped sulphur reducing archaebacteria growing optimally at 105°C, *Syst. Appl. Microbiol.*, *4*, 535 (1983).

29. Stetter, K. O., Extreme thermophile Bakterien, *Naturwissenschaften*, *72*, 291 (1985).

30. Segerer, A., Stetter, K. O., and Klink, F., Two contrary modes of chemolithotrophy in the same archaebacterium, *Nature*, *313*, 787 (1985).

31. Zillig, W., Yeats, S., Holz, I., Bock, A., Gropp, F., Rettenberger, M., and Lutz, S., Plasmid-related anaerobic autotrophy of the novel archaebacterium *Sulfolobus ambivalens, Nature, 313*, 789 (1985).

32. Stetter, K. O., Thermophilic archaebacteria occurring in submarine hydrothermal areas, in *Planetary Ecology* (D. E. Caldwell, J. A. Brierley, and C. L. Brierley, eds.), Van Nostrand Reinhold, New York, 1985, p. 320.

33. Fiala, G., Stetter, K. O., Jannasch, H. W., Langworthy, T. A., and Madon, J., *Staphylothermus marinus* sp. nov. represents a novel genus of extremely thermophilic submarine heterotrophic archaebacteria growing up to 98°C, *Syst. Appl. Microbiol., 8*, 106 (1986).

34. Baross, J. A., Lilley, M. D., and Gordon, L. I., Is the CH_4, H_2, and CO venting from submarine hydrothermal systems produced by thermophilic bacteria? *Nature, 298*, 366 (1982).

35. Baross, J. A., and Deming, J. W., Growth of 'black smoker' bacteria at temperatures of at least 250°C, *Nature, 303*, 423 (1983).

36. Baross, J. A., Deming, J. W., and Becker, R. R., Evidence for microbial growth in high pressure, high-temperature environments, in *Current Perspectives in Microbial Ecology* (M. J. Klug, and C. A. Reddy, eds.), American Society for Microbiology, Washington, D.C., 1984, p. 186.

37. Trent, J. D., Chastain, R. A., and Yayanos, A. A., Possible artefactual basis for apparent bacterial growth at 250°C, *Nature, 307*, 737 (1984).

38. White, R. H., Hydrolytic stability of biomolecules at high temperatures and its implication of life at 250°C, *Nature, 310*, 430 (1984).

39. Ahern, T. J., and Klibanov, A. M., The mechanism of irreversible enzyme inactivation at 100°C, *Science, 228*, 1280 (1985).

40. Karl, D. M., Burns, D. J., Orrett, K., and Jannasch, H. W., Thermophilic microbial activity in samples from deep sea hydrothermal vents, *Mar. Biol. Lett., 5*, 227 (1984).

41. Widdel, F., Anaerober Abbau von Fettsäuren und Benzeosäure durch neu isolierte Arten sulfat-reduzierender Bakterien, dissertation, University of Göttingen, 1980.

42. Widdel, F., and Pfennig, N., A new anaerobic, sporing, acetate-oxidizing, sulfate-reducing bacterium, *Desulfotomaculum* (emend.) *acetoxidans, Arch. Microbiol., 112*, 119 (1977).

43. Widdel, F., and Pfennig, N., Sporulation and further nutritional characteristics of *Desulfotomaculum acetoxidans, Arch. Microbiol., 129*, 401 (1981).

44. Widdel, F., and Pfennig, N., Studies on dissimilatory sulfate-reducing bacteria that decompose fatty acids. I. Isolation of new sulfate-reducing bacteria enriched with acetate from saline environments: description of *Desulfobacter postgatei* gen. nov. sp. nov., *Arch. Microbiol., 129*, 395 (1981).

45. Widdel, F., Kohring, G. W., and Mayer, F., Studies on dissimilatory sulfate-reducing bacteria that decompose fatty acids. III. Characterization of the filamentous gliding *Desulfonema limicola* gen. nov. sp. nov., and *Desulfonema magnum* sp. nov., *Arch. Microbiol., 134*, 286 (1983).

46. Klemps, R., Cypionka, H., Widdel, F., and Pfennig, N., Growth with hydrogen, and further physiological characteristics of *Desulfotomaculum* species, *Arch. Microbiol., 143*, 203 (1985).

47. Schink, B., and Zeikus, J. G., *Clostridium thermosulfurogenes* sp. nov., a new thermophile that produces elemental sulfur from thiosulfate, *J. Gen. Microbiol., 129*, 1149 (1983).

48. Zeikus, J. G., Dawson, M. A., Thompson, T. E., Ingvorsen, K., and Hatchikian, E. C., Microbial ecology of volcanic sulfidogenesis: isolation and characterization of *Thermodesulfobacterium commune* gen. nov. and sp. nov., *J. Gen. Microbiol., 129*, 1159 (1983).

49. Zeikus, J. G., and Ng., T. K., Thermophilic saccharide fermentations, in *Annual Report on Fermentation Processes*, Vol. 5 (G. Tsao, ed.), Academic Press, New York, 1982, p. 263.
50. Zeikus, J. G., Ben-Bassat, A., Ng, T. K., and Lamed, R. J., Thermophilic ethanol fermentations, in *Trends in the Biology of Fermentations for Fuels and Chemicals* (A. Hollaender, R. Rabson, P. Rogers, A. San Pietro, R. Valentine, and R. Wolfe, eds.), Plenum Press, New York, 1981, p. 441.
51. Hyun, H. H., Shen, G.-J., and Zeikus, J. G., Differential amylosaccharide metabolism of *Clostridium thermosulfurogenes* and *Clostridium thermohydrosulfuricum, J. Bacteriol., 164*, 1153 (1985).
52. Belkin, S., and Jannasch, H. W., A new extremely thermophilic, sulfur-reducing heterotrophic, marine bacterium, *Arch. Microbiol., 141*, 181 (1985).
53. Brock, T. D., Brock, K. M., Belly, R. T., and Weiss, R. L., *Sulfolobus*: a new genus of sulfur-oxidizing bacteria living at low pH and high temperature, *Arch. Microbiol., 84*, 54 (1972).
54. Belkin, S., Wirsen, C. O., and Jannasch, H. W., A new sulfur-reducing, extremely thermophilic eubacterium from a submarine thermal vent, *Appl. Environ. Microbiol., 51*, 1180 (1986).
55. Belkin, S., Wirsen, C. O., and Jannasch, H. W., Biological and abiological sulfur reduction at high temperatures, *Appl. Environ. Microbiol., 49*, 1057 (1985).
56. Jannasch, H. W., Wirsen, C. O., Molyneaux, S. J., and Langworthy, T. A., Extremely thermophilic fermentative archaebacteria of the genus *Desulfurococcus* from deep-sea hydrothermal vents, *Appl. Environ. Microbiol., 54*, 1203 (1988).
57. Zhao, H., Wood, A. G., Widdel, F., and Bryant, M. P., An extremely thermophilic *Methanococcus* from a deep sea hydrothermal vent and its plasmid, *Arch. Microbiol., 150*, 178 (1988).
58. Jones, W. J., Stugard, C. E., and Jannasch, H. W., Comparison of thermophilic methanogens from submarine hydrothermal vents, *Arch. Microbiol., 151*, 314 (1989).

23

Physiology and Biotechnology of Halophilic Anaerobes for Application to Texas Lignite

SIRIRAT RENGPIPAT and J. G. ZEIKUS* *Michigan State University, East Lansing, Michigan*

23.1 INTRODUCTION

A variety of halophilic anaerobes were enriched from hypersaline environments by selecting for bacterial strains that grew readily on inexpensive nutrients and that made either exopolysaccharides or organic acid salts and alcohols as catabolic end products during the hydrolysis of complex organic matter. One prolific species, Haloanaerobic isolate strain EO, was isolated from deep subsurface brine waters associated with an injection water filter on an offshore oil rig in the Gulf of Mexico. This species is used as a model biocatalyst to understand how chemoorganotrophic anaerobes adapt to extreme salt stress, and to understand how to bioengineer haloanaerobes for technological applications. Studies to date imply that chemotrophic haloanaerobes, unlike aerobic *Halobacterium* species, do not actively expend adenosine triphosphate (ATP) to prevent the internal accumulation of sodium or to synthesize an anabolic osmoregulant. Rather, species proton motive force and catabolism are metabolically controlled in relation to an optimal internal sodium concentration that is in dynamic equilibrium with environmental hypersalinity. The impact of this finding and the physiological properties of Haloanaerobe isolate strain EO are discussed in relation to strain improvements needed to develop an industrial fermentation of complex organic matter, such as pretreated Texas lignite in an anoxic, hypersaline bioreactor system.

23.2 OBJECTIVES

The broad goal of this research is to study the physiology of halophilic anaerobes in relation to understanding how anaerobes adapt to extreme salt stress while faced with limited chemical thermodynamic energy. The objective of the proposed research will focus on how salt influences ecophysiological function of haloanaerobes by investigating:

Current affiliation: Michigan Biotechnology Institute, Lansing, Michigan.

1. General species diversities of haloanaerobes and ecophysiological relationship between different strains that grow above 10% NaCl parameter
2. Physiological effects of salt concentration on growth and metabolism of a given species in relation to general adaptation mechanism
3. Physiological and biochemical characterization of a given species catabolism in relation to salt concentration
4. Consideration of the foregoing aspects using Texas lignite as a potential substrate

23.3 BACKGROUND

23.3.1 Ecology of Halophiles

Little work on the ecology of halophiles has been attempted, especially studies relating to haloanaerobic bacteria. Microorganisms that can grow well at above 10% NaCl are generally considered to be halophiles. A variety of environments with a salt concentration of more than 10% are found in nature (e.g., Great Salt Lake, Dead Sea, and brine ponds associated with the rocks). The deep subsurface environment of oil fields is flooded with water for secondary oil recovery and they can vary in salinity from 1 to 20% salt [2].

Since the mid-nineteenth century, it has been recognized that halophilic microorganisms could be isolated from different sources (i.e., hides, salt-preserved food, saline aquatic sediments) [9]. Halophiles were presumably noticed to be a causative agent for color changes of salts and hence were considered of economic importance.

Current knowledge of species diversity and microbial processes in hypersaline environments has been limited to studies of aerobic halophile activities in the previously mentioned environment [14,15]. In extreme hypersaline (i.e., >20% salt) aquatic environments, microorganisms appear as the only life forms that produce and decompose organic matter because both plants and animals are unable to proliferate under these extreme conditions.

In 1979, Zeikus et al. [23] initiated a research project on the microbial ecology of the anaerobic decomposition process in the Great Salt Lake (Utah). Three goups of haloanaerobes were enriched from Great Salt Lake sediment and species from each group isolated: methanol-consuming methanogens, H_2-consuming sulfate reducers, and saccharide-degrading acidogens. A prevalent anaerobe present at 10^8 cells/mL sediment, *Haloanaerobium praevalens*, was isolated from the Great Salt Lake. Oren et al. [12,13] described two other haloanaerobic species, *Clostridium lortetii* and *Halobacteroides halobius*. It is interesting that analysis of the 16S rRNA oligonucleotide sequences of haloanaerobes is different from other described eubacteria [13].

23.3.2 Physiological Adaptation of Halophiles to Salt

Physiological studies on halophilic microorganisms have emphasized the eucaryotic green algae, *Dunaliella*; the purple bacterium, *Halobacterium*; and the phototrophic bacterium, *Ectothiorhodospira* [1,14]. In order to cope with a high external salt environment, halophiles possess unique mechanisms for regulation of internal osmotic pressure. For example, *Dunaliella viridis* apparently excludes sodium ions and produces a high internal glycerol concentration for osmoregulation. *Halobacterium*, an archaebacterium, actively accumulates internal potassium ions with sodium ions remaining in high concentrations outside the cell. Halophilic eubacteria, on the other hand, produce high intracellular concentrations of betaine for osmoregulation [6].

Halobacterium and *Halococcus* species have been characterized as archaebacteria on the basis of no muramic acid residues in their cell wall, the presence of cell envelope lipids with diether linkages, and unique 16S ribosomal RNA oligonucleotide sequence patterns. They are defined as members of the family Halobacteriaceae in the 8th edition of *Bergey's Manual* [4] and possess a respirative metabolism. Halophilic bacteria are classified as moderate if they grow well at the range of 5 to 15% NaCl requirement, and as extreme if their growth requires 5 to 30% salt concentration [10]. Current interest in halophilic bacteria has focused on their ability to generate a proton motive force by photoactivation of bacterial rhodopsin, which is present in their purple membranes [19]. Presently, only three haloanaerobes have been studied: *C. lortetii* [12] and *H. praevalens* [23], both of which produce butyric and acetic acid as the major fermentation end products; and *H. halobius* [13], which form acetic acid and ethanol as major end products. The physiological mechanisms of osmoregulation in haloanaerobes have not been reported.

23.3.3 Biotechnological Attributes of Halophiles

Only limited information is available concerning the application of halophiles in industrial microbiology. Haloaerobic microbes have been suggested as having potential for producing glycerol and carotenoid-derived pigments. Haloanaerobic bacteria may have potential in microbial-enhanced oil recovery (MEOR), as well as in chemical- and fuel-producing fermentation and waste treatment processes. However, their utility in industrial microbiology remains speculation [22].

Studies on microbial exopolymer production have been limited to nonhalophilic species. *Leuconostoc mesenteroides*, a lactic acid-producing facultative anaerobe, forms copious amounts of exopolysaccharides which are used industrially [3,4]. *Xanthomonas campestris*, an aerobic species, produces a high-molecular-weight biopolymer, called xanthan [11]. Xanthans are water-soluble polymers that have been used in enhanced oil recovery processes by improving water flooding. Xanthans increase the efficiency with which water can contact and displace oil in the reservoir. Exopolymer production by halophilic anaerobes has not been reported. These organisms may be of importance for in situ MEOR as a substitute for xanthans, which are very expensive and readily biodegradable.

Organic acid-producing fermentations (e.g., acetic acid, lactic acid, etc.) are generally limited by low pH and if pH is controlled, then by species tolerance to salts of organic acids. The potential for producing concentrated organic acid salt solutions (10% by haloanaerobes has not been reported [22]. Organic solvent-producing fermentations (e.g., ethanol, butanol, etc.) are generally limited by solvent concentrations that are economically recoverable and species tolerant to solvent. Solvent-producing haloanaerobes could have higher tolerance and lower energy requirements for solvent recovery because of less process water at higher salt concentration fermenting. The processing of certain organic industrial wastes can be associated with both high salinity and anoxic conditions (e.g., spent black liquors from pulp and paper industry, oil shale rock refining wastes from the synfuel industry, and selected aspects of Texas lignite utilization). Effective biodegradation of these wastes requires salt-tolerant populations of anaerobic bacteria [22].

23.4 SIGNIFICANCE

Today we face a dwindling source of fossil fuels and will be eventually confronted with inadequate oil supplies [11,20] because only one-third of the original oil in place can be recovered by present technology. Thus interest has recently developed in the application of microorganisms and/or microbial products for the enhancement of oil production [5]. However, new and innovative biotechnology is needed to make microbial-enhanced oil recovery commercial. By its nature, oil accumulates in a variety of porous sedimentary rock and typically forms at depths where it is anoxic and associated with hypersaline connate waters [17,18]. Halophilic anaerobes have been shown to be found in oil-bearing rock areas [16]. Exopolymer producing haloanaerobes are potential candidates for in situ MEOR application because of their small size and their ability to grow on a wide variety of substrates under the existing environmental conditions of the oil reservoir. Application of these microorganisms may also be directed to novel applications for utilization of Texas lignite.

Recently, sodium chloride (rock salt) and calcium chloride have been used as highway deicing salts, which can cause serious corrosion and environmental problems. These problems include deterioration of portland cement concrete bridge decks through chloride in corrosion of reinforcing steel; corrosion of structural steel in bridge structures and other highway construction; corrosion of automobile parts; pollution of drinking water sources by sodium and chloride ions in runoff; and harm to roadside vegetation. The Federal Highway Administration (FHWA) initiated work aimed at developing substitutes for chloride salts in deicing; research conducted by Bjorksten Research Laboratories [8] identified calcium magnesium acetate (CMA) as a potentially acceptable, non-corrosive biodegradable alternative.

The limited utilization of DMA as a deicing replacement for rock salt would exceed the current chemical for acetic acid and necessitates a microbial fermentation process for production [8]. However, present biotechnology is limited because the suggested acetogenic biosystem, *Clostridium thermoaceticum*, cannot tolerate greater than 5% salt concentration. Production of CMA by fermentation seems ideally suited for a halophilic anaerobe that can produce acetic acid at 20% salt.

The sediments of hypersaline water are often described as having high numbers of anaerobic bacteria [7,9,15,21]. However, very little is known about anaerobic halophiles themselves. The first isolation of a haloanaerobe discovered by Baumgartner from Mediterranean anchovies [1], *Bacteroides halosmophilus*, was named but it unfortunately has been lost. Currently, only four species have been isolated: *H. praevalens, C. lortetii, H. halobius*, and Haloanaerobe isolate strain EO. All are gram-negative rods and require salt for growth as moderate halophiles. They are obligate anaerobes. Their G + C mol% is in the range 27 to 37 mol%, which is different from that of haloaerobes (60 to 68 mol%; see Table 23.1). Interestingly, the 16S ribosomal RNA oligonucleotide pattern of *H. halobius* showed a high degree of similarity with that of *H. praevalens* and clearly appeared to be a member of the kingdom Eubacteria.

Haloanaerobic bacteria, when considered as a natural grouping, must have evolved physiological strategies for proliferating in an extreme hypersaline environment. These studies will elaborate both on the details of species diversity and the physiology of haloanaerobic bacteria, which form organic acid and solvent. The intent is to learn how salt influences growth and metabolism of haloanaerobe. Long-term objectives relate to Texas lignite utilization.

Table 23.1 Comparison of Haloanaerobes

	C. lortetii	H. halobius	H. praevalens	Haloanaerobe isolate strain EO
Morphology	Rod	Rod	Rod	Rod
Gram stain	Negative	Negative	Negative	Negative
Size (μm)	0.5–0.6×6–10	0.5×10–20	0.5×1.5	0.4–0.7×1–1.6
t_d (h)	8	1	4 (at 25% salt conc.)	7.8–9.5
%G+C	31.5	37	27 ± 1	32 ± 0.5
pH range for growth[a]	ND	ND	6–8	5.3–8
Temperature range for growth (°C)	37–45	37–42	15–45	15–45
Salt concentration range for growth (% w/v)	6–12	9–15	5–25	6–20
Spore	+	–	–	–
Motility	+	+	–	+
Gas vacuole	+	–	–	–
Habitat rock	Dead Sea	Dead Sea	Great Salt Lake	Gas bearing
End products from glucose	Acetate, butyrate, H_2	Acetate, ethanol, CO_2/H_2	Acetate, butyrate, propionate, CO_2/H_2	Acetate, ethanol, CO_2/H_2

[a]ND, not done.

23.5 EXPERIMENTAL APPROACH

We started an extensive research program in 1984 on the isolation of haloanaerobes from hypersaline environments to obtain more details of species diversity. We have shown that anaerobic bacteria are present in deep subsurface oil-bearing sandstones from offshore oil wells in the Gulf of Mexico. A new anaerobic species, Haloanaerobe isolate strain EO [16], was isolated which grows at 6 to 20% salt concentration and produces acetic acid and ethanol as its major soluble products. Notably, Haloanaerobe isolate strain EO produces a small amount of polymer.

High salt concentration does not prevent microbial growth and activity per se, but it does select for a unique population that is adapted to growth at high salt. Generally, obligate halophiles require salt concentration about 5% to grow. It is possible that many other species of haloanaerobes are awaiting isolation and further study.

In August 1984, research studies on isolation of exopolymer producing haloanaerobes for possible use in MEOR were initiated. To be effective for in situ MEOR, an isolate must be able to produce exopolymer from inexpensive substrates and grow in an environment that is hypersaline and anoxic. With enrichment medium containing glucose as substrate, facultative anaerobes were isolated that produced polymer. The general microbial properties of these new isolates are shown in Table 23.2.

23.6 RATIONALE

Although studies on halophiles have been conducted for 50 years, most of the work has been on aerobes, emphasizing species characterization, physiological adaptation to high salt, and the ability of certain species to generate a proton motive force by photoactivation of bacterial rhodopsin. The proposed research will further fundamental biological knowledge on halophiles since only a few species of haloanaerobes have been recently isolated. Knowledge of how these species are adapted to survive in both anoxic and hypersaline environments is not available. In addition, learning how salt influences catabolism (i.e., maintenance of proton motive force and fermentation product formation and balance) will help generate a broader understanding of physiological processes in halophiles.

On an applied level, recent discussion has suggested potential utility for haloanaerobes in microbial enhanced oil recovery (MEOR), ethanol product, and noncorrosive road salt production [1,8,11]. This research will provide suitable microorganism growth

Table 23.2 Comparison of Bacteria from Two Habitats

Habitat	Gas-bearing rocks, oil-water injection filters	Great Salt Lake sediments
Isolation	*H. acetoethylicus* E1G1 (ATCC 43120)	Facultative anaerobes: GSL 5/3-1, GSL 5/3-2, GSL 5/3-3
Morphology	Rod	Rod
Gram stain	Negative	Positive
O_2 relationship	Obligate anaerobe	Facultative anaerobes
NaCl requirement (%w/v)	6–20	10
Colony appearance	No mucoid	Slightly mucoid

conditions and background data to start potential application studies. The increased knowledge gained on physiological adaptation mechanisms and the influence of salt on microbial metabolism may allow use of haloanaerobes in an industrial application, but will certainly expand our understanding on anaerobe diversity and metabolism.

We have attempted to isolate new halophiles from different hypersaline habitats. Halophilic anaerobe, Haloanaerobe isolate strain EO was isolated from green slime that clogged an injection water filter used to flood offshore oil wells. We have selected this species as a model organism to test the general hypothesis that haloanaerobes have evolved mechanisms for adaptation to high salt that are more energy efficient than those used by haloaerobes. Also, if high salt concentrations of acetic acid can be produced microbiologically, this species is a likely candidate to provide a feasibility assessment.

23.7 SPECIFIC AIMS

1. Ecological studies of haloanaerobes: attempting to isolate a new strain from hypersaline samples
   a. Selective enrichment and isolation of a new haloanaerobe
   b. Identification of isolated strain in relation to known species
2. Physiological studies and screening for the best model test haloanaerobe
   a. Characterization of the influence of external salt on growth, internal salt concentration, proton motive force, and metabolic function
   b. Characterization of the osmoregulation mechanism
   c. Characterization of how salt regulates carbon and electron flow in metabolic pathway
   d. Characterization of physiological conditions for optimal production of given species
3. Biotechnological application: attempting to isolate the best strain for:
   a. Organic acid–producing strain whose salt can replace a corrosive road salt; or
   b. Ethanol–producing tolerant strain, which could give better high yield recovery; or
   c. Polymer-producing haloanaerobe for MEOR

REFERENCES

1. Baumgartner, J. G., The salt limits and thermal stability of a new species of anaerobic halophile, *Food Res., 2*, 321–329 (1937).
2. Belyaev, S. S., Wolkin, R., Kenealy, W. R., DeNiro, M. T., Epstein, S., and Zeikus, J. G., Methanogenic bacteria from the Bondyuzhskoe oil field: general characterization and analysis of stable-carbon isotopic fractionation, *Appl. Environ. Microbiol., 45*, 691–697 (1983).
3. Brock, T. D., Smith, D. W., and Madigan, M. T., *Biology of Microorganisms*, Prentice-Hall, Englewood Cliffs, N.J., 1984.
4. Buchanan, R. E., and Gibbons, N. E., *Bergey's Manual of Determinative Bacteriology*, 8th ed., Williams & Wilkins, Baltimore, 1974.
5. Finnerty, W. R., and Singer, M. E., Microbial enhancement of oil recovery, *Biotechnol. Bioeng.*, Mar. (1983).
6. Imhoff, T. F., and Rodriquez-Valera, F., Betaine is the main compatible soluble of halophilic eubacteria, *J. Bacteriol., 160*, 478–479 (1984).

7. Goodwin, S., Lupton, F. S., Phelps, T. J., and Zeikus, J. G., Dynamics of organic matter decomposition and sulfate reduction in anoxic sediments of Great Salt Lake, Utah, *J. Gen. Microbiol.*, submitted.

8. Marynowski, C. W., Jones, J. L., Boughton, R. L., Tuse, D., Cortopassi, J. H., and Gwinn, J. E., *Process Development for Production of Calcium Magnesium Acetate*, SRI International Report, SRI, Springfield, Va., Mar. 1983.

9. Kaplan, I. R., and Friedmann, A., Biological productivity in the Dead Sea. I. Microorganisms in the water column, *Isr. J. Chem., 8,* 513-528 (1970).

10. Kushner, D. J., 1978. Life in high salt and solute concentrations halophilic bacteria, in *Microbial Life in Extreme Environments* (D. J. Kushner, ed.), Academic Press, London, 1978, pp. 317-368.

11. Moses, V., and Springham, D. G., *Bacteria and the Enhancement of Oil Recovery*, Applied Science Publishing, Barking, Essex, England, 1982.

12. Oren, A., *Clostridium lortetii*, sp. nov., a halophilic obligately anaerobic bacterium producing endospores with attached gas vacuoles, *Arch. Microbiol., 136,* 42-48 (1983).

13. Oren, A., Weisburg, W. G., Kessel, M., and Woese, C. R., *Halobacteroides halobius* gen. nov. sp. nov., a moderately halophilic obligatory anaerobic bacterium from the bottom sediments of the Dead Sea, personal communication, 1985.

14. Post, J. F., The microbial ecology of the Great Salt Lake, *Microb. Ecol., 3,* 143-165 (1977).

15. Post, J. F., Microbiology of the Great Salt Lake north arm, *Hydrobiologia, 81,* 59-69 (1981).

16. Rengpipat, S., Langworthy, T. A., and Zeikus, J. G., *Halobacteroides acetoethylicus* sp. nov., a new obligately anaerobic halophile isolated from deep subsurface hypersaline environments, *Syst. Appl. Microbiol., 11,* 28-35 (1988).

17. Taber, J. L., Research on enhanced oil recovery: past, present and future, *Pure Appl. Chem., 52,* 1323-1347 (1980).

18. Tissot, B. P., and Welte, D. H., *Petroleum Formation and Occurrence*, Springer-Verlag, New York, 1978.

19. Woese, C. R., Magrum, L. J., and Fox, G. E., Archaebacteria, *J. Mol. Evol., 11,* 245-252 (1978).

20. Zeikus, J. G., Chemical and fuel production by anaerobic bacteria, *Annu. Rev. Microbiol., 34,* 423-464 (1980).

21. Zeikus, J. G., Metabolic communication between biodegradative populations in nature, in *Microbes in Their Natural Environments*, Symposium 34, Society for General Microbiology, Cambridge University Press, Cambridge, 1983.

22. Zeikus, J. G., *Isolation and Characterization of Haloanaerobes for Use in Microbial Enhanced Oil Recovery*, research proposal prepared for Gulf Oil Company, 1983.

23. Zeikus, J. G., Hegge, P. W., Thompson, T. E., Phelps, T. J., and Langworthy, T. A., Isolation and description of *Haloanaerobium praevalens*, gen. nov. and sp. nov., an obligately anaerobic halophile common to Great Salt Lake sediments, *Curr. Microbiol., 9,* 225-234 (1983).

24

Methanogenic Fermentation by Halophilic Anaerobic Bacteria in the Hypersaline Environment of the Dead Sea

CARLOS DOSORETZ and URI MARCHAIM *MIGAL-Galilee Technological Center, Kiryat-Shmona, Israel*

24.1 INTRODUCTION

Energy is sometimes referred to as the lifeblood of our societies, for fuels are used to drive virtually all human activities. In the industrialized world, commercial fuels dominate the energy scene. Fossil fuels account for about 90% of global commercial energy requirements, and of these, oil is the most important. In 1984, commercial energy consumption throughout the world totaled 7201.6 million metric tons of oil equivalent [1,2]. The rapid growth in world energy consumption since 1950 has been disrupted in the few last years but is still high. As a result, energy systems based on renewable sources have received increased attention. The renewable materials making the more significant contribution in this aspect are, at present, biomass and hydropower. However, important advances have been made with other systems. The development of synthetic and alternative fuels to satisfy the world appetite for energy has been progressing for many years. Effort to develop and apply technology for fuels and energy from biomass and wastes has entered a self-sustaining phase, especially on examination of the scope and number of research and commercialization projects in progress. Renewable energy resources offer a vast potential source of energy supply.

Anaerobic digestion (or methanogenic fermentation) is an effective method for the generation of alternative energy, as biogas, and for the ecological treatment of organic wastes. This method has been used widely and successfully in a number of both developed and underdeveloped countries [3,4]. This treatment is affected by facultative and obligate anaerobic microorganisms, which (in the absence of oxygen) convert complex organic materials (both soluble and insoluble) into gaseous end products called "biogas." Biogas consists of a mixture of CH_4 and CO_2 (about 99% of the biogas) and traces of other gaseous products including NH_3, H_2, H_2S, H_2O, and N_2.

The anaerobic fermentation of organic residues seems attractive when the organic substrate, like wastes, or organic material, like lignite, are available in large quantities on-site, which obviates the need for expensive transportation to the fermentor. The methane produced by the system can be sold or used within the vicinity, the effluent can be used as fertilizer and/or can be fed to livestock rural area, and the use of a fermentation system provides pollution, odor, and pest control [5].

Methanogenic fermentation can be considered as a three-stage process, which requires the syntrophic interaction of metabolically different groups of bacteria. In the primary stage fermentative organisms degrade particulate and high-molecular-weight insoluble substrates extracellularly by enzymatic hydrolysis. The products of this hydrolysis are then catabolized to organic acids, carbon dioxide, and hydrogen. In the secondary stage, "acetogenic" organisms degrade higher acids to acetate, CO_2, and H_2. In the tertiary and final stage, methane is formed either by cleavage of acetate or by reduction of carbon dioxide. The overall result of this process is the generation of methane-rich gas [6,7].

Since a wide range of biochemical processes are taking place in natural systems, effective digestion of organic matter into methane requires the combined and coordinated metabolism of different kinds of carbon-catabolizing anaerobic bacteria. The substrates include organic polymers which are components of living tissue, such as cellulose, fats, proteins, pectin, starch, hemicellulose, long- and short-chain fatty acids, alcohols, ketones, and amino acids, all of which will be transformed under essentially anaerobic conditions [8].

At least four different groups of bacteria have been isolated from anaerobic digesters and these bacteria can be characterized on the basis of the substrates fermented and the metabolic end products formed. These bacteria coexist as a complex mixed population in most anaerobic ecosystems. Only a limited amount of information can be obtained about such communities using pure bacterial cultures [9]. Not all the stages or groups of bacteria above have been identified in anaerobic hypersaline environments, and our understanding of those that have been demonstrated is still incomplete.

24.2 THE METHANOGENS: PHYSIOLOGY, DISTRIBUTION, AND TAXONOMY

Methanogens are perhaps the most strictly anaerobic bacteria known (0.01 mg/L dissolved oxygen completely inhibits growth), and therefore, detailed studies require the use of stringent procedures that ensure growth in the complete absence of oxygen [9]. However, there are marked differences in oxygen sensitivity among the methanogens. Methanogenesis is extremely oxygen sensitive due to oxygen liability of certain of the unique methanogenic cofactors. The oxidation-reduction potential required for methanogenesis may be as low as -300 mV or even lower [7]. Methanogenic bacteria perform a pivotal role in anaerobic ecosystems because their unique metabolism controls the rate of organic degradation and directs the carbon and electron flow by removing toxic intermediary metabolites (such as H_2) and by enhancing the thermodynamic efficiency of intermediary metabolism [10].

The distribution and activity of methanogenic bacteria in nature are restricted to anoxic environments where associated bacteria maintain a low redox potential and produce methanogenic substrates as well as other nutrient factors [11]. Organotrophic ecosystems in which methanogens have been detected include the rumen and gastrointestinal tract of humans and animals, in particular herbivores, anaerobic digesters, landfills, and sediments (ponds, marshes, swamps, lakes, and oceans). Methanogenic bacteria have even been found inside the heart wood of living trees [12] and in hot springs (e.g., Yellowstone National Park) [9]. The methanogenic bacteria are unique among prokaryotes because they produce methane as the major product of anaerobic metabolism. However, morphologically methanogens are a diverse group of bacteria that include forms such as rods, spirilla, cocci, and various arrangements of these shapes into longer chains or aggregates [13].

The dilemma of a similar physiology but diverse morphology of methanogens was recently solved after a major revision of their taxonomy based on comparative biochemical studies of their 16S rRNA sequences, DNA sequence, cell wall, and lipids [14]. It has been proposed [15] and later largely accepted that a separate primary kingdom or urkingdom be recognized among the prokaryotes to include the methanogens, the extreme halophiles, and thermoacidophiles. This proposed urkingdom was termed the *archaebacteria*. All the remaining bacteria, cyanobacteria, and mycoplasmas would belong to the urkingdom *eubacteria*. Eukaryotic organisms belong to the *urakaryotic* urkingdom.

The archaebacteria are indeed unusual organisms. The group is now known to include three very different kinds of bacteria: methanogens, extreme halophiles, and thermoacidophiles. The extreme halophiles are bacteria that required high concentration of salt in order to survive; some of them grow readily in saturated brine. They can give a red color to salt evaporation ponds and can discolor and spoil salted fish. The extreme halophiles grow in salty habitats along the ocean borders and in inland water such as the Great Salt Lake and Dead Sea. Although the extreme halophiles have been studied for a long time, they have recently become particularly interesting for two reasons. They maintain large gradients in the concentration of certain ions across their cell membrane and exploit the gradient to move variety of substances into and out of the cell. In addition, the extreme halophiles have a comparatively simple photosynthetic mechanism based not on chlorophyll but on a membrane-bound pigment, bacterial rhodopsin, that is remarkably like one of the visual pigments [16].

24.3 METHANOGENS IN HYPERSALINE ENVIRONMENTS

Little is known on the extent of anaerobic degradation of organic matter in hypersaline environments. The sediments of hypersaline water bodies are generally anaerobic, partly as a result of biological activity in the sediments and the overlaying water, and also because of the limited solubility of oxygen in hypersaline brines. The biology of anaerobic hypersaline environments has been relatively little studied, although it is curious to note that the first bacteria ever isolated from the hypersaline environment (although not halophilic ones): *Clostridia*, causing tetanus and gas gangrene, isolated by Lortet from the Dead Sea mat in the end of the nineteenth century [17]. High salt levels cause bacterial cells to dehydrate because of osmotic pressure. Some microorganisms are more susceptible to osmotic pressure than others. *Staphylococcus aureus* is able to grow in solutions containing up to 65 g/L of NaCl, while *Escherichia coli* is inhibited at much lower levels [18].

Methanogenesis in hypersaline environments has recently attracted great attention. Although methanogenic bacteria and the extreme halophilic bacteria are both members of the *archaebacteria* kingdom, there is relatively little known about methanogens or methanogenesis in hypersaline ecosystems. However, some general observations of methanogenesis in such ecosystems as well as isolation of several pure strains have been reported. Thus a pure methanogen, *Methanococcus halophilus*, was isolated from cyanobacterial mat of the hypersaline Hamelon Pool in the Shark Bay, Australia [19]. Several strains of methanogens were isolated from hypersaline lagoons in Crimea, USSR [20], and biogenic methane has been found in the Gulf of Mexico [21]. An halophilic coccoid-shaped methanogen was isolated from the Salt Lake, Utah [22]. Another strain of halophilic methanogen was isolated from a solar saltern in San Francisco, California [23].

In all of these reports methanogenesis is ascribed to the utilization of methanol, monomethylamine, dimethylamine, trimethylamine, and methionine as a carbon and energy source. Hydrogen, acetate, and formate stimulated methanogenesis slightly or not at all in these cultures. However, methanogenic activity was reported in enrichments of mat sediments from the hypersaline Solar Lake, Sinai [24], where the predominant species enriched was a *Methanosarcina* sp., which preferentially utilized monomethylamine among H_2/CO_2, the methylated amine and acetate. In another study [25], a methanogenic coccus isolated from a solar saltern was reported to be able to grow on H_2/CO_2 or formate, while no growth was observed on trimethylamine, methanol, or acetate, despite the fact that the latter appeared to be required for growth. The optimum salt concentration of pure methanogens in most of the reports ranges from 7 to 15% (as NaCl).

24.4 METHANOGENS FROM THE DEAD SEA

We examined the sediment and mass of water from a natural hypersaline sulfur spring on the shore of the Dead Sea (in collaboration with A. Oren, the Hebrew University, Jerusalem, Israel) for the presence of methanogenic bacteria [26]. These bacteria are of interest because they are involved in the terminal dissimilation of organic carbon in nutrient-rich hypersaline environments. The sampling site chosen (Fig. 24.1) is a series of sulfur springs on the western shore of the Dead Sea lake, a few kilometers south of Ein Gedi. The water forms a number of pools on its way to the Dead Sea, and these are characterized by a rich microbial flora [27,28].

The chemical composition of the spring's water measured in the winter of 1987 (Table 24.1) was quite similar to that reported by Oren [29], measured in the spring of 1985. The average temperature of the water at the source was $39 \pm 1°C$. As shown in Table 24.1, the water was highly saline: specific gravity of about 1.12 ± 0.04 g/L and electrical conductivity of about 157.9 ± 5.3 S, both measured at $25°C$, and it contained 204.3 ± 9.3 g/L dry weight of salts. As expected, chlorine salts gave the major contribution to the salinity of these springs. We succeeded in collecting bubbles of gas, during a few hours and in several places, from the ponds. The gas was analyzed for methane by gas chromatograph, and methane was found in significant quantities (9 to 15% of the total gas phase). The methane found was ascribed to be of biogenic nature.

The enrichment procedures were in general similar to those employed elsewhere [30]. Media composition was determined in accordance to the characteristics of the natural source of inoculum, and contained 20 mM Hepes buffer 7.0, 1% of sterile water spring, 3.7 g/L KCl, and 20 g/L $MgCl_2 \cdot 6H_2O$. NaCl was added as indicated. Substrate concentration was: acetate, 100 mM; methanol, 120 mM; formate, 50 mM; methylamines, 50 mM and H_2/CO_2, 2 atm. All other components were added as reported elsewhere [30] for PB media. If not otherwise stated, NaCl concentration was 120 g/L (145 g/L total salt), substrate was trimethylamine (50 mM), and temperature $37°C$. The culture was initiated by inoculating 50 mL of suitable media, in 120-mL serum bottles, with water and sediments from the sulfur spring of the Dead Sea. The rate of methane

Fig. 24.1 Overview of the two sulfur springs on the shore of the Dead Sea, near Ein Gedi. Note the Dead Sea at the background of the lower picture.

Table 24.1 Chemical Composition of the Mass of Water of the Sulfur Springs on the Shore of the Dead Sea

Component	Value (mean ± SE)
Density$_{25}$ (g/L)	1.12 ± 0.04
EC$_{25}$[a] (S)	157.9 ± 5.3
pH	6.9 ± 0.4
Temperature ($^{\circ}$C)	37–39
Ash (g/L)	204.3 ± 9.3
Sodium (M)	1.10 ± 0.08
Potassium (M)	0.09 ± 0.01
Ammonium (M)	Not detectable
Calcium (M)	0.05 ± 0.01
Magnesium (M)	0.61 ± 0.05
Chlorine (M)	2.91 ± 0.20
Sulfide (M)	2.00 ± 0.35

[a]EC$_{25}$ is electrical conductivity (in siemens) measured at 25°C.

and carbon dioxide production and absorbance measured at 600 nm were daily recorded and after reproducible results, at each transfer, were achieved, different experiments using these cultures were started.

We obtained a stable halophilic methanogenic enrichment culture which is able to utilize trimethylamine as a preferential carbon and energy source (Figs. 24.2 and 24.3). Methanol and monomethylamine were moderately metabolized, while dimethylamine was slightly metabolized. Hydrogen/carbon dioxide, formate, and acetate cannot support growth (cannot be seen in Figs. 24.2 and 24.3 since the growth was insignificant). Interestingly, H_2/CO_2 added together with methanol caused an extension of the lag phase of growth, which resemble the partial or total inhibition observed in thermophilic strains of *Methansosarcina* sp. reported in the literature [31].

Microscopic observation showed that the enrichment culture is composed only of cocci, which resemble the pure strain *Methonococcus halophilus*. The culture is very stable and was maintained for a long period of time. The complete coincidence between production of CH_4 (Fig. 24.2) and A_{600} (Fig. 24.3) indicated that methanogens were the main species in this culture. The only exception was noted using a combination of CH_3COOH plus H_2/CO_2 as substrate, where a discrepancy appear between the rate of methane formation against the rate of absorbance increase. This finding may be based on the fact that H_2 oxidizers other than methanogens were active at low level during the earlier step of growth.

Due to the conditions of the enrichment and the presence of H_2S as reducing agent, no sulfate-reducing activity was detected. In marine and saline habitats, sulfate-reducing bacteria compete successfully with methanogenic bacteria for the available electron donnor substrates (e.g., H_2 and acetate) as long as sufficient sulfate is present [32]. Under such conditions, therefore, sulfate reducers function as the terminal oxidizers in the chain of anaerobic degradation of organic matter. This role is taken over by methano-

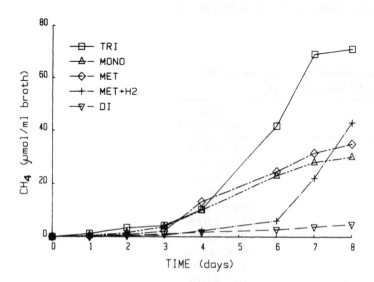

Fig. 24.2 Kinetics of methane production (μmol/mL broth) by the halophilic enrichment culture grown on several substrates. Growth temperature was 37°C, salt concentration 145 g/L (120 g/L NaCl) and pH of the media was 7.0.

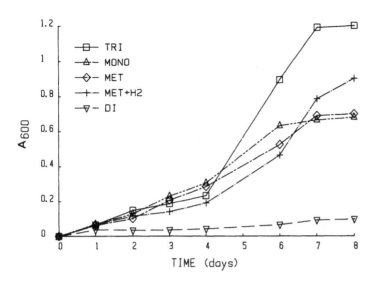

Fig. 24.3 Kinetics of bacterial growth measured as absorbance at 600 nm (A_{600}) of the halophilic methanogenic enrichment culture grown on different substrates. For details, see Fig. 24.2

gens when sulfate became depleted or its original concentration is low. This interaction is based on the fact that sulfate reduction to H_2S is thermodynamically more favorable than CO_2 reduction to CH_4.

The kinetics of CH_4 formation, CO_2 formation, and growth (measured as absorbance at 600 nm) at different salt concentrations are given in Figs. 24.4 to 24.6. Using trimethylamine as the preferential substrate, this culture grew and produced methane in

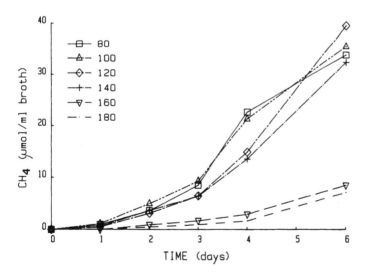

Fig. 24.4 Kinetics of methane production by the halophilic enrichment culture at different NaCl concentrations, to which 25 g/L of other salts than NaCl has been added. Trimethylamine (50 mM) was used as substrate. Other conditions are as in Fig. 24.2.

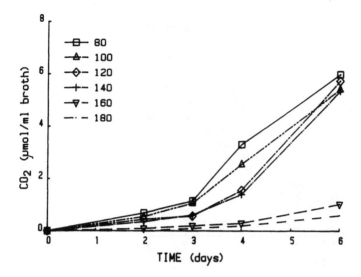

Fig. 24.5 Kinetics of carbon dioxide production by the halophilic enrichment culture at different NaCl concentrations. For conditions, see Fig. 24.4.

a wide range of salt concentrations (105 g/L, of which 80 g/L was NaCl up to 205 g/L, of which 180 g/L was NaCl). Complete inhibition was found at salt concentrations higher than 205 g/L. The optimum salt concentration is in the range of 145 to 165 g/L salt. These experiments also showed that kinetics of methane and carbon dioxide formation and kinetics of absorbance evolvement were practically identical. From the three growth temperature analyzed (Fig. 24.7), similar to those found in the natural environment, the optimum growth temperature was about 37 to 39°C, measured according to the techniques described elsewhere [30].

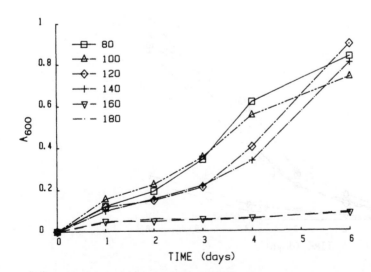

Fig. 24.6 Kinetics of bacterial growth measured as absorbance at 600 nm (A_{600}) of the halophilic methanogenic enrichment culture grown on different NaCl concentrations. For details, see Fig. 24.4.

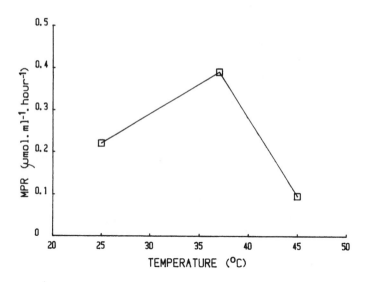

Fig. 24.7   Methane production rate (MPR) of the halophilic methanogenic enrichment culture as a function of growth temperature. For details, see Fig. 24.2.

24.5 CONCLUSIONS

The halophilic methanogenic culture described in this work was shown to be able to grow and produce methane at very high salt concentrations (2.5 to 3.0 M NaCl). According to our results, and in line with literature reports [19,20], the nutritional requirements of halophilic methanogenic bacteria appeared to be very limited. The incapability of utilizing acetate or formate as sole carbon and energy source is in line with the fact that these compounds are normally absent in ecosystems where halophilic methanogens were detected. Surprisingly, methanol, which is also absent in these environments, is a common substrate.

Although the described halophilic enrichment culture is capable of producing methane in high salt concentration, it is still limited in the substrates it can digest. To develop an "integrated-halophilic microflora" capable of degrading organic substrates in hypersaline environment to biogas it is essential to acclimatize the enriched culture to grow on other substrates. These include acetate and various organic compounds (e.g., pretreated lignite). The anaerobic fermentation of complex organic substrates, containing a high concentration of minerals compounds and toxicants, is of great importance in regard to the disposal of hazardous wastes to the environment and in regard to the purpose of generating alternative energy from lignite and other available organic materials.

As a result of the high-rate thermophilic anaerobic digestion of organic wastes, generation of methane-rich biogas and a useful digested slurry is obtained. It was shown [33] that the slurry obtained after the high-rate digestion process of cattle manure is of high value as a substitute for peat moss in horticulture and mushroom industries. If the anaerobic digestion process in hypersaline environment will succeed, together with the generation of alternative energy, it may be possible to utilize the digested slurry obtained for some commercial purposes. We hope to use our applied background in the area of high-rate anaerobic digestion to bring this very special methanogenic process from the laboratory examination stage to commercial application.

ACKNOWLEDGMENTS

The authors would like to express their thanks to Dr. A. Oren of the Hebrew University of Jerusalem for his advice and help, and to A. Kamber and Y. Gutfarb for their excellent technical assistance. This work was part of a project coordinated by Dynatech Scientific, Inc., Cambridge, Massachusetts.

REFERENCES

1. Mathews, J. T., 1986. *World Resources 1986*, World Resources Institute, Washington, D.C., and Basic Books, New York, 1986, pp. 103-119.
2. British Petroleum, *Statistical Review of World Energy*, June 1985, pp. 1-32.
3. Marchaim, U., and Criden, J., Research and development in the utilization of agricultural wastes in Israel for energy feedstock fodder and industrial products, in *Fuel Gas Production from Biomass*, Vol. 1 (D. L. Wise, ed.), CRC Press, Boca Raton, Fla., 1981, pp. 95-120.
4. Stafford, D., Hawkes, D., and Horton, D. L., *Methane Proudction from Waste Organic Matter*, CRC Press, Boca Raton, Fla., 1981.
5. Shuler, L., Utilization of farm wastes for food, in *Utilization and Recycling of Agricultural Wastes and Residues* (M. L. Shuler, ed.), CRC Press, Boca Raton, Fla., 1980, pp. 78-133.
6. Zehnder, A. J., Ingvorsen, K., and Marti, T., Microbiology of methane bacteria, in *Anaerobic Digestion 1981* (D. E. Hughes et al., eds.), Elsevier, Amsterdam, 1982, pp. 45-68.
7. Large, J. L., Methylotrophy and methanogenesis, in *Aspects of Microbiology*, Vol. 8 (J. A. Cole et al., eds.), ASM Publications, Washington, D.C., 1983, pp. 11-24.
8. McCarty, P. L., One hundred years of anaerobic treatment, in *Anaerobic Digestion 1981* (D. E. Hughes et al., eds.), Elsevier, Amsterdam, 1982, pp. 3-22.
9. Zeikus, J. G., The biology of methanogenic bacteria, *Bacteriol. Rev., 41*, 514-541 (1977).
10. Zeikus, J. G., Microbial intermediary metabolism in anaerobic digestion, in *Anaerobic Digestion 1981* (D. E. Hughes et al., eds.), Elsevier, Amsterdam, 1982, pp. 23-35.
11. Mah, R. A., Ward, D. M., Baresi, L., and Glass, T. P., Biogenesis of methane, *Annu. Rev. Microbiol., 31*, 309-341 (1977).
12. Zeikus, J. G., and Ward, J. G., Methane formation in living trees: a microbial origin, *Science, 184*, 1181-1183 (1974).
13. Mah, R. A., and Smith, M. R., The methanogenic bacteria, in *The Prokaryotes: A Handbook on Habitats, Isolation, and Identification of Bacteria*, Vol. 1, (M. P. Starr et al., eds.), Springer-Verlag, New York, 1981, pp. 948-977.
14. Blach, W. E., Fox, G. E., Magrum, L. J., Woese, C. R., and Wolfe, R. S., Methanogens: reevaluation of a unique biological group, *Microbiol. Rev., 43*, 260-296 (1979).
15. Woese, C. R., Magrum, L. J., and Fox, G. E., Archaebacteria, *J. Mol. Evol., 11*, 245-252 (1978).
16. Woese, C. R., Archaebacteria, *Sci. Am., 244*, 94-106 (1981).
17. Oren, A., *Clostridium lortetii* sp. nov., a halophilic obligatory anaerobic bacterium producing endospores with attached gas vacuoles, *Arch. Microbiol., 136*, 42-48 (1983).
18. Brock, T. D., *Biology of Microorganisms*, Prentice-Hall, Englewood Cliffs, N.J., 1970.
19. Zhilina, T. N., New obligate halophilic methane-producing bacterium, *Microbiology, 52*, 375-382 (1983).

20. Zhilina, T. N., Methanogenic bacteria from hypersaline environments, *Syst. Appl. Microbiol., 7*, 216-222 (1986).
21. Brooks, J. M., Bright, T. J., Bernard, B. B., and Schwab, C. R., Chemical aspects of a brine pool of the East Flower Garden bank, northwestern Gulf of Mexico, *Limnol. Oceanogr., 24*, 735-745 (1979).
22. Paterek, J. R., and Smith, P. H., Isolation and characterization of a halophilic methanogen from the Great Salt Lake, *Appl. Environ. Microbiol., 50*, 871-877 (1985).
23. Mathrani, I. M., and Boone, D. R., Isolation and characterization of a moderately halophilic methanogen from a solar saltern, *Appl. Environ. Microbiol., 50*, 140-143 (1985).
24. Giani, D., Giani, L., Cohen, Y., and Krumbein, W. E., Methanogenesis in the hypersaline Solar Lake (Sinai), *FEMS Microbiol. Lett., 25*, 219-224 (1984).
25. Yu, I. K., and Hungate, R. E., Isolation and characterization of an obligately halophilic methanogenic bacterium, presented at *Annual Meeting of the American Society for Microbiology*, New Orleans, La., 1983.
26. Dosoretz, C., Kamber, A., and Marchaim, U., *Methanogenic Fermentation of "Hypersaline Substrates" by Halophilic Anaerobic Fermentation*, submitted to Dynatech R&D, Cambridge, Mass., and the Houston Lighting & Power Company, Houston, Tex., 1987.
27. Oren, A., Population dynamics of halobacteria in the Dead Sea water column, *Limnol. Oceanogr., 28*, 1094-1103 (1983).
28. Oren, A., The ecology and taxonomy of anaerobic halophilic eubacteria, *FEMS Microbiol. Rev., 39*, 23-29 (1986).
29. Oren, A., *Isolation and Characterization of Novel Bacteria from Sediments of the Dead Sea*, Research Report, June 1985-May 1986, submitted to Dynatech R&D, Cambridge, Mass., and the Houston Lighting & Power Company, Houston, Tex., 1986.
30. Dosoretz, C., Origin of inhibitory phenomena in methanogenic fermentation of chicken manure, Ph.D. thesis, Tel-Aviv University, 1986.
31. Ferguson, T. H., and Mah, R. A., Effect of H_2-CO_2 on methanogenesis from acetate or methanol in *Methanosarcina* spp., *Appl. Environ. Microbiol., 46*, 348-355 (1983).
32. Pfenning, F., Widdle, and Trüper, H. G., The dissimilatory sulfate-reducing bacteria, in *The Prokaryotes: A Handbook on Habitats, Isolation, and Identification of Bacteria*, Vol. 1 (M. P. Starr et al., eds.), Springer-Verlag, New York, 1981, pp. 926-940.
33. Marchaim, U., Anaerobic digestion of agricultural wastes: the economics lie in the effluent uses, in *Proceedings of the 3rd International Symposium on Anaerobic Digestion* (R. L. Wentwort, ed.), Evans and Faulker, Watertown, Mass., 1983.

25

Microbial Formation of Methane from Pretreated Lignite at High Salt Concentrations

AHARON OREN   *The Institute of Life Sciences, The Hebrew University of Jerusalem, Jerusalem, Israel*

25.1 INTRODUCTION

During recent years the interest in the use of lower-ranked coals and lignite as energy sources has greatly increased. One of the currently investigated possibilities is the conversion of lignite material into gaseous fuels. It was suggested that biological gasification processes for coal and lignite may offer potential advantages of lower cost and higher efficiency that the various conventional thermal processes [1,2]. A process was developed by the Houston Lighting and Power Company (Houston, Texas) in collaboration with Dynatech R/D Company (Boston, Massachusetts) in which the lignaceous structure is first broken down into simple water-soluble aromatic compounds by aqueous alkali pretreatment at temperatures below 300°C [1,2]. Procedures for the microbial conversion of the resulting simple compounds into methane are under development.

One of the methods proposed for the biogasification of chemically pretreated lignite makes use of underground salt caverns as cheaply available bioreactors [2]; in this case the biological conversion of lignite breakdown products to methane has to occur at high salt concentrations, as salt will dissolve from the walls of the caverns. It was estimated that if run at 75% salt saturation, growth of the underground cavern digestor by salt dissolution would be minimized and provide 20 years of life.

The potential conversion of simple aromatic compounds, such as can be found in chemically pretreated lignite, to methane under anaerobic condition is well established [1,3]. However, the biological aspects of anaerobic breakdown of aromatic compounds to biogas under hypersaline conditions are as good as unknown; moreover, even our knowledge about methanogenesis from more easily degradable substrates at high salt concentrations is still incomplete [4].

In view of the proposed use of highly saline environments in the biogasification of lignite in underground salt caverns [2] I studied the formation of methane from chemically pretreated Texas lignite at high salt concentrations. This study shows that while a potential of methane formation from lignite breakdown products at high salt concentrations does exist, the process is not expected to occur above 16%, and methane yields obtained are very low, consistent with the conversion of only the methoxylate groups attached to the aromatic compounds to methane, while methane formation from acetate, hydrogen/carbon dioxide, and other possible breakdown products of the aromatic compounds does not occur at measurable rates.

25.2 MATERIALS AND METHODS

25.2.1 Pretreated Lignite Sample Used

A sample of pretreated Texas lignite was obtained from Dynatech R/D Company. The lignite was pretreated by aqueous alkaline hydrolysis at a relatively low temperature (below 300°C). According to the analysis performed by Dynatech, the sample contained 0.24% total solids, 0.08% total volatile solids, and 0.19% dissolved solids (passing a 0.45-μm filter). The pH of the solution was found to be 6.2. The sample was stored at 4°C. During the period in which experiments were performed, some growth of fungal mycelia was observed in the lignite sample. I did not remove the fungal growth before using a sample; however, the possibility should be taken into account that these fungi developed in the lignite preparation at the expense of substrates that would otherwise have been available for microbial fermentation and methanogenesis (see Section 25.4).

25.2.2 Testing for Toxicity of the Pretreated Lignite Preparation to Anaerobic Halophilic Bacteria

To test whether the lignite preparation contains compounds that may inhibit the growth of known anaerobic halophilic fermentative bacteria, *Haloanaerobium praevalens* DSM 2228 [5] and *Sporohalobacter marismortui* DY1 (ATCC 35420) [6,7] were inoculated in 10-mL portions of their respective standard growth media in 25-mL tubes. To these tubes portions (0, 0.2, 0.5, 1.0, and 3.0 mL) of the pretreated lignite suspension (made anaerobic by bubbling nitrogen for 15 min) were added by means of a syringe. After incubation of the tubes at 35°C the extent of growth was assessed.

25.2.3 Methane and Hydrogen Formation from Pretreated Lignite at High Salt Concentrations

To a sample of pretreated lignite, solid NaCl was added to a final concentration of 10% or otherwise, as indicated, and the suspension was further supplemented with salts in the following final concentrations: 0.1% KCl, 0.5% $MgCl_2 \cdot 6H_2O$, 0.05% $CaCl_2 \cdot 2H_2O$, 0.1% NH_4Cl, 0.02% K_2HPO_4, and PIPES buffer [piperazine N,N'-bis(2-ethanesulfonic acid)] to a final concentration of 12.5 mM, pH 6.5. In part of the experiments PIPES buffer was replaced by 0.5% $CaCO_3$. The mixture was made anaerobic by bubbling nitrogen for 15 min, whereafter cysteine·HCl was added to a final concentration of 0.5 g/L to obtain strongly reducing conditions. In part of the experiments the composition of the medium was modified as indicated. Portions of the resulting suspension were dispensed in an anaerobic glove box into 23-mL glass vials (typically 15 mL per vial) or 120-mL serum bottles (between 25 and 100 mL per bottle), which were sealed with serum stoppers and aluminum crimp seals, inoculated through the stopper by means of a syringe (inoculum as indicated, inoculum size 5% by volume), and incubated at 35°C, unless specified otherwise.

Hypersaline mud from the following sources was used as inoculum in the initial experiments:

1. A mud sample from Hamei Mazor, a hypersaline sulfur spring on the western shore of the Dead Sea near Ein Gedi [4,8].
2. Mud samples from the commercial solar salt ponds at Eilat, from a pond with a salt concentration of 8.6%.
3. Mud from the Eilat solar salt ponds, from a pond with a salt concentration of 14.5%.

Methane and hydrogen evolution by the cultures was followed by injecting 0.2-mL portions of the gas phase into a Packard model 427 gas chromatograph equipped with a thermal conductivity detector and a 180 × 0.6 cm glass column of Molecular Sieve 5A (80 × 100 mesh). Argon served as carrier gas at a flow rate of 45 mL/min. The output signal was recorded with a Goerz model RE541 Servogor recorder. Methane and hydrogen concentrations in the gas phase were calculated by reference to gas standards.

25.3 RESULTS

25.3.1 Testing for Toxicity of the Pretreated Lignite Preparation to Anaerobic Halophilic Bacteria

To test whether the lignite preparation contains compounds that may inhibit the growth of anaerobic halophilic fermentative bacteria, portions of the pretreated lignite suspension were added to cultures of the anaerobic fermentative bacteria *Haloanaerobium praevalens* and *Sporohalobacter marismortui* (0, 0.2, 0.5, 1.0, and 3.0 mL per 10 mL of culture). After overnight incubation at 35°C all tubes showed excellent growth, proving that at the concentrations employed the lignite preparation is not toxic to at least two representative anaerobic fermentative bacteria.

25.3.2 Methane Formation from Pretreated Lignite at High Salt Concentrations

To a sample of pretreated lignite solid NaCl was added to a final concentration of 15%, and the suspension was further supplemented with salts at the following final concentrations: 0.1% KCl, 0.5% $MgCl_2 \cdot 6H_2O$, 0.05% $CaCl_2 \cdot 2H_2O$, 0.1% NH_4Cl, 0.02% K_2HPO_4, and PIPES buffer to a final concentration of 12.5 mM, pH 6.5. The mixture was made anaerobic by bubbling with nitrogen, whereafter cysteine·HCl was added to a final concentration of 0.5 g/L. Portions of the resulting suspension were inoculated with mud samples from the Ein Gedi sulfur spring or from the solar salt ponds at Eilat. The vials were incubated at 35°C in the dark.

After 24 days of incubation methane was detected at concentrations of up to 1.04% in the gas phase in all vials inoculated with mud from the Eilat solar ponds; no methane was found when mud samples from the Ein Gedi sulfur spring were used. The culture that developed from the Eilat solar saltern sediment at a salt concentration of 14.5% was used as the inoculum for all subsequent experiments. This methanogenic culture could be transferred indefinitely in identical growth medium. In most cases the NaCl concentration used was lowered to 10%.

The maximum accumulation of methane observed was always in the order of 6 to 10 μl of methane per milliliter of lignite suspension, corresponding to a yield of 0.3 to 0.5% of the theoretically possible methane yield (see the discussion section).

When the gas phase above the methanogenic cultures was sampled at shorter intervals (2 to 10 days after inoculation), hydrogen was detected in addition to methane. As, at least theoretically, both the hydrogen and the methane observed can be derived from the L-cysteine added to all bottles, rather than from the lignite material, the following control experiments were performed to ascertain the source of the hydrogen and the methane:

1. To a sample of pretreated lignite, NaCl was added to a final concentration of 10%, and the suspension was further supplemented with salts at the following final concentrations: 0.1% KCl, 0.5% MgCl$_2$·6H$_2$O, 0.5% CaCO$_3$, 0.1% NH$_4$Cl, and 0.02% K$_2$HPO$_4$. Calcium carbonate was used as buffer instead of PIPES, which was used in the previous experiments. The mixture was made anaerobic by bubbling nitrogen for 15 min, and dispensed in 32-mL portions in three 120-mL glass bottles. The bottles were then treated as follows:

> *Bottle 1*: no cysteine
> *Bottle 2*: +10 mg cysteine·HCl
> *Bottle 3*: +30 mg cysteine·HCl

2. A system with cysteine but without lignite suspension was set up as follows: to 100 mL of distilled water, NaCl was added to a final concentration of 10%, and the solution was supplemented with salts to the following final concentrations: 0.1% KCl, 0.5% MgCl$_2$·6H$_2$O, 0.5% CaCO$_3$, 0.1% NH$_4$Cl, and 0.02% K$_2$HPO$_4$. A 120-mL glass bottle was filled with 100 mL of the mixture, and the mixture was made anaerobic by bubbling nitrogen, whereafter 50 mg of cysteine·HCl was added.

The bottles were sealed, inoculated (2% inoculum size by volume), and incubated at 35°C. The time course of hydrogen and methane accumulation during the incubation is shown in Figs. 25.1 and 25.2, respectively.

The following conclusions can be drawn from this experiment:

a. In all systems a transient accumulation of hydrogen was observed.
b. In the absence of cysteine less hydrogen is produced, and methane formation is delayed in time; however, cysteine alone gives rise to small quantities of hydrogen only, and no methane at all. Thus the observed methane formation can be attributed to breakdown of the lignite, not of the cysteine. Most of the hydrogen formation, too, must be due to the lignite.
c. Those cultures containing both cysteine and lignite turned black in the course of the incubation, which was probably due to the formation of hydrogen sulfide during breakdown of cysteine, which reacted with components present in the lignite preparation (possibly iron ions or other heavy metals) to give a black precipitate.

25.3.3 Effect of Salt Concentration and Temperature on Methane and Hydrogen Formation from Pretreated Lignite

Two portions of medium were prepared using pretreated lignite, one without added salt, and one in which 25 g of NaCl was added to 100 mL of lignite suspension (yielding a

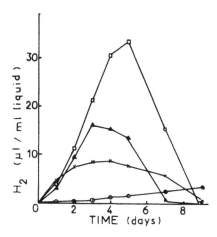

Fig. 25.1 Effect of the presence of cysteine on hydrogen evolution during anaerobic digestion of pretreated lignite. Hydrogen evolved in the gas phase was followed in cultures without added cysteine (X), with 31 mg of cysteine·HCl per 100 mL (▲), with 94 mg cysteine·HCl per 100 mL (□), and in a control experiment that contained 50 mg per 100 mL of cysteine·HCl, but no pretreated lignite (○). Cultures were incubated at 35°C.

solution of approximately 22.9% salt). To both samples salts were added as before (final concentrations 0.1% KCl, 0.5% MgCl$_2$·6H$_2$O, 0.05% CaCl$_2$·2H$_2$O, 0.1% NH$_4$Cl, 0.02% K$_2$HPO$_4$, and PIPES buffer to a final concentration of 12.5 mM), and the pH was adjusted to 6.8 to 6.9 with NaOH. The mixtures were made anaerobic by bubbling with nitrogen, whereafter cysteine·HCl was added to a final concentration of 0.5 g/L. The final pH obtained was approximately 6.4 to 6.5. Portions of the resulting suspensions were dispensed in the anaerobic glove box into 23-mL glass vials to yield 18 mL per vial

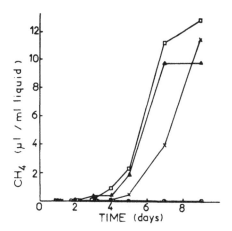

Fig. 25.2 Effect of the presence of cysteine on methane evolution during anaerobic digestion of pretreated lignite. Methane evolved in the gas phase was followed in cultures without added cysteine (X), with 31 mg of cysteine·HCl per 100 mL (▲), with 94 mg of cysteine·HCl per 100 mL (□), and in a control experiment that contained 50 mg per 100 mL of cysteine·HCl, but no pretreated lignite (○). Cultures were incubated at 35°C.

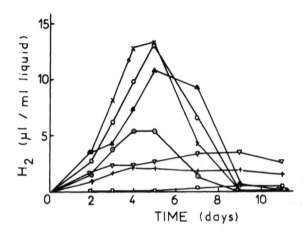

Fig. 25.3 Effect of salinity on hydrogen evolution during anaerobic degradation of pre-treated lignite inoculated with a halophilic methanogenic lignite-digesting culture. Mixtures containing lignite were prepared containing 0.8% (□), 4.5% (⊙), 8.2% (X), 11.9% (○), 15.6% (▲), 19.3% (▽), and 23.0% (+) total salt, and hydrogen evolution in the gas phase was followed during incubation at 35°C.

of the following final salt concentrations: 0.8%, 4.5%, 8.2%, 11.9%, 15.6%, 19.3%, and 23.0%. The vials were sealed, inoculated (2% inoculum size), and incubated at 35°C. The time course of hydrogen and methane accumulation during the incubation is shown in Figs. 25.3 and 25.4, respectively, and the derived optimum curve for methane formation in Fig. 25.5. It is shown that optimal hydrogen evolution occurs at 8 to 16% salt, with only low rates above 18% and below 4%. Methane formation is optimal between 8 and 12% salt, and is strongly inhibited already above 15%.

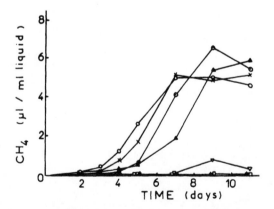

Fig. 25.4 Effect of salinity on methane evolution during anaerobic degradation of pre-treated lignite inoculated with a halophilic methanogenic lignite-digesting culture. Mixtures containing pretreated lignite were prepared containing 0.8% (□), 4.5% (⊙), 8.2% (X), 11.9% (○), 15.6% (▲), 19.3% (▽), and 23.0% (+) total salt, and methane evolution in the gas phase was followed during incubation at 35°C.

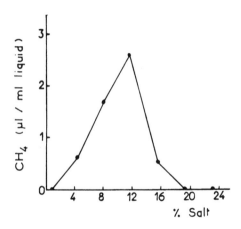

Fig. 25.5 Methane evolution during anaerobic digestion of a pretreated lignite preparation after 5 days of incubation at 35°C as a function of salt concentration. Data were derived from Fig. 25.4.

To examine the effect of temperature on hydrogen and methane formation from pretreated lignite at high salt concentrations, NaCl was added to a sample of pretreated lignite to a final concentration of 10%, and the suspension was further supplemented with salts to the following final concentrations: 0.1% KCl, 0.5% $MgCl_2 \cdot 6H_2O$, 0.5% $CaCO_3$, 0.1% NH_4Cl, and 0.02% KH_2PO_4. The mixture was made anaerobic by bubbling with nitrogen, whereafter cysteine·HCl was added to a final concentration of 0.5 g/L. Portions (25 mL) of the resulting suspension were dispensed into 120-mL glass bottles, which were sealed, inoculated, and incubated in the dark at various temperatures: 28°C, 34°C, 42°C, and 46°C. The time course of hydrogen and methane accumulation during the incuba-

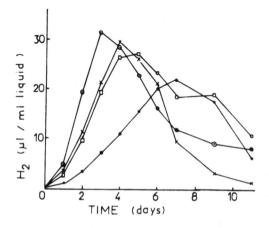

Fig. 25.6 Effect of temperature on hydrogen evolution during anaerobic degradation of pretreated lignite inoculated with a halophilic methanogenic pretreated lignite-digesting culture in the presence of 10% salt. Reaction mixtures were incubated at 28°C (●), 34°C (X), 42°C (⊙), and 46°C (□), and hydrogen evolution in the gas phase was followed during the incubation.

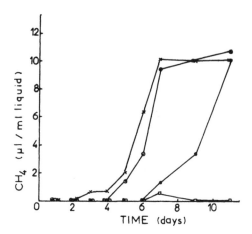

Fig. 25.7 Effect of temperature on methane evolution during anaerobic degradation of pretreated lignite inoculated with a halophilic methanogenic pretreated lignite-digesting culture in the presence of 10% salt. Reaction mixtures were incubated at 28°C (●), 34°C (X), 42°C (⊙), and 46°C (□), and methane evolution in the gas phase was followed during the incubation.

tion is shown in Figs. 25.6 and 25.7, respectively. It is shown that the optimum temperature for both hydrogen evolution and for methane formation is between 34 and 42°C. At 46°C some hydrogen was found, but no methane.

25.3.4 Elucidation of the Nature of the Direct Precursor of Methane Formation from Pretreated Lignite at High Salt Concentrations: Effect of Addition of Potential Methanogenic Substrates

To examine the nature of the immediate precursors for methanogenesis in the methanogenic cultures growing on pretreated lignite, the following experiment was performed, in which potential substrates such as hydrogen, acetate, methanol, and monomethylamine were added to a culture grown on pretreated lignite, in which methane formation had come to an end. This culture (total volume 110 mL) was divided in the anaerobic glove box into three portions of 36 mL in 120-mL glass bottles. The pH of the suspension was 6.1. The bottles were sealed and treated further as follows:

Bottle 1. The gas phase of the bottle was flushed with nitrogen. After 1 day, 0.15 mL of a 25% solution of monomethylamine (neutralized with HCl) was added and incubation was continued.

Bottle 2. 0.2 mL of a 2 M Na-acetate solution was added and the gas phase of the bottle was flushed with nitrogen. After 3 days, 0.1 mL of methanol was added to this bottle and incubation was continued.

Bottle 3. The gas phase of the bottle was first flushed with nitrogen, after which 10 mL of gas was removed and replaced by 10 mL of hydrogen.

All bottles were incubated at 35°C.

The results are shown in Fig. 25.8; methane is readily evolved from both monomethylamine and methanol, while no indication was found of any methane formation from hydrogen/carbon dioxide or from acetate.

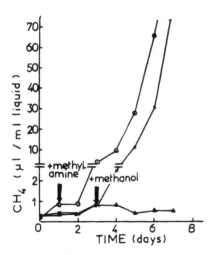

Fig. 25.8 Effect of addition of potential methanogenic precursors on methane formation by a halophilic culture of pretreated lignite degraders after the termination of methane evolution from the lignite. Cultures were incubated under nitrogen ($\odot$, X) in the presence of added acetate (X), or in its absence ($\odot$), or under nitrogen containing 12% hydrogen ($\blacktriangle$), at 35°C. After 1 day the culture under nitrogen that did not receive acetate was supplemented with methylamine ($\odot$), and after 3 days the acetate-supplemented culture was supplemented with methanol (X), and incubation was continued. The amount of methane evolved into the gas phase was determined after various periods of incubation.

25.4 CONCLUSIONS AND DISCUSSION

The main conclusions from the experiments performed are as follows.

1. Pretreated lignite can give rise to methane and hydrogen evolution in an anaerobic fermentation at high salt concentrations.

2. The optimal salt concentration for the process is between 8 and 12%; above 16% salt the reaction rates are negligible. This finding is of profound importance when planning large-scale systems for methanogenic digestion of lignite at high salt concentrations: First, the salt concentration in the reactors will have to be carefully controlled to achieve optimal reaction rates. Moreover, the optimal salt concentration values are still relatively low. It has been proposed to use underground salt caverns for the process, and in that case the process needs to be performed at much higher salinities to prevent dissolution of cavern walls, which would cause the cavern to grow and eventually collapse. It was estimated that if run at 75% of salt saturation (i.e., about 3.9 M or 23%), growth of the digestor would be minimized and provide 20 years of life [2], but this salinity seems too high for the microorganisms involved.

3. The process is mesophilic, with a temperature optimum of about 35 to 42°C for both hydrogen and methane formation (Figs. 25.6 and 25.7); at 46°C (the highest temperature tested) hydrogen evolution rates are still high (62% of the optimal rate), but methane formation was negligible.

4. Methane yields are low: typically 0.006 to 0.01 mL of methane was found to be produced per milliliter of lignite preparation with 0.24 g of total solids per 100 mL, which corresponds with 2.5 to 4 mL of methane per gram. This value should be com-

pared with a theoretically possible value of, for example, 820 mL of gas, of which 50% is methane, per gram of cellulose upon completion of a methanogenic fermentation [9].

5. Experiments in which potential methanogenic precursors were added to the fermentation mixtures after completion of methane formation showed that hydrogen and acetate are not used for methane production in this system, while methanol and mono-methylamine are readily converted to methane. This finding explains the low yields mentioned above: The methane formed is probably derived from $-O-CH_3$ groups that are abundant in the alkaline hydrolysis products formed during the lignite pretreatment (see below). The greatest part of the available carbon can at best be fermented to acetate, hydrogen, and carbon dioxide, neither of which serve as methanogenic precursors at the high salt concentrations employed.

6. During the methanogenic fermentations a transient accumulation of hydrogen was observed (in part derived from the cysteine added to achieve strongly reducing conditions). As the disappearance of this hydrogen during the later stages of the fermentation cannot be attributed to the reduction of carbon dioxide to methane, other hydrogen-consuming reactions must be postulated. I suggested previously the presence of halophilic homoacetogenic bacteria in enrichments for methanogenic bacteria from the Dead Sea [4]; whether such bacteria were responsible for the disappearance of hydrogen in the present experiments remains to be determined. An alternative explanation for the disappearance of the hydrogen is given below.

As stated above, methane yields were low, on the order of 6 to 10 μL per milliliter of lignite preparation, or 2.5 to 4 mL per gram of solids. It should be stated here that all calculations of amounts of methane formed may be slightly underestimated, for the following reasons:

1. The solubility of methane and hydrogen in the water phase was neglected, while both methane and hydrogen are somewhat soluble in water (3.5 mL of methane dissolves in 100 mL of water at $17°C$, and 1 volume of hydrogen dissolves per 50 volumes of water at $0°C$).

2. For the injection of gas samples into the gas chromatograph, simple syringes were used, not pressure-lock syringes, which might have yielded higher values when a positive pressure develops in the fermentation bottles. However, as methane and hydrogen percentages in the gas phase remained low, and as no pressure was observed to accumulate in the systems, I do not expect this underestimation to be significant.

3. The observation of growth of fungal hyphae in the pretreated lignite preparation evokes a question regarding to what extent degradable components from the mixture that are also potential methanogenic precursors were digested or incorporated by the fungi prior to the experiments and thus made unavailable for methanogenesis.

It should be stated here that the observation of fungi in the lignite digest presents us with an important question that cannot be answered as long as the proper experiments with fresh pretreated lignite preparations without the fungi have not been performed: If methanol is indeed the major precursor for methane in the system (see below), we cannot be sure whether the methanol was set free during the anaerobic fermentation at high salt concentrations or whether it was produced earlier by the activity of the fungi. In the latter case the present experiments might not support the conclusion that methane can be formed from pretreated lignite under anaerobic hypersaline conditions, as an aerobic and nonhalophilic stage was involved in generating the methanol. However, the time lag observed between the onset of the fermentative processes (as witnessed by hydrogen formation) and the first appearance of methane (generally observed after only

3 to 4 days of incubation) (Figs. 25.2, 25.4, and 25.7) strongly suggests that the methanogenic precursor (possibly methanol, see below) was indeed generated during the fermentation itself.

4. If methane can be formed only from soluble substrates formed during the chemical pretreatment of the lignite, the methane yield may better be calculated on the basis of dissolved organic matter (0.19%) rather than on the basis of total organic matter (0.24%).

Even when taking all these potential underestimations into account, yields of methane generated during pretreated lignite digestion were very low. From an easier digestible substrate such as cellulose, a theoretical yield of 820 mL of gas (methane plus carbon dioxide) can be expected per gram upon complete degradation [9]. With a variety of substrates (agricultural wastes and others) yields of 100 to 200 mL of gas per gram of substrate were typically obtained, of which 50 to 70% consisted of methane [9].

The known methanogenic bacteria can produce methane from one or more of the following precursors: hydrogen plus carbon dioxide, acetate, methanol, and methylamines. In nonhalophilic methanogenic ecosystems hydrogen/carbon dioxide and acetate are the most important methanogenic precursors, as hydrogen and acetate are among the products of most fermentative pathways. Hydrogen and acetate have also been identified among the fermentation products of isolated halophilic fermentative bacteria [5-7].

Although hydrogen was present as a potential electron donor for methanogenesis during our experiments with digestion of pretreated lignite preparations, our data suggest that it was not utilized as such. Part (but not all) of the hydrogen produced in (most) experiments presented above was derived from cysteine, which was added routinely to lower the oxidation-reduction potential of the fermentation mixtures. Cysteine can be expected to be fermented to hydrogen sulfide, ammonia, acetate, carbon dioxide, hydrogen, and possibly butyrate (note that at least theoretically, butyrate formation can serve as a hydrogen sink, and this may be an alternative explanation for the decrease in hydrogen concentrations during the later stages of the fermentation). None of these products is converted to methane under the conditions employed, and thus all methane observed must have been derived from the lignite digest itself. No assays of acetate and butyrate formation during the fermentation process were performed; thus no data are available that allow us to draw conclusions on the mechanism of the disappearance of the hydrogen.

Attempts to isolate hydrogen/carbon dioxide utilizing halophilic methanogenic bacteria from different hypersaline environments have generally failed. To the best of our knowledge, only one halophilic methanogen was described to be able to produce methane by reducing carbon dioxide with hydrogen: a methanogenic pleomorphic coccus able to grow on hydrogen and carbon dioxide or on formate, isolated from a solar saltern [10]. No growth of this organism was observed on trimethylamine, methanol, or acetate (although acetate appeared to be required for growth). It grew above 1.8 M salt (optimally at 2.7 M), and its optimal growth temperature was 35°C. The discovery of this organism was reported as an abstract at a congress in 1983; as far as I know, no subsequent data were published on this organism. There is also little direct evidence that hydrogen is an important electron donor for methanogenesis in hypersaline ecosystems in nature. One possible exception is Mono Lake (California), a hypersaline alkaline lake, in which methane formation was observed, probably derived from reduction of carbon dioxide [11].

Most studies on methanogenesis under hypersaline conditions show that at high salt concentrations the most important methanogenic precursors are more specialized substrates such as methanol, methylamines, and methionine. Hypersaline environments are usually greatly enriched in their sulfate content compared with seawater, and thus sulfate-respiring bacteria generally outcompete methanogens for available hydrogen and/or acetate. Therefore, noncompetitive substrates such as trimethylamine or dimethylsulfide can be of greater importance as methane precursors [11].

In sediments of Big Soda Lake, Nevada (87 g/L total dissolved salts, pH 9.7), methanogenesis was found to be stimulated by methanol and trimethylamine, to a lesser extent also by methionine, but hardly or not at all by hydrogen, acetate, or formate [12,13].

Using ^{14}C-labeled substrates, Zeikus [14] demonstrated high rates of microbial decomposition of simple organic compounds in Great Salt Lake sediments (e.g., in 11 days more than 20% of the label of [2-^{14}C] acetate, [U-^{14}C] glucose, and [3-^{14}C] lactate was found as $^{14}CO_2$); from these substrates no labeled methane was formed. The only substrates giving rise to the formation of methane were methylmercaptan, methionine, and methanol [14,15]. No significant methanogenesis was observed from acetate or hydrogen and carbon dioxide, and counts of methanogenic bacteria, growing in a medium with methanol, hydrogen, and carbon dioxide were as low as 10 cells/mL. In the Great Salt Lake hydrogen production is much less tightly coupled to its consumption than in freshwater environments: Hydrogen is present in the sediments (salt concentration greater than 20%) in concentrations of up to 200 μM, while methane concentrations are as low as 4 μM [14].

In the Solar Lake (Sinai), methanogenesis was demonstrated in sediments of relatively low salt concentrations (7 to 7.4%) [16], and up to 10 methanogenic bacteria per milliliter of sediment were counted. *Methanosarcina*-like organisms predominated, preferring methylamines as substrates. With hydrogen as substrate, no or little methane was produced within 3 weeks of incubation; only from the topmost layer of the core enrichments with hydrogen and carbon dioxide showed a substantial increase in methane.

Only recently halophilic methanogenic bacteria have been isolated in pure culture. Mathrani and Boone [17] isolated from a solar salt pond a methanogenic bacterium growing in salt concentrations from 1 to 3.5 M (optimum 2.1 M at 37°C). The strain grows only on methylamines and methanol. Biogenic methane formation in sediments of the Great Salt Lake was reported as early as 1978 [18]. A strain of irregular cocci, isolated from the south arm of the Great Salt Lake [19] has a similar substrate specificity; it grows on methanol, methylamine, dimethylamine, and trimethylamine, but not on hydrogen and carbon dioxide, formate, or acetate. Optimal growth was observed at 1 to 2 M NaCl and 35°C (range of growth 0.5 to 3 M salt and 25 to 40°C). The name "*Halomethanococcus mahi*" was proposed for this isolate. Another strain ("*Methanococcus halophilus*") growing on methanol or methylamine was isolated by Zhilina [20] from a cyanobacterial mat from the hypersaline Hamelin Pool (Shark Bay, Australia). This isolate also failed to grow on hydrogen and carbon dioxide or on formate, but after a long adaptation period it proved able to grow on acetate. It grew at NaCl concentrations between 1.5 and 15%, with an optimum at 7% at 26 to 36°C. Several other strains of methanogens were isolated from hypersaline lagoons in Crimea, USSR, growing in salt concentrations of up to 30%, and using methylamines as sole substrates for growth, being unable to metabolize hydrogen and carbon dioxide, acetate, or formate [21]. These strains are probably similar to "*Methanococcoides euhalobius*" [22]. A methanogenic coccus with a similar substrate specificity was also isolated from the Ein Gedi sulfur spring, an isolate growing at salt concentrations between 80 and 180 g/L, and growing best on trimethylamine and methanol, not on dimethylamine, hydrogen and carbon dioxide, or acetate, and showing slight growth on monomethylamine (C. Dosoretz, A. Kamber, and U. Marchaim, unpublished data, 1987).

The finding that methane formation in the halophilic digestion mixtures of pre-treated lignite was stimulated by methanol and methylamine, but not by hydrogen or acetate, suggests that methanol might be the direct precursor. No significant concentrations of methylamines can be expected to occur in the lignite preparations, but the occurrence of methanol among the fermentation products is not unexpected: Among the products of alkaline oxidative pretreatment of lignite [1,2] are compounds with $-O-CH_3$ groups such as syringic acid, syringaldehyde, ferulic acid, vanillic acid, and vanillin. These $-O-CH_3$ groups may have been the real precursors of the methane formed, being converted first to methanol during fermentative digestion of the aromatic rings. A variety of methoxylated aromatic compounds, such as vanillin, vanillic acid, ferulic acid, syringic acid, syringaldehyde, and 3,4,5-trimethoxybenzoic acid, can be broken down anaerobically by (nonhalophilic) bacteria [23-25]. The first step in their breakdown is the cleavage of the methoxylate groups to yield methanol [24-26].

The results of the experiments described in this report present a few interesting prospects of great academic interest on the existence of hitherto unknown types of anaerobic halophilic bacteria. Although the existence of anaerobic bacteria that ferment phenolic compounds is now well established (for reviews see, e.g., Refs. 1 and 3), we do not yet know any halophilic bacteria that perform such processes. The present study clearly suggests the existence of such bacteria. We plan to perform enrichment cultures for degraders of aromatic compounds, using the methanogenic cultures growing on lignite digests as inoculum. In addition, the present study also shows signs of the existence of homoacetogenic bacteria converting hydrogen and carbon dioxide to acetate at high salt concentrations, bacteria whose presence was suggested earlier [4]. Also, these organisms are waiting to be isolated and characterized.

ACKNOWLEDGMENTS

Support for the work described in this paper was obtained by a grant from Houston Lighting and Power Company, Houston, Texas, under a university participation program administered by Dynatech R/D Company. I thank A. P. Leuschner and A. S. Martel (Dynatech R/D Company, Boston, Massachusetts) for the sample of pretreated Texas lignite.

REFERENCES

1. Leuschner, A. P., Trantolo, D. J., Kern, E. E., and Wise, D. L., Biogasification of Texas lignite, in *Proceedings of the 13th Biennial Lignite Symposium: Technology and Utilization of Low-Rank Coal*, Vol. 1, U.S. Department of Energy, Washington, D.C., 1986, p. 216.
2. Wise, D. L., Meeting report: first international workshop on biogasification and bio-refining of Texas lignite, *Resour. Conserv., 15*, 229 (1987).
3. Berry, D. F., Francis, A. J., and Bollag, J.-M., Microbial metabolism of homocyclic and heterocyclic aromatic compounds under anaerobic conditions, *Microbiol. Rev., 51*, 43 (1987).
4. Oren, A., Anaerobic degradation of organic compounds in hypersaline environments: possibilities and limitations, in this volume, chapter nine.

5. Zeikus, J.G., Hegge, P. W., Thompson, T. E., Phelps, T. J., and Langworthy, T. A., Isolation and description of *Haloanaerobium praevalens* gen. nov. and sp. nov., an obligately anaerobic halophile common to Great Salt Lake sediments, *Curr. Microbiol., 9,* 225 (1983).

6. Oren, A., The ecology and taxonomy of anaerobic halophilic eubacteria, *FEMS Microbiol. Rev., 39,* 23 (1986).

7. Oren, A., Pohla, H., and Stackebrandt, E., Transfer of *Clostridium lortetii* to a new genus *Sporohalobacter* gen. nov. as *Sporohalobacter lortetii* comb. nov., and description of *Sporohalobacter marismortui* sp. nov., *Syst. Appl. Microbiol., 9,* 239 (1987).

8. Oren, A., Photosynthetic and heterotrophic bacterial communities of a hypersaline sulfur spring on the shore of the Dead Sea (Hamei Mazor), in *Microbial Mats: Physiological Ecology of Benthic Microbial Communities* (Y. Cohen and E. Rosenberg, eds.), American Society for Microbiology, Washington, D.C., 1989, p. 64.

9. Wise, D. L., Leuschner, A. P., Levy, P. F., Sharaf, M. A., and Wentworth, R. L., Low-capital-cost fuel-gas production from combined organic residues: the global potential, *Resour. Conserv., 15,* 163 (1987).

10. Yu, I. K., and Hungate, R. E., Isolation and characterization of an obligately halophilic methanogenic bacterium, in *Abstracts of the Annual Meeting of the American Society for Microbiology*, New Orleans, Abstract I1, 1983, p. 139.

11. Oremland, R. S., and King, G. M., Methanogenesis in hypersaline environments, in *Microbial Mats: Physiological Ecology of Benthic Microbial Communities* (Y. Cohen and E. Rosenberg, eds.), American Society for Microbiology, Washington, D.C., 1989, p. 180.

12. Oremland, R. S., Marsh, L., Culbertson, C., and DesMarais, D. J., Soda Lake III: dissolved gases and methanogenesis, *EoS, 62,* 922 (1981).

13. Oremland, R. S., Marsh, L., and DesMarais, D. J., Methanogenesis in Big Soda Lake, Nevada: an alkaline, moderately hypersaline desert lake, *Appl. Environ. Microbiol., 43,* 462 (1982).

14. Zeikus, J. G., Metabolic communication between biodegradative populations in nature, in *Microbes in Their Natural Environments*, Symposium 34 (J. H. Slater, R. Whittenbury, and J. W. T. Wimpenny, eds.), Society for General Microbiology, Cambridge University Press, Cambridge, 1983, p. 423.

15. Phelps, T., and Zeikus, J. G., Microbial ecology of anaerobic decomposition in Great Salt Lake, in *Abstracts of the Annual Meeting of the American Society for Microbiology*, Abstract I4, 1980, p. 89.

16. Giani, D., Giani, L., Cohen, Y., and Krumbein, W. C., Methanogenesis in the hypersaline Solar Lake (Sinai), *FEMS Microbiol. Lett., 25,* 219 (1985).

17. Mathrani, I. M., and Boone, D. R., Isolation and characterization of a moderately halophilic methanogen from a solar saltern, *Appl. Environ. Microbiol., 50,* 140 (1985).

18. Ward, D. M., Biogenesis of methane in Great Salt Lake sediments, in *Abstracts of the Annual Meeting of the American Society for Microbiology*, Abstract I49, 1978, p. 89.

19. Paterek, J. R., and Smith, P. H., Isolation and characterization of a halophilic methanogen from Great Salt Lake, *Appl. Environ. Microbiol., 52,* 877 (1985).

20. Zhilina, T. N., New obligate halophilic methane-producing bacterium, *Microbiology (Engl. Transl.), 52,* 290 (1983).

21. Zhilina, T. N., Methanogenic bacteria from hypersaline environments, *Syst. Appl. Microbiol., 7,* 216 (1986).

22. Obraztsova, A. Ya., Shipin, O. V., Belaev, S. S., and Ivanov, M. V., Biological characters of halophilic methanogen isolated from oil bed, *Dokl. Akad. Nauk. SSSR, 278,* 227 (1984).

23. Healy, J. B., Jr., and Young, L. Y., Anaerobic biodegradation of eleven aromatic compounds to methane, *Appl. Environ. Microbiol., 38*, 84 (1979).

24. Kaiser, J.-P., and Hanselmann, K. W., Aromatic chemicals through anaerobic microbial conversion of lignite monomers, *Experientia, 38*, 167 (1982).

25. Kaiser, J.-P., and Hanselmann, K. W., Fermentative metabolism of substituted monoaromatic compounds by a bacterial community from anaerobic sediments, *Arch. Microbiol., 133*, 185 (1982).

26. Bache, R., and Pfennig, N., Selective isolation of *Acetobacterium woodii* on methoxylated aromatic acids and determination of growth yields, *Arch. Microbiol., 130*, 255 (1981).

This article was completed in November 1987; for reasons beyond the author's control it has not been possible to update it in accordance with recent developments in the field.

26

Microbial Processing of Coal

GREGORY J. OLSON* *National Institute of Standards and Technology, Gaithersburg, Maryland*

ROBERT M. KELLY *The Johns Hopkins University, Baltimore, Maryland*

26.1 INTRODUCTION

The past few years has seen a rapidly growing interest in the potential for using microorganisms for processing of fossil fuels and minerals. Some of this interest stems from the excitement generated by the "new biotechnology" involving genetic engineering in medicine, agriculture, and biochemistry. At the same time, increased costs of conventional minerals extraction and processing, together with depletion of high-grade reserves and stricter environmental regulations have stimulated consideration of alternative technologies, including biotechnology, by the minerals and fuels industries. This need for advanced minerals processing biotechnology cannot yet be fully met by microbiologists, chemists, chemical engineers, and mining engineers since our knowledge of the full range of biotransformations of minerals, the underlying fundamental mechanisms, rate-limiting factors, and engineering studies are incomplete. Limitations in process measurement methods and lack of experimental standards are also problems.

Nonetheless, there are proven examples of commercial microbial processing of minerals. The depressed copper market has caused U.S. mining companies increasingly to utilize low-cost, low-energy hydrometallurgical processing of low-grade copper ore and waste in leaching dumps in the western United States [1]. Microbially catalyzed reactions in the dumps significantly contribute to copper recovery from many of these operations, now amounting to at least 30% of U.S. production [2]. Uranium is recovered as a result of microbially catalyzed reactions at the Denison mine in Canada [3]. Pilot plant-scale studies suggest that controlled microbial pre-leaching of host rock to assist in gold recovery by conventional cyanide treatment is near [4,5]. These example applications rely on the activities of iron-, sulfur-, and metal sulfide-oxidizing bacteria, most notably *Thiobacillus ferrooxidans*.

Some of the same organisms that are important for metal ore processing are potentially applicable to coal bioprocessing. *T. ferrooxidans*, because of its ability to oxidize

---

Current affiliation: Pittsburgh Energy Technology Center, Pittsburgh, Pennsylvania.

and solubilize pyrite (FeS$_2$), has been studied extensively in the laboratory for removal of pyritic sulfur from coal [6-10]. Other bacteria, including thermophilic *Sulfolobales* and several newly described facultative thermophiles, also hold promise in this area. Some of these organisms may be useful in removing metals from coals containing elevated metal sulfide levels [6,11-13]. Bacterial removal of organic sulfur from coal has also been reported [7,14-17]. An area of active study recently has been the solubilization of low-grade coals by certain fungi [18-22]. As yet unevaluated are microbial processes for removal of organic nitrogen and oxygen from low-rank coals.

Thus consideration of potential bioprocesses for coal upgrading not only includes removal of undesirable components associated with the coal (pyrite, metal sulfides) but also possible removal of chemical groups occurring as an integral part of the coal itself (organic sulfur, organic nitrogen, oxygen).

Much of the current interest in coal bioprocessing concerns removal of sulfur, given the recent emphasis on prevention of acid rain through a reduction of sulfur emissions from coal-fired power plants. Although several physical and chemical methods have been used or proposed for reducing sulfur emissions from coal combustion [23-25], there are associated problems of cost, efficiency, reliability, and waste disposal. Biological processes have the potential to overcome some of these coal desulfurization problems, and it is here that bioprocessing offers, perhaps, its most immediate and significant application. However, other biological processes for improving fuel quality or conversion, such as removal of oxygen, organic nitrogen, metals, or solubilization, should be investigated more fully.

The literature on coal bioprocessing has recently been reviewed or summarized by several authors. Olson and Brinckman [26,27] reviewed coal desulfurization, metals removal, and beneficiation, Kargi [28] discussed coal desulfurization, and Finnerty and Monticello [29] discussed desulfurization of fossil fuels, including coal, by microorganisms. The purpose here is to provide some perspective as to the more general issue of coal bioprocessing, with particular attention to technical challenges that prevent wide-scale implementation.

26.2 DESULFURIZATION

26.2.1 Microbial Removal of Pyrite from Coal

Much of the sulfur occurring in coal, especially in coals of the eastern United States, is in the form of the iron disulfide minerals pyrite and marcasite (pyrite is the cubic form of FeS$_2$; marcasite is orthorhombic). Removal of this mineral sulfur would significantly reduce levels of sulfur in many coals. Indeed, chemical and physical treatments for precombustion coal desulfurization have been developed.

Certain microorganisms oxidize pyrite and marcasite and have been considered as potential agents for a coal desulfurization process. Andrews and Maczuga [30] have summarized the advantages of bacterial desulfurization of coal; these include a higher pyrite removal efficiency and lower coal wastage than with physical methods, and reduced costs compared to chemical methods because microbial methods operate at ambient conditions with fewer reagents. However, microbial processes are slower, requiring days to complete.

The best known of the pyrite-oxidizing bacteria is *T. ferrooxidans*, a gram-negative iron-, sulfur-, and metal sulfide–oxidizing bacterium. This organism is acidophilic and chemoautotrophic, and is a common inhabitant of acidic environments associated with metal sulfide weathering. Indeed, this organism is important in the production of acidic coal mine drainage that results from uncontrolled pyrite oxidation during and following mining [31], and affects thousands of miles of streams in the United States.

Several investigators have studied pyrite degradation by mixed versus pure cultures of bacteria. Apel and Dugan [32] found that a mixed natural population of acid mine drainage organisms increased the rate of pyrite oxidation. Groudev and Genchev [33] reported that the presence of *Thiobacillus thiooxidans* slightly reduced the rate of pyrite oxidation by one strain of *T. ferrooxidans* but slightly increased the rate of pyrite oxidation when incubated with a different strain. They indicated that the results were "compatible" with the sulfur-oxidizing abilities of the two *T. ferrooxidans* strains. However, it is difficult to determine if the relatively small differences in pyrite oxidation rate observed were statistically significant. Andrews and Maczuga [30] noted that the lag in pyrite oxidation upon inoculation of coal slurries with *T. ferrooxidans* could be greatly reduced when *T. thiooxidans* was added to the culture. Kos et al. [14] found that mixed cultures of acidophilic bacteria from the drainage of a coal washing plant were very effective in pyrite leaching from coal.

Certain thermophilic bacteria have also been shown to remove pyritic sulfur from coal. These include *Sulfolobus acidocaldarius* [34,35], and an unidentified thermophile growing at 55°C [36]. *Sulfolobus*, a member of the archaebacteria, oxidizes pyrite, elemental sulfur, certain metal sulfides, and organic compounds at temperatures of up to 85°C. Thermophilic bacterial processes may result in faster coal desulfurization rates as associated chemical reactions are accelerated at elevated temperatures, although this has yet to be demonstrated conclusively.

Pyrite slowly oxidizes on exposure to air and water to produce acid and ferrous iron:

$$2FeS_2 + 7O_2 + 2H_2O \rightarrow 2FeSO_4 + 2H_2SO_4 \tag{1}$$

T. ferrooxidans oxidizes soluble ferrous ions to ferric ions at low pH as a source of metabolic energy:

$$4FeSO_4 + O_2 + H_2SO_4 \rightarrow 2Fe_2(SO_4)_3 + 2H_2O \tag{2}$$

Under acidic conditions, ferrous iron [reaction (2)] is relatively stable to chemical oxidation. However, the bacteria catalyze this reaction, generating ferric ions that react with additional pyrite:

$$FeS_2 + Fe_2(SO_4)_3 \rightarrow 3FeSO_4 + 2S^0 \tag{3}$$

The literature is conflicting as to whether S^0 is formed in the reaction of ferric ion with pyrite. Other metal sulfides are also oxidized and solubilized in ferric sulfate solutions:

$$MS + 2Fe^{3+} \rightarrow M^{2+} + 2Fe^{2+} + S^0 \tag{4}$$

Reaction (2) is considered the rate-limiting step in pyrite dissolution and bacteria accelerate the dissolution by a factor of up to 10^6 [37]. Elemental sulfur [reactions (3), (4)] is also oxidized by *T. ferroxidans* and other thiobacilli to sulfuric acid.

The overall biological pyrite oxidation sequence can be summarized as

$$4FeS_2 + 15O_2 + 2H_2O \rightarrow 2Fe_2(SO_4)_3 + 2H_2SO_4 \tag{5}$$

Abiotic reactions between ferric sulfate and oxygen and water produce additional acidity and various ferric hydroxy compounds. In the presence of certain cations, precipitates of iron hydroxysulfates, known as jarosites, may form [38,39]. These precipates may coat pyrite crystals and slow biological and chemical pyrite oxidation reactions.

Adsorption of T. ferrooxidans to Surfaces

The foregoing sequence of reactions has often been referred to as "indirect" microbial pyrite oxidation since it relies on biological regeneration of ferric iron, which reacts chemically with the pyrite. "Direct" attack of *T. ferrooxidans* on metal sulfides also occurs. However, much is still unknown regarding the exact mechanism of pyrite dissolution, the importance of bioadhesion in mineral degradation, and the biochemical properties of different strains of *T. ferrooxidans* that influence their rate of pyrite oxidation.

T. ferrooxidans has been shown to adsorb rapidly to particles, including pyrite glass, quartz, fluorapatite, and sulfur [40]. Bagdigian and Myerson [41] and Myerson and Kline [42] studied the adsorption of *T. ferrooxidans* on coal surfaces and found that up to 100% of the bacteria in solution were adsorbed to coal surfaces within 2 min, depending on the amount of coal present. The bacteria attached selectively and irreversibly (did not dislodge with rinsing) to pyrite, especially to massive inclusions. The pyrite has lower surface free energy than the coal matrix (as measured by mercury contact angles), which the authors suggested accounted for the selective adsorption of cells to pyrite. McCready and Zentilli [43] also showed that *T. ferrooxidans* adsorbed to pyrite and other inorganic mineral phases in finely ground coal suspensions. After 30 min of exposure, the cells could be removed from the coal but not from the inorganic minerals, indicating strong attachment to mineral phases. Keller and Murr [44] studied ferric sulfate and *T. ferrooxidans* leaching of polished pyrite crystals. Different pitting and corrosion patterns were observed with chemical and biological attack which they suggested was the result of different mechanisms of attack on the pyrite. They did not observe any significant attachment of bacteria to the pyrite. Bennett and Tributsch [45] showed that *T. ferrooxidans* attach to pyrite surfaces, with pits resulting from their activity. These authors extended this research to other metal sulfides, and suggested that during bacterial attack, surface states on the metal sulfides, $SH^{\delta-}$ groups, are produced and removed by bacterial activity [46]. In the presence of iron, Fe^{3+}, H^+, and a proposed molecular carrier removing $SH^{\delta-}$ groups from the sulfide surface all contribute to breaking chemical bonds on the metal sulfide surface, resulting in oxidation [47].

We have shown that epifluorescence microscopy coupled with fluorescent acridine orange staining can be used to quantitate the adhesion of *T. ferrooxidans* on pyrite surfaces [48]. In addition, it appears that the cell fluorescence color is indicative of cell activity [48]. Numerous investigators have attempted to correlate acridine orange fluorescence of aquatic bacteria to cellular activity, with conflicting results. However, with the acidophilic *T. ferrooxidans* and *Sulfolobus acidocaldarius*, active cells fluoresce green, whereas cells inactivated by heat treatment or anaerobic conditions or cells in prolonged stationary growth phase fluoresce red. The intracellular pH of these acidophiles, normally near neutral, may drop upon inactivation [49,50], thus affecting the acridine orange/ nucleic acid interaction. Acidophiles are known to experience a net influx of protons from their growth medium upon metabolic inhibition. This may or may not result in a drop in intracellular pH. *T. ferrooxidans* shows a drop in intracellular pH upon growth inhibition [51].

We observed that during active pyrite leaching, cells attached to pyrite surfaces fluoresced green, whereas cells in the surrounding free solution were largely red after a few days incubation. This suggests that attached cells are most active in pyrite oxidation. The tendency for *T. ferrooxidans* to adhere to or modify the pyritic component of coal led to the interesting finding that treatment of finely ground coal with *T. ferrooxidans* rendered the coal pyrite hydrophilic in minutes, making it amenable to separation by oil agglomeration [52,53]. Subsequent data have shown that to achieve 90% pyrite removal from coal by the process, the coal had to be ground to a particle about the size of 5 μm [54]. Grinding efficiencies more in line with economic considerations (ca. 10 μm) would probably restrict pyrite removal to one-third to one-half of the total using bacterial conditioning [54].

Chemical and Physical Factors Influencing Pyrite Oxidation

Different strains of *T. ferrooxidans* vary in the rates with which they oxidize iron and sulfur substrates [55,56]. However, we still do not understand the inherent properties of different strains of *T. ferrooxidans* that might account for observed differences in rates of pyrite oxidation. The lack of standard reference materials and protocols for conducting pyrite bioleaching tests make it difficult to compare data and understand properties of organisms that could be enhanced for more rapid pyrite oxidation. Nonetheless, manipulation of chemical and physical factors can enhance microbial pyrite oxidation, as shown by numerous laboratory studies.

Particle Size and Pulp Density. Intuitively, it is expected that microorganisms utilizing a solid substrate will exhibit a faster oxidation rate with greater substrate surface area, obtained chiefly by finer grinding. Numerous studies have shown that very finely ground pyrite is oxidized most rapidly with *Sulfolobus* [34], *T. ferrooxidans* [6,8,57], and mixed thiobacilli [30]. However, some engineering problems might dictate microbiological pyrite removal from coal in bioreactors not employ very finely ground coal since the processed coal may be difficult to recover [6]. Instead, continuous stirred tank reactors should use particles of at least 0.3 mm, or the use of heap leaching methods should be considered [6].

pH. *T. ferrooxidans* is an acidophilic bacterium with an optimum pH for growth in the range 2.0 to 3.5, and a pH value of 2.0 was reported as the optimum for removal of sulfur from Ohio and New Mexico coals. *Sulfolobus* also grows optimally at pH 2 to 3, coinciding with its maximum coal desulfurization activity.

Dissolved Gases. The literature is conflicting as to the effects of enrichment of CO_2 on microbial coal desulfurization, Kargi [58] found that external sources of CO_2 and air significantly increased the rate and extent of sulfur removal from coal in a batch reactor compared to a shake flask without an external CO_2 supply. However, it was not completely clear that the increased sulfur removal could be attributed to CO_2 alone since results with the batch reactor without pure CO_2 but with air were not reported. Hoffman et al. [56] showed that pyrite oxidation rates increased slightly when the partial pressure of CO_2 was raised from atmospheric pressure to 0.1 atm. Care must be taken in gas addition since hyperbaric O_2 is inhibitory to *T. ferrooxidans* [59].

Temperature. Most studies with *T. ferrooxidans* employ temperatures of 25 to 35°C. The optimum temperatures reported for coal desulfurization by this organism range from 28 to 35°C. However, both moderate (50 to 60°C) and extreme (>60°C) thermophiles have been shown to desulfurize coal at temperatures of 50 to 75°C. These organisms include *S. acidocaldarius*, which removed >90% of pyritic sulfur from a bituminous coal [34] with optimum conditions of pH 1.5 and 70°C. Adjustments of nitrogen to phosphorus and nitrogen to magnesium ratios significantly affected pyritic sulfur removal. A thermophilic, *Thiobacillus*-like bacterium removed about half of the organic sulfur and about 90% of the pyritic sulfur from a Turkish lignite coal after 25 days at 50°C [60]. Also, the hyperthermophilic archaebacteria from the genera *Pyrococcus* and *Pyrodictium* removed sulfur from a weathered, high-sulfur coal [60a]. As opposed to most desulfurization processes that generate sulfate solutions, these organisms reduce S^o to sulfide. The extent of organics removal with these organisms is not yet certain.

26.2.2 Microbial Removal of Organic Sulfur from Coal

Unlike pyrite, organic sulfur occurs as an integral part of the coal molecular matrix and is not removed by current mechanical separation methods [24]. Consequently, the potential for microbial organic sulfur removal from coal is of great interest. Several investigators have reported the removal of organic sulfur from coal ranging from <10% removal from an Indian coal (containing 6.64% sulfur, 82.8% of which was organic) with *T. ferrooxidans* and mixed natural microorganisms present in the coal [7], to about 20% removal from the same coal with a dibenzothiophene/beef extract-enriched mixed microbial population [16], to about 50% removal from a Turkish lignite (containing 2.2% organic sulfur) using a moderately thermophilic *Thiobacillus*-like bacterium (TH-1) [60]. *S. acidocaldarius* reportedly removed 44% of organic sulfur from an inorganic sulfur-depleted Pennsylvania bituminous coal containing 0.71% organic sulfur [17]; this culture was adapted to dibenzothiophene. A bacterium designated CB-1 reportedly removed up to 47% of the organic sulfur from a pyrite-cleaned coal sample; this bacterium has been demonstrated to work on a pilot scale (2500 lb/day) [15].

Unfortunately, organic sulfur in coal is generally determined by difference (total sulfur minus pyritic sulfur determinations), potentially leading to large analytical errors and making interpretations of the extent of microbial organic sulfur removal difficult. Improved methods for organic sulfur determination in coal are needed. Straszheim et al. [61] recently described a scanning electron microscopy-energy dispersive x-ray microanalysis technique for organic sulfur determination in coal. Isbister and Kobylinski [15] used this method to confirm microbial removal of organic sulfur from coal.

26.2.3 Engineering Considerations and Prospects

There have been a number of efforts addressing engineering problems that are associated with microbial coal desulfurization. These range from developing mathematical models potentially useful for design of desulfurization systems to investigation of phenemona related to system performance to detailed economic and engineering analysis related to the implementation of the technology. Since there are many that have suggested that a microbially based coal cleaning or conversion technology is not far away, the continuous interchange between scientists and engineers on this subject must take place. This is important if scientific developments are to be assessed rapidly as to their economic impact on process technology. As such, engineering studies in this area will help to rapidly commercialize microbially based coal technologies, as well as provide a framework with which to evaluate scientific advances and provide research direction.

Biological Rates

The most important consideration in the development of an actual microbial de-sulfurization process is the *biocatalyst*, that is, the particular population of microorganisms to be used. No matter how finely coal is ground, how dense the biomass concentration is, or how efficient aeration and mixing are, the rate-determining step is the rate at which the microorganisms remove the sulfur from the coal. This rate will have the largest influence on the engineering design of a microbial desulfurization process. Figure 26.1 illustrates this particular point schematically, having been discussed previously by Kargi [28].

Careful comparisons are difficult to make for coal desulfurization experiments described in the literature, because of the diversity in experimental conditions, coal type, and microbial populations. However, we have attempted to normalize these results for comparison of sulfur removal rates was made. Volumetric rates are computed simply, converting iron or pyrite leaching rates to equivalent sulfur rates where necessary through stoichiometric relationships. Surface-area-specific rates were determined by assuming that all coals studied were approximately the same density (1.5 g/cm), and coal particles were assumed to be uniform. In all but one case, internal surface area is not considered. Average particle size or, when necessary, the largest particle size reported were used for

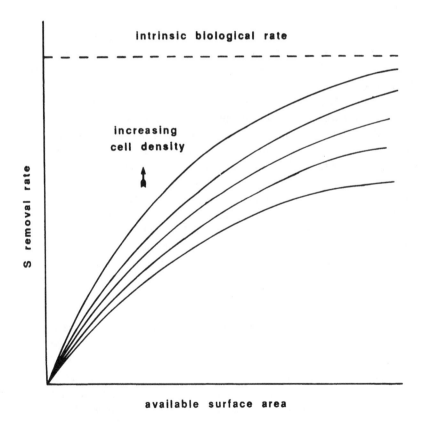

Fig. 26.1 Relationship between total available coal surface area and cell density as it affects coal desulfurization rates.

the calculation. No attempt was made to incorporate biomass concentration or inoculum size into the calculation.

The specific surface area of a coal particle a_v, was determined through the following formula:

$$a_v = \frac{6}{\phi_s D_p} = \frac{S_p}{\phi_s V_p} \qquad (6)$$

where a_v is the specific surface area of a coal particle, ϕ_s the shape factor (1.0 for a sphere, 0.73 for pulverized coal), D_p the particle diameter, S_p the surface area of sphere of equivalent diameter, and V_p the volume of a sphere of equivalent diameter. The surface area concentration in a given culture is then determined as follows:

$$a = a_v \frac{\rho_p}{\rho_c} \qquad (7)$$

where a is the surface area concentration, ρ_p the pulp density, and ρ_c the coal particle density. The surface specific rate can then be calculated from the volumetric rate as follows:

$$\text{rate}_{vol} = \text{rate}_{surface} \cdot a \qquad (8)$$

where $\text{rate}_{surface}$ is the surface specific rate and rate_{vol} is the volumetric rate.

Although these rates are very approximate estimates, it is clear that microbial desulfurization rates shown in Table 26.1 are fairly close together and, for the most part, within an order of magnitude. This is noteworthy given the diversity of experimental conditions and coal types. Rates are a strong function of available surface area, the determination of which is the weakest part of the estimates. Since in many cases the largest particle size reported was used, these rates are probably overestimates. It is worth noting that the rates reported for *S. acidocaldarius* are not significantly different for those reported at lower temperatures for *T. ferrooxidans*, although it has been frequently suggested that rates at higher temperatures would be higher.

A similar estimate was made for several reported studies of microbial coal desulfurization in different reactor configurations. These rates, shown in Table 26.2, suggest that no clear improvements can be realized through more efficient aeration and mixing. In fact, some comment on the significance of surface area on removal rate can be made by examining the work of Myerson and Kline [42]. They used a BET apparatus for measuring coal surface area to determine the dilution rate of surface area as fed to a continuously stirred tank reactor (CSTR). This surface area estimate, certainly significantly larger than would be obtained using the approach here, leads to low-surface-area specific removal rates. However, using the approach described here, their results are in line with other efforts reported in Table 26.2. The surface area available to the bacteria lies between estimates based on particle size and those measurements that incorporate surface area embedded in the porous structure of the coal. Since particle size (i.e., surface area) is a critical parameter in these systems, more attention to its determination and significance is recommended. It appears from first estimates that microbial coal desulfurization rates for a variety of coals are comparable, and that new populations of bacteria must be discovered or developed to make significant improvements.

Table 26.1 Comparison of Microbial Coal Pyrite Oxidation Rates

Coal type	Strain[a]	Pulp density (%)	Particle size	Volumetric rate ($mgS_{pyr}/L \cdot h$)	Surface rate ($mgS_{pyr}/cm^2 \cdot h$; estimated)	Ref.
Pennsylvania	ATCC-19859 TF	3%	147–417 μm	25.0	4.3×10^{-3}	58
Ohio (central)	NM isolate TF	25%	<38 μm	70.8	2.0×10^{-4}	36
Ohio (Broken Aro)	ATCC-19859 TF	10%	43–74 μm	6.7	7.2×10^{-5}	56
Illinois No. 6	ATCC-23270 TF	5%	−200 mesh	2.1	5.7×10^{-5}	88
Brogan	Canada isolate TF	10%	75–150 μm	43.4	8.8×10^{-4}	43
Subbituminous (Bulgaria)	Strain V TF	10%	<40 μm	11.7	8.5×10^{-5}	33
Pennsylvania	Strain 98-3 SA	5%	125 μm	33.0	1.7×10^{-3}	34

[a]TF, *Thiobacillus ferrooxidans*; SA, *Sulfolobus acidocaldarius*.

Table 26.2 Examples of Reactors Used for Microbial Coal Desulfurization

Reactor type	Bacteria[a]	Pulp density (%)	Particle size	Volumetric rate ($mgS_{pyr}/L\cdot h$)	Surface area (cm^2/L)	Surface rate ($mgS_{pyr}/cm^2\cdot h$)	Ref.
Airlift-recycle	SA	5%	125 μm	1.8	2.2×10^4	8.2×10^{-5}	87
Slurry pipeline	TF	25%	147–1981 μm	23.3	1.3×10^4	1.8×10^{-3}	65
CSTR	TF	15%	200 mesh	27.0	1.1×10^5	2.4×10^{-4}	42
					(7.5×10^6)[b]		
Batch	TF	3%	147–417 μm	25.0	6.4×10^3	3.9×10^{-3}	58

[a]TF, *Thiobacillus ferrooxidans*; SA, *Sulfolobus acidocaldarius*.
[b]Determined by BET measurement.

Choice of Reactor Type

Given the relatively long time it takes to desulfurize coal microbially (9 days appears to be a good estimate [10,62]), reactor types with short retention times would not be considered. As suggested by Dugan [63,64], processes more akin to waste treatment systems are more likely to be implemented. Although coal slurry pipelines have been evaluated for microbial coal desulfurization [65], these would probably be used only where significant distances between mine and user exist and only if a variety of materials and processing problems (e.g., corrosion and waste treatment) could be worked out.

It is reasonable to suggest that early microbial coal desulfurization processes for the removal of sulfur from high-sulfur coals be based on existing heap leaching systems or involve large settling ponds configured similar to a waste treatment process. As suggested by Roffman [66], land-use requirements would be considerable for the latter. Although a process based on heap leaching may be relatively simple to implement, variability in desulfurization efficiency may present problems. It is interesting to note that Bos et al. [62], in addition to an interesting discussion on the use of various reactor types for microbial coal desulfurization, report the prospect of using a tanks-in-series reactor configuration to optimize bacterial cell density and pyrite removal rate. This system was suggested after a thorough engineering and economic analysis sponsored by the Dutch National Coal Research Programme. This particular system was found to be economically feasible for the removal of pyrite from low-sulfur coals (<1% sulfur).

It is clear that the most important objective at this time is finding and/or developing microbial populations that release sulfur from coal at faster rates. While the removal of organic coal sulfur is highly desirable, significant improvements in pyrite removal rates may be just as important. Upon examining estimates shown in Table 26.1, it seems unlikely that there will be significant improvements in sulfur removal rates using those populations of bacteria that have been studied thus far.

Processing Objectives

It seems clear from the variety of studies done to this point that a particular microbial coal desulfurization technology will probably not meet a spectrum of processing objectives. However, if the specific processing objective is consistent with economic constraints, microbially based systems may be attractive.

Along these lines, there are some key considerations that arise. One of these is the ratio of pyritic to organic sulfur in the feedstock. Coals containing a high proportion of pyritic sulfur can be treated with microorganisms that are well characterized, such as *T. ferrooxidans*. Removal of both forms of coal sulfur, however, will probably require distinctly different microorganisms. If some type of continuous or semicontinuous process is envisioned, several processing stages will be required. Failure to remove a high percentage of the total sulfur content in high-sulfur coal will probably necessitate the use of flue gas desulfurization, possibly negating the economic benefits derived from the microbial process.

Scenarios have been described in which the microbial process may be attractive. The work of Bos et al. [62] shows that for the situation existing in the Netherlands, a microbially based technology is promising for upgrading low-sulfur coal. From a slightly different perspective, Dugan [63,64] has discussed the economics of microbial coal desulfurization in terms of the value added to the coal as a result of sulfur removal. In some regions of the United States, utilities pay a premium for low-sulfur coal. Dugan, using data from his coal biodesulfurization laboratory experiments, estimated that an

Appalacian coal containing 5.4% sulfur (4.2% pyritic sulfur) increased sufficiently in value to more than pay for the estimated costs of the microbially based process. When the sulfur content of coals is low, desulfurization is not a high-priority process. Dugan's work suggests that microbial precombustion removal of sulfur from certain coals is currently economically feasible. Any future environmental regulations reducing sulfur emissions from power plants would make bioprocessing more attractive.

Clearly, the motivation to use a microbially based coal desulfurization route lies in the specific circumstances surrounding a coal conversion operation. For example, since the microbial process is a wet technology, combustors that can accept a coal slurry feed are more amenable to this coal cleaning technology than are systems that require a dry coal feedstock. The decision to use a microbial coal desulfurization process must be considered in the context of an integrated process, and alternatives should be considered in the same way.

Engineering-Based Studies

Next to finding or developing microbial populations that will improve on what is currently available, the most important focus in this area is improving existing systems through manipulation of engineering parameters. This may be done on the scale of the entire process by strategic arrangement of processing steps or even on the smaller scale of improving cell-coal sulfur interaction.

Even though near-term applications of microbial coal desulfurization will probably involve less sophisticated engineering approaches, studies done under carefully controlled conditions are important to an understanding of underlying phenomena. Along these lines there have been a number of interesting investigations into the influence of engineering parameters and reactor configurations on microbial coal desulfurization rates. Table 26.2 summarizes the results from some of these studies.

In addition to experimental studies, efforts to determine specific engineering parameters and incorporate these into process models have also been reported. Myerson and Kline [42] used a steady-state bacterial mass balance model to estimate the specific growth rate of $T.$ $ferrooxidans$ on solid surface, which was found to increase with increasing dilution rate. Kargi and Weissman [67] developed a dynamic model for pyritic sulfur removal from coal which incorporated the effect of sorption coefficients, substrate concentration, pulp density, particle size, and solution and surface cell density on desulfurization rates. Efforts like these to develop process models for microbial coal desulfurization are important and useful, especially if they can be refined to include developments in understanding the underlying biological phenomena as they become available.

26.3 METALS REMOVAL

$T.$ $ferrooxidans$ can oxidize metal sulfides directly or indirectly via ferric iron generation during pyrite oxidation [equation (4)]. Consequently, metals and metalloids occurring as sulfides in coal are solubilized by this organism. Jilek and Beranova [12] found that a mixed culture of iron and sulfur oxidizing acidophilic bacteria removed up to 90% of the arsenic (and 70% of the sulfur) in a high-sulfur coal mined in Czechoslovakia (the main chemical species of arsenic in the coal was not described). $T.$ $ferrooxidans$ was used to recover certain metals, including Be, Ga, Co, Ni, and V, from coal by conversion of the insoluble metal sulfides [13]. Leachates from bioreactors containing different coals contained elevated levels of metals [14].

26.4 BIOSOLUBILIZATION OF COAL

Several investigators have recently reported that certain species of fungi, especially those that degrade lignin, can transform subbituminous or lignite coals into viscous, gummy, or liquid products [18–22]. Although this phenomenon is an interesting one, it is not clear how useful the resulting liquid product might be. Nonetheless, a fuller understanding of the process is warranted, along with assessment of process usefulness.

These studies have emphasized the oxidative biosolubilization of lignite. It is expected that lignite would be more susceptible to microbial attack than would higher-rank coals, since lignin-like polymers are identifiable in lower-rank coals [68]. However, a study of biosolubilization of several kinds of lignite and one subbituminous coal showed that only a North Dakota lignite was susceptible to rapid and extensive solubilization by certain fungi [20,21]. North Dakota lignites retain more woody character than do Gulf coast lignites [70]. Enhanced fungal solubilization of the other coals occurred following treatment with oxidizing agents.

Wilson and colleagues [20] reported that at least three different strains of fungi caused solubilization of lignite. They also reported the solubilization of leonardite, an oxidized form of lignite using a partially purified laccase preparation from a culture of a wood decay fungus, *Polyporus versicolor*. Laccase is a polyphenol oxidase enzyme involved in lignin biodegradation. Preliminary evidence was reported suggesting that oxidative pretreatment of coals affected the ability of *P. versicolor* to degrade coal.

Ether linkages are important bridging units in low-rank coals. Investigations of organisms that degrade ether linkages in model compounds might yield new pathways for coal depolymerization and solubilization. For example, a species of *Arthrobacter* converts 4-hydroxyphenoxyacetate to hydroquinone [70].

26.5 OXYGEN AND NITROGEN REMOVAL

Lignite and subbituminous coals contain much higher levels of oxygen incorporated into the coal structures than do bituminous coals. Carboxyl groups account for an estimated two-thirds of this oxygen [70]. Decarboxylase enzymes might be capable of removing this oxygen as carbon dioxide. An illustrative model compound reaction is the conversion of vanillic acid to guaiacol by strains of *Bacillus* and *Streptomyces* [71].

Nitrogen in coal occurs mainly in ring positions [23] and microbial degradation of ring nitrogen model compounds has been reported. Examples include the degradation of picolinamide [72] and nicotine [73]. Organisms such as these should be investigated for their ability to remove organic nitrogen from coal.

Studies using specific enzymes that degrade appropriate N-, O-, or S-containing compounds might also provide new information on coal chemical structure. However, the access of enzymes and microorganisms to the interior of coal particles is uncertain. Coal is porous, as shown by nitrogen and carbon dioxide adsorption studies and small-angle x-ray scattering [74]. However, pore sizes are often on the order of tens to hundreds of angstroms. Such pore sizes would exclude the entry of micrometer-sized bacteria, but not necessarily exclude small exocellular metabolites or enzymes. For example, the dimensions of the enzyme chymotrypsin (MW 22,600) are 45 × 35 × 38 Å. It is difficult to envision how a significant quantity of organic N (or organic S) could be removed from coal by microorganisms unless access to the interior of coal particles were possible. This area should be investigated further.

26.6 PROSPECTS FOR FUTURE WORK

While the economic feasibility of a microbially based coal desulfurization technology has been debated, this approach, nonetheless, presents some intriguing possibilities. When, and if, several limiting aspects of proposed processes are improved, heightened interest in this area will no doubt result. Of course, if environmental concerns about acid rain increase or if there are pressures to make wider use of coal as an energy or chemical feedstock, microbial coal desulfurization may become more attractive. In any event, several technical areas necessitate more attention.

26.6.1 Developments of Standards for Experimental Studies

One of the bigger problems in determining the effectiveness of a given microbial population for coal desulfurization is the inability to compare results obtained in different laboratories. Many such studies have been described covering a multitude of coals, microorganisms, and experimental procedures. Unfortunately, while the literature contains numerous demonstrations of the feasibility of microbially desulfurizing a given coal, current and future needs are to determine those particular microorganisms and operating conditions that are optimal.

Table 26.1 summarizes some of the numerous studies found in the literature for pyrite removal by *T. ferrooxidans*. It is clear that careful comparisons between various studies is not possible because of the different approaches used. Furthermore, as mixed cultures are considered and new isolates are found, standardization is critical to avoid spending time characterizing cultures that are ineffective relative to existing systems.

To address this problem, the National Institute of Standards and Technology is investigating prospects for developing standards and experimental procedures that will allow the comparison of results from different laboratories. One of the objectives is to be able to make comparisons of intrinsic rates of microbial coal desulfurization by particular microorganisms or mixed cultures. At present, efforts are directed at standards development for pyrite leaching that can be extended to coal. Among the factors being considered are particle size, pulp density, gas mass transfer rates, inoculum size, solution and surface cell density, total surface area, and coal sulfur forms. By establishing materials to be used as standards and by defining standard experimental procedures, indentification of the most effective strains for coal desulfurization will be possible.

26.6.2 Development and Application of Better Analytical Methodology

Unfortunately, most analytical techniques used to assess the state of a given microbial method for coal processing are indirect. For example, sulfate levels may be followed to determine the extent of coal desulfurization, although nothing is learned about the cell-surface relationship or the type of coal sulfur that has been removed. As such, more sophisticated analytical techniques are essential if a better understanding of the underlying physical, chemical, and biological phenomena is to be obtained.

Epifluorescence Microscopy

We have previously described the use of epifluorescence microscopy for examination of cells attached to iron pyrite. Because coal structure has significant microporosity, only those coals ground finely enough are amenable to this approach, but this may be the case for many microbial desulfurization processes. Figure 26.2 shows some epifluorescent photomicrographs of the attachment of *S. acidocaldarius* to coal. Work is now under way

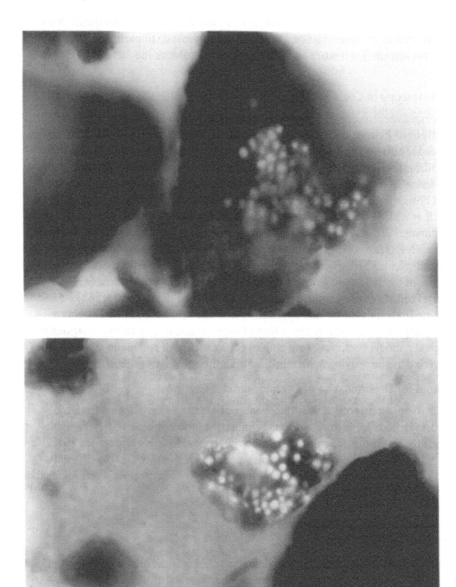

Fig. 26.2 Cells of *Sulfolobus acidocaldarius* attached to coal surfaces as shown by epifluorescence microscopy and acridine orange staining (see Ref. 48). Individual cells are approximately 1 μm in diameter.

to determine the extent to which this technique can provide qualitative and quantitative information related to the microbial coal desulfurization process. In particular, we are interested in the cell surface density and cell fluorescence color as they relate to release of sulfur.

Infrared Spectroscopy of Coals

Although infrared spectroscopy has for many years been used to study coal structure [76], the advent of Fourier transform infrared spectroscopy (FTIR), with its rapid scanning, enhanced sensitivity, and computerized data manipulation features [77], has made possible significant recent advancements in studying coal structure properties and changes during processing. Solomon and co-workers [78] investigated various maceral fractions from several coals, finding relatively discernible peak intensity differences in the aliphatic region as well as with O–C and hydroxyl groups. Within a maceral group, coupled density gradient techniques and FTIR analysis revealed an increase in oxygen functionality (in O–C bonds) with increasing density of the subspecies. The degree of oxidation of coals is readily determined by FTIR spectroscopy [79] and is useful for rapidly characterizing coal properties in the field during mining [80,81]. Mineral matter in the coal (e.g., clays, calcite, quartz) can also be determined quantitatively with FTIR techniques [82].

FTIR methods should also find application in studying changes in coal structure and mineral composition caused by the action of microorganisms, especially if organisms attacking specific model compounds are employed. As such, this technique should be useful in uncovering the underlying mechanisms of microbial-coal interactions.

Direct Methods for Determination of Organic Sulfur in Coal

It has been difficult to determine the effectiveness of organic sulfur removal from coal because direct methods for organic sulfur determination have not been widely available. New methods for direct organic sulfur determination in coal have been proposed [61, 82a] and used [15] in attempts to confirm organic sulfur removal. Wider use of techniques such as these will be important in developing effective coal organic sulfur removal methods using either biotic or abiotic approaches.

26.6.3 Discovery and/or Development of Better Microorganisms for Coal Bioprocessing

An often-mentioned problem with microbial coal processing is the relatively slow reaction rate. Thermophilic, sulfur-oxidizing organisms may improve pyrite oxidation rates, although comparisons with their mesophilic counterparts under carefully controlled conditions are not available. Different strains of *T. ferrooxidans* vary in substrate oxidation rates [83], but the underlying reasons for these differences have not been determined. As a result, we do not yet fully understand those properties of particular microorganisms that must be modified, through genetic engineering or conventional selection procedures, to improve bioprocessing rates. With pyrite removal, standard reference materials and leaching procedures will be important in making intercomparisons and evaluations of biological factors that influence sulfur oxidation rates. However, finding and/or developing more effective *biocatalyst* is paramount to the expansion of microbially based coal technologies.

Genetic manipulation of acidopphilic thiobacilli has the potential to produce strains with the desired characteristics. However, it is not clear what characteristics would result in faster sulfur or iron oxidation rates. In addition, a genetic system (mechanisms of transferring genetic information) must be developed for *T. ferrooxidans*. The unusual culturing conditions in which acidophilic thiobacilli grow and the difficulty in growing these bacteria on conventional solid media complicate the matter. Progress and prospects in this area have recently been reviewed [84,85], but no immediate effect on the current situation is expected.

Perhaps, the best prospects for breakthroughs in this area lie in discovering new groups of microorganisms that have high levels of activity toward sulfur. Candidates from groups such as the sulfur-dependent archaebacteria should be considered, although many of these microorganisms are known to be sulfur reducers. However, the ability of bacteria, such as *Pyrodictium occultum*, to convert elemental suflur to sulfide at very high rates [86] may be put to good use if an effective pretreatment of the coal sulfur can be developed. We are currently pursuing this possibility.

26.6.4 Upgrading Coal Through Attack of Specific Functional Groups

Very little work has been done investigating the selective microbial attack on certain functional groups in coal. This is, in part, the result of analytical difficulties attendant on such investigations, especially in characterization of the coal itself. Nonbiological coal liquefaction processes are often studied using model coal compounds or by controlled solubilization using chemical or physical treatments that selectively attack certain functional groups in the coal. A detailed analysis of produces formed is important in these studies, which is often useful in gaining some information concerning coal structure.

Similarly, microorganisms known to attack specific model compounds could be used to study the related occurrence in coal samples. This approach seems promising given the preliminary reports of organic sulfur from coal by dibenzothiophene-oxidizing bacteria and coal liquefaction by lignin-degrading fungi. Additional work in this area might employ organic nitrogen utilizing microorganisms to remove nitrogen from coal. Removal of oxygen from low-rank coal by enzymatic decarboxylation should also be examined carefully. Studies such as these should also contribute to our understanding of coal structure. Despite the indications that microbial model compound studies may be useful in studying coal upgrading or liquefaction, there are questions as to how accessible to enzyme attack are such functional groups in the coal matrix, and how representative are model compounds to the actual functional groups found in the coal matrix.

26.6.5 Improvements in Coal Processing Technology

Prospects for coal bioprocessing may also improve if certain improvements in coal processing are made. Better grinding technologies to achieve particle size distributions more amenable to microbial methods are desirable. Also, development of wet combustion technology so that boilers can accept a water-based coal slurry feed may make microbial methods attractive.

26.7 SUMMARY

The discussion presented here suggests that microbially based coal conversion technologies are potentially an important element of the energy industries. However, except for a few instances, economic and technical hurdles remain before their widespread use is envisioned. Improvements in sulfur removal rates and the imposition of stricter environmental standards will be the key factors influencing development of microbial desulfurization processes. Microbial coal liquefaction processes are probably further away. Nonetheless, the potential benefits from microbially based technologies should motivate continued interest and work in this area, which calls for the interaction of biologists, chemists, and engineers to solve what are clearly interdisciplinary problems.

ACKNOWLEDGMENTS

The authors would like to thank Wei-wen Su and Richard Schicho, Department of Chemical Engineering, Johns Hopkins University, for their help with calculations used in this chapter.

REFERENCES

1. Brierley, C. L., Bacterial leaching, *CRC Crit. Rev. Microbiol., 6*, 207 (1978).
2. Malouf, E. C., Present status and future prospects of biological metal extraction, presented at the *Conference on Biotechnology Applied to the Mining and Mineral Processing Industry*, Idaho National Engineering Laboratory, Idaho Falls, Idaho, Aug. 1986.
3. Wadden, D., and Gallant, A., The in-place leaching of uranium at Denison Mines, *Can. Metall. Q., 24*, 127 (1985).
4. Lawrence, R. W., and Bruynesteyn, A., Biological peroxidation to enhance gold and silver recovery from refractory pyritic ores and concentrates, *CIM Bull., 76*, 107 (1983).
5. Brunesteyn, A., Biohydrometallurgy of refractory gold ores, presented at the *Conference on Biotechnology Applied to Mining and Mineral Processing Industry*, Idaho National Engineering Laboratory, Idaho Falls, Idaho, Aug. 1986.
6. McCready, R. G. L., and Zentilli, M., Beneficiation of coal by bacterial leaching, *Can. Metall. Q., 24*, 135 (1985).
7. Chandra, D., Roy, P., Mishra, A. K., Chakrabarti, J. N., Prasad, N. K., and Chaudhuri, S. G., Removal of sulfur from coal by *Thiobacillus ferrooxidans* and by mixed acidophilic bacteria present in coal, *Fuel, 59*, 249 (1980).
8. Silverman, M. P., Rogoff, M. H., and Wender, I., Removal of pyritic suflur from coal by bacterial action, *Fuel, 42*, 113 (1963).
9. Zarubina, Z. M., Lyalikova, N. N., and Shmuk Y. I., Investigation of the microbiological oxidation of the pyrite of coal, *Izv. Akad. Nauk SSSR Otd. Tekh. Nauk Metall. Topl., 1*, 117 (1959).
10. Detz, C. M., and Barvinchak, G., Microbial desulfurization of coal, *Min. Congr. J., 65*, 75 (1979).
11. Murphy, J., Riesternberg, E., Mohler, R., Marek, D., Beck, B., and Skidmore, D., Coal desulfurization by microbial processing, in *Processing and Utilization of High Sulfur Coals* (Y. A. Attia, ed.), Elsevier, Amsterdam, 1985, p. 643.
12. Jilek, R., and Bernaova, E., Some experiences with bacterial leaching of brown coal, in *Proceedings of the International Conference on the Use of Microorganisms in Hydrometallurgy*, Pécs, Hungary, 1982, p. 167.
13. Wilczok, T., Cwalina, B., and Chrostowska, D., *Przegl. Gorn., 39*, 174 (1983) [*Chem. Abstr., 100*, 176 (1984); no. 100:970IV].
14. Kos, C. H., Poorter, R. P. E., Bos, P., and Kuenen, J. G., Geochemistry of sulfides in coal and microbial leaching experiments, in *Proceedings of the International Conference on Coal Science*, Düsseldorf, West Germany, 1981, p. 842.
15. Isbister, J. D., and Kobylinski, E. A., Microbial desulfurization of coal, in *Processing and Utilization of High Sulfur Coals* (Y. A. Attia, ed.), Elsevier, Amsterdam, 1985, p. 627.

16. Chandra, D., Roy, P., Mishra, A. K., Chakrabarti, J. N., and Sengupta, B., Microbial removal of organic sulfur from coal, *Fuel, 58*, 549 (1979).

17. Kargi, F., and Robinson, J. M., Removal of organic sulphur from bituminous coal: use of the thermophilic organism *Sulfolobus acidocaldarius, Fuel, 65*, 397 (1986).

18. Scott, C. D., Strandberg, G. W., and Lewis, S. N., Microbial solubilization of coal, *Biotechnol. Prog., 2*, 131 (1986).

19. Ward, B., Lignite-degrading fungi isolated from a weathered outcrop, *Syst. Appl. Microbiol., 6*, 236 (1985).

20. Wilson, B. W., Lewis, E., Stewart, D., and Li, S. M., Microbial benefication of coal and coal-derived liquids, in *Proceedings of the 11th EPRI Contractors Conference*, Electric Power Research Institute, Palo Alto, Calif., May 1986.

21. Cohen, M. S., and Gabriele, P. D., Degradation of coal by the fungi *Polyporus versicolor* and *Poria monticola, Appl. Environ. Microbiol., 44*, 23 (1982).

22. Strandberg, G. W., and Lewis, S. N., A method to enhance the microbial liquefaction of lignite coals, *Biotechnol. Bioeng. Symp.*, in press.

23. Montano, P. A., Granoff, B., and Padrick, T. D., Role of impurities, in *The Science and Technology of Coal and Coal Utilization* (E. D. Cooper, and W. A. Ellingson, eds.), Plenum Press, New York, 1984, p. 125.

24. Meyers, R. A., *Coal Desulfurization*, Marcel Dekker, New York, 1977, p. 5.

25. Wheelock, T. D., and Markoszewski, R., Coal preparation and cleaning, in *The Science and Technology of Coal and Coal Utilization* (E. D. Cooper and W. A. Ellingson, eds.), Plenum Press, New York, 1984, p. 47.

26. Olson, G. J., and Brinckman, F. E., Bioprocessing of coal, *Fuel, 65*, 1638 (1986).

27. Olson, G. J., Brinckman, F. E., and Iverson, W. P., *Processing of Coal with Microorganisms*, Report AP-4472, Electric Power Research Institute, Palo Alto, Calif., 1986.

28. Kargi, F., Microbiological coal desulfurization, *Enzyme Microb. Technol., 4*, 13 (1982).

29. Monticello, D. J., and Finnerty, W. R., Microbial desulfurization of fossil fuels, *Annu. Rev. Microbiol., 39*, 371 (1985).

30. Andrews, G. F., and Maczuga, J., Bacterial coal desulferization, *Biotechnol. Bioeng. Symp., 12*, 337 (1982).

31. Lundgren, D. G., Vestal, J. R., and Tabita, F. R., The microbiology of mine drainage pollution, in *Water Pollution Microbiology* (R. Mitchell, ed.), Wiley, New York, 1972, p. 69.

32. Dugan, P. R., and Apel, W. A., Microbiological Desulfurization of Coal, in *Metallurgical Applications of Bacterial Leaching and Related Microbiological Phenomena* (L. E. Murr, A. E. Torma, and J. A. Brierley, eds.), Academic Press, New York, 1978, p. 223.

33. Groudev, S. N., and Genchev, F. N., Microbial coal desulfurization: effect of the cell adaptation and mixed cultures, *Dokl. Bolg. Akad. Nauk, 32*, 353 (1979).

34. Kargi, F., and Robinson, J. M., Biological removal of pyritic sulfur from coal by the thermophilic organism *Sulfolobus acidocaldarius, Biotechnol. Bioeng., 27*, 41 (1985).

35. Kargi, F., and Robinson, J. M., Microbial desulfurization of coal by the thermophilic microorganism *Sulfolobus acidocaldarius, Biotechnol. Bioeng., 24*, 2115 (1982).

36. Murr, L. F., and Mehta, A. P., Coal desulfurization by leaching involving acidophilic and thermophilic microorganisms, *Biotechnol. Bioeng., 24*, 743 (1982).

37. Singer, P. C., and Stumm, W., Acidic mine drainage: the rate determining step, *Science, 167*, 1121 (1970).

38. Lazaroff, N., Melanson, L., Lewis, E., Santoro, N., and Pueschel, C., Scanning electron microscopy and infrared spectroscopy of iron sediments formed by *Thiobacillus ferrooxidans, Geomicrobiol. J., 4*, 231 (1985).

39. Ivarson, K. C., Ross, G. J., and Miles, N. M., Alterations of micas and feldspars during microbial formation of basic ferric sulfates in the laboratory, *Soil Sci. Soc. Am. J., 42*, 518 (1978).

40. DiSpirito, A. A., Dugan, P. R., and Tuovinen, O. H., Sorption of *Thiobacillus ferrooxidans* to particulate material, *Biotechnol. Bioeng., 25*, 463 (1983).

41. Bagdigian, R. M., and Myerson, A. S., The adsorption of *Thiobacillus ferrooxidans* on coal surfaces, *Biotechnol. Bioeng., 28*, 467 (1986).

42. Myerson, A. S., and Kline, P., The adsorption of *Thiobacillus ferrooxidans* on solid particles, *Biotechnol. Bioeng., 25*, 1669 (1983).

43. McCready, R. G. L., and Zentilli, M., Beneficiation of coal by bacterial leaching, *Can. Metall. Q., 24*, 135 (1985).

44. Keller, L., and Murr, L. E., Acid-bacterial and ferric sulfate leaching of pyrite single crystals, *Biotechnol. Bioeng., 24*, 83 (1982).

45. Bennett, J. C., and Tributsch, H., Bacterial leaching patterns on pyrite crystal surfaces, *J. Bacteriol., 134*, 310 (1978).

46. Tributsch, H., and Bennett, J. C., Semiconductor-electrochemical aspects of bacterial leaching. 1. Oxidation of metal sulfides with large energy gaps, *J. Chem. Technol. Biotechnol., 31*, 565 (1981).

47. Tributsch, H., and Bennett, J. C., Semiconductor-electrochemical aspects of bacterial leaching. 2. Survey of rate-controlling sulphide properties, *J. Chem. Technol. Biotechnol., 31*, 627 (1981).

48. Yeh, T. Y., Godshalk, J. R., Olson, G. J., and Kelly, R. M., Use of epifluorescence microscopy for characterizing the activity of *Thiobacillus ferrooxidans* on iron pyrite, *Biotechnol. Bioeng., 30*, 138 (1987).

49. Matin, A., The proton motive force and the ΔpH in spheroplasts of an acidophilic bacterium (*Thiobacillus acidophilus*), *J. Gen. Microbiol., 128*, 3071 (1982).

50. Golbourne, E., Jr., Matin, M., Zychlinsky, E., and Matin, A., Mechanisms of ΔpH maintenance in active and inactive cells of an obligately aciophilic bacterium, *J. Bacteriol., 166*, 59 (1986).

51. Cox, J. C., Nicholls, D. G., and Ingeldew, W. J., Transmembrane electrical potential and transmembrane pH gradient in the acidophile *Thiobacillus ferrooxidans, Biochem. J., 178*, 195 (1979).

52. Capes, C. E., McIlhinney, A. E., Sirianni, A. F., and Puddington, I. E., Bacterial oxidation in upgrading pyritic coals, *Can Min. Metall. Bull., 66*, 88 (1973).

53. Kempton, A. G., Mobeib, N., McCready, R. G. L., and Capes, C. E., Removal of pyrite from coal by conditioning with *Thiobacillus ferrooxidans* following oil agglomeration, *Hydrometallurgy, 5*, 117 (1980).

54. Butler B. J., Kempton, A. G., Coleman, R. D., and Capes, C. E., The effect of particle size and pH on the remvoal of pyrite from coal by conditioning with bacteria followed by oil agglomeration, *Hydrometallurgy, 15*, 325 (1986).

55. Groudev, V. I., Goudev, S. N., and Markov, K. I., Different types of *Thiobacillus ferrooxidans* mutants possessing high sulfur-oxidixing activity, *C. R. Bulg. Acad. Sci., 34*, 1549 (1981).

56. Hoffman, M. R., Faust, B. C., Panda, F. A., Koo, H. H., and Tsuchiya, H. M., Kinetics of the removal of iron pyrite from coal by microbial catalysis, *Appl. Environ. Microbiol., 42*, 259 (1981).

57. Olsen, T. M., Ashman, P. R., Torma, A. E., and Murr, L. E., Desulferization of coal by *Thiobacillus ferrooxidans*, in *Biogeochemistry of Ancient and Modern Environments*, Proceedings of the 4th International Symposium on Environmental Biogeochemistry, Canberra, 1979, p. 693.

58. Kargi, F., Enhancement of microbial removal of pyritic sulfur from coal using concentrated cell suspensions of *T. ferrooxidans* and an external carbon dioxide source, *Biotechnol. Bioeng., 24*, 749 (1982).

59. Davison, M. S., Torma, A. E., Brierley, J. A., and Brierley, C. L., Effects of elevated pressures on iron- and sulfur-oxidizing bacteria, *Biotechnol. Bioeng. Symp. Ser., 11*, 603 (1981).

60. Schicho, R. N., Brown, S. H., Olson, G. J., Parks, E. J., and Kelly, R. M., Probing coals for non-pyritic sulfur using sulfur-metabolizing mesophilic and hyperthermophilic bacteria, *Fuel, 62*, 1223 (1983).

60a. Schicho, R. N., Brown, S. H., Olson, G. J., Parks, E. J., and Kelly, R. M., Probing coals for non-pyritic sulfur using sulfur-metabolizing mesophilic and hyperthermophilic bacteria, *Fuel, 68*, 1368 (1989).

61. Straszheim, W. F., Greer, R. T., and Markuszewski, R., Direct determination of organic sulfur in raw and chemically desulfurized coals, *Fuel, 62*, 1070 (1983).

62. Bos, P., Huber, T. F., Kos, C. H., Ras, C., and Kuenen, J. G., A Dutch feasibility study on microbial coal desulferization, in *Fundamental and Applied Biohydrometallurgy*, Proceedings of the 6th Interntional Symposium (R. W. Lawrence, R. M. R. Branion, and H. E. Ebner, eds.), Elsevier, Amsterdam, 1986, pp. 129–150.

63. Dugan, P. R., Microbiological desulfurization of coal and its increased monetary value, *Biotechnol. Bioeng. Symp. Ser., 16*, 185 (1986).

64. Dugan, P. R., The value added to coal by microbial sulfur removal, in *Processing and Utilization of High Sulfur Coals* (Y. M. Attia, ed.), Elsevier, Amsterdam, 1985, p. 717.

65. Rai, C., Microbial desulfurization of coals in a slurry pipeline reactor using *Thibacillus ferrooxidans, Biotechnol. Prog., 1*, 200 (1985).

66. Roffman, H. K., Land use limitations for utilization of bacterial removal of sulfur from coal, in *Proc. Inst. Environ. Sci., 25*, 266–270 (1979).

67. Kargi, F., and Weissman, J. G., A dynamic mathematical model for microbial removal of pyritic sulfur from coal, *Biotechnol. Bioeng., 26*, 604 (1984).

68. Hayatsu, R., Winans, R. E., McBeth, R. L., Scott, R. G., Moore, L. P., and Studier, M. H., Structural characterization of coal: lignin-like polymers in coals, in *Coal Structure* (M. L. Gobarty and K. Ouchi, eds.), American Chemical Society, Washington, D.C., 1981, p. 133.

69. Loos, M. A., Roberts, R. N., and Alexander, M., Formation of 2,4-dichlorophenol and 2,4-dichloroanisole from 2,4-dichlorophenoxyacetate by *Arthrobacter* sp., *Can. J. Microbiol., 13*, 691 (1967).

70. Sondreal, E. A., and Wiltsee, G. A., The chemistry of low rank coal, *Annu. Rev. Energy, 9*, 473 (1984).

71. Crawford, R. L., and Olson, P. P., Microbial catabolism of vanillate: decarboxylation to guaiacol, *Appl. Environ. Microbiol., 36*, 539 (1978).

72. Orpin, C. G., Knight, M., and Evans, W. C., The bacterial oxidation of picolinamide, photolytic product of diquat, *Biochem. J., 127*, 819 (1972).

73. Frankenburg, W. G., Nicotine degradation *in vitro* induced by agents from tobacco seed, *Nature, 175*, 945 (1955).

74. Kalliat, M., Kwak, C. Y., and Schmidt, P. W., Small angle x-ray investigation into the porosity of coal, in *New Approaches in Coal Chemistry* (B. D. Blaustein, B. C. Backrath, and S. Friedman, eds.), American Chemical Society, Washington, D.C., 1981, p. 3.

75. Lehninger, A. L., *Biochemistry*, Worth Publishers, New York, 1970, p. 175.

76. Friedel, R. A., Retcofsky, H. L., and Queiser, J. A., *Advances in Coal Spectrometry*, Bulletin 640, U.S. Bureau of Mines, Washington, D.C., 1967.

77. Griffiths, P. R., Fourier transform infrared spectroscopy, *Science, 222*, 297 (1983).

78. Dyrkacz, G. R., Bloomquist, C. A. A., and Solomon, P. R., Fourier transform infrared study of high-purity maceral types, *Fuel, 63*, 536 (1984).

79. Fuller, M. P., Hamadeh, I. M., Griffiths, P. R., and Lowenhaupt, D. E., Diffuse reflectance infrared spectrometry of powdered coals, *Fuel, 61* 529 (1982).

80. Painter, P. C., Snyder, R. W., Pearson, D. E., and Kwong, J., Fourier transform infrared study of the variation in the oxidation of a coking coal, *Fuel, 59*, 282 (1980).
81. Fredricks, P. M., and Moxon, N. T., Differentiation of *in situ* oxidized and fresh coal using FTIR techniques, *Fuel, 65*, 1531 (1986).
82. Painter, P. C., Rimmer, S. M., Snyder, R. W., and Davis, A., A Fourier transform infrared study of mineral matter in coal: the application of a least squares curve fitting program, *Appl. Spectrosc., 35*, 102 (1987).
82a. McGowan, C. W., and Markuszewski, R., *Fuel, 67*, 1091 (1988).
83. Pichuantes, S., Cofre, G., Venegas, A., and Rodriguez, M., Studies on native strains of *Thiobacillus ferrooxidans*. 1. Growth characteristics and antibiotic suscepti-bility, *Biotechnol. Appl. Biochem., 8*, 276 (1986).
84. Woods, D. R., Rawlings, D. E., Barros, M. E., Pretorius, I. M., and Ramesar, R., Molecular genetic studies on *Thiobacillus ferrooxidans*: the development of genetic systems and the expression of cloned genes, *Biotechnol. Appl. Biochem., 8*, 231 (1986).
85. Holmes, D. S., Yates, J. R., Lobos, J. H., and Doyle, M. V., Genetic engineering of biomining organisms, in *World Biotechnology Report*, Vol. 2, Online Publications, Pinner, Middlesex, England, 1984, pp. A67–A84.
86. Parameswaran, A. K., Provan, C. N., Sturm, F. J., and Kelly, R. M., Sulfur reduc-tion by the extremely thermophilic archaebacterium, *Pyrodictium occultum, Appl. Environ. Microbiol., 53*, 1690 (1987).
87. Kargi, F., and Cervoni, T. D., An airlift-recycle fermenter for microbial desulfuriza-tion of coal, *Biotechnol. Lett., 5*, 33 (1983).
88. Risatti, J. B., and Miller, K. W., *Rate of Microbial Removal of Organic and Inor-ganic Sulfur from Illinois Coals and Coal Chars*, Final Report to the Illinois Center for Research on Sulfur in Coal, State Geological Survey, Champaign, Ill., 1986.

27

Microbial Removal of Organic Sulfur from Coal: Current Status and Research Needs

JOHN J. KILBANE II *Institute of Gas Technology, Chicago, Illinois*

27.1 INTRODUCTION

All fossil fuels contain sulfur, and when these fuels are combusted the sulfur present is released into the atmosphere, contributing to air pollution in the form of "acid rain." Current air quality standards have already placed strict limitations on the amounts of sulfurous air emissions allowed, and these regulations may well become even more stringent in the future. The current sulfur emission limit for the burning of coal is 1.2 lb of sulfur dioxide per million Btu. Furthermore, the Prevention of Significant Deterioration amendments of the Clean Air Act can prevent new emission sources from opening in those locations where local air quality is already poor, regardless of the anticipated emission levels of the proposed facility. The seriousness with which our government views the problem of sulfurous air emissions is evinced by the fact that recent versions of the Clean Air Act not only allow for fines to corporations, but also legislate that individuals who are convicted of knowingly violating environmental emission standards can be fined and/or jailed.

There is then a need and an incentive to limit sulfurous emissions resulting from the combustion of fossil fuels. The current technology focuses on sulfur removal achieved during or after combustion [1,2]. At this time there is no cost-effective technology that can desulfurize fuel prior to combustion. This is not to imply that current desulfurization technology involving sulfur removal during or after combustion is inexpensive—it is only preferable given the current alternatives. Desulfurization during combustion requires a furnace and support equipment specifically designed for the process such as fluidized-bed combustion furnaces, in which limestone is added to the fuel mix and reacts with sulfur dioxide to form gypsum. Flue gas desulfurization processes employ combinations of chemical reactions and physical separation techniques to minimize sulfurous emissions. Sulfur dioxide removal by fluidized-bed combustion is generally cheaper than flue gas desulfurization, and fluidized-bed combustion generates a solid ash in lower volume than the liquid sludge generated by flue gas desulfurization. However, the cost of desulfurization equipment and the problems associated with maintaining that equipment and disposing of the copious quantities of sulfurous wastes generated by these processes are expenses and problems that industry and the utilities would prefer to avoid. If clean low-sulfur fuel could be made available, fuel users could avoid these concerns. While there are naturally occurring sources of low-sulfur oil and coal, these high-quality fuels are not abundant enough to serve all of the world's energy needs and there is no known process for precombustion fuel desulfurization that can achieve a sufficient degree of sulfur removal in a cost-effective way. Biological processes occur under very mild reaction conditions compared to chemical reaction, so it is hoped that if suitable means of desulfurizing fuels using biological systems can be found, economically favorable precombustion fuel desulfurization processes will result.

27.2 BACKGROUND: INORGANIC SULFUR REMOVAL

It has been known for many years that acidic drainage associated with coal mines or coal storage piles is the result of microbial oxidation of sulfur in coal, which produces sulfuric acid. This microbial desulfurization of coal is almost completely limited to the inorganic forms of sulfur (pyrite) found in coal and is accomplished by the action of well-characterized bacteria: *Thiobacillus ferrooxidans* [3-5], *Thiobacillus thiooxidans* [5,6], and *Sulfolobus acidocaldarius* [7]. The microbiological removal of sulfur from coal has been shown to be capable of removing 90% or more of the inorganic sulfur, and it has been claimed that these same microorganisms are capable of removing as much as 50% of the organically bound sulfur in coal [7]. The organic sulfur in coal is generally measured by difference (total sulfur minus pyritic sulfur), which can potentially lead to large analytical errors, making difficult the interpretation of data claiming biological removal of organic sulfur. The issue of sulfur removal from organic substrates by *S. acidocaldarius* is specifically addressed in Chapter 31. In any event the microbiological removal of inorganic sulfur from coal is well documented, whereas the microbiological removal of organic sulfur from coal is certainly less complete and is less well documented. For any precombustion coal desulfurization process to obviate the need for subsequent desulfurization, greater than 90% of the total sulfur must be removed. Therefore, the focus of current research concerning the microbiological removal of sulfur from coal centers on the removal of organic sulfur.

The goal of all desulfurization processes is to remove sulfur while retaining the fuel value of the coal. Since the oxidation of inorganic sulfur can serve as a source of energy for microorganisms, it is not too difficult to understand that when species of *Thiobacillus* and *Sulfolobus* bacteria grow at the expense of the inorganic sulfur in coal, they are capable of efficient sulfur removal, yet they do not diminish the fuel value of that coal. In fact, the fuel value of coal resulting from microbiological removal of inorganic sulfur increases in terms of Btu per ton because of the loss of significant quantities of ash.

27.3 ORGANIC SULFUR REMOVAL

Unlike the metabolism of inorganic sulfur, the metabolism of organic sulfur is not known to serve as a source of energy for bacteria. Moreover, inorganic sulfur generally exists in coal in the form of discrete particles or crystals of pyrite, whereas organic sulfur occurs as an integral part of the molecular coal matrix and is not readily accessible for microbial attack. The goal, then, of developing a microbiological process for the removal of organic sulfur from coal while retaining the fuel value of that coal is a most difficult goal indeed.

Enrichment culture techniques are typically used to isolate microorganisms with desired traits, and the search for microorganisms capable of organic sulfur removal from coal is no exception. The chemical dibenzothiophene (DBT) is generally regarded as a good model compound representative of organic sulfur found in coal, and it is used as the substrate of choice in most enrichment culture experiments. There have been two reports of microorganisms that are capable of utilizing DBT as their sole source of carbon, energy, and sulfur. In 1961, Knecht [8] reported the isolation of a mixed culture of *Arthrobacter* and *Pseudomonas* species that together could metabolize DBT. He reported that sulfate was liberated from DBT by this mixed culture, but he did not report any reaction products or metabolites other than cell biomass, and there have not been any reports of continued investigations of this mixed culture. In 1976, Malik and Claus [9] reported that they were successful in isolating 20 strains of *Rhizobium* and *Acinetobacter* species that could each use DBT as their sole source of carbon, energy, and sulfur. Malik [10] reported that the metabolism of DBT by his isolates proceeds according to pathway B, outlined in Fig. 27.1. Numerous other investigators [11-16] have attempted, all without success, to isolate microorganisms that can utilize DBT. These other investigators were universally successful in isolating bacteria capable of cometabolizing DBT, that is, isolating bacteria capable of partially degrading DBT, but only when growing on an alternative carbon substrate. With one exception [15,16] all of the bacteria that can cometabolize DBT do so predominantly by pathway B, outlined in Fig. 27.1. Table 27.1 summarizes the results of research in the microbiological degradation of DBT.

If a microbiological process for coal desulfurization is to be successful, organisms that utilize pathway B cannot be used because sulfur is released only in the course of overall degradation of the substrate. If a microorganisms utilizing pathway B were to be used to achieve 90% or better desulfurization of coal, one would anticipate that only a small fraction of the original fuel value would remain. The degradation of DBT by pathway A is the only viable choice, then, if the fuel value of coal is to be retained while removing the organic sulfur. The microbial metabolism of DBT by a *Pseudomonad* species, TG232, isolated from soil by enrichment culture techniques using naphthalene and DBT as substrates has been studied at the Institute of Gas Technology (IGT). TG232 cannot grow on DBT alone, but it is capable of cometabolizing DBT to orange and red products. TG232 was grown using naphthalene as a carbon substrate, and a washed cell pellet was obtained and was resuspended in a defined salts media yielding a cell density of 10^{10} organisms per milliliter. This cell suspension was divided into two equal portions, and to one portion solid crystals of naphthalene and DBT were added, while only DBT was added to the other portion. Then both suspensions were incubated at room temperature for 24 h, at which time the cell-free supernates were obtained by centrifugation and filtration. Both supernates were reddish orange in color and were analyzed by ultraviolet/ visible spectrophotometry, gas chromatography/mass spectrometry, and by Fourier transform infrared spectroscopy (FTIR). Table 27.2 presents a summary of the identification of the metabolites of DBT produced by TG232. These results, as well as the work by Kodama et al. [14], shows that some microorganisms possess both pathways for DBT metabolism that are outlined in Fig. 27.1, and our work further shows that modifications in culture conditions can alter the metabolism of microorganisms such that pathway A is favored, resulting predominantly in sulfur specific oxidation of DBT. However, the level of sulfur-oxidizing activity exhibited by TG232 has never resulted in the release of

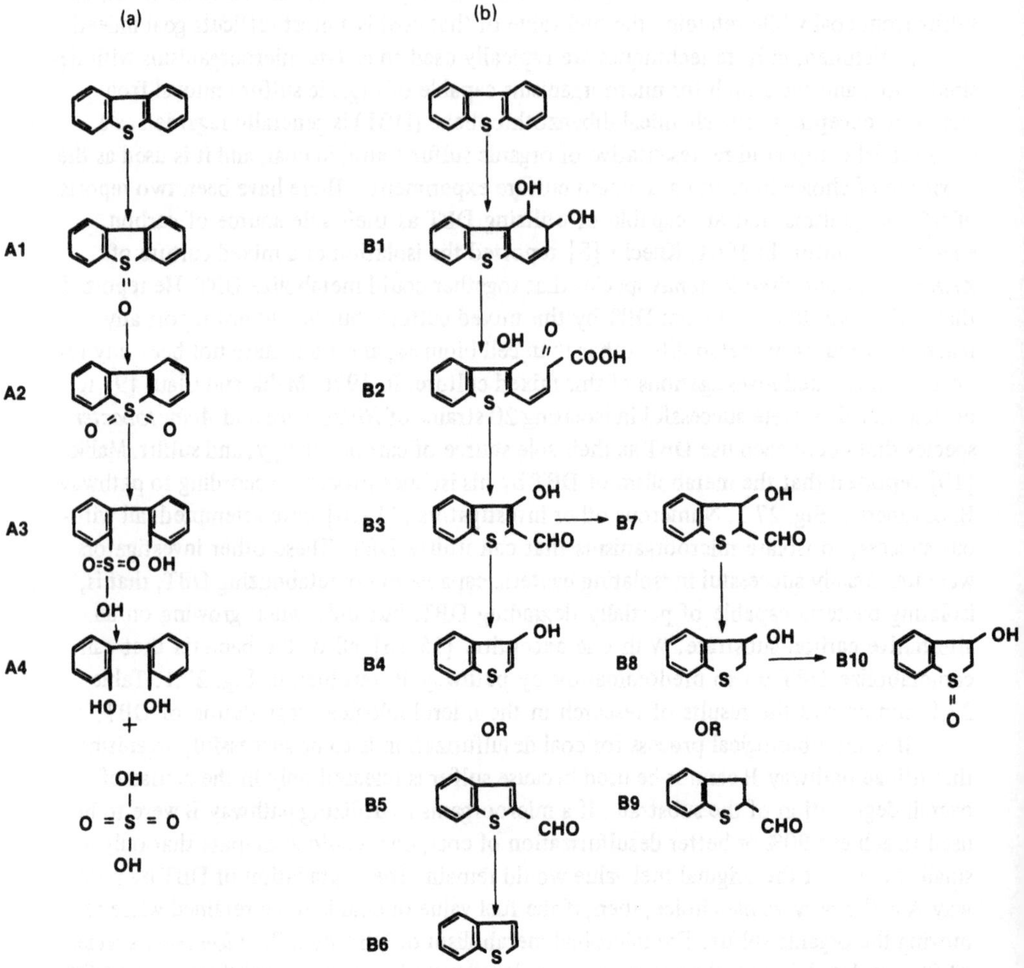

Fig. 27.1 Proposed pathway of DBT degradation.

detectable amounts of sulfate from DBT, so that while this organism possesses interesting metabolic functions relevant to the desulfurization of organic compounds, it cannot be used in its current form to desulfurize coal efficiently. There is only one research group that reports having isolated a pure bacterial culture that is capable of removing organic sulfur from coal, and that is the work of Isbister [15,16] using the *Pseudomonad* CB1.

The DBT-utilizing microorganism CB1 is reported to be capable of cometabolizing DBT by pathway A, resulting in sulfate and dihydroxybiphenyl. Furthermore, CB1 is reported to be capable of removing 40% of the organic sulfur in coal [15,16]. In any event, no microbiological process for the complete removal of organic sulfur from coal is now available; therefore, an ongoing task of research in this field is to continue efforts to isolate new strains of microorganisms that have useful desulfurization abilities and to improve the levels of desulfurizing activities present in characterized bacteria. The prospects for success in these tasks are not good, in that the vast majority of all strains of bacteria that have been isolated in the search for desulfurization of organic substrates have been

Table 27.1 Summary of the Identification of DBT Metabolites Produced by Microbial Degradation

Research group	Pathway A degradation (%)	Pathway B degradation (%)	Most complete degradation products observed
Kodama et al. [14]	8	92	A1, B3
Laborde and Gibson [12]	Minor	Major	A1, B3
Monticello et al. [13]	0	100	B3
Kilbane			
Pseudomonas strain TG232 incubated with dibenzo-thiophene	88	12	A1, B9
Pseudomonas strain TG232 incubated with a mixture of dibenzothiophene and naphthalene	32	68	A1, B6, B10
IGTS7 incubated with dibenzothiophene	100	0	A4[a]
Hou and Laskin [11]	0	100	B2
Malik and Claus [9]	0	100	B2
Isbister and Kobylinski [15]	100	0	A4

[a]IGTS7 transforms DBT to monohydroxybiphenyl rather than dihydroxybiphenyl as indicated for compound A4.

Table 27.2   GC/MS Analysis of TG232-Derived Metabolites of DBT

Compound[a]	Molecular weight	Sample	
		DBT	DBT-naphthalene
1. Dibenzothiophene	184	1	1
2. Dibenzothiophene-5-oxide plus phenoxathiin (A1)	200	0.22	0.30
3. 3-Hydroxy-2-formylbenzothiophene (B3)	178	0.002	0.21
4. Benzothiophene (B6)	134	BDL	0.016
5. Naphthalene	128	BDL	0.001
6. Three isomers of C_8H_6OS: (hydroxybenzothiophene) (B4)	150		
(a)		BDL	0.01
(b)		BDL	0.02
(c)		BDL	0.048
7. C_9H_8OS (B9)	164	0.029	0.12
8. $C_9H_8O_2S$ (B7)	180	BDL	0.067
9. C_9H_6OH (B5)	162	BDL	0.022
10. $C_{10}H_{10}OS$ or $C_9H_6O_2S$	178	BDL	0.035
11. $C_8H_8O_2S$ isomers (B10)			
(a)	168	BDL	0.033
(b)	168	BDL	0.025
12. Formula (?)	220	BDL	0.036
Total, excluding DBT		0.25	0.94

[a]The most abundant metabolites of DBT degradation were analyzed by gas chromatography/mass spectrometry. The concentration of DBT was arbitrarily set at 1.0 and the concentrations of metabolites are reported relative to the concentration of DBT. The numbers in parentheses (A1, B1, etc.) refer to structures included in the pathways presented in Fig. 27.1.

found to metabolize the carbon in the substrate preferentially and only secondarily release sulfur. That is to say: What is wanted is microorganisms that metabolize DBT via pathway A, yet nearly all of the efforts of all the researchers in the field to date have resulted in the isolation of microorganisms that metabolize DBT via pathway B. The implication is that the methods of screening isolates and of performing enrichment culture experiments may be inappropriate and in need of revision. The published reports concerning CB1 do not give any indication of a novel culture enrichment procedure and do not offer an example of how purposefully to isolate microorganisms with appropriate desulfurization abilities.

The metabolism of DBT via pathway B yields orange and red water-soluble metabolites, whereas the metabolism of DBT via pathway A does not yield any products with any appreciable visible color; therefore, the preponderance of isolates that metabolize DBT via pathway B can be seen visibly acting on DBT, whereas microorganisms that utilize pathway A cannot. Enrichment cultures that challenge bacteria to utilize DBT as their sole source of carbon and energy would be predicted to result in the isolation of bacteria that utilize pathway B because the metabolism of DBT via pathway A provides the bacteria with no carbon, as opposed to pathway B. What is provided to bacteria that metabolize DBT via pathway A is sulfur. There are several major research needs that can clearly be identified at this time. Since the rate and extent of organic sulfur removal from coal by characterized microorganisms are insufficient to warrant process development at this time, there is a need to identify new microorganisms that have sulfur-specific degradation abilities toward the full array of sulfur-containing organic molecules. Toward the isolation of new bacterial strains with appropriate desulfurization activities there is a need for a convenient and powerful strain selection technique rather than relying on laborious culture screening techniques. There also exists a need to compare the cultures and the results obtained from different laboratories concerning microbial desulfurizatic... A way is needed to normalize the results obtained by different researchers, using different microorganisms, different substrates, and different growth/reaction conditions. Existing strains of bacteria with documented desulfurization abilities are in need of strain improvement and of techniques to conveniently monitor the progress of strain improvement efforts. Finally, there is a need for better analytical techniques to assess the removal of organic sulfur from coal as the existing methods of physical/chemical analysis of sulfur by type in coal are costly, time consuming, and not particularly accurate, especially with regard to the organic sulfur in coal. All of these research needs can be served by taking advantage of the fact that all living organisms require sulfur for growth.

Enrichment cultures can be established to isolate microorganisms that possess an array of characteristics reflective of the incredible diversity present in the microbial kingdom. The challenge is to manipulate the culture conditions such that selective pressure exists favoring only those microorganisms that possess the desired trait. This is indeed a case of "survival of the fittest." Enrichment cultures are simple to establish. The compound to be degraded is supplied as the growth-limiting source of an essential nutrient, while all other growth requirements are supplied in abundance. Growth under such conditions favors those microorganisms that possess degradative activities toward the target compound such that "appropriate" microorganisms should outgrow all "inappropriate" competitors and rapidly come to dominate the culture. A balanced growth medium for an aerobic heterotrophic bacterium is presented in Table 27.3. A growth limitation can be achieved by removing, reducing, substituting, or altering any of the components listed in Table 27.3; however, in practice it is usually only the carbon source that is manipulated in enrichment culture experiments, and that also seems to be the case for enrichment culture experiments reported for the isolation of microorganisms with the ability to desulfurize DBT or the organic sulfur in coal. It would be much more to the point to establish enrichment cultures utilizing DBT as the growth-limiting source of sulfur. The

Table 27.3 Composition of a Balanced Bacterial Growth Medium

Element[a]	Relative concentration
Carbon	700
Nitrogen	80
Phosphorus	3
Sulfur	1

Source: Ref. 17.

[a]The elements oxygen and hydrogen are supplied by water, the substrate, and/or the air; no provision need be made for these elements when formulating a growth medium. All other elements required for bacterial growth (Mg, Mn, Fe, Cu, etc.) can be present in trace amounts.

DBT utilizing cultures isolated by Knecht [8], Malik and Claus [9], and Isbister [15,16] are all claimed to utilize DBT as their sole source of sulfur and/or to release detectable amounts of sulfur into the medium as a result of their metabolism of DBT, yet none of these authors have provided data documenting the ability of these cultures to utilize DBT as a sulfur source in growth experiments, nor do they report using sulfur-limited growth conditions as a means of selecting improved cultures with enhanced expression of de-sulfurization abilities.

While there are no reports, to the best of my knowledge, of sulfur limitation being using in enrichment culture experiments for the isolation of microorganisms active in DBT metabolism, there are several reports of the use of sulfur-limiting enrichment culture techniques to isolate microorganisms capable of the desulfonation of the herbicides ametryne and prometryne [18] and the desulfonation of substituted naphthalene sulfonic acids and benzene sulfonic acids [19,20]. In these instances microorganisms were found that could selectively desulfonate these substrates even though most of these cultures were incapable of utilizing these substrates as sources of carbon. This is exactly the kind of microbial activity that is needed for the removal of organic sulfur from coal, but unfortunately the structure of aromatic sulfonic acids is sufficiently different from the forms in which organic sulfur is present in coal so that these specific cultures would not be predicted to be useful in desulfurizing coal. Yet they do illustrate the potential for successfully using sulfur-limiting enrichment culture experiments.

Enrichment cultures focusing on those growth components that are required by microorganisms in relative abundance do not present any technical challenge; however, as the authors who have worked with sulfur-limited enrichment cultures point out, com-ponents, such as sulfur, that are required in very low or trace amounts present signifi-cant technical challenges. Zurrer et al. [20] report that scrupulously clean glassware was required to prevent spurious growth on contaminant sulfur, and the highest degree of purity obtainable was required for all the chemicals used to prepare growth medium. Even though these precautions were taken, their enrichments for organisms to utilize sub-stituted naphthalene sulfonates yielded 26 pure bacterial cultures on the basis of turbidity produced in the presence of naphthalene sulfonate, but when supernant fluids were examined by high-performance liquid chromatography (HPLC), only 8 cultures were observed to catalyze quantitative disappearance of the substrate. Turbidity alone, pre-sumably a measure of growth, was obviously an inadequate measure of substrate utiliza-tion. Cook and Hutter [18] report similar difficulties in that agar of all grades gave growth without added sulfur, therefore limiting the usefulness of solid media in experi-ments attempting to isolate appropriate sulfur-utilizing microorganisms. Experiments at IGT with sulfur-limited enrichment cultures have found that the source of the water used to prepare culture media can be a significant source of contaminant sulfur such that HPLC-grade water is required to prepare media to be used in quantitative sulfur utiliza-tion assays.

While the experimental details of operating sulfur-limiting enrichment cultures are somewhat problematic, especially in comparison to operating carbon-limited enrichment cultures, the successes in isolating bacteria that desulfonate aromatic sulfonic acids illustrate the usefulness of this selection technique in isolating microorganisms with sulfur-specific metabolic activities. Growth of cultures in sulfur-limited medium is useful not only for the selection of microorganisms, but the sulfur requirement of all microor-ganisms for growth can be used as the basis of a quantitative sulfur-specific bioassay by correlating bacterial growth to substrate metabolism. A detailed analysis of the bacterial metabolites of DBT such as is presented in Table 27.2 is difficult, costly, and time consuming to produce, but more important, it fails to answer a crucial question: Is any of the sulfur present in DBT available to this microorganism? If an organism possesses even a minimal ability to liberate sulfur from DBT, that organism is of interest and may ultimately be of use in developing a microbiological process for removing organic sulfur from coal, but unless the assay methods used are sufficiently sensitive, we may fail to detect any release of sulfur and discard that culture. Chemical assays for sulfate in aqueous media are generally accurate to ±5 mg/L. Yet if a microorganism were placed in a sulfur-deficient growth medium and supplied with DBT as the sole source of sulfur, any sulfate liberated would first be consumed by the bacteria to satisfy its requirements for growth; remaining sulfur might be expected to be stored by the bacteria for future use, and only when the storage capacity of the bacteria had been saturated might one expect to see sulfate accumulating in the culture medium. Therefore, by assaying only for free sulfate, microorganisms that have a significant ability to liberate sulfur from DBT might go undetected. A bioassay can detect sulfur utilization from organic substrate more easily and with greater sensitivity than can physical/chemical analyses. Moreover, a bioassay can be used to compare the results of different bacterial strains acting on different substrates. A bioassay relating sulfur metabolism of organic substrates to the rate and extent of growth observed can be a universally applicable method useful in normalizing the data obtained from every researcher in the field of microbial desulfurization of coal.

27.4 IGT SULFUR BIOAVAILABILITY ASSAY

Since all life requires some amount of sulfur for growth, one can create a situation such that by quantifying bacterial growth one can quantify the utilization of any organic or inorganic compound as a source of sulfur. In work funded by the U.S. Department of Energy, the Institute of Gas Technology developed a sulfur bioavailability assay to do just that. Inorganic sulfate is usually the form in which sulfur is supplied in bacterial growth medium, and sulfate was used to generate a standard curve relating sulfate concentration to bacterial growth for the *Pseudomonas* species TG232, which is presented in Fig. 27.2. Similarly, TG232 was tested for its ability to utilize various organic and inorganic compounds as sulfur sources in growth experiments and the results are graphically presented in Fig. 27.3. These results demonstrate how a sulfur-dependent growth assay can be used to evaluate a culture's ability to remove organically bound sulfur, but these results also illustrate that bacteria have very low requirements for sulfur. In practice it is necessary, when using the sulfur bioavailability assay, to be vigilant for possible sources of interference. First and foremost it is necessary that clean glassware and pure reagents be used; it is equally important that these experiments be performed in duplicate (at least) with all appropriate controls, including a bacterial strain known not to possess the ability to utilize organically bound sulfur. When new isolates from nature are being tested, an adaptation period of the culture to the test conditions and the growth substrate of at least one subculturing is recommended.

While the use of controls and repetition are the best way to produce reliable data using the sulfur bioavailability assay, some sources of interference can be anticipated and avoided. The easiest way to monitor growth is by optical density or turbidity, but colored products and the particulate nature of some substrates make it difficult, if not impossible, to use turbidity to monitor bacterial growth accurately. The contribution to optical density of colored products in the medium can easily be determined by obtaining a cell-

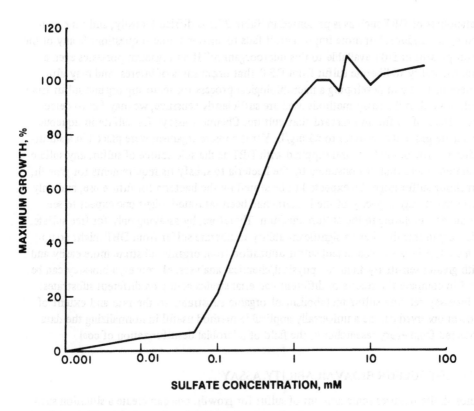

Fig. 27.2 Sulfur requirement of TG232 for growth.

free supernate and subtracting its optical density from that of the whole culture. The use
of particulate substrates such as coal precludes the use of optical density measurements as
a means of monitoring bacterial growth, but such a culture can be monitored with time
for an increase in the number of colony-forming units when dilutions of the culture are
spread onto appropriate agar plates. Alternatively, one could use a protein assay to monitor cell growth, which can be used successfully to avoid interference caused by most
colored products and by most particulate substrates. Unfortunately, coal can contribute a
rather high background in protein assays. Neither the monitoring of optical density, the
assaying for colony forming units, nor assaying for protein is completely reliable and
should only be used to gather preliminary data in a convenient fashion. Its real usefulness lies in indicating those cultures and conditions that require more careful examination. In the study of the desulfonation of aromatic sulfonic acids, Cook et al. [18,20]
found it convenient to quantify the utilization of a compound as the yield of cellular
protein per mole of sulfur. They found that 4 to 6 kg of protein per mole of sulfur was
uniformly obtained using either sulfate or an aromatic sulfonic acid as a sulfur source.
Ideally, positive results obtained using the sulfur bioavailability assay should be confirmed by using chemical/physical techniques [spectrophotometry, gas chromatography/
mass spectrometry (GC/MS), HPLC, FTIR, etc.] to document and quantify the degrada-
tion/biotransformation of the substrate and the corresponding increase in desulfurized
metabolites.

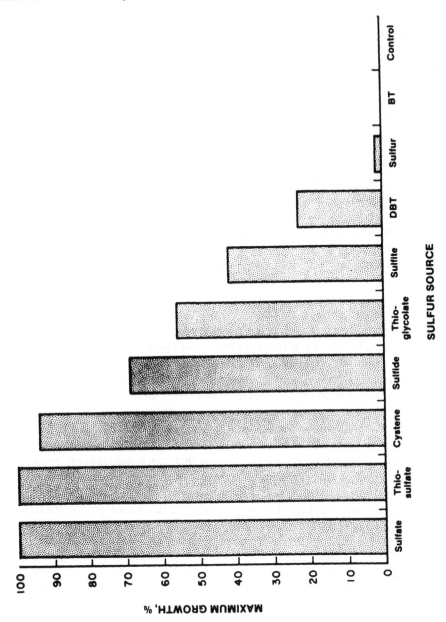

Fig. 27.3 Effectiveness of various compounds to serve as sulfur sources for the growth of TG232.

The sulfur bioavailability assay is useful because it is much more sensitive than chemical/physical assays for sulfate or other sources of sulfur. But because of its sensitivity, a culture that yields a bioassay value of 100% for utilization of organically bound sulfur may still possess a desulfurizing activity too weak to be efficient in the removal of organic sulfur from coal. As Fig. 27.2 illustrates, the sulfur requirement for maximal growth can be satisfied by 2 mM sulfate. The sulfur content of bacteria, as a percent of dry weight, is generally considered to be about 0.5%. When bacterial were grown at IGT in severely sulfur-limiting conditions—1 mM sulfate or 15 mM sulfate—the sulfur content was found to be 0.04%, 0.32%, and 3.00%, respectively. This means that the sulfur content of bacteria is not a constant value, and that bacteria have the ability to store sulfur well in excess of their growth requirements. This ability of bacteria to store sulfur in excess of their immediate needs allows for an adaptation of the sulfur bioavailability assay that can be used both as a method of assessing the sulfur utilization abilities of strains that score 100% on the bioassay using organic substrates and is also useful in strain development work.

A cell cycling experiment consists of exposing a microbial culture to two or more growth substrates or conditions in an appropriate sequence with multiple repetitions. In the context of seeking cultures with improved ability to utilize sulfur from organic substrates, an appropriate culture would first be grown under sulfur-limiting conditions until the middle of the logarithmic growth phase using a readily metabolizable carbon substrate. The culture will then be centrifuged, washed, and resuspended in a fresh sulfur-deficient growth medium containing an organic sulfur compound (a readily metabolizable carbon substrate may be present as well). The culture will then be exposed to this organic sulfur compound for a period ranging from minutes to days, after which time a portion of the culture is removed, washed, and resuspended in fresh sulfur-deficient or sulfur-limited growth medium containing a good carbon substrate, and the cycle is repeated. If the desulfurization activity possessed by the culture is inducible, one may choose to use either inducing or noninducing carbon substrates, assuming that such compounds can be identified. The use of a noninducing carbon substrate would favor the selection of microorganisms that constitutively express their desulfurization abilities. In any event, a cell cycling experiment that is carried out for several/many repetitions will provide the selective pressure needed to enrich for microorganisms with enhanced levels of desulfurization activity, while a cell cycling experiment carried out for one and one-half repetitions allows one to quantify the ability of a culture to liberate sulfur from organic substrates in excess of its immediate growth needs. That is, a cell cycling experiment of this kind allows one to measure the amount of stored sulfur that a microorganism may contain.

During the first phase of a cell cycle experiment the culture is grown in a sulfur-limited medium and therefore should have no reserve of stored sulfur. During the second phase of the experiment the culture is exposed to an organic sulfur–containing substrate, which provides an opportunity for individual cells to liberate sulfur from the organic substrate and to store that sulfur for future use. During the third phase of the experiment (the first phase of the second cycle) the culture would again be subjected to sulfur-deficient or sulfur-limiting conditions in which an abundant supply of a readily metabolizable carbon source is present. This condition allows for growth to occur in direct proportion to the amount of sulfur that the culture was able to store from the previous phase of the experiment. In this way a modified bioassay is created that allows one to quantify the desulfurization abilities of cultures that have very high desulfurization activities and would score 100% using the previously mentioned sulfur bioavailability assay. As regards strain improvement, since growth during the third phase of a cell cycling experiment occurs in proportion to the amount of sulfur stored during the previous phase, those individual cells that express the highest levels of desulfurizing ability will contain the greatest amounts of stored sulfur and will subsequently be capable of more cell divisions, such that their progeny will come to dominate the culture if a sufficient number of cycle repetitions occur. Cell cycling experiments performed for the purpose of strain improvement are usually done in conjunction with chemical mutagenesis to increase the chances of isolating microorganisms with enhanced abilities. Continuous-culture growth experiments utilizing sulfur-limited conditions can serve the same purpose as cell cycling experiments.

27.5 STRAIN IMPROVEMENT THROUGH MOLECULAR GENETICS

To put the need for strain improvement in perspective, it will be instructive to make some assumptions about microbiological desulfurizations abilities and to see how that would translate into a coal treatment process in regard to the development of a feasible microbiological process for removing organic sulfur from coal. If a desulfurization process is based on a culture that can obtain sulfur from coal to satisfy its growth requirements, but has no ability to liberate sulfur in excess of its immediate needs, then assuming typical sulfur content of 0.3% of the dry weight of the biomass, one would have to grow 10 tons (dry weight) of biomass to desulfurize 1 ton of coal that contains 3% organic sulfur. Furthermore, if one assumes that 10 g of cells can be produced per liter of culture (a liberal assumption) one requires a fermenter volume of 1 L to provide sufficient biomass to treat 1 g of coal. This example does not even consider the cell generation time/coal treatment time or the cost of a growth substrate needed to generate that biomass. It is clear that a feasible coal desulfurization process cannot be based on coupling the sulfur removal from coal to the sulfur needed for microbial growth. What is needed is a microbial culture that possesses a desulfurization activity in 1000-fold or more excess of its growth requirements.

The probability that a microorganism can be isolated from nature that possesses a very good ability to utilize organic sulfur from coal is fairly remote because there is probably no naturally occurring environment that would favor the growth of such an organism. The problem stems from the fact that the requirement of microorganisms for sulfur is very low in absolute terms, and the natural abundance and ubiquity of inorganic sulfur (the preferred form of sulfur for microbial metabolism) makes it unnecessary for an organism to utilize organic sulfur to meet its growth needs. Therefore, to obtain an organism that possesses desulfurization functions it will be necessary to establish appropriate conditions for the selection of relevant organisms and it may be necessary to alter microorganisms by mutagenesis in the event that the desired trait cannot be found in natural isolates. Moreover, to obtain a microorganism that possesses a sufficiently high level of expression of desulfurization functions to warrant the development of a microbial process for the removal of organic sulfur from coal, it will be necessary to use molecular genetics to create a microorganism with enhanced desulfurization abilities.

By using sulfur-limited continuous culture coal bioreactors and mutagenesis with 1-methyl-3-nitro-1-nitrosoguanidine (NTG), a mixed bacterial culture, IGTS7, was eventually developed that could not only utilize organically bound sulfur for growth but also possesses a specificity for the oxidation of carbon-sulfur bonds. Bioreactors were established using Illinois No. 6 coal and inoculated with bacteria derived from soils obtained from coal storage sites and with activated sludge obtained from an oil refinery wastewater treatment plant. The same batch of coal was retained within the bioreactor for the duration of the experiment by using -9 + 14 mesh coal and an inclined nonmixed sedimentation tube containing several wiers/baffles, from which the bioreactor effluent was withdrawn at relatively slow flow rates. The flow rates were adjusted according to the ability of the bacteria to respond to the sulfur-limitation challenge. Typically, hydraulic retention times averaged 72 h. The necessity of retaining the same batch of coal with the bioreactor is because coal contains both organic and inorganic forms of sulfur. Some of the inorganic forms of sulfur can be leached from the coal and are preferred over organically bound forms of sulfur as sources of sulfur for bacterial growth. The leachable forms of sulfur are not easily leached/removed from coal, but rather can be completely removed in an aqueous situation only with extremely long leaching times. The bioreactors were supplied on a continuous basis with a sulfur-deficient mineral salts solution of the following composition: 4 g of $K_2 HPO_4$, 4 g of $Na_2 HPO_4$, 2 g of $NH_4 Cl$, 0.2 g of $MgCl_2 \cdot 6H_2 O$, 0.001 g of $CaCl_2$, and 0.001 g of $FeCl_3$ per liter. Additionally, 50 mM glucose and 50 mM sodium benzoate were supplied as carbon sources. On a periodic basis, bacteria were obtained from the effluent, mutagenized by exposure to NTG, and returned to the bioreactors. Bacteria obtained from the bioreactors were routinely evaluated using the sulfur bioavailability assay. In this way the mixed bacterial culture, IGTS7, was developed/isolated and shown to possess the ability to utilize the sulfur present in DBT and related compounds.

The metabolites of DBT degradation/utilization by IGTS7 were examined by gas chromatography/mass spectroscopy (GC/MS); these results are presented in Table 27.4, which also lists for comparison the metabolites of DBT degradation by the *Pseudomonas* culture TG232. TG232 produces all the metabolites of DBT originally identified by Kodama et al. [14], as well as metabolites representing further degradation by that pathway. In contrast, IGTS7 metabolizes DBT predominantly to monohydroxybiphenyl, and the metabolites of pathway B pictured in Fig. 27.1 cannot be detected. IGTS7 differs from TG232 not only in the pathway of DBT degradation but also in the amount of DBT metabolized, with IGTS7 showing a 60-fold increase compared with TG232. This GC/MS analysis also reveals that very little dihydroxybiphenyl is detected as a result of DBT metabolism by IGTS7. Dihydroxybiphenyl is the end product of the proposed 4S pathway of sulfur-specific metabolism (pathway A of Fig. 27.1). The abundance of monohydroxybiphenyl suggests that an alternative pathway, similar but not identical to the 4S pathway, is used by IGTS7. It is also interesting to note that the product of the chemical reaction of DBT with molten caustic (the gravimelt process) is monohydroxybiphenyl [21].

Although research on organic sulfur removal at IGT was focused on the development/isolation of useful microbial cultures and not on optimizing a process for the desulfurization of coal, a sulfur-limited continuous culture bioreactor eventually led to the isolation of a microbial culture capable of desulfurizing organic substrates, and it is instructive to examine the changes in the sulfur content of coal from that bioreactor. A sulfur-by-type analysis of the untreated coal and the coal obtained from the bioreactor after 212 days is presented in Table 27.5. Sulfur determinations were performed according to two procedures developed at IGT as modifications of ASTM Method D2493-84. The ASTM method calls for each coal sample to be split into two portions: One portion is used to determine total sulfur, and the other portion is progressively subjected to chemical treatments to determine sulfate and pyritic sulfur. Organic sulfur is then calculated

Table 27.4 GC/MS Analysis of Bacterial-Derived Metabolites of DBT[a]

Compound	Mol wt		TG-232[b]	IGTS7[b]
Dibenzothiophene		184	1	1
(A1) Dibenzothiophene-5-oxide	200		0.30	1.8
plus phenoxathiin		200	–	–
(A4)* Dihydroxybiphenyl[c]		186	BDL	0.033
(A4)* Hydroxybiphenyl[c]		170	BDL	59
(B3) 3-Hydroxy-2-formylbenzothiophene	178		0.21	BDL
Biphenyl		154	BDL	0.001
(B6) Benzothiophene		134	0.016	BDL
Naphthalene		128	0.001	BDL
(B4) Three isomers of C_8H_6OS (hydroxybenzothiophene):				
No. 1		150	0.01	BDL
No. 2		150	0.02	BDL
No. 3		150	0.048	BDL
(B9) C_9H_8OS		164	0.12	BDL
(B7) $C_9H_8O_2S$		180	0.067	BDL
(B5) C_9H_6OS		162	0.022	BDL
$C_{10}H_{10}OS$ or $C_9H_6O_2S$	178		0.035	BDL
(B10) $C_8H_8O_2S$ isomers				
Isomer (a)	168		0.033	BDL
Isomer (b)	168		0.025	BDL
Formula (?)		220	0.036	BDL
Total excluding DBT			0.94	60.8

[a]The most abundant metabolites of DBT degradation were analyzed by GC/MS. The concentration of DBT, present throughout as a saturated aqueous solution, was arbitrarily set at 1.0, and the concentrations of metabolites were reported relative to the concentration of DBT. The numbers in parentheses (A1, B1, etc.) refer to structures included in the pathways presented in Fig. 27.3.
[b]BDL, below detection limit.
[c]Hydroxybiphenyl is the terminal metabolite in a modified A/4S pathway.

by summing the amount of sulfate and pyritic sulfur and subtracting this from the value obtained for total sulfur. IGT follows this procedure but also determines the amount of sulfide. A further extension of this ASTM procedure is to take the residue that remains after sulfate, sulfide, and pyritic sulfur have been chemically removed and perform a total sulfur determination on that residue. This allows for an addditional and more direct determination of the organic sulfur in coal and provides an internal control for the accuracy of the analysis of each sample. The data in Table 27.5 show that the organic sulfur content of the coal calculated from the ASTM testing procedure was 0.39% for the day 212 bioreactor sample, compared with 2.43% for the control samples. However, when the organic sulfur content of the day 212 sample was directly determined by a modification of the ASTM testing procedure, the amounts were found to decrease from 2.25% to 0.205%, which represents a removal value of 91%.

An additional experiment was performed to acquire information about the sulfur content of coal samples resulting from bioprocessing. Coal biodesulfurization yields a mixture of bacteria and coal that must be separated before the coal can be analyzed to determine its sulfur content. Desulfurizing bacteria are known to be associated with coal

Table 27.5 Sulfur Analysis of Illinois Coal: Bioreactor Samples

Sample	Untreated control −9/+44 mesh	Untreated control −9/+44 mesh	Untreated control +9 mesh	Untreated control +9 mesh	Bioreactor sample day 212	Bioreactor sample day 212
Sulfide	0.017	0.014	0.010	0.011	0.017	0.012
Sulfate	0.42	0.41	0.34	0.34	0.11	0.090
Pyritic	0.23	0.21	0.21	0.18	0.093	0.10
Organic						
Determined	2.17	2.20	2.31	2.33	0.21	0.20
Calculated	2.33	2.37	2.51	2.52	0.37	0.41
Total						
Determined	3.00	3.00	3.06	3.05	0.59	0.61
Calculated	2.84	2.83	2.86	2.86	0.43	0.40

particles, and coal surfaces may well adsorb sulfur-containing molecules. Therefore, a potential complicating factor in the analysis of the sulfur content of coal is the incomplete removal of bacterial cells and adsorbed molecules from coal samples. An experiment was performed to detect and quantify this potential source of interference in coal analysis.

Coal samples obtained from biodesulfurization experiments are processed by differential centrifugation and by boiling to yield clean coal samples suitable for analysis. Coal particles are generally larger and denser than bacterial cells, so coal samples are first shaken vigorously to dissociate bacteria that may be attached to coal particles. The aqueous suspension is then centrifuged at 1000g for 5 min. The supernate, which contains bacteria, is carefully removed by aspiration. The pellet, which contains coal, is resuspended in water and transferred to a beaker, placed over a flame, and brought to a boil. The boiled coal suspension is allowed to cool, transferred to a centrifuge tube, shaken vigorously, and then centrifuged at 1000g for 5 min. The supernate is removed, and the pellet/coal sample is dried and then analyzed for sulfur content.

To test the efficacy of this process for the removal of bacterial cells and adsorbed molecules from coal particles, a sample of washed coal was divided into two portions. One portion was analyzed for its sulfur content by performing an elemental analysis and determining the sulfur-to-silicon ratio, and the other portion was Soxhlet extracted with toluene for 24 h prior to being analyzed for its sulfur content. The toluene extraction was designed to exhaustively remove any bacteria that remained associated with coal particles as well as any organic molecules adsorbed to coal surfaces. The coal samples showed no significant differences in their sulfur content as a result of toluene extraction, as illustrated in Table 27.6.

The results shown in Table 27.6 are significant for several reasons. First, these data show that the coal sample preparation procedure is successful in separating coal from biomass and components of the growth medium. Second, using x-ray diffraction to perform an elemental analysis of the coal samples and then reporting the sulfur content as a sulfur-to-silicon ratio is a fundamentally different way of analyzing the sulfur content of coal that avoids some of the possible limitations of the ASTM procedure. The ASTM procedure determines values for sulfur based on the weight percent of the total coal sample; however, the weight of the coal samples could change during bioprocessing and therefore bias the result of the sulfur determination. For example, the coal could have undergone

Table 27.6 Sulfur Analysis of Toluene-Extracted Coal Samples[a]

Non-de-ashed Illinois No. 6 control	Non-de-ashed Illinois No. 6 day 212 bioreactor sample
S/Si (original) = 33.72	S/Si (original) = 3.91
S/Si (extracted) = 33.42	S/Si (extracted) = 3.75
% retained = 100.9	% retained = 104.3

[a]X-ray diffraction analyses of an original and a toluene-extracted portion of the Illinois No. 6 non-de-ashed control and experimental coal (day 212 sample) were used to determine sulfur-to-silicon elemental ratios.

extensive oxidation and increased in weight due to the addition of oxygen. Thus even if the sulfur remained unchanged, the ASTM procedure would reveal an apparent decrease in the sulfur content. On the other hand, the sulfur/silicon ratio would not be influenced by the oxidation of the coal. The data in Table 27.6 reveal that the total sulfur content of the coal calculated from the sulfur/silicon ratio decreased from 33.42 for the control sample to 3.75 for the bioreactor sample, representing a decrease of 89%, which agrees very closely with the ASTM results. Moreover, the coal samples obtained from the bioreactor on day 212 were also analyzed at Iowa State University using the electron microbeam technique. An electron microbeam can be focused on extremely small regions or particles of coal. If a mineral-free region of coal is analyzed, the elemental analysis can be used as a direct measurement of the organic sulfur in coal [22]. This technique determined that the biotreated coal samples have a decrease in organic sulfur of 80%. The mixed culture, IGTS7, has been shown to be capable of removing organic sulfur from coal; however, the rate of sulfur removal is too slow to be used in a commercial process.

To develop strains of bacteria with a sufficiently high level of desulfurization activity, it will be helpful, if not essential, to use the tools of molecular genetics. Once an organism is found that can efficiently desulfurize organic substrates and is genetically stable for this trait (i.e., does not give rise to spontaneous derivatives that lack desulfurization abilities), mutagenesis experiments can be conducted. The desulfurizing *Pseudomonas* species CB1 would appear to be a candidate for genetic analysis and strain improvement. However, CB1 has been reported to be genetically unstable as regards desulfurization (J. D. Isbister, personal communication), which reduces the possibility of meaningful mutagenesis experiments and makes any genetic analysis of this strain difficult. Work is currently in progress to isolate a pure culture from the mixed culture IGTS7 that possesses the desulfurization trait. Either chemical mutagenesis or transposon-mediated mutagenesis could be used, but transposon-mediated mutagenesis would be the method of choice because it has the advantage of inserting a readily identifiable segment of DNA into the mutated gene, which can subsequently be of great value in identifying the gene of interest. Moreover, since transposons often encode selectable functions, such as resistance to antibiotics, by using transposon-mediated mutagenesis, one can readily identify and isolate only those members of a culture that have experienced a mutagenic event. This streamlines subsequent screening of colonies to identify mutants whose desulfurization ability is affected. On the other hand, when chemical mutagens are used, there is no way to identify or enrich those cells that have experienced a mutagenic event.

In any event, once a strain (or a collection of strains) is available that is mutated in a desulfurization-related function, cloning can begin. If a cloned genetic segment that encodes an intact desulfurization function is introduced into a strain that is mutant in that function, the gene introduced should "complement" the mutant gene, and the resulting strain should be restored to proficiency as regards desulfurization. Having identified genes that encode desulfurization functions, one can attempt to alter the regulation and the level of expression of these genes and ultimately obtain strains with increased desulfurization activity. That dramatic increase in metabolic activity can be achieved by a program of strain improvement is illustrated by the example of penicillin production [23] where more than a 1000-fold increase in yields has been achieved. Techniques for altering the level of expression of genes are well known, so the rate-limiting step in this strain improvement scheme is the identification and cloning of genes that encode desulfurization functions.

One can also use genetic engineering techniques to aid in the isolation of strains with enhanced levels of desulfurization functions without having to clone those genes first, or even have any detailed information about the number of functioning of those genes whatsoever. A cell's requirement for sulfur can be increased by introducing genes into that cell, which encode for proteins that contain a high proportion of the sulfur-containing amino acids cysteine and methionine. A plasmid that encodes such a protein could be introduced into a cell which has a known ability to desulfurize organic substrates. Then the expression of this sulfur-rich protein would increase the amount of sulfur needed per cell so that the selective pressure favoring those cells with increased expression of desulfurization activity would be intensified. A gene that encodes for a sulfur-rich protein could be chemically synthesized to maximize the cysteine and methionine content of the protein; then that artificial gene could be inserted into a broad host range plasmid vector in such a way that the gene was inducible but would be expressed at very high levels when induced. Such a plasmid would be of general utility in strain improvement experiments involving diverse species of bacteria that possess desulfurization abilities.

27.6 SUMMARY

The removal of organic sulfur from coal is the focus of ongoing microbiological research. This work is still in its infancy and many research needs can easily be identified. There is a need to isolate new strains that have sulfur-specific degradation abilities toward the full array of sulfur-containing organic substrates, and to enhance the abilities of strains already identified. But most of all, there is a need for new methodologies that will allow researchers to proceed more directly toward their goals, quantify their progress, and be able to compare their results meaningfully with others. Utilizing the fact that all living organisms require sulfur for growth may provide the basis to satisfy these research needs. Techniques for culture isolation, strain improvement, and a sulfur bioavailability assay based on this fact have been proposed in the hope that they will prove to be useful.

ACKNOWLEDGMENTS

The support of the U.S. Department of Energy, Contract DE-AC22-85PC81201, to IGT for the study of "Microbial Removal of Organic Sulfur from Coal" is gratefully acknowledged. Special thanks are extended to Bruce Solka for help in interpreting mass chromatograms, and to Kee Rhee, Dharamvir Punwani, and John Conrad for critical reading of the manuscript.

REFERENCES

1. Working Panel of the European Federation of Chemical Engineers Steering Committee on Aspects of Chemical Engineering in the Environment, *Desulfurization Techniques*, ISBN 0901001651, Society of Chemical Industry, London, 1980.
2. Wheelock, T. D., and Markuszewski, R., Coal preparation and Cleaning, in *The Science and Technology of Coal and Coal Utilization* (B. R. Cooper and W. A. Ellingson, eds.), Plenum Press, New York, 1984, pp. 47-113.
3. Silverman, M. P., and Lundrgren, D. B., Studies on chemosynthetic iron bacterium *Ferrobacillus ferrooxidans*, *J. Bacteriol.*, *78*, 321-326 (1959).
4. Silverman, M. P., Rogoff, M. H., and Wender, I., Removal of pyritic sulfur from coal by bacterial action, *Fuel*, *42*, 113-124 (1963).
5. Beier, E., Removal of pyrite from coal using bacteria, in *Proceedings of the First International Conference on Processing and Utilization of High Sulfur Coals*, Columbus, Ohio, Oct. 13-17, 1985.
6. Hoffmann, M. R., Faust, B. C., Panda, F. A., Koo, H. H., and Tsuchiya, H. M., Kinetics of the removal of iron pyrite from coal by microbial catalysts, *Appl. Environ. Microbiol.*, *42*, 259-271 (1981).
7. Kargi, F., and Robinson, J. M., Microbial oxidation of dibenzothiophene by the thermophilic organism *Sulfolobus acidocaldarius*, *Biotechnol. Bioeng.*, *26*, 687-690 (1984).
8. Knecht, A. T., Jr., Dissertation, Order No. 621235, Louisiana State University, 1961.
9. Malik, K. A., and Claus, D., *5th International Fermentation Symposium*, Berlin, 1976.
10. Malik, K. A., Microbial removal of organic sulfur from crude oil and the environment: some new perspectives, *Process Biochem.*, *13*, 10 (1978).
11. Hou, C. T., and Laskin, A. I., Microbial conversion of dibenzothiophene, *Dev. Ind. Microbiol.*, *17*, 351 (1976).
12. Laborde, A. L., and Gibson, D. T., Metabolism of dibenzothiophene by a *Beijerinkia* species, *Appl. Environ. Microbiol.*, *34*, 783 (1977).
13. Monticello, D. J., Bakker, D., and Finnerty, W. R., Plasmid mediated degradation of dibenzothiophene by *Pseudomonas* species, *Appl. Environ. Microbiol.*, *49*, 756-760 (1985).
14. Kodama, K., Nakatani, S., Umehara, K., Shimizu, K., Minoda, Y., and Yamada, K., Microbial conversion of petrosulfur compounds: isolation and identification of products from dibenzothiophene, *Agric. Biol. Chem.*, *34*, 1320-1324 (1970).
15. Isbister, J. D., and Kobylinski, E. A., Microbial desulfurization of coal, in *Processing and Utilization of High Sulfur Coals*, Vol. 9 in *Coal Science and Technology Series* (Y. A. Attia, ed.), Elsevier, Amsterdam, 1985, p. 627.
16. Isbister, J., Biological removal of organic sulfur from coal, in *Biological Treatment of Coals Workshop*, U.S. Department of Energy, Office of Fossil Energy, and the Bioprocessing Center, Idaho National Engineering Laboratory, Washington, D.C., 1986, p. 18.
17. Cook, A. M., Grossenbacher, H., and Hutter, R., Isolation and cultivations of microbes with biodegradative potential, *Experientia*, *39*, 1911-1918 (1983).
18. Cook, A. M., and Hutter, R., Ametryne and prometryne as sulfur sources for bacteria, *Appl. Environ. Microbiol.*, *43*, 781-786 (1982).
19. Thurnheer, T., Kohler, T., Cook, A. M., and Leisinger, T., Orthanilic acid and analogues as carbon sources for bacteria: growth physiology and enzymic desulphonation, *J. Gen. Microbiol.*, *132*, 1215-1220 (1986).
20. Zurrer, D., Cook, A. M., and Leisinger, T., Microbial desulfonation of substituted naphthalene sulfonic acids and benzene sulfonic acids, *Appl. Environ. Microbiol.*, *53*, 1459-1463 (1987).

21. Nowak, M. A., Chemistry of molten caustic desulfurization of organics, paper presented at the *4th Annual Coal Preparation, Utilization and Environmental Control Contractors Conference*, Pittsburgh, Aug. 8-11, 1988.

22. Straszheim, W. E., Greer, R. T., and Markuszewski, R., Direct determination of organic sulfur in raw and chemically desulfurized coals, *Fuel, 62*, 1070-1075 (1983).

23. Queener, S. W., and Lively, D. H., Screening and selection for strain improvement, in *Manual of Industrial Microbiology and Biotechnology* (A. L. Demain and N. A. Solomon, eds.), American Society for Microbiology, Washington, D.C., 1986, p. 156.

28

Developments in the Biological Suppression of Pyritic Sulfur in Coal Flotation

ANTHONY S. ATKINS *Staffordshire Polytechnic, Staffordshire, England*

28.1 INTRODUCTION

Sulfur, mainly in the inorganic form of iron pyrites (FeS_2), is a common detrimental contaminant of coal. Figure 28.1 illustrates the distribution by sulfur content of coal in the United Kingdom in 1982. It shows the average sulfur content of the coal to be 1.5%, and virtually no low-sulfur coal (<0.5% sulfur) is available in the United Kingdom. The burning of fossil fuels is associated with the production of harmful sulfur dioxide emissions, which are implicated in the formation of acid rain [1,2]. Within the United Kingdom alone, combustion of power station coal releases approximately 2.3 million metric tons of sulfur dioxide per annum and is the major source of sulfur dioxide emissions [3]. Figure 28.2 indicates the U.K. sulfur dioxide emission compared to Europe and the USSR in 1982.

Acid rain became recognized as a major pollutant in the later 1970s, due to extensive publicity concerning its effects on the environment, particularly regarding the acidity of lakes and the devastation of forests. It is not a new phenomenon; as far back as 1852 the effects on cattle and grazing land of a corrosive rain in the area around Manchester and in Wales was reported by a British chemist, Robert Angus Smith [4]. The effects of this rain were reported in Sweden in the 1920s by the formation of acid lakes, although it was not until the 1950s that investigations into the causes behind the acidification were begun with the introduction of the European Air Chemistry Network, which, coupled with Eastern Europe, began accumulating data on windborne pollutants [5]. Analysis of these data illustrated the localized pollution problems generated by short-range transportation of these pollutants, which subsequently led to the development of the "tall stacks" policy [5]. The implementation of this policy has alleviated localized pollution problems by creating a much larger dispersion plume, making it possible for the pollutant emitted by one country to be deposited in others.

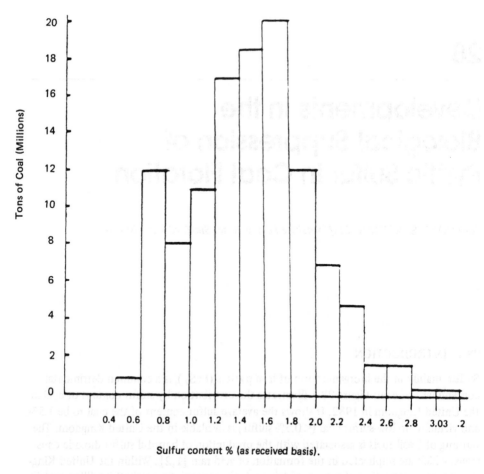

Fig. 28.1 Sales of British coal by sulfur content, 1987–1983.

The major pollutants—sulfur and nitrogen oxides—and their involvement in the formation of acid rain were highlighted by the 1982 Stockholm conference [6]. The published effects of acid rain are numerous and include the destruction of aquatic life and forestry [7], the destruction of buildings by the dissolution of calcium carbonate, and the effect on public health through increased toxic metal concentration in domestic water supplies, causing osteomalacia and related bone diseases [8,9]. Fossil fuels (notably coal) contain sulfur in various forms—elemental, pyritic (in the mineralogical form of pyrite/marcasite FeS_2), organic (disulfides), and sulfate (calcium sulfate), although the elemental and sulfate constituents exist in negligible quantities and can be ignored with respect to desulfurization [10].

As a result of the increased consumption of coal, the total emissions of sulfur dioxide in Europe may increase by up to a third over the next 20 years. Atmospheric modeling has predicted that sulfur dioxide depositions in Europe will have risen from 57 million metric tons in 1982 to a projected 74 million tons by 2002, with a corresponding effect on acidification [11,12]. Because of the serious, costly, and cumulative environmental damage that appears to be attributable to sulfur dioxide, and the implications for human health and welfare, there is considerable pressure, based on sound and sensible

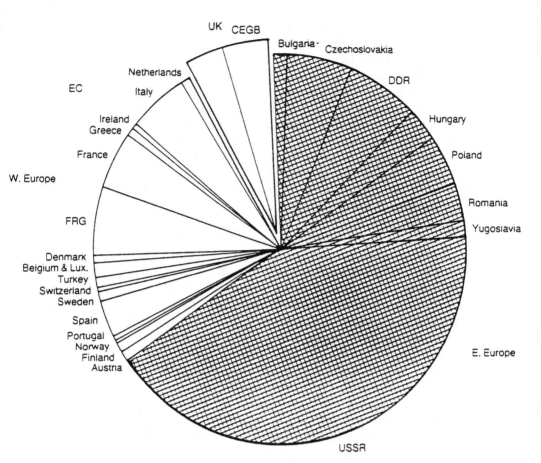

UK SULPHUR DIOXIDE EMISSIONS 1982

UK sulphur dioxide (SO₂) emissions	Million tons SO₂ per annum	% UK SO₂ emissions
overall	4.0	100
from power generation by CEGB	2.6	65
from coal fired generation	2.3	57

(TOTAL EMISSIONS 61.4 MILLION TONS SO₂ PER ANNUM)

Fig. 28.2 Estimated sulfur dioxide emissions in Europe and the USSR, 1982.

arguments, for restrictions to be placed on the sulfur content of the coal used for power generation. In view of the dearth of indigenous low-sulfur coal, this implies using some form of treatment to reduce its sulfur content before combustion. The possibilities include a more efficient use of energy, resulting in lower fuel consumption and the alternative use of nuclear energy with its potential environmental hazards.

The type of strategy adopted will depend on economic, technical, and political considerations, but it would seem quite logical that countries with large coal reserves containing relatively high amounts of sulfur will opt for fuel cleaning and/or emission control. The present strategy in the United Kingdom is to develop both pressurized fluidized-bed combustion (PFBC) and flue gas desulfurization (FGD) systems for the control of sulfur dioxide emissions. At present the PFBC system is still in prototype development in the United Kingdom (80 MW at Grimethorpe, Yorkshire, U.K.) and with a scaleup to the typical Central Electricity Generating Board (CEGB) capacity of 2000 MW, it would be unrealistic to assume a commissioning date before the year 2000 [13]. The application of the FGD system, of which there are several types, would require the retrofitting of possibly 12 power stations to meet anticipated European Economic Community legislation and will require both extensive capital expenditure of between £120 and £160 million per station and an annual operational cost of £35 million per station [1,14]. Both systems (assuming the application of the limestone/gypsum FGD system) will also have to solve the environmental waste disposal problems associated with the residues from this type of technology.

Obviously, the most feasible route to investigate would be to lower the sulfur content of the coal prior to combustion. A major source of sulfur is pyrite, and as this can occur in a finely disseminated form, conventional gravity concentration equipment has unfortunately limited application for desulfurization processes [3,15,16]. In the United Kingdom, 12 to 14% of the run-of-mine (ROM) output (1983: 110 million metric tons salable, 67 million tons dirt) is a fine coal less than 0.5 mm in size (of which 70% is less than 63 μm and 30% is less than 10 μm); however, through mechanization the amount of fines (less 0.5 mm) may be up to 25% of the ROM [17], and this coal is amenable to treatment by a novel bioflotation technique.

Conventional flotation is also ineffective as a separation technique, as pyrite appears to have similar flotation properties to that of coal and is consequently concentrated by flotation [18,19]. Other workers have shown that significant amounts of pyrite can be removed from coal by bacterial leaching followed by oil agglomeration, but the fundamental drawbacks of long operational time (days) and fine feed size ($<$53 μm) have prevented full-scale implementation [20-22].

To overcome these obstacles a bioflotation process has been developed in which pyrite can be suppressed during flotation by preconditioning in the presence of the bacterium *Thiobacillus ferrooxidans* and/or extracellular components prior to flotation. Dogan et al. [23] achieved pyritic removal in finer size ranges after a conditioning time of 4 h. In the current work the surfaces are sufficiently modified to suppress the pyrite after 2 min of conditioning prior to conventional flotation [24]. Further development of this system into a commercial process is constrained by the large quantities of active cells that are required and can normally only be produced after 500 h of growth under selective conditions. To relieve this constraint, an investigation into the suppression mechanism was conducted to establish an alternative biological suppressant that could be produced in sufficiently large quantities for commercial usage [25,26].

Progressive improvement in the bioflotation system has shown that not only *T. ferrooxidans* but also other heterotrophic bacteria and certain yeasts can suppress pyritic sulfur in heterogeneous systems [27]. Further refinement has also shown that by-products from the food-processing industry can be used [26]. This chapter outlines the development of such a bioflotation system for use in conventional coal preparation plants; the results for both synthetic and medium- to high-sulfur coals are shown.

28.2 MATERIALS AND METHODS

28.2.1 Growth of *Thiobacillus ferrooxidans*

The culture of iron-oxidizing *T. ferrooxidans* used throughout the study was grown in a specially designed air-sparged fermentation vessel (pachuca) as shown in Fig. 28.3. The pachuca was inoculated using 100 mL of bacterial liquor (2% v/v), and this culture being maintained on a modified 9K salt medium [28] at pH 2.0, supplemented with pyrites (-53 μm d_{50} 6 μm) as the energy source at 6% pulp density. The operating temperature was maintained at 35°C and the 5-L working volume was readjusted daily for evaporation losses with pH 2.0 distilled water. As a result of sulfuric acid production, the operating pH of 2.0 was allowed to fall naturally to approximately 1.0, at which point it leveled off, signifying the end of the growth cycle.

The growth cycle of the culture was monitored every 48 h by measuring the total and soluble iron release using a Shimadzu AA646 atomic absorption spectrophotometer and also measuring the pH. The end of the growth cycle occurred after approximately 500 h, after which the bacteria were recovered using a differential centrifugation technique. The cell concentration was determined by absorbance at 600-nm wavelength using a Pye Unicam PU 8600 ultraviolet/visible spectrophotometer. Storage at pH 2.0 and 25°C for up to 4 weeks appeared to have no detrimental effect on the bacteria.

28.2.2 Heterotrophic Bacteria, Yeasts, and Filamentous Fungi

The bacteria tested were *Echerichia coli* and *Pseudomonas maltophila* stock cultures. Bacteria were grown on nutrient agar and harvested by flooding with distilled water and agitation. Enumeration was by plate counting. The yeasts tested included the baker's yeast *Saccharomyces cerevisiae* and the fodder yeast *Candida utilis*. Spores of the filamentous fungus *Aspergillus niger* were also tested. Growth of the yeasts and the filamentous fungus was on oxytetracycline-yeast extract agar. Again harvesting was performed by flooding with distilled water, and enumeration in the resultant microbial suspensions was by direct counting using a haemocytometer (improved Nebauer chamber).

28.2.3 Biological Agents

One of the commercial biological agents (A) was obtained as a waste by-product from the food-processing industry. The remaining two products (B and C), again from the food industry, were used in the trials at a dilution of 10:1 in order to assist in dispersion.

28.2.4 Materials

Pyrite

The main metalliferous pyrite sample was obtained from Capper Pass and Sons Limited, North Humberside, Hull, U.K. The head sample of +50 mm was comminuted and

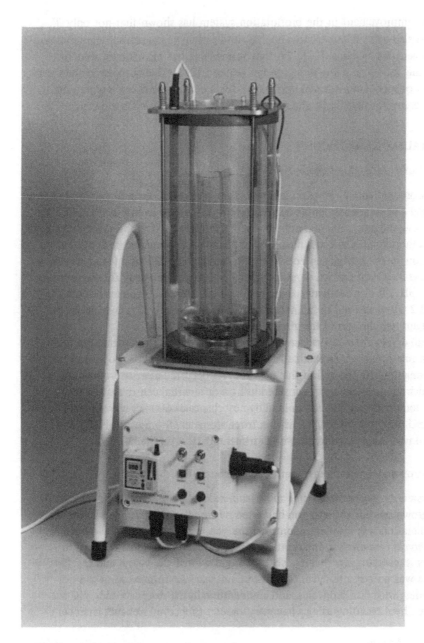

Fig. 28.3 Air-sparged fermentation vessel.

screened to produce the following size fractions (in μm): -500 + 425, -425 + 300, -300 + 212, -212 + 106, -106 + 53, and -53. The Tharsis pyrite was obtained from a massive pyrite deposit, Tharsis, Rio Tinto, Spain, and was used after screening to -212 + 106 μm. Additional metalliferous pyrites for zeta-potential measurements were supplied by Cardiff University.

Coal/Pyrite Mixture and Coal Samples

A run-of-mine (ROM) low-sulfur coal sample was obtained from the High Hazle seam, Nottinghamshire, U.K. To ensure complete liberation of the pyritic constituents in simulated coal flotation experiments, a synthetic feed consisting of low-sulfur coal and pyrite was prepared to produce a 10.75% and 5.03% sulfur coal, respectively (dry basis). The synthetic feed size consists of $-300 + 106 \mu$m coal with $-212 + 106 \mu$m of pyrite, which was the optimum size for Hallimond tube tests [19,27]. An American high-sulfur coal sample of undisclosed origin and previously sized was also received from Century Oils. Two samples of coal with a total sulfur content of 3.8% and 4.98%, respectively, were tested from a surface mining operation at Ellerbeck, Leicestershire, U.K., from different parts of the Lower Tops seam. The coal was comminuted using a Sturtevant Jaw Crusher and an End Running Mill to produce about 60% below 63 μm, and the size distribution is shown in Table 28.1.

28.2.5 Pyrites Reactivation

After bacterial suppression of the pyrites, three activation methods were tested in both the Hallimond tube and the Leeds cell. Reactivation technique 1 included membrane filtration of the sink product (81.1% feed), drying at 60°C for 1 h and resuspension in the Hallimond tube in pH 2.0 distilled water, after preconditioning in the orbital incubator at 100 rpm for 2 min in 15 mL of distilled water (pH 2.0) containing 73 μL of Centifroth SR1. Reactivation technique 2 involved reactivation of the bacterially treated sink product (89.3% of the feed) in 15 mL of pH 2.0 distilled water containing 73 μL of Centifroth SR1 for 2 min. This suspension was then transferred to the Hallimond tube, made up to 230 mL with pH 2.0 distilled water and refloated for 10 min. Reactivation technique 3 involved incubation of the bacterially treated sink product (84.7% of the

Table 28.1 Size Distribution of Ellerbeck Coals

Particle size (μm)	Ellerbeck coal A (3.8% sulfur)		Ellerbeck coal B (4.98% sulfur)	
	Percent retained	Cumulative percent	Percent retained	Cumulative percent
$-500 + 425$	0.11	100.00	0.08	100.00
$-425 + 300$	0.36	99.89	0.49	99.92
$-300 + 212$	0.51	99.53	1.30	99.43
$-212 + 106$	12.46	99.02	8.65	98.13
$-106 + 75$	14.30	86.56	14.64	89.48
$-75 + 63$	13.33	72.26	9.65	74.84
$-63 + 53$	15.52	58.93	2.94	65.19
$-53 + 38$	9.41	43.41	20.86	62.25
$-38 + 22$	11.72	34.00	16.43	41.39
$-22 + 10$	11.20	22.28	13.51	24.96
$-10 + 6$	4.43	11.08	4.79	11.45
$-6 + 4$	3.17	6.65	3.42	6.66
$-4 + 2.4$	2.00	3.48	1.99	3.24
-2.4	1.48	1.48	1.25	1.25

initial feed) in 15 mL of pH 2.0 distilled water containing 100 μL of 0.1 M copper sulfate solution 13 and 73 μL of Centifroth SR1 for 2 min in a conical flask rotating at 100 rpm in an orbital incubator. The contents were then made up to 230 mL with pH 2.0 distilled water, transferred to the Hallimond tube, and refloated for 10 min. The procedure for re-activation techniques 2 and 3 was scaled up and repeated in the Leeds cell.

28.2.6 Analysis

Analysis for sulfur content, carbon, and volatiles was by using a Leco SC 132 sulfur de-terminator and a MAC 400 Leco proximate analyzer according to BS 1016 Part 3 [29]. To obtain a direct comparison all analytical results were converted to a dry basis. The proximate analyses of all samples were carried out in triplicate according to BS 1016 Part 3 [29] for coals and showed excellent repeatability. However, the ash contents (especial-ly in the cases of synthetic mixture containing 10.75% sulfur) determined are artificially high, due to the oxidation of iron. The increase could approach 37.5% of the pyritic sulfur content of the sample; consequently, the ash values are correspondingly influenced, as is the fixed carbon value obtained by difference. The pyritic sulfur content of the coal was determined by analyzing the total and nonpyritic iron content using the digestion procedure outlined in BS 1016 Part 3 [29] on a representative dry sample milled to -53 μm. The total and nonpyritic iron analysis was determined by atomic absorption spectro-photometer and the pyritic sulfur calculated from the ideal stoichiometric formulas for pyrite.

28.2.7 Flotation Trials

Hallimond Tube

The Hallimond tube was used in trials for initial feasibility in view of the large number of experiments that were conducted and also because it was more economical as to bacteria and materials than were standard laboratory cells. Investigations involving flotation of pyrites in the Hallimond tube were performed at 2% pulp density, as limited by the characteristics and dimensions of the apparatus (4.6% pyrites, $-212 + 106$ μm in a volume of 230 mL). The Hallimond tube and ancillary equipment are illustrated in Fig. 28.4. A detailed discussion of the operation of the Hallimond tube and flotation theory is given in Pryor [30].

Initial flotation tests were carried out at pH 2.0 on a solids concentration of 2% coal or pyrite suspension using 73 μL of the nonphenolic flotation oil in 230 mL of water, as this represented normal growth conditions for the bacteria. The conditioning period in the presence of the collector was 2 min and an airflow rate of 200 mL per minute was used. Flotation time was 10 min, but in tests demonstrating the influence of *T. ferrooxidans*, heterotrophic bacteria, and *S. cerevisiae* at 3.25×10^{10} cells per gram of pyrite, a further short preconditioning time of 2.5 min was introduced prior to collector addition and flotation.

Where alternative microorganisms were tested, the same cell densities were used during treatment. In an experiment that illustrated the effect of cell disintegration on suppression, the same density of cells of *T. ferrooxidans* were disintegrated in a Braun cell homogenizer (Braun Industries, West Germany) for 1 and 5 min with constant cooling with CO_2. Cell demage was checked microscopically. All conditioning steps were con-ducted in 15-mL volumes of water adjusted to pH 2.0 in the controls, or 15 mL of bac-teria, yeast, or biological suspension. The products were dewatered by filtration on cellu-

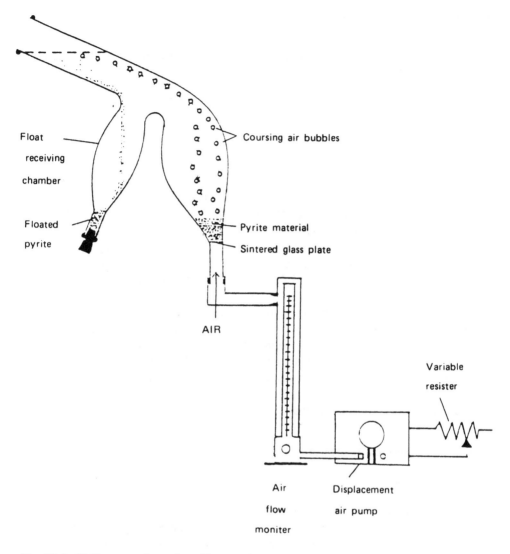

Fig. 28.4 Hallimond tube and ancillary equipment.

lose nitrate membrane filters (Whatman No. 42, 47 mm diameter, 0.45 μm pore size). The percentage by mass was calculated for all products and where appropriate, proximate analysis conducted. All tests were performed in triplicate unless otherwise indicated.

Modified Denver, Denver, and Leeds Cell

The flotation tests were all conducted at the relatively low solids concentration of 2% in order to conserve samples. The optimum conditions for pyrite flotation were standardized previously and were retained for coal flotation—hence the necessary high reagent dosage using the synthetic coal/pyrite mixtures and pyrite [24]. Although previous research has shown that the suppression is effective over the pH range 1 to 11, the tests concerned with the synthetic mixtures and pyrite were conducted at pH 2.0 since this represented normal growth conditions for *T. ferrooxidans*.

Table 28.2 Summary of Flotation Parameters

	Modified Denver cell	Denver cell	Leeds cell
Capacity (mL)	600	2500	3500
Impeller speed (rpm)	1900	1200	1500

All the tests were at least duplicated and the maximum error was 4% in the modified Denver cell, while in the Leeds and Denver standard laboratory cells the minimum error was greater than 4% even though the latter cells are some six times the capacity of the modified cell. In all cases the error is associated mainly with loss of sample in the cell. A summary of the flotation parameters is given in Table 28.2 and the cell operating parameters were based on commercial flotation plant data supplied by the Technical Headquarters, British Coal, Bretby, U.K. (private communication).

The conditioning period for the flotation reagent was 2 min, and for biological conditioning a further period of 2.5 min was used with the cell operating at low speed. The flotation time was 3 min for all tests. The modified Denver cell incorporates an extended quiescent zone and has been described in detail previously [31]. All products were dewatered using standard laboratory vacuum filter apparatus with Whatman No. 541, 185-mm filter papers. The percentage by mass was calculated for all products, and where appropriate, sulfur and proximate analyses were conducted.

28.2.8 Electrokinetic Measurements

Sample Preparation

Five samples of pyrite were used for zeta-potential measurements, three metalliferous and two extracted from coal. Each sample was crushed to $-106\,\mu m$ and then milled in a Spex type 27-28 eccentric ball mill (Glen Creston, 16 Dalston Gardens, Stanmore) for a period of 20 min to produce a d_{50} of 6 μm. Size analysis was performed by a Malvern 2600/3600 Particle Analyser. All samples were nitrogen sparged and stored in airtight containers.

Apparatus

The zeta potential of each sample at various pH values was determined by calculating the particle mobility using a modified Rank Brothers MK II particle electrophoresis apparatus fitted with a flat-type cell (Rank Brothers, Bottisham, Cambridgeshire, U.K.) and operated using blacked platinum electrodes (Fig. 28.5).

Theory

The formula used to relate particle mobility to zeta potential was a rationalized version of the Helmholtz-Smoluchowski equation [32,33] for water at 25°C.

Fig. 28.5 Electrophoresis apparatus and ancillary equipment.

$$\text{Zeta potential} = 12.83 \times 10^5 \times U \quad \text{(V)} \tag{1}$$

where U is the particle mobility.

$$U = \frac{Vel}{E} \quad (m^2/s \cdot V) \tag{2}$$

where Vel is the particle velocity (m/s) and E is the applied field strength (V/m).

$$E = \frac{V_{op}}{L} \tag{3}$$

where V_{op} is the constant operating voltage (V) and L is the interelectrode distance (m).

$$L = RKA \quad (m) \tag{4}$$

where R is the electrical resistance between electrodes (Ω), K the specific conductance of electrolyte (Ω^{-1}/m), and A the cross-sectional area of cell (m^2). Substituting (4) into (3) yields

$$E = \frac{V_{op}}{RKA} \quad (V/m) \tag{5}$$

Substituting (5) into (2) gives

$$U = Vel \times \frac{RKA}{V_{op}} \quad (m^2/s \cdot V) \tag{6}$$

Substituting (6) into (1), we obtain

$$\text{zeta potential} = 12.83 \times 10^8 \times Vel \times \frac{RKA}{V_{op}} \quad (mV) \tag{7}$$

This equation was used throughout all experimentation to calculate the zeta potential.

Calibration

The interelectrode distance L was calculated prior to each run by determining the RKA values using a conductance cell in conjunction with the modified electrophoresis apparatus. R was determined by calculating the voltage drop across a switchable standard resistor (V_r) connected in series with the cell and the voltage drop across the cell (V_c). To achieve this a total potential of 20 V dc (V_t) was used, this being supplied from the instrument's power supply and measured independently using an AVO Mk VI meter (number 2 in Fig. 28.5). The voltage drop across the resistor r (V_r) was measured similarly (number 3 in Fig. 28.5). The voltage drop across the cell (V_c) was obtained by difference:

$$V_c = V_t - V_r \quad (V) \tag{8}$$

R was then determined using the relationship

$$R = r \frac{V_c}{V_r} \quad (V) \tag{9}$$

A 10-W resistor of 4560 Ω (measured) was used.

The electrolytes used throughout all experimentation consisted of twice-distilled water containing various quantities of potassium chloride, sulfuric acid, and potassium hydroxide for pH regulation. The quantities of potassium chloride/potassium hydroxide were varied to obtain K values in the range 7×10^{-3} to $3 \times 10^{-2-1}$ Ω/cm, due to the operating voltage limitations. The specific conductivity (K) of each solution was determined using a Model MC-1 Mark V electrolytic conductivity test, fitted with a Philips type PW 9510/60 immersible conductance probe (cell constant = 0.75) (Kent Electronic Instruments Limited, Chertsey, Surrey, U.K.). The measuring cell and the holder were cleaned thoroughly and rinsed twice with double-distilled water. The cell was further flushed with a sample of the electrolyte to be tested. The sample was placed into the cell and the temperature noted, the corresponding value of which was set on the temperature correction dial. The selector switch was set to the anticipated range and the specific conductance rapidly determined (to avoid cell polarization) by establishing the central position on the balance indicator.

The cross-sectional area of the cell was determined by direct measurement using the micrometer fixings on the apparatus (number 4 in Fig. 28.5). This was repeated at various intervals along the cell length to ensure uniformity. All cell measurements were performed dry. To determine the velocity of an observed particle by means of measuring the time required for it to traverse one graticule, the length of the graticule has to be established. This was achieved by observing a stage micrometer (available from Rank Brothers) mounted in the same plane as the flat cell. To measure accurately the velocity of the particle due solely to electrophoresis, the particle must be observed in the stationary layer as the walls of the cell are charged in the presence of solvent leading to the streaming of the oppositely charged solvent toward the appropriate electrode near the walls.

The calculation for the stationary level is based on the Komagata equation [34]:

$$\frac{s}{d} = 0.5 - \left(0.833 + \frac{32d}{\pi^5 \ell}\right)^{1/2} \tag{10}$$

where s is the distance from wall to stationary level (μm), d the internal width of cell (μm), and ℓ the internal height of cell (μm).

Operation

Thirty milligrams of sample were conditioned in 30 mL of electrolyte for 3 min [35]. The suspension was poured into the flat cell and the platinum electrodes carefully inserted to ensure that no air was trapped. The cell was placed into the surrounding water bath and allowed to attain equilibrium to 25°C for 2 min. With the cell connected, R was determined as previously described. The internal thickness was measured, the stationary layer calculated, and the corresponding adjustments made on the instrument to allow observations of the particles in the stationary layer. This procedure was rapidly carried out to prevent sedimentation of the pyrite.

A dc potential (V_{op}) ranging from 8 to 60 V was applied; the value selected depended on K but was minimized to prevent electrode polarization and the development of thermal overturns. The time taken for each particle to traverse a graticule was measured accurately using a digital timer. This measurement was repeated in the reverse direction using the reverse-polarity switch on the hand set (number 5 in Fig. 28.5). Approximately 30 measurements were taken to provide a statistically accurate mean velocity for use in 7. At all voltages, a constant check on electrode polarization was maintained by measuring the current through the cell and by visual observation of the electrodes.

28.3 RESULTS AND DISCUSSIONS

28.3.1 Optimization of Flotation Parameters

Effect of Different Flotation Oils on Pyrite Flotation

The effect of various commercial collector reagents on the floatability of the Capper Pass pyrites was assessed in the Hallimond tube at pH 2.0 and 2% pulp denisty (-212 + 106 μm). In bacterially conditioned samples, the pyrite was initially treated with 15 mL of a suspension of *T. ferrooxidans* at a cell density of 1×10^{10} cells/mL (harvested during the late exponential phase of growth from a Pachuca reactor) in an orbital incubator rotating at 100 rpm at 35°C for 2.5 min. After a further 2-min conditioning period with added collector, the subsequent flotation trials were conducted at ambient room temperature.

An initial dosage of 73 μL of Centifroth SR1 was used to float the Capper Pass pyrites, and control tests (no bacterial treatment) were conducted to assess the influence of the bacteria on the flotation of the pyrites. Figure 28.6 illustrates the effect of various collectors on pyrites flotation in the control tests and the corresponding suppression effects of the bacteria. The graph also indicates that the natural floatability (no collector additions) at pH 2.0, 2% pulp density, was 17%. The graph illustrates that the addition of the respective flotation reagents at doses of 73 μL in 230 mL markedly improved the

Pyrite flotation

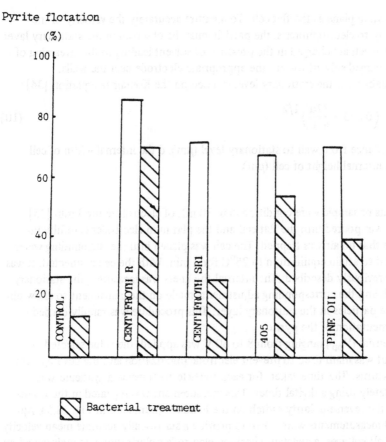

Fig. 28.6 Influence of collector frother type on Capper Pass pyrite (-212 + 106 μm) flotation and amenability to bacterial suppression at pH 2.0 in the Hallimond tube. (From Ref. 24.)

flotation of the pyrites, achieving 72% float in the chemical control with Centifroth SR1. The bacterial pretreatment suppressed the pyrites in the case of Centifroth SR1 and resulted in only 13% of the pyrites floating. Centifroth SR1 was subsequently selected for all further trials, since it exhibited the desired investigative criteria of high collector activity in the control float and low activity in the bacterially pretreated pyrites.

Effect of Conditioning Time on Pyrite Flotation

The influence of collector conditioning time on the Capper Pass pyrites flotation at pH 2.0 and 2% pulp density is illustrated in Fig. 28.7. The collector activity appeared to be instantaneous, with a recovery of 60% within a few seconds of conditioning. The graph illustrates that increasing the conditioning time to 1 min resulted in a slight improvement in flotation yield to 71%, although the statistical significance (2 × SE difference) was marginal. Increasing the conditioning time to 10 min resulted in no further increase in flotation yields. Subsequent flotation tests were conducted with a 2-min conditioning period, which was selected to ensure sufficient collector activity and reproducibility of results.

Pyrite flotation

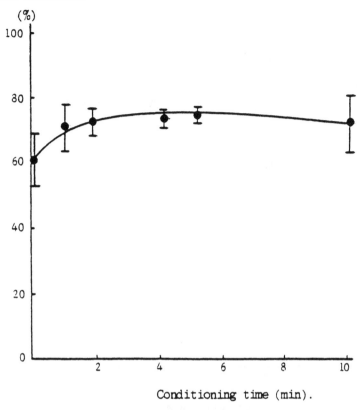

Fig. 28.7 Influence of collector frother conditioning time on Capper Pass pyrite (-212 + 106 μm) in the Hallimond tube at pH 2.0. (Values shown represent mean of three replicate trials ± SEM.) (From Ref. 24.)

Effect of Collector Dosage on Pyrite and Coal Flotation

The influence of collector dosage (Centifroth SR1) on the Capper Pass pyrites at pH 2.0, 2% pulp density (- 212 + 106 μm) during flotation together with the low-sulfur coal (-300 + 106 μm) under similar conditions is illustrated in Fig. 28.8. The graph indicates that in the case of the Capper Pass pyrites, a collector dosage of 40 mg produced a pyritic yield of 75%. However, a further increase to 100 mg of collector produced no significant improvement in yield, and similarly, in the test with the low-sulfur coal, a dosage of 30 mg produced the maximum yield. From these results, the optimum collector dosage of Centifroth SR1 of 60 mg (73 μL) was selected for further trials to compromise between flotation reagent economies and providing a conservative amount of collector in the system.

Effect of Various Bacterial Pretreatments on Pyrite Flotation

Figure 28.9 shows the effect of various pretreatments of the -212 + 106 μm pyrite on flotation. The natural floatability under standard conditions was 84.5% of the feed. After a 2.5-min pretreatment in membrane-filtered bacterial liquor from the fermenter as a preconditioning solution, 17.4% of the pyrite feed floated. This significant reduction in the

Pyrite flotation

(%)

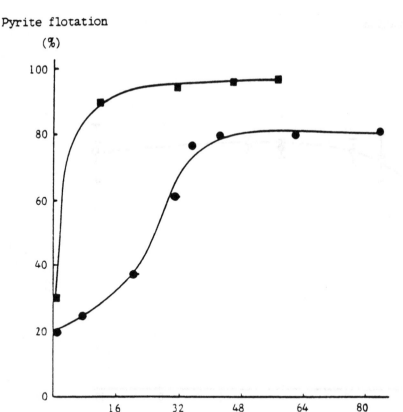

Collector dosage (μLiters)

Fig. 28.8 Influence of collector frother dosage (centifroth SR1) on Capper Pass pyrite (−212 + 106 μm), •; and High Hazles seam coal, Nottinghamshire (−300 + 212 μm), ■; in the Hallimond tube at pH 2.0. (From Ref. 24.)

amount of pyrite floated was reduced even further in the presence of 1.0×10^{10} cells/mL in 15 mL of distilled water at pH 2.0 to a level of 7.7% flotation of the feed. The best suppression was obtained with 15 mL of a filtered liquor supplemented with bacteria to a density of 1.0×10^{10} cells/mL, which demonstrated 4.0% pyrite flotation. All further experiments were performed with the bacterial suspension in distilled water at pH 2.0 so that the experimental components could be standardized. However, the fact that bacterial liquors can be used directly from the pachuca reactor indicates the possibility of continuously producing effective suppressive solutions without the need for selectively centrifuging and resuspending pure bacteria in distilled water. This would overcome the difficulty encountered with surface-active bacteria in that they are associated predominantly with the mineral particles until the end of the exponential phase of growth, at which time they are released from the particle surface rapidly into the supernatant [36,37].

Pyrite flotation

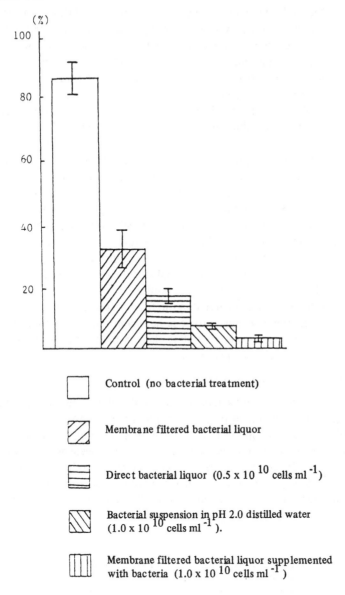

Fig. 28.9 Comparison of different pretreatments on subsequent pyrite flotation in the Hallimond tube at pH 2.0. (Results represent mean of three replicate trials ± SEM.)

Effect of Bacterial Conditioning Time on Pyrite Flotation

Table 28.3 shows the effect of bacterial conditioning time on pyrite flotation. Most treatment times appeared to produce similar levels of pyrite suppression, and adequate activity was apparent even at the shortest time tested, of 10 s contact. Subsequent trials were conducted with a 2.5-min conditioning time since this time period was experimentally manageable and represents a realistic plant holding period of a few minutes. The adopted short conditioning time of 2.5 min appeared to be very satisfactory in terms of pyrite suppression, and this is significantly shorter than other reported bacterial conditioning times. One report indicated that a conditioning time of 1 to 3 days was needed for pyrite suppression during oil agglomeration of coal [38]. Other workers reported a time of 15 min [23] and they suggested that the process was one of bioadsorption. The results obtained in this study are in agreement with this proposal of using a novel bioflotation system since conditioning times as short as 10 s were effective in reducing pyrite flotation.

Figure 28.10 illustrates for comparison purposes the leaching of -53 μm Capper Pass pyrite at 2% and 6% pulp density in the presence of actively growing *T. ferrooxidans*. The 2% pulp density was leached to 80% in 240 h of incubation with the bacterium, while the 6% pulp density was leached to only 63% in 340 h of incubation. Control experiments that were not inoculated and included 10 mM sodium azide showed very little leaching. Also shown in the graph are the pH profiles resulting from the leached pyrite. Although it appears that the leaching of liberated pyrite from coal is feasible, a residence time of about 250 h and greater would appear to be rather inappropriate for commercial-scale treatment at the mine or at the power generating plant. Although this might be a way of desulfurizing coal from coal stock piles heavily contaminated with pyrites.

Effect of Bacterial Density on Pyrite Flotation

Figure 28.11 illustrates the influence of bacterial density during the 2.5-min conditioning period prior to flotation. The amount of pyrite floated was reduced to 38.7% in the presence of 2.6×10^{10} cells/g pyrite in the 15 mL of conditioning volume containing 4.6 g of pyrite. Much greater reductions in the amount of pyrite floated were achieved with 6.5×10^{11} cells/g pyrite at 12% and 3% flotation of the feed, respectively. An optimum cell count of 3.25×10^{10} cells/g pyrite was chosen as the normal operating density, as this level of bacteria showed good suppression at a density that was realistically attainable from the Pachuca reactor in sufficiently large volumes.

Table 28.3 Effect of Bacterial Preconditioning Time on Pyrite Flotation (-212 $+$ 106 μm) in a Hallimond tube at pH 2.0[a]

Conditioning time (min)	Pyrite floated (%)
0 (control no bacterial treatment)	91.04 (±3.61)
0.17	11.03 (±1.13)
2.50	6.39 (±0.32)
10.00	2.64 (±0.20)
30.00	3.78 (±0.29)
240.00	2.08 (±0.28)

Source: Ref. 25.
[a]Results represent mean of three replicates ±SEM.

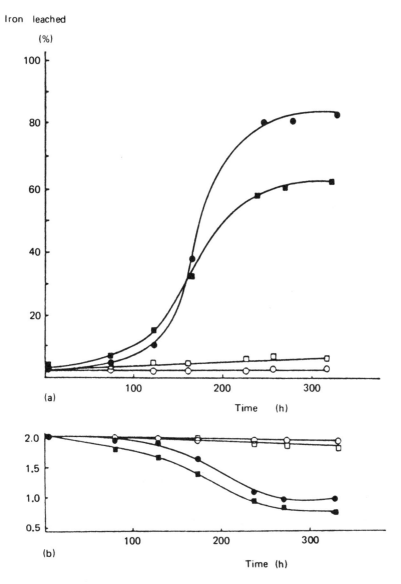

Fig. 28.10 (a) Leach profiles for Capper Pass pyrite (-53 μm) using *T. ferrooxidans* inoculated at 1 × 10⁹ cells/mL per pulp density in 9K salts medium; (b) pH profiles during leaching of Capper Pass pyrite. ●, 2% pulp density in the presence of bacteria; ○, 2% control; ■, 6% pulp density in the presence of bacteria; □, 6% control. (From Ref. 25.)

Effect of Temperature on Pyrite Flotation

Figure 28.12 shows the effect of temperature on pyrite flotation in the range 10 to 40°C. There was no appreciable effect of temperature on the natural floatability of the pyrite, and similarly, there was little influence of temperature on the high degree of suppression brought about by the bacterial presence at the optimum cell density. This is significant because a process dependent on metabolic energy would be much more temperature dependent than was indicated in this study and thus an adsorptive mechanism for this

Pyrite flotation

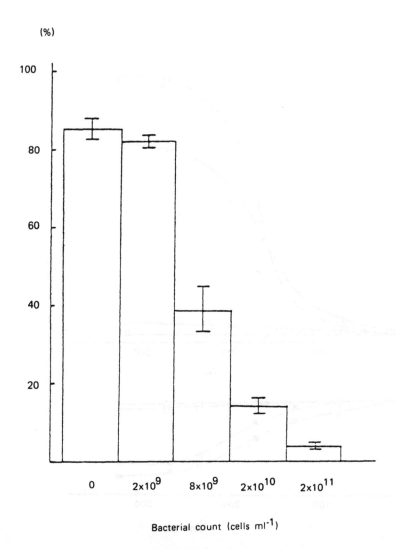

Fig. 28.11 Effect of bacterial density on pyrite flotation in the Hallimond tube using Capper Pass pyrite ($-212 + 106$ μm) at pH 2.0. (Values represent mean of three replicate trials ±SEM.) (From Ref. 25.)

process is suggested. Flotation operating temperatures above 30 to 40°C are not envisaged because of collector frother breakdown, as they are oil based.

Effect of pH on Pyrite Flotation

Figure 28.13 shows the effect of pH on pyrite flotation in the range 2.0 to 11.0. The amount of Capper Pass pyrite floated was reduced from 85% to 19% as the pH was increased from 2.0 to 5.0 in control tests without the addition of bacteria. Thus pH above 5.0 appeared to reduce the amount of pyrite floated. However, the degree of suppression

Pyrite flotation

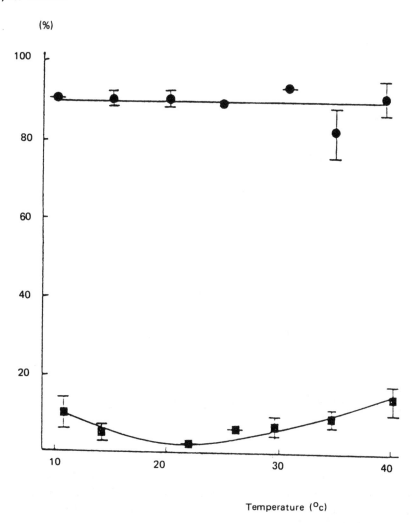

Fig. 28.12 Effect of temperature on pyrite flotation in the Hallimond tube at pH 2.0 using Capper Pass pyrite (−212 + 106 μm). ●, Controls (no bacterial pretreatment); ■, pyrite flotation after 2.5 min of preconditioning with 1.0×10^{10} cells/mL *T. ferrooxidans*. (Results are mean of three replicate trials ±SEM.) (From Ref. 25.)

of pyrite floated in the presence of the bacteria was consistent at greater than 98%. Thus Capper pass pyrite suppression seems to be achievable by increasing the pH to >5.0. In the case of Tharsis pyrite, however, flotation did not appear to be dependent on pH to this degree. It is likely that all the different types of pyrite present in a coal seam would need to be investigated individually to determine if bacterial preconditioning is a necessary prerequisite to prevent pyrite flotation, or whether pH alone could be used to control it. Although some pyrites could probably be suppressed by pH control, we know from experience that in most coal preparation plants operating in the pH range 6.5 to 7.5 pyrite flotation is a problem in these plants (Technical Headquarters, British Coal, Bretby,

Pyrite flotation

(%)

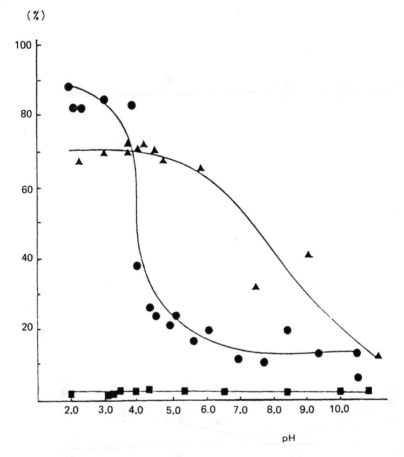

Fig. 28.13 Influence of pH on pyrite flotation in the Hallimond tube at 20°C. ■, After bacterial preconditioning for 2.5 min at a cell density of 1.0×10^{10} cells/mL; ●, control float using Capper Pass pyrite ($-212 + 106\ \mu$m); ▲, control float using Tharsis pyrite ($-212 + 106\ \mu$m). (From Ref. 25.)

private communication). The fact that pH had no influence on the suppressive activity of the bacteria is very important, since although leaching is dependent on the environmental pH being in the range 1.0 to 2.0 for the acidophillic *T. ferrooxidans*, the suppression effect was equally effective at all the pH values tested in the range 1.0 to 11.0.

Effect of Particle Size on Pyrite and Coal Flotation

The influence of screened size fractions on pyrites and coal flotation in both the Hallimond tube and the Leeds cell is illustrated in Table 28.4. Size fractions were prepared as follows: $-500 + 425\ \mu$m, $-425 + 300\ \mu$m, $-300 + 212\ \mu$m, $-212 + 106\ \mu$m, $-106 + 53\ \mu$m, and $-53\ \mu$m. Pyrites flotation in the presence of Centifroth SR1 was more effective in the Leeds cell than in the Hallimond tube at all size fractions tested, and this was probably due to the better kinetic characteristics of the Leeds cell. The larger size fraction of $-500 + 300\ \mu$m did not appear to float well in either apparatus. Particles in the

Table 28.4 Influence of Particle Size on Flotation of Capper Pass Pyrites and High Hazles Seam Coal, Nottinghamshire, at pH 2.0

Particle size (μm)	Pyrites flotation (%)		Coal flotation (%)	
	Hallimond tube	Leeds cell	Hallimond tube	Leeds cell
−500 + 425	1.9	8.4	96.4	92.4
−425 + 300	5.2	42.3	97.1	93.4
−300 + 212	77.8	81.3	95.5	92.8
−212 + 106	89.4	90.6	97.6	93.0
−106 + 53	85.6	87.5	93.0	90.7
−53	59.1	84.4	86.7	58.8

Source: Ref. 24.

size range −300 + 53 μm were floated efficiently at a level of 80 to 90% in both the Hallimond tube and the Leeds cell, whereas −53 μm floated in the Leeds cell, producing an 84.4% float, while in the Hallimond tube only 59.1% of the feed sample appeared to float.

In the tests on untreated pyrites illustrated in Table 28.4, the floatability of the coal in the Hallimond tube and the Leeds cell generally appeared to be greater than that of the pyrites. For similar collector dosages, and in particular in the size range −500 + 300 μm, most results reached 90% floated or greater. In the size range −53 μm, both coal and pyrites flotation was poorer than in the larger size range. In general, coal flotation appeared to be more efficient in the Hallimond tube than in the Leeds cell. In contrast, however, the opposite appeared to be the case for pyrites.

Effect of Particle Size on the Bacterial Suppression of Pyrite in Flotation

The influence of size and the bacterial suppression effects on pyrite flotation are shown in Fig. 28.14. Under control conditions (no bacterial treatment) the larger size fractions of −500 + 300 μm resulted in poor flotation, achieving a maximum of 11% yield in the range −300 + 212 μm. The reason for this high degree of natural suppression is attributed to the launder design, which provides an extended quiescent zone. Previous work has shown that the larger size ranges float well, achieving an 80% float in the −300 + 212 μm fraction [25]. Below 212 μm, around 80% of the pyrite floated under control conditions. Since the pyritic contaminant in coal fines flotation was typically in this size range, the results indicate a significant proportion of pyritic material being concentrated in the clean coal product.

The influence of bacterial pretreatment on the screened fractions of pyrite is also illustrated in Fig. 28.14. Poor flotation in the larger size range (−500 + 300 μm) was obtained under control conditions. However, after bacterial treatment, significant suppression of the floatable material was achieved, resulting in a pyritic float of only a few percent. The most dramatic biosuppression effect was observed in the range −212 + 106 μm where the floatable pyrite was reduced from 84% to 2%. This effect was further illustrated in the −106 + 53 μm and −53 μm size fractions, where bacterial treatment resulted in a reduction from 83 and 75% to 37 and 25%, respectively.

To further reduce the quantity of pyrite floatable in the −106 μm range, it was assumed that insufficient quantities of bacteria of bacterial products were present due to

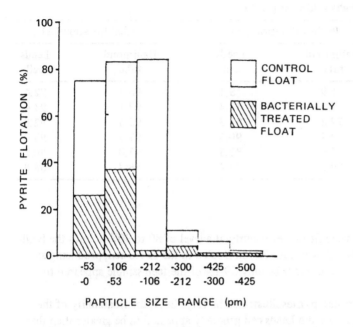

Fig. 28.14 Comparison of the effect of particle size on pyrite on conventional and bacterially treated flotation (dosage 3.26 × 10¹⁰ cell/g). (From Ref. 31.)

the high surface area-to-mass ratio. The dosage rates were optimized for particles in the size $-212 + 106$ μm, while in further experiments four times the original bacterial dosage $(1.30 \times 10^{11}$ cell/g of solids) was applied to the $-106 + 53$ μm and -53 μm size fractions. The results are shown in Fig. 28.15. These illustrate that the amount of floatable material in the fractions above was reduced from 83 and 75% to 5 and 8%, respectively.

Effect of Pyrite Reactivation

In conventional coal flotation, the tailings from the flotation cells are dewatered by a number of techniques before disposal in a prescribed manner on to surface tips. In this novel coal treatment system using a bioflotation stage, it would be beneficial to concentrate the pyrites from the waste tailings as a possible commercial source of sulfur values and to minimize problems of environmental cost envisaged with surface tipping. To establish the feasibility of recovery, the Capper Pass pyrite ($-212 + 106$ μm) was suppressed by a bacterial pretreatment at 1×10^{10} cells/mL in 15 mL of distilled water at pH 2.0 and 2% pulp density, as described previously. The float and tailing products were subjected to the three alternative reactivation techniques described previously.

Table 28.5 compares the results obtained in each case in the Hallimond tube and the Leeds cell. Method 1, using membrane filtration, was not carried out with the Leeds cell because the filtration was considered unpractical on a commercial scale. Direct addition of further collector frother effectively reactivated the suppressed pyrites without any need for dewatering. In the Hallimond tube, the initial bacterially treated floats ranged from 10.7 to 18.9% of the feed pyrites from the three experiments.

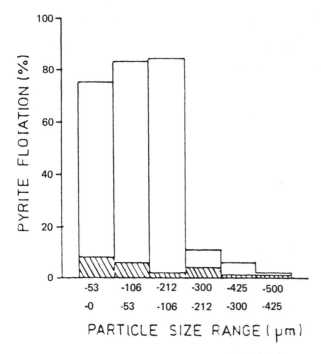

Fig. 28.15 Effect of increased bacterial dosage (1.304×10^{11} cell/g) on pyrite in the $-106 + 53$ μm and -53 μm size fractions in comparison to Fig. 28.14. (From Ref. 31.)

28.3.2 Coal Pyrite Mixtures

The flotation results on the coal/pyrite synthetic feed are given in Table 28.6. Under control conditions (no bacterial treatment), the cell reduced the sulfur content from 10.75 to 6.25% (by mass), and the effect of bioflotation was to reduce the sulfur content to 2.7%. A reduction in floats of 11% resulted from the bioflotation and was accompanied by a reduction in ash from 7.7% to 3.6%. Most of this reduction may be associated with pyrite loss, although proximate analysis shows that the associated loss of fixed carbon to discard as a result of bacterial treatment is 2.6%.

Table 28.5 Comparison of Pyrites Floats After Bacterial Suppression and Reactivation Treatment in a Hallimond Tube and Leeds Cell (Capper Pass Pyrites, $-212 + 106$ μm; pH 2.0)

	Pyrites flotation			
	Hallimond tube		Leeds cell	
Reactivation technique	Float 1 (bacteria)	Float 2 (reactivate)	Float 1 (bacteria)	Float 2 (reactivate)
1	18.9 ±2.9	78.9 ±1.1	a	a
2	10.7 ±0.3	81.5 ±3.2	27.3 ±3.0	87.3 ±3.5
3	15.3 ±1.1	84.7 ±0.4	27.4 ±3.2	87.5 ±6.2

Source: Ref. 24.
[a]Not tested.

Table 28.6 Effect of Bacterial Treatment on the Separation, of a Pyrite/Coal Mixture in a Flotation Cell (Percent by Mass, Dry Basis)

	Float		Tails	
	Control	Bacterially treated	Control	Bacterially treated
Product	91.60	80.60	–	–
Discard	–	–	8.40	19.40
Sulfur	6.25	2.74	44.01	38.79
Ash[a]	7.69	3.64	66.44	56.51

Source: Ref. 31.
[a]See Section 28.2.6.

28.3.3 Comparison of *Thiobacillus ferrooxidans*, Heterotrophic Bacteria, and Yeasts in the Suppression of Pyrite in Flotation

Figure 28.16 shows the influence of cultivation time on the subsequent suppressive activity of the *T. ferrooxidans* cells in suspension in water at pH 2.0 at a cell density of 3.25×10^{10} cells/g of pyrite, in suppressing flotation in the Hallimond tube [27]. Also shown is the effectiveness of the liquor following a 2-h sedimentation period for solids removal prior to experimentation (this constituted the lixivant).

The concomitant leach profile is shown in Fig. 28.17, which represents the batch growth curve. Two periods of exponential growth were indicated, which suggests that the culture was either a mixture of two strains or that two nutrient sources were utilized sequentially. The bacteria exhibited two periods of suppressive activity, which may relate to the respective stationary phases of the two strains. The lixiviant only showed activity in the final stationary phase. Electron microscope studies show two morphological bacterial types which could not be isolated and grown on solidified 9K ferrous sulfate media. The fact that no growth was apparent on nutrient agar suggests that they were not heterotrophic contaminants. Although the cells harvested during the first period of stationary phase showed about 80% suppression of flotation while suspended independently from the liquor in distilled water adjusted to pH 2.0 with 0.1 m H_2SO_4, the suspension of the bacterial cells in distilled water may have caused some cell lysis, releasing cellular components that may have adsorbed to the pyrite. Activity shown by the lixiviant during stationary phase was probably also related to the presence of cell lysis products.

It seems probable, therefore, that the activity of cell harvests could be stimulated by cell lysis, and this was confirmed by increasing the suppressive activity of the cells following disintegration in a Braun Cell Homogeniser (Braun Industries, West Germany). Activity was increased by 11.2% over controls which showed 70.1% suppression after 1 min of disintegration and 17.6% after 5 min to almost 100% suppression. Also, by suspending inactive cells in water at pH 2.0 the suppressive activity could be enhanced from 30 to 85% within a further 100 h of storage probably by cell death and lysis. One major disadvantage with this desulfurization method is that relatively large quantities of *T. ferrooxidans* are required to permit administration of the required bacterial dosage of 3.25×10^{10} cells/g of pyrite.

Table 28.7 shows the levels of pyrite suppression that are attainable by microorganisms other than *T. ferrooxidans*. The two bacteria used showed almost identical levels of suppression as *T. ferrooxidans*, while the spores of *A. niger* and yeast cells at compar-

Pyrite

floated (%)

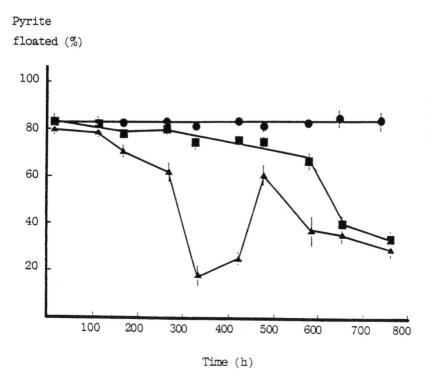

Time (h)

●; Control (water adjusted to pH 2·0 with 0·1M H₂SO₄).

■: Lixiviant (this constituted liquor from the Pachuca reactor after sedimentation for 2 hours).

▲: Suspension of cells of *Thiobacillus ferrooxidans* in water at pH 2·0 at a cell density of 3·25 × 10¹⁰ cells per gram of pyrite.

(Results of three replicate trials = S:E:M.)

Fig. 28.16 Suppressive activity of lixiviant and cells of *T. ferrooxidans* harvested at various times during batch growth.

able cell densities of 3.25×10^{10} cells/g of pyrite in water at pH 2.0 showed slightly inferior levels of suppression, but nevertheless were in the region of 90% or better. Quite clearly, the activity of microbial cells in the suppression of the flotation of pyrite was not restricted to *T. ferrooxidans* and the fact that these other cell types can be propagated in much shorter times than *T. ferrooxidans* could introduce a significant advantage in their use.

Table 28.7 also shows the effect of various microorganisms on the Nottinghamshire coal yield when the coal was floated independently of the pyrite. While *T. ferrooxidans* caused no loss in yield with 90.7% coal floated in relation to the float yield of 89.8% achieved in the control trials, the presence of the other bacteria resulted in the yield falling to 84.4 and 79.1% with *E. coli* and *P. maltophila*, respectively. The yeasts and the fungal spores caused no reduction in coal yield.

Table 28.8 shows the comparative coal desulfurization by *T. ferrooxidans* and *S. cerevisiae*. The yeast *S. cerevisiae* was chosen for comparison to *T. ferrooxidans* in preference to the heterotrophic bacteria on potential health hazard grounds and also because

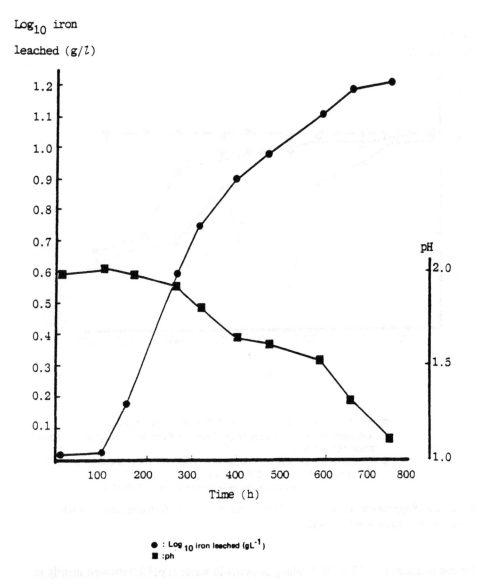

Log$_{10}$ iron leached (g/l)

Fig. 28.17 Progress on iron leach from -53 μm by *T. ferrooxidans* and the associated pH profile. (From Refs. 27 and 39.)

this organism may well be commercially available as a by-product from the brewing industry. Both *T. ferrooxidans* and *S. cerevisiae* were very effective in reducing the pyrite component from the synthetic coal/pyrite mixture.

The sulfur content of the floats of untreated control mixtures was 11.4%, whereas after bacterial and yeast pretreatment this was reduced to 2.2 and 1.4%, respectively. The ash content of the floats was also dramatically reduced as a result of biological pretreatment, while fixed carbon and volatiles were increased. Associated with these improvements in the coal quality, there was an unfortunate reduction in the coal yield from 89.1% to 73.6% and 71.0% after treatment with *T. ferrooxidans* and *S. cerevisiae*, respectively. It is worthy of note, however, that the control floated coal contained 21.1% pyrite.

Table 28.7 Pyrite Suppression in a Hallimond Tube Using a Variety of Microorganisms in Distilled Water Adjusted to pH 2.0[a]

Organism	Pyrite floated (%)	Coal floated (%)
Control	90.0 ± 0.7	89.8 ± 2.2
T. ferrooxidans	5.0 ± 0.3	90.7 ± 1.1
E. coli	4.2 ± 0.9	84.4 ± 1.0
P. maltophila	5.7 ± 0.5	79.1 ± 0.9
S. cerevisiae	10.7 ± 0.9	90.0 ± 0.7
C. utilis	8.3 ± 0.9	89.7 ± 1.1
A. niger (spores)	6.9 ± 0.4	88.6 ± 1.2

Source: Ref. 27.
[a]Results represent mean of three replicate trials ±SEM.

It was originally postulated that the suppression activity was specific to *T. ferrooxidans* because of its implication in iron sulfide oxidation. The significant advance reported here is that the suppression of pyrite during the simulated flotation was not restricted to *T. ferrooxidans* and that fungal spores, yeast cells, and other heterotrophic bacteria showed similar activity toward suppression at similar cell densities. Although treatments with *T. ferrooxidans* and *S. cerevisiae* reduced the coal yields slightly from 89.1% in control tests to around 70%, this may be improved upon by reformulation of the flotation oil.

28.3.4 Development of *Thiobacillus ferrooxidans*, the Yeast *Saccaromyces cerevisiae*, and Biological Depressant in Simulated Coal Flotation

Table 28.9 shows the comparative coal desulfurization by *T. ferrooxidans*, *S. cerevisiae*, and an aliquot of 115 μL of a biological depressant. All three agents were very effective in reducing the pyrite component from the synthetic coal/pyrite mixtures. The sulfur content of the floats of untreated control mixtures was 11.4%, whereas after pretreatment with either bacteria, yeast, or the biological suspension, this was reduced to 2.2%, 1.4%, and 2.5%, respectively. The ash content of the treated floats was also dramatically reduced while fixed carbon and volatiles were increased. Associated with these benefits, however, there was reduction in the coal yield from 81.9% to 73.6%, 71.0%, and 75.5% after treatment with *T. ferrooxidans*, *S. cerevisiae*, and the biological suspension, respectively. It is notworthy, however, that the control floated coal contained 21.1% pyrite.

Table 28.8 Comparative Coal Desulfurization with *T. ferrooxidans* and *S. cerevisiae* (Percent by Mass, Dry Basis)

	Control	Bacteria	Yeast
Product	89.1	73.6	71.0
Sulfur	11.4	2.2	1.4
Ash	13.1	3.2	1.7
Fixed carbon	52.3	56.3	59.9
Volatiles	35.0	40.4	37.9

Source: Ref. 27.

Table 28.9 Comparative Coal Desulfurization with *T. ferrooxidans* and *S. cerevisiae* (3.25 × 10^{10} cells/g pyrite) and 115 μL) Biological Suspension, Hallimond Tube at 2% Pulp Density of a Synthetic Coal/Pyrite Mixture Containing 10.75% Sulfur (Percent by Mass, Dry Basis)

	Control	*T. ferrooxidans*	*S. cerevisiae*	Biological depressant
Product	89.1	73.6	71.0	75.5
Sulfur	11.4	2.2	1.4	2.5
Ash	13.1	3.2	1.7	4.3
Volatiles	35.0	40.4	37.9	36.7
Fixed carbon	52.3	56.3	59.9	59.0

Source: Ref. 26.

Although it was originally postulated that pyrite suppression was specific to *T. ferrooxidans*, it has been shown clearly here that suppressive activity was also apparent, not only in other microbial cell types, but in a suspension of biological suppressant from the food-processing industry also. Although the coal yields were slightly inferior following treatments, the relatively small losses are probably outweighed by very significant reductions in the sulfur content, from 11.4% to around 2.5% or lower, and this was also complemented by an advantageous reduction in the ash content of the coal.

A similar test using a synthetic mixture of coal/pyrite at 5.03% total sulfur was repeated using 1.25 mL of biological agent and 208 μL of nonphenolic flotation oil in a 2.5-L Denver Cell at a solid concentration of 2%, at pH 2.0, and the results are shown in Table 28.10. The results confirm that the sulfur content of the coal could be reduced by 66% accompanied by a reduction of the ash content from 6.65% to 3.01% and a small decrease of coal yield from 98.56% to 92.29%, respectively, as a result of pyrite suppression.

28.3.5 Biological Agent

Table 28.11 illustrates the effect of the three biological agents, A, B, and C, and the bacteria *T. ferrooxidans* upon flotation of Capper Pass pyrite in the Hallimond tube [40]. Biological agent A proved to be the most effective suppressant at low dosages, reducing the pyrite in the float product from 90.7% control to 22.6% with the addition of only

Table 28.10 Synthetic Mixture of Coal/Pyrite Mixture Containing 5.03% Sulfur (Percent by Mass, Dry Basis)

	Head analysis	Floats	
		Control	Biological depressant
Product	–	98.56	92.29
Sulfur	5.03	4.77	1.72
Ash	11.04	6.65	3.01
Volatiles	35.11	36.46	36.95
Fixed carbon	53.49	56.89	60.04

Source: Ref. 26.

Table 28.11 Comparison of the Various Biological Agents on the Flotation of Capper Pass Pyrite in a Hallimond Tube (Solids Concentration 2%, pH 2.0, Particle Size −212 + 106 μm, Percent by mass, Dry Basis)

			Biological agent				
	A		B		C		*T. ferrooxidans* 4.4 × 10⁷ cell/μL
Dosage (μL)	Pyrite float (%)	Dosage (μL)	Pyrite float (%)	Dosage (μL)	Pyrite float (%)	Dosage (μL)	Pyrite float (%)
0	90.7 ± 1.3	0	90.7 ± 1.3	0	90.7 ± 1.3	0	90.7 ± 1.3
50	22.6 ± 1.0	500	51.0 ± 1.3	500	42.8 ± 1.4	5,000	90.2 ± 0.9
100	13.1 ± 0.9	1,000	13.4	1,000	18.8 ± 0.9	15,000	28.3 ± 0.5
125	10.9 ± 0.8	1,250	12.3	1,250	14.2 ± 1.4	25,000	21.6 ± 0.9
150	9.7 ± 0.5	1,500	12.5 ± 0.5	1,500	11.4 ± 1.0	35,000	15.1 ± 0.9
200	8.7 ± 0.2	2,000	13.4 ± 1.1	2,000	7.5 ± 2.6	45,000	11.4 ± 0.8

50 µL. On increasing the dosage the float product was further reduced to 8.7% at the final dosage of 200 µL. In general, the effect of biological agents B and C on pyrite were similar to that of A, in that they also suppressed flotation. The addition of 500 µL of each suppressant reduced the pyrite in the float to 51.0% and 42.8%, respectively. The floats were correspondingly reduced to 13.4% and 7.5% by addition of 2000 µL. In the case of *T. ferrooxidans*, Table 28.11 shows that very high cell concentrations and dosages were required, to produce comparable suppressions of pyrite flotation.

Table 28.12 illustrates the quantitative effect of various biological agents on the zeta potential of Capper Pass pyrite (30 mg). All three agents reduced the zeta potential from +14.6 mV to below −10.0 mV. All dosages are quoted in absolute (undiluted) values. The isoelectric point was obtained using biological agent A at a dosage of 0.33 µL. The potential was reduced to a negative value of −17.2 mV at 20 µL. Further increases in the agent dosage resulted a reversal of the zeta potential trend. The effects of biological agents B and C on the zeta potential are generally similar. The isoelectric point was obtained at dosages of 0.031 µL and 0.046 µL for B and C, respectively. Zeta potential became negative and continued to decrease with increasing dosage. *T. ferrooxidans* was much less effective on the zeta potential in comparison with the other agents.

The comparative effects of treating various size fractions of an American coal with A and *T. ferrooxidans* are illustrated in Table 28.13. The head analysis of each size fraction and the results of control flotation are also shown for further comparison. In the size range −500 + 150 µm, there are significant reductions in the sulfur content of the float under normal flotation conditions (control), while in the finer size fractions, no significant reduction was obtained under these conditions. Treatment with A reduced the sulfur content to approximately 4% and 5% for the +150 and −150 µm, respectively, while the bacteria produced a product having corresponding values of around 6% throughout. All treatments resulted in a marked reduction in the ash value of the float. Treatment with both the A and bacteria resulted in reduction in coal recovery of around 6% overall. The product is reduced by between 4% and 10%.

28.3.6 Comparison of Laboratory Flotation on Capper Pass Pyrite Using the Biological Depressant

Previous work conducted in a modified Denver Cell [31] to conserve samples was used to optimize the flotation and biological depressant dosage of 100 µL and 350 µL, respectively. The flotation dosage was established previously to optimize pyrite flotation, hence the unnecessarily high reagent dosage in relation to coal flotation [24]. The flotation and biological depressant dosage in the Denver and Leeds cell was increased proportionally to the cell volumes.

A comparison of the performance of several laboratory cells was conducted using the representative size fractions of Capper Pass pyrite to establish the effect of pyrite suppression using the biological agent. The tests were conducted with a solids concentration of 2% at pH 2.0 and the results are illustrated in Table 28.14. The results of the Hallimond tube were given for comparative purposes since the kinetics of the tube are not amenable for the treatment of fine particle sizes or for scaling. However, the small capacity of the cell enables conservation of materials and reagents while providing comparative assessments of floatability.

Table 28.14 illustrates that in most of the size fractions, except for the size range below 106 µm in the Hallimond tube and Leeds cell, almost total suppression of the pyrite was achieved. The biological depressant dosage was originally optimized in the

Table 28.12 Comparison of the Various Biological Agents on the Zeta Potential of Capper Pass Pyrite (Dosages Shown are Absolute Values)

	Biological agent						
	A		B		C		T. ferrooxidans 4.4 × 10⁷ cell/μL
Dosage (μL)	Zeta potential (mV)	Dosage (μL)	Zeta potential (mV)	Dosage (μL)	Zeta potential (mV)	Dosage (μL)	Zeta potential (mV)
0	+14.6	0	+14.6	0	+14.6	0	+14.6
0.1	+8.7	0.01	+10.4	0.01	+11.8	1	+14.6
0.3	+1.2	0.03	+1.5	0.03	+8.8	3	+14.5
0.6	-7.2	0.06	-11.7	0.06	-9.7	6	+14.4
1	-11.1	0.10	-13.6	0.10	-12.1	10	+14.3
2	-13.6	0.20	-15.4	0.20	-13.4	20	+14.1
4	-14.8	0.40	-16.3	0.40	-14.6	40	+13.6
6	-15.3	0.60	-16.6	0.60	-15.2	60	+13.1
10	-15.5	1	-16.6	1	-16.0	100	+11.9
20	-17.2	2	-17.0	2	-17.1	200	+8.0
40	-12.8	4	-17.6	4	-18.7	400	+5.6
60	-11.9	6	-18.1	6	-19.7	600	+4.6
100	-11.0	10	-18.6	10	-20.7	1000	+3.9

Source: Ref. 40.

Table 28.13 Effect of Bacterial and Biological Preconditioning in Relation to Conventional Flotation Using a High-Sulfur American Coal (Percent by Mass, Dry Basis)

Size fraction (μm)	Control				Biological agent A				T. ferrooxidans				Head analysis			
	−500+250	−250+150	−150+106	−106+75	−500+250	−250+150	−150+106	−106+75	−500+250	−250+150	−150+106	−106+75	−500+250	−250+150	−150+106	−106+75
Float																
Product	50.60	65.70	66.10	63.80	52.80	53.60	54.90	56.30	50.20	52.00	50.90	56.60	–	–	–	–
Ash	12.94	18.70	26.89	23.00	11.01	10.84	14.57	16.00	11.30	13.96	16.73	16.67	48.84	47.76	49.67	49.28
Fixed carbon	52.83	48.74	48.37	46.91	53.44	53.86	52.05	51.34	54.59	52.20	50.43	51.26	22.62	24.56	23.18	24.01
Volatiles	34.23	32.50	24.74	30.09	35.55	35.30	33.38	32.66	34.11	33.84	32.84	32.07	28.54	27.68	27.15	26.77
Sulfur Total	6.39	6.88	10.15	10.49	3.95	3.93	4.58	5.64	4.95	6.18	6.45	6.00	11.30	9.45	10.41	11.23
Pyritic	–	–	–	–	–	–	–	–	–	–	–	–	8.47	4.42	5.90	7.63
Coal recovery[a]	75.62	85.74	87.83	85.61	76.99	78.97	79.66	80.51	76.39	81.35	80.35	80.79				
Sink																
Product	49.40	34.30	33.90	36.20	47.72	46.40	45.10	43.70	49.80	48.00	49.10	43.40				
Ash	77.55	83.41	84.51	83.26	78.88	79.55	80.71	80.76	78.41	80.75	80.77	80.83				
Fixed carbon	2.46	0.00	0.36	1.51	0.00	0.36	0.27	1.30	0.93	0.04	0.11	1.05				
Volatiles	19.99	16.59	15.13	15.23	21.12	20.09	19.02	17.94	20.66	19.21	19.12	18.12				
Total sulfur	16.98	13.70	12.63	15.13	16.90	15.10	16.95	17.90	16.73	14.10	15.00	16.28				

Source: Ref. 40.

[a]Percentage coal recovery = % floats $\left(\dfrac{\text{combustibles in floats}}{\text{combustibles in feed}}\right)$.

Table 28.14 Comparison of Laboratory Flotation Cells on Capper Pass Pyrite (Solids Concentration 2%, pH 2, Percent by Mass, Dry Basis)

Size (μm)	Hallimond tube float		Modified Denver cell float		Denver cell float		Leeds cell float	
	Control	Biological depressant	Control	Biological depressant	Control	Biological depressant	Control	Biological depressant
-500 + 425	12.27	0.02	3.55	0.00	4.93	0.00	2.52	1.59
-425 + 300	4.06	0.47	5.34	0.10	4.63	0.00	23.95	1.87
-300 + 212	76.08	0.05	13.67	0.25	22.91	0.00	77.48	1.90
-212 + 106	98.44	8.60	73.20	0.62	81.20	1.77	94.38	3.64
-106 + 53	98.45	96.79	77.75	1.55	95.85	6.92	96.39	75.02
-106 + 53*	98.45	10.12	–	–	–	–	96.39	6.94
-53	89.99	83.00	77.33	37.08	90.81	58.85	92.81	88.44
-53*	89.99	13.2	77.33	14.72	90.81	21.01	92.81	29.95
	Dosage (μL), flotation Reagent/biological depressant							
Standard	75/115		100/350		417/1460		583/2042	
*	75/1000		100/1200		417/6500		583/22,000	

Source: Ref. 26.

modified Denver Cell for particles in the size range $-212 + 106$ μm. Since the surface area increases with decreasing particle size, the biological dosages were increased in the modified Denver Cell to achieve a reasonable suppression of approximately 85% for sizes below 53 μm without adversely affecting coal flotation.

As indicated in Table 28.15, size analysis of the -53 μm size fraction shows that 15% of the particles are below -6 μm, and therefore a higher dosage would be required to achieve complete suppression. The results indicate that the Leeds Cell suppressed only 75% of the $-106 + 53$ μm size fraction for the standard biological dosage. Using the increased dosage of biological depressant, the pyrite float was 14.72% in the modified Denver Cell, 21.01% in the Denver Cell, while in the Leeds Cell floated only 29.95% pyrite even with an increase of 3.5 times the dosage. Consequently, in all future work using the biological agent the Denver Cell is preferred. Another important aspect of these results is that by using the modified Denver Cell, which incorporates an increased quiescent zone, a reduction of approximately 15% of the pyrite can be achieved by mechanical effects. This would suggest that a mechanical modification to existing industrial flotation cells could make a significant contribution to improved desulfurization of coal.

28.3.7 Ellerbeck Coals

Preliminary investigation using the biological agent in a 2.5-L standard laboratory Denver Cell was conducted on 3.8% and 4.98% sulfur Ellerbeck coals, respectively. The dosage of 208 μL of nonphenolic and 1.25 mL of biological agent was optimized, and all flotation tests on the coals were conducted at pH 5.5, which represented typical operational pH in coal flotation circuits. Table 28.16 summarizes the flotation test work on the Ellerbeck coals. The table shows that the total sulfur content of the recovered float was reduced by about 17% and 23%, respectively, with a corresponding reduction of 19% and 23% of the ash content. Petrographic studies indicate that some of the pyrite was distributed in the coal meceral with an average grain size below 10 μm in diameter. The size distribution of the Ellerbeck coals presented in Table 28.1 illustrates that approximately 11% of the coal is below 10 μm in diameter and consequently, further communition would be required for further pyrite liberation. The work on actual coals is preliminary and the results are presented merely to complement the development work on the biological depressant. The results suggest that coals containing small pyritic grain sizes are less effectively treated by biological processes.

Table 28.15 Size Distribution of -53-μm Capper Pass Pyrite (Malvern 2600/3600 Particle Analyzer)

Size distribution (μm)	Percent retained	Cumulative percent
$-53 + 38$	12.1	100.00
$-38 + 22$	21.1	87.9
$-22 + 10$	32.8	66.7
$-10 + 6$	18.5	33.9
$-6 + 4$	11.3	15.4
$-4 + 2.4$	3.5	4.1
-2.4	0.6	0.6

Source: Ref. 26.

Table 28.16 Flotation Test on Ellerbeck Coals of 3.8% Sulfur (A) and 4.98% Sulfur (B) Using a Biological Depressant (Percent by Mass, Dry Basis)

	Head analysis		Float			
			Control	Biological	Control	Biological
	A	B	A	A	B	B
Product	–	–	78.84	64.69	84.00	67.06
Total sulfur	3.80	4.98	3.28	2.47	3.56	2.91
Pyritic sulfur	2.28	3.86	2.36	1.73	2.46	1.75
Nonpyritic sulfur	1.52	1.12	0.92	0.74	1.10	1.16
Ash	25.27	14.98	11.16	8.58	8.45	6.85
Volatiles	31.00	35.40	35.40	35.79	36.75	36.94
Fixed carbon	43.73	49.62	53.55	55.63	54.80	56.21

Source: Ref. 26.

Table 28.17 Comparative Coal and Pyrite Flotation, Tharsis and Capper Pass (−212 + 106 μm), in a Hallimond Tube at 2% Pulp Density in the Presence of the Formulated Blend of Flotation Oil and Biological Depressant

	Float (percent by mass, dry basis)		
Amount of flotation agent added (μL)	Coal	Tharsis pyrite	Capper pass pyrite
Control[a]			
0	60.0	29.3	4.9
10	83.4	29.3	4.8
50	85.8	35.6	5.2
100	88.1	37.7	8.5
150	99.0	38.3	14.0
250	98.7	52.4	18.6
500	98.9	62.5	86.7
1000	99.9	75.7	91.1
Floation agent containing biological suppressant			
0	61.4	–	–
10	70.0	21.7	2.2
50	69.9	23.1	4.8
100	70.1	27.9	6.5
150	88.2	20.3	4.0
250	88.4	19.9	5.5
500	90.3	23.7	7.1
1000	97.1	30.7	11.2

Source: Ref. 26.
[a]Control tests were performed with the blend containing water substituted for the biological suppressant.

Table 28.18 Effect of Conditioning Time on Pyrite Flotation (Capper Pass −212 + 106 μm), 2% Pulp Density in a Hallimond Tube at pH 2.0 Following Treatment with 1000 μL of the Formulated Flotation Oil Blend Containing Biological Depressant

	Pyrite float (% by mass, dry basis)	
Conditioning time (s)	Controls (blend containing water)	Treated samples (blend containing biological depressant)
10	95.7	11.8
20	96.9	12.3
30	97.1	13.1
60	98.9	13.7
120	99.3	13.0

28.3.8 Formulation of Flotation Oil Incorporating Biological Depressant

Table 28.17 shows the comparative coal and pyrite flotation (Tharsis and Capper Pass) when the biological depressant was incorporated in a newly formulated flotation oil. The control flotation oil mixture contained water instead of the biological depressant complement. Flotation of the coal was slightly reduced in the presence of the biological depressant in the addition range 10 to 1000 μL of the product tested. At an addition of 1000 μL, control floats were 99.9%, whereas in the blend containing the biological depressant, 97.1% coal flotation was achieved. Flotation of Capper Pass pyrite in controls increased above additions of 150 μL of the blend containing water to a 91.1% at 1000 μL. In contrast, an addition of 1000 μL of the formulated blend containing the biological depressant, only 11.2% of the pyrite floated. Reduction in flotation of the Tharsis pyrite was also obtained as illustrated in Table 28.17. Table 28.18 shows the conditioning time for effective pyrite suppression following the addition of 1000 μL of the formulated product. Suppression was achieved in the shortest conditioning time of 10 s.

28.4 CONCLUSIONS

There have been many processes outlined for the desulfurization of coal [1,41,42] ranging from magnetic separation, flue gas desulfurization, selective flotation, and oil agglomeration techniques. Although leaching, as opposed to surface conditioning, is another possible method for removing pyritic contaminants in coal, it is not practical for commercial operation because of the long retention times needed. For example, residence times for bacterial leaching of pyrites in coal have ranged from 10 days [43] and 6 days [44] to 5 days [20]. The bacterial surface conditioning of pyrites before flotation took a mere 2.5 min. This would appear to be rather more appropriate to commercial treatment at the mine or power generating plant, particularly as pyrites are conceived of as a potential valuable sulfur source. In contrast, during bacterial leaching processes it is converted to sulfuric acid of a concentration that is too low to be of any use and high enough to be objectionably corrosive. This 2.5-min conditioning period is also significantly shorter than those used by other workers. In fact, it has been shown that this conditioning time can be reduced to 10 s, although 2.5 min was chosen as the standard test time to ensure adequate mixing and also to conform to realistic plant operating procedures.

It has also been demonstrated that a pyrites-rich tailings product could be achieved after bacterial treatment of coal/pyrites mixtures before flotation. It is also likely that this pyritic sink product could be reactivated to produce a pyritic concentrate. The encouraging result of the reactivation study could have important implications in the processing of commercial sulfides, as in selective separation of sulfides, as well as in the concentration of pyrites and removal of pyritic contaminants from coal. Continued development of this work was limited to the laboratory scale due to the problems of producing sufficient quantities of bacteria. This is highlighted in Table 28.11, where 4.3×10^{11} cells are required to treat 1 g of pyrite to an acceptable level of suppression. The large quantities of *T. ferrooxidans* required coupled with the extended production times will undoubtedly restrict the possible application of this technique on an industrial scale.

The experimental results also demonstrated that the suppression of pyrite in simulated flotation was not restricted to *T. ferrooxidans*, heterotrophic bacteria, and that the yeast *S. cerevisiae*, together with a commercial biological agent from the food-processing industry, can produce similar effects. The use of an off-the-shelf biological suppression agent dramatically simplifies the application to coal desulfurization in existing flotation circuits, particularly with a residence time as low as 10 s.

The effects of each of these biological suppressants on the zeta potential confirm the results obtained in the microflotation trials. All three biological agents completely reversed the potential on the pyrite surface from electropositive to electronegative, while the bacteria merely reduce the positive potential. These results suggest that the active negatively charged components of the biological agent and the similar component of the gram-negative *T. ferrooxidans* adsorp onto the pyrite, reducing the potential. Once the pyrite surface has been rendered electronegative, collector adsorption cannot occur due to either electrostatic repulsion or simply to the nonavailability of any "free sites" at the surface of the pyrite.

Suppressant A is a waste by-product from the food-processing industry and relatively cheap, and consequently all further coal flotation trials were performed using this product. It was found that a critical dosage exists at which no further improvement in sulfur reduction could be made. Zeta-potential measurements also indicated a characteristic dosage of A which the reducing potential trend was reversed. The coal flotation trials illustrate that biological agent A can effect a much larger sulfur reduction than can *T. ferrooxidans* without a significant difference in coal recovery values.

Experimentation of various particle sizes of pyrite in different laboratory cells indicate that mechanical modification to produce an increased quiescent zone can make a useful contribution to the suppression of pyrite and its implication in the modification of existing industrial flotation cells. Preliminary investigations using the biological agent on Ellerbeck coals with a sulfur content of 3.8% and 4.98% showed a reduction of 17% and 23% of the total sulfur content, respectively. Petrographic studies, however, indicate that further communition to liberate the pyrite in this particular case is required to further reduce the pyritic sulfur content of the coals.

It is of some significance that the biological suppressant can now be incorporated in the reagent to effect pyrite suppression. However, it must be recognized that no single flotation reagent incorporating the suppression agent could suffice for all coals, and a degree of variability in the formulation of such a product is likely to be desirable. This new approach will overcome the difficulties of managing a process involving living cells and the unreliability inevitably associated with such a process.

If scaled up, the biological reagent dosages used in the laboratory test work would approximate to a cost of approximately 5 cents per metric ton of coal treated. It seams feasible that coals with high levels of contaminating pyrite which is liberated during coal cutting and crushing could be treated by a specially formulated flotation oil containing the biological suppressive agent, so that existing plant could be used to assist in the reduction of SO_2 during combustion.

ACKNOWLEDGMENTS

The financial support of a co-funded British Coal/Science and Engineering Research Council grant is gratefully acknowledged. The author also acknowledges the cooperation and assistance of Mr. J. Morris, Surface Environmental Group, Headquarters Technical Department, British Coal. Special thanks are due also to Dr. C. C. Townsley, Dr. S. I. Al-Ameen, and Mr. A. J. Davis of the research team, and also to several engineers and scientists within British Coal for their support and assistance. The cooperation provided by Mrs. P. Hawthorne and the technical assistance of Mr. E. W. Bridgwood in the preparation of this manuscript are also gratefully acknowledged.

REFERENCES

1. *Acid Rain*, ISBN 0-902543-77-6, Central Electricity Generating Board, London, Aug. 1984.
2. Owens, R. V., and Owens, R., *Acidification of the Environment Including Acid Rain*, ISBN 946-924-00-7, Pyramid Publications, 1983.
3. *Acid Rain*, OR Bulletin, National Coal Board, London, Mar. 1985.
4. Ministry of Health, *Mortality and Morbidity During the London Fog of December 1952*, Reports on Public Health and Medical Subjects No. 95, Her Majesty's Stationery Office, London, 1954.
5. Ellison, J., and Waller, R. E., A review of sulphur oxides and particulate matter as air pollutants with particular reference to effects on health in the United Kingdom, *Environ. Res., 16*, 302-325 (1978).
6. Hinrichsen, D., A beautiful death, *Sweden Now*, 21-23 (1983).
7. Franks, J., Acid rain, *Chem. Br.*, June, 504-509 (1983).
8. Acid rain estimated at $5 billion annually, *Min. Eng.*, Sept., 1105 (1985).
9. Acid rain may be a cause of bone disease, *The Guardian*, Sept. 14, 1985, p. 2.
10. Bolton, T., Atkins, A. S., and Proudlove, P., The sulphur, ash relationship in coal seams and its implications in mine planning, in the United Kingdom. *Min. Sci. and Tech., 7*, 265-275 (1988).
11. Highton, N. H., and Chadwick, M. J., The effects of changing patterns of energy use on sulphur emissions and depositions in Europe, *Ambio, 11* (6), 324-329 (1982).
12. EMEP, *The Cooperative Programme for Monitoring and Evaluation of Long Range Transmission of Air Pollutants in Europe*, ECE United Nations, 1981.
13. *Energy and Environment* (Environmental Data Services Report 31), *117*, Oct. 11, 1984.
14. *Energy and Environment* (Environmental Data Services Report 30), *117*, Oct. 9, 1984.
15. Meyers, R. A., Hamersma, J. W., Land, J. S., and Kraft, M. L., Desulphurisation of coal, *Science, 177*, 1188 (1972).
16. Zimmermann, R. E., Economics of coal desulphurisation, *Chem. Eng. Prog., 62* (10), 61-66 (1966).
17. MacGregor, I., The role of technology in the mining industry, *Min. Technol.*, Jan. 3-6 (1986).

18. Chapman, W. R., and Rhys-Jones, D. C., The removal of sulphur from coal, *J. Inst. Fuel, 28*, 102 (1955).

19. Leonard, J. W., and Mitchell, O. R. (eds.), *Coal Preparation*, American Institute of Mining, Metallurgical, and Petroleum Engineers, New York, 1968.

20. Dugan, P. R., and Apel, W. A., Microbial desulphurisation in coal, in *Metallurgical Applications of Bacterial Leaching and Related Phenomena* (L. E. Murr, A. E. Torma, and J. A. Brierley, eds.), Academic Press, New York, 1978, pp. 223-250.

21. Andrews, G. F., and Maczuga, J., Bacterial removal of pyrite from coal, *Fuel, 63*, Mar., 297-302 (1984).

22. Capes, C. E., McIlhinney, A. E., Sivianni, A. F., and Puddington, I. E., Bacterial oxidation in upgrading pyritic coals, *Can. Inst. Min. Metall. Bull., 66*, 88-91 (1973).

23. Dogan, M. Z., Oxbayoglu, G., Hicyilmaz, C., Serikaya, M., and Ozcengiz, C., Bacterial leaching versus bacterial conditioning and flotation in desulphurisation of coal, *15th Congress International de Metallurgie*, Cannes, June 11, 1985, Vol. 2, pp. 304-313.

24. Atkins, A. S., Davis, A. J., Townsley, C. C., Bridgwood, E. W., and Pooley, F. D., Production of sulphur concentrates from the bio-flotation of high pyritic coals, in *Proceedings of the International Conference, Sulphur 85*, 1985, pp. 83-104.

25. Townsley, C. C., Atkins, A. S., and Davis, A. J., An investigation of suppression of pyritic sulphur during flotation tests using the bacterium *Thiobacillus ferrooxidans, Biotechnol. Bioeng.*, in press.

26. Atkins, A. S., Townsley, C. C., and Al-Ameen, S. I., Application of a biological sulphur depressant to the desulphurisation of coal in froth flotation, Minprep 87, *International Symposium on Innovative Plant and Processes for Mineral Engineering*, Doncaster, Yorkshire, England, Apr. 1987, pp. 1-13.

27. Townsley, C. C., and Atkins, A. S., Comparative coal fines desulphurisation using the iron-oxidising bacterium *Thiobacillus ferrooxidans* and the yeast *Saccharomyces cerevisia* during simulated froth flotation, *Process Biochem.*, Dec., 188-191 (1986).

28. Silverman, M. P., and Lundgren, D. G., Studies on the chemoautrophic iron bacterium *Thiobacillus ferrooxidans*. 1. An improved medium and a harvesting procedure for securing high cell yields, *J. Bacteriol., 77*, 642-647 (1959).

29. *British Standard Method for the Analysis and Testing of Coal and Coke*, BS 1016, Part 3, 1965.

30. Pryor, E. J., *Mineral Processing*, 3rd ed., Elsevier, Amsterdam, 1965.

31. Atkins, A. S., Bridgwood, E. W., Davis, A. J., and Pooley, F. D., A study of the suppression of pyrite sulphur in coal froth flotation by *Thiobacillus ferrooxidans*, in *Coal Preparation*, Gordon and Breach, New York, 1987.

32. Instruction manual supplied with electrophoresis apparatus, Rank Brothers, Bottisham, Cambridge, 1984.

33. Lyklema, J., and Overbeek, J. T. G., On the interpretation of electrokinetic potentials, *J. Colloid Sci., 16*, 501-512 (1961).

34. Black, A. P., and Smith, A. L., Suggested method for calibration of Briggs microelectrophoresis cell, *J. Am. Water Works Assoc., 58*, Apr., 445-454 (1966).

35. Elgillani, D. H., and Fuerstenau, M. C., Mechanisms involved in cyanide depression of pyrite, *Trans. Am. Inst. Min. Metall. Pet. Eng., 241*, Dec., 437-445 (1968).

36. McGoran, C. J. M., Duncan, D. W., and Walden, C. C., *Can. J. Microbiol., 15*, 135 (1969).

37. Atkins, A. S., Studies on the oxidation of ferrous sulphides in the presence of bacteria, Ph.D. thesis, University College, Cardiff, Wales, 1976.

38. Capes, C. E., McIlhinney, A. E., Sirianni, A. F., and Puddington, E. I., *Can. Inst. Min. Metall. Bull., 66*, 3 (1986).

39. Atkins, A. S., Pooley, F. D., and Townsley, C. C., Comparative mineral sulphide leaching, *Process Biochem., 21*, 3 (1986).

40. Davis, A. J., and Atkins, A. S., A comparison between *Thiobacillus ferrooxidans* and biological by-products in the desulphurisation of coal fines in flotation, *Workshop for the Biological Treatment of Coal*, U.S. Department of Energy, Office of Fossil Energy, and the Bioprocessing Center, Idaho National Engineering Laboratory, Washington, D.C., July 1987.

41. Argaval, J. C., Gilberti, R. A., Irminger, P. F., Petrovic, L. F., and Sareen, S. S., *Min. Congr. J., 40*, Mar. (1975).

42. Miller, J. K., *Flotation of Pyrite from Coal*, Pilot Plant Study 7822, U.S. Bureau of Mines, Washington, D.C., 1973.

43. Torma, A. E., and Murr, L. E., *Desulphurisation of Coal by Microbial Leaching, Final Report on Project EMD 2-67-3319*, 1981, pp. 1–58.

44. Detz, C. M., and Barvinchak, G., Microbial desulphurisation of coal, *Min. Congr. J.*, July, 78–83, 86 (1979).

Microbe-Assisted Pyrite Removal from Hard Coal with Due Consideration of Ensuing Alterations of the Organic Coal Substance

ERNST BEIER   *DMT Fachhochschule Bergbau, Bochum, West Germany*

29.1   INTRODUCTION

This chapter deals with pyrite removal assisted by *Thiobacillus ferrooxidans*. As other reactions are also of some relevance in this context, a distinction is made between the following reaction types: (1) purely chemical reaction, (2) biochemical reaction, and (3) chemobiochemical reaction sequence. During our experiments on *purely chemical reaction*, no air-induced oxidation of the pyrite in a sterile medium had taken place at the end of 13 days [1]. Oxidation was brought about, however, by means of iron(III) salts under formation of either iron(II) sulfate and elementary sulfur:

$$FeS_2 + Fe_2(SO_4)_3 = 3FeSO_4 + 2S \tag{1}$$

or iron(II) sulfate and sulfuric acid:

$$FeS_2 + 7Fe_2(SO_4)_3 + 8H_2O = 15FeSO_4 + 8H_2SO_4 \tag{2}$$

As U. Zagberg reported in his thesis (1977, Fachhochschule Bergbau, Bochum), this occurred at equal rates in air and nitrogen. Reaction (1) is preferred in most cases. But even if the final product is that of equation (2), elementary sulfur is generated probably as an intermediate product which, however, need not be precipitated but rather, may remain as a colloid in solution. One drawback of the purely chemical reaction with iron(III) salts is that the oxidation potential of the solution drops very low, due to the formation of iron(II) salts. Theoretically, oxidation of these salts by the oxygen in air would be possible by increasing the pH value of the solution. In that case, however, the basic salts of the trivalent iron would be precipitated, which would then no longer be available for pyrite oxidation and moreover, might cause some congestion of the coal pores.

By "biochemical reaction" is understood a reaction where, by direct contact between pyrite crystals and bacteria and due to the catalytic effect of enzymes, oxidation of the pyrite is brought about, yielding, ultimately, iron(III) sulfate and sulfuric acid. A study implemented by Studiengesellschaft Kohlegewinnung Zweite Generation highlights the distribution of pores [2] in hard coals. As can be concluded from this study, a network of macropores (>0.5 μm) that is permeable by thiobacilli will normally be inadequate to make the pyrite accessible to bacteria.

By "chemobiochemical reaction sequence" should not be understood a biochemical reaction but rather a chemical reaction of the pyrite with iron(III) sulfate under the formation of iron(II) sulfate followed by a microbe-supported retro-oxidation of iron(II) sulfate to give iron(III) sulfate. The chemobiochemical reaction sequence eliminates the weak point of the purely chemical reaction, where for the reasons mentioned, no retro-oxidation of the bivalent iron takes place. Such retro-oxidation is brought about, even at low pH values, by *T. ferrooxidans*. The thiobacillus penetrates the pores to the degree that its size permits. On its way, whenever it comes into contact with pyrite crystals, it oxidizes them. For the rest, *T. ferrooxidans* represents the medium oxidizing iron(II) salts. Iron(III) salts diffuse into the micropores, which are inaccessible to *T. ferrooxidans*, oxidize the pyrite, and then diffuse jointly with the iron(II) ions generated from the pyrite toward the *T. ferrooxidans*, where they are reoxidized. Should colloid sulfur be generated at the pyrite crystal, that sulfur may migrate toward the *T. ferrooxidans* as well as being oxidized by it and giving sulfuric acid.

When studying pyrite removal from hard coals one should also consider oxidation of the organic coal substance. Since iron salts soluble in water function as catalysts [3] in coal oxidation, such a catalytic process necessarily also happens during pyrite removal. Theoretically, oxygen adsorption by coal from the air, CO_2 discharge into the air, or increase in oxygen content of the coal could be used as a unit of coal oxidation. However, due to the fact that all three parameters are difficult to identify during pyrite removal from coal, we selected different parameters to characterize oxidation (i.e., changes in the volatile content and in dilatation).

Pyrite removal by microbe attack is brought about both by suspension and by seepage washout. Beyer [4] notes that depending on the conditions of suspension washout, reactor sizes of total volume between 20 and 70,000 m^3 were calculated for removing 80 or 90% pyrite from an output of 1000 tons of coal per day. These volumes are on the order of magnitude of big-city gas tanks and could, of course, be distributed among several reactors (which, by the way, should be made of anticorrosive material). Another method is leaching by seepage on coal heaps, which requires much space but much less cost for apparatus. Whereas the problem with suspension leaching consists in maintaining the entire coal mass in suspension, leaching by seepage may give rise to complications by accumulating ultrafines and ensuing congestion of the biopulp seepage path or, alternatively, channel formation in the coal. To determine the scope of the latter phenomenon, experiments on the seepage speed of water through coal heaps and columns of different grain sizes were run. When trying to determine the influence of grain size on the seepage rate, one should bear in mind that some size degradation goes on while bacteria are decomposing the pyrite [1]. The experimental results were entered in the design and cost estimations of pyrite removal plants utilizing the seepage concept. Such calculations were carried out jointly with the UHDE engineering company in Dortmund.

29.2 EXPERIMENTATION AND RESULTS

29.2.1 Dependence of the Degree of Pyrite Removal on the Pyrite Content of Coals

In his work with Bergbau-Forschung, Essen, Guntermann [5] refers to 20 different coal types. The coals were subjected to coarse crushing and then ground in a disk mill to finer than 0.5 mm. The size fraction between 0.1 and 0.5 mm was screened off the mill discharge and used for the experiments on pyrite removal. All tests were carried out with *T. ferrooxidans* BF 219 of Bergbau-Forschung [6]. Twenty grams of each coal sample was placed in three 500-mL Erlenmeyer flasks and mixed with 150 mL each of a modified nutritive medium according to Leathen. Two of these preparations were inoculated with 15 mL each of a preliminary preparation (redox potential 630 mV, pH value 1.8), whereas the third preparation served to control sterility. Then the preparations were shaken at a frequency of 100 times per minute at 30°C for a period of 27 days. During the shaking tests pH values were monitored, and as soon as they had risen due to basic substances diffused from the coal, they were readjusted to 1.8 by the addition of sulfuric acid. Disulfide determination of both the fresh and the pyrite-free coals took place according to an internal Bergbau-Forschung procedure [7,8]. The decomposition of pyrite as determined aftrer 27 days of treatment was plotted against the initial pyrite content of the coal (Fig. 29.1). The fact that a regression line of a correlation coefficient of 0.954, definition of 91%, and residual standard deviation of 0.14 can be drawn through the points is indicative of the proportionality of the decomposed pyrite volume to the initial pyrite content of the coal or, in other words, of the independence of the degree of pyrite removal from the initial pyrite content.

The phenomenon may become more understandable by way of an example. If we have two spherical coal grains of identical size into which lead an identical number of pores, allowing the passage of iron ions but no bacteria, these iron ions will diffuse in a similar way into either of the grains. Suppose in the first coal grain that every fourth pore leads to pyrite, whereas in the second grain this is just every twelfth pore (pyrite of an identical type), In this case the pyrite of the first coal grain will be oxidized as fast as in the second one (i.e., the extent of pyrite removal is identical for both grains, although the first contains three times as much pyrite as the second). This applies, of course, on the provision that an adequate oxygen volume for retro-oxidation of the iron (II) salts formed during pyrite oxidation be available. At this point we will note one effect whose details will be dealt with later in the chapter. Whenever the iron (III) salts do not meet any pyrite in the pores to form iron (II) ions, the oxidation potential of the solution remains effective and leads to relatively strong oxidation of the organic coal substance.

The foregoing reconfirms a working hypothesis [1] established in 1985. Guntermann was able to back the earlier measured data as well (Fig. 29.1). When transferring the Guntermann data straight to a system of coordinates of identical scales for abscissa (pyrite content of the feed coal) and ordinate (pyrite decomposition), one obtains a line steepness of tan $\alpha = 0.675$. This means that during the 27-day test duration the coals lost 67.5% of their pyrite, on average. The author calculated a rate of advance of 3 μm/day [1] for the grain size considered by Guntermann. By "rate of advance" is to be understood the speed at which the pyrite reactants—thiobacilli or iron (II) ions—penetrate from outside into the coal grains to oxidize the pyrite. The calculation based on the foregoing

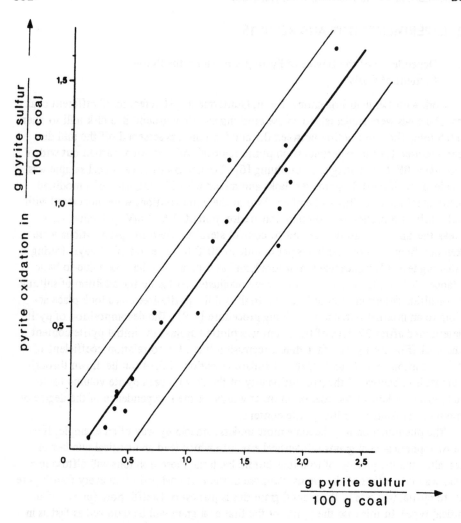

Fig. 29.1 Dependence of the depyritization from the content of pyrite sulfur of coals (27 days). (From Ref. 5.)

rate of advance yields a pyrite removal efficiency of 69.1% for the Guntermann experiments.

29.2.2 Pyrite Removal Efficiency as a Function of Pyrite Genesis

Guntermann [5] determined the origin of pyrite (syngenetic or epigenetic) by microscopy and associated the five coal types easiest to identify to any one of the categories. Contrary to expectations, the epigenetic pyrite (pyrite removal efficiency $\eta = 56\%$) was not more easily accessible to oxidation than was the syngenetic pyrite ($\eta = 65\%$). This may be attributed to the fact that normally the bacteria or iron(III) ions get to both of the pyrite types through similar pore systems. To determine the dependency of pyrite removal efficiency on the coal type, the coals were classified, after reflectance measurements and volatile analysis, into three groups: (1) gas flame coals (10 samples), (2) gas coals (4 samples), and (3) coals of less than 30% volatile matters (i.wf) (3 samples). Pyrite removal efficiencies (η) obtained were as follows:

η gas flame coals   57%
η gas coals   66%
η coals <30% volatiles   61%

Given the reduced number of coal samples, the results obtained lead only to the conclusion that there are apparently no conspicuous rank-dependent differences in the accessibility of coals to pyrite removal.

29.2.3 Dependence of Pyrite Removal Efficiency on Bulk Configuration and Grain Size

Prerequisite for an economic removal of pyrite by seepage is a sufficient seepage rate of the biopulp through the coal heap. Wuch [9] heaped coals of 1 to 2 mm and 0.5 to 15 mm grain sizes up in cone shapes to examine the cone stability with central sprinkling, distribution of the biopulp (in the model test: water) over the cone cross section, and changes in seepage velocity over time. Whereas a cone consisting of a narrow size range of coals (1 to 2 mm) was flattened considerably by the action of water sprinkling, the shape of a cone made up by coarser coals (0.5 to 15 mm) was essentially unchanged.

To measure the distribution of seepage liquid over the entire cone base, Wuch arranged nine funnels underneath that base (one in the center and four pairs at varying distances from the center). The upper circular face of the nine funnels equaled 19% of the surface of the coal base. This means that if distributed uniformly over the cone, 19% of the water should gather in the funnels and 81% should run off. As a matter of fact, a constant volume of 18% of the water collected in the funnels over the entire duration of the test. At the beginning of the tests much more water seeped through the central part of the cone than through its peripheral zones; after 100 h of seepage, however, the distribution had changed considerably in favor of the peripheral zones (Fig. 29.2). The latter phenomenon is attributed to the fact that for one thing, water impacting centrally flushes small grains preferably into the cavities around the cone axis, thus making water passage there difficult; second, visible channels form at the surface and invisible channels probably form below the surface, through which water gets into the peripheral zones without having earlier been involved in the seepage process.

To determine the dependency of seepage rate on coal size, tests were run in glass tubes of 15 mm diameter into which were placed 10-cm-high columns of one coal type but of different size ranges. The tubes were covered and a layer of glass wool was put underneath them. On top of the coal a 1.00-m-high water column was placed and kept constant. As shown on Fig. 29.3, the seepage rate did not change over time for the size ranges 63 to 90 μm and 90 to 125 μm, whereas it slowed down for the range 125 to 180 μm. This is probably due to the fact that the coal grains moved and thereby reduced the cavity volume, which meant more resistance to water passage. Since the laboratory is in the immediate neighborhood of a railroad track, it cannot be excluded that railroad traffic oscillations contributed to that movement of grains. Investigations into the dependency of seepage rate on time and bulk configuration will be pursued.

On Fig. 29.4 has been plotted the apparent seepage rate against grain size, which in this case is the arithmetic mean of the grain size limit involved. The apparent seepage rate was calculated from the volume flow $\dot{V}$ of the water and from the tube cross

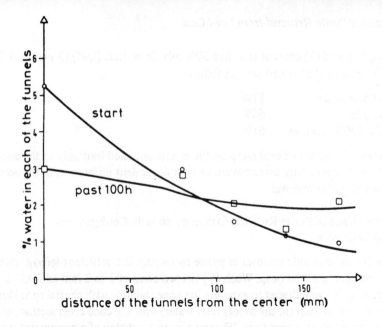

Fig. 29.2 Distribution of water about a cone of coal.

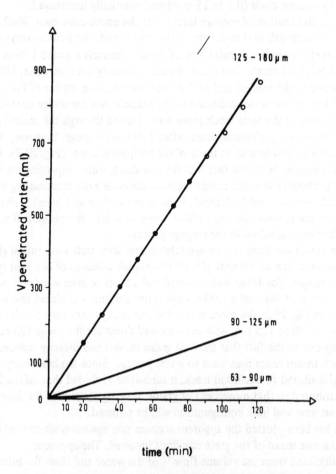

Fig. 29.3 Penetration of water through coals at different size of grains.

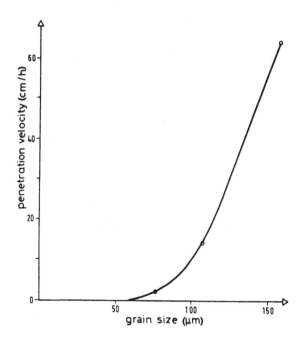

Fig. 29.4 Dependence of the penetration velocity from the grain size.

section. As the free cross section (i.e., not filled up by coal) is substantially smaller than the tube cross section, a much higher effective seepage speed results. For design calculation, however, the apparent seepage rate is more important than the effective rate. For the plant layout, these apparent seepage rates should enter the calculation as being the maximum attainable seepage rates.

From the foregoing rates can be derived the maximum admissible heights of coal heaps to be subjected to seepage. In the last section we discuss a circular heap containing 11,800 tons or 20,000 m³ of coal [10]. Treatment involved 240,000 m³/day of biopulp flow volume. When building up a coal heap its surface coverage (A) should be such as to ensure the aforementioned volume flow ($\dot{V}$) in due consideration of the seepage rate (v) developing:

$$\dot{V} = Av \qquad (3)$$

The size of the coal was assumed to be 0.5 mm [11,12]. In the absence of accurate data for 0.5 mm, the data available for the smaller size range in general were extrapolated, whereby we found an apparent seepage rate of 3 m/h. By inserting this rate into equation (3), we obtain a cross section of 3300 m² for the coal heap. Given a volume of 20,000 m³, we obtain for a heap of constant cross section 3300 m² an admissible height of 6 m. Due to the uncertainty inherent in extrapolation, we stayed with a height of 3 m.

When running their suspension experiments, Rinder and Beier [7] and Rostalski [8] found the pyrite removal efficiency to be strongly dependent on grain size. On the other hand, Beyer [13] did not observe such clear dependency during his percolation experiments. The results are summarized in Table 29.1 and some of them also in Fig. 29.5. Contrary to the indications in the original literature, the table does not give size ranges but rather, the arithmetic means from the biggest and smallest grains of one size range.

Table 29.1 Pyrite Removal Efficiency as a Function of Grain Size

	Pyrite removal efficiency, η (%)					
	Measured					
Grain size (mm)	27 days after suspension (Ref. 13)	15 days after suspension (Ref. 8)	70 days after percolation (Ref. 14)	Calculated: 15 days at v value:		
				3 μm/day	4 μm/day	6 μm/day
0.02	95.8	–	–	100.0	100.0	100.0
0.055	95.7	–	–	99.3	100.0	100.0
0.10	–	–	–	83.4	93.6	99.9
0.125	–	–	73	73.8	85.9	97.8
0.25	–	62.8	–	44.9	56.1	73.8
0.35	78.1	–	–	–	–	–
0.375	–	–	87	–	–	–
0.50	–	–	–	24.6	31.8	44.9
0.70	–	–	84	–	–	–
0.75	65.6	43.3	–	–	–	–
1.00	–	–	–	12.9	16.9	24.6
1.075	–	–	67	–	–	–
1.50	40.1	34.5	–	–	–	–
1.625	–	–	68	–	–	–
2.50	–	32.5	–	–	–	–
5.00	–	–	–	2.7	3.6	5.3
8.50	–	27.2	–	1.6	2.1	3.1
10.00	–	–	–	1.3	1.8	2.7

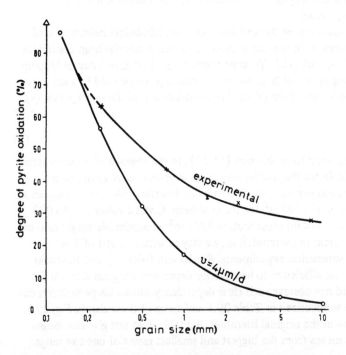

Fig. 29.5 Dependence of the depyritization of the grain size.

Moreover, the table contains the pyrite removal efficiencies, which were calculated under the assumption of spherical coal grains which do not exhibit pores permeable to thiobacilli. Furthermore, it was assumed that pyrite decomposition progresses at constant velocity v from the outside toward the grain center. As seen in the table, the extent of time-related pyrite removal is strongly dependent on grain size. To provide some preliminary figures on the progress of pyrite removal over time, neglecting temporarily the complex chemical and physical mechanisms involved, we computed only the extent of pyrite removal brought about within a given period provided that all the coal grains are spherical and that pyrite oxidation takes place solely from the exterior toward the interior at a constant rate of $v = ds/dt$. The figures computed were then compared with the measurements, and it turned out that there was one specific size range within which the figures computed for a decomposition rate of 4 μm/day correlated well with the data found by measurement. For the smaller size ranges, however, the measured values are lower, and for the coarser fractions they were higher (Fig. 29.5). These results induced us to derive the following hypothesis for the various size fractions (with the figures just giving a rough classification).

Size Range 0.1 to 1 mm

The biopulp is assumed to contain a sufficient number of bacilli to completely wrap the surface of all the grains present. Starting from a bacillar thickness of 1 μm and a grain diameter of 1 mm, we need a bacillar volume of 0.6% of the coal volume. The number of pores permeable to thiobacilli, referred to the average number of grains, is quite restricted, so it can be neglected. The pyrite within the grains is oxidized by iron (III) salts diffusing from their surface deeply into the pores. The job of the thiobacilli on the grain surface is, in addition to direct oxidation of the surface pyrite, to oxidize the iron (II) ions diffusing toward the exterior. The average diffusion rate is identical for all equivalent iron ions. Along with the progression of ions into the depth of the coal grain to meet the pyrite crystals, the depyritization rate will slow down. With increasing "traveling" distance the residence of the iron (III) ions on the coal substance will extend such that a steadily growing proportion of iron (III) ions oxidates the coal substance rather than the pyrite.

Size Range <0.1 mm

The assumption made for the 0.1 to 1 mm range of complete coverage of the grain surface by a bacillar layer no longer applies to very small grains. Emanating, as before, from a bacillar thickness of 1 μm, the required volume of bacilli to wrap up all the 8-μm-diameter coal grains would be equal to the coal volume. Even for grains of 20 μm diameter, we arrive at a bacillar volume equal to 33% of the coal volume. Thus complete coverage of the grain surface is logically impossible; nor will the aforesaid pyrite removal rate of $v = ds/dt$ be attained as is the case with coarser grains entirely covered by thiobacilli.

Size Range >1 mm

It was found by calculation that to wrap up spherical coal grains of 1 mm diameter by thiobacilli, biomass equaling approximately 0.4% of the coal mass is required. Let us assume that such a large biomass is also used for smaller and coarser grain sizes. While being inadequate to cover all the surface of smaller grains, one layer of biomass around the bigger grains would not absorb all of that mass. Nevertheless, as mentioned earlier, one should bear in mind that unlike the smaller ones, the coarser grains exhibit pores permeable to bacilli. It is just because of such pores that the bigger grains are subjected to more thorough pyrite removal than one would have expected from microbe attack exclusively from the outer surface.

Studies on this side and—as far as the author is informed—those carried out by other authors so far considered only the size range up to 10 mm. On the other hand, it is with fractions beyond 10 mm that cleats in the coal play a critical part during depyritization. Cleats are, indeed, spaces in which not only can microbes move and ions diffuse but which also function as passages (although at low speed) for the biopulp. Coarse grains exhibiting cleats therefore may behave similarly to smaller ones without cleats. From the preceding we see that probably (in contrast to general belief) the durations required for pyrite removal from cobbles are not by some orders of magnitude longer than those for ultrafine coals. It may even be that less surface is required to remove pyrite from cobbles than is needed for smalls since the former allow much faster biopulp seepage rates. Tests on cobbles will begin soon. If they are encouraging, the potential for biological pyrite removal will definitely improve.

Size Degradation During Depyritization

In addition to the grain size of feed coal, some attention should be given to size degradation during pyrite removal. At this juncture one should distinguish between mechanical degradation (abrasion) occurring during suspension leaching, on the one hand, and chemical degradation or decomposition, on the other hand, caused by the dissociation of pyrite from the grain and occurring with any and all methods of pyrite removal. Table 29.2 shows to what extent size degradation is brought about either by shaking the coal in a sterile nutritive solution or by thiobacilli cultures. Similar conditions were applied to samples 1–3, and samples 5 and 6. The coal (No. 5) from the Albert seam was so brittle that chemical grain decomposition, if any would have been masked by the unusually strong abrasive effect. In all the remaining cases, particularly for No. 4, chemical decomposition by the dissociation of pyrite was considerable. By the way, such decomposition (cracking) is, indeed, stronger than it appears from the results since the fragments that form, let us say, by degradation of the 0.5-mm range to half its size, remain in the initial size range of 0.5 to 0.2 mm (i.e., 0.5 to 0.1 mm).

Size reduction, while basically promoting pyrite removal, is not as beneficial as was expected because with the comminution of coarse coal grains, part of the bacteria-

Table 29.2 Size Reduction During Pyrite Removal from Shaken Coal

		Size Percentage (mm)					
				Treated coal			
		Feed coal		Sterile		By bacilli	
Sample	Seam	0.2–0.5	0.1–0.5	<0.2	<0.1	<0.2	0.1
1	Wilhelm	100		7		15	
2	Plassh. I	100		17		28	
3	Plassh. II	100		18		26	
4	Wilhelm	100		11		80	
5	Albert		100		72		70
6	Zollverein		100		2		19

Source: Ref. 13 (for samples 1, 2, 3), Ref. 7 (for sample 4), Ref. 5 (for samples 5, 6).

permeable pores is lost. To illustrate the phenomenon, let us compare a big lump of coal with a Swiss cheese run through by fermentation channels. These channels represent, so to speak, the macropores that allow bacteria to pass through a big coal lump. Once the cheese is cut into progressively smaller cubes, one will soon obtain cubes containing not channels but, at best, depressions left over from the channels. In analogy, with progressive reduction of a coal grain having pores larger than 0.5 μm, one will obtain an increasing number of grains no longer containing pores permeable to bacteria, not to mention that pores are weak points where grains tend to break most easily.

29.2.4 Dependence of the Pyrite and Sulfur Removal Efficiencies on the Biopulp pH Status

The microbes used for pyrite removal will attack pyritic sulfur exclusively; they are ineffective for organic sulfur. This means that a coal whose sulfur is 50% pyritic and organic in nature can have at best 50% of its sulfur removed by the action of thiobacilli. In many cases the expected equal rate of decrease was observed in pyritic and total sulfur [7,13-15]. The figures entered in Fig. 29.6 were taken from a study by Golomb [15], according to whom the decrease in pyritic and total sulfur correlated over the entire pH range examined, from 1.5 to 2.5. An important fact for assessing such a statement is that all of the coal samples were subjected to secondary treatment with hydrochloric acid

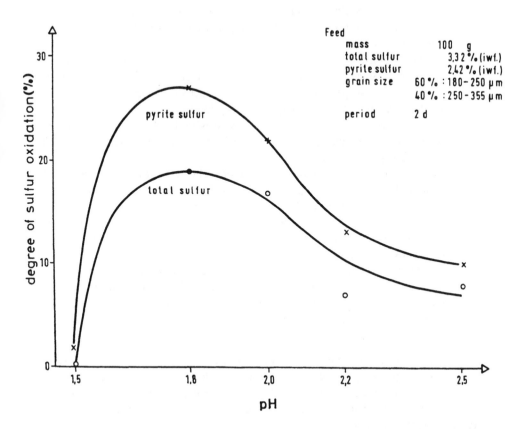

Fig. 29.6 Dependence of the degrees of depyritization and desulfurization of the pH value.

after pyrite removal and before sulfur analysis. The highest desulfurization efficiencies were attained at pH values between 1.7 and 2.0. In some cases the reduction of total sulfur was either less important than that of pyritic sulfur [7,13] or even increased [13]. This can be explained by the fact that the sulfur stemming either from pyrite or from the biopulp precipitates in a modified form on the coal [7,13].

Beyer et al. [7] measured 0.1% (by mass) elementary sulfur for an initial pH value of 3.6, and 0.7% (by mass) for a pH of 1.9. This sulfur definitely originates from pyrite. Either pyrite or biopulp is a susceptible contributor of sulfate, which precipitates on the coal as basic iron salts. The increase in total sulfur of the coal during pyrite removal is explained by sulfur being released from solution [13]. Such sulfur releases at pH values beyond 2.0 (see Ref. 1, p. 598) are the reason for impaired pyrite removal efficiency during percolation experiments, due to pore congestion [15].

29.2.5 Oxidation of Organic Coal Matter During Pyrite Removal

Open-air storage of any type of coal and many other materials causes their uninterrupted oxidation [16]. As oxidation is catalyzed by iron salts soluble in water [17], it must not be neglected during pyrite removal. Oxygen absorption by the coal, changes in oxygen content, and finally, the release of carbon oxides are, however, difficult to identify for the following reasons:

1. Air oxygen is absorbed not only by carbonaceous matter, but also by pyrite (this is the reason for aeration of the biopulp) and by dead bacteria. Oxygen absorption by the coal could be calculated simply by deducting the oxygen absorbed by pyrite and dead bacteria from the oxygen volume released by the aeration air. Since the oxygen loss from aeration air during laboratory tests according to the suspension method is in an order of magnitude of just some tenths of a percent and the oxygen absorption by pyrite and dead bacteria escapes precise identification, it is impossible with the given state of the art to determine oxygen absorption by means of the carbonaceous matter to anything like an acceptable accuracy.

2. The CO_2 originating from the oxidation of coal escapes from complete experimental identification since part of it is assimilated immediately by the bacteria.

3. A rising oxygen content in the coal alone is not appropriate evidence of oxidation having taken place, as it may be that the coal released the majority of oxygen that it had taken in, in the form of carbon oxides (in the case of open-air storage, up to 89% [16]) and water. But even if we determined how much oxygen was absorbed by the coal, we would still have to break such oxygen absorption down into that going into the organic coal substance and/or into the mineral constituents. The latter, in turn, is difficult to determine since during pyrite removal, some of the mineral constituents become dissociated from coal either directly or after oxidation [1].

As it was too difficult to determine the parameters above, which, moreover, were of low purity, we stopped trying. Basing on previous studies [18] having revealed some change in volatile content of the coals and (for coking coals) in dilatation behavior after open-air storage, we used those parameters to check whether the coal substance had altered during pyrite removal.

Guntermann [5] determined the volatiles of 16 coals prior to and after a 27-day pyrite removal treatment. Alterations were found to be between +0.2% for high-rank coal and –3.4% for low-rank coal. The results are plotted in Fig. 29.7. Guntermann made a regressive linear calculation of the values yielded after extended open-air storage [18], referring them to 27 days; the results have also been plotted on the diagram. The regression lines for normal storage and pyrite removal show roughly identical inclinations, so that one should subject highly similar mechanisms.

The fact that the absolute alterations during pyrite removal of high-volatile coals are more than 10 times as important is during open-air storage can certainly be attributed to the catalytic effects occurring with pyrite removal, but also to the linear conversion to short-term values of values associated with extended storage periods. The fact that the oxidation speed of coal stored in an open space declines over time was neglected for this study [3]. Beside volatiles, changes in dilatation were pursued as an alternative approach to define coal oxidation. Needless to say, that is applicable only to coals exhibiting some dilatation.

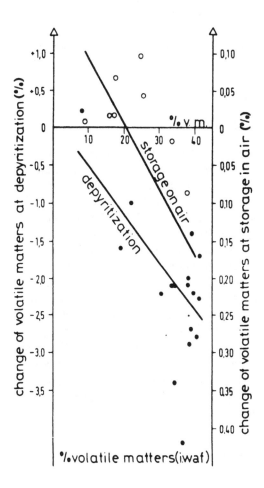

Fig. 29.7 Change of volatile matters of coals at depyritization and at storage on air in 27 days.

In earlier experiments a decreasing dilatation behavior for coals stored between 50 days and 4 years between 20 and 60°C at relative humidity contents from 0 to 97% and oxygen contents from 21 to 100% by volume has always been observed [18]. Guntermann [5], during his pyrite removal experiments run at Bergbau-Forschung in Essen, found a decrease in dilatation from 17% initially toward zero after 27 days. This result is also indicative of considerable oxidation of the coal during pyrite removal. The critical issue of alterations of caloric power inherent with depyritization of coal is being studied.

29.2.6 Tracing the Mineral Matter Dissociated from the Coal

During the symposion held in Columbus on high-sulfur coals the question had been raised as to where the mineral matter coming free from the coals during pyrite removal would go [1]. Beyer et al. [7] determined reductions in the ash content of coals up to 7.5% (by mass); Guntermann [5] found an average ash reduction of 2.5% (by mass) for 22 coal samples, with the smallest reduction of 0.1% (by mass) for a low-ash coal (initial ash content 2.5% by mass) and a maximum of 11.4% (by mass) for middlings of initially 42.7% (by mass) (wf). Provided that the mineral matter released by the coal is virtually dissolved in the biopulp, it passes along with the waste slurry into the settling tanks either to be precipitated or to flow off with the clean water. As the volumes involved are considerable, one should pay more attention to them in the future. It is certain, though, that part of the sulfuric acid formed during pyrite removal is neutralized by the mineral matter so that less lime is needed in the settling tank.

29.2.7 Design of Plants from the Biotechnical Pyrite Removal from Hard Coals

Biotechnical pyrite removal is conceived either as suspension leaching or as bulk leaching of heaped-up material. While it is true that suspension leaching does the job faster than bulk leaching, the author believes that the equipment expenditure for the former is unjustifiably high. He therefore designed, jointly with F. Keil, of UHDE, Dortmund, and D. Koch [11] and P. Schaper [12], both diploma candidates of the WBK Mining College, two plants for seepage leaching. One of the plants has been laid out for leaching in a coal ditch [11] and the other for heap leaching [12]. Figure 29.8 shows the block diagram applying to both plants. Figure 29.9 shows ditch leaching and Fig. 29.10 heap leaching. Biopulp treatment will be similar for both types of leaching. As is visible in Fig. 29.8, the circulating biopulp is sprinkled over the coal, then collected at the base of the coal heap, aerated, and recycled for sprinkling, and so on. As the waste pulp becomes enriched in iron salts and sulfuric acid, part of it is skimmed off continuously and replaced by fresh water which had previously been used for elution of the pyrite-free coal. Then lime milk is used to precipitate a calcium sulfate/iron(III) hydroxide mixture from the waste pulp. As the representation of details would go beyond the scope of this chapter, only the main data regarding the two design jobs will be given; dimensions and costs will be indicated only for the cheaper heap leaching alternative.

Prevailing Conditions

Power plant capacity	50 MW$_{el}$
Coal	
Calorific value (gross)	29,310 kJ/kg
Grain size	0.5 mm
Pyritic sulfur	0.6% (by mass)

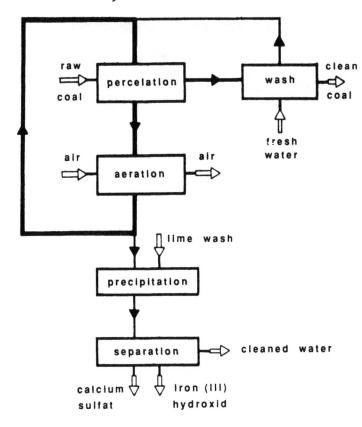

Fig. 29.8 Depyritization of coal during percolation. (From Ref. 19.)

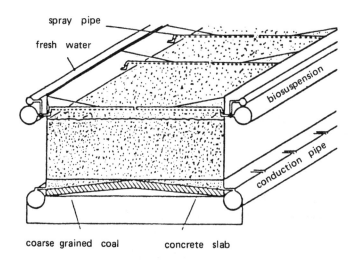

Fig. 29.9 Percolation of coal in a ditch.

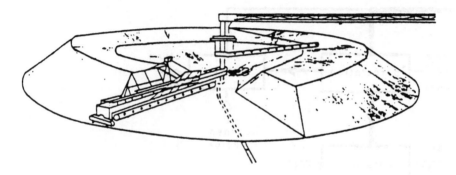

Fig. 29.10   Percolation of coal on a ring heap.

Pyrite removal efficiency	80%
Pyrite oxidation	
By O_2 in biopulp (direct)	50%
By Fe (III) in biopulp	50%
Consumption of dissolved O_2	
For pyrite oxidation	75%
For coal oxidation	25%
Iron content of the waste pulp	$4 \ kg/m^3$

Calculated Parameters

Coal feed	422 t/day
Time requirement for pyrite removal	26 days
Time requirement for coal on storage heap	28 days
Coal mass on storage heap	11,800 tons
Coal volume ($\rho = 0.6 \ ton/m^3$)	$20,000 \ m^3$
Circular coal heap	
Diameter	100 m
Height	3 m
Volume flows	
Biopulp	$240,000 \ m^3$/day
Waste pulp	$360 \ m^3$/day
Fresh water	$3000 \ m^3$/day
Air	$70,000 \ m^3$/day
Space requirement of the entire plant	$30,000 \ m^2$
(For comparison: for ditch leaching)	$47,000 \ m^2$)

Costs

Capital requirement	
Installation of storage facility (coal heap),	
including totality of transport facilities	10,600,000 DM
Aeration and precipitation plant	3,200,000 DM
	13,800,000 DM

Operational Costs

Electric current	1,110,000 DM/yr
Ca(OH)$_2$	210,000 DM/yr
Fresh water	280,000 DM/yr
Wastewaster	280,000 DM/yr
Wages (eight workers)	540,000 DM/yr
Repairs, etc.	760,000 DM/yr
Insurances	290,000 DM/yr
	3,470,000 DM/yr

Assuming one-third of the capital requirements to be raised by proper funds, linear depreciation over 8 years, repayment of loans within 3 years, and interests on loans of 8%, the average cost of pyrite removal over a period of 8 years will amount to DM 54 per ton of coal. This figure does not take into account any alteration of the coal mass and quality during pyrite removal, nor that costs may have to be incurred to get rid of mineral matter which, beside pyrite, might have been detached from the coal and gotten into the biopulp.

For comparison we quote the costs of the dry-type absorption method [20], which has been designed for flue gas desulfurization using lime and water to be fed separately to the column. The process has an SO$_2$ emission-reducing effect similar to that of pyrite removal. The costs of the latter process depend strongly on the sulfur content of the coal and amount to 70 DM per ton of coal on average, which is slightly above pyrite removal costs. When comparing the alternative it should be borne in mind that dry-type absorption is by now an introduced technology, whereas calculations related to pyrite removal still represent laboratory experiments.

29.3 CONCLUSIONS

The main part in pyrite oxidation in the presence of *T. ferrooxidans* is probably played by a chemobiochemical reaction sequence in which iron (III) salts oxidize the pyrite, with iron (II) salts formed all the while being subjected to retro-oxidation by the oxygen in the air, via the *T. ferrooxidans*. The pyrite removal efficiencies attained within a given period for different coals depend neither on their pyrite content nor their rank nor on the syn- or epigenesis of the pyrite. For those grains which due to their fineness probably no longer contain pores permeable to bacteria, pyrite removal efficiencies correlate well with experimental results on the assumption of a 3- or 5-μm/day advance rate of pyrite oxidation from the outside toward the center. Coarser grains will always contain pores permeable to bacteria. Thiobacilli will penetrate the pores and be available in the center of grains as oxidation catalysts for the iron (II) ions. Oxidation of the iron (II) ions therefore does not require their being diffused from the grains as is the case with ultrafine grains. Initial investigations have been indicative of the likelihood of biologic pyrite removal also being applicable to relatively coarse coals (cobbles) as they exhibit cleats and pore accessible to the thiobacilli.

The oxygen needed for pyrite oxidation is supplied to the coal via the biopulp. The speed with which the biopulp seeps through the coal depends strongly on the grain size. A seepage rate of 0.64 m/h was measured, for example, for the size range between 125 and 180 μm. To ensure a preestablished pyrite removal efficiency in a stated period, one needs a minimum pulp flow, which, however, can be attained only if the coal is not heaped up beyond an admissible height.

During pyrite removal the organic coal matter is oxidized as well. Lacking accuracy of oxygen balances prompted investigations into changes in the quality of the coals. As with open-air storage, the volatile content of low-rank coals was reduced during a 27-day pyrite removal treatment (to a maximum of 3.5% by mass), whereas for high-rank coals it went up to (0.2% by mass). The dilatation behavior of coking coal went down from 17% initially to zero. Investigations into possible changes in the caloric value of depyritized coals have not been completed. Pyrite removal will dissociate other mineral matter from the coal. For 22 different coals "de-ashing" between 0.1 and 11.4% (by mass), 2.5% (by mass) on average, was observed. The mineral matter is passed along with the waste pulp into either the settling tank or the clean water. The exact whereabouts still need to be delineated.

Two designs for pyrite removal plants associated with a 50-MW_{el} power station were completed, one for leaching of coal heaps and the other for ditch leaching. Leaching of heaped-up coal turned out to be more economical. The design utilized a heap diameter of about 100 m at a 3-m bulk height. The costs involved amounted to DM 54 per ton of coal, which roughly equals the costs of flue gas desulfurization of equivalent sulfur removal efficiency. The work will be carried on by extensive investigations into the seepage of biopulp through heaped-up coal of various grain sizes, the diffusion of iron ions within the coal grains, and pyrite removal from coarse coals.

ACKNOWLEDGMENTS

The work for the theses by A. Guntermann und U. Rostalski mentioned in this paper was carried out at Bergbau-Forschung in Essen, the work by D. Koch and P. Schaper, at UHDE in Dortmund. The work by Guntermann and Rostalski was included in the project BMFT-PbE (038 4220), entitled "Desulphurization of Coals by Microbes," sponsored by the Federal Ministry of Research and Technology. J. Klein, H. G. Ebner, and especially, M. Beyer discussed the task and the implementation of relevant work with diploma candidates A. Guntermann and U. Rostalski and the author, and supervised progress. Sincere appreciation is extended to them for their cooperation. Thanks are also extended to F. Keil of UHDE, who acted as a consultant to diploma candidates D. Koch and P. Schaper.

REFERENCES

1. Beier, E., Removal of pyrite from coal using bacteria, in *Proceedings of the First International Conference on Processing and Utilization of High Sulfur Coals*, Columbus, Ohio, 1985.
2. *Final Report on the Project BMFT 03 E-6153-A (Properties of Hard Coals)*, Studiengesellschaft Kohlegewinnung Zweite Generation, Essen, West Germany, Nov. 1984, p. 524.
3. Beier, E., Gas transfer of hard coals and other materials during several decades of open-air storage, *Erdoel Kohle*, *38*(3), 127–129 (1985).
4. Beyer, M., *Microbial Desulphurization of Coals*, Bergbau-Forschung, Bochum, West Germany, Mar. 1987.
5. Guntermann, A., Investigations into the suitability of different coals for pyrite re-

moval by microbes and into the oxidation of the carbonaceous matter during pyrite removal, thesis, WBK Fachhochschule Bergbau, Bochum, 1985.

6. Beyer, M., Ebner, H. G., and Klein, J., Bacterial desulfurization of German hard coal, *6th International Symposium on Biohydrometallurgy*, Vancouver, British Columbia, Canada, Aug. 21-24, 1985.

7. Beyer, M., Ebner, H. G., Assenmacher, H., and Frigge, J., Elemental sulphur in microbiologically desulphurised coals, *Fuel, 66*, Apr., 551-555 (1987).

8. Rostalski, U., Investigations into the chemical and microbiological pyrite oxidation of coals in the percolator, thesis, WBK Fachhochschule Bergbau, Bochum, West Germany, 1987.

9. Wuch, G., Model tests on pyrite removal from heaped-up coal, thesis, WBK Fachhochschule Bergbau, Bochum, West Germany, 1979.

10. Ruhrkohle manual, Essen, West Germany, 1984, p. 264.

11. Koch, D., Design of a plant for the biotechnical pyrite removal from hard coals (ditch leaching), thesis, WBK Fachhochschule Bergbau, Bochum, West Germany, 1986.

12. Schaper, P., Design of a plant for the biotechnical pyrite removal from hard coals (leaching of a coal heap), thesis, WBK Fachhochschule Bergbau, Bochum, West Germany, 1986.

13. Rinder, G., and Beier, E., Microbiologic pyrite removal from coals in a suspension, *Erdoel Kohle, 36*(4), 170-174 (1983).

14. Beyer, M., Microbial removal of pyrite from coal using a percolation bioreactor, *Biotechnol. Lett., 9*(1), 19-24 (1987).

15. Golomb, M., Oxidation of bivalent iron by *Thiobacillus ferrooxidans* as a function of the pH value of the nutritive solution, thesis, WBK Fachhochschule Bergbau, Bochum, West Germany, 1975-1976.

16. Beier, E., The incidence of moisture, iron salts and microorganisms on atmospheric oxidation of coal pyrite, *Glueckauf Forschungsh., 34*(1), 24-32 (1973).

17. Beier, E., Ferrous compounds as catalysts of coal oxidation, *Bergbau, 21*(4), 81-88 (1973).

18. Beier, E., The oxidation of hard coal in the open air, *Mitt. Westfaelische Berggewerkschaftskasse, 22*, Nov., 1-105 (1962).

19. Beier, E., Desulphurization process for coal-fueled power stations, *Glueckauf Forschungsh., 42*(6), 275-280 (1981).

20. Schultess, W., Coal-based firing systems between 1 and 50 MW, *Energie, 37*, 17-23 (1985).

30

Transformations of Dibenzothiophene by Axenic Cultures of *Sulfolobus acidocaldarius* and Other Bacteria: A Critique

STEVEN KRAWIEC *Lehigh University, Bethlehem, Pennsylvania*

30.1 INTRODUCTION

The principle of microbial infallibility states that there exist no naturally occurring organic compound that is not degraded by some microorganism under suitable conditions [1] . The principle appears to assert that specific microorganisms can degrade specific parts of organic structures. Should this claim be true, exposure of properly prepared coal to a judicious selection of microorganisms could yield a modified product with qualitites more suitable for its utilization as a commercial source of energy.

30.2 THE CARBON CYCLE AND ITS LIMITS

Autotrophs shunt carbon from the inorganic realm to the organic realm, and heterotrophs reverse the process (Fig. 30.1). The autotroph converts an inorganic substrate, CO_2, to organic products collectively represented by the general formula $(CH_2O)_n$. This material is the generic substance of living matter. Because autotrophs generate organic material, they are commonly called synthesizers or producers. Heterotrophs, in a contrasting process called mineralization, convert $(CH_2O)_n$ to CO_2. Mineralization is the antithesis of carbon fixation and heterotrophs are often called degraders or consumers. These roles in the carbon cycle are so basic that their significance is sometimes exaggerated. While any component of the pool of inorganic carbon might ultimately be transformed into organic carbon, not all forms of inorganic carbon are immediately susceptible to biological transformation. Indeed, the vast bulk of inorganic carbons must first be transformed by some chemical or physical processes. Clearly, such materials as petrological carbonates and graphites are not directly substrates for biological reactions. Thus the microbiological aspects of the carbon cycle are confined solely to CO_2 as an inorganic substrate and, presumably, any organic substrate.

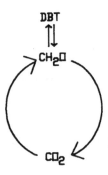

Fig. 30.1 Carbon cycle. Autotrophs "fix" CO_2 into organic matter, $[CH_2O]_n$; hetero-trophs oxidize organic matter to CO_2. DBT, a refractory compound, may or may not be in a biological equilibrium with organic matter in general.

30.2.1 Infallibility and Recalcitrance

The principle of microbial infallibility states that "no naturally occurring *organic* compound exists which is not degradable by some microorganism" [1]. This assertion appears crafted to eliminate obvious confounding circumstances. Plastics (and other xenobiotics), for example, are excluded because their occurrence is not "natural." While inorganic carbons other than CO_2 and synthetic organics are readily eliminated from consideration, there remain organic compounds that are resistant to microbiological transformation. Some examples of naturally occurring organic materials that have a remarkable persistence are the cortex of bacterial endospores, chitin, porphyrins, terpenoid resins, and acacia wood [2]. Such materials which are impervious to microbiological degradation are said to be "recalcitrant." But recalcitrance may merely represent a quantitatively slower transformation rather than a wholly different state in which no microbiological processing whatsoever occurs. If recalcitrance were a qualitatively distinct phenomenon and the organic compounds that were recalcitrant were completely inert to microbial transformation, recalcitrant compounds would constitute a "carbon trap." During each stoichiometric turn of the carbon cycle, some portion of the organic constituents would be entrapped in forms that were biologically unreactive. Because such compounds were not degraded, they would, if continuously synthesized, accumulate, and in the course of time, would become an ever-increasing proportion of all extant organic molecules (Fig. 30.2). An obvious question for microbiologists who are interested in coal processing is whether coal is properly considered a recalcitrant compound.

30.2.2 Coal: A Natural Substrate

Organic molecules are commonly considered the carbon compounds of living beings. These are the "naturally occurring" compounds that are typically perceived as the substrates of consumers or degraders. Interpretations of what might be the organic matrix of coal [3] depict structures that are vastly more complex than typical biological macromolecules, which are linear polymers of a limited array of monomeric units. Also, the organic sulfur contents of coals [4-7] are often much higher than the sulfur contents of biomass [8].

Coals are clearly not conventional organic compounds that readily enter the carbon cycle. They may be recalcitrant molecules. Alternatively, they may be molecules which

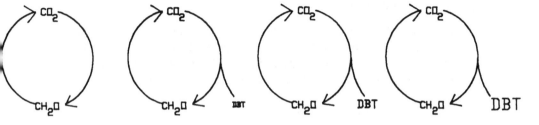

Fig. 30.2 Carbon trap. If an organic compound (e.g., DBT) is synthesized by organisms but not degraded by some organism(s), the compound will accumulate as a consequence of continuing activity in the carbon cycle.

were first produced by "synthesizers" and which were, subsequent to their generation, transformed by nonbiological processes. Heat, pressure, sulfurous degassing of the earth, and the presence of sulfates in the water column, among other conditions, may have contributed to conversions that protect coals from present-day microbiological degradative processes.

30.2.3 Additional Constraining Factors

Another provision of the principle of microbial infallibility is that organic molecules function as heterotrophic substrates under "suitable" conditions. The refractory nature of recalcitrant molecules may not be some special feature inherent in the molecule; instead, the persistence may represent environmental conditions. As a consequence, the status of being recalcitrant became an ambiguous one: It is not known whether a compound is genuinely refractory or whether the appropriate environment for degradation has not been established.

Even with all these limiting conditions presented, the expectations for heterotrophic performance in the carbon cycle are often too great. Mineralization of a complex organic molecule may be mediated by a consortium of organisms functioning in sequence or synergistically. Furthermore, the organisms that participate in the mineralization process can be quite varied, including multicellular organisms as well as unicellular ones, and eukaryotes as well as prokaryotes. Fungi, which are eukaryotic and often coenocytic—a variant form of multicellularity, constitute a kingdom of microbes. Fungi are typically powerful degraders. Another microbial kingdom, one very distinct from fungi, is that of the bacteria. This group of microbes is notable for its metabolic versatility. Collectively, eukaryotic microbes and prokaryotic microbes may have the capacity to mineralize most organic substrates. Determining exactly which microbe mediates what transformation during mineralization is an important activity in understanding completely the events associated with the carbon cycle.

30.3 SULFUROUS MOITIES IN COAL

Coal is derived from fossilized vegetable matter that lived predominantly in the Carboniferous period (and to lesser extents in Permian, Jurassic, Cretaceous, and Tertiary periods). Its composition may reflect the structure at the time of its deposition, or the structure may reflect any subsequent modification.

30.3.1 Inorganic Sulfurous Components

Inorganic sulfur is often prevalent in coal in the form of pyrite or iron disulfide (FeS_2) and marcasite [4-7]. The latter is a mineral which shares the same composition as pyrites but which differs in its crystalline organization. Oxidation of the former by microorganisms is well established and will not be considered in this presentation.

30.3.2 Organic Sulfurous Components

The abundance of organic sulfur in coal is greater than the abundance of sulfur in extant biomass. Characteristically, sulfur in the organic matter of contemporary organisms is less than 1% by weight [8]. While organic sulfur is more prevalent in coal than in present-day organisms, some of the forms in which it appears in coal are quite conventional. Organic sulfur occurs in coal in three forms [4-7]: sulfhydryls (–SH; or mercaptans), moieties that are also present in proteins, coenzymes, enzyme inhibitors, and vitamins; sulfides (–S–) and disulfides (–S–S–), linkages that are again present in proteins and vitamins; and thiophenes (C_4R_4S). The occurrence of thiophenes is limited; they were first recognized (in 1883) as coal tar derivatives. In addition to being found in fossilized materials, the natural occurrence of thiophenes is restricted to some plant pigments. Because sulfhydryls, sulfides, and disulfides are commonplace, it is not surprising that they readily enter into enzymatic reactions. The obverse of this premise implies that thiophenes would be less susceptible to enzymatic transformation. Thiophenes, in a sense, are properly regarded as the "tough nut to crack" among the organic sulfurous forms of coal.

30.3.3 DBT: A Model Compound

Dibenzothiophene is often used as a model compound for the thiophenic entities in coal. At the same time, it should be noted that there is negligible, if any, dibenzothiophene per se in coal. (Were dibenzothiophene present in coal as simple, free molecules, it could be flushed from the coal matrix with a hot organic solvent and much of the commercial concern about removal of organic sulfurs from coal would be resolved.) Thiophenes exist in coal as derivatized benzothiophenes and derivatized dibenzothiophenes. This circumstance is important in understanding the adequacy of dibenzothiophene as a model compound. Only the thiophene ring is representative of the organization of substituted benzothiophenes and dibenzothiophenes in coal; the benzenoid components are not. Accordingly, degradations of dibenzothiophenes that involve the peripheral parts of the molecule are irrelevant to consumption of the derivatized structures in coal. Furthermore, such degradations typically do not proceed to the thiophene ring, where desulfurization would occur.

30.3.4 Vehicles

For dibenzothiophene to be transformed by a microorganism, the microorganism must encounter the molecule. This obvious condition is simple in conception; however, it is a confounding circumstance in experimentation. A critical feature of experiments with model compounds is how the molecule is presented or delivered in the experimental system. Commonly, dibenzothiophene is dispersed in the aqueous medium as a heterogeneous suspension of "needlelike" crystals or dissolved in an immiscible light oil or paraffin [9]. In the latter circumstance, the organic solvent is sometimes called a "vehicle." With vigorous agitation, the vehicle bearing dibenzothiophene and the aqueous medium can form an emulsion. An alternative delivery system is to condense dibenzothiophene onto an inert surface (e.g., glass beads) [10] and to add the glass beads to the medium that supports growth of the bacterium. Whatever delivery system is employed, it is important to relate the method of delivery to the manner in which a bacterium might encounter thiophenes in coal. Such is done infrequently. Dibenzothiophene may, to some extent, be an adequate model compound with respect to its chemical structure. It does not appear to be a meaningful physical model.

30.4 INTERESTS IN DESULFURIZATION

Selective modification of the coal matrix, specifically desulfurization of thiophene components, by bacteria is not merely a commercially valuable prospect; it is an intriguing *biological* prospect as well. Such a transformation bespeaks an understanding of a phenomenon that is sufficiently complete to allow willful alteration of an enormously complex organic substrate. The manipulation and commercial gain is not so impressive; the understanding that emerges is.

For the transformation to be commercially valuable, specific biological features are desirable. The process should be selective or limited. In particular, only covalent bonds connecting sulfur atoms to the coal matrix should be severed. Disruption of covalent bonds between carbons, carbon and oxygen, or carbon and hydrogen would reduce the energy value of coal. Furthermore, the process should be regulated or controlled. Such features may allow augmentation of the biological activity and, in turn, maximization of the desulfurization.

During combustion of coal, the sulfurous components are oxidized along with the carbonaceous components. The former are released as oxides which precipitate as acids that are damaging to the environment. This environmental burden has been deemed unacceptable; thus coals must be desulfurized prior to combustion, during combustion, or afterward. Physical and chemical desulfurization techniques are presently prohibitively expensive. The appeal of microbial desulfurization is that the activity may not be costly. Unlike physical and chemical desulfurizations, microbial desulfurizations are not energy intensive; they do not require elevated pressures or extremely high temperatures. The aspects of microbial desulfurizations that presently contribute to increased cost are the long residence time for the biological process, the anticipated volume of the reactions, and the limited extent of desulfurization. As the biological understanding of desulfurization activities increases, the desired features of the commercial process can be optimized.

30.5 HETEROTROPHIC TRANSFORMATIONS OF THIOPHENES

Organisms abound that can transform dibenzothiophene (DBT) [11]. The truth of this claim can be adequately demonstrated with a technique that was designed to exhibit the microbial reactivity of compounds that are insoluble in water [12]. The procedure is as follows. Bacteria from a suitable environment (e.g., a coal storage site) are distributed onto an agar surface and allowed to establish extremely minute colonies. Then an aerosol of DBT in a volatile organic solvent is sprayed onto the surface of the agar. The solvent evaporates and the DBT precipitates onto the agar surface. Because DBT is insoluble on the surface of the aqueous gel, it appears as a "frost" or translucent film. If the organisms in the developing bacterial colony have the capacity to transform DBT from a hydrophobic substrate to a more polar, hydrophilic product, zones of clearing will appear around the colony. Some of the degradation products of DBT have brilliant colors; accordingly, not only are the rings clear, but some are brightly colored [11].

30.5.1 Transformations Not Involving Desulfurization

Organisms that carry out such transformations are varied but the predominant types are free-living, aerobic, gram-negative rods commonly found in soils. Pseudomonads and related organisms are especially prevalent. Pseudomonads are known specifically for their capacity to metabolize an enormous variety of low-molecular-weight compounds [13]. For example, *Pseudomonas cepacia* has the capacity to use any one of more than 90 distinct compounds as the sole source of carbon. Some of the substrates are strinkingly different from simple, straight-chain carbohydrates and amino acids. Extrachromosomal genetic elements sometimes confer the peculiarly extensive ability of pseudomonads to degrade various low-molecular-weight, heterocyclic, organic substrates. The genetic determinants specifying the enzymatic machinery for the degradation of toluene, camphor, naphthalene, *n*-octane, salicylate, and xylene by *P. putida* reside on distinct plasmids (i.e., the TOL, CAM, NAH, OCT, SAL, and XYL plasmids, respectively [14].

Microbial Identities

Pseudomonads that have been specifically identified as having the capacity to degrade DBT or benzothiophene are *P. abikonensis* [15], *P. jianii* [15], *P. aeruginosa* [16,17], *P. alcaligenes* [17], *P. stutzeri* [17], and *P. putida* [17]. The first two species do not appear on the current Approved Lists of Bacterial Names [18], meaning that the organisms do not satisfy the criterion of having been "adequately described." Among the organisms that have a sufficient description, all fall into a common taxonomic group established by rRNA hybridization [13]. Notably, the *P. alcaligines* and *P. putida* strains harbor a 55-megadalton plasmid which also confers the capacity to oxidize naphthalene and salicylate [20]. Organisms with a Dbt⁻ phenotype lack the plasmid and have no capacity to oxidize naphthalene or salicylate [20].

　　　Bacteria other than pseudomonads are known to share the metabolic capacity that allows partial consumption of DBT. Included among the other genera are *Acinetobacter* [17] (essentially ellipsoid pseudomonads), *Beijerinckia* [20], and *Rhizobia* [17]. The last two genera are also gram negative, aerobic, and (more or less) rod shaped; the former is capable of "free-living" N_2 fixation, while the latter performs N_2 fixation most often in conjunction with leguminous plants. Also, the last two genera are present in families that are closely related to the Pseudomonadaceae. *Beijerinckia* are included among "other bacteria" which are assigned to no established family; nonetheless, they are classified in *Bergey's Manual of Systematic Bacteriology* as being in the "gram-negative, aerobic rods and cocci" along with pseudomonads, *Rhizobia*, and *Acinetobacter*.

Biochemical Pathway of Transformation

Independent investigations have, collectively, established a pathway by which DBT is degraded. The sequence of conversions is depicted in Fig. 30.3. Dibenzothiophene is oxidized to a dihydroxy, dihydro derivative, which, in turn, is converted to a dihydroxy-dibenzothiophene [20]. Subsequently, an interdiol cleavage generates a series of water-soluble, colored reactants. A principal intermediate is 4-[2-(3-hydroxy)thionaphthenyl]-2-oxo-3-butenoic acid [20,21], which is orange in visible light. Also produced is *trans*-4-[2-(3-hydroxy)thionaphthenyl]-2-oxo-3-butenoic acid [20,21], which is red in color.

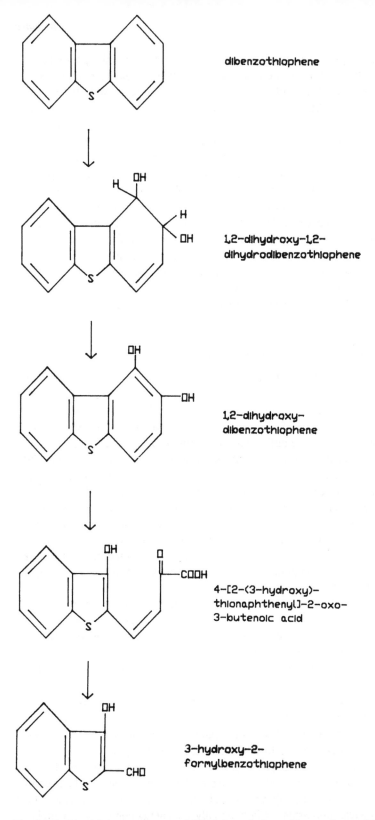

dibenzothiophene

1,2-dihydroxy-1,2-
dihydrodibenzothiophene

1,2-dihydroxy-
dibenzothiophene

4-[2-(3-hydroxy)-
thionaphthenyl]-2-oxo-
3-butenoic acid

3-hydroxy-2-
formylbenzothiophene

Fig. 30.3 Established pathway of DBT degradation. Observations with a variety of bacteria [15,17,20] have established the sequence of reactions by which DBT is metabolized. The sulfur atom is not reactive in this pathway.

The latter compound is rapidly converted to 3-hydroxy-2-formyl benzothiophene [20, 21], which is yellow. In addition to these compounds, which appear to represent a progressive degradation of DBT, a condensation product is generated as well. In particular, 3-oxo[3'-hydroxythionaphthenyl-(2)-methylene] dihydrothionaphthene [21], a purple compound, is formed. These colored compounds are, presumably, those seen on plates of elective cultures in petri dishes exposed to an aerosol containing DBT.

Two features are especially notable. The initial additions that generate the dihydroxy, dihydro derivative of DBT occur on the periphery of the benzenoid component. The sequential oxidation and interdiol ring cleavage contribute to the increased hydrophilicity of the successive products. These reactions and the subsequent ones are *excluded* with derivatized DBT in coals because the reaction sites are the sites at which substitutions occur. Stated another way, the degradation of DBT as a low-molecular-weight organic substrate consumed by some pseudomonads and related bacteria is irrelevant to the degradation of DBTs which are fully incorporated by covalent linkages into the coal matrix. The second and more significant feature is that *no desulfurization* occurs. The final product in the established sequence 2-hydroxy,3-formylbenzothiophene retains the thiophene moiety.

30.5.2   Coal Bug Number 1 and Desulfurization of DBT

A scientist from Atlantic Research Corporation exposed soil samples from her backyard to elective conditions that favored the growth of organisms that could degrade DBT [22, 23]. Many of the methodological particulars remain vague, presumably because the company wishes to retain its commercial advantage through exclusivity. Nonetheless, the published record establishes that the initial capacity of the organism to consume DBT was augmented through a regimen of mutagenesis. Another detail is that consumption of DBT is dependent upon stimulation (or "induction") of the culture with sodium benzoate. Desulfurization requires the presence of the organism; the presence of the fluids in which the organism grew is insufficient for the desulfurization process to occur. This observation suggests that the activity is associated with the organism as a whole or, at least, its surface rather than extracellular enzymes. The organism need not be exhibiting robust growth for the desulfurization to occur; during authentic *coal* desulfurization (as opposed to desulfurization of a model compound), fully grown, induced cultures of coal bug number 1 (CB-1) are harvested and introduced to coal slurries.

CB-1 has been described as a pseudomonad that differs from other organisms of its genus. However, this assertion must be approached with caution. Many species within this genus are distinguished by a limited number of phenotypic characteristics. Mutagenesis of CB-1 may have altered phenotypic manifestations to such an extent that only molecular techniques will now provide reliable measures of similarity to established species.

Experiments with DBT have yielded an intriguing result: Not only is DBT transformed by CB-1, it is desulfurized. The presumed pathway (Fig. 30.4) is oxidation of DBT to DBT-5-oxide, a compound that is further oxidized to the sulfone, DBT-5-dioxide. In turn, the dioxide is further oxidized to sulfate and 2,2'-dihydroxybiphenyl. Such a reaction has features that are of striking biological and commercial consequence. The biological interest arises from a reaction that is directed at the sulfur of the thiophene component rather than any of the many other covalent linkages. The commercial value is that the sulfur is removed while diminishing only minimally the energy content of the molecule.

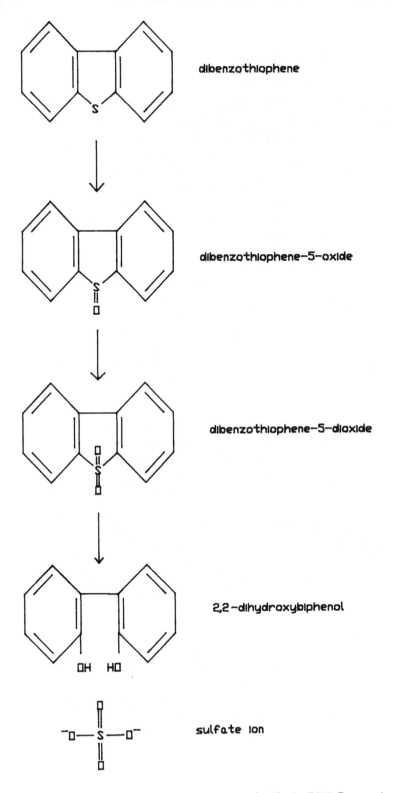

dibenzothiophene

dibenzothiophene–5–oxide

dibenzothiophene–5–dioxide

2,2–dihydroxybiphenol

sulfate ion

Fig. 30.4 Proposed pathway of oxidation of sulfur in DBT. Progressive preferential oxidation of sulfur and limited oxidation of carbon leads to the release of sulfate from an organic frame.

30.5.3 Investigations with *Sulfolobus acidocaldarius*

Taxonomy of the Organism: Its Distinctiveness

Sulfolobus acidocaldarius, first isolated in 1970 [24], is a member of the archaebacteria. All extant organisms are considered to belong to one of three major evolutionary groups: the eukaryotes, the prokaryotes, or the archaebacteria [25-27]. Analysis of ribosomal RNA sequences indicates that there are two major realms within the archaebacteria: the methanogenic organisms and the sulfur-dependent bacteria [28]. (The former group includes, as well, extreme halophiles and the *Thermoplasma* genus.)

Although organisms presently consigned to the archaebacteria have been known for more than 75 years [29], the archaebacteria were not recognized as constituting a third evolutionary distinct cellular form until the past decade. That these organisms are fundamentally different from eukaryotes and prokaryotes is supported by a variety of biochemical and genetic criteria, including composition of the cell wall, presence of distinctive metabolic cofactors, resistance to differentiating antibiotics, unique nucleotide sequences in ribosomal RNAs, and so forth [27]. The notion that the archaebacteria are a wholly distinctive evolutionary line from the primitive "protogenote," the presumed progenitor of all living forms, has achieved prominence. Nonetheless, other microbiologists contend that the archaebacteria, while distinctive, are merely a variant form of prokaryotes [13]. By ecological and morphological criteria, the archaebacteria appear similar to eubacteria [30].

Natural Habitat

S. acidocaldarius was originally isolated from an "extreme" environment—an acidic, sulfur-rich, miniature caldera in Yellowstone National Park [24,31,32]. The name chosen for the organism reflects, in part, these origins. "Acido-," in the species-specific name, refers to the pH optimum, which is 2.0 to 3.0. The organism tolerates a minimum pH of 0.9 and a maximum of 5.8. "Caldarius" reflects the site of isolation and suggests a hot environment. The optimum temperature for growth is 70 to 75°C; the organism can survive at a minimum of 55°C and a maximum of 85°C. The hot spring is also rich in sulfur, a circumstance readily recalled by the "Sulfo-" prefix of the genus name. The sulfur-rich environment contains elemental sulfur and volatile forms of sulfur. Whether this environment also contains organic sulfurs in any appreciable concentration was not stated in the original or, to my knowledge, subsequent descriptions of the habitat.

Structure

The organism is a lobe-shaped, colloidal-sized, unicellular organism with a relatively simple cytoplasmic organization. The lobes are a prominent feature, a circumstance that is, once more, reflected in the genus name of the organism. The irregular shape is evident in electron micrographs of these organisms, which are grown in a liquid medium (see Weiss [33] for electron micrographs). The lobed character of the organism is more pronounced when the specimens are taken from colonies grown on a solid substrate. The size and organization support the original (and possibly erroneous) conception that the organism is a conventional prokaryote.

Trophy

Microorganisms can be differentiated by their trophic (feeding) modes. Autotrophs are self-feeders that obtain their organic molecules by absorbing CO_2, fixing it to a receptor molecule, and then reducing it to an organic product. The source of "reducing power" for such syntheses by nonphotosynthetic autotrophs is a reduced inorganic molecule (e.g., H_2S). The alternative trophic mode is heterotrophy. Heterotrophs consume organic molecules obtained from other organisms. Through constructive metabolism (anabolism), organic molecules are transformed into a variety of compounds from which additional biomass is elaborated. Through destructive metabolism (catabolism), organic molecules are degraded and a portion of the released energy is conserved in a biologically useful form. (Furthermore, a toxic compound may be destroyed through catabolism with or without conservation of energy.)

S. acidocaldarius can exist as either an chemoautotroph or a chemoheterotroph. (The prefix "chemo-" distinguishes these organisms from "photoautotrophs" and "photoheterotrophs," which transform radiant energy into a biologically useful form.) As an autotroph, *S. acidocaldarius* can oxidize (among other sulfurous materials) hydrogen sulfide, elemental sulfur, or pyrite. The last of these substrates may reveal why the organism can be grown as colonies on particles of coal suspended in a presumably inert carbohydrate matrix. As a heterotroph, the organism responds to a limited set of organic substrates [24,31,32]. For the 98-3 strain, substrates include the disaccharide sucrose but neither of its constituent monosaccharides, glucose and fructose. The monosaccharide galactose is also not a substrate, nor is the disaccharide lactose.

All strains are able to grow on ribose and none grow on fructose. Only one isolate, 129-1, grows on sucrose, lactose, glucose, galactose, or ribose. Typically, the isolates grow on a limited array of amino acids, including glutamic acid, glutamine, alanine, and aspartic acid. However, strain 129-1, which grows on a wider array of carbohydrates, grows poorly on the first three of these amino acids and not at all on aspartic acid. All strains grow on protein-based complex media. But no isolates have been found that utilize phenylalanine, histidine, proline, and leucine. The sum of this information suggests that the organism is fastidious with respect to organic substrates, is possibly not a robust heterotroph, but is capable of heterotrophic growth, particularly in rich complex media. With conventional bacteria, "facultative autotrophs" function as autotrophs unless an organic substrate is present. With *S. acidocaldarius*, the "mood" is that the organism exhibits a more vigorous life cycle as an autotroph. This sense is supported by a few observations: The report of the original habitat described an environment capable of supporting autotrophic growth; the report was silent on the potential of the environment to support heterotrophic growth; the array of organic substrates is extremely limited and the inability of some monosaccharides to support growth—even monosaccharides that are constituents of disaccharides that do support growth—is suggestive. [*Note:* The weak response of *S. acidocaldarius* to commonplace organic compounds (some of which are present in all living organisms) [8] may contribute to a bias. The basic notion which underlies the bias is that DBT is very different in its chemical organization from the few organic structures that serve as heterotrophic substrates for *S. acidocaldarius*. Accordingly, the prospect that DBT would also serve as an organic growth substrate seems remote.]

DBT, a planar heterocyclic compound, is vastly different in its structure from the few conventional organic compounds that are known to function as heterotrophic substrates for *S. acidocaldarius*. There is nothing in the prior metabolic description of the organism which suggests that DBT might be a metabolic substrate. The bias does not exclude the prospect that DBT would be transformed by *S. acidocaldarius*; rather, any transformation would have to represent some activity other than catabolism. An immediately evident alternative was that DBT is a toxic material and that transformation of DBT could represent conversion of a noxious material to an innocuous one (Fig. 30.5).

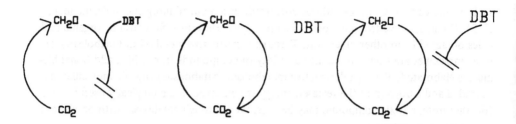

Fig. 30.5 Inhibition of the carbon cycle by DBT and the release of inhibition by degradation of DBT or by resistance to inhibition. DBT may be a toxic compound for some heterotrophic organisms and may interfere with metabolic activity. The effects of toxicity may be avoided by destroying the toxic compound or by becoming resistant to the compound.

Strains

Three strains of *S. acidocaldarius* have been used in recent experimentation in the Department of Biology at Lehigh University. The strains were obtained from W. Zillig of the Max-Planck-Institut fur Biochemie, West Germany. The first of these, 98-3, is supposedly identical to a strain commonly circulated among laboratories in the United States. However, unverified oral reports shared at professional meetings indicate that this strain has been irradiated at some locations. Accordingly, investigators should be cautious in assuming that some strains are equivalent to natural isolates. DSM 639 is a separate isolate which is thought, on the basis of the organization of its RNA polymerase, to be similar in character to 98-3. Finally, the B12 strain (isolated in Beppu, Japan) contains a fairly small genetic element [34] which may exist extrachromosomally and which is seemingly associated with a virus. This episome/virus contains approximately 13,000 base pairs [35,36] or enough genetic information to encode approximately 13 or so conventional genes. Because the B12 strain has an extrachromosomal genetic element (ECE) [37], because the ECE may increase the overall genetic potential, and because ECEs are known in some organisms to increase the degradative potential of the organism, *S. acidocaldarius* B12 was favored in the experiments described below.

Maintenance

Individual samples of all three strains, grown on "Shivvers' salts" [38] (see below) supplemented with 0.1% yeast extract, were introduced into equal volumes of glycerol and stored at -196°C in liquid nitrogen. Such samples were demonstrated to retain viability for at least 6 months. The B12 strain was maintained continuously at thermophilic temperatures through weekly serial transfers. B12 was so maintained in six media: Shivvers' salts plus 0.1% yeast extract; Shivvers' salts plus 0.1% yeast extract plus 2% v/v pristane plus 500 μg DBT/mL; "new salts" (see below) plus 0.1% yeast extract; new salts plus 0.1% yeast extract plus 2% pristane plus 500 μg DBT/mL; new salts plus 0.1 mM MgSO$_4$ plus sucrose; and new salts base plus 1.0 mM MgSO$_4$ plus sucrose plus 2% v/v pristane plus 500 μg DBT/mL. An important circumstance to note is that maintenance in the presence of DBT may represent a selective condition.

Media

All organisms require a standard set of "principal atomic components," including C, H, N, O, P, and S, a series that is sometimes called the macroconstituents, and, in addition, Na, Mg, Cl, K, and Ca, a series that is sometimes called the microconstituents [8,39]. (More exacting measures of mineral dependence have extended the second series to include a variety of additional less commonly recognized elements.) All the essential elements must obviously be provided in whatever aqueous media microorganisms are cultured. The liquid media used for the growth of microorganisms are commonly divided into two general groups: the complex and the defined. The former have no specific chemical formula. Rather, some organic material (e.g., yeast, beef parts, hay) is used as the source of an extract that can be dehydrated and subsequently reconstituted by the addition of water. The advantages of such media are convenience and the abundant provision of "components" that promote bacterial growth. The disadvantage is that the chemical composition of the components is, more often than not, not known. The defined medium has a precise chemical composition; the identity and abundance of every component is specified. Defined media are adequate for robust microorganisms; they work less well with microorganisms that are designated as being "fastidious."

Often a "salts base" is defined for a particular organism. The salts base is a solution of the microconstituents and some macroconstituents. The macroconstituent that is typically omitted from the macroconstituent series is the carbon source. In the salts base, the concentrations of partial salts of a weakly dissociating acid can be adjusted to achieve control of pH in the medium. By establishing a specific salts base, the effect of supplementing the salts base with a specific carbon source (e.g., sucrose or DBT) can be determined. Furthermore, other individual components, say SO_4^{2-}, can be omitted to determine if a supplement like DBT can serve as an alternative source of sulfur.

Shivvers' Salts Medium. The salts medium developed by those that first isolated and characterized *S. acidocaldarius* has the following composition [38]:

20.00 mM $(NH_4)SO_4$
2.00 mM KH_2PO_4
1.00 mM $MgSO_4 \cdot 7H_2O$
0.48 mM $CaCl_2 \cdot H_2O$

The pH of this medium is adjusted to be 3.0 by the addition of 1 N H_2SO_4. Trace elements are provided by the adding to every 100 mL of culture 1 mL of a solution having the following component concentrations:

20.00 mg/L $FeCl_3 \cdot 6H_2O$
4.50 mg/L $Na_2B_4O_7 \cdot 7H_2O$
0.22 mg/L $ZnSO_4 \cdot 7H_2O$
0.05 mg/L $CuCl_2 \cdot H_2O$
0.03 mg/L $NaMo_4 \cdot 2H_2O$
0.024 mg/L $CoCl_2 \cdot 6H_2O$
1.80 mg/L $MnCl_2 \cdot 4H_2O$

Inappropriateness of a Medium for Desulfurization Studies. The Shivvers' salt medium supports rapid proliferation of *S. acidocaldarius*. Rapid proliferation, or an approximation of maximum growth rate, is a circumstance frequently sought when designing a medium. Maximum growth rate allows an organism to achieve as great a presence in an environment as is possible. Under such conditions, the organism has exploited the environment or is adapted to the environment as effectively as possible. By this criterion, Shivvers' salts is a well-constructed base. Shivvers' salts medium contains many ionic forms that contain sulfur. Furthermore, the pH is adjusted with H_2SO_4. As a consequence, inorganic sulfur is supplied from many sources and to an extent that is not precisely established. This circumstance is essentially inconsistent with an attempt to define the dependence of *S. acidocalderius* on specific sulfur sources.

   Experimental Approach to Establishing New Salts Base.   The process of characterizing the metabolic potential of a microorganism is served by knowing the effect of each of the chemical components in the growth medium for that organism. Accordingly, a medium was sought that provides for rapid and extensive growth of *S. acidocaldarius*, maintains an established pH, and permits measurement of the sulfate concentration throughout the period of incubation. A straightforward, empirical approach was pursued. The experimental design was as follows: Cultures of *S. acidocaldarius* known to be growing exponentially on conventional Shivvers' salts plus 0.1% yeast extract were divided into six equivalent volumes; incremental concentrations of some component of the new salts medium were added and the effects on rate and extent of growth were measured during the course of 2 days. Repetition of this experimental protocol provided well-controlled conditions in which to assess responsiveness of *S. acidocaldarius* to additional media components (e.g., organic and inorganic sulfur-containing compounds).

   Composition of New Salts Base.   The composition of the new salts base was as follows:

   20.00 mM NH_4Cl
   1.00 mM $MgCl_2 \cdot 7H_2O$
   0.48 mM $CaCl_2 \cdot H_2O$

A source of sulfur was either omitted to create a sulfur-free new salts base or, alternatively, added in the form of 1.00 mM $MgSO_4 \cdot 7H_2O$ to create a new salts base + $MgSO_4$. The pH was adjusted to 3.0 with phosphate buffer. Trace elements were added in the same manner as with the Shivvers' salts medium.

Technical Reference

Details of methods of assessing mass by optical density, microscopic enumeration, or a protein assay that is accurate at low pH, quantitatively measuring the presence of DBT by spectrophotometry and high-performance liquid chromatography, fractionating cultures, disrupting cells, and measuring SO_4^{2-} concentration have been presented elsewhere [40].

Heterotrophic Response of *S. acidocaldarius* and Control Organisms to DBT

In this section of the critique, observations obtained in Department of Energy- and Pennsylvania Energy Development Authority-supported project directed by the late Bland S. Montenecourt and the author will be emphasized. Subsequently, the observations will be compared with those obtained by other investigators.

   During the investigation, *S. acidocaldarius* exhibited, at a minimum, three DBT-related "states." The first of these was DBT sensitivity, a condition associated with long-term growth in the absence of DBT. Such organisms were designated as being "naive." The data will demonstrate that for naive organisms two aspects of the growth potential, but neither a third nor a fourth aspect, were impaired by the presence of DBT. The two affected states were the initial ability to grow and the subsequent rate of growth; the unaffected conditions were the continuation of exponential growth and the extent of growth. Extensive subculturing of the organism in the presence of DBT yields a variant that shows neither a lag nor a reduced growth rate when introduced into a medium containing DBT. Such organisms were considered to be DBT adapted. Continued subculturing in the presence of DBT produced a third condition, DBT resistance. *S. acidocaldarius* in this state exhibited no adverse response to DBT after being cultured either in the presence or the absence of DBT.

The Naive or Sensitive Organism.

Batch Cultures in a Complex Medium. Knowledge of the growth characteristics of *S. acidocaldarius* in "standard conditions" of batch culture served as a reference for all subsequent interpretations. Figure 30.6 depicts the growth profile of *S. acidocaldarius* in Shivvers' salts plus 0.1% yeast extract. The extent of growth is presented as the logarithm of 1000 times the optical density at 420 nm; elapsed time is also presented on an arithmetic scale. The growth behaviors in batch culture of strains 98-3, DSM 639, and B12 of *S. acidocaldarius* are largely equivalent. These observations are in very good agreement with the results from batch culture obtained by other investigators. The generation time or doubling time between hours 19 and 38 for the 98-3 culture was 9.6 h. The maximum optical density, 0.710, corresponds to approximately 2×10^8 cells/mL. *S. acidocaldarius*

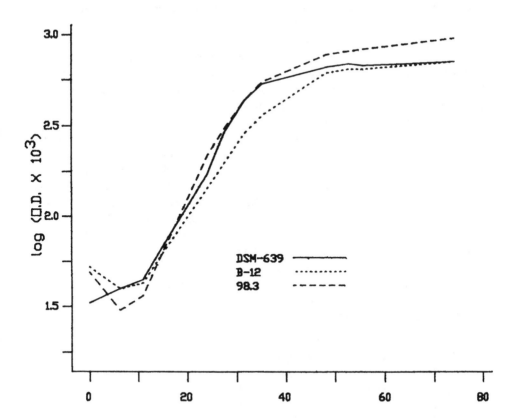

Fig. 30.6 Growth profile of *S. acidocaldarius* in batch culture. Strains 98-3, Dsm 639, and B12 were grown on new salts base plus 0.1% yeast extract. The organisms exhibited a conventional bacterial gorwth profile. (See the text for details.)

B12 growing on new salts base plus 0.1% yeast extract had a generation time in an early growth experiment of 14.3 h; the relative extent of growth was 0.933. The longer generation time indicates that the growth rate is somewhat less in the new salts base. It should be noted that no inorganic source of sulfur exists in the medium. Under these growth conditions, *S. acidocaldarius* obtains sulfur from components in the yeast extract. During growth of *S. acidocaldarius* on complex media, a pattern of neither SO_4^{2-} consumption nor SO_4^{2-} production could be discerned. Rather, the measurable SO_4^{2-} fluctuated. Similar observations have been made in other laboratories but have not been recorded in the literature.

Continuous Culture in a Complex Medium. A chemostat with a 5000-mL capacity was constructed for the continuous culture of *S. acidocaldarius* B12. The growth medium contained new salts base plus 0.03% yeast extract. Initially, the flow rate was set at 30 mL/h. The flow rate divided by the volume of the active culture yields the dilution rate, which is numerically equivalent to the instantaneous growth rate [41] (i.e., the growth rate expressed in terms of the base *e*). From this value, the generation time or doubling time can readily be calculated. The inoculum grew exponentially (with a generation time of approximately 20 h) to an optical density of approximately 0.400 at 420 nm. Then a steady state with a generation time of 115 h (4.8 days) was maintained through the twentieth day of the exercise. (Maintaining organisms with an extended generation time is an especially stringent test of the capacity of a chemostat to maintain steady state [41].) After 20 days, the flow rate was adjusted to 165 mL/h, a condition that required *S. acidocaldarius* to grow with a generation time of 21 h.

Batch Cultures in a Defined Medium. Growth *S. acidocaldarius* B12 in the new salts base plus 1.0 mM $MgSO_4$ plus 0.1% sucrose is much less robust than when yeast extract is present in the medium. A generation time of approximately 25 h was observed during exponential phase. The extent of growth during stationary phase fluctuated between 0.850 and 0.930 (i.e., no less than with an equivalent weight of yeast extract). Omission of 1.0 mM $MgSO_4$ prevented growth.

The foregoing observations demonstrate heterotrophic growth of *S. acidocaldarius* in increasingly stringent conditions, conditions that range from complex to fully defined. In the latter circumstance, all concentrations of substrates, including the carbon source and the sulfur source, are known. The growth is not robust under any of these situations. In comparison, some thermophiles (e.g., *Bacillus stearothermophilus*) grow with a generation time of less than 15 min; others (e.g., *Thermus aquaticus*) grow more laggardly with a generation time of 90 min or so [42]. In any event, *S. acidocaldarius* exhibits reproducible patterns of heterotrophic growth that are consistent with the characteristic phases of batch growth of conventional bacteria.

Growth of *S. acidocaldarius* in a chemically defined medium allows determination of what substrates can serve as sources of carbon and sulfur. The following list of controlled conditions is useful in determining the role of DBT:

1. New salts base + sucrose
2. New salts base   + $MgSO_4$
3. New salts base + sucrose + $MgSO_4$
4. New salts base   + DBT
5. New salts base + sucrose   + DBT
6. New salts base +   + $MgSO_4$ + DBT
7. New salts base + sucrose + $MgSO_4$ + DBT

The first set of conditions establishes that a sulfur source is required for growth. The second set of conditions establishes that a carbon source is required. The third set of conditions establishes that sucrose and magnesium sulfate are, respectively, appropriate sources of carbon and sulfur. The fourth set of conditions tests whether DBT can serve as a source of both sulfur and carbon. The fifth and sixth sets of conditions test, respectively, whether DBT can serve as a sulfur source or a carbon source. Conditions 4, 5, and 6 were examined to a limited extent. Growth was never observed under these circumstances. The seventh set of conditions measures the effect of DBT on *S. acidocaldarius* in a medium demonstrated to have the capacity to support heterotrophic growth. Several outcomes can be envisioned: DBT has no effect; DBT stimulates growth; DBT inhibits growth.

Some Effects of DBT on the Heterotrophic Growth of S. Acidocaldarius. DBT does not have an adverse effect on batch cultures of *S. acidocaldarius* in exponential phase. In particular, strain 98-3 in Shivvers' salts plus 0.1% yeast extract and no additional supplements grew with a generation time of 7.1 h. The addition of 2% v/v pristane had either no effect or a slightly stimulative one. The generation time under these conditions was 6.3 h. The addition of 500 μg DBT/mL in 2% v/v pristane also had either no effect or a slightly stimulative one. The generation time in the presence of DBT and pristane was 5.6 h.

Growth medium composed of new salts base plus 0.1% yeast extract was prepared and supplemented with 2% v/v pristane plus various concentrations of DBT. Medium of this character was then inoculated with *S. acidocaldarius*, which had no prior exposure to DBT. Figure 30.7 depicts the results. The control culture, which lacked either pristane or DBT, had a lag phase of approximately 10 h and a generation time of 19.6 h during the exponential phase. The culture, which contained 2% v/v pristane plus zero milligrams DBT/mL, had a slightly shorter lag and a generation time of 16.5 h during the exponential phase. The presence of 2% v/v pristane plus 250 μg DBT/mL extended the lag to approximately 40 h; the generation time during exponential growth was 21 h, a time that is not appreciably greater than that of the control. The higher concentration of DBT (i.e., 500 μg/mL) caused the lag to be extended further and produced an increased generation time. In particular, the generation time between hours 80 and 195 of incubation was 77 h. The results from this experiment were representative of the growth behaviors of *S. acidocaldarius*. Some variation exists. For example, in an experiment that was nearly identical in its design, the organism in the control culture had a generation time of 14.2 h during exponential phase, the organism growing in the presence of 2% v/v pristane had a generation time of 15.2 h, and the organism growing in the presence of 2% v/v pristane plus 500 μg DBT/mL had a average generation time of 181 h.

Experiments of similar design were performed in which sucrose was substituted for yeast extract. More specifically, *S. acidocaldarius* which had been grown on sucrose but which had no previous exposure to DBT was inoculated into a variety of media composed of new salts base plus 1.0 mM $MgSO_4$ plus 0.1% sucrose. The control culture had no additional supplements, while the two experimental cultures contained 2% v/v pristane or 2% v/v pristane plus 500 μg DBT/mL. The control culture had a lag of approximately 20 h and then grew with a generation time of 11.5 h during exponential phase. The inoculum introduced into the medium containing 2% v/v pristane did not exhibit exponential growth until a lapse of slightly greater than 6 days. Then the culture grew

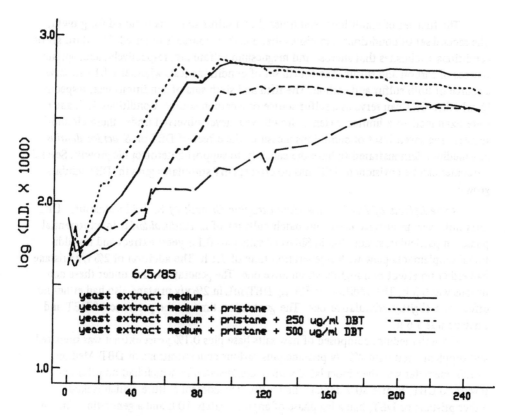

Fig. 30.7 Growth profile of *S. acidocaldarius* sensitive to DBT. Increasing concentrations of DBT have increasingly adverse effects on cultures of *S. acidocaldarius* which have had no prior exposure to the compound. (See the text for details.)

with a generation time of 32.6 h. The lag for the culture in the presence of 2% v/v pristane plus 500 μg/mL was more extreme; it lasted approximately 215 h. The generation time during exponential phase ranged between 38 and 81 h; the overall generation time was 62 h.

Exposure of Lag Phase Cultures of Escherichia coli *and* Salmonella typhimurium *to DBT.* Two questions about the sensitivity of *S. acidocaldarius* to DBT merit attention. One is whether the responsiveness to DBT is commonplace. The other is whether pristane serves as a sink for DBT and as such protects an organism to some extent from any adverse effects of DBT. These concerns were approached by growing *Escherichia coli* and *Salmonella typhimurium* in the presence of DBT and the presence of DBT dissolved in pristane. To have the test organisms be especially susceptible to any toxic effects of DBT, the cultures that were used as sources of inocula were maintained in stationary phase for lengthy periods of time. Long lags in freshly inoculated cultures correspond with having maintained organisms in stationary and are an indication that cells in the inoculum are not robust. For the analysis, the concentrations of pristane, 2% v/v, and DBT, 500 μg/mL, were those that had been used in previous experiments. The growth medium was a standard chemically defined medium, a circumstance chosen because it imposes the greatest metabolic demands on the organism. Such a circumstance is appropriate when attempting to measure adverse effects. For a fully controlled experiment, four conditions were selected:

1. Glucose salts
2. Glucose salts + 2% v/v pristane
3. Glucose salts + 2% v/v pristane + 500 μg DBT/mL
4. Glucose salts   + 500 μg DBT/mL

In essence, the patterns of growth of either *E. coli* or *S. typhimurium* in any of these four conditions were indistinguishable in the four conditions. Lags of considerable length occurred, but the lags were not extended by the presence of DBT. As is the ordinary circumstance, the lags seemingly are a reflection of the immediately preceding growth conditions of the inoculum. The following table summarizes the generation times during exponential phase for the two organisms.

Organism	Growth condition	Generation time (min)
E. coli	1	60
	2	70
	3	72
	4	70
S. typhimurium	1	62
	2	60
	3	65
	4	63

The preeminent conclusion that arises from these experiments is that DBT is not noxious with respect to *E. coli* or *S. typhimurium*. This conclusion precludes an evaluation of whether pristane serves as a sink for DBT, which, in effect, diminishes the exposure of an organism to DBT. However, it can be inferred that pristane plays no such role with respect to *S. acidocaldarius*. Increased concentrations of DBT in a constant amount of pristane have increased effects on the length of lag and generation time for this organism. Pristane with its similarity to phytane seems to have the desired effect of exposing *S. acidocaldarius* to DBT rather than protecting the organism from the compound.

Adaptation of S. acidocaldarius *to the Presence of DBT.* Because the prolonged lags were an impediment to the progress of the experimental investigation, a protocol was adopted to circumvent the problem. In particular, stock cultures of *S. acidocaldarius* were serially transferred in the presence of pristane plus DBT. When such stocks were used as inocula, no extended lags were observed. This circumstance is exhibited in Fig. 30.8. In this particular experiment, the B12 strain in the presence of the new salts base plus 0.1% yeast extract grew during exponential phase with an average generation time of 19.9 h; B12 in the same medium supplemented with 2% v/v pristane grew with a generation time of 18.8 h; and B12 in the same medium supplemented with both 2% v/v pristane and 500 μg DBT/mL grew with an average generation time of 18.4 h. No lag

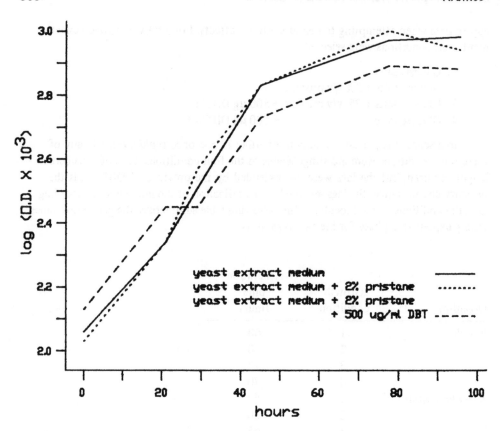

Fig. 30.8 Growth profile of *S. acidocaldarius* adapted to DBT. Subculturing in the presence of DBT generates inocula which exhibit no adverse affects from the presence of DBT. In some cultures of "adapted" *S. acidocaldarius*, a diminution of DBT occurs. (See the text for further discussion.)

whatsoever was evident. [*Note*: Reduction in the amount of DBT present in media that had supported the growth of *S. acidocaldarius* occurred while using DBT-adapted stocks as a source of the inoculum (see Section 30.5.4).]

Insensitivity to DBT. DBT appears to affect adversely growth of *S. acidocaldarius*. Serial transfer of the organism to maintain a "DBT-adapted" inoculum is a sufficient procedure to contend with the adverse effects of DBT. But this regimen is also a selective procedure for the isolation of DBT-insensitive or DBT-resistant cells. Organisms that were indifferent to the presence of DBT were first recognized by the loss of the ability "to transform" DBT (see below). The growth behavior of such organisms is depicted in Fig. 30.9. Two aspects are notable: The rate of growth is very rapid and the extent of growth has increased. The average generation time of strain B12 in the presence of new salts base plus 0.1% yeast extract plus 2% v/v pristane plus either 0 μg DBT/mL, or 250 μg DBT/mL, or 500 μg DBT/mL, or 1000 μg DBT/mL was, respectively, 9.3 h, 8.9 h, 8.9 h, and 9.9 h. The average generation time for the control culture that lacked both pristane and DBT was 8.5 h. The extent of growth was approximately 150% of that seen with the original strains grown in Shivvers' salts plus 0.1% yeast extract (see Fig. 30.6).

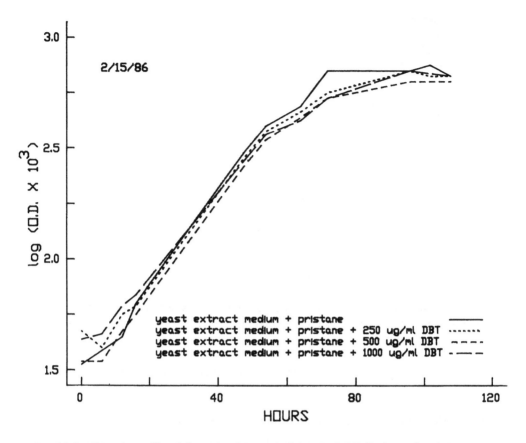

Fig. 30.9 Growth profile of *S. acidocaldarius* resistant to DBT. Prolonged maintenance of *S. acidocaldarius* in the presence of DBT yields cultures which grow rapidly and extensively in the presence of all concentrations of DBT examined. Such cultures appear indifferent to the presence of DBT and exhibit no transformation of DBT.

In the extreme, the insensitivity could arise from the environment (i.e., from the manner in which the inoculum was maintained) or, alternatively, the insensitivity could be inherent in the organism. The latter possibility would, presumably, arise from mutation (s). An experiment was conducted to determine if the DBT insensitivity persisted in the absence of exposure to DBT (Fig. 30.10). An inoculum of presumptively DBT-insensitive *S. acidocaldarius* cells were inoculated into media (new salts base plus 0.1% yeast extract) with and without 2% v/v pristane plus 500 μg DBT/mL. The rates of growth were monitored to note the effects, if any, of DBT. After 1 week, inocula were taken from the culture *lacking* DBT. The inocula were again introduced into media with and without pristane plus DBT. At the conclusion of another week, the inoculation regime was repeated. In all stances, the inoculum proliferated as rapidly and as extensively in the presence of DBT as in its absence, despite the fact that the organisms in the inoculum had been perpetuated in the absence of DBT. Stated another way, the indifference to DBT could not be originating due to the presence of DBT in the environment because the environment was constructed to lack DBT. The interpretation is that the in-

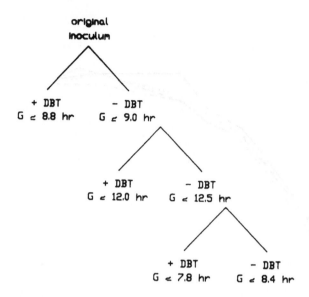

Fig. 30.10 Resistance to DBT. A regimen of culturing resistant strains of *S. acidocaldarius* in the absence of DBT yields progeny that retain their insensitivity.

difference to DBT is a genetic character and is perpetuated independently of environmental stimuli.

30.5.4 Diminution of DBT Concentration by *S. acidocaldarius* B12

During a 4-month period from September 1985 through December 1985, there was a diminution in the amount of DBT that could be recovered from cultures which had supported the growth of *S. acidocaldarius*. Representative results (from one of four experiments) are presented in Table 30.1. No product of DBT degradation was observed by HPLC or spectrophotometry. In a strict sense, there is no evidence of a "transformation" of a substrate, DBT, into identifiable products. Rather, there is a decrease in the amount of DBT recovered and measured from cultures that have supported the growth of *S. acidocaldarius*.

This intriguing observation requires that more experimentation be performed and that care be exercised with interpretations. Care must be taken to avoid an artifact. For the diminution to be meaningful, the ability to recover and to measure DBT must be adequately demonstrated. Accurate measurement of the concentration of DBT by HPLC has

Table 30.1 Diminution of DBT by *S. acidocaldarius* B12

Culture and phase of growth	HPLC measurement (μg DBT/mL)
Unionoculated	408 ± 140 (N = 11)
Two days, late lag–early exponential	457
Four days, late exponential	30
Ten days, stationary	2

been shown through standard curves. The data for recovery of DBT from 11 uninoculated cultures (Table 30.1) demonstrate a high standard deviation. This observation signals the need to use some statistical technique to assure sufficient sampling. No such techniques were used during this project. Nonetheless, the observed diminution is credible. The extents of the decreases in recoverable DBT in experimental cultures that were 4, 6, or 10 days old exceed the decreases that were observed with equally old uninoculated cultures. Stated another way, the decreases attributable to the variable, the presence of *S. acidocaldarius* for varying periods of time, appear greater than the decreases attributable to experimental error. Two other circumstances are inconsistent with an artifactual origin of the observation. In the four experiments, the amount of DBT recovered always decreased with time. Sampling errors alone might not yield such a distinctive pattern. Also, the phenomenon of decreased recoverable DBT was confined to a limited period of time (i.e., 4 months). It seems unlikely that there were no confounding sampling problems during the first year of the project, that there was then a phase for 4 months when the problem always existed, and that this phase was followed by the final period of the project during which sampling problems were again absent. A simpler interpretation is that there existed for a time a DBT-adapted subculture of *S. acidocaldarius* that could transform DBT. But this interpretation needs to be demonstrated through more experimentation.

30.5.5 Additional Investigations of the Capacity of *S. acidocaldarius* to Desulfurize DBT

The results from the investigation performed at Lehigh University do not exclude some intriguing speculations. But confidence in any interpretation requires that the observations on which they are based be both reproduced and confirmed. However, results from separate laboratories that allow comparisons are in conflict. Fikrit Kargi has claimed that *S. acidocaldarius* grows as a heterotroph at the expense of DBT [43]. However, the data that he has provided are not definitive. Kargi's conclusions are based solely on the detectable increase in SO_4^{2-}. He observed that a culture which contained DBT as the sole carbon source and which was seeded with a 5% inoculum of *S. acidocaldarius* 98-3 exhibited an increase in SO_4^{2-} after 10 days and through 20 days of incubation. The observation is provided without also providing adequate standard curves (achieved by adding incremental amounts of SO_4^{2-} to stock medium) to demonstrate the reliability of the assay. Data in successive figures of a single report on the appearance of SO_4^{2-} from 300 mg DBT/L appear to be in disagreement. Also, the sulfate that is reported to be produced appears, in some instances, to exceed the amount that could be obtained from the DBT added to the cultures. The doubling time for the appearance of SO_4^{2-} is on the order of 192 h, which is vastly different from most observed generation times of *S. acidocaldarius* and the increase in SO_4^{2-} is modest (approximately 25 to 40%), whereas the increase in biomass in actively growing cultures is usually enormous (a 5% inoculum returning to the cell concentration of the parent culture would increase 2000% .)The changes in SO_4^{2-} concentration do not appear to reflect directly the growth behaviors of *S. acidocaldarius*, the organism cited by Kargi as being responsible for desulfurization of DBT. The investigator only claims that SO_4^{2-} appears as a consequence of the presence of *S. acidocaldarius*, not its growth. If growth is not required, the effects of varied inocula sizes becomes relevant in confirming a role for the organism. Kargi indicates that the mediator of desulfurization was *S. acidocaldarius* 98-3 and that the organism was grown on glucose prior to inoculation into medium containing DBT. Yet the investigator who first characterized *S. acidocaldarius* claims that strain 98-3 does not grow on glucose. Microscopic evaluations of Kargi's cultures were performed in the Department of Biology at Lehigh University; the examination revealed rod-shaped organisms in addition to lobe-shaped ones. Despite these deficiencies, the reproducible appearance of $SO_4{}^{2-}$ during a characteristic period of incubation merits further attention.

Duane Skidmore and collaborators (at Ohio State University) claim that *S. acidocaldarius* desulfurizes coal to such an extent that sulfur must be released from thiophenes as well as from other compounds [44]. But such assertions are based on inferences rather than measurements. Changes in the abundance of thiophenes in coals were not directly measured; such is understandable since technologies for such measurements are not commonly available. The conclusion was based, instead, on the changes in total sulfur contents of coal, the changes in the inorganic sulfurs of coals, and the calculation that organic sulfurs must have been removed from coals as well. However, unless the assumptions on which these inferences are founded are verified, the conclusion cannot be regarded as established [45]. Skidmore's published accounts emphasize the engineering process; the biological particulars remain vague. As such, the basis of the conclusion that "[the investigators] believe that the microbial process is capable of removing 50-60% of the 'organic' sulfur" needs to be documented.

Kathleen W. Miller and J. Bruno Risatti, working at the Illinois State Geological Survey, found by observing (once more) the production of sulfate that *S. acidocaldarius* had the ability to "convert 10-15% of the sulfur in DBT to sulfate" [10]. However, there are several limitations in Miller's and Risatti analysis. DBT was condensed onto the surface of sterile 3-mm beads, and then bacterial growth medium was added to flasks containing the beads. No data were provided which indicated the success of this method in delivering DBT to cultures inoculated with *S. acidocaldarius*. Also, no information is provided about the viability of the organism during the incubation with DBT or at the end of the experiment. Given that there is limited biochemical activity against DBT, the object of the experimentation, it is imperative that the organism be shown to be biochemically active against a known substrate. Negative results are always tenuous; they require special attention to retain credibility. The investigators attributed some loss of DBT to sublimation, although neither information about vapor pressures nor experiments pertaining to this hypothesis were reported. Furthermore, the role of emulsifiers that might be produced by *S. acidocaldarius* was not developed. An intriguing result is the observation that as much as 15% of the DBT present in a culture may be removed by *S. acidocaldarius*. This circumstance is reminiscent of Isbister's work with CB-1 [21,22]. There a minimal capacity to desulfurize DBT was incremented through mutagenesis and selection. The observations by Miller and Rissati offer the prospect that *S. acidocaldarius* has a capacity that can be exploited.

An attempt to reproduce our observations yielded ambiguous results: Diminished amounts of DBT were recovered from cultures of adapted *S. acidocaldarius*; unfortunately, the control experiments did not yield straightforward results. Hence no useful interpretation can be derived. None of these investigations, to say the least, is definitive. Yet each contains a significant element of promise. An adequate measure of the capacity of *S. acidocaldarius* to desulfurize DBT might be achieved by adding ^{35}S-labeled DBT to steady-state cultures of the organism maintained in a defined medium in a chemostat. Experiments of this character are about to be performed at Lehigh University.

30.6 ABSENCE OF A MUTAGENIC CAPACITY IN DBT

The hydrophobicity, planarity, and size of DBT suggest that it may function as an intercalating mutagen. Comparison of the chemical structure of DBT and 2-aminofluorene, a mutagenic standard, reveals striking similarities between the compounds (Fig. 30.11). While the similarities are notable, chemical relatedness is not a good predictor of mutagenicity. Accordingly, experiments were performed.

The Ames test correlates the appearance of specific types of mutations in *Salmonella typhimurium* with the presence of specific compounds [46-48]. If specific sorts of changes occur more frequently in the presence of a stimulus (e.g., the presence of DBT), and the extent of the change increases with increasing concentrations of the stimulant, the stimulant can be identified as a mutagen that causes a particular type of genetic change.

Five strains of *S. typhimurium* obtained by the late Bland Montenecourt of Lehigh University from Charlotte Witmer of the Department of Pharmacology and Toxicology at UMDNJ Rutgers Medical School were used: TA97, TA98, TA100, TA102, and TA104. The first two are especially useful for identifying intercalating mutagens.

Standard mutagens [i.e., sodium azide (99%), crotonaldehyde (98%), and 2-aminofluorene (98%)] were obtained from Aldrich Chemical Company. Because the sterilizing agent ethylene oxide is a mutagen, petri dishes treated with ethylene oxide were not used. Instead, petri dishes sterilized by exposure to gamma irradiation were used; these were obtained from Thomas Scientific. Nutrient broth number 2 prepared by Oxoid U.S.A. is demonstrably low in mutagen content; accordingly, it was used in the experiments. Standard procedures were followed in the conduct of the experiments.

The data generated with control mutagens (sodium azide, crotonaldehyde) in the absence of S9 liver homogenate show good agreement between the observed and expected

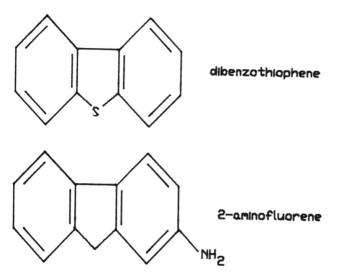

dibenzothiophene

2-aminofluorene

Fig. 30.11 2-Aminofluorene and dibenzothiophene. 2-Aminofluorene, an established mutagen, and DBT bear structural similarities.

numbers of induced mutations. Stated another way, the single concentration of established mutagen produced effects of the magnitude anticipated.

The data obtained with *S. typhimurium* strains TA97, TA98, TA100, and TA104
demonstrate that DBT is not a mutagen. (The results obtained with strain TA102 are uninformative because the control experiments with this strain did not conform to expectations.) The results with strains TA97 and TA98 are particularly relevant because results
with these strains, as noted earlier, are pertinent to frame shift mutations. This type of
mutation is the sort that occurs by intercalation, the mechanism of mutagenesis envisioned for DBT.

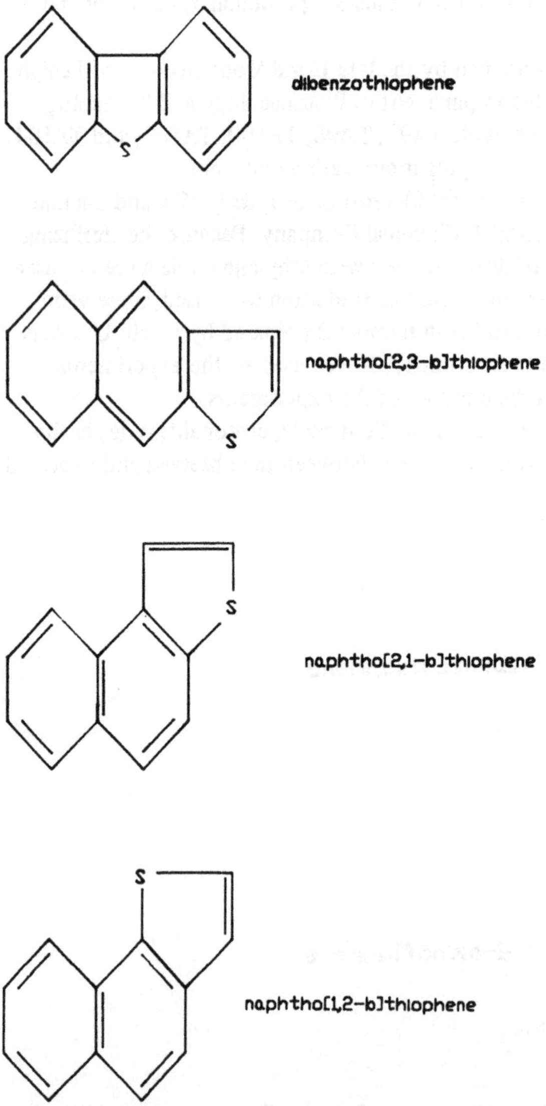

dibenzothiophene

naphtho[2,3-b]thiophene

naphtho[2,1-b]thiophene

naphtho[1,2-b]thiophene

Fig. 30.12 DBT and isomers of DBT. DBT and its isomers have been tested for mutagenicity [49]. Only naphtho[1,2-*b*] thiophene had any mutagenic activity.

Other investigators have demonstrated that DBT either alone or processed with S9 liver homogenates has no mutagenic potential [49,50]. S9 liver homogenates contain a series of enzymes that sometimes transform nonmutagenic compounds into mutagenic ones. Such transformations are particularly relevant to the effect of compounds on complex organisms where metabolic conversions by liver enzymes might occur; they are not especially informative about the effects of potentially mutagenic compounds on free-living bacteria.

Napththo[1,2-*b*] thiophene, an isomer of DBT, has a mild mutagenic potential [49]. Naphtho[2,1-*b*] thiophene is very similar to the mutagenic compound. It differs only in the position of the sulfur in the three fused rings that constitute this polycyclic aromatic sulfur heterocycle. Similarly, naphtho[2,3-*b*] thiophene shares a common composition and some common structural features; it, too, lacks any mutagenic potential. The structures of these compounds are shown in Fig. 30.12.

Seemingly, DBT and its isomers have either no potential or a minimal prospect of enhancing metabolic degradation of DBT through stimulation of mutations that might generate a degradative capacity. However, these conclusions are based on experiments with standard strains of *S. typhimurium*. Growth experiments with *S. typhimurium* showed no effect of the presence of DBT in a medium with or without pristane. This observation reveals an ambiguity. It is not evident whether DBT is innocuous or whether it was not taken up by the test strain. By contrast, increasing concentrations of DBT in a constant amount of pristane had progressively more adverse effects on the growth of naive *S. acidocaldarius*. This observation suggests that DBT is noxious and possibly internalized by *S. acidocaldarius*; as a consequence, there is the prospect that the genetic material of *S. acidocaldarius* could be exposed to DBT and possibly affected by it.

30.7 GENETIC INSTABILITY IN THE ARCHAEBACTERIA

A legitimate interpretation of the preceding results pertaining to the response of *S. acidocaldarius* to DBT is that the phenotypic character of the organism is not stable. Because the environment in which the organism was subcultured was constant, the differences in phenotype, if not artifactual, would appear to have a genetic origin. Culturing in the presence of 2% v/v pristane plus 500 μg DBT/mL may be a sufficiently powerful form of selection to account for the apparent shift from DBT sensitivity to DBT adaptation and then from DBT adaptation to DBT resistance. A second possibility is that the organism may exhibit an *inherent* genetic variability. Manifestation of such variability may have been augmented by the selective effect of the culture conditions.

The rate at which mutations occur in halobacteria, another archaebacterium, may be as high as 10^{-4} to 10^{-2} mutant per Colony Forming Unit [51]. these rates were estimated on the basis of phenotype changes in gas vacuole formation and pigment synthesis. Chromosomal rearrangements that are a manifestation of a series or repeated nucleotide sequences are thought to be the origin of these changes. As many as 500 different repeated nucleotide sequences may be present in the genome of a halobacterium species [52]; the extent of repetition of a distinct sequence may vary from a single repeat to numerous repeats. On the one hand, the repeated sequences tend to be clustered; on the other hand a single 5-kilobasepair segment of a genome may harbor one or more copies of 10 or more distinct sequences. The sizes of the repeated sequences are not known [52].

Rearrangements may occur without any recognizable change in the phenotype. Hybridization patterns establish that genomic organizations in the halobacteria quickly diverge during binary fission. As an individual cell passes through 34 generations, it will generate approximately 1.7×10^{10} division products. Experimentation with halobacteria led to the interpretation that only 3×10^7 of the progeny preserve the genomic organization that was present in the original progenitor cell [51].

Similar repeated elements have not been seen in *Thermoplasma* [52], nor in methanogens [53]. A highly repetitious nucleotide sequences is known to occur in enteric eubacteria [54], yet their genomes are remarkably stable [55]. Repeated sequences are shared among species of halobacteria and, similarly, repeated sequences are shared among enterics. That sequences are shared among species and even genera of a family suggests that in addition to promoting intraspecific rearrangements (or having the apparent potential to do so), the sequences may permit interspecific exchanges.

The relevance to *Sulfolobus* of the intriguing suggestions about "genomic rearrangements of astonishingly high frequency ... which may possess uniquely archaebacterial features" must remain speculative [56]. The mechanisms that foster chromosomal instability in halobacteria are not known to exist in *Sulfolobus*. The absence of repetitive sequences in *Thermoplasma* is uninformative since the extent of DNA-DNA homology between *T. acidophilum* and *S. acidocaldarius* has been shown to be less than 0.1%, an amount that could not be distinguished from nonspecific associations and experimental error [57]. What is known is that an important phenotype, diminution of DBT, is not stable and is not readily attributable to artifact. Whether this variability is an expression of genotypic instability remains to be demonstrated.

30.8 ACCESS OF BIOLOGICAL MEDIATORS OF DESULFURIZATION TO THIOPHENES IN COAL

A perplexing problem associated with microbial desulfurization of organic compounds in coals pertains to how bacteria might encounter derivatized DBT which is incorporated into the coal matrix. When DBT is used as a model compound, provision is often made for a "vehicle" that delivers DBT to the organism in an accessible form. The physical features of solubilized DBT or crystalline DBT appear to be vastly different from the physical features of DBT in coal.

The width of a DBT molecule is on the order of 1 nm. A type of coal that has frequently been used in organic desulfurization studies is 200 mesh; the largest cross-sectional dimension of coal of this size is 75 μm. If DBT were uniformly distributed throughout a spherical coal particle, approximately one ten-thousandth of the molecules would be in the outer 1-nm shell. Molecules in the outer shell would conceivably be oriented such that they could be subject to microbial desulfurization. This rendering of desulfurization is not consistent with the extents of organic desulfurization which have been reported. Accordingly, some assumptions pertinent to desulfurization need to be re-examined and others need to be identified.

In the preceding description, the abundance of molecules in a spherical particle with a diameter of 75 μm was considered. Spheres have the most efficient ratio of surface to volume. Also, the surface-to-ratio volume decreases as the diameter increases. In other words, large spheres represent the least favorable conditions for exposing components that might be the substrates for desulfurization. Typical coal particles are somewhat irregular in shape and have a cross-sectional dimension less than the maximum. Nonetheless, it seems unlikely that irregularity of shape and slightly reduced size are sufficient to provide the high extent of accessibility that the high degrees of organic desulfurization would seem to require. Additional surface could be achieved through cracks, clefts, and pores, although the last of these has dimensions that would exclude microbes and possibly macromolecules like extracellular enzymes. In these considerations, surface exposure of a relatively small substrate present in a large solid is assumed to be the limiting condition. (For this reason, it is moot as to whether the biological mediator of desulfurization is an enzyme [or series of enzymes] on the surface of the bacterium, an enzyme released across the surface, transport across the surface, or some other phenomenon.) The issue presently does not have a resolution because there is insufficient knowledge of what might constitute availability of substrate in conditions of solid-state chemistry. Furthermore, such other recently recognized phenomena as liquefaction may have some relevance. Until there is a more complete understanding of how a substrate is exhibited, some of the value of observations with model compounds must be considered more relevant to microbial capacities of degradation than to coal desulfurization.

30.9 BROAD SUBSTRATE SPECIFICITY FOR SOME DEGRADATIVE ENZYMES

Enzymes and enzyme systems are characterized, in part by specificity for particular substrates; however, the specificity has some latitude [58]. Discrimination among commonplace organic compounds contributes to regulation of reactions; compounds are not inadvertently metabolized. Such is especially useful when substrates are being partitioned between various energy-generating catabolic activities and biomass-generating anabolic activities. In other circumstances, a broad compatability between an enzyme and a spectrum of substrates is evolutionarily advantageous. An example would be conversion of all members of a class of noxious compounds into innocuous products through a single type of transformation that is common to all molecules. Similar to this circumstance is the dehalogenation of a variety of chlorinated compounds (e.g., monochloroacetic acid, dichloroacetic acid, monochloropropionic acid, or dichloropropionic acid, by individual dehalogenases [59]. (Interestingly, recent studies on the genetic origins of some dehalogenases have shown them to be associated with large plasmids [60] and transposable elements [61].)

A similar circumstance would be advantageous for genuine desulfurization of compounds which have, in common, a thiophene ring. Derivatized DBTs in coal vary widely with regard to the substitutions that occur on the benzenoid components. Having different degradative enzymes for each substituted form of DBT would be costly. By contrast, a desulfurizing enzyme with a specificity toward the sulfur of thiophenes and an unresponsiveness toward any other features of the substrate molecule would be efficient. Notably, these might be the features of an enzyme that transforms a compound because it is toxic rather than nutritional.

30.10 CONCLUSIONS

The microbial transformation of DBT is consistent with the principle of microbial infallibility. Several aspects of DBT utilization are noteworthy. DBT is degraded through more than one metabolic sequence, and one of the pathways is known to achieve authentic desulfurization. Furthermore, the complete transformation resulting in desulfurization is sometimes mediated by an individual microbial species. These significant observations need to be characterized more thoroughly. The genetic and biochemical bases of the transformations need to be extensively characterized so that full control of the transformation can be achieved. Further, the relationship between the physical and chemical character of relevant substrates in natural materials and desulfurizing bacteria needs to be understood. In sum, there have been striking observations with bacterial desulfurization of dibenzothiophenes. But understanding presently lags behind observation.

ACKNOWLEDGMENTS

Investigation of the response of *S. acidocaldarius* to DBT by the late Dr. Bland S. Montenecourt and the author has been supported by contracts from the Department of Energy and the Pennsylvania Energy Development Authority. Dr. Montenecourt is remembered with respect for her commitment, determination, and courage. Special thanks are extended to Ms. Diane Dutt, who carefully, competently, and seemingly tirelessly performed most of the experiments. Alexander Szewczak, who at the time of the project was an undergraduate "energy research intern," also merits genuine thanks for his analysis of the mutagenic protential of DBT.

REFERENCES

1. Alexander, M., Nonbiodegradable and other recalcitrant molecules, *Biotechnol. Bioeng., 15*, 611 (1973).
2. Alexander, M., Recalcitrant molecules, fallible micro-organisms, in *Microbial Ecology: A Conceptual Approach* (J. M. Lynch and N. J. Poole, eds.), Blackwell Scientific, Oxford, 1979.
3. Whitehurst, D. D., A primer on the chemistry and constitution of coal, in *Organic Chemistry of Coal* (J. W. Larsen, ed.), American Chemical Society, Washington, D.C., 1978.
4. Kargi, F., Microbiological coal desulfurization, *Enzyme Microb. Technol., 4*, 13 (1982).
5. Kargi, F., Microbial methods for desulfurization of coal, *Trends Biotechnol., 10*, 293 (1986).
6. Malik, K. A., Microbial removal of organic sulphur from crude oil and the environment: some new perspectives, *Process Biochem., 10*, 10 (1978).
7. Monticello, D. J., and Finnerty, W. R., Microbial desulfurization of fossil fuels, *Annu. Rev. Microbiol., 39*, 371 (1985).
8. Morowitz, H. J., *Energy Flow in Biology: Biological Organization as a Problem in Thermal Physics*, Academic Press, New York, 1968.
9. Kiyohara, H., Nagao, K., and Yana, K., Rapid screen for bacteria degrading water-insoluble, solid hydrocarbons on agar plates, *Appl. Environ. Microbiol., 43*, 454 (1982).
10. Miller, K. W., and Risatti, J. B., Rates of microbial removal of organic and inorganic sulfur from Illinois coals and coal chars, *Annual Report to the Illinois Coal Development Board*, Center for Research on Sulfur in Coal, Champaign, Ill., 1986.
11. Ochman, M., and Klubek, B., Isolation of thiophene utilizing bacteria from soil and strip-mine spoil, *Abstr. Annu. Meet. Am. Soc. Microbiol., 86*, 285 (1986).
12. Kiyohara, H., Nagao, K., and Yana, K., Rapid screen for bacteria degrading water-insoluble, solid hydrocarbons on apar plates, *Appl. Environ. Microbiol., 43*, 454 (1982).
13. Kreig, N. R., *Bergey's Manual of Systematic Bacteriology*, Williams & Wilkins, Baltimore, 1984.
14. Broda, P., *Plasmids*. W.H. Freeman, San Francisco, 1979.

15. Yamada, K., Minoda, Y., Kodama, K., Nakatani, S., and Akasaki, T., Microbial conversion of petro-sulfur compounds. I. Isolation and identification of dibenzothiophene utilizing bacteria, *Agric. Biol. Chem., 32*, 840 (1968).

16. Sagardia, F., Rigau, J. J., Martinez-Lahoz, A., Fuentes, F., Lopez, C., and Flores, W., Degradation of benzothiophene and related compounds by a soil *Pseudomonas* in an oil-aqueous environment, *Appl. Environ. Microbiol., 29*, 722 (1975).

17. Monticello, D. J., Bakker, D., and Finnerty, W. R., Plasmid-mediated degradation of dibenzothiophene by *Pseudomonas* species, *Appl. Environ. Microbiol., 49*, 756 (1985).

18. Skerman, V. R. D., McGowan, V., and Sneath, P. H. A., Approved lists of bacterial names, *Int. J. Syst. Bacteriol., 30*, 225 (1980).

19. Monticello, D. J., Bakker, D., Schell, M., and Finnerty, W. F., Plasmid-borne Tn5 insertion mutation resulting in accumulation of gentisate from salicylate, *Appl. Environ. Microbiol., 49*, 761 (1985).

20. Laborde, A. L., and Gibson, D. T., Metabolism of dibenzothiophene by a *Beijerinckia* species, *Appl. Environ. Microbiol.., 34*, 783 (1978).

21. Kodoma, K., Nakatani, S., Umehara, K., Shimizu, K., Minoda, Y., and Yamada, K., Microbial conversion of petro-sulfur compounds. III. Isolation and identification of products from dibenzothiophene, *Agric. Biol. Chem., 34*, 1320 (1970).

22. Isbister, J., Biological removal of organic sulfur from coal, in *Biological Treatment of Coals Workshop*, U.S. Department of Energy, Office of Fossil Energy, and the Bioprocessing Center, Idaho National Engineering Laboratory, Washington, D.C., 1986.

23. Isbister, J. D., and Kobylnski, E. A., Microbial desulfurization of coal, in *Processing and Utilization of High Sulfur Coals*, Vol. 9 of Coal Science and Technology Series (Y. A. Attia, ed.), Elsevier, Amsterdam, 1985.

24. Brock, T. D., Brock, K. M., Belly, R. T., and Weiss, R. L., *Sulfolobus*: a new genus of sulfur-oxidizing bacteria living at low pH and high temperature, *Arch. Mikrobiol., 84*, 54 (1972).

25. Fox, G. E., Stackebrandt, E., Hespell, R. B., Gibson, J., Maniloff, J., Dyer, A., Wolfe, R. S., Balch, W. E., Tanner, R. S., Magrum, L. J., Zablen, L. B., Blakemore, R. Gupta, R., Bonen, L., Lewis, B. J., Stahl, D. A., Luehrsen, K. R., Chen, K. N., and Woese, C. R., The phylogeny of prokaryotes, *Science, 209*, 457 (1980).

26. Stackebrandt, E., and Woese, C. R.. Evolution of bacteria, in *Molecular and Cellular Aspects of Microbial Evolution* (M. J. Carlile, J. F. Collins, and B. E. B. Mosley, eds.), 32nd Symposium of the Society for General Microbiology, 1981.

27. Woese, C. R., Archaebacteria, *Sci. Am., 244* (6), 98 (1981).

28. Yang, D., Kaine, B. P., and Woese, C. R., The phylogeny of archaebacteria, *Syst. Appl. Microbiol., 6*, 251 (1985).

29. Buchanan, R. E., and Gibbons, N. E., *Bergey's Manual of Determinative Bacteriology*, 8th ed., Williams & Wilkins, Baltimore, 1974.

30. Carlile, M., Prokaryotes and eukaryotes: strategies and successes, *Trends Biochem. Sci., 7*, 128 (1982).

31. Brock, T. D., *Thermophilic Microorganisms and Life at High Temperatures*, Springer-Verlag, New York, 1978.

32. Brock, T. D., Extreme thermophiles of the genera *Thermus* and *Sulfolobus*, in *The Prokaryotes* (M. P. Starr, H. Stolp, H. G. Truper, H. Balows, and H. G. Schlegel, eds.), Springer-Verlag, New York, 1981.

33. Weiss, R. A., Subunit cell wall of *Sulfolobus acidocaldarius, J. Bacteriol., 118*, 275 (1974).

34. Stetter, K. O., and Zillig, W., *Thermoplasma* and the thermophilic sulfur-dependent archaebacteria, in *The Bacteria*, Vol. 8, *Archaebacteria* (C. R. Woese and R. S. Wolfe, eds.), Academic Press, New York, 1985.

35. Yeats, S., McWilliam, P., and Zillig, W., A plasmid in the archaebacterium *Sulfolobus acidocaldarius, EMBO J., 9*, 1035 (1982).
36. Martin, A., Yeats, S., Janekovic, D., Reiter, W-D., Aicher, W., and Zillig, W., SAV1, a temperate u.v.-inducible DNA virus-like particle from the archaebacterium *Sulfolobus acidocaldarius* isolate B12, *EMBO J., 3*, 2165 (1984).
37. Reanney, D., Extrachromosomal elements as possible agents of adaptation and development, *Bacteriol. Rev., 40*, 552 (1976).
38. Shivvers, D. W., and Brock, T. D., Oxidation of elemental sulfur by *Sulfolobus acidocaldarius, J. Bacteriol., 114*, 706 (1973).
39. Green, D. E., and Goldberger, R. F., *Molecular Insights into the Living Process*, Academic Press, New York, 1968.
40. Krawiec, S., and Montenecourt, B. S., Investigation of the potential of *Sulfolobus acidocaldarius* to transform dibenzothiophene, *Final Report of Contract DE-AC22-83PC63046*, Department of Energy, Pittsburgh Energy Technology Center, Pittsburch, Pa., 1986.
41. Kubitschek, H. E., *Introduction to Research with Continuous Culture*, Prentice-Hall, Englewood Cliffs, N.J., 1970.
42. Mohr, P. W., and Krawiec, S., Temperature characteristics and Arrhenius plots for nominal psychrophiles, mesophiles, and thermophiles, *J. Gen. Microbiol., 121*, 311 (1980).
43. Kargi, F., and Robinson, J. M., Microbial oxidation of dibenzothiophene by the thermophilic organism *Sulfolobus acidocaldarius, Biotechnol. Bioeng., 26*, 687 (1984).
44. Murphy, J., Riestenberg, E., Mohler, R., Marek, D., Beck, B., and Skidmore, D., Coal desulfurization by microbial processing, in *Processing and Utilization of High Sulfur Coals*, Vol. 9 of Coal Science and Technology Series (Y. A. Attia, ed.), Elsevier, Amsterdam, 1985.
45. Spiro, C. L., Wong, J., Lytle, F. W., Greegor, R. B., Maylotte, D. H., and Lamson, S. H., X-ray absorption spectroscopic investigation of sulfur sites in coal: organic sulfur identifications, *Science, 226*, 48 (1984).
46. Ames, B. N., McCann, J., and Yamasaki, E., Methods for detecting carcinogens and mutagens with the *Salmonella*/mammalian microsome mutagenicity test, *Mutat. Res., 31*, 347 (1975).
47. Maron, D. M., and Ames, B. N., Revised methods for the *Salmonella* mutagenicity test, *Mutat. Res., 113*, 173 (1983).
48. Gerhardt, P., *Manual of Methods for General Bacteriology*, American Society for Microbiology, Washington, D.C., 1981.
49. PelRoy, R. A., Stewart, D. L., Tominaga, Y.-I., Iwao, M., Castle, R. N., and Lee, M. L., Microbial mutagenicity of 3- and 4-ring polycyclic aromatic sulfur heterocycles, *Mutat. Res., 117*, 31 (1983).
50. McFall, T., Booth, G. M., Lee, M. L., Tominaga, Y., Pratap, R., Tedjamulia, M., and Castle, R. N., Mutagenic activity of methyl-substituted tricyclic and tetracyclic aromatic sulfur heterocycles, *Mutat. Res., 135*, 97 (1984).
51. Sapienza, C., Rose, M. R., and Doolittle, W. F., High-frequency genomic rearrangements involving archaebacterial repeat sequence elements, *Nature, 299*, 182 (1982).
52. Sapienza, C., and Doolittle, W. F., Unusual physical organization of the *Halobacterium* genome, *Nature, 295*, 384 (1982).
53. Bollschweiler, C., Kuhn, R., and Klein, A., Non-repetetive AT-rich sequences are found in intergenic regions of *Methanoccus voltae* DNA, *EMBO J., 4*, 805 (1985).
54. Stern, M. J., Ames, G. F. L., Smith, N. H., Robinson, E. C., and Higgins, C. F., Repetitive extragenic palindromic sequences: a major component of the bacterial genome, *Cell, 37*, 1015 (1984).

55. Krawiec, S., Concept of a bacterial species, *Int. J. Syst. Bacteriol., 35*, xx (1985).
56. Doolittle, W. F., Genome structure in archaebacteria, in *The Bacteria*, Vol. 8, *Archaebacteria* (C. R. Woese and R. S. Wolfe, eds.), Academic Press, New York, 1985.
57. Christiansen, C., Freundt, E. A., and Vinther, O., Lack of deoxyribonucleic acid: deoxyribonucleic acid homology between *Thermoplasma acidophilum* and *Sulfolobus acidocaldarius, Int. J. Syst. Bacteriol., 31*, 346 (1981).
58. Hulbert, M. H., and Krawiec, S., Cometabolism: a critique, *J. Theor. Biol., 69*, 287 (1977).
59. Hardman, D. J., and Slater, J. H., Dehalogenases in soil bacteria, *J. Gen. Microbiol., 123*, 117 (1981).
60. Hardman, D. J., Gowland, P. C., and Slater, J. H., Large plasmids from soil bacteria enriched on halogenated alkanoic acids, *Appl. Environ. Microbiol., 51*, 44 (1986).
61. Slater, J. H., Weightman, A. J., and Hall, B. H., Dehalogenase genes of *Pseudomonas putida* PP3 on chromosomally located transposable elements, *Mol. Biol. Evol., 2*, 557 (1985).

31

Use of *Sulfolobus acidocaldarius* for Microbial Desulfurization of Coal

FIKRET KARGI *Washington University, St. Louis, Missouri*

Removal of sulfur compounds from coal has become a promising area of research in recent years due to acid rain problems [1,2]. Among various methods of desulfurization of coal, biological methods offer unique advantages over physical and chemical methods [2-4]. Microbial methods of desulfurization of coal require low capital and operating costs, operate under mild conditions (T = 25 to 70°C, P = 1 atm), and result in removal of finely disseminated inorganic and organic sulfur compounds [3]. Despite a major disadvantage of being a slow process, due to aforementioned advantages, microbial desulfurization methods have been given considerable attention. The two most widely used organisms for coal desulfurization are *Thiobacillus ferrooxidans* and *Sulfolobus acidocaldarius*. *T. ferrooxidans* is a mesophilic (T = 25 to 35°C), acidophilic (pH = 2 to 4), and obligate autotrophic organism capable of oxidizing pyritic sulfur only [4-9]. The organism was reported to remove more than 90% of pyritic sulfur over an incubation time period of 4 to 6 days [8]. *S. acidocaldarius* is a thermophilic (T = 50 to 80°C), acidophilic (pH = 1 to 5), and facultative autotrophic organism capable of oxidizing both pyritic sulfur and some organic carbon and sulfur compounds [10,11]. *Sulfolobus* species have distinct advantages over *Thiobacillus* species in removing sulfur compounds from coal. Being a facultative autotroph, *Sulfolobus* may oxidize some organic sulfur compounds present in coal. Also, operation of desulfurization process at high temperatures and low pH values reduce the chance of contamination of the reaction medium [3].

Sulfolobus-type organisms have been used for pyritic sulfur removal from coal by Detz and Barvinchak [4]. *Sulfolobus acidocaldarius* (strain 98-3), originally isolated by Brock [10], was used extensively by the author for pyritic sulfur removal from bituminous coal [12-14]. Kinetics of this oxidation, effect of process variables (coal pulp density, particle size, cell density), and environmental parameters (T, pH) on the rate and extent of pyritic sulfur removal have been investigated and conditions were optimized [14]. A mathematical model was developed and kinetic parameters were determined [15-16].

S. acidocaldarius was also proven to be able to oxidize dibenzothiophene (DBT, a refractory organic sulfur compound) in a specially designed carbon-free nutrient medium [17]. The kinetics of sulfate release from DBT oxidation was investigated [17]. Oxidation was found to be inhibited by DBT concentration above [DBT] > 500 mg/L [17]. Organisms capable of oxidizing DBT were used for organic sulfur removal from inorganic sulfur-free coal samples [18]. Nearly 40% of organic sulfur was removed by *S. acidocaldarius* from inorganic sulfur-free coal samples, within 4 weeks of incubation time. The organisms were also capable of oxidizing thianthrene and thioxanthene in carbon-deficient nutrient media [19].

S. acidocaldarius was shown to be very effective in nearly complete removal of pyritic sulfur and partial removal of organic sulfur from coal. The rate of pyritic sulfur removal and the rate and extent of organic sulfur removal from coal by *S. acidocalaldarius* need to be improved for an economically feasible process. Enzymology of organic sulfur compound oxidations is not known and needs to be elucidated for more effective organic sulfur removal from coal. Cell-free enzymatic treatment of coal may result in improved sulfur removal rates.

A two-stage conceptual process was developed on the basis of our experimental studies [20]. The first stage was for inorganic sulfur, and the second stage was for organic sulfur removal. Preliminary economic analysis of the process indicated a fixed capital cost of $70 × 10^6 and an operating cost of $17 per ton of coal to process 120 tons of coal per hour, which is the coal supply of a 250-MW power plant [20].

REFERENCES

1. Eliot, R. C. (ed.), *Coal Desulfurization Prior to Combustion*, Noyes Data Corporation, N.J., 1978, pp. V–VI, 33–42.
2. Kargi, F., Microbiological coal desulfurization, *Enzyme Microb. Technol.*, *4*, 13 (1982).
3. Kargi, F., Microbiol desulfurization of coal, in *Advances in Biotechnological Processes*, Vol. 3 (A. Mizrahi and A. L. van Wezel, eds.), Alan R. Liss, New York, 1984, pp. 241–272.
4. Detz, C. M., and Barvinchak, G., Microbial desulfurization of coal, *Min. Congr. J.*, *7*, 75 (1979).
5. Silverman, M. P., Rogoff, M. H., and Wender, I., Removal of pyritic sulfur from coal by bacterial action, *Fuel*, *42*, 113 (1963).
6. Silverman, M. P., Mechanism of bacterial pyrite oxidation, *J. Bacteriol.*, *94*, 1046 (1967).
7. Dugan, P. R., and Apel, W. A., Microbiological desulfurization of coal, in *Metallurgical Applications of Bacterial Leaching and Related Microbiological Phenomena* (L. E. Murr, A. E. Torma, and J. A. Brierley, eds.), Academic Press, New York, 1978, pp. 223–250.
8. Hoffmann, M. R., Faust, B. C., Panda, F. A., Koo, H. H., and Tsuchiya, H. M., Kinetics of the removal of iron pyrite from coal by microbial catalysis, *Appl. Environ. Microbiol.*
9. Kargi, F., Enhancement of microbial removal of pyritic sulfur from coal using concentrated cell suspension of *T. ferrooxidans* and an external carbon dioxide supply, *Biotechnol. Bioeng.*, *24*, 749 (1982).
10. Brock, T. D., Brock, K. M., Belly, R. T., and Weiss, R. L., *Sulfolobus*: A new genus of sulfur oxidizing bacteria living at low pH and high temperature, *Arch. Microbiol.*, *84*, 54 (1972).

11. Brock, T. D., The genus *Sulfolobus* in *Thermophilic Microorganisms and Life at High Temperatures*, Springer-Verlag, New York, 1978, pp. 118-179.

12. Kargi, F., and Robinson, J. M., Removal of sulfur compounds from coal by the thermophilic organism *S. acidocaldarius, Appl. Environ. Microbiol., 44*, 878 (1982).

13. Kargi, F., and Robinson, J. M., Microbial desulfurization of coal by thermophilic organism *S. acidocaldarius, Biotechnol. Bioeng., 24*, 2115 (1982).

14. Kargi, F., and Robinson, J. M., Biological removal of pyritic sulfur from coal by the thermophilic organism *Sulfolobus acidocaldarius, Biotechnol. Bioeng., 27*(1), 34-40 (1985).

15. Kargi, F., and Weissman, J. G., A dynamic mathematical model for microbial removal of pyritic sulfur from coal, *Biotechnol. Bioeng., 26* (6), 604-612 (1984).

16. Kargi, F., and Weissman, J. G., Kinetic parameter estimation in microbial desulfurization of coal, *Biotechnol. Bioeng.* (1987).

17. Kargi, F., and Robinson, J. M., Microbial oxidation of dibenzothiophene by the thermophilic organism *S. acidocaldarius, Biotechnol. Bioeng., 26* (7), 687-690 (1984).

18. Kargi, F., and Robinson, J. M., Removal of organic sulfur from bituminous coal by the thermophilic organism *S. acidocaldarius, Fuel, 65*, 397-399 (1985).

19. Kargi, F., Biological oxidation of thianthrene, thioxanthene and dibenzothiophene by the thermophilic organism *S. acidocaldarius, Biotechnol. Lett., 9*(7), 474-482 (1987).

20. Kargi, F., Microbial methods for desulfurization of coal, *Trends Biotechnol., 4*(11), 293-297 (1986).

32

Bacterial Desulfurization of Lignites

CELAL F. GOKCAY and REYHAN YURTERI *Middle East Technical University, Ankara, Turkey*

32.1 CHARACTERISTICS OF TURKISH LIGNITES

Control of air pollution originating from sulfur oxide emissions has gotten increasing attention in Turkey during the past decade. Consequent upon serious episodes of air pollution in the capital city, Ankara, and in several other Anotolian cities, it is now clear that the prime cause of pollution is the burning of high-sulfur Turkish lignites for space heating. Furthermore, the Transboundary Air Pollution Control International Act stipulates that sulfur oxide emissions from power plants must be lowered drastically as we approach the end of the century.

As a means of combating air pollution in Ankara, some tentative measures, such as the use of imported low-sulfur bituminous coal or Siberian natural gas, have begun. However, it is generally believed that in the long term the nation will almost certainly have to rely on its own energy resources, such as lignites. Control of sulfur oxide emissions from power plants is currently being tried by employing expensive flue gas desulfurization facilities, which in turn adds to the already high cost of electricity.

Generally, two forms of sulfur are recognized in coals; inorganic and organic. Inorganic sulfur falls into two categories, disulfides and sulfates. Sulfate sulfur, which is often quantitatively negligible in coals, remains in the ash during combustion and does not contribute to sulfur oxide pollution. The other form of inorganic sulfur, iron disulfide, commonly referred to as pyritic sulfur, appears in two crystalline forms, pyrite and marcasite, with a chemical formula of FeS_2. Pyritic sulfur may be present as large particles or as microscopic crystals in the coal matrix, or may simply be disseminated throughout the organic matter.

Organic sulfur is spread throughout the hydrocarbon matrix of coals and is found in the form of thiophanes, arylsulfides, cyclic sulfides, aliphatic sulfides, and as aryl and aliphatic thiols [1]. Organic and pyritic forms constitute the combustible fraction of sulfur and contribute to air pollution.

Table 32.1 Analyses of Beypazari-Cayirhan
Lignites (-80 Mesh)

Moisture content (% of coal)	25.0
Ash content (% of coal)	22.26
Low heating value (kcal/kg)	3264.0
High heating value (kcal/kg)	3576
Total sulfur (% of coal)	4.22
Pyritic sulfur (% of coal)	1.48
Organic sulfur (% of coal)	2.23
Sulfate sulfur (% of coal)	0.51

Turkish lignites are characterized by their high organic sulfur and low caloric values. For example, among the 243 discrete lignite reserves in the country, with 3.2 billion tons of proven reserve, the Afsin-Elbistan field is the largest and accounts for over half the national reserves. The Afsin-Elbistan lignite contains around 4.4% sulfur, of which over 50% is organic. Furthermore, the caloric value of this lignite is extremely low, averaging around 2500 kcal/kg.

Nallihan-Cayirhan lignites were studied because of their proximity to Ankara and for their significant reserves, which ranked them third in significance in the country. Chemical analyses of a sample of Beypazari-Cayirhan lignites, with a particle size of -80 mesh, are shown in Table 32.1.

32.2 OVERVIEW OF MICROBIAL DESULFURIZATION

One method of controlling sulfur oxide emission is by desulfurization of coals prior to combustion. Among the methods available, bacterial desulfurization prior to coal combustion seems promising for the future in many respects. For example, Detz and Barvinchak compared the cost of bacterial desulfurization resulting in a 60% reduction in total sulfur content with the costs of other precombustion desulfurization and flue gas desulfurization processes and rated microbial desulfurization as the cheapest process, at $10 to $14 per ton [2]. Second best was flue gas desulfurization, with a desulfurization cost of $16 per ton. Bacterial desulfurization seems preferable for its ease of application. However, the sulfur content of microbially desulfurized coal reported by these workers was comprised almost totally of pyritic sulfur. The pyritic sulfur in coals is known to be amenable to bacterial desulfurization.

There have been numerous reports of bacterial coal desulfurization in the literature, and the bacteria involved in the oxidation of pyritic sulfur in coal almost invariably belong to the genus *Thiobacillus*. For example, Chandra et al. [3], Detz and Barvinchak [2], Dugan and Apel [4], Hoffmann et al. [5], Silverman et al. [6], and Zaburina and Ashmed [7] have all reported involvement of lithoautotrophic, acidophilic thiobacilli in bacterial desulfurization of coals. On the other hand, Chandra et al. [8] have reported bacterial removal of organic sulfur in an Indian coal by a thiophene-enriched heterotrophic isolate. Recently, similar findings have been reported on organic sulfur removal from coals by a thiophene-enriched heterotrophic thermophile, *Sulfolobus acidocalderius* [9]. Gokcay and Yurteri [10] have also reported bacterial removal of organic sulfur from lignites by a mixotroph formerly identified as *T. ferrooxidans* TH1. Further evidence of organic sulfur removal by thermophilic *S. acidocalderius* has recently been presented by Chen and Skidmore [11] together with a kinetic model of sulfur removal.

Silverman et al. [12] have reported that *T. thiooxidans* is not active in coal desulfurization, and Dugan and Apel [4] have shown that the two species *T. ferrooxidans* and *T. thiooxidans* can best desulfurize coals collectively. However, the most commonly cited bacterium in coal desulfurization is *T. ferrooxidans.*

T. ferrooxidans, unlike its counterpart *T. thiooxidans*, is known for its ability to oxidize pyritic sulfur. Both strains can oxidize elemental sulfur and other reduced inorganic sulfur species. The chemolithotrophic *T. ferrooxidans* is reported to be metabolically active between 25 and 35°C and at a pH range between 2.0 and 3.6, as reported by Silverman and Lundgren [13]. Organics in the medium are noted for their inhibitory effect on this bacterium. Concurrently, isolation of acidophilic heterotrophic bacteria in acidic colliery effluents and mine drainages was attributed to this effect. Groudev and Genchev [14] reported that the rate of coal desulfurization was increased when a particular *T. ferrooxidans* strain was adapted to coal or pyrite prior to desulfurization.

The exact mechanism of bacterial desulfurization is still poorly understood. It is postulated that two mechanisms are responsible for desulfurization. The direct contact mechanism assumes intimate physical contact between the bacteria and sulfide minerals under aerobic conditions [15]. As a result, oxidation of the sulfide mineral is achieved according to the following reaction:

$$FeS_2 + H_2O + (7/2)O_2 \rightarrow FeSO_4 + H_2SO_4$$

The direct mechanism was also favored by Beck and Brown [16], who showed that a significant amount of carbon dioxide uptake occurred in the absence of ferric ions. Berry and Murr [17] produced electron micrographs of *T. ferrooxidans* attached to mineral surfaces as direct evidence of a contact mechanism.

Silver [18] and others [19,20] have proposed an enzymatic mechanism following Michael-Menten kinetics, with ferrous iron acting as the substrate. Kelly and Jones [21] modified this model by introducing end product inhibition to the model. However, Michael-Menten kinetics does not readily explain experimental findings of Hoffmann et al. [5], who showed that at different pulp densities (solids concentration) of coal, different specific desulfurization rates are obtainable. Conversely, if the Michael-Menten kinetics would hold, the specific desulfurization rate would be constant at every level of solids concentration provided that the specific particle surface area is not rate limiting.

In the indirect mechanism the role of bacteria is considered not to attack pyrite directly but to catalyze aerobic oxidation of ferrous iron in solution to the ferric state. The ferric ion in solution then oxidizes pyrite to the ferrous state and additional acidity is produced due to protons released in this reaction [2,4,6].

Experimental results indicated that the removal of pyritic sulfur from coal in batch systems is governed by first-order reaction kinetics as described by the following equation:

$$\frac{d(FeS_2)}{dt} = -k(FeS_2)$$

The relative ease of coal desulfurization follows the order bituminous > subbituminous > lignite, depending on the neutralizing capacity of coals. Silverman et al. [6] report that lignites are resistant to bacterial oxidation without acid pretreatment. These workers indicate that acid pretreatment is needed to remove the buffering effect of the gang material.

The optimum pH for bacterial desulfurization is reported to lie between 2.0 and 3.0, although growth is sustained between pH 1.5 and 4.0 [4]. The initial pH of the medium is considered important as the lag period in batch cultures increases with increasing pH. An initial pH of 1.5 is reported by Hoffmann et al. [5] as being completely inhibitory to the organism during coal desulfurization. Detz and Barvinchak [2] suggest that the culture of the organism should be maintained below pH 2.5 to prevent possible precipitation of iron hydroxides and sulfate complexes onto the coal surfaces, thus hindering bacterial contact.

Microbial coal desulfurization may be carried out effectively in two temperature ranges. The mesophilic range 23 to 37°C is suitable for *T. ferrooxidans* and *T. thiooxidans* [22], while the thermophilic *T. ferrooxidans* TH1 grows best in the range 40 to 60°C [23] and *S. acidocalderius* at 55 to 80°C [4,9]. The temperature quotient, Q10, for *T. ferrooxidans* is reported as 1.8 for ferrous iron oxidation.

For culturing of *T. ferrooxidans* the 9K liquid medium developed by Silverman and Lundgren [13] is usually chosen. More recently, Hoffmann et al. [5] developed a low-phosphate, low-sulfate medium known by the acronym LOPOSO. These workers claim that the medium is superior to 9K. For culturing of the thermophilic TH group, a liquid medium described by Brierley et al. [23] is recommended. In addition to the basal salts, the TH group requires an outside organic carbon and sulfur source, which is usually supplied by the yeast extract in the medium.

Hoffmann et al. [5] and many others report that the particle size of coal affects the desulfurization rate. Generally, the smaller the particle size, the higher the reaction rate. Finally, Detz and Barvinchak [2], and Dugan and Apel [4] report that the pulp density of coal affects the rate of desulfurization, 20% pulp density being most suitable in the majority of cases.

32.3 EXPERIMENTAL PROCEDURES

The common method of testing microbial desulfurization is by using shake flask cultures. Although this test system is far from able to simulate commercial-size reactors, the ease of handling and the high rates obtained make it ideal for desulfurization studies. It is quite probable that the sulfur removal rates obtained in shake flasks would be higher than those that would be obtained in commercial-size reactors. Nevertheless, shake flask culture systems usually represent the highest attainable rates and they are of additional merit for this effect.

The shake flask culture technique was also used in the present study. The 500-mL conical flasks contained varying amounts of pulverized lignite, whose chemical analysis is shown in Table 32.1, 100 mL of 9K liquid medium devoid of iron, and an appropriate inoculum. The control flasks lacked inoculum but contained 1% mercury chloride in addition, to stop any bacterial activity. The lignite samples were washed with acidified water (pH 2.5) and dried prior to experimentation. During incubation samples were drawn from flasks, and pH and sulfate were determined in the filtered aliquotes according to APHA *Standard Methods* [24]. At the end of a 25-day incubation period, residual total and pyritic sulfur were determined in the filtered coal samples by ISO standard analysis procedures (Refs. 25 and 26, respectively). Organic sulfur was calculated by subtracting pyritic and sulfate sulfur from total sulfur. If not otherwise stated, the initial pH of the medium was 3.0 and temperature was 35°C at the mesophilic range.

The strains of *T. thiooxidans* and *T. ferrooxidans* (strain OSU 380) were cordially provided by P. R. Dugan from the type culture collection at Ohio State University. The strain of *T. ferrooxidans* TH1 was kindly provided by J. A. Brierley from the New Mexico State Bureau of Mines. For stock culturing these strains were grown on 3K medium, a modification of 9K medium containing one-third the amount of iron. The TH1 organism was grown on a liquid culture described by Brierley et al. [23].

The biomass concentration during the experiments was determined by filtering off the entire solids content on a 0.45-μm membrane filter and extracting protein from filter into IN NaDH by boiling. Protein concentration was then determined in the alkali fraction by the classical Folin-Ciocalteu method. Insoluble matter was removed by centrifugation and the color produced was read in a spectrophotometer against a reagent blank developed from control flasks receiving the same treatment.

The most probable number (MPN) method utilizing two tubes per dilution was used [27] for the enumeration of viable cells. The MPN method involved decimal dilutions of a 1-mL sample in tubes containing the 3K solution. Inoculated tubes were then incubated for 10 days. A reddish color and pH drop were taken as evidence of growth, and these tubes were scored as positive. The most probable numbers were estimated from the combination of positive and negative scores by consulting a chart in Ref. 27.

32.4 DESIGN OF DESULFURIZATION EXPERIMENTS

Several factors are reported to affect the rate and extent of bacterial desulfurization. Among these parameters, temperature (T), particle size (D_p), pulp density (P), and initial pH are generally considered most important.

One way of determining the relative magnitude of the effect of each parameter and of their combination is by construction of a proper experimental design. For example, the linear first-order model that can be developed by performing a 2^3 factorial experimental design would be

$$Y = b_0 + b_1 X_1 + b_2 X_2 + b_3 X_3 + b_{12} X_{12} + b_{13} X_{13} + b_{23} X_{23} + b_{123} X_{123}$$

where Y represents the cumulative response and the b values are coefficients representing the magnitude of the associated parameter(s), as described in Table 32.2. The variables and levels suggested in the factorial design are given in Table 32.3. The coefficients of the variables are then determined by solving the design matrix according to the following equation:

$$|b| = (\{X\}^T \{X\})^{-1} (\{X\}^T \{Y\})$$

Table 32.2 Coefficients of the Linear First-Order Model

Variable	Coefficient
	b_0
x_1 (T, °C)	b_1
x_2 (P, %)	b_2
x_3 (D_p, μm)	b_3
x_{12} (T, P)	b_{12}
x_{13} (pH, T, D_p)	b_{13}
x_{23} (P, D_p)	b_{23}
x_{123} (T, P, D_p)	b_{123}

Table 32.3 Variables and Levels of Factorial Design

	Temperature (°C)	Pulp density (%)	Particle size (µm)	pH
Upper level (+)	35	20	<177	3.0
Lower level (−)	25	10	<53	2.0

32.5 DESULFURIZATION OF LIGNITES BY MESOPHILIC THIOBACILLI

32.5.1 Effect of *Thiobacillus* Strains

The effect of *Thiobacillus* strains on coal desulfurization is controversial. For example, Dugan and Apel [4], after researching with an American coal using *T. ferrooxidans* and *T. thiooxidans*, concluded that neither *T. ferrooxidans* nor *T. thiooxidans* alone was effective in sulfur removal from coal. Others [5,14] have reported that *T. thiooxidans* has no noticeable effect on desulfurization.

The effect of *Thiobacillus* species on lignite desulfurization was tested using *T. ferrooxidans* and *T. thiooxidans* acting alone or as a mixture. The change in sulfate and protein concentrations with time in the test flasks are shown in Figs. 32.1 to 32.4.

The specific growth rates and desulfurization rates were calculated from slopes of semilog plots in Figs. 32.1 and 32.2 and Figs. 32.3 and 32.4, respectively, as summarized in Table 32.4. It is evident from these data that there is no marked effect of strain on the rate and extent of desulfurization in 3% pulp density. However, in 10% slurry the *T. ferrooxidans* strain is clearly superior over *T. thiooxidans* and the mixture. Hence *T. ferrooxidans* was the sole organism used in the remainder of this study.

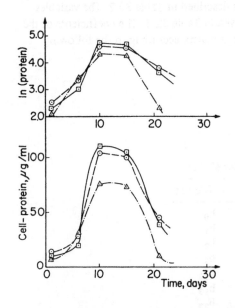

Fig. 32.1 Protein analyses of samples from flasks containing different species of thiobacilli in 3% coal slurry. ▫, *T. ferrooxidans*; ⊙, *T. thiooxidans*; △, mixed culture.

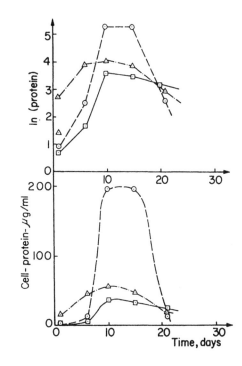

Fig. 32.2 Protein analyses of samples from flasks containing different species of *Thiobacilli* in 10% coal slurry. ⊙, *T. ferrooxidans*; ▫, *T. thiooxidans;* △, mixed culture.

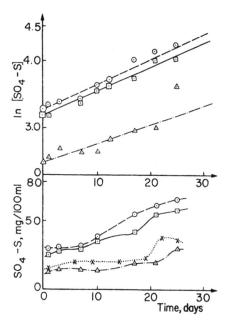

Fig. 32.3 Sulfate release in 3% coal slurries inoculated with different species of *Thiobacilli*. ⊙, *T. ferrooxidans*; ▫, *T. thiooxidans*; △, mixed culture; X, control.

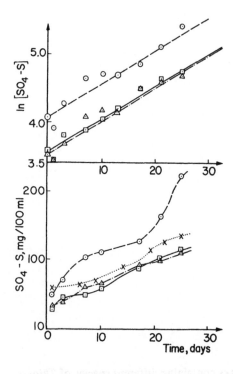

Fig. 32.4 Sulfate release in 10% coal slurries inoculated with different species of *Thio-bacilli*. ⊙, *T. ferrooxidans*; ▣, *T. thiooxidans*; △, mixed culture; X, control.

Table 32.4 Effect of Different *Thiobacillus* Strains on Lignite Desulfurization (Initial pH = 3, 35°C, 3K Basalt Salts, 53 < D_p < 177 μm, 7-Day 3K-Grown Inoculum)

Pulp density (%)	Inoculum	Rate constant, k (h^{-1})	Maximum specific growth rate (h^{-1})	Total sulfur removal (%)
3	*T. ferrooxidans*	0.0016	0.018	29.5
3	*T. thiooxidans*	0.00142	0.013	25.2
3	Mixed culture	0.00119	0.010	13.0
10	*T. ferrooxidans*	0.00216	0.029	42.0
10	*T. thiooxidans*	0.00216	0.018	30.2
10	Mixed culture	0.00208	0.005	25.4

32.5.2 Effect of Nutrient Medium

Hoffmann et al. [5], studying the effect of nutrients on coal desulfurization by *T. ferrooxidans*, concluded that classical 9K medium (3K medium in this study) basal salts are somewhat inhibitory to this organism in the concentrations supplied and proposed the low-phosphate, low-sulfate LOPOSO medium for improved results. In the present study, the effect of the nutrient supply on the microbiological desulfurization of lignites was investigated with 3% and 10% slurries. Concurrently, as deduced from the sulfate release curves shown in Fig. 32.5 and rates tabulated in Table 32.5, a supply of 9K (or 3K) basalt salts to the bacteria is by no means justified in the desulfurization of this lignite. The probable source of supply of the necessary nutrients were the minerals present in the lignites and/or carryovers with the inoculum. Based on these results, nutrient salts were thenceforth removed from the experiment flasks.

32.5.3 Effect of Pulp Density

The effect of increasing pulp density (% solids) had an increasing effect on the specific rate of sulfur removal by *T. ferrooxidans* up to 20%, as deduced from Table 32.6. The increase in rate with the increase in pulp density can be explained by the fact that a higher solids concentration would yield a greater coal surface area exposed to microorganism attack. However, this reasoning does not seem to hold in the pesent study, as the specific rate of desulfurization drops with pulp density values above 15%, as shown in Fig. 32.6 and in its loagarithmic version in Fig. 32.7. The percent sulfur removal also drops in pulp densities exceeding 15%, as indicated in Table 32.7 and in Fig. 32.8. A linear relationship cannot be observed in Fig. 32.7 for the case of 25% pulp density, suggesting a different order

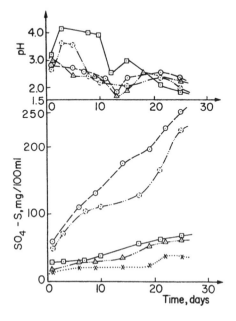

Fig. 32.5 pH values and sulfate release in the flasks supplemented and unsupplemented with nutrient. T = 35°C, pH_i = 3.0. ⊙, +N, 10%; ⊙, -N, 10%; ◻, +N, 3%; △, -N, 3%; X, control.

Table 32.5 Total Sulfur Removal in Samples Supplemented and Unsupplemented with
3K Basal Salts ($35°C$, Initial pH $= 3.0, 53 < D_p < 177 \, \mu m$)

Pulp density (%)	3K Basal salts	Total sulfur removal (%)	Rate constant, k (h^{-1})
3	+	29.5	0.0016
3	−	36.5	0.00204
10	+	45.0	0.00222
10	−	42.0	0.00216
3 (control)	−	16.6	0.00145
10 (control)	−	17.7	0.00147

in that case. One explanation for such behavior may be the product inhibition suggested
by Kelly and Jones [21].

The most probable number (MPN) of cells measured in 15% pulp density follows
the typical batch growth curve as shown in Fig. 32.9. The cell concentration increases
from 10^6–10^7 cells/mL to 10^{11}–10^{12} cells/mL by the eighth day of incubation. However, from Fig. 32.7 it is understood that sulfur removal proceeds at the same initial rate
until the twenty-fifth day, although growth stops after the eighth day. The protein release curves in Fig. 32.2 also support the MPN data. This suggests that some of the sulfur
removal obtained by bacteria may not be growth-associated phenomena after all and that
thiobacilli are required to initiate the reaction.

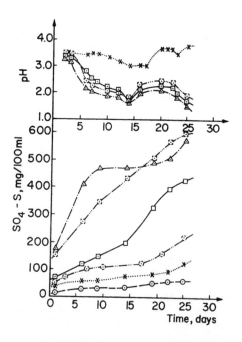

Fig. 32.6 Sulfate concentration and pH versus time plots as a function of pulp density
for fixed particle size (53 $\mu m < D_p < 177 \, \mu m$). T = $35°C$, pH$_i$ = 3.0. ⊙, 3%; ⊙, 10%;
⊡, 15%; ⊡, 20%; △, 25%; X, 20% control.

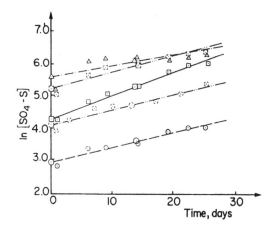

Fig. 32.7 Logarithm of sulfate concentration versus time for the coal slurry concentrations of 3, 10, 15, 20, 25% by weight. ⊙, 3%; ⊙, 10%; ▫, 15%; ▫, 20%; △, 25%.

Table 32.6 Desulfurization Rates at Different Pulp Densities

Pulp density (%)	Leach rate (mg/L·h)	First-order rate constant (h^{-1})	Maximum specific growth rate[a] (h^{-1})
3	1.08	0.00204	0.018
10	3.75	0.00216	0.029
15	4.33	0.00316	
20	10.41	0.00217	
25	19.58	0.00217	

[a]Calculated from Figs. 32.1 and 32.2.

Table 32.7 Percent Sulfur Removal at Different Pulp Densities [Initial pH = 3.0, Total sulfur (S_t) = 4.22, Pyritic sulfur (S_p) = 1.48]

Pulp density (%)	S_t removal (%)	S_p removal (%)
3	38.8	81.7
3 (control)	16.8	—
10	42.6	87.1
10 (control)	15.1	—
15	53.5	97.9
15 (control)	1.9	—
20	46.2	97.3
20 (control)	9.5	—
25	37.2	84.5
25 (control)	5.2	—

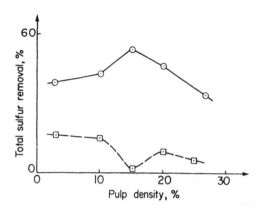

Fig.32.8 Percent of total sulfur removal versus pulp density. □, Uninoculated control; ⊙, inoculated.

32.5.4 Effect of Particle Size

One other method of increasing total particle surface area exposed to microbial attack is to lower the particle size as much as possible. Concurrently, specific sulfur rates and percentages were in accord with this reasoning, as shown in Figs. 32.10 and 32.11 and as summarized in Tables 32.8 and 32.9. The highest rate and extent of desulfurization coincided with the -270 mesh size. However, unlike the pulp density effect, an increase in rate of sulfur removal with decreasing particle size did not reverse the results beyond a critical particle size.

32.5.5 Effect of Inoculum Size

In a growth-associated phenomenon, initial concentration of microorganisms normally should not influence the specific reaction rate constant, although the effect would be obvious on the total amount of reactants changed in a finite time interval. It is evident from the data presented in Table 32.10 that this assumption does not necessarily hold in

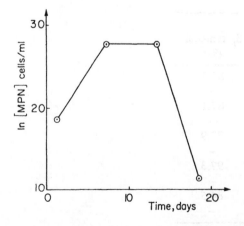

Fig. 32.9 Logarithm of most probable number of cells versus time curve.

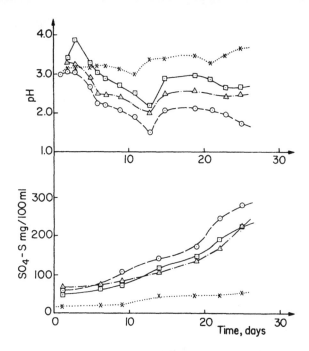

Fig. 32.10 Curves showing sulfate release with time from coals of different particle size and pH versus time. $T = 35°C$, $pH_i = 3.0$, pulp density = 10%. $\triangle$, -80 mesh; $\square$, -150 mesh; $\odot$, -270 mesh; X, control.

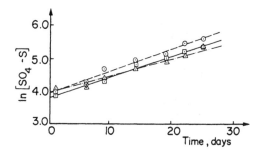

Fig. 32.11 Semilogarithmic plot of sulfate concentration versus time. Particle size: $\odot$, -270 mesh; $\square$, -150 mesh; $\triangle$, -80 mesh.

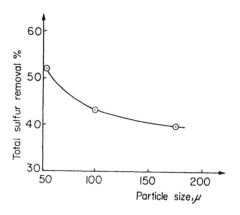

Fig. 32.12 Change of total sulfur removal percent with particle size of coal.

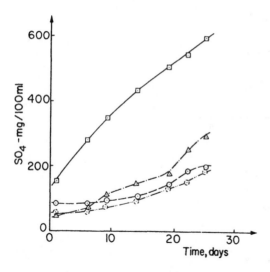

Fig. 32.13 Curves showing sulfate release versus time at 25°C and 35°C. ☐, 20% Pulp density, 53 μm < D_p < 177 μm, 35°C; △, 10% pulp density, D_p < 53 μm, 35°C; ⊙, 20% pulp density, 53 μm < D_p < 177 μm, 25°C; ⊙, 10% pulp density, D_p < 53 μm, 25°C.

Table 32.8 Rate of Sulfur Removal at Different Particle Sizes

Particle size (mesh)	Removal rate (mg/L·h)	Rate constant, k (h⁻¹)
−80	1.0	0.00225
−150	1.08	0.00275
−270	1.5	0.00294

Table 32.9 Percent Sulfur Removal with Different Particle Size

Particle size (mesh)	Initial S (%)		Final S (%)		Total sulfur removal (%)	Pyritic sulfur removal (%)
	S_t	S_p	S_t	S_p		
−80	4.22	1.48	2.60	0.13	38.30	91.20
−150	4.30	1.40	2.40	0.06	44.10	95.70
−270	4.32	1.32	2.00	0.02	53.70	98.50
−270 (control)	4.32	1.32	3.50		18.90	

Table 32.10 Effect of Initial Inoculum Size on Bacterial Lignite Desulfurization (53 μm $< D_p < 177$ μm, 10% Pulp Density, 35°C)

Inoculum size (cells/mL)	Total sulfur removal (%)	Rate constant, k (h^{-1})
10^8	42.6	0.00216
$>10^8$	50.0	0.00240
Control	18.0	

the case of sulfur removal by *T. ferrooxidans*. For the sulfur removal rate in flasks inoculated with concentrated cells ($>10^8$ cells/mL) was greater than in flasks inoculated with the same amount of unconcentrated inoculum (10^8 cells/mL). Therefore, it can be concluded that increasing the cell number in sulfur removal has a dual effect: increasing the observed plus specific rates of desulfurization.

32.5.6 Effect of Temperature

The van't Hoff rule states that chemical reaction rates generally double with each 10°C rise in reaction temperature. Microorganisms also obey this rule, and the relationship between temperature and reaction rate is represented mathematically by the Arrhenius equation. However, with thiobacilli the increase is not always twofold, as indicated by temperature quotient values (Q_{10}) of less than 2. For example, in the present study, the desulferization rates of lignites by thiobacilli increase approximately 1.3 times when the temperature is raised from 25°C to 35°C (see Table 32.11, Fig. 32.13). Activation energies (E_a) calculated from Arrhenius plots of natural logarithms of reaction rate coefficients (ln k) versus reciprocal temperature in kelvin, shown in Fig. 32.14, were somewhat low, averaging around 5 kcal/g mol. Conversely, if the reaction rate doubled with each 10°C rise, the activation energy should be around 12 kcal/g mol.

Table 32.11 Effect of Temperature on the Reaction Rate and Removal Efficiency

Temperature (°C)	Pulp density (%)	Particle size (mesh)	Total sulfur removal (%)	Rate constant, k (h^{-1})	Q_{10}	Activation energy, E_a (cal/g mol)
25	20	-80	13.6	0.00164	1.34	-5961
35	20	-80	46.2	0.00217		
25	10	-270	31.2	0.00232	1.26	-4570
35	10	-270	53.7	0.00294		

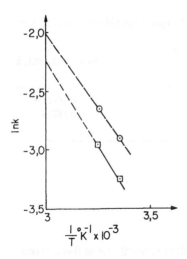

Fig. 32.14 Arrhenius plot of rate of sulfate release versus reciprocal temperature (I/K).
□, 10% Pulp density, $D_p < 33$ μm; ⊙, 20% pulp density, 53 μm $< D_p < 177$ μm.

32.6 COMBINED EFFECTS OF VARIOUS PARAMETERS ON THE DESULFURIZATION OF LIGNITES BY THE MESOPHILIC *T. FERROOXIDANS*

The effect of selected individual parameters on the desulfurization efficacy of *T. ferrooxidans* was discussed in Section 32.5 and the best combination affecting sulfur removal can be picked up among these results. Accordingly, the combination yielding the highest rate and percent of sulfur removal is the combination giving *T. ferrooxidans* as the inoculum, 15% pulp density, initial pH 3.0, and 35°C incubation temperature (see Table 32.7) with a particle size of 53 μm $< D_p < 177$ μm. Under these conditions *T. ferrooxidans* could remove 53.5% of the total sulfur present in this lignite sample within 25 days of incubation at a specific desulfurization rate of 0.00316 h^{-1}.

Although many studies have been carried out assuming that these parameters are independent, no information on the combined effects of these seems to exist in the literature. To test the interdependence of the parameters, the experimental design described earlier was carried out. While some of the experimental protocol of the preceding section was being used, some additional experiments were also needed to evaluate the design. To randomize the experimental conditions, the data were obtained in the sequence indicated by the "run order" in Table 32.12. The levels of variables used in the 2^3 factorial design were as summarized in Table 32.13.

The collective results of the factorial experimental design are summarized in Table 32.12. As shown in the table, slightly more than 53.5% of the total sulfur removal could be obtained in the combination corresponding to run number 1. The calculated coefficients of the linear first-order model for the experimental design, as shown in Table 32.14, indicate that temperature, in the range selected, is the most important factor affecting sulfur removal from lignites. In addition, particle size and pulp density are important variables in the order given and should be considered for an efficient desulfurization process. Conversely, the combined effects of these parameters were found to be negligible over the range studied.

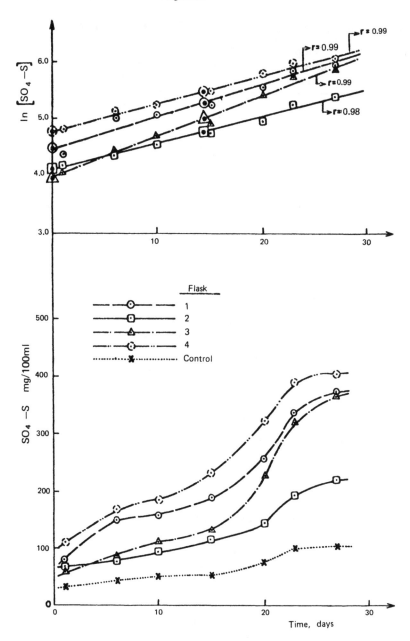

Fig. 32.15 Sulfate release versus time and log-sulfate concentration versus time plots with *T. ferrooxidans TH1*, 50°C, 10% pulp density. Flask contents are summarized in Table 32.16.

Table 32.12 Composition of Samples and Corresponding Sulfur Removal Data

Run order	Temperature ($^\circ$C)	Pulp density, P (%)	Particle size, D_p (μm)	Initial pH	Sulfur removal (%)
6	25	10	<53	3.0	25
5	35	10	<53	2.0	50
2	25	20	<53	3.0	34
1	35	20	<53	2.0	55
8	25	10	<177	2.0	16
7	35	10	<177	3.0	40
4	25	20	<177	2.0	14
3	35	20	<177	3.0	46

Table 32.13 Variables and Levels of the Factorial Design

Variable	Temperature ($^\circ$C)	Pulp density (%)	Particle size (μm)	pH
Upper level (+)	35	20	<177	3.0
Lower level (+)	25	10	<53	2.0
Coded form of the levels	$x_1 = \dfrac{T - 30}{5}$	$x_2 = \dfrac{P - 15}{5}$	$x_3 = \dfrac{D_p - 115}{5}$	$x_4 = \dfrac{pH - 2.5}{0.5}$

Table 32.14 Coefficients of the Linear First-Order Model

Variable	Coefficient	Value of coefficient
	b_0	35.00
x_1 (T, $^\circ$C)	b_1	12.75
x_2 (P, %)	b_2	2.25
x_3 (D_p, μm)	b_3	-6.0
x_{12} (T, P)	b_{12}	0.5
x_{13} (pH, T, D_p)	b_{13}	1.25
x_{23} (P, D_p)	b_{23}	-1.25
x_{123} (T, P, D_p)	b_{123}	1.5

Several trials were made using the first-order linear equation to calculate any possible higher removal efficiency, as shown in the following numerical example, but removal higher than 55% could not be obtained.

Example: Run number 1: T = 35°C, P = 20%, D_p = 53 μm, pH = 2.0. Calculated: $x_1 = 1, x_2 = 1, x_3 = -1, x_4 = -1$. Linear first-order model:

$$Y = b_0 + b_1 X_1 + b_2 X_2 + b_3 X_3 + b_{12} X_{12} + b_{13} X_{13} + b_{23} X_{23} + b_{123} X_{123}$$

Solution:

$$Y = 35.0 + 12.75 (1) + 2.25 (1) - 6.0(-1) + 0.5 (1)(1) + 1.25 (1)(-1)$$
$$- 1.25 (1)(-1) + 1.50(1)(1)(-1)$$

From this point on, the value of Y is calculated as 55%, as in Table 32.12.

32.7 DESULFURIZATION OF LIGNITES WITH THE THERMOPHILIC *THIOBACILLUS* TH1

Around 50% of the sulfur content of Nallihan-Cayirhan lignites are organic in character, as understood from Table 32.1. From the treatment in previous sections it is clearly evident that the sulfur removed from this lignite by the mesophilic *T. ferrooxidans* is almost entirely pyritic in character. However, removal of pyritic sulfur alone is not sufficient to lower the sulfur content of these lignites to 0.7% or below, which is the generally accepted figure in U.S. practice [4].

Although there is no standard regarding the sulfur content of lignites in Turkey, U.S. practices are generally accepted in this country. Therefore, it is essential to lower the combustible sulfur content of this lignite, which is the sum of pyritic and organic sulfur, by around 80% to meet the current clean air requirements. The 55% removal, which accounts for almost the entire pyritic content of the lignite, attained by the mesophilic *T. ferrooxidans* is not adequate in this respect, and further lowering of the sulfur content by at least another 25% is necessary.

The thermophilic *T. ferrooxidans* TH1 is a thermophile noted for its inability to assimilate and thus reduce sulfate or to fix carbon dioxide. The bacterium meets its carbon and reduced sulfur needs from the yeast extract supplied in the growth medium. Furthermore, it can oxidize reduced sulfur and iron compounds for energy.

Brierley et al. [23,28] have proposed the medium described in Table 32.15 for culturing of the TH group. In the desulfurization tests undertaken with this organism at

Table 32.15 Medium Composition for the TH Group

Component	g/L
$(NH_4)_2 SO_4$	0.4
$K_2 HPO_4$	0.4
$MgSO_4 \cdot 7H_2 O$	0.4
Yeast extract	0.02
$FeSO_4 \cdot 7H_2 O$	27.8[a]

[a]In experiments, iron was omitted.

Table 32.16 Composition of Test Flasks Inoculated with *T. ferrooxidans* TH1

Flask	Mineral salts	Yeast extract
1	−	+
2	−	−
3	+	+
4	+	−

50°C and using several modifications of this medium, including no nutrient case, as summarized in Table 32.16, appreciable amounts of organic sulfur removal were observed in addition to the pyritic sulfur removed in excess of 90% by this organism. Falling still short of the 80% barrier, the organism proved promising, with its ability to remove around 70% of the combustable sulfur in lignites at a comparable specific desulfurization rate of 0.00298 h^{-1} (Table 32.17) in the presence of mineral salts and yeast extract.

Table 32.17 Percent Sulfur Removal and Removal Rates by the TH1 Organism (50°C, 5 mL Inoculum 10^{10} cells/mL, 53 μm < D_p < 177 μm)

Flask	S_p rate constant (h^{-1})	Percent removal of sulfur compounds[a]			
		S_t	S_c	S_p	S_o
1	0.00226	59.7	67.9	91.2	52
1 (control)		20.1			
2	0.00195	34.8	40.4	92.5	5.8
2 (control)		17.0			
3	0.00298	62.0	71.7	94.6	56.5
3 (control)		24.0			
4	0.00197	58.0	66.0	90.0	50.0
4 (control)		13.0			

[a]S_t, total sulfur; S_c, combustible sulfur; S_p, pyritic sulfur; S_o, organic sulfur.

32.8 APPRAISAL OF BACTERIAL LIGNITE DESULFURIZATION

The figures in the literature related to rates of bacterial desulfurization are not in agreement. The differences arise mainly from the variety of experimental conditions involved, especially with relation to the type of coal tested. For example, Kargi [29], experimenting with 11% total sulfur containing American refuse coal and with 3.3% pulp density, in a batch-type stirred tank reactor with carbon dioxide supplied, obtained sulfur removal rates as high as 5×10^{-3} mg S removal/cm^2·h, as expressed in his original units. Comparable figures calculated by him for the data of Dugan and Apel [4] and Silverman et al. [6] were 1.4×10^{-4} (20% coal slurry, approximately equal mixtures of 100 to 200 mesh) and 1.9×10^{-4} mg S removed/cm^2·h, respectively.

The rate of desulfurization obtained with the mesophilic *T. ferrooxidans* in this study was calculated in Kargi's units, considering the data in Table 32.6 and the density of lignite (d) as 1.2 g/cm^3. Accordingly, the specific surface area (A) of the lignite for the average particle size (D$_p$) of 115 μm is calculated from the following equation [30].

$$A = \frac{6}{d \cdot D_p}$$

as

$$A = \frac{6}{(1.2 \text{ g/cm}^3)(115 \times 10 \text{cm}^{-4})} = 434.78 \quad \text{cm}^2/\text{g}$$

In 3% to 25% pulp densities, corresponding sulfur removal rates are calculated as in the following example. For 3% pulp density,

$$\frac{1.08 \text{ mg/L·h}}{434.78 \text{ cm}^2/\text{g} \times 30} = 8.28 \times 10^{-5} \text{ mg/cm}^2\text{-h}$$

The calculated sulfur removal rates for the 3, 10, 15, 20, and 25% pulp densities were 8.28×10^{-5}, 8.6×10^{-5}, 6.6×10^{-5}, 1.2×10^{-4}, and 1.8×10^{-4} mg S removed/cm^2·h, respectively. It can be seen that the sulfur removal rates obtained in this study are on the same order of magnitude as those of Dugan and Silverman, but 25 to 50 times slower than those of Kargi. However, after replotting Kargi's data semilogarithmically it is found that the specific rate of desulfurization is $2.2 \times 10^{-2} \text{h}^{-1}$, which is around 7 times faster than that observed in this study with mesophilic *T. ferroxidans* and 7.4 times faster than the thermophilic TH1. The specific desulfurization rate calculated from Hoffmann's data is also of the same order of magnitude as in this study [i.e., 2.1×10^{-3} h^{-1} (1% slurry)]; however, the maximum specific desulfurization rate reported by Detz and Barvinchak, 6×10^{-3} h^{-1} is the highest in the literature.

It can be concluded that the rate of sulfur removal in lignites is comparable with the rates of bituminous coal desulfurization, and even higher in several instances. The observable sulfur removal rate expressible as mg S removed/cm^2·h can be improved further by increasing pulp density, inoculum size, and carbon dioxide supply—in short, by increasing cell population. Furthermore, increasing the sulfur content of lignites should have an increasing effect on the removal rate. Increasing effects on the sulfur removal rate of the decrease in particle size and increase in temperature are obvious. Kargi suggests the use of stack gas for the supply of carbon dioxide in commercial-scale reactors.

It can be deduced from Table 32.6 that the specific growth rate of *T. ferrooxidans* increases with increasing pulp density, the highest observed being 0.029 h^{-1}. By a crude approximation this figure may also be used to estimate the specific growth rate at 15%

pulp density, assuming that the desulfurization rate/specific growth ratio has the average value of 11.1 as deduced from previous figures. The calculated specific growth rate is then $0.035 \ h^{-1}$.

According to MacDonald and Clark [19], the specific growth rate of *T. ferrooxidans* on pure pyrite is $0.145 \ h^{-1}$, which is about fourfold faster than the case of 15% pulp density in this study. On the other hand, the calculated specific growth rate of $0.035 \ h^{-1}$ is 3.5 times faster than the figure reported by Hoffmann et al. [5].

It is this author's view that going to commercial-scale pilot-plant studies, especially with high pyritic sulfur lignites, is now probably justified. Only then will there be reliable data on the cost of desulfurization for comparison with the other alternative desulfurization techniques. There is also evidence that the observable desulfurization rate may be higher in percolator-type reactors, which are also easier to apply to commercial scale.

REFERENCES

1. Attar, A., Chemistry, thermodynamics and kinetics of sulphur in coal-gas reactions: a review, *Fuel, 57*, 210 (1978).
2. Detz, C. M., and Barvinchak, G., Microbial desulfurization of coals, *Min. Congr. J., 8*, 75-82 (1979).
3. Chandra, D., Roy, P., Mishra, A. K., Chakabarti, J. N., Parasad, N. K., and Chaudhuri, S. G., Removal of sulphur from coal by mixed acidophilic bacteria present in coal and by *thiobacillus ferrooxidans, Fuel, 59*, 249-252 (1980).
4. Dugan, P. R., and Apel, W. A., Microbiological desulfurization of coal, in *Metallurgical Application of Bacterial Leaching and Related Microbiological Phenomena* (E. L. Murr, A. E. Torma, and J. A. Brierley, eds.), Academic Press, New York, 1978.
5. Hoffmann, M. R., et al., Kinetics of the removal of iron pyrite from coal by microbial catalysis, *Appl. Environ. Microbiol., 42*, 259-271 (1981).
6. Silverman, M. P., Rogoff, M. H. and Wender, I., Removal of pyritic sulfur from coal by bacterial action, *Fuel, 42*, 113-124 (1963).
7. Zarubina, Z. M., and Ashmed, D., *Izv. Akad. Nauk SSSR Otd. Tekh. Nauk.*, Metallurgica Toplivo., *1*:117-119 (1959.
8. Chandra, D., Roy, P., Mishra, A. K., Chakabarti, J. N. and Segupta, B., Microbial removal of organic sulphur from coal, *Fuel, 58*, 549-550 (1979).
9. Kargi, F., and Robinson, J. M., Removal of organic sulfur from coal, *Fuel, 65*, 397-399 (1986).
10. Gokcay, C. F., and Yurteri, R. N., Microbial desulfurization of lignites by a thermophilic bacterium, *Fuel, 62*, 1223-1224 (1983).
11. Chen, C. Y., and Skidmore, D. R., Microbial coal desulfurization with thermophilic microorganisms, research paper presented at the *INEL Workshop: Coal Bioconversion*, Washington, D.C., July 8-10, 1987.
12. Silverman, M. P., Rogoff, M. H. and Wender, I., Bacterial oxidation of pyrite material in coal, *Appl. Microbiol., 9*, 491-496 (1961).
13. Silverman, M. P., and Lundgren, D. G., Studies on the chemoautotrophic iron bacterium *Ferrobacilus ferrooxidans, J. Bacteriol., 78*, 326-331 (1959).
14. Groudev, S. N., and Genchev, F. N., Microbial coal desulfurization: effect of the cell adaptation and mixed cultures, *C.R. Acad. Bulg. Sci., 32*(3), 353-355 (1979).
15. Sato, M., Electrochemical study of oxidation of sulfide minerals at $25°C$, *Abstracts, 40th Annual Meeting, Society of Economic Geologists*, p. 110A (1959).

16. Beck, J. V., and Brown, D. G., Direct sulfide oxidation mechanism in the solubilisation of sulfide ores by *Thiobacillus ferrooxidans, J. Bacteriol.*, *96*, 1433–1434 (1968).

17. Berry, V. K., and Murr, L. E., Direct observation of bacteria and quantitative studies of their catalytic role in the leaching of low-grade, copper bearing waste, in *Metallurgical Application of Bacterial Leaching and Related Microbiological Phenomena* (E. L. Murr, A. E. Torma, and J. A. Brierley, eds.), Academic Press, New York, 1978.

18. Silver, M., Metabolic Mechanism of iron oxidizing *Thiobacillus*, in *Metallurgical Application of Bacterial Leaching and Related Microbiological Phenomena* (E. L. Murr, A. E. Torma, and J. A. Brierley, eds.), Academic Press, New York, 1978.

19. MacDonald, D. B., and Clark, R. H., The oxidation of aqueous ferrous sulphate by *Thiobacillus ferrooxidans, Can. J. Chem. Eng.*, *48*, 669–676 (1970).

20. Dugan, P. R., and Lundgren, D. G., *J. Bacteriol.*, *89*, 825 (1970).

21. Kelly, D. P., and Jones, C. A., Factors affecting metabolism and ferrous iron oxidation in suspensions and batch cultures of *Thiobacillus ferrooxidans*, in *Metallurgical Application of Bacterial Leaching and Related Microbiological Phenomena* (E. L. Murr, A. E. Torma, and J. A. Brierley, eds.), Academic Press, New York, 1978.

22. Torma, A. E., The role of *Thiobacillus ferrooxidans* in hydrometallurgical processes, in *Advances in Biochemical Engineering*, Vol. 6 (T. K. Ghose, et al., eds.), Springler-Verlag, Berlin, 1977.

23. Brierley, J. A., and Le Roux, N. W., A facultative thermophilic *Thiobacillus*-like bacterium, *Conference Reports on Bacterial Leaching*, Gesselschaft für Biotechnologische Forschung, 1977, pp. 55–56.

24. APHA, AWWA, WPCF, Standard Methods, 13th ed., 1971, p. 337.

25. ISO, *Coal-Coke Determination of Total Sulfur, Eschka Method*, No. 334, 1975-01-15, International Standard Organization, 1975.

26. ISO, *Determination of Forms of Sulfur*, No. 157, 1975-01-05, International Standards Organization, 1975.

27. Finstein, M. S., *Pollution Microbiology*, Marcel Dekker, New York, 1972.

28. Brierley, J. A., Norris, P. K. and Le Roux, N. W., Characteristics of a moderately thermophilic and acidophilic iron-oxidizing *Thiobacillus, Eur. J. Appl. Microbiol. Biotechnol.*, *5*, 291–299 (1978).

29. Kargi, F., Enhancement of microbial removal of pyritic sulfur from coal using concentrated cell suspension of *Thiobacillus ferrooxidans* and an external carbon dioxide supply, *Biotechnol. Bioeng.*, *24*, 749–752 (1982).

30. Emmett, P., *Catalysis: Fundamental Principles*, Vol. 1, Reinhold, New York, 1944.

33

Removal of Sulfur from Assam Coals by Bacterial Means

D. CHANDRA   *Indian School of Mines, Dhanbad, India*

A. K. MISHRA   *Bose Institute, Calcutta, India*

33.1 INTRODUCTION

There is a substantial reserve of tertiary coals in Assam which are of good quality, particularly with regard to a characteristically low ash content, usually combined with a good coking character. Despite this, these coals cannot, in general, be utilized properly, as they contain a high amount of sulfur. It has been established that sulfur in coal generally occurs in three different forms: (1) iron disulfide (pyrite or marcasite), (2) inorganic sulfate (generally of iron, calcium, aluminum, etc.), and (3) organically bound sulfur. The amount of these various forms of sulfur in coal depends on the source of the sample. Generally, the major contribution to the total sulfur content of coal sample is from pyritic and organic sulfur. Assam coals are unique in having a high organic sulfur content (2 to 6%) but little pyritic sulfur.

The combustion of coal is the major source of sulfur dioxide pollution. During combustion the sulfur in coal (except for inert inorganic sulfate) is oxidized to produce sulfur dioxide. When burned sulfur-bearing coal gives off sulfurous smoke, causing air pollution, corrosion of the metallic parts of the boiler, and so on. When subjected to high-temperature carbonization, sulfur-containing coal produces a gas which when used for domestic requirements needs purification. The resulting residue or coke also retains some sulfur, called fixed sulfur, which renders the coke unsuitable for smelting operations if it exceeds a certain specific limit. When coal seams in mines are exposed to the atmosphere, pyrite is gradually oxidized. As a result, mine water becomes acidic, causing various environmental hazards, including severe damage to vegetation, soil erosion, and water pollution. A good, economical method of desulfurization of coal at its source is needed so that the various hazards arising from the sulfur compounds in coal can be avoided.

Attempts have been made to desulfurize coal by solvent refining electrophoretic separation, treatment with various gases, acid peroxide treatment, oxidation by ferric chloride, and so on. But so far these methods have failed to lead to a commercially feasible process. The purpose of the present investigation was to remove sulfur from Assam coals by bacterial means under laboratory-controlled conditions. When we began our study a literature survey showed that only two teams of workers—one Soviet, one American—had attempted to desulfurize coal by bacterial means. Both teams claimed partial success in removing pyritic sulfur using a strain of *Thiobacillus ferrooxidans* [10, 11,15]. However, none attempted to remove either sulfate or organic sulfur.

33.2 DEVELOPMENT OF TECHNIQUES

For the purpose of this investigation an iron-oxidizing strain of *T. ferrooxidans* was used. The culture was obtained as *Ferrobacillus ferrooxidans* (NCIB-2580) from the National Chemical Laboratory in Poona, India. *F. ferrooxidans* is now universally accepted to be *T. ferrooxidans*, so we have used the name *T. ferrooxidans* throughout this chapter. To start with, considerable difficulties were encountered in growing the bacteria. First, it was found that the bacteria were extremely slow growing, and during growth, usually in 9K

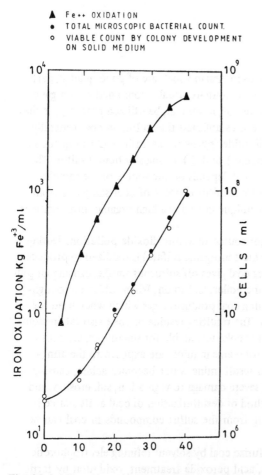

Fig. 33.1 Growth and iron oxidation of the bacterium *Thiobacillus ferrooxidans*.

liquid medium, it was difficult to secure cells free of iron salts due to their precipitation. Second, there was no way to grow the bacteria in a solid medium. Hence, to proceed with the investigation, we attempted to overcome the aforementioned two problems in the following manner.

1. Precipitation in the liquid medium was avoided by changing the acidity of the usual culture medium from pH 3.0 to pH 1.3. The bacterial strain of *T. ferrooxidans* was grown at pH 1.3 by stepwise adaptation. It was found that this acidity had no inhibitory effect on growth (i.e., increase in cell numbers against time, quantitatively expressed as generation time) and ferrous iron oxidation. The generation time, determined from the growth curve (Fig. 33.1) and the iron oxidation rate, was found to be in good agreement with those of other workers.

2. *T. ferrooxidans* is autotrophic in nature (i.e., solely dependent on inorganic nutrition and inhibited by most organic compounds). Because of this, it was difficult to grow on a solid agar medium (agar is an organic compound used as a solidifying agent in microbiological work). There was not a single report of culturing this bacterium on a solid medium except for the membrane filter technique developed by Tuovinen and Kelly [12], which is not only costly but also laborious and has limitations. Fortunately, it was possible to grow the bacterial strain on a solid medium (using agar, agarose, and a new solidifying agent, a carrageneen). This method of growing bacteria will be of great help in the future in industrial exploitation of microbial metal leaching processes, as well as in removing pyritic sulfur from coal.

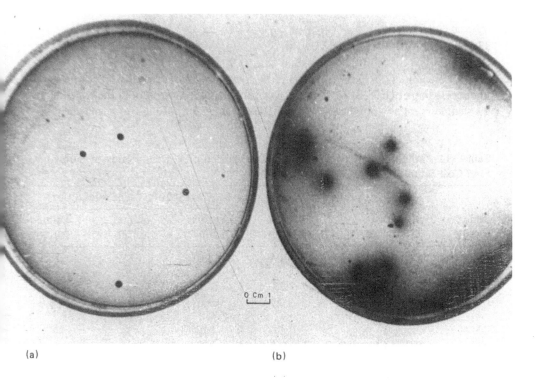

(a)   (b)

Fig. 33.2 Colonies of *T. ferrooxidans* on solid medium: (a) carrageneen solidified medium; (b) agar solidified medium.

Table 33.1 Effect of Ultraviolet Irradiation
of the Survival of *T. ferrooxidans* (Dose inci-
dent 50 ergs/mm^2/sec)

Time of irradiation (sec)	Percent survival (%)
0	1×10^2
5	4×10^1
10	1×10^1
20	1×10^{-1}
40	$8 \times \cdot 10^{-3}$
80	1×10^{-4}
160	3×10^{-5}

With this bacterial strain of *T. ferrooxidans* adapted to growth on a solid medium
(Fig. 33.2) the possibility of increasing the efficiency (the rate of pyrite oxidation) by
mutation was explored. This is the most important and valuable tool for manipulating a
specific function of a microbe by changing its genetic makeup in industrial microbiology.
Ultraviolet irradiation was applied to get an efficient mutant but as yet without success.
However, it was found that ultraviolet irradiation had a marked killing effect on this bacterium (Table 33.1), and hence sunlight would be detrimental in this microbial leaching
system.

33.3 EXPERIMENTAL PROCEDURE

33.3.1 Coal Sample

Four coal samples (J/IV, J/VI, B-5, and B-60) from Assam were used. In the experiments
the samples were powdered to −240 B.S.S. (less than 65 μm). Total sulfur and the dis-

Table 33.2 Total Sulfur Content and the Distribution of Various Forms of Sulfur in
Four Coal Samples

Sample	Colliery	Total sulfur	Distribution (% of total sulfur) Pyritic sulfur	Sulfate sulfur	Organic sulfur
J/IV	Jeypore IV seam, Assam, India	6.04	48.0	13.9	38.1
J/VI	Jeypore VI seam, Assam, India	2.29	9.0	5.0	86.0
B-5	Baragoloi 5-ft (1.5-m) seam, Assam, India	6.64	1.0	16.2	82.8
B-60	Baragoloi 60-ft (18.2-m) seam, Assam, India	3.50	25.4	9.1	65.5

tribution of different forms of sulfur content of the coal samples were estimated according to standard procedures [4]. Total sulfur content and the distribution of various forms of sulfur in the coal samples are shown in Table 33.2.

33.3.2 Treatment of Coal with Bacteria

For the treatment of coal with bacteria, a slurry of 10% with acidified water (pH 2.0) was made and 50 mL was dispensed in 100-mL Erlenmeyer flasks. The flasks were then inoculated with a 5-mL cell suspension (10^9 cells/mL) and incubated at 28°C on a rotary shaker (200 rpm). After prescribed periods of time the coal was filtered, washed with water, dried, and analyzed. The filtrate was used to estimate soluble iron. The uninoculated samples were treated similarly to serve as controls.

33.4 RESULTS

A sample of coal (J/IV) with a high pyritic sulfur content (Table 33.2) (pyritic sulfur 48.0% of the total sulfur content) was selected for bacterial treatment. The idea was to check if the technique was good enough to "take up" at least pyritic sulfur from Assam coal. The sample was treated with a strain of *T. ferrooxidans* for 3, 6, 9, and 12 days, with results shown in Table 33.3. It will be seen from Fig. 33.3 that the total sulfur content decreased progressively with the lengths of time of bacterial activity. About 44% sulfur was removed after 9 days of treatment with bacteria, whereas the control removed 27%. It was assumed for the sake of simplicity that bacteria were taking up only pyritic sulfur. The desulfurization from the control sample indicated that some pyrite oxidizers of the same *Thiobacillus* type were present in the coal sample and/or that some chemical oxidation of pyrite was occurring. So another experiment was performed with a coal sample (J/VI) having a high organic sulfur content (Table 33.2) (86% of the total sulfur content). The sample was treated with bacteria for a period of 3, 6, 9, and 12 days, with the results shown in Fig. 33.4 and Table 33.4. It appeared that for the same duration there was no difference between the control and the corresponding sample treated with bacteria. In other words, the bacteria were not effective in leaching out sulfur from the sample. It was thought that the bacterial treatment was not effective because of the low total sulfur content (2.29%) or the high organic sulfur content of the sample. To verify these results another coal (B-5) was taken which had not only a high total sulfur content

Table 33.3 Effect of Control and Bacterial Treatment on Sample J/IV for Various Time Periods

Treatment	Total sulfur content of the treated sample	Percent reduction of total sulfur from the original coal (total content of the original coal, 6.04%)
Three days with bacteria	4.0	33.7
Six days with bacteria	3.8	37.0
Nine days with bacteria	3.4	43.7
Nine days without bacteria (control)	4.4	27.1
Twelve days with bacteria	3.3	45.3

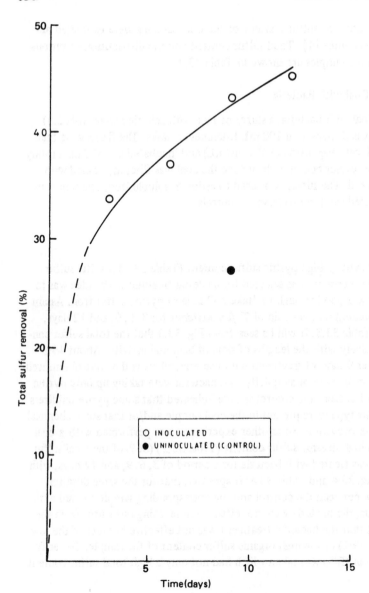

Fig. 33.3 Percent removal of total sulfur with time of treatment (sample J/IV).

(6.6%) but also a high organic sulfur content (83% of the total sulfur content). The coal was treated with *T. ferrooxidans* for periods of 6, 11, 16, 21, and 26 days. While carrying out the experiments, an opportunity was taken to determine the effect of bacterial treatment on various forms of sulfur. The distribution of sulfur in the original coal is shown in Table 33.2.

The effect of bacterial treatment on total sulfur content and different forms of sulfur is shown in Table 33.5. From the results it was observed that:

1. The control acted as a desulfurizing agent for all forms of sulfur (Table 33.5 and Figs. 33.5 to 33.8).

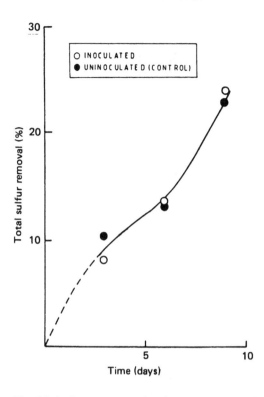

Fig. 33.4 Percent removal of total sulfur with time of treatment (sample J/VI).

Table 33.4 Effect of Control and Bacterial Treatment on Sample J/VI for Various Time Periods

Treatment	Total sulfur (%)	Percent of removal of total sulfur from the original coal (total sulfur content in the original coal, 2.29%)
Three days with bacteria	2.10	8.2
Three days without bacteria (control)	2.05	10.4
Six days with bacteria	1.98	13.5
Six days without bacteria (control)	1.99	13.1
Nine days with bacteria	1.74	24.0
Nine days without bacteria (control)	1.76	23.1

Table 33.5 Effect of Bacterial Treatment on Different Forms of Sulfur on Sample B-5

Particulars of sample taken	Total sulfur (%)	Percent removal of total sulfur from the original coal	Sulfate sulfur (actual) (%)	Percent removal of sulfate from the original coal	Pyritic sulfur (actual) (%)	Percent removal of pyritic sulfur from the original coal	Organic sulfur (actual) (%)	Percent removal of organic sulfur from original coal
Original coal	6.642		1.073		0.069		5.500	
Treated 6 days without bacteria (control)	6.065	8.68	0.585	45.48	0.046	33.33	5.434	1.20
Treated 6 days with bacteria	5.719	13.89	0.418	61.04	0.035	49.27	5.266	4.25
Treated 11 days without bacteria (control)	5.881	11.46	0.412	61.60	0.046	33.33	5.423	1.40
Treated 11 days with bacteria	5.567	16.18	0.379	64.68	0.035	49.27	5.153	6.31
Treated 16 days without bacteria (control)	5.812	12.50	0.376	64.95	0.046	33.33	5.390	2.00
Treated 16 days with bacteria	5.499	17.21	0.324	69.80	0.035	49.27	5.130	6.73
Treated 21 days without bacteria (control)	5.753	13.38	0.376	64.95	0.046	33.33	5.331	3.07
Treated 21 days with bacteria	5.490	17.34	0.320	70.18	0.023	66.66	5.157	5.23
Treated 26 days without bacteria (control)	5.663	14.74	0.344	67.94	0.035	49.27	5.384	2.11
Treated 26 days with bacteria	5.496	17.25	0.300	72.04	0.012	82.60	5.184	5.74

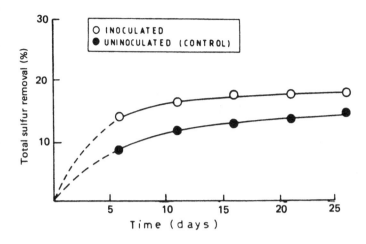

Fig. 33.5 Percent removal of total sulfur with time of bacterial treatment (sample B-5).

2. The bacteria were also active in removing various forms of sulfur. For any particular form of sulfur, bacteria-treated samples always desulfurized more than the corresponding control (Figs. 33.5 to 33.8). But the difference in desulfurizing potentiality between the bacteria-treated sample and corresponding control remained almost constant, as shown by the almost parallel curves. This was particularly true for the removal of total sulfur, sulfate sulfur, and organic sulfur (Figs. 33.5, 33.6, and 33.8). For example, (a) a sample treated with bacteria for 26 days removed 17% of the total sulfur, whereas the control removed 15% of the total sulfur during the same period (Table 33.5); (b) a sample treated for 26 days removed 72% of the sulfate sulfur, whereas the corresponding control removed 68% of the sulfate sulfur (Table 33.5); and (c) a sample treated for 26 days leached 6% of the organic sulfur, whereas the control removed 2% of the organic sulfur (Fig. 33.8).

3. The control removed about 49% of the pyritic sulfur when the coal was treated for 26 days (Table 33.5 and Fig. 33.7), but the bacterial action had a greater, more pronounced effect in the removal of pyritic sulfur. *T. ferrooxidans* were very sensitive in the removal of pyritic sulfur. Even when the coal sample contained only 0.069% pyritic sulfur, about 83% of this minute quantity of pyritic sulfur was removed when the coal was treated with bacteria for 26 days (Table 33.5). It should be pointed out that this technique would be very useful in countries where it is a great problem to remove pyritic sulfur from coal to control atmospheric pollution.

4. The bacteria appeared to have a limited effect on removing organic sulfur. About 6 to 7% of organic sulfur was removed when the coal was treated for 11 to 16 days (Table 33.5 and Fig. 33.7). Although there was no further improvement for total sulfur, sulfate sulfur, or organic sulfur after 6 days of bacterial treatment, there was a further fall in pyritic sulfur content after 21 days of bacterial treatment (Figs. 33.5 to 33.8).

From the foregoing results it appeared that the assumptions made earlier were not true. Despite the coal having a high organic sulfur content, the bacteria were quite active in removing pyritic sulfur (although the latter was present in minute quantities) from the coal sample used in the experiment. Further, it was presumed that the bacterial action was inhibited in the previous sample (J/VI) due to other factors. It was noted from the

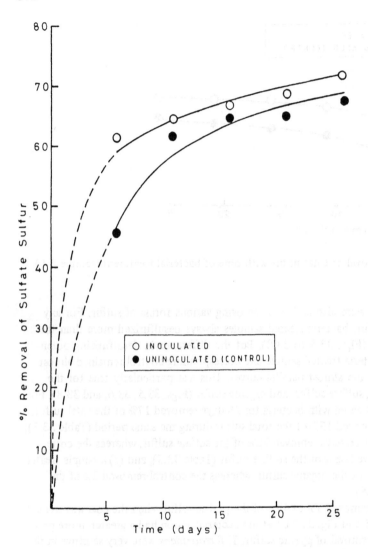

Fig. 33.6 Percent removal of sulfate sulfur with time of bacterial treatment (sample B-5);

literature that carbonate prevents the action of bacteria on coal. Therefore, the calcite contents of the coal samples used for bacterial treatment were measured. It was found that only sample J/VI contained calcite. So it was probable that calcite in the sample might have inhibited bacterial action in the sample.

33.4.1 Microbes in Coal as Desulfurizing Agent

In all the experiments on coal, it was observed that coal itself served as a desulfurizing agent when it was used as a control. For example, in sample B-5 the control itself removed 49.27% of pyritic sulfur (Table 33.5). In other words, the results indicated that raw coal contained microbes that were capable of removing pyritic sulfur from coal.

To prove this assumption a sample of coal (J/IV) was sterilized by autoclaving at a pressure of 15 psi for 15 min to make it free, as much as possible, from bacteria that

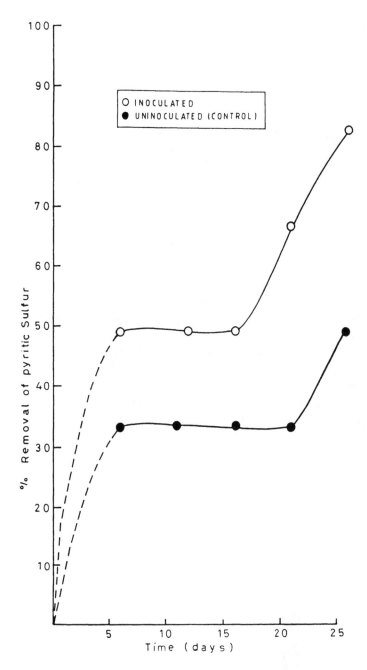

Fig. 33.7 Percent removal of pyritic sulfur with time of bacterial treatment (sample B-5).

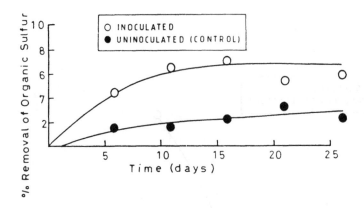

Fig. 33.8 Percent removal of organic sulfur with time of bacterial treatment (sample B-5).

were inherent in the coal. Then the sterilized coal was treated with and without addition of bacteria for 15 days at a temperature of 28°C at pH 2.0. For comparison, unsterilized coal (i.e., the raw coal retaining the inherent bacteria) of the same sample was treated with and without addition of bacteria for the same length of time (15 days), the same temperature (28°C), and the same pH (2.0). It was observed (Fig. 33.9) that unsterilized, uninoculated coal (i.e., raw coal with inherent bacteria) showed greater reduction of pyritic sulfur than did sterilized noninoculated coal (i.e., raw coal devoid of inherent bacteria). In other words, unsterilized, noninoculated coal (i.e., raw coal), because of the in-

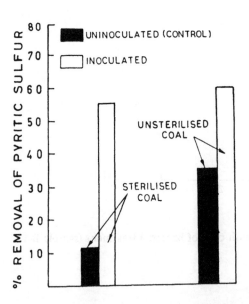

Fig. 33.9 Removal of pyritic sulfur from the sterilized (almost free from inherent bacteria) and unsterilized coal (with inherent bacteria) (Sample J/IV; time of incubation, 15 days; temp., 28°C; pH, 2.0).

herent bacteria, removed a greater percentage of pyritic sulfur than did sterilized coal, as the inherent bacteria were lacking in the sterilized coal.

This observation has considerable significance for future industrial use of the leaching technique. For example, if ever bacterial leaching technique is to be used, say, for leaching of pyrites, additional expenses for producing bacteria for the leaching could be avoided. It is most likely that the coals used in previous experiments could immediately be used as the source of bacteria.

33.4.2 Effect of Nutrient

To see the effect of nutrient on bacteria, a sample of coal (J/IV) was sterilized and from the sterilized sample, three subsamples were drawn and treated as follows:

1. Sample A was treated for 100 h in much the same way as samples B and C but without bacteria or nutrient (in other words, this sample was used as a control).
2. Sample B was treated with bacteria for 100 h.
3. Sample C was treated with bacteria and a nutrient for 100 h.

The results are shown in Fig. 33.10. It will be seen that the sample with the nutrient had the greater effect on desulfurization.

33.4.3 Effect of Surfactant

Like nutrient, the addition of a surfactant (Tween-80, 100 ppm) increased the removal of pyritic sulfur (Fig. 33.11).

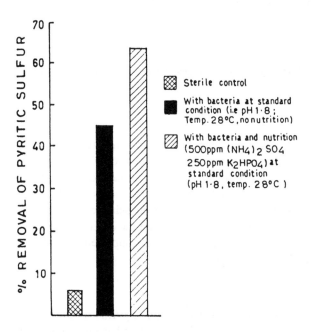

Fig. 33.10 Effect of nutrient on the removal of pyritic sulfur by *T. ferrooxidans* (sample J/IV, sterilized, treated for 100 h).

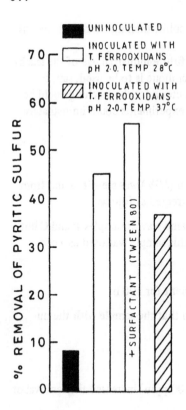

Fig. 33.11 Effect of surfactant and temperature on the removal of pyritic sulfur after 100 hours of incubation of the sterilized coal sample (J/IV) with *T. ferrooxidans*.

33.4.4 Effect of Temperature

A higher temperature (37°C) was found to have an adverse effect on the removal of pyritic sulfur (Fig. 33.11).

33.4.5 Effect of Calcite

Earlier it was mentioned that sample J/VI contained a significant amount of calcite, which inhibited the bacterial action on pyrite. To verify this assumption, the coal was treated with dilute HCl to remove calcite. This acid-pretreated coal was then subjected to bacterial action. The coal samples were also treated by the bacteria under different conditions. Eighty-five percent of the pyritic sulfur was removed from acid-treated coal (Fig. 33.12), whereas only 60% of the pyritic sulfur was removed from the original sample that had not been treated with acid (Fig. 33.12). The results therefore supported the assumption that calcite inhibited bacterial action on pyrite.

33.4.6 Effect of pH

An attempt was made to determine the effect of pH on removal of pyritic sulfur with *T. ferrooxidans*. For this purpose sterilized coal (5% w/w) in acidified water with varying pH (2.0 to 3.5) was inoculated with *T. ferrooxidans* and incubated at 28°C in a shaker (150 rpm). The optimum pH for coal pyrite oxidation, as estimated by the release of

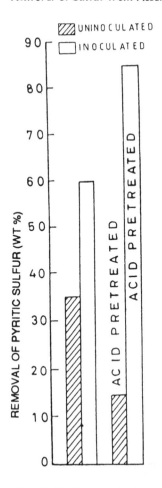

Fig. 33.12 Removal of pyritic sulfur without and with pretreatment of acid (HCl) (sample J/VI; unsterilized, inoculated with *T. ferrooxidans*; Time of incubation, 15 days; temp., 28°C; pH, 2.0).

soluble iron, was around 2.0 to 2.5. A higher pH (3.0 to 3.5) was detrimental for pyrite oxidation (Fig. 33.13). The maximum pyrite oxidation was about 7 mg iron/L per hour. A lag was observed at pH 3.0 and 3.5. At 200 h 90% pyrite was oxidized at pH 2.5, whereas only 47% oxidation occurred at pH 3.5 during this period.

33.4.7 Effect of Pulp Density

To determine the effect of pulp density, coal suspensions (100 mL) at different initial concentrations and at pH 2.0 were used. The iron leach rate increased with increased coal concentration and the rate was maximum at 20% pulp density (16.5 mg iron/L per hour; Fig. 33.14). At 2.5% pulp density the iron leach rate was only 3.0 mg iron/L per hour, which increased linearly with the increase in pulp density up to 10%, where the rate was 13.0 mg iron/L per hour. Almost 90% oxidation of coal pyrite was observed at 5% or 10% pulp density, whereas only 60% oxidation occurred at 20% pulp density after 200 h.

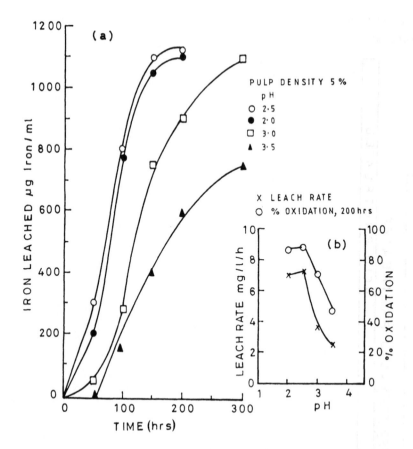

Fig. 33.13 Effect of pH on oxidation of pyrite of a coal sample (J/IV) by *T. ferrooxidans*.

33.4.8   Removal of Organic Sulfur

From the foregoing results it became evident that the strain *T. ferrooxidans* is capable of removing organic sulfur to a limited extent. Therefore, it was essential to try another line of action in removing organic sulfur from coal. The presence of thiophenol and thiophene-like organic sulfur compounds in coal has been established indirectly [5]. Thus, assuming the presence of thiophene in coal, a mixed bacterial culture was isolated from dibenzo-thiophene (DBT). For this purpose both sterilized and unsterilized coal samples were treated with DBT-enriched bacterial culture grown in a DBT medium (Table 33.6). The sterilized coal sample removed about 20% of the organic sulfur (Table 33.7). The removal of organic sulfur was slightly better in the case of sterilized coal samples, as perhaps there was no interaction with inherent bacteria of the *Thiobacillus* type present in the unsterilized (i.e., raw) coal. The results definitely proved the efficacy of the DBT-enriched culture in attacking organic sulfur and successfully removing it.

33.4.9   Isolation and Purification of Microbes

An attempt was made to isolate and identify the various microbes grown in DBT to evaluate the potentiality of the DBT-isolated strains to remove forms of sulfur other than or-

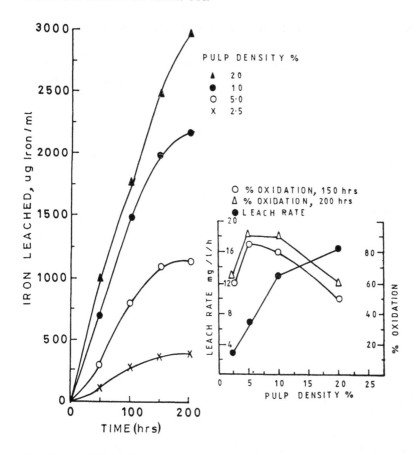

Fig. 33.14 Effect of pulp densities on oxidation of pyrite of a coal sample by *T. ferro-oxidans.*

ganic sulfur from coal. A number of soil samples were inoculated in 50 mL of medium containing DBT and incubated at 28°C in a shaker (150 rpm). Transfers were made weekly. After several such transfers, the best growing cultures were selected by both visible turbidity and microscopic observations. The cultures were streaked on nutrient agar plates. On the basis of the foregoing search, seven bacteria were isolated and designated as B, N, S, T, V, W, and X, B, N, T, and V belonged to the *Bacillus* group, S and W were species of *Micrococcus*, and X was a strain of *Pseudomonas*.

Table 33.6 Composition of the Medium Containing DBT Used

NH_2CL	500 mg
$MgCl_2$	250 mg
KH_2PO_4	250 mg
DBT	1.0 g
Water	1000 mL
pH	7.5

Table 33.7 Effect of Bacterial Treatment on the Removal of Organic Sulfur from a Coal (B-5)

Sample	Total sulfur (%)	Organic sulfur (%)	Removal of organic sulfur (% of organic sulfur)
Original coal < 240 BSS	6.68	5.53	—
Unsterilized coal treated under same conditions without adding bacteria (control samples)	6.56	5.44	1.6
Sterilized coal without adding bacteria (control sample)	6.54	5.42	2.0
Sterilized coal treated with bacterial culture	5.40	4.43	19.9
Unsterilized coal treated with bacterial culture	5.62	4.62	16.5

33.4.10 Nutrient Broth Medium

Since dibenzothiophene is a costly chemical, the possibility of using a cheaper medium without DBT was explored. Four different strains of *Bacillus*—B, N, T, and V—were grown in the DBT as well as in the standard bacteriological medium (nutrient broth) without DBT. The results obtained are shown in Table 33.8. The composition of the nutrient broth is shown in Table 33.9. It will be seen that the strains grown in nutrient broth medium are more efficient in removing total and organic sulfur from coal than are those grown in DBT medium.

Table 33.8 Removal of Organic Sulfur by DBT-Isolated Strains Grown in Different Media for 10 Days at 28°C and pH 7.0

Original coal (B-5) treated with strain:	Medium[a]	Total sulfur[b] (%)	Percent reduction	Organic sulfur[c] (%)	Percent reduction
Bacillus T	NB	4.210	36.60	3.661	30.81
	DBT	4.414	33.52	3.781	28.52
Bacillus B	NB	3.911	41.09	3.374	36.29
	DBT	4.378	34.08	3.961	25.12
Bacillus N	NB	4.152	37.46	3.540	33.08
	DBT	4.449	32.99	3.894	26.39
Bacillus V	NB	4.120	37.95	3.660	30.81
	DBT	4.555	31.42	3.877	26.51

[a]NB, nutrient broth; DBT, dibenzothiophene.
[b]6.64 in original coal.
[c]5.29 in original coal.

Table 33.9 Composition of the Nutrient Broth Medium Used

Peptone	10 g
Beef extract	10 g
NaCl	5 g
Water	1000 mL
pH	7.2

33.4.11 Effect of pH on the Removal of Organic Sulfur from Coal by the Bacterial Species

To show whether pH had any effect, an experiment was conducted at two different pH values: 7.0 and 8.5. The results are shown in Table 33.10. It is apparent from the result that organic sulfur removal is effective in the pH range 7.0 to 8.5. This result is in favor of the commercial feasibility of the process because maintenance of a constant pH is an extra burden in any industrial processes. Therefore, pH tolerance over a range is rather advantageous.

33.4.12 Removal of Different Forms of Sulfur from a Coal Sample

Since Coal India Ltd. plans to mine a 60-ft seam in the Baragoloi Colliery, an attempt was made to treat the coal sample (B-60) with individual strains isolated from DBT-enriched culture. The results show that all the strains isolated from a DBT culture were capable of removing pyritic, sulfate, and organic sulfur from this coal, and about 45% of the total sulfur was removed (Table 33.11). The sample had an original total sulfur content of 3.5%, which after treatment with bacteria decreased to 1.93%.

Table 33.10 Effect of pH on the Microbial Removal of Organic Sulfur Treated in Nutrient Broth with 10% Coal (w/v) at 28°C for 10 Days

Original coal (B-5) treated with strain:	pH	Total sulfur[a] (%)	Percent reduction	Organic sulfur[b] (%)	Percent reduction
Bacillus T	7.0	4.71	29.06	3.99	24.42
	8.5	4.69	29.36	3.94	25.51
Bacillus B	7.0	4.73	28.76	3.96	25.08
	8.5	4.76	28.30	3.94	25.61
Bacillus N	7.0	4.75	28.37	4.00	24.38
	8.5	4.69	29.23	3.94	25.50
Bacillus V	7.0	4.65	29.89	3.90	26.14
	8.5	4.85	26.95	4.11	22.17
Pseudomonas X	7.0	4.62	30.48	3.92	25.81
	8.5	4.87	26.71	4.14	21.66

[a]6.64 in original coal.
[b]5.29 in original coal.

Table 33.11 Removal of Different Forms of Sulfur by DBT-Isolated Strains in Nutrient Broth (pH 7.2)[a]

Original coal (B-60) treated with strain:	Total sulfur[b] (%)	Percent reduction	Sulfate sulfur[c] (%)	Percent reduction	Pyritic sulfur[d] (%)	Percent reduction	Organic sulfur[e] (%)	Percent reduction	Final pH of the medium[f]
Bacillus T	1.95	44.25	0.17	46.88	0.06	93.48	1.72	24.75	2.1
Bacillus B	1.99	43.14	0.19	40.95	0.04	96.51	1.75	23.36	2.2
Bacillus N	1.93	44.85	0.19	40.95	0.10	88.31	1.64	28.30	2.5
Bacillus V	2.06	41.14	0.19	40.95	0.03	96.05	1.83	19.85	2.5
Pseudomonas X	2.06	41.14	0.15	52.50	0.07	92.30	1.84	19.61	3.0
Micrococcus W	2.04	41.71	0.17	46.88	0.17	80.89	1.85	19.21	2.2
Micrococcus S	2.15	38.57	0.16	47.81	0.17	80.89	1.93	15.22	3.0

[a]Coal sample was sterilized before treatment.
[b]3.50 in original coal.
[c]0.32 in original coal.
[d]0.89 in original coal.
[e]2.29 in original coal.
[f]Original pH was 7.

Table 33.12 Properties of Coal Before and After Leaching with Bacteria (Baragoloi 5-ft Seam)

Treatment	As received			
	Moisture	Ash	BS swelling number	Volatile matter (dry, ash free)
Untreated sample	4.02	18.36	3	49.19
Treatment with *Bacillus* strains T, B, N, V grown in nutrient broth medium	2.76	9.90	3	46.23
Treatment with *Bacillus* strains T, B, N, V grown in DBT medium	2.10	9.76	2-1/2	45.38

33.4.13 Quality of Coal After Bacterial Treatment

The coal sample (B-5) was treated in the usual manner by the bacteria isolated from DBT, by growing them either in nutrient broth or in DBT medium at neutral pH (7.0 to 7.2). In two bacteria-treated samples, there was a reduction of about 50% ash content (Table 33.12); also, there was a marginal decrease in volatile matter content, suggesting an improvement in rank.

33.5 CONCLUSIONS

Earlier it was believed that organic sulfur is so intimately bonded in the chemical structure of coal that it will be extremely difficult to get rid of. However, the results of Chandra et al. [2] definitely show that a portion of the organic sulfur comes out very easily. The authors kept the coal exposed to the atmosphere and organic sulfur was found to decrease with time due to weathering. It also appears that part of the organic sulfur is also linked with a thiophene-like structure in coal, which may be removed using DBT-grown bacteria. However, to remove organic sulfur completely from coal it is necessary to grow bacteria (as grown on DBT) on any other sulfur-linked organic compounds that may be present in the coal.

We should mention that, so far, all work on desulfurization of coal has been "groping in the dark," because it is not known how sulfur is linked in the structure of coal. Therefore, for efficient and effective removal of organic sulfur from coal, it is essential that a separate research project be undertaken to determine the distribution of sulfur in the coal structure. In this connection, it may also be pointed out that so far there is no standard method of direct determination of organic sulfur in coal. Total sulfur minus the sum of pyritic and sulfate sulfur is taken to be the organic sulfur content [i.e., organic sulfur = total sulfur – (pyritic sulfur + sulfate sulfur)]. Therefore, an attempt should be made to find a satisfactory direct method of organic sulfur determination. However, with our present knowledge of coal desulfurization, the total sulfur content may be reduced to the tolerance limit of utilization for most coals.

It is very much easier to desulfurize coals other than Assam coals. This is because Assam coals are abnormal [3] and contain high levels of organic sulfur. They are, therefore, far more difficult to desulfurize.

The prospects of commercial success of desulfurization will become much brighter if the acidic water remaining after desulfurization of coal could be utilized either for leaching of sulfide ores [7] or for producing ammonium sulfate (fertilizer) by reacting with ammonia [1,8]. The acidic water may also be used for solubilization of rock phosphates for the preparation of phosphoric acid [1,9]. The volume or pH of the acidic solution may be controlled by adding either pyrite or pyritiferous shales [1,6]. Besides microbial desulfurization of Assam coals several chemical and physicochemical processes of desulfurization of coal have been evolved [2,13]. An attempt is being made to combine the bacterial process with the physicochemical and/or chemical method for desulfurization of coal to improve the efficiency of removal of sulfur from coal.

REFERENCES

1. Chandra, D., Feasibility of beneficiation of pyritiferous shales by bacterial leaching on a laboratory scale, and solubilisation of rock phosphate by microbial means, *Report on the PPCL-Sponsored Research Project*, Indian School of Mines, Dhanbad, India, 1982.
2. Chandra, D., Chakrabarti, J. N., and Swamy, Y. V., Autodesulphurization of coal, *Fuel, 61*, 204-205 (1982).
3. Chandra, D., Ghose, S., and Choudhuri, S. G., On certain abnormalities in the chemical properties of Tertiary coals of Upper Assam and Arunachal Pradesh, *Fuel, 63*(9), 1318-1323 (1984).
4. Chowdhury, A. N., *Methods of Testing for Coal and Coke*, 1350 (Part III), Indian Standards Institution, 1969, pp. 5-13.
5. Iyengar, M. S., Guha, S., Beri, M. L., and Lahiri, A., Oxidation of abnormal coals, in *Proceedings of the Symposium on the Nature of Coal* (A. Lahiri, ed.), Council of Scientific and Industrial Research, India, 1959, pp. 206-214.
6. Mishra, A. K., Roy, P., Roy Mahapatra, S. S., and Chandra, D., Low grade pyrites and their possible beneficiation by *Thiobacillus ferrooxidans*, *Proc. Indian Natl. Sci. Acad., B50*(5), 519-524 (1984).
7. Roy Mahapatra, S. S., Roy, P., Mishra, A. K., and Chandra, D., Leaching of sulfide ores by *Thiobacillus ferrooxidans*, *Proc. Indian Natl. Sci. Acad., B51*(1), 85-95 (1985).
8. Roy Mahapatra, S. S., Mishra, A. K., Chandra, D., Pandalai, H. S., Ghosh, P. K., and Banerjee, P. K., Oxidation of pyrite from pyritiferous shales using *Thiobacillus ferrooxidans, Indian J. Exp. Biol., 23*, 42-47 (1985).
9. Roy Mahapatra, S. S., Mishra, A. K., and Chandra, D., Solubilisation of rock phosphate by *Thiobacillus ferrooxidans, Curr. Sci., 54*(5), 235-237 (1985).
10. Silverman, M. P., Rogoff, M. H., and Wender, I., Bacterial oxidation of pyritic minerals in coal, *Appl. Microbiol., 9*, 491-496 (1961).
11. Silverman, M. P., Rogoff, M. H., and Wender, I., Removal of pyritic sulfur from coal by bacterial action, *Fuel, 42*, 113-124 (1963).
12. Tuovinen, O. H., and Kelly, D. P., Studies on growth of *Thiobacillus ferrooxidans, Arch. Microbiol., 88*, 285-298 (1973).
13. Venkata Swamy, Y., Chandra, D., and Chakravarti, J. N., Removal of sulphur from Indian coals by sodium hydroxide, *J. Inst. E. (London), 57*, 433, 438-443 (1984).
14. Yamda, K., Minoda, Y., Kodama, K., Nakatani, S., and Akasaki, T., Microbial conversion of petro-sulfur compounds. I. Isolation and identification of dibenzothiophene utilizing bacteria, *J. Agric. Biol. Chem., 32*, 840-845 (1968).
15. Zarubina, Z. M., Lylikova, N. N., and Shmuk, Ye, I., Investigation of microbiological oxidation of coal pyrite, *Izv. Akad. Nauk SSSR Otd. Tekh. Nauk Metall. Topl., 1*, 117-119 (1959).

34

Microbial Coal Desulfurization with Thermophilic Microorganisms

CHARLES C. Y. CHEN and DUANE R. SKIDMORE   *The Ohio State University, Columbus, Ohio*

34.1 INTRODUCTION

Coal is one of the major energy sources in the United States, contributing about 22 to 24% of domestic consumption. It is abundant, with a resource base of about a trillion tons, estimated to be able to last 400 to 4000 years [1]. It is relatively cheap; at a price of $22 to $40, a ton of typical bituminous coal generates about 24 million Btu of energy. Although the current oil glut discourages coal production, the coal production rate is still at a level close to 900 million tons per year [2]. In 1984, 56% of the coal was consumed by power generation [2]. However, sulfur and nitrogen oxides released in coal combustion brought some problems to the coal industry, especially in the eastern part of the country.

Through complex chemical reactions, SO_2 and SO_3 are converted to sulfuric acid and nitrogen is converted to nitrogen oxides which precipitate as acid rain. To control the acid rain problem, the Clean Air Act and its amendments were passed. These laws set an emission limit of 1.2 lb SO_2 per million Btu, which is equivalent to 0.7% sulfur for a typical coal [3]. The impact of these laws on the coal industry and utility power plants is pronounced. A large fraction of coal used for power generation is produced east of the Mississippi River, while 62% of low-sulfur coal reserves are west of the Mississippi [4]. Many coal mines have been shut down because their coals have a high sulfur content. Power plants in the east had to pay extra transportation costs to ship low-sulfur coal from the west. The alternatives to shipping coal from the west are numerous. Sulfur and nitrogen can be removed before combustion, or SO_2 can be eliminated during or after combustion. At least 50 precombustion coal cleaning techniques are under development, based on biological, physical, and chemical principles [5]. The published literature lends credibility to the allegations that biological cleaning methods are feasible [3,6]. For example, the cost of microbial coal desulfurization was estimated to be $10 to $14 per ton, compared with \$20 to \$30 per ton for chemical leaching [3]. Numerous review articles about microbial coal desulfurization are available in the literature [6-8].

In this chapter we review work, unpublished and published, performed in the Chemical Engineering Department at the Ohio State University. The objective of the program was to develop a commercial process for microbial coal desulfurization using a thermophilic microorganism, *Sulfolobus acidocaldarius*. The effort started in 1978 with *Thiobacillus ferrooxidans* and shifted in 1981 when *S. acidocaldarius* was chosen as being a more active bacterium for coal desulfurization than the mesophilic *T. ferrooxidans* and another thermophile, *S. brierleyi*. Then the effects of coal types and particle size, yeast extract, temperature, coal slurry pulp density, CO_2 and O_2 enrichment, reactor scaleup, surfactant addition, stirring and aeration rates, and other process variables were investigated. In addition, fundamental studies were undertaken to describe biological, chemical, and physical processes.

34.2 CHARACTERISTICS OF *SULFOLOBUS*

In 1972, Brock et al. [9] reported the isolation and characterization of a new genus of sulfur-oxidizing bacteria able to grow at high temperatures and low pH. This genus was designated *Sulfolobus* gen. nov. and characterized as follows: (1) generally, spherical cells producing frequent lobes; (2) facultative autotroph with growth on sulfur or on a variety of simple organic compounds; (3) unusual cell wall structure devoid of peptidoglycan; (4) acidophilic—pH optimum of 2 to 3 with a range from 0.9 to 5.8; and (5) thermophilic—with a temperature optimum of 70 to 75°C. A more detailed description is given in *Bergy's Manual of Determinative Bacteriology* [10].

The organisms of this genus are widespread and can readily be isolated from both acid soils and acid hot springs rich in sulfur. Different species have been isolated from thermal spring areas in the United States, Italy, Iceland, New Zealand, Dominica, and El Salvador. These species are *S. acidocaldarius*, *S. brierleyi*, *S. solfotaricus*, and *S. ambivalens* [9,11,12]. *S. acidocaldarius* was first isolated from an acid thermal region of Yellowstone National Park by Brock et al. [9]. The species is able to grow on ferrous sulfate, elemental sulfur, pyrite, yeast extract, and some organic compounds. Among the other species in the family, *S. brierleyi* has also been applied to effect mineral leaching [13]. *S. acidocaldarius* and *S. brierleyi* differ in many ways: *S. brierleyi* ls larger and is unable to use yeast extract as an energy source.

34.3 CHARACTERISTICS OF SULFUR IN COALS

Sulfur is found in coal in three forms: pyritic, sulfate, and organic sulfur. Definitions of the sulfur forms are based on standard leaching experiments specified by the American Society of Testing and Materials (ASTM) [14]. Pyritic sulfur is the nitric acid-leachable part and sulfate sulfur is the hydrochloric acid-soluble part of the sulfur. Organic sulfur, then, balances the total sulfur content determined by combustion [14]. Pyritic sulfur generally exists as iron pyrite (FeS_2) and in coal is a major form of sulfur. For example, most of Ohio's major coal seams contain more than 50% of total sulfur as pyrite [15]. The pyrite in coal has a wide range of grain-size distributions—from a few micrometers to 300 μm, depending on the type of coal [16]. For most coals, the content of sulfate sulfur in fresh coal is very low. The sulfate sulfur is in the form of mineral salts, some of which could also contribute to SO_2 emission through combustion. Organic sulfur is poorly defined and very little direct information is available about its structure in coal. However, sulfide, disulfide, thiols, and thiophenes have been identified in the liquids derived from solvent extraction of coal [17]. Organic sulfur is thought to be part of the matrix of coal structure, and in any event, is the most difficult part of the sulfur to remove from coal.

Generally, microorganisms of the genera *Thiobacillus* and *Sulfolobus* remove sulfate and pyritic sulfur at a significant rate, and a recent report showed that some uncharacterized mesophiles removed organic sulfur [18]. Isbister and Kobylinski reported that 18 to 47% of organic sulfur was removed from various coals by a microorganism created by mutagen treatments of naturally occurring microorganisms [19]. Reduction of organic sulfur was also observed in the authors' laboratory [20]. Some investigators observed the oxidation of dibenzothiophene (DBT) by *S. acidocaldarius* [21]. However, a study is under way to establish this point more solidly.

34.4 MICROBIAL COAL DESULFURIZATION BY *S. ACIDOCALDARIUS* AND *S. BRIERLEYI*

In this study the more active species of the thermophilic genus *Sulfolobus* were selected for further development of a process for microbial coal desulfurization. Finely divided coal was suspended in a mineral salt medium and inoculated with the thermophilic bacteria. Sulfate release was measured to determine the rate and extent of bacterial release of sulfur from coal. A mineral salt medium was prepared with the following composition (g/L): $MgSO_4 \cdot 2H_2O$, 0.5; $(NH_4)_2SO_4$, 0.5; KH_2PO_4, 0.5; yeast extract, 0.02% (w/v); the pH was adjusted to 2.0 with H_2SO_4. The pH was adjusted to such a low level to prevent minerals precipitation. The coal was Clarion 4-A coal sampled from the McArthur Mine in Vinton County, Ohio. The coal contained 2.54% pyritic sulfur, 0.16% sulfate, and 0.63% organic sulfur. For the reaction, coal was ground to below 200 mesh (< 74 μm) and suspended in the mineral salt medium to make a 5% slurry by weight. The cultures of *S. acidocaldarius* and *S. brierleyi* were obtained from Dr. Brierley at the New Mexico Institute of Mining and Technology. The microorganisms showed growth in the previously described medium with elemental sulfur, pyrite, and ferrous sulfate as the energy sources [9,22-24].

Laboratory experiments on microbial sulfur leaching were carried out in round-bottomed three-necked flasks kept in a water bath with the temperature at or near 70°C. The flask, containing 800 mL of medium, was inoculated with 15 to 20 mL of stock culture. Continuous aeration with house air was provided to supply oxygen and carbon dioxide for microbial activity. Since the air evaporated some moisture, a condenser was attached to the flask to condense the moisture and to prevent dehydration of medium in the flask. Sulfate was measured by the turbidimetric method specified by the ASTM [25].

Figure 34.1 shows the history of sulfate release from coal during a typical leaching experiment. Both *S. acidocaldarius* and *S. brierleyi* removed sulfur from coal. The lag phase present in this experiment is a result of subculturing from the stock culture to the coal slurry. The results indicated that *S. acidocaldarius* removed sulfur from coal at a higher rate than that of *S. brierleyi*. Later experiments conducted at 70°C with pyrite and elemental sulfur as the energy sources showed that *S. acidocaldarius* oxidized both elemental sulfur and pyrite faster than did *S. brierleyi* [4]. Consequently, *S. acidocaldarius* was chosen as the only microorganism for subsequent study. Uninoculated controls showed very slight sulfate release, which was presumed to represent solubilization of mineral sulfate and chemical oxidation of pyritic sulfur from the coal.

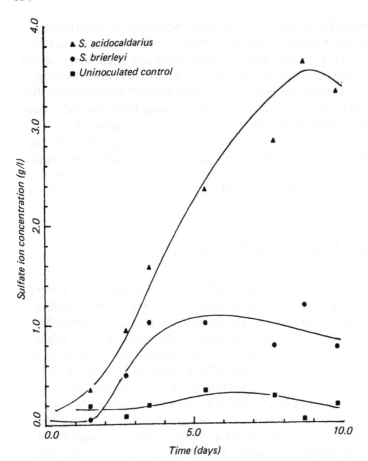

Fig. 34.1 Sulfate release curves for *S. acidocaldarius* and *S. brierleyi* on 5% Clarion No. 4A coal (−74 μm) slurries, aerated and maintained at 70°C with initial pH 2.0.

35.4.1 Effects of Particle Size

Clarion No. 4-A coal in sub-200-, 150-, and 65-mesh size ranges were prepared for the leaching study. The sizes in micrometers for 200, 150, and 65 mesh are 74, 90, and 240, respectively. Experimental procedures were similar to those just described. Again, sulfate concentration in the solution was determined as the measure of sulfur removal. Figure 34.2 shows the sulfate release curves for the experiments with different coal particle sizes. Similar results have been reported in the literature [7], where finer coal particle sizes yielded higher sulfur removal rates and greater amounts of sulfur removal.

To compare the sulfur removal rates for different particle sizes quantitatively, the concept of first-order kinetics has been adopted. Some researchers [3] found that the rate of sulfate release followed first-order kinetics, where the rate expression was defined as

$$\frac{d(FeS_2)}{dt} = -k(FeS_2) \tag{1}$$

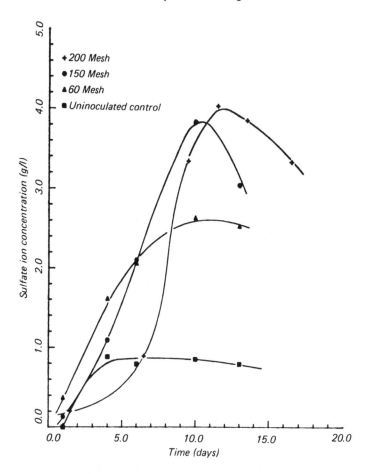

Fig. 34.2 Sulfate release curves on 5% slurries of sub-200-, 150-, and 65-mesh Clarion No. 4A coal, aerated and maintained at 70°C with initial pH 1.5. (From Ref. 9.)

where FeS_2 is the concentration of leachable pyrite in coal and k is the first-order rate constant. The equation may be integrated to yield the form

$$\ln(FeS_2) = -kt + C \tag{2}$$

Since (FeS_2) is $FeS_2)_0$ at $t = 0$, $C = \ln(FeS_2)_0$ and $\ln[(FeS_2)_0/(FeS_2)] = kt$. A logarithmic plot of the data yields the rate constant. In the following calculation, the FeS_2 in equations (1) and (2) was replaced by S, which is the total leachable sulfur to accommodate the possibility that other forms were removed, and it was also assumed that sulfate is the only net or final product of sulfur oxidation. The rate constant evaluated this way for the experiment with sub-200-mesh coal is 2.9 and 3.7 times the rate constants for the 150- and 65-mesh data, respectively. As indicated in Fig. 34.2, the extent of sulfur removal also increased with decreasing coal particle size. Table 34.1 summarizes the calculated rate constants and $t_{0.9}$, the time required for 90% leachable sulfur removal for different mesh sizes. The $t_{0.9}$ values are calculated from the equation $t_{0.9} = (\ln 10)/k$.

To promote interpretation of the data, some theories of microbial oxidation of pyrite are discussed here. First, the rate of sulfate production is a function of pyrite surface

Table 34.1 Sulfur Leaching Rate Constants and $t_{0.9}$ for Different Mesh Sizes, Ohio Clarion No. 4A Coal, at $70°C$ and Initial pH 1.5

	Coal particle size		
Mesh size	Maximum size (μm)	Rate constant (day^{-1})	$t_{0.9}$ (days)
65	240	0.1983	11.6
150	90	0.2527	9.1
200	74	0.7455	3.1

area or concentration [3]. The two can vary differently since the pyrite grain size may vary. Other investigators found that bacterial attachment is not necessary for pyrite oxidation but is prevalent with *Sulfolobus*-type organisms [26]. As the coal particle size decreases, more sulfur surface is exposed to the reactive solution and made available for microbial attachment and oxidation. Comparison of the sulfur removal rates obtained in this study to those from the literature for the mesophile *T. ferrooxidans* shows that *S. acidocaldarius* is several times faster. For example, the $t_{0.9}$ observed in this study for -74-μm (sub-200-mesh) coal was 3.1 days, while the $t_{0.9}$ reported by Detz and Barvinchak for a different coal with *T. ferrooxidans* was 16 days [3].

34.4.2 Effect of Temperature on Sulfur Removal

The goal of this study was to determine temperature effects on microbial leaching of coal-derived sulfur by the thermophile. The microorganism *S. acidocaldarius* has optimum growth in the range 70 to $75°C$ [9]. However, in microbial leaching processes, oxidation of pyrite and leaching of sulfur forms in general may involve purely chemical reactions independent of microbial mediation, as suggested by Silverman [27]. The rates of purely chemical reactions are higher at higher temperatures. On the other hand, operating the process at higher temperatures increases the cost and may diminish the economic advantages that the process enjoys compared with other desulfurization methods. This study was therefore undertaken to optimize temperature in the thermophilic process.

A coal mixture collected from the Columbus and Southern Ohio Electric Power Plant in Pickaway County, Ohio, was designated as Ohio coal mixture No. 1 and was used for this study. The sample was a mixture of Ohio coals from the Sands, King Quarries, Benedict, and Crooksville mines in southeastern Ohio. The coal was pulverized, with 30% -400 mesh and 59.9% -200 mesh (cumulative). A more common pulverized coal specification would have shown weight percent -74 μm. Again, the experiments were performed in a round-bottomed three-necked flask with 5% coal slurry. Sulfur removal from coal by *S. acidocaldarius* was investigated at temperatures ranging from 67 to $83°C$. To compare the sulfate production rates at different temperatures, the maximum rates were evaluated. The maximum was defined as the maximum slope on the sulfate release curve. Figure 35.3 shows the plot of maximum sulfate release rate versus temperature. The maximum occurred at $72°C$ with a sulfur removal rate of 10.21 mg/L per hour. This value is larger than the 4.5 mg/L per hour reported by Kargi and Robinson [29]. However, due to the differences in coals and in particle-size distributions, a direct comparison is subject to interpretation. Table 34.2 shows the maximum rates and the extents of sulfur removal at different temperatures. The data indicated that total sulfur removal of up to

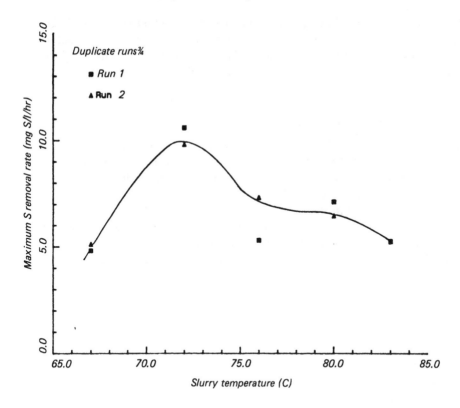

Fig. 34.3 Maximum rates of total sulfur removal versus slurry temperatures on Ohio coal mixture No. 1 under standard conditions. (From Ref. 28.)

60% was obtained. Sulfur form analyses were not available for this work. In later studies, information was provided to show the capability of the thermophile to remove organic sulfur as well as sulfate and pyritic forms.

34.4.3 Effects of Yeast Extract Supplementation

Increased growth of *S. acidocaldarius* had been observed in a medium where both yeast extract and sulfur were present [9]. However, the presence of yeast extract was shown to have no effect or to have slightly detrimental effects on rates and extents of sulfur oxida-

Table 34.2 Maximum Rates and Extents of Sulfur Removal from Ohio Coal Mixture No. 1 at Different Temperatures

Temperature (°C)	Maximum rate of sulfur removal (mg/L/h)	Extent of sulfur removal (% total S)
67	4.96	—
72	10.21	59.39
76	6.33	50.28
80	7.92	43.37
83	6.50	41.71

tion [9]. Some investigators showed that yeast extract promoted the growth of pili in the outer cell structure of the bacteria [30]. Without yeast extract, the cells did not have pili and were not able to attach to the substrate. In contrast to the foregoing, Kargi and Robinson [29] reported a higher sulfur oxidation rate in the absence of yeast extract supplementation.

34.4.4 Effects of Coal Slurry Density

Since sulfate production rates depend on the amount of coal surface area exposed, higher concentrations of coal particles in the slurry corresponded to higher absolute rates. However, if the rates were normalized to the amounts of sulfur remaining, the rates at higher pulp densities were lower. Figure 34.4 shows sulfate release curves at different slurry densities. The coal involved in the experiment was the Ohio Clarion No. 4A coal described previously. Other researchers showed a linear relationship between the sulfur removal rates and the coal pulp densities for densities lower than 15% [6]. This result means that the specific sulfur removal rate (the rate per unit amount of sulfur available)

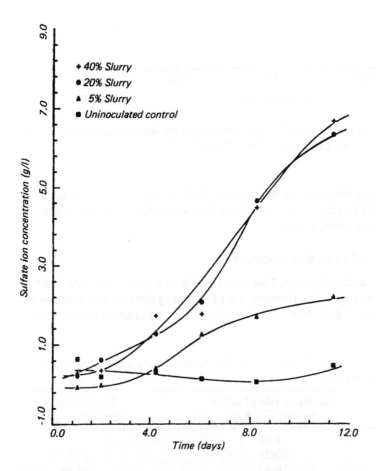

Fig. 34.4 Sulfate release curves for varying weight percent slurries of Clarion No. 4A coal (−74 μm), with yeast extract supplementation, under standard conditions. (From Ref. 31.)

is constant. Detz and Barvinchak [3] found drastically reduced rates of desulfurization with slurries denser than 20% coal by weight, using *T. ferrooxidans*. Some researchers suggested that the organic compounds solubilized from coal could be toxic to the bacteria [32]. It is more likely that lower specific rates can be attributed to reduced mass transfer rates and loss of uniform particle distributions due to poor mixing. Kargi and Robinson [6] reported that the lower rates at higher pulp densities could be attributed to particle agglomeration or to limited effective transfer rates of O_2 and CO_2.

34.4.5 Effects of Basal Salt Addition

Some basal salts, such as ammonium sulfate, potassium phosphate, magnesium sulfate, and calcium chloride, are essential for microbial growth. However, part of the mineral matter in coal could be solubilized and provide necessary chemical nutrients, in whole or in part. In most published reports on *T. ferrooxidans*, the 9K medium was considered adequate as a nutrient supplement for microbial leaching. The medium, containing $(NH_4)_2SO_4$, KCl, K_2HPO_4, $MgSO_4 \cdot 7H_2O$, $Ca(NO_3)_2$, and $FeSO_4 \cdot 7H_2O$, was first proposed by Silverman and Lundgren [33]. In this study, the basal salts solution described in Section 34.2 was used instead of the 9K medium.

The experiments were conducted with Kentucky No. 9 coal. The coal was received from Babcock and Wilcox Co. of Alliance, Ohio. The total sulfur content of the coal was 4.42%, with 2.88% pyritic sulfur. The coal was finely ground with only 15% by weight greater than 200 mesh (74 μm). "Single-strength basal salts" refer to the concentrations described in Section 34.4. Concentrations of sulfate ion, ferrous iron, and ferric iron ion in the solution were determined. Table 34.3 shows the maximum sulfate release rates versus number of doses of added basal salts. Both the rates of sulfur removal and extents of sulfur removal increased with increasing amounts of basal salts at low basal salt concentration. When basal salt concentration got higher, sulfur removal rates were finally limited and decreased. High sulfate concentration associated with a high dose of basal salts caused problems such as precipitation of jarosite to reduce significantly the apparent rates and extents of sulfur removal.

Lacey and Lawson [34] reported that the precipitation of basic ferric sulfate onto the coal surface hindered the access of bacteria to reaction sites. The precipitation also negated the beneficial effects on original removal. Sulfate in the precipitate can decompose upon combustion and form sulfur dioxide pollutants.

One example of mineral sulfate precipitate is hydronium jarosite:

$$3Fe_2(SO_4)_3 + 12H_2O \rightarrow 2HFe_3(SO_4)_2(OH)_6 + 5H_2SO_4 \tag{3}$$

Table 34.3 Rates and Extents of Sulfur and Iron Release in 5% Slurry of Kentucky No. 9 Coal at 72°C and Initial pH 2.0

Doses of salts	Max. $\dfrac{dS}{dt}$ (mg S/L·h)	Extent of S removal (g S/L)	Max. $\dfrac{d[Fe]}{dt}$ (mg/L·h)	Extent of Fe removal (mg/L)
0	4.86	0.713	4.86	538
1	14.60	1.123	6.38	756
2	6.57	0.870	3.04	441

In the presence of potassium ions, potassium jarosite, $KFe_3(SO_4)_2(OH)_6$, is formed preferentially to hydronium jarosite [35]. Potassium, one of the ions necessary for microbial growth, is present in the basal salts medium, and presumably is leached from coal in the absence of salts. Lower potassium concentration and lower pH are desired to reduce jarosite precipitation. Reprecipitation of sulfur in other mineral compounds may also be observed. For example, gypsum was detected on one microbially leached coal.

34.4.6 Effects of Airflow Rates

Airflow in some runs provided turbulence for particle mixing and for improved mass transfer of oxygen and carbon dioxide. At higher air rates, bubble holdups were higher, and consequently the interfacial area between the gas phase (air) and the slurry phase was larger. Larger interfacial areas yielded higher transfer rates for O_2 and CO_2. Higher gas velocity also induced turbulence, which enhanced mixing. Better mixing meant a thinner film and lower diffusion resistance at the gas-liquid interface for O_2 and CO_2. A higher gas transfer rate would result. Since mixing and aeration requirements contributed most of the operating cost of the process, the data interrelating air rates and sulfur removal rates were crucial for estimating operating costs and evaluating overall process economics.

A total of 18 batch leaching runs defined the effects of airflow rates on microbial coal desulfurization in the authors' laboratory. The airflow rate ranged from 2 to 7 standard cubic feet per hour (SCFH). Experiments were conducted in the flasks described previously. Each batch vessel was charged with 40 g of a coal. The coal involved in the study, designated as Ohio coal mixture No. 2, was a mixture from the Benedict, Natter, and Boyle mines in southeastern Ohio. The coal was coarser than Ohio coal mixture No. 1, and had 12%-250 mesh and 46% -200 mesh (cumulative). The raw coal contained 2.12% pyritic sulfur, 0.27% sulfate, and 1.29% organic sulfur as reported by the Black Rock Test Lab in Morgantown, West Virginia. Some of the flasks were maintained at a total volume of 800 mL, while others maintained at 1000 mL (lower slurry density). No other physical agitation was provided in any of the leaching experiments. Observation showed that some coal particles tended to settle in the vessels with higher pulp densities (lower slurry height).

The total sulfur removal rates obtained from the flasks maintained at 800 mL were slightly lower. It is not clear whether the greatest contribution to lower rates was insufficient agitation and particle agglomeration or insufficient mass transfer for O_2 or CO_2. The increased height of the larger volume slurry provided additional time and opportunity for mass transfer. Lower pulp density also prevented agglomeration of fine particles, which could decrease the surface area available for sulfur leaching.

The maximum sulfate production rates were used to evaluate the effect of airflow rates. Figure 34.5 shows a plot of maximum rate of sulfate production versus airflow rate. The maximum rate increased rapidly with initial airflow rate increases, but only a slight upward trend occurred above 4 SCFH in the 800-mL flasks.

The data presented above provide only qualitative information about mixing and mass transfer needed in a process, since the results were correlated with overall airflow rates. The overall performance of mixing and mass transfer is governed by the hydrodynamics, which, in turn, is defined by the design of the reactor, airflow rate, and physical properties of the slurry, such as viscosity, surface tension, particle size and shape, and temperature. The most critical part of this type of reactor is the air sparger. The design

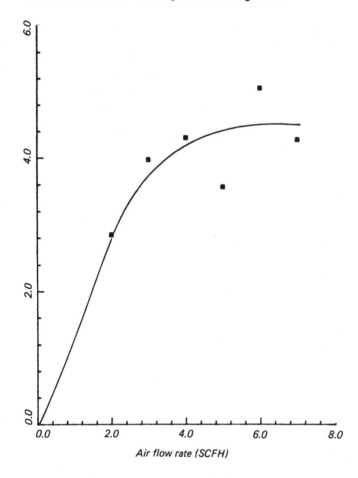

Fig. 34.5 Maximum rates of total sulfur removal versus airflow rates on Ohio coal mixture No. 2 under standard conditions. (From Ref. 20.)

of the air sparger will undoubtedly play an important role in determining the gas bubble size, bubble size distribution, and bubble holdup. These will, in turn, decide the interfacial areas between air and slurry, which will affect the overall mass transfer rates. For example, if a porous plate is used as the sparger, the bubble size will be much smaller than that generated by a perforated plate. Finer bubbles yield a higher interfacial area for mass transfer. Consequently, higher mass transfer rates could be obtained. In an air-agitated vessel, mixing is determined by air velocity, shape of the vessel, and physical properties of the slurry. Airflow rate alone, therefore, is not adequate to provide quantitative estimates for scaleup of aeration and mixing effects separately.

34.4.7 Effects of Elevated Pressure

One of the problems introduced by substitution of a thermophile for a mesophile is that the O_2 and CO_2 supplies are limited since the solubility of gas in a liquid is lower at elevated temperatures. To improve gas solubility, the partial pressure of the gas can be increased. Partial pressure can change with composition or with total pressure. However, higher total pressure could kill some microorganisms; yet others would be barotolerant.

Sulfur-oxidizing bacteria have been isolated from seawater samples taken at the Galapagos Rift, where the hydrostatic pressure is 280 atm [36].

An experiment was conducted at elevated pressure to study the effects of total
pressure on the desulfurization of coal using the thermophile *S. acidocaldarius*. The pressure vessel for the experiment was an Autoclave Engineers stirred 1-gallon vessel, model
HT 80016. The vessel was fitted with a glass insert to prevent corrosion of its interior. A
heater jacket provided the heat to maintain the temperature at desired levels. The operating pressure was 3 atm and the pulp density of the coal slurry was 5%. The coal was
Ohio coal mixture No. 1 described in Section 34.4.3. Figure 34.6 shows a schematic diagram of the experiment apparatus.

The maximum desulfurization rate of the run at a pressure of 3 atm was 3.8 mg
S/L per hour, which was only 42% of the average rate of runs at 1 atm (9.0 mg S/L per
hour). The first-order rate constant for the 3-atm run (0.0059 per day) was 35% of the
rate constant for the 1-atm run (0.017 per day). Inhibition of microbial growth or metabolism by a high oxygen tension may be responsible for the observed reduction in rate.

34.4.8 Oxygen and Carbon Dioxide Enrichment at Barometric Pressure

Oxygen is utilized by the microorganisms as an electron acceptor during the aerobic
process of coal desulfurization. Oxidation of sulfur supplies the energy needs of the bacteria. The rate-limiting step in oxygen transfer could be the step where dissolved oxygen
is transferred from the bulk solution to the bacteria. The solubility of oxygen in water
at a pressure of 1 atm decreases from 8.5 ppm to 5.7 ppm when the temperature rises
from 25°C to 70°C [37]. Increased concentrations of electrolytes such as sulfate, in

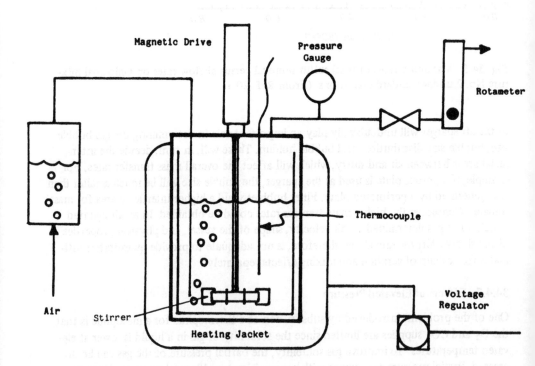

Fig. 34.6 Apparatus for elevated pressure experiment.

water, reduce oxygen solubility [38]. When the oxygen concentration in the medium is higher than a critical value, the cell can obtain sufficient oxygen for its respiratory processes. On the other hand, when the oxygen concentration is even higher and above a certain value, toxic effects may occur.

Detz and Barvinchak [3] addressed the issue of optimum dissolved oxygen concentration during the microbial desulfurization of coal. When utilizing *T. ferrooxidans*, they found that in all cases the bulk concentration of dissolved oxygen in the slurry was in equilibrium with the partial pressure of oxygen in the feed gas. They determined that the optimum desulfurization rate occurred at 8 ppm dissolved oxygen.

The effect of carbon dioxide enrichment on coal desulfurization by *T. ferrooxidans* was observed by Detz and Barvinchak [3]. They concluded that the concentration of dissolved CO_2 in equilibrium with the CO_2 at its natural atmospheric partial pressure was sufficient for bacterial desulfurization. Kargi reported that sulfur removal rates were higher when the gas supply contained 23% carbon dioxide [39]. However, the result may not be due to CO_2 enrichment since he used a concentrated inoculum and a coal with an extra-high sulfur content (11%). Hoffmann et al. [40] found that the pyrite leaching rate was increased by 31%, with the partial pressure of CO_2 increased from natural atmospheric values to 0.1 atm when using *T. ferrooxidans*. Improvement of growth of *S. acidocaldarius* under enhanced CO_2 conditions has also been described in the literature [9].

Experiments were conducted in the authors' laboratory to determine if enrichment of the gas sparged into the coal slurry would increase the rate or extent of sulfur removal. Oxygen molar fractions of 0.14, 0.21, 0.30, and 0.40 were investigated in combination with carbon dioxide molar fractions of 0.000032, 0.007, 0.10, and 0.33. The experimental results are summarized in Table 34.4.

In noninoculated control runs without gas enrichment, the maximum sulfur removal rate, presumably by a chemical mechanism, of 1.3 mg S/L per hour was measured.

Table 34.4 Results of Gas Enrichment Experiments

Run	Partial pressure (atm)			Maximum sulfur removal rate (mg S/L·h)
	CO_2	O_2	N_2	
1	0.00021	0.14	0.86	7.7
2	0.00024	0.40	0.60	8.0
3	0.00028	0.30	0.70	9.7
4	0.00032	0.21	0.79	8.5
5	0.00032	0.21	0.79	9.0
6	0.00032	0.21	0.79	9.4
7	0.007	0.21	0.79	11.1
9	0.10	0.14	0.76	10.1
10	0.10	0.19	0.71	11.1
11	0.10	0.19	0.71	12.6
12	0.10	0.21	0.69	10.5
13	0.10	0.30	0.60	10.4
14	0.10	0.40	0.50	9.3
16	0.33	0.21	0.46	9.5
17	0.33	0.30	0.37	10.1
18	0.33	0.40	0.27	8.9

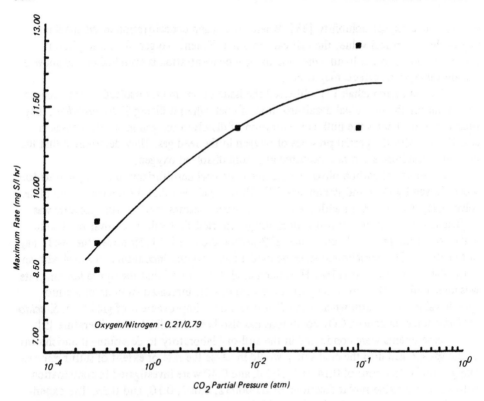

Fig. 34.7 Maximum rates of total sulfur removal versus CO_2 partial pressure in the CO_2-enriched air on Ohio coal mixture No. 1 under standard conditions.

The maximum rates obtained from inoculated runs under the same condition were 8.5, 9.0, and 9.4 mg S/L per hour. This indicates that the microbial leaching rate was about seven times the chemical leaching rate.

In one set of experiments, the partial pressure of carbon dioxide was varied and the ratio of oxygen to nitrogen was held constant at 0.21/0.79. As indicated in Fig. 34.7, the addition of carbon dioxide to a partial pressure of 0.007 atm increased the rate by 23%. At a carbon dioxide partial of 0.10 atm, the rate was 11.9 mg S/L per hour, a 32% increase over that of atmospheric levels of carbon dioxide. As the partial pressure of carbon dioxide increased, the partial pressures of oxygen and nitrogen decreased, since the total pressure was maintained at 1 atm.

The effect of carbon dioxide partial pressure was analyzed on the basis of the same oxygen partial pressure. Figure 34.8 summarizes the effect of carbon dioxide partial pressures on sulfur removal rates. In this figure, maximum sulfur removal rates were plotted against partial pressure of carbon dioxide, with oxygen partial pressure as the parameter. Experiments were conducted with the oxygen partial pressure at 0.21 atm and the partial pressure of carbon dioxide varied. Nitrogen made up the remaining gas to give a total pressure of 1 atm (runs 4, 5, 6, 7, 12, and 16 in Table 34.4). The maximum rate was observed at the carbon dioxide partial pressure of 0.007 atm. For experiments with oxygen partial pressure held constant at 0.30 or 0.40 atm, the variation of carbon dioxide partial pressure showed little effect on the sulfur removal rate, as indicated in Fig. 34.8. At oxygen partial pressure of 0.14 atm (run 1 and 9), sulfur removal increased from 7.7

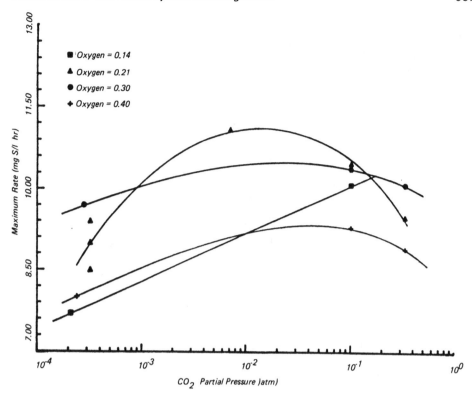

Fig. 34.8 Maximum rates of total sulfur removal versus CO_2 partial pressure, at varying O_2 partial pressure in the enriched air, on Ohio coal mixture No. 1 under standard conditions.

mg S/L to 10.1 mg S/L per hour. In general, the rate reached a maximum and then decreased as the carbon dioxide increased.

Several reports about the effects of gas enrichment on microbial coal desulfurization are available for comparison. Detz and Barvinchak [3] reported that carbon dioxide enrichment showed no effect on microbial leaching of coal with *T. ferrooxidans*. However, Hoffmann et al. [40] observed an increased rate with carbon dioxide enrichment, which parallels our findings. The *Sulfolobus* species is a facultative autotroph. Without supplementation of organic carbon source, the microbe fixes carbon dioxide, just as the *Thiobacillus* species does. An increase in carbon dioxide partial pressure simply increases the transfer rate of carbon dioxide from the gas to the slurry and eventually to the cells for better growth. However, the reason for the decreased rates observed at carbon dioxide partial pressures higher than 0.30 atm is unknown, although the effect is commonly observed in facultative autotrophs.

A similar analysis was made to investigate the effects of oxygen partial pressure. The results are shown in Fig. 34.9, where the maximum sulfur removal rates are plotted versus oxygen partial pressure, with carbon dioxide partial pressure as the parameter. With oxygen enrichment of the air, the sulfur removal rate reached a maximum and then decreased, as indicated by the curve with carbon dioxide partial pressure smaller than 0.00032 atm. Addition of oxygen to the air simply diluted carbon dioxide to give partial pressures smaller than the ambient carbon dioxide partial pressure of 0.00032

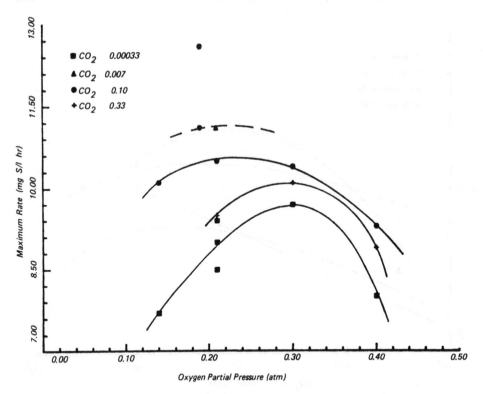

Fig. 34.9 Maximum rates of total sulfur removal versus O_2 partial pressure at varying CO_2 pressure in the enriched air on Ohio coal mixture No. 1 under standard conditions.

atm. Consequently, the effect on the sulfur removal rate is probably due to the dilution of carbon dioxide rather than the enrichment of oxygen. Similar trends were observed at different levels of carbon dioxide partial pressure, where the rates increased with increasing oxygen partial pressure and then declined.

An explanation of the experimental results is possible based on the role of oxygen in sulfur oxidation and in other microbial activities. In the microbial oxidation of sulfur to sulfate, oxygen serves as the terminal electron acceptor. At oxygen partial pressures lower than 0.30 atm, the sulfur removal rate increased with increasing oxygen partial pressure due to the higher oxygen transfer rate to the cells. At an 0.14 atm oxygen partial pressure, the solubility of oxygen is approximately 3.8 ppm, which seems to be lower than the critical oxygen concentration. The reduced sulfur removal rate at a high oxygen concentration may be due to inhibition by high oxygen tensions. It was reported earlier that the introduction of high oxygen tensions reduced the growth rate of *Thiobacillus* strain TH3 [41].

34.4.9 Surfactants and Microbial Coal Desulfurization

A surfactant might aid particle wetting and dispersion, prevent agglomeration of particles, expose more surface area, enhance the attachment of microorganisms to the solid substrate, increase microbial growth rate, and eventually promote coal desulfurization. Some researchers reported that surfactants might have an effect on the oxidation of pyrite in coal by *T. ferrooxidans*. Surfactants produced by selected cells made the process of slurrying coal in recycled culture easier than slurrying coal in water [8]. Wakoa et al. [42] showed that the nonionic surface-active agents Tween-20 and sugar ester accelerated pyrite oxidation. Since *T. ferrooxidans* organisms adhere to coal via setting, Wakoa et al. concluded that the surface-active agents had a physical-chemical effect on the relationship between pyrite and the cells.

In the authors' laboratory, the effects of surfactants on sulfur removal from coal were investigated [43]. The experiments were conducted in a series of round-bottomed three-necked flasks described previously. Mechanical agitation was introduced to promote particle mixing and mass transfer of gases. The surfactants tested were a series of condensates of ethylene oxide with hydrophobic bases formed by condensing propylene oxide with propylene glycol. The surfactants were L62, L43, L44, P103, and F127 with HLB number 7, 9, 12, 16, and 22, respectively, from the Pluronic series manufactured by BASF Wyandotte Corporation.

The Kruskal-Wallis test was applied to analyze the experimental results of 55 batch runs. In 5% coal slurry, Pluronic surfactant P103 slightly enhanced microbial coal desulfurization, while L62 inhibited desulfurization. No significant increases or decreases of sulfur removal rate or extent of sulfur removal were observed for the surfactants L43, L44, and F127. This means that unlike *Thiobacillus* species, the surfactants did not significantly affect the sulfur removal from coal. Either concentrations were too low at 3.5 g/L, or the mechanism of sulfur removal from coal or the mechanism of cell attachment on coal for *Sulfolobus* species is different from those for the *Thiobacillus* species.

34.4.10 Summary of the Effects of Process Parameters on Various Coals

Various effects on the rates and extents of sulfur removal are mentioned in previous sections. However, not all the studies were on a single coal. For better comparison, rates and extents of sulfur removal are summarized in Table 34.5. The rate is characterized by the maximum sulfate production rate and the first-order rate constant. The extent of sulfur removal is quantitified by percent of total sulfur removal based on sulfate release.

A direct comparison of the data among different coals is difficult since the observed effect of one variable on one coal was actually a combination of the effects of many other variables. Coal particle size strongly affected sulfur removal rates from coal. The optimum rate of sulfur removal from Clarion No. 4A coal was obtained at a size of 100%–200 mesh (<74 μm), while the data from Ohio coal mixture No. 1 were obtained at a size of 60%–200 mesh.

The extent of sulfur removal depends significantly on the sulfur forms in coal. *Sulfolobus* species is able to remove pyritic sulfur and some organic sulfur, depending on the coal types and particle size. Clarion No. 4A coal had 76% of its total sulfur content as pyritic, while Ohio coal mixture No. 2 had 58% of its total sulfur content as pyritic. Kentucky No. 9 coal, which had a particle size of 73%–200 mesh, finer than the two Ohio coal mixtures, had 65% of total sulfur content as pyritic. The data in Table 34.5 indicate that the extent of sulfur removal were near the percentage of pyritic sulfur. Sulfur form analyses were not available for the microbially leached coals.

Table 34.5 Effects of Process Parameters on Various Coals

Coal type	Variable	Range (optimum)	Rates and extent of sulfate removal at optimum conditions		
			Maximum rate (mg S/L·h)	Rate constant (day^{-1})	Percent total sulfur removed
Clarion No. 4A	Size	−65 to −200 mesh (−200)	14.87	0.7455	73
Ohio mixture No. 1	Temperature	67–83°C (72)	10.21	0.5533	59
	CO_2	0.00021–0.33 atm (0.1)	10.50	0.4080	58
	O_2	0.14–0.40 atm (0.30)	9.70	0.4320	60
Ohio mixture No. 2	Air rate	2–7 SCFH (4)	5.06	0.1629	51
Kentucky mixture No. 9	Basal salts	0–2 doses (1)	14.60	0.4176	64

Table 34.6 Sulfur Forms in Microbially Leached Coals

Coal type	Sulfur form	Percent as sulfur	
		Before leaching	After leaching
Kentucky No. 9	Total sulfur	3.87	1.79
	Sulfate	0.17	0.46
	Pyritic	1.95	0.15
	Organic	1.75	1.18
Ohio Clarion No. 4A	Total sulfur	8.31	1.83
	Sulfate	0.18	0.34
	Pyritic	6.80	0.14
	Organic	1.33	1.35

34.5 SULFUR FORMS IN MICROBIALLY LEACHED COAL

Leaching experiments were conducted to investigate the removal of different sulfur forms from coal. As stated in Section 34.4.10, the extent of sulfur removal depends strongly on particle size. To eliminate the effects of particle size, coal samples with a narrow size distribution were prepared for microbial leaching experiments. The sample contained only the portion of 270 to 325-mesh particles separated from a Kentucky No. 9 coal. Another coal sample with a particle size of 100% - 200 mesh was prepared from a run-of-mine, high-sulfur Ohio Clarion No. 4A coal. Sulfur forms were determined before and after microbial leaching following the standard procedurds specified by the ASTM as closely as possible [14].

The data in Table 34.6 show organic sulfur removal. Calculation of percentage of sulfur removal was not based on an "ash-free basis" since although some ash was removed during the leaching process, ash analyses were not available for the coals. Further investigation is planned to confirm the results.

34.6 ATTACHMENT OF *S. ACIDOCALDARIUS* CELLS TO COAL

S. acidocaldarius cells have been seen to attach to a solid substrate surface. Shivvers and Brock [44] reported that the bacteria attached to the surface of elemental sulfur particles in the late exponential phase and in the stationary phase. Weiss [45] found that pili enable the cell to attach to sulfur crystals. Other researchers [24,46,47] observed that a *Sulfolobus*-like bacterium adsorbed on pyrite surface preferentially. Kargi and Robinson [48] reported that most of the cells of *S. acidocaldarius* attached to the coal surface during a batch leaching experiment. However, quantitative studies on the rate and extent of cell attachment to coal or pyrite, such as those performed on *T. ferrooxidans*, are not available in the literature [49,50].

The objective of the adsorption experiment was to study the rate and extent of *S. acidocaldarius* cell attachment to coal and some other particles. Cells were grown in the medium described previously, where they grew with sublimed elemental sulfur as the energy source; supplemented with yeast extract. Cell harvest occurred in 3 to 5 days. The culture was first centrifuged at 1000g for 10 min to remove the solid particles. Then the cells were separated from the fluid by centrifugation at 4000g and resuspended in a basal salt solution without an energy source. Cell counts were achieved by measuring the total

protein in solution calibrated by visual cell counts in a phase-contrast microscope
equipped with a Petroff-Hausser counting chamber. Protein was analyzed by the modified
Bradford method [51].

Activated carbon, pyrite, Kentucky No. 9 coal, Ohio coal (a mixture of Nos. 6 and
12), a physically cleaned Ohio No. 6 coal, and a European coal from Sulzer Brothers, Co.,
Switzerland, were tested for cell attachment. Particles sized to a narrow range at 270 to
325 mesh (45 to 53 μm) were used. One gram of the solid particles was mixed with 5 mL
of the cell solution in a series of test tubes and agitated vigorously. Then the cell densities
in the solution were measured to determine by difference the number of cells attached to
the solid particles. Experiments with different initial cell densities were conducted to determine the equilibrium relationships.

The major finding about the rate of adsorption is shown in Fig. 34.10. As indicated
by the data, cell attachment is a very fast process. Equilibrium was reached in less than 1
min of contact between the cells and the solid particles. Comparison of these data with
those for *T. ferrooxidans* published in the literature are difficult since the earlier mixing
conditions were not mentioned. If the mixture of solid particles and cell solution is not
well agitated, the particles tend to settle and diffusion of the cells from the bulk to the
particle surface becomes the rate-limiting step.

Table 34.7 gives the amounts of pyritic and organic sulfur in the coals involved in
the cell attachment study. Total pyrite surface area is determined not only by total
amount of pyritic sulfur, but also by the pyrite grain size. Information about pyrite grain
size in these coals was not available. The results of the experiments with different initial
cell densities on the coals are shown in Fig. 34.11. In this figure the number of attached

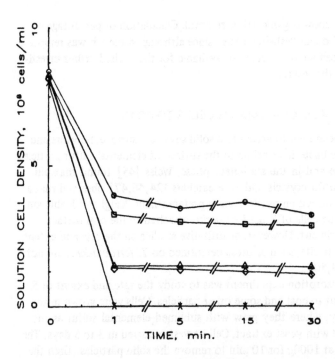

Fig. 34.10 Cell adsorption on different particles. △, Kentucky No. 9; ◊, Sulzer coal; ○,
Ohio No. 6; □, Ohio coal mixture No. 1; X, activated carbon.

Table 34.7 Sulfur Contents in Coals for Cell Attachment Study

	Sulfur contents	
	Percent pyritic	Percent organic
Kentucky No. 9	2.68	1.49
Ohio coal mixture	2.12	1.29
Ohio No. 6	0.68	1.37
Sulzer	0.12	0.54
Pyrite	53.40	—

cells was plotted against the equilibrium cell density in the solution. Contacting time between cell solution and coal particles was 30 min. Desorption experiments were also conducted to show the reversibility of cell attachment. Cell attachment to activated carbon, Sulzer coal, and pyrite was shown to be irreversible. For reversible attachment, the Langmuir adsorption isotherm, although not theoretically applicable, fit the data well, as indicated by the solid lines drawn through the data.

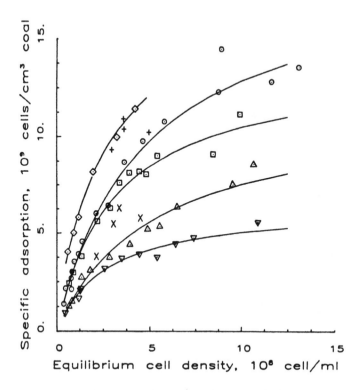

Fig. 34.11 Adsorption isotherms for the cell adsorption on coals and pyrite. ◊, Sulzer coal; ○, Kentucky No. 9; □, pyrite; △, Ohio coal mixture No. 1; ▽, Ohio No. 6; +, desorption data from Kentucky No. 9 coal; x, desorption data from Ohio coal mixture No. 1.

A different set of adsorption experiments was carried out to determine the available pyrite surface area on the coal and on the mineral pyrite. Instead of microbial cells, a pyrite-selective absorbate called polyacrylic acid/xanthate (PAAX) was used. Details about this polymer are available in the literature [52]. The amount of adsorbed PAAX can qualitatively define the pyrite surface available.

As indicated in Fig. 34.12, the isotherms showed the relative amounts of pyrite surface on the coals and other solids. The sulfur analyses shown in Table 34.7 are consistent with the trend in Fig. 34.12. However, comparison of Fig. 34.11 with Fig. 34.12 shows that the Sulzer coal, a coal with little pyritic sulfur, adsorbed more cells than did another high-sulfur coal. This result indicates that the broader surface properties (surface area, surface charge, and hydrophobicity) may play more important roles in cell attachment than does pyrite selectivity alone.

As proposed by some researchers, the adsorption-desorption of cells is a major part of the model [53]. Information about cell attachment is important in modeling the continuous leaching process. In a continuous process, the feed coal must contact the process liquor in the fermentor to allow cells to attach to the sulfur sites. Further study would better define surface properties of the solid particles and promote insight into the mechanism of cell attachment.

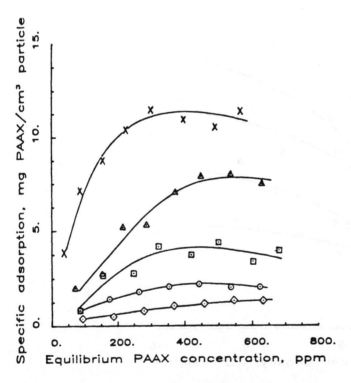

Fig. 34.12 Adsorption isotherms of PAAX on coals and pyrite. △, Kentucky No. 9 coal; ▫, Ohio coal mixture No. 1; x, pyrite; ⊙, Ohio No. 6; ◊, Sulzer coal.

34.7 KINETICS OF SULFUR LEACHING FROM COAL

All three forms of sulfur—sulfate, pyritic, and organic sulfur—are removed by leaching with the microorganisms *S. acidocaldarius*. Oxidation of pyrite by acidophilic bacteria is a well-explored subject, as is sulfate dissolution. However, the mechanism for the removal of organic sulfur, if different from that for the other forms, is not clear since the structure of the organic sulfur compounds in coal is not well defined.

Silverman [25] proposed two mechanisms for the oxidation of pyritic sulfur by *T. ferrooxidans*: direct and indirect mechanisms. In the direct mechanism, direct contact between the cell wall and pyrite is necessary for oxidation of pyrite. The microorganisms oxidized the pyrite into ferrous iron and sulfuric acid through enzymes in the cell wall. Oxidation of ferrous iron into ferric iron by the bacteria also occurred, as described in the following equations:

$$2FeS_2 + 7O_2 + 2H_2O \xrightarrow{bacteria} 2FeSO_4 + 2H_2SO_4 \tag{4}$$

$$2FeSO_4 + (1/2)O_2 + H_2SO_4 \xrightarrow{bacteria} Fe_2(SO_4)_3 + H_2O \tag{5}$$

In the indirect oxidation mechanism, pyrite is oxidized by ferric iron to yield ferrous iron and elemental sulfur. Then the ferrous iron and elemental sulfur are oxidized by the bacteria to give ferric iron and sulfuric acid. The ferric iron can oxidize pyrite to produce more ferrous iron and elemental sulfur. The reactions are summarized as follows:

$$FeS_2 + Fe(SO_4)_3 \rightarrow 3FeSO_4 + 2S \tag{6}$$

$$2S + 3O_2 + 2H_2O \xrightarrow{bacteria} 2H_2SO_4 \tag{7}$$

Following reaction (7) is reaction (5), in which ferrous iron is oxidized to ferric iron. Simultaneous action of direct and indirect mechanisms is reported by other researchers [6].

The goal of the study in the authors' laboratory was to determine the relationship between the rate of each reaction and the cell density in the coal slurry. In the direct mechanism, if the reaction rate of (4) is much greater than that of reaction (5), the products released into the solution are ferrous iron (Fe^{2+}) and sulfate (SO_4^{2-}). In the indirect mechanism, release of ferrous iron and sulfate is also observed if reaction (5) has the lowest rate. However, if the rate of reaction (5) is higher than that of reaction (4), the net products of pyrite oxidation are ferric iron and sulfate. The rate of reaction (4) is determined by the available pyrite surface or the number of bacteria attached to the pyrite surface. On the other hand, the rate of reaction (5) is governed by the concentration of ferrous iron and the number of cells that are not in direct contact with the pyrite in coal.

Experiments were conducted in a series of mechanically stirred round-bottomed three-necked flasks equipped with condensers. The flasks were aerated with laboratory air and the condensers served to remove the moisture carried away from the slurry by the air. The flasks were kept in a water bath at 76°C, which with evaporative and condensate cooling gave a slurry temperature of 72°C. The inoculum was obtained from a stock culture of *S. acidocaldarius*. The stock culture was prepared by growing the cells in a medium with elemental sulfur as the energy source and enriched with yeast extract. The stock culture was transferred every 5 days since the cells attached to the sulfur particles through longer incubation, as indicated in the literature [44]. The stock culture was first centrifuged at 1000*g* for 10 min to remove the sulfur particles. The cells were harvested by centrifugation at 4000*g* for 20 min and resuspended in the basal salt solution for inoculation. Cell density in the inoculum was determined as described previously. Concentrations of ferrous iron, ferric iron and sulfate ions, and sulfur contents were followed through each batch run. Detailed analytical methods are available in the literature [27]. Kentucky No. 9 coal with a size range of 270 to 325 mesh was used in this study. One coal sample was thoroughly washed with basal salts solution at 72°C to remove soluble iron. Both washed and unwashed samples were subjected to microbial leaching experiments. The initial cell density was 6.81×10^7 cells/mL for the unwashed coal and 7.94×10^7 for the prewashed coal.

Figure 34.13 shows the concentration profiles of ferrous iron, ferric iron, and sulfate in a batch run with prewashed coal. Figure 34.14 shows the results from the unwashed coal. Figure 34.13 indicates that the products released from pyrite leaching were ferrous iron and sulfate. The ferrous iron was converted into ferric iron and ferrous iron

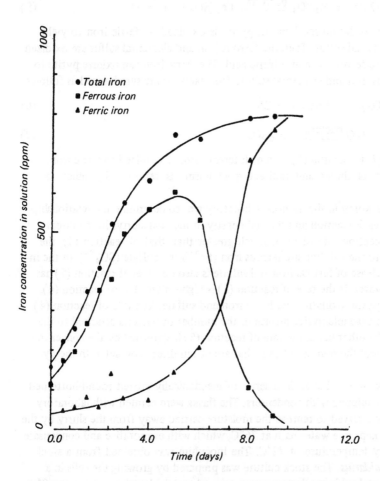

Fig. 34.13   Iron release in batch leaching of a prewashed −270 + 325 mesh Kentucky No. 9 coal.

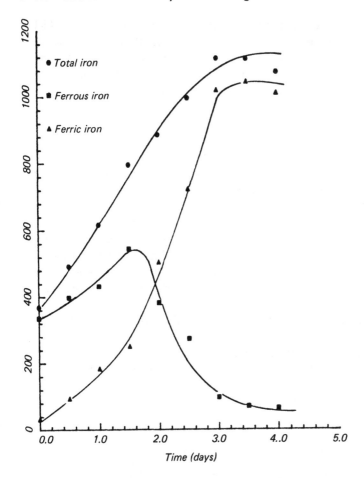

Fig. 34.14 Iron release in batch leaching of an unwashed −270 + 325 mesh Kentucky No. 9 coal.

concentration dropped at the point where 76% of the leachable iron was released. However, in Fig. 34.14, the ferrous iron started decreasing at the point where only 60% leachable iron was released. Comparison of Figs. 34.13 and 34.14 indicates that a higher cell density and higher ferrous iron concentration in the solution, the rate of reaction (5) exceeded the rate of reaction (4) at lower conversion levels of pyrite.

The results suggest that at low cell density, most of the cells attached to the pyrite surface and reaction (4) was the dominant reaction. The cells switched from pyrite oxidation to ferrous ion oxidation when the pyrite had substantially disappeared. However, at a high cell density ($> 10^7$ cells/ml), pyrite surfaces were occupied by bacteria, while other cells not in contact with the pyrite were attached elsewhere or were free floating. These cells oxidized ferrous iron produced by the direct mechanism or the elemental sulfur produced by the indirect mechanism. Consequently, the rate of reaction (5) exceeded the rate of reaction (4) and ferric iron became the major product of pyrite leaching. Information obtained in this study is qualitative rather than quantitative. To define the leaching rate in a more specific way, the surface area of pyrite on the coal and the total surface area of the coal itself are necessary.

Information provided by the kinetic study is very useful in modeling the microbial leaching system. A well-developed model can be used in the design of the reactor and the control of the process. The release of ferric iron is of more than casual importance since the reprecipitation of jarosite occurs at high levels of ferric iron concentration. More work in this area would be desirable to improve process design and control.

34.8   SEMICONTINUOUS OPERATION

This set of experiments was designed to study the possibility of semicontinuous operation of a microbial leaching process. A total of 10 flasks with solids densities ranging from 5 to 15% were employed. The experiments proceeded in batch mode until successive sulfate concentrations indicated that sulfur oxidation had slowed. It was assumed that the bacteria had reached the stationary phase at this point. Then a volume of slurry was extracted from the flask and replaced with additional, fresh coal-water-acid slurry. The flasks were mechanically agitated and aerated to ensure good mixing and mass transfer. Samples were taken and analyzed for sulfur content each day so that a direct comparison was possible between average residence time of the coal in the slurry and the extent of sulfur removal.

Sampling error made the experiments very difficult. The sulfur contents of the samples varied substantially and made comparisons between average residence time and sulfur removal difficult. The variability was influenced by the sample size for sulfur analysis. In the analysis with a LECO sulfur analyzer, only 50 mg of coal was used. A sample contained particles of different sizes. The finer particles achieved higher sulfur removal for the same residence times. Consequently, the sample with more finer coal would

Table 34.8   Summary of Data from Semicontinuous Operation

Day	Residence time (days)	Weight of coal in flask (g)	Weight added (g)	Weight removed (g)	Percent sulfur
0	0	275	–	–	–
10	10	275	50	38	1.5
11	9.3	287	50	45	1.4
12	8.7	292	50	47	1.5
13	8.2	295	27	55	1.5
14	8.4	267	50	44	1.6
15	8.0	273	50	51	2.0
16	7.6	272	66	61	1.7
17	7.0	277	41	58	1.9
18	6.9	260	58	58	1.7
19	6.4	260	60	55	2.1
20	6.1	265	55	46	1.8
21	6.0	274	46	64	2.4
22	6.0	256	64	9	1.9
23	5.8	311	9	81	2.1
24	6.6	239	81	59	2.1
25	5.7	261	–	58	1.8
26	5.7	203	–	157	3.2

contain less sulfur than would a sample with more coarser coal. Part of the results are shown in Table 34.8. The raw coal contained 3.68% sulfur.

The data in Table 34.8 were obtained with 6% average solids density. Sulfur varied from 1.4 to 2.4% during the run. The same level of sulfur was attained in the same average residence time even though the average solids densities differed. In this experiment, 1032 g of coal was treated, with an average residence time of 6.65 days and a 47% sulfur removal rate on average. Experiments with higher solids densities showed lower levels of sulfur removal at longer residence times [3].

34.9 PILOT-SCALE OPERATION

Experiments were performed in 55-gallon drums fitted with polypropylene liners. Agitation was accomplished by means of rotating, with mixing motors mounted on the side of each drum adjusted so that the shaft was at a 15° angle. The impellers were three-blade marine propellers. Both the shaft and the impellers were made of stainless steel. The coal pulp density of the slurry ranged from 5 to 45%. Temperature control was achieved by supplying steam through a heating coil inside the drum. The drums were aerated with 1/8-in. stainless tubing from the bottom of the drums. Sulfate concentration was determined every day as a measure of microbial activity. At the end of each batch leaching experiment, samples were taken from the top, middle, and bottom of the drums for sulfur analysis. The slurry remaining in the drum was filtered and dried for size distribution determinations after each batch run, to determine the effect of mechanical stirring on coal particle size.

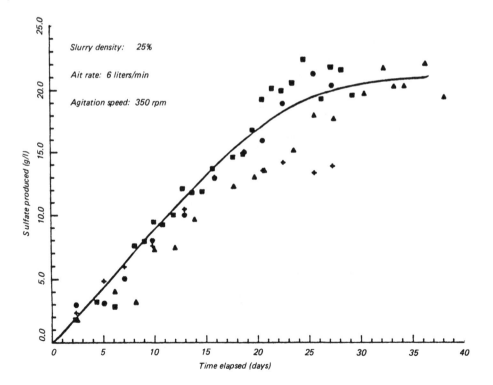

Fig. 34.15 Reproducibility of the results from pilot-scale experiment on Ohio coal mixture No. 1 under standard conditions. ■, ●, ▲, and + are duplicate runs. (From Ref. 54.)

Figure 34.15 shows the sulfate release for different runs at 25% slurry density. In these runs the air rate was 6 L/min and the stirring rate was at 350 rpm. The maximum sulfur removal rate for the run at 5% slurry density was 5.14 mg S/L per hour. This value is smaller than the average of those from 1-L runs, which were about 9 mg S/L per hour, as shown in Section 34.4.8. This suggested that mixing and mass transfer were not adequate in the 55-gallon drums as operated here.

34.10 PROCESS FLOWSHEET

Figure 34.16 shows a flowsheet proposed for microbial leaching based on work performed in this laboratory using the thermophilic microorganism *S. acidocaldarius*. The coal is finely divided (–200 mesh) to expose the sulfur surface to attack by the microorganisms. Then the acidified coal slurry is processed at elevated temperature (about 70°C). Cleaned coal is separated from the slurry, rinsed, and dried. Process liquor is treated before recycle. Sulfate is precipitated in the treatment and separated from the leach liquor and rejected to waste. Nutrients are to be added where necessary for balanced growth of the microorganisms. The process should be built near a power plant to gain certain advantages. Coal is ground at the plant site to avoid the necessity of shipping fine coal. The cheap, waste heat in the flue gas and boiler water recycled from the plant helps maintain reactor temperature. Carbon dioxide generated by combustion provides a carbon source for better microbial growth.

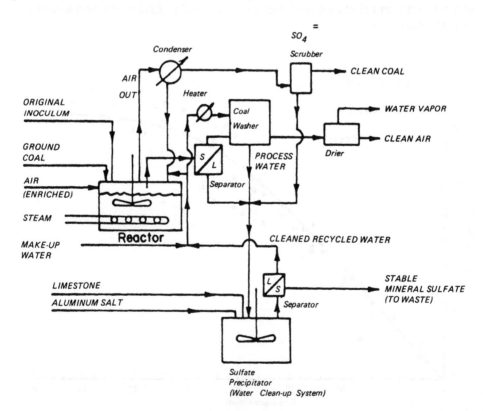

Fig. 34.16 Schematic of continuous process flowsheet for microbial coal desulfurization.

34.11 CONCLUSIONS AND RECOMMENDATIONS

The past few years of effort at the Ohio State University have yielded information pertinent to the development of a process for microbial desulfurization. The effects of process variables, such as coal types, coal particle size, pulp density of the coal slurry, temperature, pressure, gas composition, nutrient addition, and surfactants were studied experimentally. The mechanisms of chemical reactivity and of cell attachment to the coal were investigated. Finally, to test the feasibility of commercial operation, semicontinuous and continuous leaching were conducted on an expanded scale.

Further work is planned to improve understanding of the process and to promote commercialization. The key to commercial operation is cost. More detailed work will improve definitions of capital and operating costs and sharpen estimates of time to commercialization. The cost should be estimated based on more fundamental terms. Mass transfer, kinetics, and equilibrium values are of considerable value for estimation of the cost of a scaled-up process. Information about heat transfer and mixing in a larger-scale reactor would be advantageous to improve ranges of estimates and to attract investors. Siting at a user plant seems justified on a cost basis. Estimates based on continuous work in commercially sized reactors would be beneficial. A continuous leaching experiment on an extended time scale will provide more insight into process feasibility. To improve the efficiency of mass transfer, a new type of reactor might be developed in this application. For example, the air-lift fermenter has gained attention recently [5].

REFERENCES

1. Westerstrom, L., Though not a record year, the coal industry performs well in 1983, *Min. Eng., 36*, 191 (1984).
2. Galuszka, P., 1911-1986: Coal's rise and fall and rise, *Coal Age,* 47 (1986).
3. Detz, C. M., and Barvinchak, G., Microbial desulfurization of coal, *Min. Congr. J., 65*(7), 75 (1979).
4. Beck, B., Master's thesis, The Ohio State University, 1982.
5. Merritt, P. C., Advance coal cleaning processes sought for superclean coal, *Coal Age,* June (1986).
6. Kargi, F., Microbial desulfurization of coal, *Adv. Biotechnol. Processes, 3,* 241 (1984).
7. Dugan, P. R., and Apel, W. A., Microbial desulfurization of coal, in *Metallurgical Application of Bacterial Leaching and Related Microbiological Phenomena* (L. E. Murr, A. E. Torma, and J. A. Brierley, eds.), Academic Press, New York, 1978.
8. Andrews, G. F., and Maczuga, J., Bacterial coal desulfurization. *Biotechnol. Bioeng. Symp., 12,* 337 (1982).
9. Brock, T. D., Brock, K. M., Belly, R. T., and Weiss, R. L., *Sulfolobus*: a new genus of sulfur-oxidizing bacteria at low pH and high temperature, *Arch. Mikrobiol., 84,* 54 (1972).
10. Buchanan, R. E., and Gibbons, N. E., *Bergy's Manual of Determinative Bacteriology*, 8th ed., Williams & Wilkins, Baltimore, 1974.
11. Segerer, A., Stetter, F. O., and Klink, F., Two contrary modes of chemolithotrophy in the same archaebacterium, *Nature, 313,* 787 (1985).
12. Zillig, W., Yeqats, S., Holz, I., Bock, A., Gropp, F., Rettenberger, M., and Lutz, S., Plasmid-related anaerobic autotrophy of the novel archaebacterium *Sulfolobus ambivalens, Nature, 313,* 789 (1985).
13. Berry, V. K., and Murr, L. E., Direct observations of bacteria and quantitative studies of their catalytic role in the leaching of low grade copper bearing waste, in

Metallurgical Applications of Bacterial Leaching and *Related Microbial Phenomena* (L. E. Murr, A. E. Torma, and J. A. Brierley, eds.), Academic Press, New York, 1978.

14. ASTM, *Standard Method for Forms of Sulfur in Coal*, ASTM Annual Book, D2492, 05.05, American Society for Testing and Materials, Philadelphia, 1985, p. 350.

15. Kneller, W. A., and Maxwell, G. P., Size, shape, and distribution of microscopic pyrite in selected Ohio coals, in *Processing and Utilization of High Sulfur Coals* (Y. A. Attia, ed.), Elsevier, Amsterdam, 1985, pp. 41–65.

16. Calton, R. W., Image analysis of pyrite in Ohio coal: relation between pyrite grain-size distribution and pyritic sulfur reduction, in *Processing and Utilization of High Sulfur Coals* (Y. A. Attia, ed.), Elsevier, Amsterdam, 1985.

17. Eliot, R. C. (ed.), *Coal Desulfurization Prior to Combustion*, Noyes Data Corporation, N.J., 1978, pp. V–VI, 33.

18. Chandra, D., Roy, P., Mishra, A. K., Chakrabarti, J. N, and Sengupta, B., Microbial removal of organic sulfur from coal, *Fuel, 58*, 549 (1979).

19. Isbister, J. D., and Kobylinski, E. A., Microbial desulfurization of coal, in *Processing and Utilization of High Sulfur Coals* (Y. A. Attia, ed.), Elsevier, Amsterdam, 1985, pp. 627–641.

20. Murphy, J., Master's thesis, The Ohio State University, 1984.

21. Kargi, F., and Robinson, J. M., Microbial oxidation of dibenzothiophene by the thermophilic organism *Sulfolobus acidocaldarius, Biotechnol. Bioeng., 26*, 687 (1984).

22. Brierley, C. L., Thermophilic microorganisms in extraction of metals from ores, *Dev. Ind. Microbiol., 18*, 273 (1977).

23. Brierley, C. L., Leaching of chalcopyrite ore using *Sulfolobus* species, *Dev. Ind. Microbiol., 21*, 435 (1977).

24. Brock, T. D., Cook, S., Petersen, S., and Mosser, J. L., Biochemistry and bacteriology of ferrous iron oxidation in geothermal habitats, *Geochim. Cosmochim. Acta, 40*, 493 (1976).

25. ASTM, *Standard Test Methods for Sulfate Ion in Water*, ASTM Annual Book, D516, 11.01, American Society for Testing and Materials, Philadelphia, 1986, p. 700.

26. Murr, L. E., and Berry, V. K., Direct observations of selective attachment of bacteria on low-grade sulfide ores and other mineral surfaces, *Hydrometallurgy, 2*, 11 (1976).

27. Silverman, M. P., Mechanism of bacterial pyrite oxidation, *J. Bacteriol., 94*, 1046 (1967).

28. Frye, R., Masters thesis, The Ohio State University, 1985.

29. Kargi, F., and Robinson, J. M., Microbial desulfurization of coal by thermophilic microorganism *Sulfolobus acidocaldarius, Biotechnol. Bioeng., 24*, 2115 (1982).

30. Millonig, G., De Rosa, M., Gambacorta, A., and Bu'Lock, J. D., Ultrastructure of an extremely thermophilic acidophilic microorganism, *J. Gen. Microbiol., 86*, 165 (1975).

31. Marek, D., Masters thesis, The Ohio State University, 1983.

32. Gruder, S., and Genchev, F., Microbial coal desulfurization: Effects of the cell adaptation and mixed cultures, *C.R. Acad. Bulg. Sci., 32*, 353 (1979).

33. Silverman, M. P., and Lundgren, D. G., Studies on the chemoautotrophic iron bacterium *Ferrobacillus ferrooxidans, J. Bacteriol., 77*, 642–647 (1959).

34. Lacey, D. T., and Lawson, F., Kinetics of the liquid-phase oxidation of acid ferrous sulfate by the bacterium *Thiobacillus ferrooxidans, Biotechnol. Bioeng., 12*, 29 (1970).

35. Brown, J. B., A chemical study of some synthetic potassium–hydronium jarosites, *Can. Mineral., 10*, 696 (1970).

36. Karl, D. M., Wirsen, C. O., and Jannasch, H. W., Deep sea primary production at the Galapagos hydrothermal vents, *Science, 207,* 1345 (1984).
37. Emmert, R. E., and Pigford, R. L., Gas adsorption and solvent extraction, in *Chemical Engineering Handbook* (R. H. Perry, C. H. Chilton, and S. D. Kirpatrick, eds.), McGraw-Hill, New York, 1963.
38. Bailey, J. E., and Ollis, D. F., *Biochemical Engineering Fundamentals,* McGraw-Hill, New York, 1977.
39. Kargi, F., Enhancement of microbial removal of pyritic sulfur from coal using concentrated cell suspension of *T. ferrooxidans* and external carbon dioxide supply, *Biotechnol. Bioeng., 24,* 749 (1982).
40. Hoffmann, M. R., Faust, B. C., Panda, F. A., Koo, H. H., and Tsuchiya, H. M., Kinetics of the removal of iron pyrite from coal by microbial catalysis, *Appl. Environ. Microbiol., 42,* 259 (1981).
41. Davison, M. S., Torma, A. E., Brierley, J. A., and Brierley, C. L., Effects of elevated pressure on iron-sulfur oxidizing bacteria, in *Biotechnol. Bioeng. Symp., 11,* 603 (1979).
42. Wakao, N., Mishina, M., Sakuai, Y., and Shiota, H., Bacterial pyritic oxidation. II. The effect of various organic substances on release of iron from pyrite by *Thiobacillus ferrooxidans, J. Appl. Microbiol., 29,* 177 (1983).
43. Barbara, A., Master's thesis, The Ohio State University, 1985.
44. Shivvers, D. W., and Brock, T. D., Oxidation of elemental sulfur by *Sulfolobus acidocaldarius, J. Bacteriol., 114,* 706 (1973).
45. Weiss, R. L., Attachment of bacteria to sulphur in extreme environments, *J. Gen. Microbiol., 77,* 505 (1973).
46. Berry, V. K., and Murr, L. E., Bacterial attachment to molybdenite: an electron microscope study, *Metall. Trans., 6B*(1), 448 (1975).
47. Murr, L. E., and Berry, V. K., An electron microscope study of bacterial attachment to chalcopyrite: microstructural aspects of leaching, in *Extractive Metallurgy of Copper,* Vol. 2 (J. C. Yannopoulos and J. C. Agarwal, eds.), The Metallurgical Society of AIME, New York, 1976.
48. Kargi, F., and Robinson, J. M., Biological removal of pyritic sulfur from coal by the thermophilic organism *Sulfolobus acidocaldarius, Biotechnol. Bioeng., 27,* 41 (1985).
49. Myerson, A. S., and Kline, P., The adsorption of *Thiobacillus ferrooxidans* on solid particles, *Biotechnol. Bioeng., 25,* 1669 (1983).
50. DiSpirito, A. A., Dugan, P. R., and Tuovinen, O. H., Sorption of *Thiobacillus ferrooxidans* to particulate material, *Biotechnol. Bioeng., 25,* 1163 (1983).
51. Peterson, G. L., Determination of total protein, in *Methods in Enzymology,* Vol. 91, *Enzyme Structure,* (C. H. W. Hirs and S. N. Timasheff, eds.), 1983, Part I, p. 95.
52. Attia, Y. A., and Fuerstenau, D. W., Feasibility of cleaning high sulfur coal fines by selective flocculation, *CIM 14th International Mineral Processing Congress,* Toronto, Ontario, Canada, Oct. 17-23, 1982.
53. Kargi, F., and Weissman, J. G., A dynamic mathematical model for microbial removal of pyritic sulfur from coal, *Biotechnol. Bioeng., 26,* 604 (1984).
54. Harsh, D., Master's thesis, The Ohio State University, 1985.

35

Microbiological Studies on High-Sulfur Eastern Canadian Coals

RONALD G. L. MC CREADY *Canadian Centre for Mineral and Energy Technology (CANMET), Ottawa, Ontario, Canada*

35.1 USE OF THE SABA PROCESS FOR COAL BENEFICIATION

35.1.1 Introduction

Eastern Canadian coals can vary from 4 to 12% in their pyritic sulfur content; generally, the pyrite is disseminated throughout the carbon matrix as 1 to 2-μm crystals. Thus to beneficiate these coals, they must be finely ground and the carbon matrix is recovered using fine-particle technology.

Over the past decade there has been increasing pressure from environmental groups and government agencies for coal producers to reduce the sulfur content of their coals in order to reduce atmospheric pollution on combustion. Many research groups have investigated the biological desulfurization of coal and there is evidence that such a process may be the most economical approach to this problem. In this chapter we document an extensive study on the biological desulfurization of eastern Canadian coals from various locations within the Maritime provinces of Canada.

At present, the method used, the SABA (spherical agglomeration–bacterial adsorption) process, is most successful in the beneficiation of Minto coal from New Brunswick; the process does not appear to be suitable for the beneficiation of Cape Breton coals. In the future, the process may be modified to accomplish beneficiation of Cape Breton coals, but such modification will require an appreciation and understanding of the differences in the chemistry of the iron sulfide and mineral ash components among the various coals.

Several factors that adversely affect the economic application of the SABA process to coal beneficiation require additional research. As optimal beneficiation is obtained with finely ground coal slurries, research is required on the economics of fine grinding and new means of grinding to produce consistent, known particle sizes within the resultant coal slurries. Second, research is required on the surface chemistry of the various coal components; if the surface charge of the mineral ash could be altered to prevent adsorption of the thiobacilli, fewer cells per unit mass of coal would be required, as they would bind specifically to the iron sulfides, rendering them hydrophilic. Third, research is required to develop a more economical means of recovering the carbonaceous material from the fine coal slurries. At present the oil consumption in the spherical agglomeration step is excessive and a new technology is required for coal recovery for direct combustion or production of coal-water mixtures. Finally, research is required on the separation of the mineral ash from iron sulfides in the process tailings to allow the use of the iron sulfides as a substrate to grow additional thiobacilli for the process. The cost of the medium could be greatly reduced by replacing the $FeSO_4$ with the iron sulfides removed from the coal. Thus further laboratory studies are required to optimize the SABA process, and once these objectives have been achieved, pilot scale and economic evaluations of the process could be conducted on various Maritime coals.

Cape Breton coal-waste dumps are estimated to contain from 300,000 to 500,000 tons of coal, with mineral ash contents of up to 40% and pyrite contents up to 10% w/w. With the increasing cost of mining, increased emphasis on preventing environmental pollution, and development of new coal-cleaning technologies, there has been renewed interest in recovering the combustible material from these dumps. As the pyrite and ash content of the waste coal is so high, new technologies are required to produce a salable, recovered commodity. One approach may be to bacterially leach the pyrite and ash from the waste coals prior to the physical separation and coal recovery process.

35.1.2 Specific Conclusions

The microbiological studies conducted on application of the bacterial adsorption-spherical agglomeration process for the removal of pyrite from Cape Breton coals have lead to the following conclusions.

1. Iron-depleted thiobacilli bind exclusively to the pyrite and iron-containing ash components of ground coal suspensions.
2. There are differences among the bacterial strain in their ability to bind to the pyrite and ash components. At present, the original Elliot Lake thiobacillus isolate produces the best levels of pyrite rejection of the bacterial strains tested.
3. Iron-depleted cells are more efficient in rejecting pyrite from finely ground coal suspensions than are non-iron-depleted cells. Therefore, the iron-depletion step of the process is essential.
4. As a large portion of the indigenous pyrite in Cape Breton coals is in the crystal size 1 to 2 μm, these coals will have to be finely ground to reduce their pyrite content by any physical process.
5. Studies on the mineralogy and chemistry of both the iron sulfides and the mineral ash components of various Maritime coals are required to explain their differences in reactivity to the SABA process.
6. Cape Breton coals are amenable to bacterial leaching to remove pyrite and mineral ash components, and larger-scale studies should be initiated to evaluate this process as a means of reducing the pyrite and ash content of the coal-waste dumps.
7. During bacterial leaching the leach solution chemistry will have to be monitored continuously until the leach dump attains chemical equilibrium to prevent the destruction of the inoculated thiobacilli by ciliate and zooflagellate predators.

35.1.3 Background

Eastern Canadian coals are considered high-sulfur coals, as they contain 0.1 to 0.5% sulfate, 2 to 12% pyritic sulfur, and 0.5 to 1.5% w/w organic sulfur. Coincident to their high iron sulfide content, these coals also contain 4 to 25% w/w inorganic mineral ash. Current environmental regulations restrict the combustion of coal to coals containing ≤ 3% total sulfur. Therefore, as Maritime coals contain an average of about 1% organic sulfur, prior to combustion, the total inorganic sulfur content must be reduced to less than 2% w/w. If stricter environmental regulations are imposed in the future, the inorganic sulfur may have to be totally removed from the coal prior to combustion.

The SABA process was developed on a laboratory scale using Minto coal. Initial studies resulted in greater than 90% rejection of the pyritic sulfur and greater than 75% rejection of the inorganic mineral ash components using finely ground Minto coal. The current study was initiated to assess application of the SABA process to other Maritime coals of different physical and chemical characteristics and to optimize the chemical and biological parameters of the process.

35.1.4 Objectives

1. Isolate and culture indigenous *Thiobacillus ferrooxidans* associated with stockpiles of coal waste using standard microbiological techniques.
2. Conduct experiments using ^{32}P-labeled thiobacilli to determine the site of attachment of the bacteria to the components of coal slurries.
3. Optimize conditions for pyrite and ash rejection in the coal beneficiation process.
4. Assess the feasibility of reducing the pyrite and ash content of coal within coal-waste dumps by bacterial leaching, thereby allowing reclamation of the dumps.

For simplicity, the results and discussion are divided into five sections:

1. Isolation and cultivation of indigenous *T. ferrooxidans* from Cape Breton coal wastes
2. Selective adsorption of ^{32}P-labeled *T. ferrooxidans* in finely ground coal suspensions
3. Studies on iron sulfide and ash rejection using the SABA process and various Maritime coals
4. Bacterial leaching of iron sulfides from coal
5. Ancillary basic studies on nutrition, fimbriae, plasmids, and predator organisms

35.2 ISOLATION AND PURIFICATION OF *T. FERROOXIDANS* CULTURES FROM COAL-WASTE WATER SAMPLES

35.2.1 Growth Medium

The thiobacilli were maintained on Tuovinen and Kelly's medium of the following composition: KH_2PO_4, 0.4 g; $MgSO_4 \cdot 7H_2O$, 0.4 g; $(NH_4)_2SO_4 \cdot 7H_2O$, 0.4 g; and $FeSO_4 \cdot 7H_2O$, 33.3 g per liter of distilled water, pH adjusted to 2.0 with concentrated H_2SO_4.

35.2.2 Procedure for Determining the Culture Purity of *Thiobacillus* Isolates

After successive transfers of the *T. ferrooxidans* isolates on Tuovinen and Kelly's medium, an aliquot of each culture was streaked onto Nutrient Agar (Difco) plates and incubated at 23 to 24°C for 24 and 48 h. As *T. ferrooxidans* will not grow in the presence of organic acids or sugars, any evidence of growth on these plates would have indicated culture contamination by acidophilic heterotrophs.

35.2.3 Results: Isolation of *T. ferrooxidans* Indigenous to Cape Breton Coals

Water samples were collected from five sites on Cape Breton Development Corporation (CBDC) property: lagoon No. 14, pH 2.5; Donnely bog, pH 3.0; final plant effluent, pH 3.5; CBDC wash plant discharge, pH 2.5; and lagoon No. 9, pH 2.5. *T. ferrooxidans* organisms were isolated from all locations, but the strains isolated from lagoon No. 14 and from the wash plant discharge have the fastest rates of growth and iron oxidation of the five isolates. The five isolates are identified as Cape Breton *T. ferrooxidans* No. 1 through No. 5, their numerical order corresponding to their sites of isolation listed above.

35.3 SELECTIVE ADSORPTION OF ^{32}P-LABELED *T. FERROOXIDANS* IN FINELY GROUND COAL SLURRIES

The attachment of *T. thiooxidans* to elemental sulfur granules has been observed both microscopically and by the use of ^{14}C-labeled cells [8,32,42]. Similarly, using scanning electron microscopy, Murr and Berry [29] reported that thermophilic, sulfur-oxidizing *Caldariella* were specifically attached to pyrite or chalcopyrite intrusions in a low-grade copper ore and were not bound to other mineral phases. In 1973, Bennett and Tributsch [4] reported that *T. ferrooxidans* bind to pyrite crystals along fracture lines, dislocations, and crystal imperfections and that sulfide oxidation was initiated at these sites.

In 1980, Kempton et al. [17] proposed a novel process for the removal of pyrite from finely ground coal prior to combustion. *T. ferrooxidans* organisms were harvested, resuspended in iron-free medium, and incubated for 2 h to deplete the cells of endogenous Fe^{2+}. The authors postulated that when added to the coal suspension, the iron-depleted *T. ferrooxidans* adsorbed to the pyrite surfaces, creating a hydrophilic surface on the particles, which could then be separated from the hydrophobic coal particles by oil agglomeration. The proposed mechanism, allowing separation of the pyrite and coal, has been questioned, as there was no direct evidence of the binding of *T. ferrooxidans* to the pyrite particles. In the present chapter we report on the distribution of ^{32}P-labeled *T. ferrooxidans* after 30 min incubation in finely ground coal suspensions.

35.3.1 Materials and Methods

A strain of *T. ferrooxidans* isolated from abandoned pyritic uranium mine tailings near Elliot Lake, Ontario was grown on Tuovinen and Kelly's [37] medium at pH 2.3. Stock cultures were prepared in 250-mL Erlenmeyer flasks, each containing 100 mL of medium and grown at 30°C on a gyratory shaker for 72 h. Stock cultures remain active at 4°C for at least 1 week.

Radioactively ^{32}P-labeled cells were obtained by preparing 3 L of medium containing 10 μCi of ^{32}P. The medium was inoculated with 300 mL of stock culture and incubated at 30°C on a gyratory shaker at 150 rpm for 72 h. The cells were harvested by centrifugation at 10,000 rpm for 10 min. The cell pellet and sedimented $Fe(OH)_3$ was resuspended in $FeSO_4$-free medium at pH 2.3 and centrifuged at 1000 rpm for 5 min to separate the $Fe(OH)_3$ from the cells. The cell suspension was decanted from the $Fe(OH)_3$ pellet and centrifuged at 10,000 rpm to pellet the cells. The distribution of the ^{32}P in the phosphorus-containing cellular components was determined by chemical fractionation of the cells using the modified Schmidt-Thannhauser technique described by Myers and McCready [30].

For studies on the distribution of ^{32}P-labeled cells in coal suspensions, the cells were harvested as described previously, but the $Fe(OH)_3$-free cell suspension was added to 1 L of $FeSO_4$-free medium at pH 2.3 and incubated at 30°C for 2 h to deplete the cells of endogenous Fe^{2+}. Cell numbers in the final suspension were determined by direct count using a Petroff-Houser counting chamber under a phase-contrast microscope. After the 2-h incubation, 100 mL of coal suspension (35 g of coal) was added to the 1 L of iron-free cells and incubated for 30 min at 30°C on a gyratory shaker at 150 rpm. The coal was then recovered from suspension by the spherical agglomeration technique described by Capes et al. [6]. After the initial screening to recover the agglomerated coal, a 100-mg aliquot of coal was spread on a 45-mm, 0.2-μm Millipore filter which was placed in a scintillation vial. Twenty milliliters of scintillation fluid was added and the activity of ^{32}P determined by scintillation counting. The coal remaining on the screen was washed with 500 mL of deionized water, and an additional 100-mg aliquot of coal was spread on a Millipore filter and counted as described previously.

The mineral ash components and the pyrite that passed through the screen during coal recovery were separated from the aqueous phase of the tailings by filtration of an aliquot (10 mL) of the tailings suspension through an 8-μm Millipore filter. A second 10-mL aliquot was filtered, and the solids retained on the filter were washed with 100 mL of $FeSO_4$-free medium to remove loosely bound thiobacilli prior to drying and scintillation counting of the filter and solids.

The distribution of ^{32}P in the cellular components and the distribution of the cells in the coal suspensions were determined using the scintillation fluid and techniques described by McCready and Din [26]. Radioactive counts were determined using the LKB Rackbeta Scintillation Counter and all counts were corrected for background radioactivity.

Two samples of coal indigenous to Cape Breton were used in this study. Their characteristics are presented in Table 35.1. Brogan coal is from a surface outcrop of the

Table 35.1 Properties of the Coals Used (Percent by Weight, Dry Basis)

	Brogan coal	Metallurgical coal
Ash	16.34	4.5
Total sulfur	6.95	2.21
Inorganic sulfur (sulfate)	1.13	0.30
Pyritic sulfur	4.24	1.01
Organic sulfur	1.59	0.91
Particle-size distribution:	$10\% > 14.5~\mu m$	$10\% > 18.5~\mu m$
	$25\% > 10.5~\mu m$	$25\% > 13.0~\mu m$
	$50\% > 7.3~\mu m$	$50\% > 8.7~\mu m$
	$75\% > 5.1~\mu m$	$75\% > 5.8~\mu m$
	$90\% > 3.7~\mu m$	$90\% > 4.1~\mu m$

Harbor coal seam, which is being strip-mined at Point Aconi, Nova Scotia. The metallurgical coal is a blended coal product produced at the Victoria Junction Coal-Wash Plant operated by the Cape Breton Development Corporation in Sydney, Nova Scotia.

35.3.2 Results

The percent assimilation and intracellular distribution of ^{32}P in *T. ferrooxidans* grown on Tuovinen and Kelly's [37] medium containinn 10 μCi of ^{32}P are presented in Table 35.2. As the percent assimilation of the available ^{32}P was low (1.8%), the medium was modified to increase the specific activity of the PO_4^{3+}. Both the KH_2PO_4 and the $MgSO_4 \cdot 7H_2O$ were reduced by 75%, to 0.1 g/L, and the experiments were repeated (see Table 35.3). Medium modification resulted in an increased percent assimilation of the available ^{32}P to 10% and an apparent increase of the radioactive label in the DNA.

Previous studies (Myers and McCready [30]) have shown that isotope labeling of microbial cells is most stable when the major portion of the isotope is incorporated into the nucleic acids. From the data in Table 35.3, between 89 and 92% of the ^{32}P is incorporated into the nucleic acids. Therefore, modified Tuovinen and Kelly's medium was used to prepare labeled cells for subsequent experiments on the adsorption and distribution of *T. ferrooxidans* in finely subdivided coal.

To follow the procedure of Kempton et al. [17], the ^{32}P-labeled cells were suspended in 1 L of $FeSO_4$-free medium and incubated at 30°C and 150 rpm for 2 h to deplete the cells of endogenous Fe^{2+}. During the 2-h incubation of labeled cells in the $FeSO_4$-free medium, no apparent cellular lysis or loss of the ^{32}P label was observed. A 5-mL aliquot of the cell suspension was filtered through a 0.22-μm Millipore filter at 15-min intervals, washed with $FeSO_4$-free medium, dried, and the radioactive counts determined for each aliquot of cells. The radioactive counts were within ±3% of each other over the 2-h incubation period, and total cell counts, by microscopy, were almost constant over the 2-h period.

After depletion of the endogenous iron, 35 g of coal was added in 100 mL of suspension to the 1 L of $FeSO_4$-free culture and incubated at 30°C and 150 rpm for an additional 30 min. The coal was then recovered by spherical agglomeration and screening. Aliquots of the different suspension constituents were analyzed, the coal was then washed with 500 mL of deionized water, and an aliquot of the washed coal and the wash

Table 35.2 Assimilation and Intracellular Distribution of ^{32}P in *T. ferrooxidans* Grown on Tuovinen and Kelly's Medium at 30°C for 72 Hours

Cellular fraction	Percent (^{32}P)
Acid soluble	8.9
Lipid	0.6
RNA	41.5
DNA	49.0
Percent of available ^{32}P assimilated	1.8

Table 35.3 Assimilation and Distribution of ^{32}P by *T. ferrooxidans* Grown on Modified (0.1 g KH_2PO_4 and $MgSO_4 \cdot 7H_2O$ per Liter) Tuovinen and Kelly's Medium at 30°C for 72 Hours

Cellular fraction	Percent ^{32}P	
	Culture 1	Culture 2
Acid soluble	5.9	8.2
Lipid	2.1	2.1
RNA	15.5	20.6
DNA	76.5	68.7
Percent of available ^{32}P assimilated	10	10

water were analyzed. The distribution of the labeled cells after contact with the Brogan coal is presented in Table 35.4.

About 20% of the ^{32}P-labeled cells were found adsorbed to either the inorganic mineral ash components or the pyrite associated with the coal. The cells reported initially with the coal were either loosely bound or had been entrapped in the hydrocarbon phase used in the oil agglomeration step. These cells were easily removed from the coal by washing and the radioactivity initially associated with the coal was redistributed into the tailings and aqueous phase after the water wash. In contrast, the cells that were adsorbed to the tailings solids were not removed by subsequent washing. The majority (70%) of the cells remained in the aqueous phase.

In the experiments with metallurgical coal, which had much lower concentrations of pyritic sulfur and ash, the number of cells added per gram of coal was reduced substantially. From Table 35.5, the percentage of cells binding to the tailings solids increased to between 63 and 88% and the remainder of the cells were found in the aqueous phase of the tailings after washing the coal with water. The radioactivity initially associated with the coal was due to cells being trapped in the agglomerates; these were easily removed by washing with water.

Table 35.4 Distribution of ^{32}P-Labeled *T. ferrooxidans* (5.5×10^{11} cells/g of Coal) in Brogan Coal after 30 Minutes' Contact Time and Oil Agglomeration

Phase	Percent distribution of cells	
	Experiment 1	Experiment 2
Initial separation		
Coal after screening	43.3	14.4
Tailings after screening	20.3	21.8
Screening water phase	36.4	63.8
After a water wash		
Coal	Nil	Nil
Tailings in wash water	10.1	4.1
Wash water	32.5	7.9

Table 35.5 Distribution of ^{32}P-Labeled *T. ferrooxidans* $(6.6 \times 10^5$ Cells/g of Coal) in Metallurgical Coal After 30 Minutes' Contact Time and Oil Agglomeration

	Percent distribution of cells	
Phase	Experiment 1	Experiment 2
Initial separation		
Coal after screening	32.5	8.2
Tailings after screening	62.9	87.7
Screening water phase	4.6	4.1
After a water wash		
Coal	Nil	Nil
Wash water	30.5	8.2

35.3.3 Discussion

A major problem in the isotope labeling of viable bacteria is obtaining assimilation of sufficient radioactive isotope by the cells so that counting of emissions is practical while avoiding any deleterious effects on the cells. Myers and McCready [30] reported no lethal effects from ^{32}P labeling of *Serratia marcescens* using 10 μCi of ^{32}P per 300 mL of medium. Lea et al. [23] calculated that 4×10^3 roentgens of beta emission per cell is required to kill *Escherichia coli*. The emission rate for the amount of isotope used in our experiments (10 μCi of ^{32}P per 3 L of medium) is less than 0.1 roentgen per second. As this emission is distributed throughout the total volume of the medium, it should have no lethal effect on *T. ferrooxidans* during growth.

Fuerst and Stent [15] state that the rate of ^{32}P assimilation is proportional to the specific activity of the growth medium. "Hence the bacteria in a high specific activity medium incorporate initially many more atoms of ^{32}P into their phosphorylated constituents per unit of time than those in a low specific activity medium." During growth on Tuovinen and Kelly's medium, the *T. ferrooxidans* cells assimilated only 1.8% of the available ^{32}P. By reducing the total phosphate content of the medium, thereby increasing the specific activity of the medium, the cells were able to assimilate 10% of the available ^{32}P.

For the radioactive label to indicate the presence of a viable bacterial cell, the major portion of the label must be incorporated into a stable intracellular cell component. The data presented in both Tables 35.2 and 35.3 show that the majority of the ^{32}P was incorporated into the cellular nucleic acids. Thus, in the coal suspension studies, the presence of ^{32}P in a particular phase indicated the presence of intact bacterial cells and was not the result of labeled capsular or external membranous material being sloughed off and adsorbed to a particular coal constituent.

Hoffman et al. [16] reported increased growth yields by growing *T. ferrooxidans* on a low-phosphate, low-sulfate medium. Comparing the data from Tables 35.2 and 35.3 for the percent of ^{32}P incorporated into DNA, the increased incorporation of ^{32}P in the low-phosphate medium indicates an increased growth rate. As DNA is synthesized only during cell division, an increased incorporation of ^{32}P over 72 h of growth can occur only by having an increased rate of cell multiplication.

The results presented in Tables 35.4 and 35.5 indicate that the ^{32}P-labeled *T. ferrooxidans* organisms adsorb to the pyrite and inorganic mineral phases of the ash. They were associated with the hydrophobic coal initially but were easily removed with a water wash. Although the percentage of cells associated with the coal varied considerably between experiments, the percentage of cells associated with the solid phase of the tailings after screening are similar for each coal type. Coal is very heterogeneous in its composition, and this heterogeneity is evident from the data presented in Tables 35.4 and 35.5. The variance in the percentage of cells associated with the coal is not believed to be a biological phenomenon but is believed to be due to physical entrapment of the bacteria within the coal agglomerates during the spherical agglomeration process. As the cells were easily removed from the coal by a water wash, there was no physical attachment of the cells, via fimbriae or glycocalyx material, to the coal per se. In contrast, washing the solid phase of the tailings with water caused no loss of radioactivity, suggesting that the cells are tenaciously attached to the mineral constituents.

DiSpirito et al. [10] have shown that *T. ferrooxidans* organisms adsorb to glass beads, sulfur particles, pyrite, fluorapatite, and quartz particles and that maximum sorption of the cells to these particulates occurred within 30 to 40 min.

In the present study the cells were incubated with the finely ground coal suspensions for 30 min prior to recovery of the coal by oil agglomeration. The cells were not bound to the coal, as they were removed by a water wash; in contrast, they are not removed from the mineral ash and pyritic components of coal by washing with water. The results presented in this chapter are consistent with the mechanism postulated by Kempton et al. [17]. The iron-depleted thiobacilli adsorb to the pyrite and mineral ash components of finely ground coal and thereby facilitate the recovery of a coal product, with a reduced ash and pyrite content, by a hydrophobic reagent using the spherical agglomeration technique.

35.4 STUDIES ON THE IRON SULFIDE AND ASH REJECTION FROM COAL SLURRIES USING THE SABA PROCESS AND VARIOUS MARITIME COALS

35.4.1 Materials and Methods

The majority of the coal samples used in these studies were supplied by the Cape Breton Development Corporation in Sydney, Nova Scotia. Samples of Minto coal were supplied by both the Chemistry Division of the National Research Council and the New Brunswick Coal Corporation. The St. Rose coal sample was supplied by the Geology Department of Dalhousie University.

Ground coal slurries were prepared either by ball milling or by grinding, in the Attritor, sufficient coal and water to produce a 30% w/v slurry. Coal particle-size analyses were determined initially by the Chemistry Division of the National Research Council, but the majority of the particle-size determinations were made using an optical micrometer in a phase-contrast microscope at a magnification of 400X or 1000X.

The procedure of Kempton et al. [17] was used for testing the beneficiation effect of the SABA process on the various Maritime coals. All coal constituent analyses were performed by the Atlantic Regional Laboratory of the National Research Council (NRC), or were performed, under contract, by the Atlantic Coal Institute.

35.4.2 Results

As the bacteria specifically bind to the pyrite and ash components of the coal suspension, studies were initiated to assess the SABA process with Cape Breton coals. As the chloride

Table 35.6 Sulfur Component and Ash Content of Brogan Coal (Mean Particle Size 5.4 µm) After SABA Treatment with Cape Breton *T. ferrooxidans*

Sample	Total sulfur (%)	Inorganic sulfur (%)	FeS$_2$ sulfur (%)	Percent FeS$_2$ rejected	Organic sulfur (%)	Ash (%)	Percent ash rejected
Ground coal	6.67	1.01	4.07	–	1.68	16.34	–
Agglomerated coal	6.48	0.35	4.67	-13.0	1.46	15.32	6.3
SABA No. 1	5.89	0.40	3.78	8.0	1.71	7.48	54.0
SABA No. 2	6.69	0.62	4.44	-22.0	1.64	8.47	49.0

content of the Cape Breton coals is higher than that of the Minto coal, indigenous thiobacilli were tested, as they should have a greater tolerance to chloride.

In the initial experiments (Table 35.6) the ash rejection following bacterial adsorption was good; however, the pyrite rejection was poor. The experiments were therefore repeated using the Elliot Lake *Thiobacillus* isolate, which had been used in developing the process (Table 35.7).

Although the pyrite rejection was better with the Elliot Lake strain, it was not as high as had been obtained with the Minto coal. It was learned from discussions with Dalhousie geologists and P. Hacquebard that about 20% of the pyrite in Cape Breton coals occurs as 1 to 2-µm crystals which are finely disseminated throughout the coal organic matrix. A sample of coal ground to 90% < 1 µm obtained from Magstone Developments Ltd. was tested for pyrite rejection by spherical agglomeration and the SABA process (Table 35.8).

Desulfurization of Prince Coal by Bacterial Adsorption Followed by Spherical Agglomeration

A series of experiments to assess the pyrite rejection by the SABA process were carried out with Prince coal (Tables 35.9 and 35.10). As the pyrite rejections were low in comparison to previous studies [17], the process was reassessed using a sample of Minto coal (Table 35.11). From the data in Table 35.11, the pyrite rejection was much lower than expected for a Minto coal sample. However, Kempton et al. [17] had found low percentages of pyrite rejection on initial exposure of iron-depleted *T. ferrooxidans* to a coal slurry. Therefore, the experiment was repeated, but the tailings from the first batch of

Table 35.7 Sulfur Component Content of Brogan Coal (Mean Particle Size 5.4 µm) After SABA Treatment with the Elliot Lake *T. ferrooxidans*

Sample	Total sulfur (%)	Inorganic sulfur (%)	FeS$_2$ sulfur (%)	Percent FeS$_2$ rejected	Organic sulfur (%)
Ground coal	6.76	1.01	4.07	–	1.68
Agglomerated coal	6.48	0.35	4.67	-13.0	1.46
SABA No. 1	5.46	0.64	3.11	23.6	1.72
SABA No. 2	5.47	0.61	3.25	20.1	1.60

Table 35.8 Sulfur Component Content of Finely Ground (90% <1 μm) Brogan Coal After SABA Treatment with the Elliot Lake *T. ferrooxidans*

Sample	Total sulfur (%)	Inorganic sulfur (%)	FeS$_2$ sulfur (%)	Percent FeS$_2$ rejected	Organic sulfur (%)
Ground coal	6.95	1.15	4.24	–	1.59
Agglomerated coal	3.07	0.42	1.60	62	1.05
SABA coal	3.55	0.69	2.58	39	0.29

Table 35.9 Sulfur Component and Ash Content of Prince Coal (Mean Particle Size 7.3 μm) Following Various Desulfurization Treatments (2.3 × 10^8 Cells/g of Coal)

Sample	Total sulfur (%)	Inorganic sulfur (%)	FeS$_2$ sulfur (%)	Percent FeS$_2$ rejected	Organic sulfur (%)	Ash (%)	Percent ash rejected
Ground coal	2.21	0.3	1.01	–	0.91	4.50	–
Agglomerated coal	2.02	0.12	0.91	9.9	1.03	2.56	43.1
SABA No. 1 (not Fe depleted)	2.10	0.19	0.88	12.9	1.02	2.64	41.3
SABA No. 2	2.10	0.19	0.85	15.8	1.14	2.45	45.5
SABA No. 3	2.10	0.19	0.84	16.8	1.08	2.71	39.8

Table 35.10 Sulfur Component and Ash Content of Prince Coal Ground to 80% < 1 μm After SABA Treatment with Elliot Lake *T. ferrooxidans* (2 × 10^8 Cells/g of Coal)

	Total sulfur (%)	Inorganic sulfur (%)	Organic sulfur (%)	FeS$_2$ sulfur (%)	Percent FeS$_2$ rejected	Ash (%)
Ground coal	4.27	0.60	1.10	2.57	–	18.00
Agglomerated coal	3.52	0.20	1.10	2.22	13.6	14.40
SABA coal	3.55	0.29	1.26	2.00	22.2	12.86

Table 35.11 Sulfur Component and Ash Content of Ball-Milled Minto Coal (Mean Particle Size 7.3 μm) After SABA Treatment with the Elliot Lake *T. ferrooxidans* (1.47 × 10^8 Cells/g of Coal)

	Total sulfur (%)	Inorganic sulfur (%)	Organic sulfur (%)	FeS$_2$ sulfur (%)	Percent FeS$_2$ rejected	Ash (%)
Ground coal	5.94	0.36	2.16	3.42	–	10.4
Agglomerated coal	5.55	0.11	2.04	3.40	0.6	6.28
SABA No. 1	5.55	0.12	2.00	3.43	Nil	6.65
SABA No. 2	4.64	0.16	1.72	2.76	19.3	6.70

Table 35.12 Sulfur Component and Ash Content of Ball-Milled Minto Coal After SABA Treatment with Elliot Lake *T. ferrooxidans* (3.73×10^7 Cells/g of Coal), Tailings Cycled Through Five Additional Batches of Coal (Mean Particle Size 7.3 μm)

	Total sulfur (%)	Inorganic sulfur (%)	Organic sulfur (%)	FeS_2 sulfur (%)	Percent FeS_2 rejected	Ash (%)	Percent ash rejected
Ground coal	5.94	0.36	2.16	3.42	–	10.4	–
Agglomerated coal	4.52	0.36	2.00	2.16	36.8	4.74	54.4
SABA No. 1	3.82	0.31	1.15	2.36	31.0	6.64	36.2
SABA No. 2	3.58	0.28	1.32	1.98	42.1	6.90	33.7
SABA No. 3	3.42	0.29	1.13	2.00	41.5	7.15	31.2
SABA No. 4	5.98	0.08	2.30	3.60	-5.2	7.11	31.6
SABA No. 5	6.13	0.08	2.41	3.64	-6.4	7.14	31.3
SABA No. 6	6.08	0.08	2.36	3.64	-6.4	7.12	31.5

coal were utilized to treat a second batch of coal; this procedure was repeated for an additional four batches of coal slurry (Table 35.12).

Previous studies on Minto coal had shown pyrite and ash rejections greater than 70%. Thus the results shown in Tables 35.11 and 35.12 were of concern. Kempton (personal communication) had observed a gradual decrease in the efficiency of pyrite and ash rejection using bacterial cultures that had been maintained on synthetic laboratory media for several years. A decision was made to reassess the SABA process and attempt to explain the variations in process efficiency with coals and to determine the effects of slurry age, coal storage prior to treatment, and other parameters that could affect the process.

Studies were reinitiated on Minto coal to assess whether the results observed previously were due to inherent properties of the coal or whether the bacterium had undergone a change due to continuous cultivation on a synthetic medium. Table 35.13 shows the effect of slurry age when a stored Minto coal sample was freshly ground and subsequently treated by spherical agglomeration and the SABA process at various time intervals. Maximum pyrite and ash rejection were observed with both processes when the slurry had been aged 24 h. However, in this series of experiments the percent carbon recovery was poor. As a result, a study of the percent v/w Varsol/coal was initiated to determine the optimal Varsol concentration required for fine coal particle recovery by spherical agglomeration. Table 35.14 shows maximum pyrite and ash rejection combined with maximum carbon recovery when 60% v/w Varsol was used, and that these results were independent of the pH of the slurry.

Kempton et al. [17,28] had reported consistent pyrite rejections of about 70% using *T. ferrooxidans* grown on a defined medium with Minto coal. As the level of pyrite rejection in Table 35.13 was much lower than this value and the data of Table 35.14 indicated that a suboptimal Varsol concentration had been used for coal recovery, the experiment was repeated using a fresh sample of Minto coal obtained directly from the mine.

Table 35.15 shows the effect of aging a fresh coal sample in liquid suspension. With spherical agglomeration, some chemical or physical change in the slurry component surface properties appears to be required before good pyrite and ash rejections are observed using fresh coal. In contrast, the slurry age does not appear to affect the beneficia-

Table 35.13 Comparison of the Effect of Coal Suspension Age on Pyrite and Ash Rejection and Carbon Recovery Between Spherical Agglomeration and the SABA process; Stored Minto Coal Ground to Between 3 and 28 μm and Treated with 7.2×10^8 Cells/g of Coal and Recovered Using 36.5% Varsol

Time (hours)	Procedure	FeS_2 (%)	Ash (%)	Percent FeS_2 rejected	Percent ash rejected	Percent carbon recovered
0	Raw coal	3.69	22.25			
	Agglom.	3.22	12.43	13	45	75
	SABA	2.82	8.12	24	64	97
	SABA	2.77	10.34	25	54	86
24	Agglom.	3.08	12.89	17	42	82
	SABA	2.20	9.38	40	58	86
	SABA	2.08	9.46	44	58	84
96	Agglom.	3.22	15.07	13	32	89
	SABA	2.86	15.95	22	28	87
	SABA	3.12	16.26	15	27	91

tion observed using the SABA process, and the levels of FeS_2 and ash rejection are very similar to those observed in the initial studies on Minto coal using *T. ferrooxidans* cultivated on a synthetic medium. Therefore, the ineffectiveness of the SABA process on Cape Breton coals is not a result of microbial changes during laboratory cultivation but is due to differences in the physical or chemical properties between the coals. Therefore, studies of the mineralogy and chemistry of the iron sulfides and mineral ash content of the various Maritime coals are required in order to understand and overcome the difficulties in coal beneficiation.

Table 35.14 Effect of Varsol/Coal Ratio on Pyrite and Ash Rejection and Carbon Recovery During Spherical Agglomeration of a Freshly Ground Stored Minto Coal Sample (Particle Size 5–10 μm)

		Total sulfur (%)	FeS_2 (%)	Percent FeS_2 rejected	Ash (%)	Percent ash rejected	Percent carbon recovery
Raw coal		4.64	3.46	—	13.46	—	
Percent Varsol	pH						
20% v/w	2.0	3.80	2.01	42	8.35	38	61
30% v/w	2.0	3.56	1.99	43	7.12	47	81
40% v/w	2.0	3.40	1.63	53	5.24	61	93
60% v/w	2.0	3.00	1.59	54	4.76	65	96
60% v/w	2.0	2.96	1.58	54	5.09	62	97

Table 35.15 Comparison of the Effect of Coal Suspension Age on Pyrite and Ash Rejection and Carbon Recovery Between Spherical Agglomeration and the SABA Process; Minto Coal Ground to Between 5 and 10 μm and Treated with 2.88 $\times$ 10^{10} Cells/g of Coal and Recovered Using 60% v/w Varsol

Time (h)	Procedure	FeS_2 (%)	Ash (%)	Percent FeS_2 rejected	Percent ash rejected	Percent carbon recovered
0	Raw coal	1.43	5.92			
	Agglom.	1.56	3.13	Nil	47	89.2
	SABA	0.44	2.02	69	66	100
	SABA	0.59	1.69	59	71	100
24	Agglom.	1.40	2.99	2	49	98
	SABA	0.45	1.60	68	73	98
	SABA	0.66	1.66	54	72	98
48	Agglom.	1.30	2.88	9	51	100
	SABA	0.57	1.73	60	71	96
	SABA	0.48	1.68	66	72	97
72	Agglom.	1.12	2.75	22	54	96
	SABA	0.53	1.82	63	69	99
96	Agglom.	1.04	2.68	27	55	100
	SABA	0.44	1.77	69	70	100

35.4.3 Discussion

Brogan Coal

Strain differences between *T. ferrooxidans* were noted in their ability to cause pyrite rejection using the SABA process on Brogan coal. The Cape Breton thiobacillus No. 4 resulted in substantial ash rejection but had little or no effect on pyrite rejection compared to spherical agglomeration alone (Table 35.5). In contrast, the Elliot Lake thiobacillus resulted in 20 to 24% pyrite rejection (Table 35.7), which is the highest pyrite rejection obtained to date with ball-milled Cape Breton coals. The strain differences observed in particulate binding may be due to differences in the fimbriae or cellular envelopes of the different strains. Fimbriae are known to mediate bacterial cell binding to solid surfaces and have been found on thiobacillus cells [9,20]. Alternatively, the differences in cellular binding may reflect differences in the capsular or cell envelope materials that may be necessary for surface adsorption [22].

When Brogan coal was ground to greater than 90% less than 1 μm, 62% pyrite rejection was obtained by spherical agglomeration, while only 39% pyrite rejection was obtained using the SABA process. The low pyrite rejection rate obtained with the SABA experiment may be due to cellular lysis caused by the large volume of Varsol required to agglomerate the finely divided coal. To obtain 95% coal recovery, the suspension was agglomerated twice and a total of 30 mL of Varsol was required to recover 35 g of coal. In the other experiments, only 11 mL of Varsol was required and the cells were in contact with the Varsol for only 5 min. Tuovinen [39] has shown that aromatic and aliphatic mineral floatation reagents are inhibitory to *T. ferrooxidans*.

Prince Coal

SABA treatment of Prince coal ground to a mean particle size of 7.3 μm resulted in about 16% rejection of the pyrite and 40 to 45% rejection of the ash. SABA treatment resulted in about 27% greater pyrite rejection than spherical agglomeration alone.

The low levels of pyrite rejection appear to reflect the fact that about 20% of the pyrite in Cape Breton coals is finely disseminated 1 to 2-μm crystals which would be freed from the coal matrix only by grinding the coal to 1 to 2-μm particle size [14].

Experiments with finely ground Prince coal (Table 35.10) were not as successful as were experiments with finely ground Brogan coal (Table 35.8). Bacterial adsorption prior to agglomeration resulted in 22.2% of the pyrite being rejected, whereas agglomeration only resulted in 13.6% pyrite rejection. There must be chemical and physical differences between these two coals; the components of the Prince coal appear to be non-amenable to cellular binding and agglomeration compared to the results obtained with the Brogan coal.

Minto Coal

The most recent critical reassessments of the SABA process using Minto coal as the test substrate have indicated a number of previously unrecognized parameters which require further study and assessment as to their role in coal beneficiation. Comparing the data of Tables 35.11 to 35.13 to the data of Table 35.15, it is apparent that coal storage and the age of finely ground coal in aqueous suspension is one critical factor that must be considered during coal beneficiation. The coal used in the experiments of Tables 35.11 to 35.13 was obtained from E. Capes of NRC in Ottawa and had been stored for a considerable period of time prior to being shipped to Halifax. Thus some undetermined amount of surface chemical oxidation of the coal components had occurred prior to these investigations. To take this hypothesis one step further, the coal suspensions utilized in Tables 35.11 and 35.12 were prepared by ball milling the coal in water for 24 h. The coal slurry was stored for 10 days prior to being used in the experiments of Table 35.11. In contrast, the coal was treated immediately after grinding for the experiments of Table 35.12 and the percent pyrite rejections in SABA Nos. 2 and 3 of Table 35.12 are double the value obtained for SABA No. 2 of Table 35.11. As the same stored coal sample was used in both experiments and the experimental conditions were as similar as possible, the only variable that could have produced the observed difference was the age of the coal slurry.

Similarly, levels of pyrite rejection were observed in Table 35.13 with the 24-h coal slurry as were observed in Table 35.12 with the coal slurry that had been ball-milled for 24 h. In contrast, the coal used in the experiments in Table 35.15 was obtained as a freshly mined sample and was treated immediately after a 30-min grind in the Attritor. The level of pyrite rejection using the SABA process is consistent with time and is very similar to the values reported by Kempton et al. [17,18,19].

A rather surprising observation in regard to spherical agglomeration is presented in Table 35.15. It appears the the efficiency of this process is dependent on either surface oxidation or some alteration in the surface chemistry, or component surface electrostatic charges are required to obtain an acceptable rejection of pyrite.

The data of Table 35.14 indicate the uneconomical concentrations of Varsol are required to obtain fairly high levels of pyrite rejection combined with high percentage carbon recovery. For oil agglomeration to be utilized on an industrial scale, the percent w/w oil to coal is generally considered uneconomical if greater than 10 to 15% oil is required. Obviously, some other mechanism of carbon recovery must be developed for the economical beneficiation of finely subdivided coal.

35.5 BACTERIAL LEACHING OF IRON SULFIDES FROM COAL

In 1947, Colmer and Hinkle [7] showed that acid mine drainage was the result of an oxidative attack of the iron-mineral sulfides by *T. ferrooxidans* to produce sulfuric acid. Ashmead [3] proposed the use of these organisms to desulfurize coal, but the subsequent research effort was concentrated on the metabolism and physiology of these extreme acidophiles.

A limited number of studies have been reported on the bacterial leaching of pyrite from coal [12,35]. However, the bacterial oxidation of pyrite is a self-limiting first-order reaction, probably due to the precipitation of $Fe(OH)_3$ and elemental sulfur, which progressively inhibits intimate contact between the bacterial cells and the sulfide-mineral surface. Consequently, the best pyrite oxidation experiments with coal reported to date required 5 days to remove 97% of the pyrite [13] using mixed cultures of *T. ferrooxidans* and *T. thiooxidans*. Although bacterial leaching is too slow a process for reducing the sulfur content of all the coal mined, it does appear to be a feasible process for the reclamation of waste-coal dumps.

35.5.1 Materiala and Methods

A large bulk sample of run-of-the-mill Brogan coal was obtained from the Cape Breton Development Corporation. This coal had an average total sulfur content of 6.93%, which was composed of 1.6% inorganic sulfur, 4.24% pyritic sulfur, and 1.61% organic sulfur and an average mineral ash content of 16.34%. The coal sample was dry sieved to produce aliquot samples in the following size ranges: 0.045 to 0.075 mm, 0.075 to 0.15 mm, 0.15 to 0.3 mm, 0.3 to 0.6 mm, 0.6 to 0.85 mm, 0.85 to 1.7 mm, and 6.35 to 12.7 mm (1/4 to 1/2 in.).

A strain of *T. ferrooxidans* isolated from uranium mine tailings near Elliot Lake, Ontario, was maintained on a modified Tuovinen and Kelly's [37] medium containing (per liter): KH_2PO_4, 0.05 g; $MgSO_4 \cdot 7H_2O$, 0.05 g; NH_4NO_3, 0.4 g, and $FeSO_4 \cdot 7H_2O$, 33.3 g. The medium was adjusted to pH 2.3 with H_2SO_4 and sterilized by autoclaving.

Shake-Culture Studies

Two Fernbach flasks containing 900 mL of $FeSO_4$-free medium were prepared and 100 g of sized coal was added to both flasks. The control flask was brought up to 1 L of liquid by adding 100 mL of sterile water acidified to pH 2.3 and containing 0.2 g of sodium azide. The test flask was inoculated with 100 mL of a 96-h culture of *T. ferrooxidans*. The flasks were incubated on a New Brunswick gyratory shaker at 200 rpm, 30°C.

Monitoring Ferric Iron Release During Leaching

The ferric iron produced from the bacterial oxidation of pyrite was determined spectrophotometrically at 410 nm using the assay procedure of Schnaitman et al. [34].

Bacterial Column Leaching of Coal

The 1/4- to 1/2-in. coal fraction was packed into a 3 in. × 24 in. column. The 1.5 kg of coal within the column was washed with water to remove any soluble salts and then the water was replaced with 500 mL of $FeSO_4$-free modified Tuovinen and Kelly medium.

The medium was inoculated with 100 mL of a 96-h culture of *T. ferrooxidans*. The column effluent was recycled by means of a peristaltic pump at 132 mL/h. When the ferric iron concentration increased to 7 to 8 mg/mL, the solution was collected and replaced with fresh $FeSO_4$-free modified Tuovinen and Kelly medium.

Analysis of the Heavy Metal Ions Released During the Bacterial Leaching of Coal

Aliquots of the column effluent were saved with each addition of fresh medium, and the aqueous phase of the shake cultures were collected when the experiments were terminated. The heavy metal ion content of all leachates was determined using a Beckman Atomic Absorption Analyzer and standard analytical procedures.

Analysis of the Coal

The percent sulfur of the various sulfur compounds in the coal samples was determined by the Atlantic Research Laboratory of NRC using the ASTM procedure.

35.5.2 Results

Shake-Culture Leaching

Based on an average pyritic-sulfur content of 4.24%, approximately 4 g of iron should be released by complete oxidation of the pyrite contained in 100 g of coal. Preliminary experiments indicated that only 50 to 55% of the iron released during bacterial leaching was derived from pyrite; the balance of the iron is from the inorganic ash components within the coal. Thus, all the shake-culture leaching experiments were terminated when the iron content of the leach solution reached a concentration of 8 mg/mL or 8 g/L. The iron content of the leach solutions versus time for the various shake-culture experiments on the sized coal fractions are presented in Fig. 35.1.

Initially, the coal was recovered after leaching by pouring the shake-culture contents through a 0.045-mm (300-mesh) screen. The coal was then washed with water, dried, and weighed. This procedure was suitable for coal particles greater than 0.3 mm, but below this size, large quantities of coal were lost unless the filtrate was subjected to oil agglomeration.

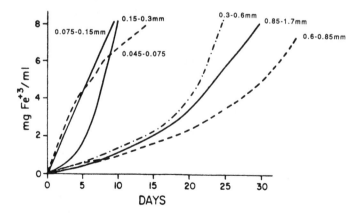

Fig. 35.1 Iron concentration of the shake-culture leach solution versus time for Brogan coal samples of various particle sizes.

Table 35.16 Bacterial Shake-Culture Leaching of 100 g of Brogan Coal Screened to
>0.85 mm and <1.7 mm for 29 Days

	Raw coal	Azide control	Leached coal
Sulfur (%)			
Total	6.76	5.52	3.60
FeS_2	4.07	3.45	1.11
Inorganic	1.01	0.13	0.19
Organic	1.68	1.94	2.30
Ash (%)	17.02	6.78	6.84
Coal recovered (g)			
Screening		77.31	
Agglomeration		4.17	85.19
Carbon recovered (%)		93.7	99.7

The results for the bacterial leaching of coal ground to between 0.85 and 1.7 mm
are presented in Table 35.16. After 29 days of bacterial leaching, the pyrite content was
reduced by 72.7% and the ash content was reduced by 59.8%. At this particle size, >99%
of the original carbon was recovered. The micrbial leaching of coal particles ranging between 0.6 and 0.85 mm resulted in a 95.5% reduction in the pyrite content after 57 days
(Table 35.17). Again, the recovery of carbon was >97%.

The leaching of coal particles ranging from 0.3 to 0.6 mm for 26 days resulted in a
87.7% reduction in the pyrite content and 69.1% reduction in the ash content (Table
35.18). The recovery of carbon was similar to that in the two previous experiments
(97.9%).

With decreasing particle size, an increase was observed in the ash and pyrite content
of the smaller fractions. As the increase in the ash and pyrite content increased the total
iron content of the coal sample, monitoring the soluble iron content of the leachate resulted in the experiments being terminated too early. As a result, the reduction in the
pyrite and ash content of the smaller-particle-sized samples is not as complete as that of
the larger sizes. Furthermore, the continuous abrasion, combined with the physical
rupturing of coal particles due to the oxidation of pyrite, resulted in a reduction in
particle size. This reduction in particle size resulted in a poor recovery of the carbon by

Table 35.17 Bacterial Shake-Culture Leaching of 100 g Brogan Coal Screened to >0.6
mm and <0.85 mm for 57 Days

	Raw coal	Azide control	Leached coal
Sulfur (%)			
Total	6.93	5.70	3.03
FeS_2	4.24	3.15	0.19
Inorganic	1.1	0.10	0.12
Organic	1.6	2.45	2.72
Ash (%)	16.34	8.04	4.72
Coal recovered (screening) (g)		86.4	81.7
Carbon recovered (%)		96.8	97.8

Table 35.18 Bacterial Shake-Culture Leaching of 100 g of Brogan Coal Screened to >0.3 mm and <0.6 mm for 26 Days

	Raw coal	Azide control	Leached coal
Sulfur (%)			
Total	6.93	5.77	3.2
FeS$_2$	4.24	3.53	0.52
Inorganic	1.1	0.12	0.16
Organic	1.6	2.12	2.52
Ash (%)	16.34	8.44	5.05
Coal recovered (screening (g)		85.3	82.1
Carbon recovered (%)		95.0	97.9

screening on a 0.045-mm (300-mesh) screen. These effects can be seen in the data presented in Table 35.19. After 10 days of leaching, only 37.4% of the pyrite was removed and only a 25.7% reduction in the ash content of the coal was observed. As many of the particles were of less than 0.045-mm diameter following leaching, only 59% of the carbon was recovered by screening.

As the particle size of the coal subjected to bacterial leaching was reduced, the ash and pyrite contents increased and the recovery of carbon became more difficult, even when the oil agglomeration technique was used. From Table 35.20 it is obvious that the microbial cell components or their secreted metabolites are masking or altering the surface of the carbonaceous particles, rendering them nonamenable to agglomeration or coalescence during oil agglomeration. This fact becomes even more evident from the data presented in Table 35.21.

The heavy metal content of the leachates from these coal leaching experiments was determined. The concentrations of the various metals were: manganese 11 to 28 ppm; Zn, 11 to 100 ppm; Co, 0.8 to 2.4 ppm; Ni, 1.8 to 5.1 ppm; and vanadium, 0.1 to 0.2 ppm. In general, the concentration of the various metals in the leachate increased with decreasing particle size of the coal sample.

Table 35.19 Bacterial Shake-Culture Leaching of 100 g of Brogan Coal Screened to >0.15 mm and <0.3 mm for 10 Days

	Raw coal	Azide control	Leached coal
Sulfur (%)			
Total	6.96	6.54	5.93
FeS$_2$	4.76	4.33	2.98
Inorganic	0.27	0.21	0.22
Organic	1.93	2.00	2.73
Ash (%)	17.34	15.73	12.89
Coal recovered (screening) (g)		57.05	55.65
Carbon recovered (%)		58.6	59.6

Table 35.20 Bacterial Shake-Culture Leaching of 100 g of Brogan Coal Screened to >0.075 mm and <0.15 mm for 11 Days

	Raw coal	Azide control	Leached coal
Sulfur (%)			
Total	7.01	6.67	5.87
FeS$_2$	4.76	4.19	2.72
Inorganic	0.27	0.35	0.89
Organic	1.98	2.13	2.26
Ash (%)	25.03	24.25	23.79
Coal recovered (agglomeration) (g)		86.6	66.6
Carbon recovered (%)		87.6	68.9

Column Leaching Study

The leaching of iron from the 1.5 kg of Brogan coal contained within the column is presented in Fig. 35.2. After 110 days of leaching, 47.2 g of iron has been removed from the coal. Accompanying this loss of iron, 62.3% of the pyrite and 68.2% of the ash has been removed from the upper surface of the coal column (Table 35.22). Leaching of the column was continued for 269 days or about 9 months, at which time the column was dismantled and the coal analyzed as 10-cm fractions with depth (Table 35.23).

The organic sulfur content of the coal after leaching seemed excessively high, and it was postulated that the apparent high value for organic sulfur may have resulted from incomplete oxidation of sulfides, resulting in the deposition of elemental sulfur or the formation of insoluble jarosites within the column. The coal was extracted with carbon disulfide and subsequently washed with acetone and absolute alcohol to remove any residual CS$_2$. The total sulfur content of the coal was redetermined and the organic sulfur content recalculated (Table 35.24).

The rate of leaching of several metal ions from the coal is presented in Fig. 35.3. High concentrations were leached initially but the concentration declined rapidly over a 40-day period and then leveled off at the concentration range 0.5 to 4.0 ppm. The cumulative amounts of each metal leached from the coal are presented in Fig. 35.4. From these

Table 35.21 Bacterial Shake-Culture Leaching of 100 g of Brogan Coal Screened to >0.045 mm and <0.075 mm for 14 Days

	Raw coal	Azide control	Leached coal
Sulfur (%)			
Total	10.07	9.23	6.83
FeS$_2$	6.39	5.92	3.53
Inorganic	1.51	0.47	1.11
Organic	2.17	2.84	2.19
Ash (%)	25.23	22.44	26.61
Coal recovered (agglomeration) (g)		58.5	17.64
Carbon recovered (%)		61.8	18.1

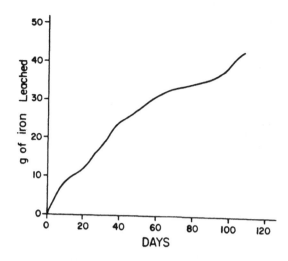

Fig. 35.2   Release of iron due to bacterial leaching of 1.5 kg of Brogan coal in a 7.6 X 61 cm column.

Table 35.22   Analysis of Brogan Coal After 110 Days of Column Leaching at 22 to 23°C

	Initial coal	Leached coal	Percent removal
Sulfur (%)			
Total	6.93	4.05	41.5
FeS_2	4.24	1.60	62.3
Inorganic	1.1	0.29	73.6
Organic	1.6	2.16	-35.0
Ash (%)	16.34	5.20	68.2

Table 35.23   Pyrite and Ash Removal Versus Depth in the Column: Leaching with *T. ferrooxidans* at pH 2.3 for 9 Months at 22 to 23°C

Sample	Total sulfur (%)	FeS_2 (%)	Percent FeS_2 leached	Organic sulfur (%)	Ash (%)	Percent ash leached
Raw coal	6.60	4.58	—	1.04	15.29	—
Leached coal						
0–10 cm	4.47	2.07	54.8	2.22	8.03	47.5
10–20 cm	4.39	2.43	46.9	1.62	6.89	55.0
20–30 cm	4.60	2.17	52.6	2.24	8.77	43.0
30–40 cm	4.82	1.80	60.7	2.84	8.22	46.0
40–50 cm	5.34	2.94	35.8	2.23	8.93	42.0

Table 35.24 Analyses of Coal Fractions from the Leaching Column After Extraction with Carbon Disulfide and Washing with Acetone and Ethanol; Change in Organic Sulfur Content

Leached coal sample	Total sulfur	Percent organic sulfur
0–10 cm	3.83	1.58
10–20 cm	2.85	0.08
20–30 cm	2.93	0.57
30–40 cm	4.06	2.08
40–50 cm	3.85	0.74

data one can calculate that 112 g of Mn, 58.7 g of Zn, 14 g of Ni, 5.6 g of Co, and 2 g of vanadium will be released from each metric ton of Brogan coal.

35.5.3 Discussion

On a theoretical basis, the rate of microbial oxidation of pyrite should increase with decreasing particle size. In general, this is true; however, this study has illustrated several factors that will deter the use of microbial leaching of finely powdered coal to remove the pyrite. Because of differences in the relative densities of the components of coal, the

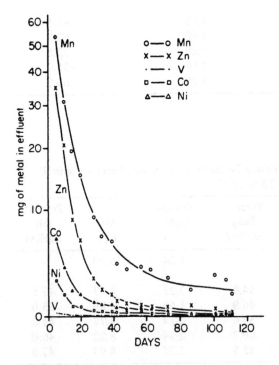

Fig. 35.3 Metal ion concentration in the effluent versus time during the bacterial column leaching of 1.5 kg of Brogan coal.

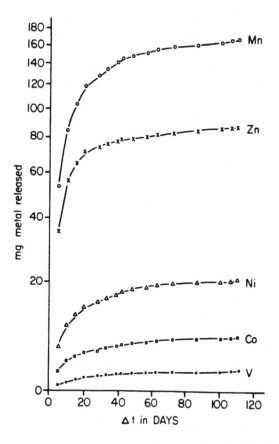

Fig. 35.4 Cumulative amounts of metal ions released versus time during the bacterial column leaching of 1.5 kg of Brogan coal.

small-sized fractions become preferentially enriched in pyrite and ash during dry screening of the large coal sample. Dugan and Apel [13] have also observed this phenomenon in their studies. Second, during microbial leaching there is a reduction in particle size due to interparticle abrasion and the physical rupturing of particles due to the molecular expansion resulting from the oxidation of S^{2-} to SO_4^{2-} [40]. This phenomenon is presented clearly in the data of Table 36.19. The initial size of the coal was greater than 0.15 mm, but only 58 to 59% of the carbon could be recovered on a 0.045-mm screen after leaching. Therefore, about 40% of the particles had degraded to a size of less than 0.045 mm. A third factor against the bacterial leaching of finely subdivided coal is the fact that the coal is difficult to recover, and this may be due to the absorption of cellular components or metabolites to the surface of the carbon particles, which prevents their agglomeration by oil.

The data presented clearly indicate that the pyrite and ash content of coal ground to greater than 0.3 mm can be reduced substantially with good recoveries of the carbonaceous material. However, such a process takes time and would require a continuously stirred tank reactor to provide aeration of the medium and suspension of the particles.

The most feasible and economical approach to desulfurizing coal waste dumps is the use of heap leaching. This technique has been used successfully for years to recover copper from low-grade ores [24,31].

Our column leaching experiments probably simulate large-scale industrial conditions, although the latter might be more effective. The spreading of the solution by sprinkler systems over finger dumps rather than a trickle of solution from a peristaltic pump would increase the bathing of the particles by the liquid, and would ensure the maximum solution aeration that is required for the process. Conversely, the possible effects of seasonal temperature variations may have to be evaluated with respect to recovery rates; our experiments were run at a constant temperature. For the Maritimes, evaporation should not be a problem. The washing effect of rains would presumably have the beneficial effect of removing any precipitated salts, thus enhancing the attack of sulfide particles by bacteria and restoring permeability reduced through the molecular volume increase of altered products.

The problem of elemental sulfur precipitation within the leaching column or dump due to the partial oxidation of sulfide may be overcome by inoculation of the leach dumps with a mixed culture of *T. ferrooxidans* and *T. thiooxidans*. As *T. thiooxidans* specifically oxidizes elemental sulfur to sulfuric acid, the addition of this organism to the inoculum should prevent the accumulation of elemental sulfur within the dump. Jarosite precipitation may be reduced substantially by modification of the leaching medium. Doddema [11] reported jarosites precipitate during growth of *T. ferrooxidans* on pyrite when the concentration of sulfate, ferric iron, and sodium or ammonium exceeded 100, 50, and 75 mM, respectively. Under the present experimental conditions the precipitation of sulfur could have inhibited the oxidation of pyrite by preventing intimate contact between *T. ferrooxidans* and the sulfide mineral phases. These studies should now be carried out on a larger scale (e.g., 500- to 1000-kg leaching tests) to give a better indication of the feasibility of reclaiming waste-coal dumps using the bacterial-heap leaching concept.

35.6 ANCILLARY BASIC STUDIES ON NUTRITION, FIMBRIAE, PLASMIDS, AND PREDATORS OF *T. FERROOXIDANS*

35.6.1 Nutrition

During the ^{32}P-labeling study, the K_2HPO_4 and $MgSO_4$ contents of Tuovinen and Kelly's medium were reduced to 25% of their original concentration. Over the past year further modifications of the medium have been tried, and at present stock cultures of the various *Thiobacillus* isolates are maintained and grow well on a medium of the following composition adjusted to pH 2.3:

Component	g/L
K_2HPO_4	0.1
$MgSO_4$	0.1
NH_4NO_3	0.4
$FeSO_4$	33.3

These changes have reduced the cost of the chemicals required for the medium and have resulted in a slightly faster growth rate of the various isolates. Further, the amount of ferric hydroxysulfate precipitate that normally accumulates during growth has been greatly reduced.

35.6.2 Fimbriae and Plasmids

To optimize the SABA process, an understanding of the mechanism of cellular binding to the iron sulfide mineral phases is required. Preliminary studies have shown that the environmental isolates of *T. ferrooxidans* used in this project produce fimbriae and contain a varying number of intracellular plasmids. (See fig. 35.5).

Fimbriae are known to facilitate cellular binding to solid surfaces in other microorganisms and are postulated to have a similar role in the thiobacilli. Studies are required to confirm this postulate and to investigate cultural conditions or genetic alteration techniques to cause an increased production of fimbriae per cell if they do facilitate cellular binding to solid substrates. Similarly, the intracellular plasmids found in these organisms are postulated to confer heavy metal tolerance to the organisms. However, recent studies

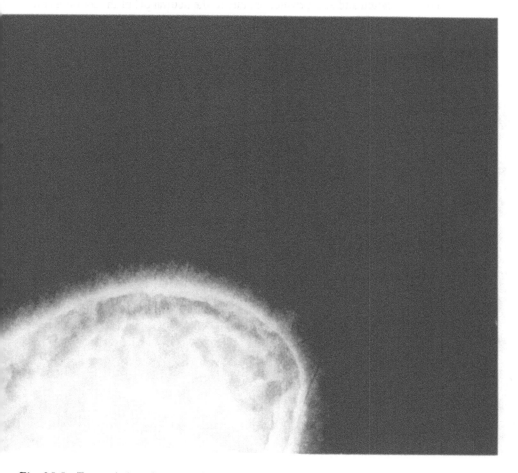

Fig. 35.5 Transmission electron micrograph of the polar end of a *T. ferrooxidans* cell. f, Flagellum; p, pilus or fimbrium. (Magnified 170,000X.)

by the molecular genetics section of the NRC in Ottawa have shown that these plasmids do not contain genes that encode for heavy metal tolerance.

Alternatively, the plasmids may encode for specific fimbrial proteins required for specific metal sulfide binding capabilities. Therefore, studies are required to test this hypothesis, and if it is correct, it may be possible to transfer the thiobacillus plasmid for FeS_2 binding to a faster-growing heterotrophic organism. If this were the case, this pyrite-binding heterotroph could then be utilized in the SABA process. If this is possible, it would have several advantages over the present technique:

1. Large numbers of cells could be obtained in 18 to 24 h, compared to the 96 h required to grow *T. ferrooxidans*. Simultaneously, this time saving would greatly reduce the operating costs of the project.
2. In contrast to the present process, in which the coal must be in a slurry at pH 2.3, the use of heterotrophic organisms would allow the process to be operated at a neutral pH, thereby reducing the cost of process effluent treatment.
3. As the heterotrophic organism would have the pyrite binding capability but not the pyrite oxidizing capability, the effluent waste would not be subject to bacterial oxidation and acid production, due to the neutral pH of the solution and the lack of thiobacilli in the effluent.

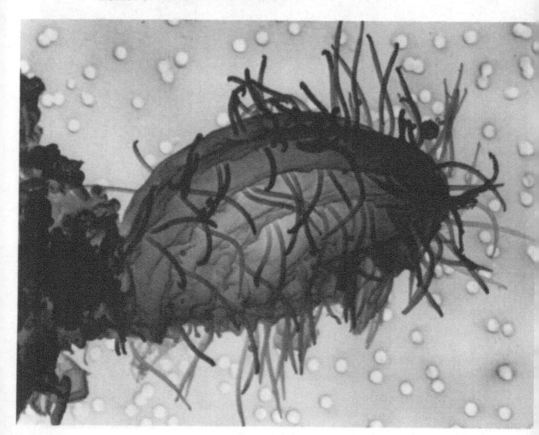

Fig. 35.6 Scanning electron micrograph of the ciliate predator of *T. ferrooxidans*.

35.6.3 Protozoal Predators of Thiobacilli

During the course of these studies, a zooflagellate predator of *T. ferrooxidans* was found quite by accident. We were maintaining a 20-L continuous culture of *T. ferrooxidans*, and on one occasion we added nonsterile medium to the culture. When we left that evening the cell population was about 10^8 mL^{-1}, but the following morning the cell count was reduced to 10^2 mL^{-1} and large numbers of a zooflagellate were observed in the culture. Preliminary studies showed that this predator was active over a pH range of 2.3 to 7.0, but in the presence of low cell numbers the organisms encysted and could not be activated to the vegetative state.

Attempts to reisolate the zooflagellates have been unsuccessful to date, but we have isolated a ciliate that acts as a predator over the pH range 4.0 to 7.0. Further, in an attempted coal leaching experiment in Sydney, Nova Scotia, about 2 weeks after starting the experiment, pyrite oxidation suddenly ceased. A leach solution sample was obtained and analyzed. The pH of the solution had risen from 2.3 to 4.0 due to acid consumption by alkaline coal components, and the solution was infested with a predator ciliate that had almost completely eliminated the *T. ferrooxidans* from the leach solution. Collaborative studies on these predators and assessment of their potential utilization to control thiobacillus populations in mine wastes and effluents have been initiated with colleagues at the Ontario Research Foundation. (See fig. 35.6).

35.7 ADSORPTION STUDIES RELATED TO USE OF THE SABA PROCESS

35.7.1 Introduction

^{32}P-labeled *T. ferrooxidans* adsorption studies on the various components of coal have reinforced the conclusions made in the earlier sections of this chapter. For the SABA process to be economically feasible, further research is required to chemically modify the surfaces of the mineral ash components of the coal to prevent, or at least reduce, cellular adsorption by these materials. In contrast, the data presented in this and subsequent sections suggest that the sulfur and ash content of stockpiled coal wastes could be reduced substantially, thereby allowing their reclamation and utilization for thermal power generation. Further, if partial pyrolysis of coal proves to be the preferred means of utilizing coal, the inorganic sulfur content of the combustible char produced can be reduced to very low levels in a reasonable period of time by microbial oxidation.

Over the past decade, with the increasing cost of petroleum, interest has again shifted to coal as an energy source. Environmental awareness has necessitated a reduction in industrial pollutant release, and thus a means of reducing the sulfur content of coal has been sought over the last several years.

Previous studies on the SABA process have shown it to be effective in reducing the sulfur content of Minto coals but much less effective when applied to the Cape Breton coals. Studies were required on the bacterial adsorption of the individual components of coal to obtain a better understanding of the mechanisms of the process.

Preliminary studies on the bacterial leaching of pyrite from coal suggested that the process may be suitable for the reclamation of wash plant-rejected coal wastes, but further studies were required to optimize and pilot plant the process.

35.7.2 Objectives

1. Assess the adsorption of ^{32}P-labeled *T. ferrooxidans* to the individual com-
    ponents of coal.
2. Optimize the rate of bacterial oxidation of pyrite from coal.
    a. Develop an optimal medium for growth and iron oxidation while mini-
        mizing the formation of inorganic precipitates.
    b. Assess the effect of surfactants of the rate of pyrite oxidation.

35.8 MATERIALS AND METHODS OF ADSORPTION

35.8.1 Adsorption of ^{32}P-Labeled *T. ferrooxidans* to Various Components of Run-of-Mine Coal

A strain of *T. ferrooxidans*, isolated from abandoned pyritic uranium mine tailings near
Elliot Lake, Ontario, was grown on Tuovinen and Kelly's [37] medium at pH 2.3. Stock
cultures were prepared in 250-mL Erlenmeyer flasks, each containing 100 mL of medium,
and grown at 30°C on a gyratory shaker for 72 h. Stock cultures remain active at 4°C for
at least 1 week.

Radioactively ^{32}P-labeled cells were obtained by preparing 3 L of medium containing 10 μCi of ^{32}P. The medium was inoculated with 300 mL of stock culture and incubated at 30°C on a gryatory shaker at 150 rpm for 72 h. The cells were harvested by
centrifugation at 10,000 rpm for 10 min. The cell pellet and sedimented Fe(OH)$_3$ was
resuspended in FeSO$_4$-free medium at pH 2.3 and centrifuged at 1000 rpm for 5 min to
separate the Fe(OH)$_3$ from the cells. The cell suspension was decanted from the Fe(OH)$_3$
pellet and was centrifuged at 10,000 rpm to pellet the cells. The distribution of the ^{32}P
in the phosphorus-containing cellular components was determined by chemical fractionation of the cells using the modified Schmidt-Thannhauser technique described by Myers
and McCready [30].

For studies on the distribution of ^{32}P-labeled cells in charcoal, pyrolyzed char,
pyrite, or sand suspensions, the cells were harvested as described previously, but the
Fe(OH)$_3$-free cell suspension was added to 1 L of FeSO$_4$-free medium at pH 2.3 and
incubated at 30°C for 2 h to deplete the cells of endogenous Fe^{2+}. Cell numbers in the
final suspension were determined by direct count using a Petroff-Houser counting chamber under a phase-contrast microscope. After the 2-h incubation, charcoal, pyrolyzed
char, pyrite, or sand was added to the 1 L of iron-free cells and incubated for 30 min at
30°C on a gyratory shaker at 150 rpm. The charcoal and char were then recovered from
suspension by the spherical agglomeration technique described by Capes et al. [6]. After
pyrite and sand were recovered by screening, the initial screening to recover the test material, a 100-mg aliquot of the solid was spread on a 45-mm, 0.2-μm Millipore filter which
was placed in a scintillation vial. Twenty milliliters of scintillation fluid was added and
the activity of ^{32}P determined by scintillation counting. The material remaining on the
screen was washed with 500 mL of deionized water, and an additional 100-mg aliquot of
the solid was spread on a Millipore filter and counted as described previously.

The distribution of ^{32}P in the cellular components and the distribution of the cells
in the solid suspensions were determined using the scintillation fluid and techniques
described by McCready and Din [26]. Radioactive counts were determined using the
LKB Rackbeta Scintillation Counter, and all counts were corrected for background radioactivity.

35.8.2 Development of a Minimal Medium

Previous studies have shown that *T. ferrooxidans* grows well on 0.1 mM PO_4^{3+}, but minimal concentrations of NH_4^+ and Mg^{2+} had not been determined for this organism. Media with varying ammonium and magnesium contents were prepared (see Table 35.25) to assess the best growth conditions for the various strains of *T. ferrooxidans*. Growth of the various strains was also compared to the growth and iron-oxidizing capability of a standard culture of *T. ferrooxidans* (ATCC 23270).

35.8.3 Bacterial Leaching of Pyrite from Coal

A large bulk sample of run-of-mine Brogan coal was obtained from the Cape Breton Development Corporation. This coal had an average total sulfur content of 6.93%, which was composed of 1.6% inorganic sulfur, 4.24% pyritic sulfur, and 1.61% organic sulfur and an average mineral ash content of 16.34%. The coal sample was dry sieved to produce aliquot samples in the following size ranges: 0.3 to 0.6 mm, 0.6 to 0.85 mm, 0.85 to 1.7 mm, and 6.35 to 12.7 mm (1/4 to 1/2 in.).

A strain of *T. ferrooxidans* isolated from uranium mine tailing near Elliot Lake, Ontario, was maintained on a modified Tuovinen and Kelly's [37] medium containing (per liter): KH_2PO_4, 0.1 mM; $MgSO_4 \cdot 7H_2O$, 0.5 mM; NH_4SO_4, 0.5 mM; and $FeSO_4 \cdot 7H_2O$, 120 mM. The medium was adjusted to pH 2.3 with H_2SO_4 and sterilized by autoclaving.

Shake-Culture Studies

Two Fernbach flasks containing 900 mL of $FeSO_4$-free medium were prepared and 100 g of sized coal was added to both flasks. The control flask was brought up to 1 L of liquid by adding 100 mL of sterile water acidified to pH 2.3 and containing 0.4 g of sodium azide. The test flask was inoculated with 100 mL of a 96-hr culture of *T. ferrooxidans*. The flasks were incubated on a New Brunswick gyratory shaker at 150 rpm, 30°C.

Monitoring Ferric Iron Release During Leaching

The ferric iron produced from the bacterial oxidation of pyrite was determined spectrophotometrically at 410 nm using the assay procedure of Schnaitman et al. [34].

35.8.4 Bacterial Column Leaching of Coal

The 1/4- to 1/2-in. coal fraction was packed into a 3 in. × 24 in. column. The 1.5 kg of coal within the column was washed with water to remove any soluble salts and then the water was replaced with 500 mL of $FeSO_4$-free modified Tuovinen and Kelly medium. The medium was inoculated with 100 mL of a 96-h culture of *T. ferrooxidans*. The

Table 35.25 Composition of the Various Growth Media Tested (mM)

Medium	$[PO_4^{3+}]$	$[NH_4^+]$	$[Mg^{2+}]$	$FeSO_4]$
A	0.1	0.5	0.5	120
B	0.1	0.5	0.1	120
C	0.1	1.0	0.05	120
D	0.1	0.5	0.025	120
E	0.1	0.5	0.05	120

column effluent was recycled by means of a peristaltic pump at 132 mL/h. When the ferric iron concentration increased to 7 to 8 mg/mL, the solution was collected and replaced with fresh $FeSO_4$-free modified Tuovinen and Kelly medium.

Analysis of the Heavy Metal Ions Released During the Bacterial Leaching of Coal

Aliquots of the column effluent were saved with each addition of fresh media, and the aqueous phases of the shake cultures were collected when the experiments were terminated. The heavy metal ion content of all leachates was determined using a Beckman Atomic Adsorption Analyzer and standard analytical procedures.

Analysis of the Coal

The percent sulfur of the various sulfur compounds in the coal samples was determined by the Atlantic Coal Institute using the ASTM procedure.

35.8.5 SABA Study with *Escherichia coli*

A mucoidal strain of *E. coli* was grown on a glucose-mineral salts medium for 24 h at 30°C and 150 rpm. The cells were harvested by centrifugation, resuspended and washed in 0.1 M $PO_4{}^{3+}$ buffer, and finally, suspended in 900 mL of 0.1 M $PO_4{}^{3+}$ buffer at pH 6.8. Cell counts were determined microscopically using the Petroff-Houser Cytometer.

One hundred milliliters of a Minto coal suspension was added to the culture and incubated at 30°C and 150 rpm for 30 min. Coal was then recovered by the spherical agglomeration process outlined by Kempton et al. [17]. The procedure utilized to treat coal with *T. ferrooxidans* was identical to that described by Kempton et al. [17].

35.8.6 Bacterial Leaching of Pyrolyzed Char

Pyrolysis of coal produces a char that has a very porous structure. Due to the increased porosity, *T. ferrooxidans* organisms adsorb to the char surface and will be in intimate contact with the pyrite and pyrrhotite crystals associated with the char. Due to the strong adsorption of *T. ferrooxidans* to pyrolyzed char, it was postulated that the organisms may be able to leach the inorganic sulfur from the char more readily than from the intact coal matrix.

A leaching study was initiated using the procedure outlined for shake-culture leaching of pyrite from coal, except that the 100 g of coal was replaced by 20 g of a low-temperature (400°C), low-pressure (50 torr) pyrolyzed coal char. The rate of pyrite and pyrrhotite leaching was monitored by measuring the Fe^{3+} produced with time.

35.9 RESULTS OF ADSORPTION STUDIES

35.9.1 Adsorption of ^{32}P-Labeled *T. ferrooxidans* to the Various Components of Run-of-Mine Coal

After 96 h of growth on modified Tuovinen and Kelly's medium, the *T. ferrooxidans* cells assimilated 10.8% of the available [32]P. Analysis on the various cellular components indicated that greater than 80% of the radioactive label was incorporated into the nucleic acids (Table 35.26).

Adsorption of ^{32}P-labeled *T. ferrooxidans* cells to a sample of Brogan coal ground to between 5 and 10 μm indicated that the bacteria do not adsorb to the carbon matrix but that 25 to 30% of the cells bind to the pyrite and mineral ash components of the coal

Table 35.26 Assimilation and Distribution of ^{32}P by *T. ferrooxidans* Grown on Modified Tuovinen and Kelly's Medium at 23°C for 96 Hours

Cellular fraction	Percent ^{32}P
Acid soluble	12.6
Lipid	4.0
RNA	60.9
DNA	22.5
Percent available ^{32}P assimilated	10.8

and the remaining 65.70% of the cells remain in the aqueous phase following oil agglomeration and washing with acidified water (Table 35.27).

In contrast to the adsorption of *T. ferrooxidans* to coal, the cells are adsorbed quite strongly to the surface of animal charcoal. Following oil agglomeration, about 18% of the cells were found adsorbed to the charcoal; after washing vigorously with acidified water, 16.7% of the cells remained bound to the charcoal (Table 35.28).

As the bacteria apparently adsorbed tenaciously to the animal charcoal, an alternative form of carbon matrix was chosen, and the experiments were repeated. Table 35.29 presents the data for the adsorption of ^{32}P-labeled *T. ferrooxidans* to a sample of pyrolyzed coal char. In this case, about 43% of the cells adsorbed to the char initially, and few (about 10% of the bound cells) were removed by vigorous washing.

The adsorption of ^{32}P-labeled *T. ferrooxidans* to the surface of acid-washed pyrite was also determined (Table 35.30). Initially, about 30% of the available bacterial cells bound to the surface of the pyrite. However, after a vigorous water wash, only 5 to 6% of the cells remained bound to the pyrite surface.

The final coal component tested for the adsorption of bacterial cells was the inorganic ash component. Acid-washed silica sand was chosen as the best material, as the

Table 35.27 Distribution of ^{32}P-Labeled *T. ferrooxidans* (5.5×10^{11} Cells/g of Coal) in Brogan Coal After 30 Minutes' Contact Time and Oil Agglomeration

Phase	Percent distribution of cells	
	Experiment 1	Experiment 2
Initial separation		
Coal after screening	43.3	14.4
Tailings after screening	20.3	21.8
Screening water phase	36.4	63.8
After a water wash		
Coal	Nil	Nil
Tailings in wash water	10.1	4.1
Wash water	32.5	7.9

Table 35.28 Distribution of ^{32}P-Labeled *Thiobacillus ferrooxidans* (5.5×10^{11} Cells/g) After 30 Minutes' Exposure to Animal Charcoal in Suspension and Charcoal Recovery by Oil Agglomeration

| | Percent distribution of cells | |
Phase	Experiment 1	Experiment 2
Initial separation		
Charcoal after screening	17.7	18.2
Screening water phase	82.3	81.8
After a water wash		
Charcoal	16.6	16.8
Wash water	1.1	1.4

Table 35.29 Distribution of ^{32}P-Labeled *T. ferrooxidans* (5.5×10^{11} Cells/g) After 30 Minutes' Exposure to Pyrolyzed Coal Char in Suspension and Char Recovery by Oil Agglomeration

| | Percent distribution of cells | |
Phase	Experiment 1	Experiment 2
Initial separation		
Char after screening	42.5	44.3
Screening water phase	57.5	55.7
After a water wash		
Char	37.7	38.1
Wash water	4.8	6.2

Table 35.30 Distribution of ^{32}P-Labeled *Thiobacillus ferrooxidans* (1.5×10^{12} Cells/g After 30 Minutes' Contact with 100-Mesh Pyrite in Suspension and Recovery of FeS_2 by Screening on a 300-Mesh Screen

| | Percent distribution of cells | |
Phase	Experiment 1	Experiment 2
Initial separation		
FeS_2	30.5	31.1
Screening water phase	69.5	68.9
After a water wash		
FeS_2	5.8	6.0
Wash water	24.7	25.1

Table 35.31 Distribution of ^{32}P-Labeled *T. ferrooxidans* (5.5×10^{11} Cells/g) After 30 Minutes' Contact with 100-Mesh Acid-Washed Ottawa Silica Sand in Suspension and Recovery of the Sand on a 300-Mesh Screen

	Percent distribution of cells	
Phase	Experiment 1	Experiment 2
Initial separation		
Sand	30.4	30.1
Screening water phase	69.6	69.9
After a water wash		
Sand	29.8	29.7
Wash water	0.6	0.4

major constituent of the inorganic ash associated with maritime coals are silicate minerals. The data presented in Table 35.31 indicate that 30% of the available bacterial cells bind to the silica sand initially, and that the adsorption to the silica is quite tenacious, as very few cells were removed by a vigorous water wash.

35.9.2 Development of a Minimal Medium for *T. ferrooxidans*

One of the major problems observed in the past in regard to the recovery of metals by bacterial leaching has been to provide sufficient inorganic nutrients to support good bacterial growth, but simultaneously to prevent the development of inorganic precipitates which can coat the pyritic surfaces and reduce their rates of oxidation.

The ^{32}P isotope labeling study indicated that *T. ferrooxidans* utilized about 10% of the available phosphate in a medium containing 0.1 g of K_2HPO_4 per liter. Thus the initial phosphate concentration was reduced from 0.568 mM to 0.1 mM. Good growth was obtained with medium containing 0.1 mM $PO_4{}^{3+}$, and a series of media containing various concentrations and combinations of Mg^{2+} and $NH_4{}^+$ were tested to determine the optimal level of inorganic nutrients required by *T. ferrooxidans*. Optimal growth and iron oxidation were observed on a medium containing 0.1 mM $PO_4{}^{3+}$, 0.5 mM Mg^{2+}, and 0.5 mM $NH_4{}^+$ at pH 2.3 (Fig. 36.7).

35.9.3 Use of Heterotrophic Bacteria in the SABA Process

Attachment of bacteria to solid surfaces can occur via cellular fimbriae (pili) or via glycocalyx (polysaccharide cellular capsules) structures, which surround many bacterial cells. The data presented in Table 35.32 compare the pyrite and ash rejection observed when coal was treated by oil agglomeration, the SABA process with *T. ferrooxidans*, and the SABA process with *E. coli*. Treatment of the coal with either bacterial genus resulted in greater pyrite rejection than was observed by oil agglomeration alone. The best pyrite and ash rejection were observed when *T. ferrooxidans* was used in the SABA process. The results do indicate, however, that other heterotrophic organisms, which are known to bind to solid substrates, may be utilized in the SABA process.

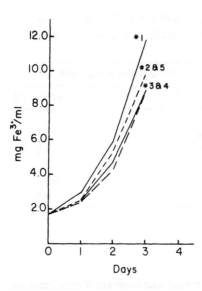

Fig. 35.7 Growth of various isolates of *T. ferrooxidans* from acidic drainage on the low phosphate medium at 28°C and 150 rpm.

35.9.4 Effect of a Surfactant on the Leaching of Pyrite

Numerous reports over the last two decades have suggested that the addition of surfactants to the medium may increase the rate of bacterial iron oxidation by *T. ferrooxidans*. As a follow-up to the previous study on bacterial shake-culture leaching of pyrite from coal, we examined the effect of a surfactant (Tween-20) on the rate of pyrite oxidation using Brogan coal. Three different-sized fractions of coal were comparatively leached; however, the surfactant had no apparent effect on the rate of pyrite leaching (Table 35.33).

Table 35.32 Comparison of the SABA Process Using *T. ferrooxidans* and *E. coli* as the Active Agents Relative to Oil Agglomeration on Minto Coal Ground to 5 to 10 μm

Procedure	FeS_2 (%)	Ash (%)	Percent FeS_2 rejected	Percent ash rejected	Percent carbon recovered
Nil	7.24	25.42			
Oil agglomeration[a]	5.12	8.63	41.3	66.0	94.9
SABA	1.91	5.56	78.9	78.1	88.7
(*T. ferrooxidans*)	2.06	6.05	77.1	76.2	91.7
Nil	4.01	13.50			
SABA	1.71	5.87	57.3	56.5	92.1
(*E. coli*)	1.43	4.62	64.3	65.7	94.7

[a]All oil agglomerations utilized 11 mL of Varsol, except the last experiment with *E. coli*, in which 18 mL of Varsol was used.

Table 35.33 Effect of a Surfactant (Tween-20) on the Bacterial Leaching of Pyrite and Ash from Coal Incubated at 30°C, 150 rpm for 25 Days

Sample	Total sulfur	FeS_2 (%)	Percent FeS_2 leached	Organic sulfur (%)	Ash (%)	Percent ash leached
Raw coal	6.52	3.83	—	1.56	13.78	—
		Without Tween Added				
0.35–0.6 mm	3.28	0.80	79	2.27	4.96	64
0.6–0.85 mm	3.78	1.10	71	2.45	5.35	61
0.85–1.7 mm	3.89	1.41	63	2.24	4.68	66
		With Tween Added				
0.35–0.6 mm	3.53	0.98	74	2.33	4.87	65
0.6–0.85 mm	3.60	1.14	70	2.24	4.59	67
0.85–1.7 mm	3.98	1.40	63	2.34	4.64	66

35.9.5 Bacterial Trickle Leaching of 1 kg of Brogan Coal (>1/2 in. to <1 in.) in Laboratory Columns

Sized samples of Brogan coal were loaded into two columns and trickle leached with the low-phosphate medium inoculated with *T. ferrooxidans*. After 1 year of continuous leaching, the columns were dismantled in 10-cm segments and the sulfur forms and ash content of each segment were analyzed (Table 35.34). In two identical columns the

Table 35.34 Pyrite and Ash Removal Versus Depth in the Columns: Leaching of Brogan Coal with *T. ferrooxidans* for 1 Year at 22 to 23°C

Sample	Total sulfur	FeS_2 (%)	Percent FeS_2 leached	Organic sulfur (%)	Ash (%)	Percent ash leached
Raw coal	6.55	4.26	—	1.04	17.98	—
No. 1 coal						
0–10 cm	3.66	1.51	64.5	1.84	8.61	52.1
10–20 cm	3.79	1.77	58.4	1.85	6.09	66.1
20–30 cm	3.55	1.38	67.6	1.98	9.22	48.7
30–40 cm	3.14	0.95	77.7	1.99	6.96	61.3
40–45 cm	3.61	1.47	65.5	1.95	6.68	62.8
Average values			66.7			58.2
No. 2 coal						
0–10 cm	3.77	1.12	73.7	2.23	7.14	60.3
10–20 cm	3.33	0.88	79.3	2.23	6.61	63.2
20–30 cm	3.17	0.95	77.7	2.03	5.70	68.3
30–40 cm	3.73	1.29	69.7	2.22	8.97	50.1
40–50 cm	3.69	1.46	65.7	2.05	7.06	60.7
50–55 cm	3.48	1.51	64.5	1.80	6.43	64.2
Average values			71.7			61.1

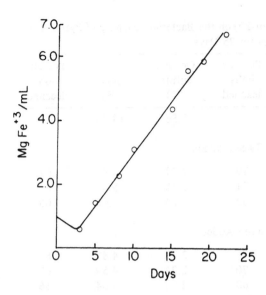

Fig. 35.8 Growth and iron oxidation observed during the leaching of pyrolyzed coal char with *T. ferrooxidans* at 28°C and 150 rpm.

average value for pyrite removal was 69% and that for ash removal was 59% after 1 year of bacterial leaching.

35.9.6   Bacterial Leaching of Pyrolyzed Coal Char

Since the ^{32}P-labeled *T. ferrooxidans* adsorption study had indicated a strong binding of the cells to pyrolyzed coal char, and physical analyses of the char had indicated a porous, spongelike structure in the char particles, it was postulated that the inorganic sulfur associated with the char may be amenable to bacterial leaching. The rate of iron oxidation was linear after the initial lag phase of growth (Fig. 35.8) and the bacterial metabolic activity greatly reduced both the pyrite and pyrrhotite content of the char (Table 35.35).

Table 35.35   Effect of Pyrolysis (400°C and 50 Torr) and Bacterial Leaching on the Sufur Forms of the Residual Char Obtained from Prince coal

Sample	Total sulfur (%)	SO_4-S (%)	FeS_2 (%)	Pyrrhotite (%)	Organic sulfur (%)
Washed Prince coal	4.78	0.56	2.68		1.30
Pyrolyzed char	5.49	0.30	3.14	0.69	1.36
Leached char	2.32	0.11	0.67	0.02	1.52

35.10 DISCUSSION OF ADSORPTION STUDIES

The ^{32}P-labeled *T. ferrooxidans* adsorption study indicated areas in which further research is required to develop the SABA process into a feasible and economical process. Although the studies with run-of-mine coal indicated that bacteria do not bind to the carbon matrix of coal, the cells tenaciously bind to heat-treated animal charcoal or to pyrolyzed coal char. Unfortunately, the bacteria also bind to the silicate, thereby reducing the number of cells available for binding to pyrite; this observation indicates that the potential for recycling batches of cells to treat greater amounts of coal will be substantially reduced, as 30% of the total cell population will be lost due to adsorption to the mineral ash components. DiSpirito et al. [10] reported that 27 to 30% of a cell population was lost due to adsorption to elemental sulfur, pyrite, silica sand, or small glass beads. Therefore, the specificity of binding exclusively to potential metabolites is not exhibited by *T. ferrooxidans*.

For the SABA process to be commercially feasible, research is required to find an organic or inorganic reagent that will specifically mask the surfaces charges of the coal mineral ash components. This would allow exclusive binding of the *T. ferrooxidans* to the pyrite particles, thereby making them hydrophilic, allowing their separation from the hydrophobic carbon matrix via spherical agglomeration.

In contrast to the previous study on the binding of *T. ferrooxidans* to coal (McCready and Legallais [27]), the present study on the binding of cells to pyrite indicates that the organisms could be removed by vigorous washing. This result is also in contrast to the findings of DiSpirito et al. [10] and may be due to physical abrasion of the cells from the pyrite during screening or to washing the pyrite with wash water not adjusted to pH 2.3.

To prevent the precipitation of ferric hydroxysulfate or jarosites during the bacterial leaching of pyrite from coal, the inorganic nutrient levels must be minimized to prevent these precipitates, but must be at sufficient concentrations to allow optimal growth and iron oxidation by *T. ferrooxidans*. The laboratory study indicates that the media commonly utilized for bacterial leaching [36,38] contain excessive amounts of the inorganic ions required by *T. ferrooxidans*. The data presented in Fig. 35.7 indicate that all the *T. ferrooxidans* isolates grow well on 0.1 mM PO_4^{3+}, 0.5 mM Mg^{2+}, and 0.5 mM NH_4^+. Kos et al. [21] reported that 0.1 mM PO_4^+ and 0.5 mM NH_4^+ were the optimal concentrations required for the bacterial leaching of pyrite from coal by *T. ferrooxidans*. These concentrations of bacterial nutrients allow optimal growth while substantially reducing the amount of ferric hydroxysulfate observed in pure cultures. Therefore, this level of nutrients should reduce the amount of ferric hydroxysulfate and jarosite precipitation within a leaching dump and thus alleviate the plugging that has been observed by these precipitates and reduce the masking of pyritic surfaces by the precipitates, which generally reduce the rate of pyrite oxidation in a bacterial leaching situation.

Kempton (personal communication, 1983) had observed that mucoidal strains of heterotrophic bacteria such as *E. coli* and *Aerobacter aerogenes* resulted in slightly lower pyrite and ash rejection than he observed using *T. ferrooxidans* in the SABA process. The results presented in Table 35.32 confirm Kempton's findings and suggest that the adsorption of bacterial cells to solid surfaces is dependent on electrostatic charge rather than a chemotactic response to a nutrient. Numerous recent studies [5] have shown that the initial adsorption of cells to solid surfaces is by electrostatic attraction followed by the secretion of a binding substance (glycocolyx material) by which the cells anchor themselves to the surface.

Over the last two decades a controversy has existed as to whether the addition or cellular production of surfactants aided the bacteria in the oxidation of sulfur or sulfide minerals to sulfuric acid [1,2,33]. The data presented in Table 36.33 indicate that the addition of the surfactant Tween-20 had no effect on the oxidation of pyrite associated with coal. In contrast, Wakao et al. [41] reported that the addition of Tween-20, protein, nucleic acids, yeast extract, or peptone caused the desorption of the bacterial cells from pyrite surfaces and increased the rate of oxidation.

The column leaching study on coal sized from >½ in. to <1 in. confirmed the previous conclusion that bacterial heap leaching may be the most economical means of recovering combustible coal from stockpiled waste dumps. Continuous bacterial leaching for 1 year resulted in the reduction of pyrite content by 67 to 72% and simultaneously reduced the mineral ash content of the coal by about 60%.

Should partial pyrolysis of coal to produce gases, tar, and a combustible char become the selected means of utilizing coal, the inorganic sulfur content of the char may be reduced substantially by bacterial oxidation of the pyrite and pyrrhotite. The data presented in Fig. 36.8 and Table 35.35 indicate that the inorganic sulfur content of a pyrolyzed char was reduced by 80% in only 21 days of leaching using *T. ferrooxidans*. In contrast, to obtain a similar reduction in the inorganic sulfur content of coal, leaching had to be continued for 59 days [25] or more. Thus, during vacuum pyrolysis, the carbon matrix of the coal develops a spongelike structure which apparently makes the internally located pyrite and pyrrhotite particles accessible to the organisms, thereby allowing their rapid bacterial oxidation and solubilization.

ACKNOWLEDGMENTS

The preceding work was supported by research contracts 09 SC 3115-2-6301 and 9 SC 31028-4-48034 From the National Research Council of Canada while the author was on staff with the Department of Biology, Dalhousie University, Halifax, Nova Scotia. The author would like to thank Mr. L. C. Johnson and Dr. C. E. Capes for their assistance in obtaining coal samples, and Dr. Stirling Whiteway, Dr. C. E. Capes, and Mr. D. Forgeron for the analyses of coals. I would also like to thank Mr. W. J. D. Stone for supplying the finely ground coal samples, Ms. A. M. Coyle and Mr. S. M. Parikh for their excellent technical assistance, and Dr. M. Zentilli, Geology Department, Dalhousie, for the heavy metal analyses during the leaching project.

REFERENCES

1. Agate, A. D., Korczynski, M. S., and Lundgren, D. G., Extracellular complex from the culture filtrate of *Ferroobacillus ferrooxidans, Can. J. Microbiol., 15*, 259–264 (1969).
2. Agate, A. D., and Vishniac, W., Transport characteristics of phospholipids from thiobacilli, *Bacteriol. Proc., 50* (1970).
3. Ashmead, D., The influence of bacteria in the formation of acid mine waters, *Colliery Guardian, 190*, 694–698 (1955).
4. Bennett, J. C., and Tributsch, H., Bacterial leaching patterns on pyrite crystal surfaces, *J. Bacteriol., 134*, 310–317 (1978).
5. Berkley, R. C. M., Lynch, J. M., Melling, J., Rutter, P. R., and Vincent, B., *Microbial Adhesion to Surfaces*, Ellis Horwood, Chichester, West Sussex, England, 1980.
6. Capes, C. E., McIlhinney, A. E., Sirianni, A. F., and Puddington, I. E., Bacterial oxidation in upgrading pyritic coals, *CIM Bull., 66*, 88–91 (1973).

7. Colmer, A. R., and Hinkle, M. E., A role of microorganisms in acid mine drainage: a preliminary report, *Science, 106,* 253-256 (1947).
8. Cook, T. M., Growth of *Thiobacillus thiooxidans* in shaken cultures, *J. Bacteriol., 88,* 620-623 (1964).
9. DiSpirito, A. A., Silver, M., Vose, L., and Tuovinen, O. H., Flagella and pili of iron-oxidizing thiobacilli isolated from a uranium mine in northern Ontario, Canada, *Appl. Environ. Microbiol., 43,* 1196-2000 (1982).
10. DiSpirito, A. A., Dugan, P. R., and Tuovinen, O. H., Sorption of *Thiobacillus ferrooxidans* to particulate material, *Biotechnol. Bioeng., 25,* 1163-1168 (1983).
11. Doddema, H. J., Partial microbial oxidation of pyrite in coal followed by oil-agglomeration, presented at the *International Symposium on Biohydrometallurgy,* Cagliari, Italy, May 1983.
12. Dugan, P. R., Microbiological removal of sulfur from pulverized coal blend, presented before the *3rd Symposium on Coal Preparation, NCA/BRC Coal Conference and Expo IV,* Louisville, Ky., 1977.
13. Dugan, P. R., and Apel, W. A., Microbial desulfurization of coal, in *Metallurgical Applications of Bacterial Leaching and Related Microbiological Phenonomena* (L. E. Murr, A. E. Torma, and D. A. Brierley, eds.), Academic Press, New York, 1978, pp. 223-250.
14. Hacquebard, P. A., 1982. *Composition, Origin and Geology of Coal,* text prepared for a geology course at Dalhousie University.
15. Fuerst, C. R., and Stent, G. S., Inactivation of bacteria by decay of incorporated radioactive phosphorus, *J. Gen. Physiol., 40,* 73-90 (1956).
16. Hoffman, M. R., Faust, B. C., Panda, F. A., Koo, H. H., and Tsuchiya, H. M., Kinetics of the removal of iron pyrite from coal by microbiol catalysis, *Appl. Environ. Microbiol., 42,* 259-271 (1981).
17. Kempton, A. G., Moneib, N., McCready, R. G. L., and Capes, C. E., Removal of pyrite from coal by conditioning with *Thiobacillus ferrooxidans* followed by oil agglomeration, *Hydrometallurgy, 5,* 117-125 (1980).
18. Kempton, A. G., Butler, B. J., and Levadoux, W., *Bacterial Desulfurization of Coal,* NRC Project 102-13, Contract 081-030/1-6616.
19. Kempton, A. G., and Butler, B. N., 1982. *Bacterial Desulfurization of Coal,* NRC Project 201-06, Contract 05x81-00130.
20. Korhonen, T. K., Nurmiaho, E. L., and Tuovenin, O. H., Fimbriation in *Thiobacillus* A-2, *FEMS Microbiol. Lett., 3,* 195-198 (1978).
21. Kos, C. H., Bijleveld, W., Grotenhuis, T., Bos, P., Kuenen, J. G., and Poorter, R. P. E., Composition of mineral salts medium for microbial desulfurization of coal, *International Symposium on Biohydrometallurgy,* Cagliari, Italy, May 1983.
22. Laishley, E., Bryant, R., and Costerton, J. W., Glycocalyx attachment of *Thiobacillus albertis* to elemental sulfur, *Abstr. Can. Soc. Microbiol. Ann. Meet.,* 1983, p. 79.
23. Lea, D. E., Haines, R. B., and Bretscher, E., The bactericidal action of x-rays on neutrons and radioactive radiations, *J. Hyg. (Cambridge), 41,* 1-16 (1941).
24. Malouf, E. E., Copper leaching practice, in *Inplace Leaching and Solution Mining,* MacKay School of Mines, Reno, Nev., 1975.
25. McCready, R. G. L., 1984. *Microbiological Studies on High-Sulfur Coals,* ARL Technical Report 50, NRCC 23601.
26. McCready, R. G. L., and Din, G. A., Active sulfate transport in *Saccharomyces cerevisiae, FEBS Lett., 38,* 361-363 (1974).
27. McCready, R. G. L., and Legallais, B. P., Selective adsorption of ^{32}P-labelled *Thiobacillus ferrooxidans* in finely ground coal suspensions, *Hydrometallurgy, 12,* 281-288 (1984).
28. Moneib, N. A. M., Study of biological desulfurization of coal, M.Sc. thesis, University of Waterloo, 1978.

29. Murr, L. E., and Berry, V. K., Direct observations of selective attachment of bacteria to low-grade sulfide ore and other minerbal surfaces, *Hydrometallurgy*, 2, 11-14 (1976).
30. Myers, G. E., and McCready, R. G. L., Non-lethal assimilation and distribution of radioactive phosphorus in *Serratia marcescens*, *Can. J. Microbiol.*, 10, 317-322 (1964).
31. Rudershausen, C. G., Copper solution at Old Reliable, in *Inplace Leaching and Solution Mining*, MacKay School of Mines, Reno, Nev., 1975.
32. Schaeffer, W. I., Holbert, P. E., and Umbreit, W. W., Attachment of *Thiobacillus thiooxidans* to sulfur crystals, *J. Bacteriol.*, 85, 137-140 (1963).
33. Schaeffer, W. I., and Umbreit, W. W., Phosphatidylinositol as a wetting agent in sulfur oxidation by *Thiobacillus thiooxidans*, *J. Bacteriol.*, 85, 491-493 (1963).
34. Schnaitman, C. A., Korczynski, M. S., and Lundgren, D. G., Kinetic studies of iron oxidation by whole cells of *Ferrobacillus ferrooxidans*, *J. Bacteriol.*, 99, 552-557 (1969).
35. Silverman, M. P., Rogoff, M. H., and Wender, I., Removal of pyrite sulfur from coal by bacterial action, *Fuel*, 42, 113-124 (1963).
36. Silverman, M. P., and Lundgren, D. G., Studies on the chemoautotrophic iron bacterium *Ferrobacillus ferrooxidans*. 11. Manometric studies, *J. Bacteriol.*, 78, 326-331 (1959).
37. Tuovinen, O. H., and Kelly, D. P., Studies on the growth of *Thiobacillus ferrooxidans*. I. Use of membrane filters and ferrous iron agar to determine viable numbers and comparison with $^{14}CO_2$-fixation and iron oxidation as measures of growth, *Arch. Mikrobiol.*, 88, 285-298 (1973).
38. Tuovinen, O. H., and Kelly, D. P., Studies on the growth of *Thiobacillus ferrooxidans*. V. Factors affecting growth in liquid culture and development of colonies on solid media containing inorganic sulfur compounds, *Arch. Microbiol.*, 98, 351-364 (1974).
39. Tuovinen, O. H., Inhibition of *Thiobacillus ferrooxidans* by mineral floatation reagents, *Eur. J. Appl. Microbiol. Biotech.*, 5, 301-304 (1978).
40. Wadsworth, M. E., Physico-chemical aspects of solution mining, in *Inplace Leaching and Solution Mining*, MacKay School of Mines, Reno, Nev., 1975.
41. Wakao, N., Mishino, M., Sakurai, Y., and Shiota, H., Bacterial pyrite oxidation. 111. Adsorption of *Thiobacillus ferrooxidans* cells on solid surfaces and its effect on iron release from pyrite, *J. Gen. Appl. Microbiol.*, 30, 63-77 (1984).
42. Waksman, S. A., *Principles of Soil Microbiology*, 2nd ed., Williams & Wilkins, Baltimore, 1932.

36

New Filamentous Bacteria for Coal Desulfurization

KWANGIL LEE* and TEH FU YEN *University of Southern California, Los Angeles, California*

M. LUISA BLAZQUEZ *Complutense University of Madrid, Madrid, Spain*

36.1 INTRODUCTION

Since the energy crisis in the 1970s, interest in coal as an energy source for production of electric power has increased. However, the feasibility of coal use for electric energy purposes has decreased because of recent stringent environmental regulations.

One of the main concerns regarding environmental regulations is associated with the sulfur content of coal. During combustion, sulfur is converted almost entirely to gaseous products that cause environmental pollution and material corrosion. Acid mine drainage at mine sites is often related to the weathering of exposed pyritic materials and chemolithotropic soil bacteria. These problems have created new interest in developing processes for removing sulfur, prior, during, or after combustion of coal.

Sulfur present in coal consists of inorganic and organic compounds. Inorganic sulfur occurs predominantly as ferrous sulfide (FeS_2) in its mineral forms of pyrite and marcasite, whereas organic sulfur compounds are part of the molecular configuration of coal, believed to be in the form of monosulfide (−S−), disulfide (−S−S−), thiol (−SH), and heterocyclic compounds [1,2].

There are several methods for coal desulfurization [2,3]. Chemical desulfurization processes operate at high temperatures (100 to 400°C) and pressures (100 to 800 psi). Therefore, such methods demand high levels of energy consumption and sophisticated techniques to become practical. Physical methods such as flotation and magnetic separation appear to be more sound economically than do chemical methods. However, the former have some disadvantages, due to finely scattered pyrite and poor organic sulfur removal, which may result in energy losses.

Numerous papers have revealed that a significant portion of inorganic and organic sulfur is removed from coal by microorganisms. The importance of biological desulfurization from coal can hardly be overemphasized, perhaps because microbial removal of sulfur compounds before combustion requires low capital and operating costs and is

Current affiliation: Cal Science Engineering, Cypress, California

feasible, yielding a negligible loss of the initial energetic value of coal. The microorganisms involved in sulfur removal from coal are generally characterized by their unique capability to obtain metabolic energy from reactions involving sulfur. Bacteria such as *Thiobacilli*, *Beggiatoa*, and *Sulfolobus* are sulfur-oxidizing organisms that use aerobic-respiratory oxidation of sulfur and sulfide to sulfate as a source of energy [4].

Thiobacillus ferrooxidans, an acidophilic autotroph microorganism, has so far been the most widely used organism for the removal of pyritic sulfur from coal [5,6]. However, it is a comparatively slow process due to the long residence time required, and oversized reactors are needed [7]. In addition, *Thiobacillus* species are ineffective in removing organic sulfur. Recently, Kargi and Robinson explored biological sulfur removal from coal by *Sulfolobus acidocaldarius*, a thermophilic organism [8-10]. Surprisingly, little is known about the ability of *Beggiatoa* to remove sulfur from coal. *Beggiatoa* is a filamentous gliding bacterium that can oxidize sulfide to elemental sulfur and then to sulfate [11]. It is also reported that *Beggiatoa* inters mutalistically with plant roots [12]. Therefore, *Beggiatoa* may play an important role in a plant-soil ecosystem at coal mine sites.

Recently, we have attempted to study *Beggiatoa* species for sulfur removal from coal. *B. alba* ATCC 33555, known as a sulfur-oxidizing filamentous bacteria, was tested for its use in coal desulfurization. *Thiobacilli* and *Sulfolobus* were also tested.

36.2 MATERIALS AND METHODS

36.2.1 Coal Samples

North Dakota lignite, one of the major low-ranking coals for electric power plants in the United States, was tested in our laboratory and found to have a low pyritic sulfur content (0.25%) and a relatively high organic sulfur content (1.7%). Since the preparation of very fine particles reduces the economic value of coal, proper particle size must be attained. The results of the study of complex sulfides related to pulp density showed that a reasonable bioleaching rate was achieved at 10% of pulp density using *T. ferrooxidans* (Fig. 36.1). Therefore, coal particles with 10% of pulp density below 100 mesh size (<150 μm) were used throughout this study.

36.2.2 Microorganisms

Three pure cultures obtained from the American Type Culture Collections were used in this study.

1. *T. ferrooxidans* ATCC 19859 are acidophilic, flagellated, non-spore-forming, mesophilic, aerobic, and gram-negative rods (0.5 × 1 to 2 μm). A mineral salts medium having the following composition was used: 3.0 g/L $(NH_4)_2SO_4$, 0.1 g/L KCl, 0.5 g/L $KH_2PO_4 \cdot 3H_2O$, 0.5 g/L $MgSO_4 \cdot 7H_2O$, 0.01 g/L $Ca(NO_3)_2 4H_2O$, 700 mL of deionized water, and 300 mL of a 14.74 % w/v $FeSO_4 7H_2O$ solution. The pH was adjusted to 2.5 with H_2SO_4.

2. *S. acidocaldarius* ATCC 33909 is a thermophilic (60 to 90°C), acidophilic, and facultative autotrophic organism. A mineral salts medium having the following composition was used: 3.0 g/L $(NH_4)_2SO_4$, 0.28 g/L KH_2PO_4, 0.5 g/L $MgSO_4 7H_2O$, 0.07 g/L $CaCl_2 \cdot 2H_2O$, and 0.02 g/L $FeCl_3 \cdot 6H_2O$. The pH was adjusted to 2.5 with H_2SO_4 and incubated at 72°C.

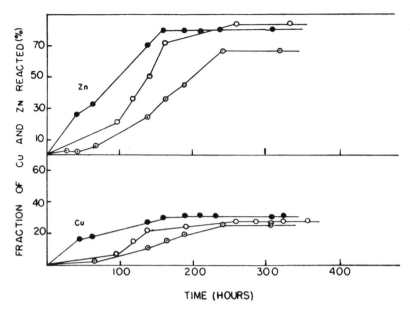

Fig. 36.1 Impact of pulp density on bioleaching. (From Ref. 13.)

3. *B. alba* ATCC 33555 is a filamentous gliding bacterium that can oxidize sulfide to elemental sulfur as a deposit. *B. alba* was grown in the following mineral salts medium: 200 mg/L NH_4Cl, 10 mg/L K_2HPO_4, 5 mL tap water, 500 mg/L sodium acetate, and 1 mM Na_2S. The pH was adjusted to 7.5 with 0.1 N HCl. After 3 weeks the culture growing in this medium at 25°C was adopted to pyrite. A medium having 2.5 g of oil shale was sterilized and inoculated with a culture previously concentrated by centrifugation (0.505 g wet weight of cells).

36.2.3 Coal Desulfurization Experiments

Cultures of *T. ferrooxidans*, *S. acidocaldarius*, and *B. alba* were routinely maintained. Sulfur-oxidizing bacteria were inoculated to flasks containing their respective mineral media containing 10% coal pulp density. For the inoculation with *B. alba*, the culture medium was centrifuged and 1.5 g wet weight of cells was used for 300 mL of medium. Control flasks were used to determine nonbiological sulfur removal from lignite.

36.2.4 Analysis

Conventional chemical procedures used for sulfur analysis of coal samples were adapted from ASTM. The Eschka method was used to measure the total sulfur content of lignite samples [14]. Prior to precipitation of $BaSO_4$ during the Eschka method, 1 mL of sample was collected and analyzed by ion chromatography (Dionex 2000I). The ASTM D2494-79 procedure was used for sulfate and pyritic sulfur analyses and the ISO 1171 procedure to determine the ash content. Sample liquids were taken to measure soluble sulfate concentration by ion chromatography every 48 h. After bioleaching, the collected coal samples were filtered and then washed with 0.1 N HCl and deionized water to extract soluble sulfate adsorbed in coal particles. Then the samples were dried overnight under vacuum at 60°C.

36.3 RESULTS AND DISCUSSION

The experiments described above were performed to study (1) the ability of *B. alba* ATCC 33555 to remove sulfur from coal, and (2) the desulfurization of lignite by the sulfur-oxidizing bacteria *T. ferrooxidans* and *S. acidocaldarius*. Total soluble sulfate concentrations of the respective sulfur-oxidizing bacteria are compared in Fig. 36.2. The maximum dissolution of sulfur was reached after 6 days for *T. ferrooxidans*, 16 days for *S. acidocaldarius*, and 10 days for *B. alba*. Total sulfur removal from lignite was measured by ion chromatography and the Eschka method. The data are presented in Fig. 36.3. Table 36.1 shows data for sulfur removal from lignite by sulfur-oxidizing bacteria. Although ion chromatography and the Eschka method yielded different results, both results showed that *S. acidocaldarius* was the best sulfur-removing bacteria for lignite. The new bacteria species, *B. alba* showed significant sulfur removal from lignite (Fig. 36.4). Total sulfur removal from lignite by *Beggiatoa* (36.8% of initial total sulfur) was more than that of *T. ferrooxidans* (14% of initial total sulfur). In addition, *B. alba* showed 6% initial organic sulfur removal from lignite. Therefore, *B. alba* was a significant sulfur-oxidizing bacteria for coal desulfurization.

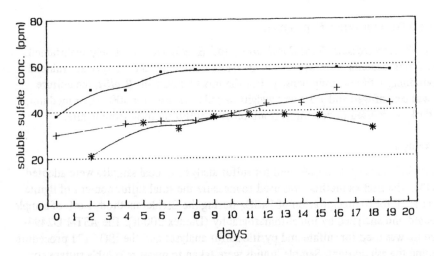

Fig. 36.2 Soluble sulfate concentration with lignite by sulfur-oxidizing bacteria.

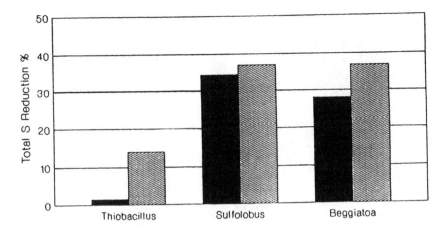

Fig. 36.3 Total sulfur reduction from lignite by sulfur-oxidizing bacteria.

Table 36.1 Sulfur Compositions of Lignite Desulfurized by *T. ferrooxidans, S. acidocaldarius,* and *B. alba*

	North Dakota lignite (%)	Percent removal		
		T. ferrooxidans	*S. acidocaldarius*	*B. alba*
Ash	19.90	23.0	36.8	10.6
Total sulfur	2.80	14.0	37.0	36.8
Pyritic sulfur	0.85	100.0	87.4	68.4
Sulfate sulfur	0.25	16.5	84.7	84.7
Organic sulfur	1.70	0.0	6.0	6.0

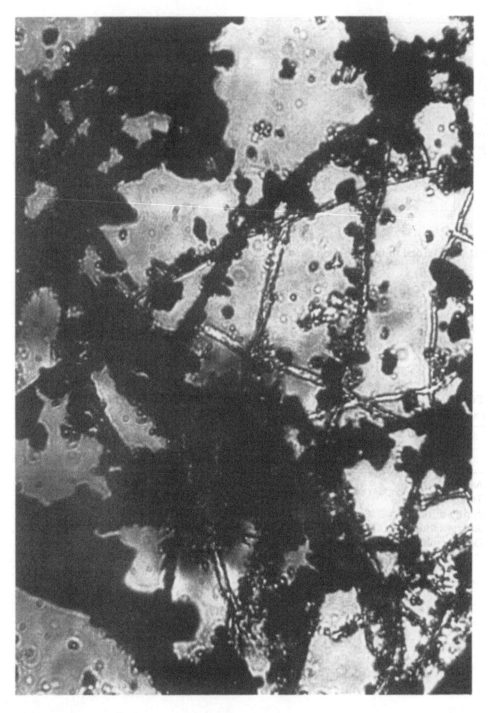

Fig. 36.4 *B. alba* with coal particles (400X).

36.4 CONCLUSIONS

Initial pyritic sulfur present in lignite was completely removed by *T. ferrooxidans*, and *S. acidocaldarius* was the most effective bacterium for total sulfur removal from lignite. *B. alba* significantly removed sulfur from lignite; more than 68% of initial pyritic sulfur was removed by this new sulfur-oxidizing bacteria. In terms of total sulfur removal, *B. alba* (about 37% of total sulfur removal from lignite) was more effective than *T. ferrooxidans* (14% of total sulfur removal).

   In its natural habitat, *B. alba* is known to remove sulfur. This strongly suggests that these filamentous bacteria are feasible for biological desulfurization of coal. Furthermore, a mutually favorable interaction between the *Beggiatoa* species and plant roots may be helpful for the soil-plant ecosystem. The use of *Beggiatoa* species for desulfurization in both industry and soil-plant ecosystems will depend on basic knowledge of its activity and survival in various conditions.

   Ion chromatography was found to be a useful analytical tool for measuring the total sulfur content of coal. However, further work is required to optimize this method.

ACKNOWLEDGMENT

M. Luisa Blazquez would like to acknowledge the receipt of bilateral travel support in a program aiding a visiting scientist from Spain at the University of Southern California.

REFERENCES

1. Nishioka, M., Leeand, M. L., Castle, R. N., Sulphur heterocycles in coal-derived products, *Fuel, 65*, 390–396 (1986).
2. Wheelock, T. D. (ed.), *Coal Desulfurization: Chemical and Physical Methods*, ACS Symposium Series 64, American Chemical Society, Washington, D.C., 1977.
3. Meyers, R. A., *Coal Desulfurization*, Marcel Dekker, New York, 1977.
4. Kelly, P., Sulphur bacteria first again, *Nature, 326*, 830 (1987).
5. Silverman, M. P., Rogoff, M. H., and Wender, I., Bacterial oxidation of pyritic materials in coal, *Appl. Microbiol., 9*, 491–496 (1961).
6. Silverman, M. P., Mechanisms of bacterial pyrite oxidation, *J. Bacteriol., 94*, 1046–1051 (1967).
7. Hoffmann, M. R., Faust, B. C., Panda, F. A., Koo, H. H., Tsuchiya, H. M., Kinetics of the removal of iron pyrite from coal by microbial catalysis, *Appl. Environ. Microbiol., 42*, 259–271 (1981).
8. Kargi, F., and Robinson, J. M., Removal of sulfur compounds from coal by the thermophilic organism *Sulfolobus acidocaldarius, Appl. Environ. Microbiol., 44*, 878–883 (1982).
9. Kargi, F., and Robinson, J. M., Biological removal of pyritic sulfur from coal by the thermophilic organism *Sulfolobus acidocaldarius, Biotechnol. Bioeng., 27*, 41–49 (1985).
10. Kargi, F., and Robinson, J. M., Removal of organic sulfur from bituminous coal, *Fuel, 65*, 397–399 (1986).
11. Mezzino M. J., Strohl, W. R., and Larkin, J. M., Characterization of *Beggiatoa alba, Arch Microbiol., 137*, 139–144 (1984).
12. Joshi, M. M., and Hollis, J. P., Interactions of *Beggiatoa* and rice plant: detoxification of hydrogen sulfide in the rice rhizosphere, *Science, 195*, 179–180 (1977).

13. Ballester, A., Gonzalez, F., Blazquez, M., Mier, J. L., and Munoz, J., Spanish sulphides: attack by bioleaching, *Biohydrometallurgy International Symposium*, Coventry, England, 1987.

14. Karr, C., *Analytical Methods for Coal and Coal Products*, Vol. 1, Academic Press, New York, 1978.

Index